Pecos Pueblo Revisited

Papers of the
Peabody Museum of
Archaeology and Ethnology
Harvard University
Cambridge, Massachusetts, U.S.A.

Volume 85

Pecos Pueblo Revisited

The Biological and Social Context

Edited by
Michèle E. Morgan

Foreword by
Matthew Liebmann and Christopher Toya

With contributions by
Lane Anderson Beck, Karin S. Bruwelheide, Patricia Capone, Richard L. Jantz, Corina M. Kellner, Steven A. LeBlanc, Katherine Moore, Michèle E. Morgan, Douglas W. Owsley, Christopher Ruff, Margaret Schoeninger, Katherine Spielmann, Robert G. Tague, Christy G. Turner, Katherine E. Weisensee

Peabody Museum of
Archaeology and Ethnology
Harvard University
Cambridge, Massachusetts, U.S.A.

2010

Cover design by Mary Sweitzer
Editorial direction by Joan K. O'Donnell
Copyediting by Jane Kepp
Composition by Donna Dickerson
Production management by Donna Dickerson
Proofreading by Janice Herndon
Manufactured in the United States of America by Edwards Brothers, Inc.

Peabody Museum of Archaeology and Ethnology, Harvard University
Cambridge, Massachusetts 02138

♾ The paper used in this publication meets the minimum requirements of the American National Standard for Information Sciences—Permanence of Paper for Printed Library Materials, ANSI Z39.48-1992.

Cover illustration by Singleton Peabody Moorehead from A. V. Kidder, *Pecos, New Mexico: Archaeological Notes,* Papers of the Robert S. Peabody Foundation for Archaeology 5: fig. 22 (Andover, MA, 1958).

ISBN: 978-0-87365-213-1

Library of Congress Cataloging-in-Publication Data

Pecos Pueblo revisited : the biological and social context / edited by Michèle E. Morgan ; with contributions by Lane Anderson Beck ... [et al.].
p. cm. -- (Papers of the Peabody Museum of Archaeology and Ethnology ; v. 85)
Includes bibliographical references.
ISBN 978-0-87365-213-1 (pbk. : alk. paper)
1. Pecos National Monument (N.M.)--Antiquities. 2. Pueblo Indians--Antiquities. 3. Human remains (Archaeology)--New Mexico--Pecos National Monument. 4. Excavations (Archaeology)--New Mexico--Pecos National Monument. 5. Social archaeology--New Mexico--Pecos National Monument. 6. Ethnoarchaeology--New Mexico--Pecos National Monument. I. Morgan, Michèle E., 1962- II. Beck, Lane A.
E99.P34P43 2010
978.9004'974--dc22

2009034156

Contents

Figures

Tables

Foreword

Matthew Liebmann and Christopher Toya

The people of Pecos Pueblo, Jemez Pueblo, and Harvard University will forever be connected through a series of remarkable journeys that have taken place over the past two centuries. At the time Harvard was founded in 1636, Spanish documents tell us that Towa-speaking people regularly trekked the 60-plus miles between the pueblos of Jemez and Pecos on foot, carrying news and information from one village to the other. More than 200 years later, in 1838, the last permanent residents of Pecos left their ancestral home, traveling west to take up residence with their Jemez cousins. In 1910, a young Harvard graduate student named Alfred V. Kidder first visited Pecos Pueblo. Little did he know that this journey would link Pecos, Jemez, and Harvard in a unique and enduring relationship. Five years later Kidder began excavations at the pueblo, shipping the remains of the individuals he unearthed back across the country to Harvard's Peabody Museum of Archaeology and Ethnology (PMAE).

The skeletal material from Pecos was housed at the PMAE until 1999, when representatives from Jemez Pueblo traveled to Cambridge to take formal possession of their ancestors in order to return them to the place of their original burial. At the same time that the human remains left Massachusetts to cross the country one last time, hundreds of Jemez people left their pueblo to retrace the steps of their forebears, walking and running over ancient mountain trails on a pilgrimage back to Pecos. The two processions met just outside of Pecos—one coming from the east, the other from the west—to form a single column as they approached the ancient village. Later that day the remains of 1,922 individuals were reinterred by their descendants at Pecos National Historical Park.

Although the physical journey of the Pecos individuals from New Mexico to Cambridge and back again may have concluded on May 22, 1999, their intellectual journey continues to the present day. *Pecos Pueblo Revisited* represents another series of footsteps on the generations-long route to a better understanding of the lives of the people of Pecos. In the chapters collected between these covers, Michèle E. Morgan has assembled a series of important new insights regarding the inhabitants of Pecos. Through these studies of their diet, relationships with neighboring peoples, demography, patterns of activity, health, and disease, we now have a better grasp of the deep history of Pecos than ever before. The contributors are to be commended for their careful, thoughtful, and respectful analyses, as well as their commitment to the study of the ancestors of the modern Jemez people.

Matthew Liebmann is Assistant Professor, Department of Anthropology, Harvard University. Christopher Toya is Traditional Cultural Properties Manager for the Pueblo of Jemez Department of Resource Protection.

The series of journeys connecting Pecos, Jemez, and Harvard will not end with the closing pages of this book, however. As Dr. Morgan so rightly notes, although this volume represents the culmination of decades of research on the archaeology of Pecos, the publication of these studies, along with the comprehensive data sets included in the appendixes, will allow future researchers to carry out further analyses, venturing anew on journeys to places yet unknown.

As many of the authors note, the studies presented here are a direct outcome of the Native American Graves Protection and Repatriation Act (NAGPRA). In the past, NAGPRA has been characterized by some as a law that contributes to the termination of archaeological investigations, ultimately resulting in a diminution of knowledge regarding the human past. And while it is true that the remains detailed in this volume are no longer available for primary analysis, it is equally true that these studies would not have taken place in the absence of NAGPRA. *Pecos Pueblo Revisited* thus stands as a testament to the constructive aspects of the act, contributing to a richer understanding of the history of Pecos than we have ever had before. This is not to suggest that repatriation is ever an easy or painless process. It is a lengthy negotiation marked by difficult and frank discussions. To be sure, there remain varied opinions regarding the appropriateness both of studying human remains and of repatriating them. Dr. Kidder's excavations of nearly 2,000 burials at Pecos, while consistent with the professional standards of his day, continue to be a source of consternation to many of the Jemez people. In the end, however, those of us who have worked in the context of NAGPRA have learned that the process is positive and edifying more often than not. This volume is further evidence of the productive potential of NAGPRA.

Pecos occupies a unique and cherished place in the tribal history of the contemporary residents of Jemez Pueblo, and many of the elements introduced into Jemez culture as a result of the 1838 migration from Pecos remain in practice as living traditions at Walatowa (Jemez Pueblo) today. Nowhere is this more conspicuous than at the main plaza every August 2, when the feast of St. Porcingula, patron saint of the Pecos Pueblo mission, is celebrated. In the shrine at the west end of the plaza you will find the Pecos Governor, a tribal official who still carries the cane of office that was conveyed to the Pecos people from the King of Spain in the seventeenth century. If you're lucky, you might even catch a glimpse of the famous Pecos Bull, surrounded by members of the Pecos Eagle Watchers' Society, who continue to make pilgrimages to the sacred landscape surrounding Pecos. As a result of these regular visits by the Jemez people, Pecos should not be considered "abandoned." Rather, it remains inhabited by their ancestors and patiently awaits the return of more Towa-speaking inhabitants in the future.

In 1925 Dr. Kidder recognized the importance of the connections between Pecos and Jemez, noting that it was vital "to bring the history of the Pecos people down to the present day." With the publication of this volume, Dr. Morgan and her colleagues have accomplished this goal. We are grateful to all the contributors to *Pecos Pueblo Revisited* for making it possible for us to learn more about the people of Pecos, and we look forward to continuing the journey and strengthening the ties of scholarship and friendship among the people of Pecos, Harvard, and Jemez in the future.

Acknowledgments

The idea for this book developed out of conversations with Douglas Owsley and Richard Jantz shortly after the physical repatriation of the Pecos Pueblo human remains and associated funerary objects from the Peabody Museum of Archaeology and Ethnology (PMAE) to the Pueblo of Jemez in New Mexico. The original goal was to assemble the primary data on the PMAE human remains and present current interpretations of those data in a way that would be useful to future researchers.

Members of the Pueblo of Jemez who were involved in the repatriation, particularly Pete Toya, Benny Shendo Jr., Vincent Toya, and Franklin Toya, enriched this work by sharing their knowledge, history, and insight during our consultation visits in Massachusetts and New Mexico. I am grateful to them for their interest and and for their ongoing support for the Native American Graves Protection and Repatriation Act (NAGPRA) inventory study that underlies this volume.

I owe a special debt of gratitude to a number of people at the PMAE. Rubie Watson and William Fash, past and present directors, respectively, of the PMAE, along with the PMAE Publications Committee, were crucial in their endorsement of the publication of this volume. David Pilbeam was an early and important supporter of my research and the editorial process. Steven LeBlanc enthusiastically encouraged me to expand my original idea to include as much primary data as possible in a publishable format. He offered valuable advice and encouragement, along with material support, throughout the development of the volume.

Many thanks to Laura MacLatchy, Joan Merriman, and the two anonymous reviewers, who provided constructive comments on parts or all of the manuscript. I also warmly thank my fellow contributors to the volume for their interest, expertise, and generosity. It has been a pleasure and an honor to work with so many outstanding scholars in the field.

Matthew Liebmann and Christopher Toya continue to research and preserve the material and cultural traditions of the Jemez Pueblo people. I thank them for sharing their perspective on the enduring ties that connect Pecos Pueblo, Jemez Pueblo, and Harvard University.

This book could not have been published without the persistent work of Joan O'Donnell and Donna Dickerson of the Peabody Museum Press. I am grateful to them for their editorial and production oversight. Jane Kepp did an outstanding job copyediting the volume with thematic and stylistic insight. Eliza McClellan prepared the maps, and Janice Herndon proofread the volume. Jessica Desany, Genevieve Fisher, Patricia Kervick, and India Spartz assisted with images and copyright issues. My thanks to all of them for helping to see this volume to press.

Finally, I thank my husband, Robert, my children, James, Jason, and Elizabeth, my father, Jason, and my in-laws, Al and Sue, for their loving support in this and all my endeavors.

1 Introduction

Michèle E. Morgan

In May 1999, the Pueblo of Jemez, New Mexico, formally repatriated human remains and associated funerary objects originally from Pecos Pueblo and associated sites. The human remains and associated artifacts had been curated at the Peabody Museum of Archaeology and Ethnology (PMAE) of Harvard University and at the Robert S. Peabody Museum of Andover, Massachusetts. The Pueblo of Jemez reinterred them at Pecos National Historic Park, New Mexico, along with human remains and associated funerary objects repatriated from other museums and institutions, in a well-documented event (Tarpy 2000). The reinterment provided closure for the descendants of Pecos Pueblo. At the same time, it ended the possibility of any additional scientific study of the human and cultural remains. The preceding 80 years of anthropological research on the Pecos human remains represent a milestone in the development of our understanding of what can be learned from skeletal remains and how they inform us of the human condition in the past.

Pecos Pueblo has figured prominently in Southwestern archaeology for more than 100 years (Elliott 1995), with initial description by Adolph Bandelier in the late nineteenth century (figs. 1.1, 1.2, 1.3) and extensive excavations by Alfred V. Kidder between 1915 and 1929 (fig. 1.4). The pueblo's influence on the peoples, cultures, and landscapes of the Southwest, however, extends back in time for more than 650 years. Strategically located on a small mesa overlooking the Pecos River, Pecos Pueblo marked a crossroads between the Glorieta Mountains to the west and the Great Plains to the east. The inhabitants, known historically as the Cicuye, became dominant in the region in the late fifteenth century, and Pecos Pueblo emerged as a key settlement as Plains Indian groups from the east developed increasing trade with eastern Pueblo Indians and as Spaniards forayed north into the southwestern United States in the late 1500s. A complex suite of factors, including warfare, disease, and an increasingly demanding Spanish presence, severely stressed the people of Pecos Pueblo in the eighteenth century. In 1838, the few remaining Cicuye left Pecos and trekked 60 miles west to join the Pueblo of Jemez. Pecos Pueblo traditions continue at the Pueblo of Jemez, and members of the Pueblo of Jemez regularly visit Pecos National Historic Park.

Just as Pecos Pueblo played an important role in the history of the Southwest, so the excavations there played an important role in the development of American archaeology. The excavations by A. V. Kidder (fig. 1.5) set a new standard for fieldwork and interpretation, as reviewed by Beck (2006). Pecos became a mecca for visiting archaeologists who wanted to learn new methods and share ideas, and the annual conference of Southwestern archaeologists is called the Pecos Conference in memory of the origin

FIGURE 1.1. Ruins of the fourth mission church at Pecos Pueblo. Photographed by George C. Bennett for Adolph Bandelier in 1880. (Photographic Archives, 2004.29.1536, Peabody Museum of Archaeology and Ethnology, Harvard University.)

of the meeting. Along with other innovations, Kidder recognized that skeletal remains were a valuable source of information. He enlisted the expertise of Earnest A. Hooton (fig. 1.6) to study the human remains, which were transferred to the PMAE, Harvard University, between 1915 and 1929. This resulted in Hooton's major work, *The Indians of Pecos Pueblo* (1930), and many subsequent studies, which are discussed in this volume. Because it has been so thoroughly studied by a number of prominent biological anthropologists, the Pecos sample is also a significant population for use in comparative studies of fossil hominins and recent populations alike.

The impetus for this volume arose near the conclusion of the PMAE's inventory of the human remains from the Pecos Valley in its collections, which was mandated by the Native American Graves Protection and Repatriation Act (NAGPRA). The PMAE took this endeavor very seriously. While documenting the collections in accordance with NAGPRA regulations, Michèle Morgan, then the PMAE's repatriation staff osteologist, realized that although Hooton's measurements remained extremely valuable, inaccuracies in the transcription and assessment of the provenience, age, sex, and skeletal pathologies of individuals compromised interpretations of the data. Moreover, no subsequent researcher had reexamined a majority of the human skeletal remains. The painstaking review of field notes and verification of associated funerary objects, provenance, and so forth, in collaboration with the Robert S. Peabody Museum of Andover, has yielded a vastly improved database.

It was also recognized that additional information could be recovered that had not been recorded by Hooton or subsequent researchers. The PMAE approached this opportunity by arranging and supporting research visits in 1998 by Douglas Owsley, Richard Jantz, Karin Bruwelheide, and others from the Smithsonian Institution and the University of Tennessee, Knoxville, to study the burials from the Pecos mission church. Lane Beck also studied the Pecos human remains in 1998 to collect additional pathology data. These efforts resulted in a new accumulation of data about the soon-to-be-repatriated historic collection.

These new data, together with insights into the biological and social contexts of Pecos Pueblo from scholars who have made significant scientific contributions through study of the Pecos Pueblo human and material remains, are presented in this mono-

FIGURE 1.2. View of the Southern House (foreground) and Pecos mission church. Photographed by George C. Bennett for Adolph Bandelier in 1880. (Photographic Archives, 2004.29.1531, Peabody Museum of Archaeology and Ethnology, Harvard University.)

FIGURE 1.3. View of the court of the Northern House at Pecos Pueblo. Photographed by George C. Bennett for Adolph Bandelier in 1880. (Photographic Archives, 2004.29.1537, Peabody Museum of Archaeology and Ethnology, Harvard University.)

FIGURE 1.4. Visit to Pecos Pueblo by the board members of the School of American Research to observe A. V. Kidder's research, 1915. Left to right: Mrs. Percy Jackson, Jesse L. Nusbaum, unknown, E. L. Hewett, unknown, Alfred V. Kidder, Percy Jackson, H. R. Fairclough, Charles F. Lummis, unknown, unknown, Judge J. R. McFie, unknown. (A. V. Kidder Papers, 65-18-10/100077.1.5, Peabody Museum of Archaeology and Ethnology, Harvard University.)

graph. Many of the contributors have generously allowed their primary data to be included in appendixes for future analysis. Although the human remains have now been reinterred, scientists will continue to benefit from the data these scholars have provided. In addition to the raw data presented in Appendixes A–K, the PMAE NAGPRA inventory data and many photographic materials relating to Pecos Pueblo and the Kidder excavations are curated at the PMAE.

Scholars from many disciplines have researched the material culture, language, rituals, social networks, economy, health, and biology of the people of Pecos Pueblo and their relatives and descendants, as well as the influence of Bandelier and Kidder on Southwestern archaeology. It is not our goal in this volume to summarize all that is known about Pecos Pueblo and the people who dwelled there. Rather, our objectives are twofold: to review some of the most significant findings from Pecos Pueblo in the context of current Southwestern archaeological and osteological perspectives and to present an improved series of data and new interpretations of the behavior and biology of the inhabitants of Pecos Pueblo.

In what follows, the physical, cultural, and sociopolitical contexts of Pecos Pueblo are explored first, in chapters by Patricia Capone and Katherine Spielmann, who summarize current thinking on external influences that shaped the initiation and evolution of Pecos Pueblo. All the archaeological sites referred to in these and subsequent chapters are shown in figure 1.7. In chapter 2, Capone reviews the landscape ecology of the Pecos Valley and highlights the strategic location of Pecos Pueblo. She discusses the cultural and historical settings of Pecos Pueblo and its nearest neighbors, ending not with the abandonment of Pecos in 1838 but with the repatriation of Pecos collections to the Pueblo of Jemez in 1999.

In chapter 3, Spielmann examines the social and political interactions of Pecos Pueblo with other Pueblo groups and with Plains groups in the Rio Grande area. She explores the history of occupation in Pecos Valley and posits the likely origin of the founders of Pecos Pueblo. She compares two ritual systems, one centered on pueblos to the north and the other maintained by pueblos to the south, that appear to have coexisted at Pecos and at other sites near the interface of these cultural practices. Drawing on both historical sources and archaeological evidence, Spielmann also examines the relations of the Cicuye with the Plains Indians and with Spanish missionaries and soldiers beginning in 1540–1541.

FIGURE 1.5. Alfred V. Kidder at Pecos Pueblo in 1915. (A. V. Kidder Papers, 65-18-10/100077.1.3, Peabody Museum of Archaeology and Ethnology, Harvard University.)

FIGURE 1.6. Earnest A. Hooton in his office at the Peabody Museum, January 1950. (Courtesy Harvard University News Office, negative no. F941.)

The subsequent eight chapters focus on different approaches to studying the human remains that were excavated by Kidder and his associates between 1915 and 1929 and curated at the PMAE until the 1999 repatriation. These chapters reveal what can be learned about the lifestyles of individuals and populations when skeletal data are interpreted within a detailed archaeological and ethnographic context. Specific aspects addressed include diet, health, level of physical activity, fertility, demography, mobility, social networks, biological variation, and conflict.

This volume does not include a comprehensive assessment of health and disease at Pecos Pueblo, with the exception of the mission church burials reported in chapter 11 by Karin Bruwelheide and her colleagues. Data collected during the NAGPRA inventory are summarized in chapter 4 and presented in Appendixes A–D. A detailed analysis of the pathology data on the Pecos Pueblo human remains collected by Beck in 1998 could not be included in this volume, but a preliminary analysis is offered in chapter 8.

In chapter 4 I present the data resulting from the NAGPRA inventory of the human remains from the upper Pecos Valley sites that were excavated by Kidder and his colleagues. I describe the efforts undertaken to verify and establish context for every individual, as well as problems encountered in estimating and reporting the minimum number of individuals for such a large collection. My primary contribution is a series of lengthy appendixes (A–D) presenting improved data about the archaeological context and physical condition of all the human remains that were curated at the PMAE. In chapter 4, basic demographic and pathological data for 1,127 individuals from Forked Lightning and Pecos Pueblo are presented in a series of summary tables and are compared with Hooton's synthesis of nearly 600 individuals (Hooton 1930). Temporal trends during the occupation of Pecos Pueblo are also noted.

The demographic data I offer accord well with the age and sex assignments presented by Katherine Weisensee and Richard Jantz (on the basis of Hooton's measurements) in chapter 5 and by Christopher Ruff (chapter 9), Robert Tague (chapter 10), and Bruwelheide and colleagues (chapter 11). Occasional discrepancies occur, however. Because the human remains

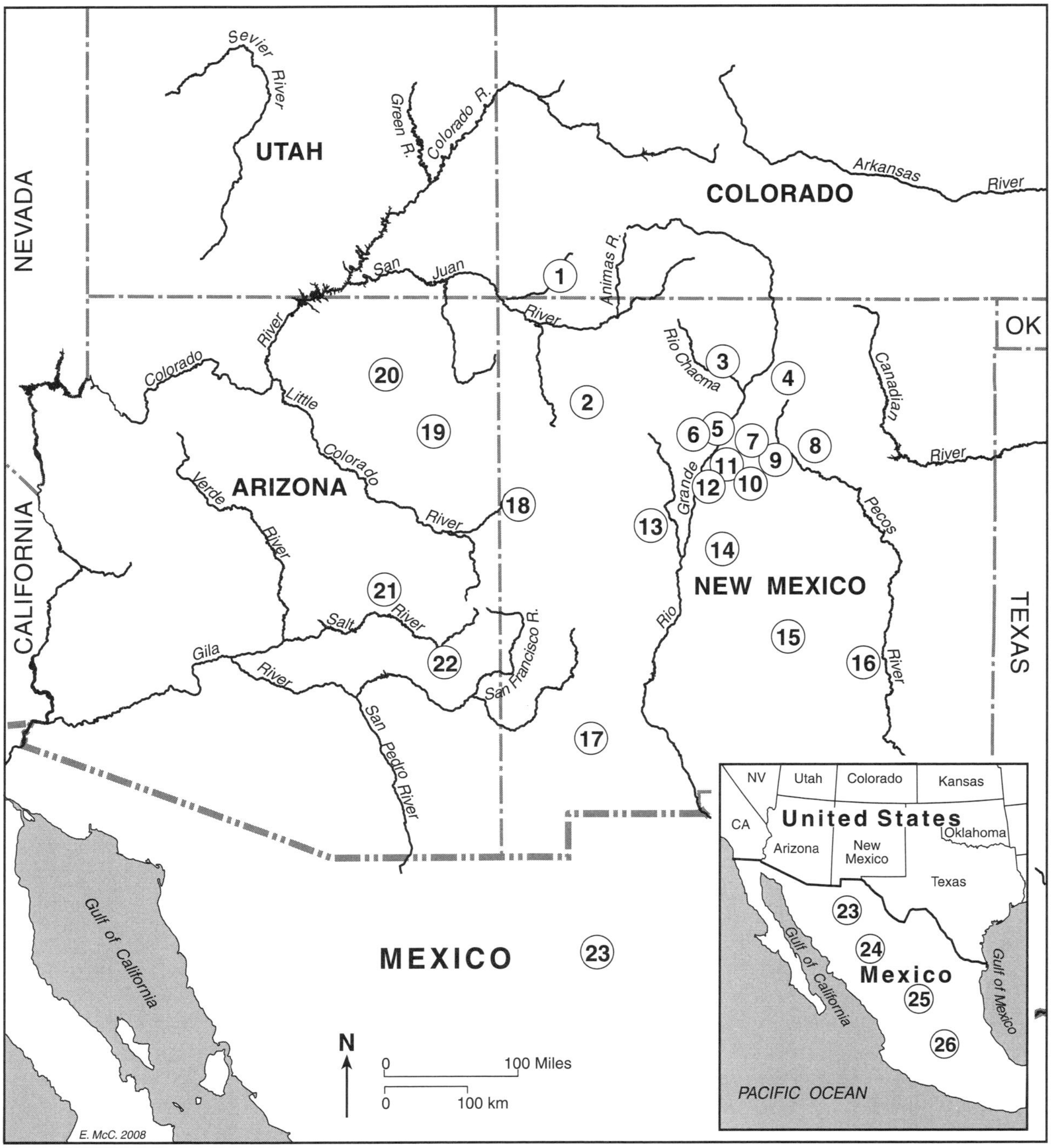

FIGURE 1.7. The southwestern United States and Mexico, showing principal sites discussed in this volume: 1, Mesa Verde; 2, Chaco Canyon (Pueblo Bonito); 3, Chama area (Sapawe); 4, Taos area (Old Picuris); 5, Puye (Pajarito Plateau); 6, Jemez area (Kwasteyu); 7, Arroyo Hondo and Pindi Pueblos; 8, Tecolote; 9, Pecos Pueblo area (Arrowhead Ruin, Dick's Ruin, Forked Lightning, Hobson-Dressler, Loma Lothrup, Pecos Pueblo, Rowe Pueblo); 10, Galisteo Basin (San Cristóbal); 11, San Marcos; 12, Tijeras area; 13, Pottery Mound, Kuaua; 14, Salinas Pueblo area (Gran Quivira); 15, Capitán area (Sierra Blanca); 16, Henderson; 17, Mimbres Valley (Harris, NAN Ranch Ruin, Swarts Ruin); 18, Hawikku; 19, Antelope Mesa (Awatovi); 20, Kayenta ancestral Pueblo; 21, Grasshopper Pueblo; 22, Point of Pines; 23, Casas Grandes; 24, Hidalgo del Parral; 25, Zacatecas; 26, Cuicuilco, Tlatelolco.

could not be reexamined once work on this volume began, these differences could not be resolved. The data sets used by the authors are presented in the appendixes.

The contributors of chapters 5 and 6 examine metric and nonmetric traits in order to address questions of relatedness both internal and external to the Pecos community. The two studies offer insights into patterns of regional movements of people, culture, and language. In chapter 5, Weisensee and Jantz reexamine a subset of Hooton's cranial and postcranial metric and nonmetric data (Hooton's data are presented in Appendix E) in order to address several questions regarding biological and population stasis and change at Pecos Pueblo over time. Their work reaffirms Hooton's claim that Pecos Pueblo comprised a heterogeneous population, and it provides additional details about the pattern of this observed heterogeneity. Further, they illustrate the continued utility of well-documented data sets for new types of analyses and approaches.

In chapter 6, Christy Turner and Steven LeBlanc present dental nonmetric data on Pecos Pueblo inhabitants (included in Appendix F) that Turner collected in the late 1980s. They compare these with similar data collected on 15 other Southwestern population groups. Turner and LeBlanc use these data along with published cranial nonmetric data to investigate the relationships of linguistic and cultural similarities and differences with biological markers. They also offer their assessment of the factors primarily responsible for shaping the observed genetic relationship of Pecos Pueblo to its neighbors. Another significant outcome of this study is a clearer understanding of the possibilities and limitations of applications of dental and cranial nonmetric data to regional population studies.

Diet and health are the focuses of the next two chapters. In chapter 7, Corina Kellner and colleagues present new isotopic data and reassess previously collected data (presented in Appendix G). They compare Pecos Pueblo with three other Southwestern sites: Gran Quivira, Grasshopper Pueblo, and Antelope Creek. Interesting differences in access to dietary resources, including maize, meat, and native plants, are presented. A startling outcome is the apparent relative consistency of diet at Pecos Pueblo over hundreds of years, despite significant changes in trading relationships and occupancy by Spanish colonists.

Lane Beck presents statistics on the incidences of various diseases and pathological conditions at Pecos Pueblo in chapter 8. By comparing survey and detailed data that she collected with Hooton's 1930 data, she demonstrates how different excavation, curation, and analytical practices and approaches bias available pathological data. Although a comprehensive analysis of paleopathological data is still in progress, Beck's findings suggest that the incidence of pathology at Pecos Pueblo increased during the historic period, relative to earlier periods, a change that may be partially attributed to the selective abandonment of Pecos Pueblo by healthy people.

Biomechanical analyses of the limbs and pelves of Pecos individuals provide insights into the physical demands of life, work, and childbearing on Pecos inhabitants. Christopher Ruff and Robert Tague both focused their Ph.D. dissertation studies on Pecos Pueblo skeletal anatomy. They have continued to analyze their primary Pecos data as a comparative population standard and to explore new research questions about Pecos. In chapter 9, Ruff examines change over an age series of males and females, focusing on both the upper and lower limbs (data are presented in Appendix H), in order to address questions regarding sex differences in behavior, dimorphism, growth, and aging. In chapter 10, Tague compares the pelvic anatomy of Pecos Pueblo females with that of females from five other North American populations (data are presented in Appendixes I–K) and suggests that Pecos Pueblo females had relatively greater obstetrical constraints than other populations.

In the final analytical chapter, Bruwelheide, Owsley, and Jantz present an in-depth analysis of the individuals recovered from the mission church constructed at Pecos in the early eighteenth century. These human remains were not included in Hooton's 1930 tome. Bruwelheide and colleagues assess the demography, health, and biological affinities of individuals from the church and compare them with the same variables for individuals recovered from the rest of the site. Drawing on historical sources (Kessell 1987; Levine 1999; Levine and LaBauvre 1997), they consider disease, violence, and the increasing Spanish presence as factors that both influenced the composition of the church burials and affected the eventual abandonment of the site by the Pecos Indians in 1838.

The concluding chapter presents a brief integrative synthesis of the preceding chapters. Links between chapters are explored, and apparent discrepancies between lines of evidence are noted. These discrepancies yield questions for future studies of Pecos and other Rio Grande sites.

Pecos Pueblo has had a tremendous influence on the cultural continuity and change, sociopolitical structure and economy, human biology, and landscape ecology of its region. Although Cicuye occupation of the pueblo ended in 1838, Pecos Pueblo, now a national historic park, remains vital to Native Americans, scholars, and students and is an impressive educational resource for visitors.

2 Pecos Pueblo

The Natural, Cultural, and Historical Settings

Patricia Capone

Pecos Pueblo lies in the valley of the Pecos River, which extends nearly 800 miles from the Sangre de Cristo Mountains all the way to the Gulf of Mexico (figs. 2.1, 2.2). Pecos Pueblo's location is prominent—relatively near the headwaters of the Pecos River in the upper Pecos Valley at the geographical and cultural crossroads between the Great Plains and the Rio Grande Puebloan Southwest. The upper Pecos Valley is bounded by the Tecolote Range of the Sangre de Cristos to the north and east and by Glorieta Mesa to the south and west. The valley forms a pass that serves as a corridor between the southern Great Plains and the Rio Grande Valley.

Pecos Pueblo rises in masonry and adobe in the shadow of Glorieta Mesa (fig. 2.3). The site is currently administered through the National Park Service as part of Pecos National Historical Park, which comprises some 6,000 acres. Pecos Pueblo and associated occupations appear to extend over approximately 365 of those acres (Head and Orcutt 2002:2).

THE NATURAL SETTING

The geological setting of Pecos Pueblo offers ample raw materials and soil suitable for agriculture. In particular, the sedimentary series, including the Magdalena group and Glorieta sandstone, was used in abundance for architectural purposes and for ceramic production, especially between A.D. 1200 and 1600. The local geological signature of Pecos ceramics is readily identifiable even in sites far from the pueblo during this time frame.

At 6,900 feet above sea level, Pecos Pueblo's elevation is high for year-round occupation but offers access to a variety of altitudinal zones over short distances. Elevation largely dictates Pecos's environmental setting. The flora and fauna are characterized as those of the Upper Sonoran life zone (Bailey 1913). Vegetation consists of mixed woodlands to coniferous forest. The vegetation present at Pecos National Historical Park today differs from the plant life that was extant when the pueblo was occupied, especially because the last 50 years of emphasis on ranching and grazing have promoted a greater abundance of grass. Paleobotanical study at nearby Rowe Pueblo offers an idea of the plants that might have been in use at Pecos Pueblo, including amaranth, chenopodium, corn, juniper, piñon, and wild tobacco (Cordell 1998). Relative to other Rio Grande sites, Pecos Pueblo is situated at a high elevation and experiences a low mean annual temperature.

Aspects of climate relating to water and temperature are relevant to the agricultural orientation of Pecos Pueblo. Annual average precipitation today measures 40–50 cm (16–20 inches). Mountain snowmelt is a significant factor in supplying the Pecos River and therefore in irrigation potential. The Pecos

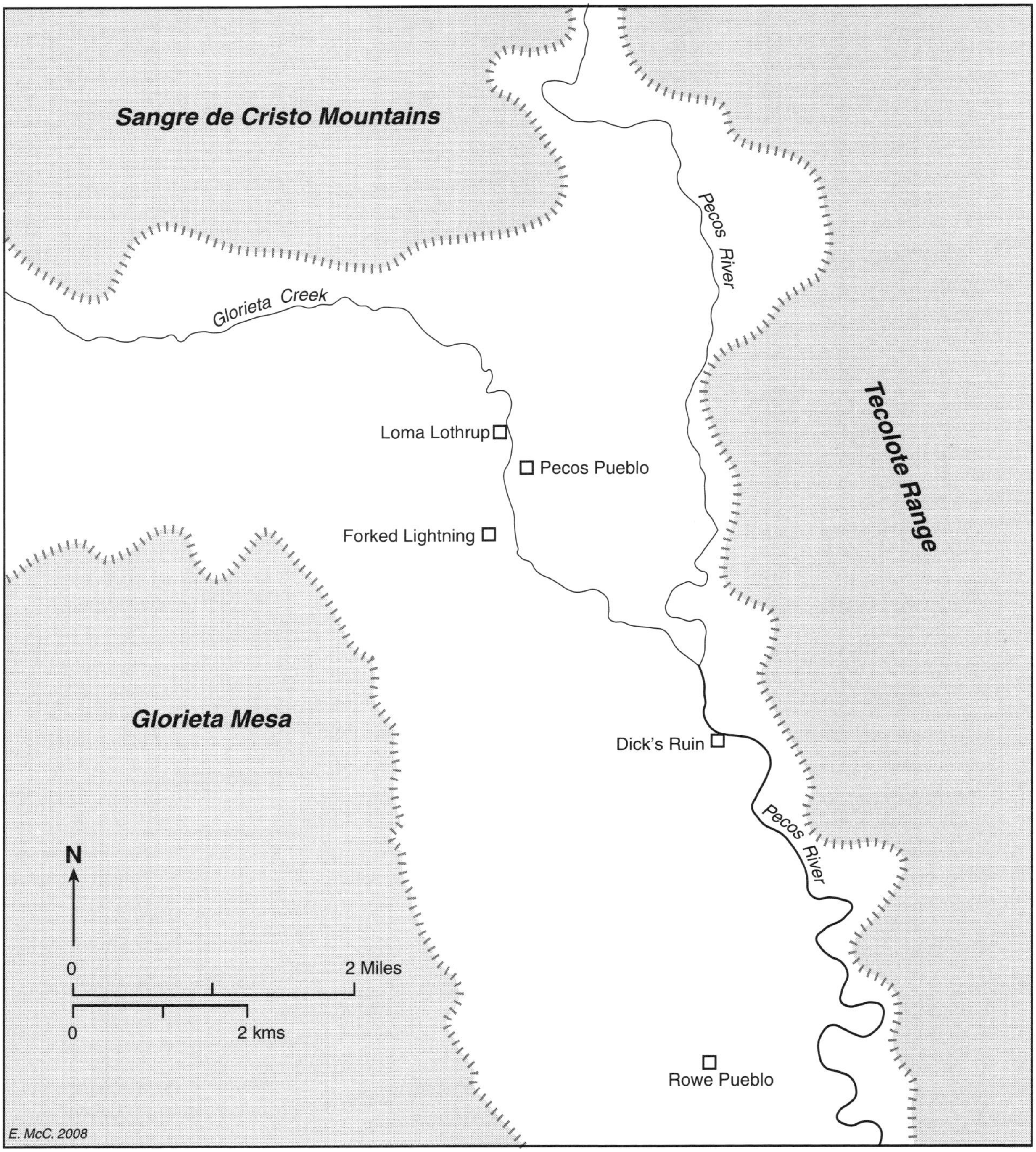

FIGURE 2.1. The Pecos Valley, showing the locations of Pecos Pueblo and associated sites.

River forms the major source of water. The average growing season is approximately 127 days. This figure is so close to the minimum growing season for corn (120 days) that it has been suggested that maize farming at Pecos Pueblo might not have afforded a dependable agricultural strategy (Head and Orcutt 2002:16; Spielmann 1991a:73).

The seasonal distribution of rainfall is a critical factor in agricultural sustainability. Reconstructions of precontact climate suggest that most of the annual rainfall came during the summer growing season, as it does today, and that this pattern remained relatively stable throughout the occupation of Pecos Pueblo. Yet because the balance of agricultural sustainability

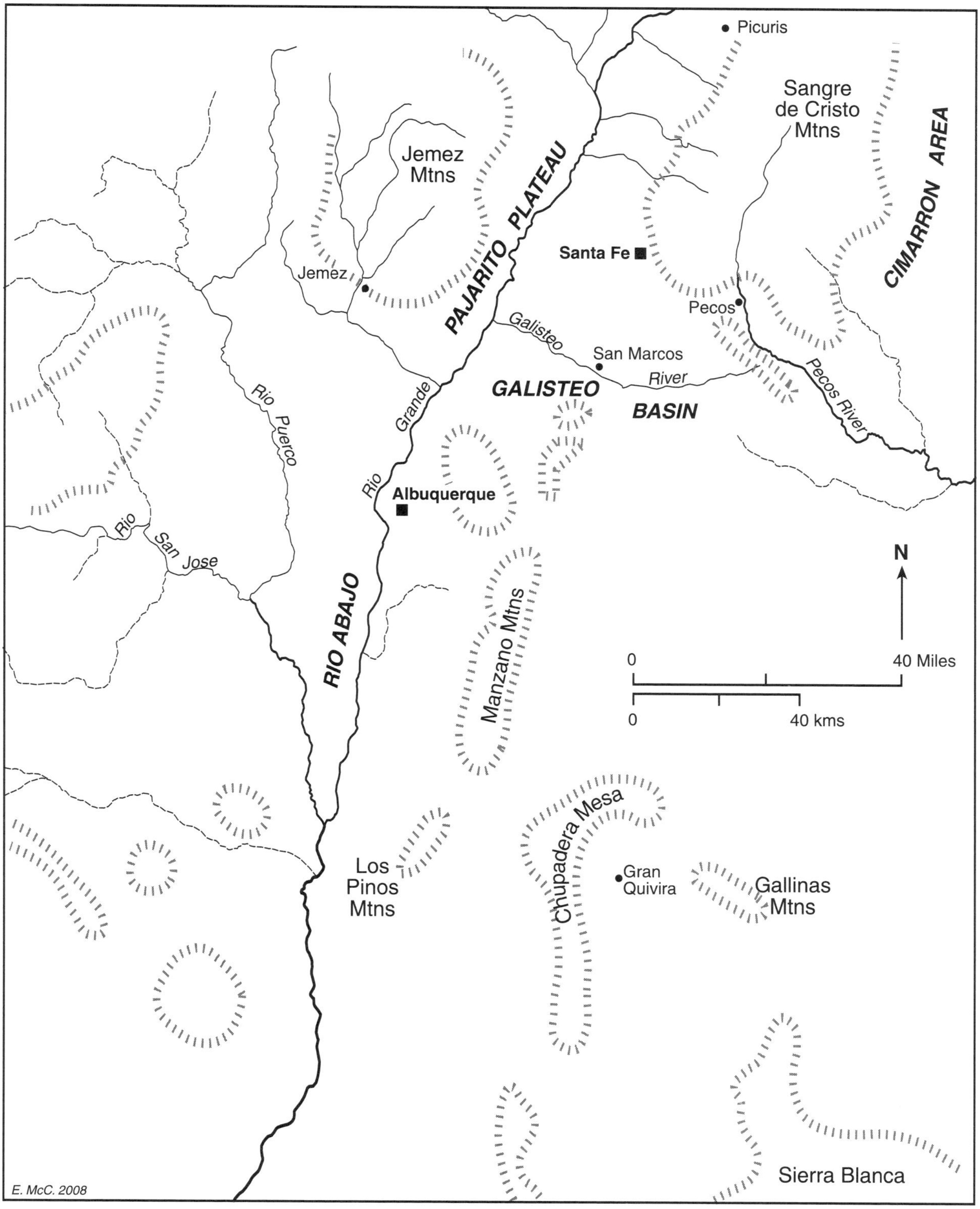

FIGURE 2.2. The northern Rio Grande area, showing the locations of Pecos Pueblo and major landforms.

FIGURE 2.3. Aerial view of Pecos Pueblo, about 1929. (Photograph by Charles Lindbergh, courtesy Palace of the Governors Photo Archives [NMHM/DCA], 130328.)

would have been so delicate, owing to the marginally adequate growing season and rainfall, coping strategies for dry or cold years might have been necessary. Environmental challenges evidenced by tree-ring data suggest that areas northwest of the Rio Grande Valley were subject to an even more variable climate, especially in seasonal rainfall patterning, from A.D. 1250 to 1450, while the Pecos area and much of the Rio Grande Valley remained more stable. This may be a significant factor to consider for Pecos Pueblo and its neighbors (Rose, Dean, and Robinson 1981).

THE CULTURAL AND HISTORICAL SETTINGS

From a cultural historical perspective, the ruins of Pecos Pueblo stand as a monument to the gateway that flourished for hundreds of years between the agricultural Pueblo Southwest and the hunting-and-gathering southern Plains Indian cultures. Pecos Pueblo also remains a vital center in the practices and memories of contemporary Pueblo and Plains people.

Pecos Pueblo is classified as having been Towa speaking; its descendants at the Pueblo of Jemez testify to this. Today, Towa is spoken only at the Pueblo of Jemez, where most of the approximately 3,000 residents are fluent Towa speakers. Pueblo peoples and their relationships to one another have been considered from a number of perspectives, most commonly that of linguistic groupings. Among nearly 20 Pueblo groups in New Mexico, there are six linguistic groupings: Keres, Piro, Tewa, Tiwa, Towa, and Zuni (see Levine 1999:75 for a sample map). Tewa, Tiwa, and Towa represent the Kiowa-Tanoan language family, which is a close linguistic relative to Kiowa, a Plains Indian language spoken by the tribe of the same name. Judging from archaeological and historical evidence, Towa appears to have differentiated from its closest language affiliates, Tiwa-Tewa, prior to historic documentation. Possibly it developed in the Largo-Gallina area in Rio Arriba County of north-central New Mexico sometime between A.D. 1000 and 1275 (Hale and Harris 1979:177). Jemez people describe their origins in the Rio Arriba County area through oral history (Sando 1982).

Although such attempts at categorization can lead to oversimplification, some key points are usefully taken from this framework. Linguistic grouping not only implies a shared language family but also can indicate other shared relationships such as those entailed by sociopolitical or religious societies. The case of Pecos Pueblo and its Towa-speaking relation, the Pueblo of Jemez, is all the more remarkable. The

TABLE 2.1
CHRONOLOGICAL GUIDE TO OCCUPATION OF THE UPPER PECOS VALLEY

Period	Dates	Expression
Preceramic (Paleoindian and Archaic)	11,500 B.C.–A.D. 600	Scant occupation
Developmental	A.D. 600–1200	Initial sedentary settlements, pithouses to adobe: Forked Lightning, Rowe
Coalition	A.D. 1200–1325	Sedentary pueblos: Rowe, Forked Lightning, Black-on-White House, Loma Lothrop, Dick's Ruin
Classic	A.D. 1325–1600	Arrowhead, Forked Lightning, Hobson-Dressler, Loma Lothrop, Pecos Pueblo
Mission	Approx. A.D. 1620–1680	Pecos Pueblo mission
Post-Revolt, colonial	A.D. 1680–1838	European encroachment on Pecos Pueblo land and relocation of residents

Source: Wendorf and Reed 1955.

formal link between the two, grounded in the linguistic relationship and reflected in concomitant social, political, and religious relationships, laid a robust foundation for the eventual repatriation of the human remains excavated from Pecos Pueblo in the early twentieth century.

The formation of Pecos Pueblo and the aggregation of population there are topics of debate. Did neighboring villagers, who were long resident in the Pecos Valley, join together in a consolidated center at Pecos Pueblo? Did immigrants from another part of the Southwest contribute to the growth of population? Or did Pecos Pueblo result from some combination of these factors? For this volume, with its focus on biological aspects of the Pecos population, these questions of ancestral ties and behavioral patterns are pertinent and may be addressed by new perspectives on the data presented throughout the book. The cultural historical framework I present here sets the stage for considering some of these questions and perhaps eliciting new ones. The framework generally employed for ancestral Pueblo contexts in the Santa Fe–northern Rio Grande region is distinct from the well-known Southwestern Basketmaker-Pueblo framework. The upper Pecos Valley archaeological sequence follows Developmental–Classic terminology as detailed in table 2.1.

The upper Pecos Valley appears not to have been favorable for human occupation during the Paleoindian and Archaic periods. Archaeological evidence for occupation during these periods is scant, with just two isolated finds from the Paleoindian period and only a few sites with Archaic remains (Head and Orcutt 2002:5). Human occupation increased during the Developmental period (A.D. 600–1200), with initial evidence for sedentary settlements in pithouse villages (pithouses are shallowly dug, sometimes stone-lined dwellings) and possibly later evidence for settlements in adobe structures at the sites of Forked Lightning (Kidder 1958) and Rowe (Cordell 1998).

For the subsequent Coalition period (1200–1325), significant changes are evident. Diversity is seen in that not all sites conform to a specific set of characteristics. However, the existence of a number of sedentary pueblos of at least 50 rooms each indicates a population increase, agricultural intensification, craft intensification, and development of long-distance exchange networks. It has been suggested that the Coalition period be divided into two phases—an early, or "Pindi," phase, from 1200 to 1250, and a late phase, from 1250 to 1325 (Cordell 1998; Wendorf and Reed 1955).

Notable among Coalition period sites is Black-on-White House, which underlies the subsequent Pecos Pueblo occupation of the Classic and historic periods. Other Coalition sites within a five-mile radius include Dick's Ruin, Forked Lightning, Loma Lothrop, and Rowe Pueblo. Forked Lightning was the largest of the near neighbors during this period. Slightly farther afield is Tecolote Pueblo, also of considerable size (Cordell 1998). Coalition sites show no evidence of an overall site plan. Masonry architecture forms both residential rooms and kivas (semisubterranean ceremonial structures) at some sites, such as Rowe, whereas others, such as Dick's Ruin and Forked Lightning, consist of coursed adobe. Architectural materials may be temporal indicators of early or late occupation. At Forked Lightning, however, that may not be the case, because it was occupied throughout the period but retained coursed adobe. A recent notable study of Rowe Pueblo (Cordell 1998) detailed the current understanding of the late Coalition period from the perspective of settlement aggregation.

Throughout the Coalition period, ceramics were abundant and were characterized by white wares, black-on-white wares, and undecorated wares. These families of wares tend to be low-fired (wood in open pit) earthenware ceramics. White wares usually have a slip of fine white clay covering the surface. Black-on-white wares are similar to white wares with the addition of a decoration painted in black over the white slip surface. Evidence for long-distance exchange of ceramics, turquoise, obsidian, and California shell with southern Plains groups begins to appear at this time (Lintz 1991).

During the Classic period (1325–1600), significant shifts in settlement pattern toward aggregated centers led to the formation of Pecos Pueblo. Arrowhead Ruin, Forked Lightning, Hobson-Dressler, Loma Lothrop, and Pecos are major sites of this period. The Classic also is viewed as a time of diversification in material culture, including painted murals and glaze-decorated ceramics and their associated production and exchange complexes. Architecture grew in scale and formality. Planned, multistory pueblos were set around plazas and kivas. By the mid-fifteenth century, Pecos Pueblo became the sole aggregation center of the upper Pecos Valley. Evidence suggests that populations of the previously mentioned large sites relocated to Pecos. Specific reasons for this shift are unknown, but factors such as defensive strategy and sociopolitical or religious developments have been offered (Levine 1999:8).

The economic and sociopolitical implications of the population shifts are discussed further by Spielmann in chapter 3. Briefly, the factors contributing to these changes are topics of debate. At Rowe Pueblo, for which the results of intensive survey and excavation have been synthesized (Cordell 1998), evidence shows that the local population resided at the pueblo throughout the Developmental and Coalition periods. There appears to be no evidence for immigration during either period. In contrast, an early synthetic study of the Pecos Valley by A. V. Kidder (1958) suggested that the dispersal of population out of the Four Corners area around A.D. 1200 might have contributed directly or indirectly to the increase in population at Pecos. Although others since Kidder have pondered this possibility (Stanislawski 1983), no direct evidence supports it. For this volume in particular, questions relating to the increase and aggregation of population in the Pecos Valley and the biological implications of those trends are noteworthy.

Following a number of exploratory visits to Pecos by Spanish representatives in the mid- to late sixteenth century, sustained contact in the form of missionization began in 1617. Between 1617 and 1680, two mission churches were built, and Pecos became part of the Spanish mission empire. The second church, which was completed by 1629 and was very large, must have required substantial labor. Conflict over religious ideas, labor demands, and other aspects of the mission program have been documented (Kessell 1979). Nuances of the intercultural relationships extant at Pecos have not been well explored beyond the documentary evidence, but works by Levine (1999) and Spielmann (1991a) have created a good basis for future study. Pecos participated in the Pueblo Revolt of 1680 and remained autonomous until the reconquest of 1692. Two additional churches were built after the reconquest. Disease, incursions by Plains raiders, drought, civil difficulties, and subsequent out-migrations all likely contributed to the relocation of the remaining Pecos residents to the Pueblo of Jemez in 1838 (Adams and Chavez 1956:213; Levine 2004:95).

Mexican and United States expansion into New Mexico over the subsequent decades found the site of Pecos Pueblo hospitable. Farming and ranching operations developed nearby. For American settlers on the Santa Fe Trail, a new trading post (1858), built largely from the ruins of Pecos Pueblo, became a welcome oasis (Head and Orcutt 2002:11). The area later was an important local battleground in the Civil War, and the trading post served as a regional Union Army headquarters. Following the development of a major railroad line through Glorieta Pass in 1878, further changes took place in the upper Pecos Valley, with logging, mining, increased farming, ranching settlement, and tourism, until the National Park Service acquired the area in the 1960s. In 1965 the immediate area surrounding Pecos Pueblo was established as Pecos National Monument, and in 1990 an expanded area was redesignated Pecos National Historical Park.

REPATRIATION

Pecos's role in the intellectual history of American archaeology, in groundbreaking scholarship, and in the national reckoning of the Native American Graves

Protection and Repatriation Act (NAGPRA) has been singularly important for anthropology and its relevance to American society. This has been true since the earliest investigations at Pecos in the nineteenth century, the documentation of which served as pivotal evidence in raising consciousness about protecting American antiquities. Kidder's development of the stratigraphic method at Pecos Pueblo in the 1920s provided the basis for broad regional strategies of cultural-chronological synthesis and laid the foundation for the field. In the late twentieth century, far-reaching studies in bioanthropology and medicine based on data gained from study of Pecos Pueblo, paired with reexamination and debate over study ethics and NAGPRA, emerged as an avenue of national reconciliation.

A summary of previous studies at Pecos provides the foundation for understanding its roles. In 1880, early in the practice of archaeological anthropology in the United States, Adolph Bandelier conducted the first systematic archaeological study of the upper Pecos Valley. He contributed descriptive notes, maps, and drawings that documented the site, and he offered the first formal classification of a culture sequence (Bandelier 1881). He reported on the great extent of looting and vandalism at Pecos, which raised concern for American antiquities. Eventually his reports featured prominently in the U.S. Congress's ratification of America's first cultural preservation law, the 1906 Antiquities Act. Archaeology at Pecos became part of the drive for the preservation of natural resources and, through the legacy of the Antiquities Act, part of the underpinning of contemporary society's desire to balance preservation, study, and ethical considerations of multiple interests in America's past.

Pecos became a focus of major archaeological investigation under Alfred V. Kidder, of Phillips Academy in Andover, Massachusetts, from 1915 to 1929. Most of the resulting artifact collections became part of the Robert S. Peabody Museum at Phillips Academy. The large group of human remains (nearly 2,000 individuals) joined the osteological collections at the Peabody Museum of Archaeology and Ethnology (PMAE), Harvard University, and was studied by the prominent biological anthropologist Earnest Hooton. This collection was the subject of Hooton's important 1930 publication, *The Indians of Pecos Pueblo: A Study of Their Skeletal Remains.*

Kidder's work at Pecos was among the most groundbreaking in the young field of American archaeology and focused the field for many years to come. His main reason for choosing Pecos was its long history of occupation, lasting well into the historic period. He also chose it for its historical role as a major trade center, which he thought might allow for the development of a regional correlation for his classificatory and chronological schemes and for a broad synthesis of the diachronic development of cultures in the Southwest (Kidder 1931). His contributions in these areas were substantial (Kidder 1924, 1931, 1932, 1958) and have been widely celebrated (Cordell and Fowler 2005; Willey and Sabloff 1993). Intellectual exchange among Southwestern archaeologists, as inspired by Kidder, continues today through the annual Pecos Conference of Southwestern Archaeology, first convened by Kidder at Pecos in 1927.

One of Kidder's early chronological schemes—the one used by Hooton in his 1930 work—is compared with current interpretations (Powell and Benedict 2002) in table 2.2. Among the developments that led from Kidder's scheme to the current ceramic chronology were tree-ring data, regional chronologies for neighboring areas, analytical studies relating to production sources, and the ceramic probability dating of the large-scale survey of Pecos National Historical Park that was conducted beginning in 1995 (Powell and Benedict 2002). Some of the key ceramic characteristics that are involved in deriving these intricate chronologies are vessel form, raw materials, design style, and type of surface treatment and decoration, such as slip, glaze, and paint.

Various important works have focused on Pecos since those early days, including the previously mentioned studies of aggregation and mutualistic interaction among the Puebloan groups and between them and nomadic hunter-gatherers from the southern plains (Spielmann 1991a). Biological anthropological studies of the large collection of human remains have also been important. Among these are the medically inspired study of aging and osteoporosis that was so significant for understanding the causes and prevention of the disease (Ruff 1991), as well as the essays presented herein. Following the National Park Service's adoption of the site in the 1960s and the addition of associated property in the 1990s, a robust body of data and study has been generated from specialists under the auspices of public management (e.g., Head and Orcutt 2002; Levine, Norcini, and Foster 1994).

Kidder's and others' accomplishments at Pecos set the tone for many decades of significant research based on the site and paved the way for the Pecos archaeological collections to become a superior candidate for early application of the Native American Graves Protection and Repatriation Act of 1990. On May 17, 1999, Native American human remains representing 1,922

TABLE 2.2
CERAMIC STAGE CHRONOLOGY FOR PECOS PUEBLO

Ceramic stage	Description	Chronology presented in Hooton 1930 (A.D.)	Current chronology (A.D.)[a]
Black-on-White	Black-on-white corrugated ware	800–950	ca. 1175–1425
Glaze 1	Red- or yellow-slipped ware with thin black glaze	950–1100	ca. 1325–1425
Glaze 2	Transitional, yellow-slipped ware with thicker black glaze; red-filled decoration appears	1100–1200	ca. 1400–1450
Glaze 3	Light-colored slip, thick blackish glaze lines of poorer quality, red-filled figures	1200–1350	ca. 1425–1490
Glaze 4	Light-red-colored slip, black glaze, overall good quality, figures filled with dark red	1350–1500	ca. 1475–1515
Glaze 5	Thick rims, poor-quality whitish slip, heavy glaze, figures filled with red pigment	1500–1650	ca. 1515–1700
Glaze 6	Poor-quality light-colored slip, thick glaze, no red filling	1650–1700	ca. 1625–1700
Late or modern	Dull paints, slip yellowish	1700–1838	ca. 1700–1838

[a] Dates are from Hooton 1930, Powell and Benedict 2002, and M. Powell, personal communication 1999.

individuals were repatriated from the PMAE and from the Robert S. Peabody Museum of Phillips Academy to the Pueblo of Jemez, home of the modern-day descendants of the people of Pecos. The remains had been excavated as part of Kidder's investigations at Pecos Pueblo and at the neighboring sites of Forked Lightning, Rowe Pueblo, Dick's Ruin, and Loma Lothrop. More than 1,000 funerary objects from the Robert S. Peabody Museum and 23 funerary objects from the PMAE were also repatriated. The funerary objects included antler tools; bead bracelets; bone awls, beads, button, flutes, knives, tubes, and whistles; a brush; buffalo hair; ceramic fragments, ladles, and vessels; chipped stone implements; a clay ball; a concretion; cordage; a crystal; effigies; faunal bone implements; faunal remains; flint chips; fossils; hammerstones; medicine bundle contents; moccasins; obsidian pieces; paint raw materials; pebbles; pieces of copper ore and lead ore; pipes; polishing stones; projectile points; sandals; shell fragments, tinklers, ornaments, necklaces, and pendants; stone axes, drills, knives, palette, pendants, scrapers, and shaft straighteners; textiles; turquoise pendants; wooden and copper crosses; worked shell; and wrappings.

On May 22, 1999, these Native American human remains, along with others from Pecos repatriated from the Maxwell Museum of Anthropology at the University of New Mexico and the Museum of Indian Arts and Culture in Santa Fe, were reinterred as part of a large ceremony held at Pecos National Historical Park (figs. 2.4, 2.5). Even in repatriation, Pecos continues to chart new territory—the repatriation of human remains from the PMAE is the largest to date, and the remains were among the first to be reinterred on federal property under NAGPRA. Many of the repatriated funerary objects, which are under the control of the Pueblo of Jemez, are housed at Pecos National Historical Park through a collaborative relationship.

The reinterment was the culmination of the consultation and research mandated by NAGPRA, which lasted for several years. The essential requirement of NAGPRA is that in order for repatriation of Native American human remains to take place, cultural affiliation must be demonstrated. A finding of cultural affiliation means that a relationship of shared group identity must be reasonably traced historically or prehistorically between a present-day Indian tribe or Native Hawaiian organization and the human remains

FIGURE 2.4. Journey from the Pueblo of Jemez to Pecos Pueblo as part of the repatriation ceremony, May 1999. (Copyright © 1999 *The Santa Fe New Mexican,* reprinted with permission. All rights reserved.)

of an earlier group. In the case of Pecos, most of the human remains in collections were Native American, with a small number of non–Native Americans from the historic period. Non–Native American human remains are not subject to repatriation under NAGPRA. Demonstrating a relationship of shared group identity between the Pueblo of Jemez and Native American human remains from Pecos Pueblo was relatively straightforward. The multitude of evidence for shared group identity between the Pueblo of Jemez and individuals buried at Pecos Pueblo made a strong NAGPRA case. Geographical, kinship, biological, archaeological, ethnographic, linguistic, oral traditional, and historical data all supported a finding of cultural affiliation. Historical documents—ethnohistoric, ethnographic, and linguistic—related the last Pecos residents directly to Jemez and even described the Pecos governmental officials who were to be represented at the Pueblo of Jemez (Levine, Norcini, and Foster 1994; Sando 1982). The official document with the finding of cultural affiliation is on file with the PMAE and the National Park Service (Peabody Museum of Archaeology and Ethnology 1998).

Nevertheless, questions of identity are complex, and the NAGPRA consultation process, which in this case was supported by the National Park Service, rightly requires that everyone with potential for cultural affiliation be given an opportunity to participate. The potential for cultural affiliation of the Pecos remains with other Pueblo and Plains groups, based on a variety of evidence, was explored through consultation and research at the PMAE and the Robert S. Peabody Museum. Ultimately, the process involved the following 11 groups: the Apache Tribe of Oklahoma, the Comanche Tribe, the Hopi Tribe, the Jicarilla Apache Tribe, the Kiowa Tribe, the Mescalero Apache Tribe, the Navajo Nation, the Pueblo de Cochiti, the Pueblo of Jemez, the Pueblo of Santo Domingo, the Pueblo of Zuni, and the Wichita and Affiliated Tribes. There was particular concern that the potential for some Pecos individuals to be of Plains tribal descent be explored on the basis of oral tradition, historical and ethnohistorical evidence for seasonal Plains encampments just outside the pueblo, and potential archaeological evidence for Plains material culture associated with human remains. Each of these lines

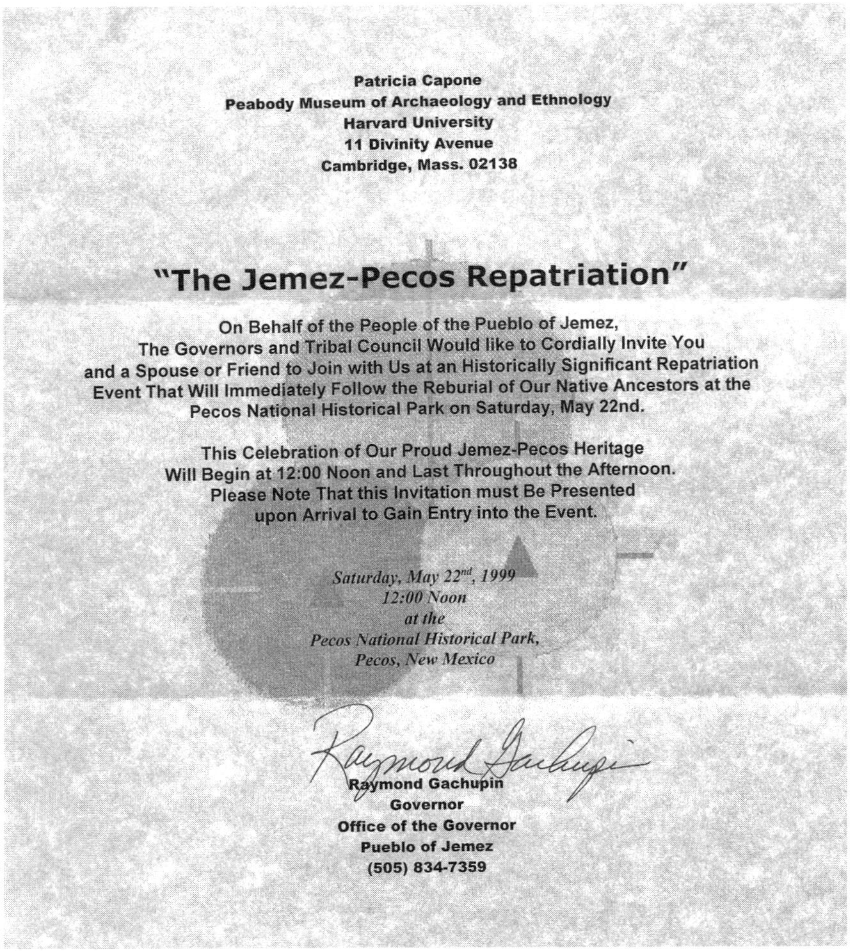

Patricia Capone
Peabody Museum of Archaeology and Ethnology
Harvard University
11 Divinity Avenue
Cambridge, Mass. 02138

"The Jemez-Pecos Repatriation"

On Behalf of the People of the Pueblo of Jemez,
The Governors and Tribal Council Would like to Cordially Invite You
and a Spouse or Friend to Join with Us at an Historically Significant Repatriation
Event That Will Immediately Follow the Reburial of Our Native Ancestors at the
Pecos National Historical Park on Saturday, May 22nd.

This Celebration of Our Proud Jemez-Pecos Heritage
Will Begin at 12:00 Noon and Last Throughout the Afternoon.
Please Note That this Invitation must Be Presented
upon Arrival to Gain Entry into the Event.

Saturday, May 22nd, 1999
12:00 Noon
at the
Pecos National Historical Park,
Pecos, New Mexico

Raymond Gachupin
Governor
Office of the Governor
Pueblo of Jemez
(505) 834-7359

FIGURE 2.5. Invitation to attend the repatriation ceremony at Pecos National Historical Park, May 1999.

was explored, including a detailed morphometric analysis of selected human remains for potential Plains traits and concurrent examination of funerary objects for Plains characteristics, which were performed at the request of potentially affiliated groups. In the end, the preponderance of evidence illustrated that if an individual had been buried in Pecos Pueblo itself and not outside its surrounding wall, middens, and plazas, then the individual was of Pecos Pueblo identity. Strongest among this evidence were the bioanthropological data and the historical accounts that all outsiders, including Plains individuals, had to vacate the pueblo before sunset (Kessell 1979).

The physical repatriation was welcomed as a homecoming by the descendants of Pecos Pueblo and their relatives. The repatriating institutions and Native American partners in the consultations joined with the Pecos descendants to welcome the ancestors back to Pecos Pueblo and to commemorate their reinterment. The events included a several-day, 60-mile walk by members of the Pueblo of Jemez from Jemez to Pecos—a reverse migration just prior to the reinterment. Pueblo members generously shared their story with the mainstream media and with schoolchildren by conducting programs in Cambridge, Massachusetts, public schools to recount their ancestors' homecoming and explain the message of their culture.

The story of the human remains from Pecos, representing some of the ancestry of the Pueblo of Jemez, is a poignant part of American history, the history of science, and the field of anthropology. It is one that will continue to be considered from many perspectives for years to come.

3 The Sociopolitical Context of Pecos

Katherine Spielmann

The occupation of Pecos Pueblo and the activities that took place there are in part unique to that time, place, and people, but they also relate in part to processes and activities common to contemporaneous populations elsewhere in the eastern Pueblo world. In this chapter I discuss some of the more prominent commonalities between Pecos inhabitants and their contemporaries in the larger Rio Grande area. I focus particularly on the history of occupation in the Pecos Valley, relations with Plains Indians, the Pecos people's decisions concerning two large-scale, competing ritual systems, and their experience with Spanish missionization.

RIO GRANDE IMMIGRATION

Our understanding of the history of ancestral Pueblo occupation in the Pecos Valley is enhanced by placing this history in the larger context of occupation and change in the eastern Pueblo area as a whole. That context is one of patchily distributed eastern populations and massive migration across the western Pueblo region. In the east (the Rio Grande Valley and vicinity), an archaeological understanding of occupational history requires a close reading of data concerning the relative contributions of local and immigrant populations to the sizable aggregations that began in the mid-1200s. Here, I briefly summarize relevant data.

Until the A.D. 1200s, much of the eastern Pueblo area was sparsely occupied. Dean and colleagues have estimated that perhaps 5,000 people lived in the entire northern Rio Grande area, from the Galisteo Basin to the Colorado border—roughly 13,000 square kilometers (Dean, Doelle, and Orcutt 1994:fig. 4.9; and see fig. 2.2). The majority of this population was concentrated on the Pajarito Plateau, while populations east of the Rio Grande were much smaller (see Crown, Orcutt, and Kohler 1996; Spielmann 1996a). The eastern foothills of the Sangre de Cristo Mountains (Glassow 1980) and the Taos area (Adler and Dick 1999; Crown, Orcutt, and Kohler 1996) on the north, and the valleys of the Sierra Blancas to the south (Kelly 1984, 1991), supported dispersed pithouse, jacal, and rockshelter habitations. In between, substantial jacal settlements characterized the Chupadera Mesa portion of the Salinas area (Caperton 1981; Chamberlin 2008). Given the available data, however, the Galisteo Basin, Tijeras Canyon, and the Pecos Valley exhibit no evidence of permanent habitations between around A.D. 900 and 1150.

To the west and north of the Rio Grande Valley, beginning in the 1100s but increasing in tempo into the 1200s, massive population movements took place that profoundly affected the subsequent demographic and cultural histories of the eastern Pueblo area. Ancestral Pueblo populations abandoned many portions of the Southwest that they had occupied for centuries; people who had had long-term ties to particular places and landscapes moved well outside their homelands. One gains an appreciation for the scale of population movement during the 1200s by contemplating the abandonment of only a single, relatively populous area, that of Mesa Verde. The pre-migration population estimate for the central Mesa Verde area—which encompasses approximately 5,800 square kilometers, less than half the area of the northern Rio Grande—ranges between 12,000 and 14,000 people, many of whom abandoned the Mesa Verde area only in the later decades of the 1200s (Wilshusen 2002).

Explanations for migration on this scale are diverse and not mutually exclusive. Tree-ring data indicate a marked disruption in rainfall patterns from 1250 to 1450 in the western Pueblo area but not in the Rio Grande, where annual precipitation patterns remained stable (Dean 1996). Ahlstrom and coauthors (1995) defined an environmental gradient from the Mesa Verde area to the Rio Grande, and Kohler and Van West (1996) argued that the Mesa Verde area was not productive enough to support aggregated populations. The Rio Grande region enjoyed greater precipitation and was less affected by the Great Drought, the decline in groundwater levels, and floodplain erosion than the Mesa Verde area. Intercommunity hostilities have also been implicated in the large-scale migrations of the thirteenth century (Haas and Creamer 1996; Kuckelman 2002; LeBlanc 1998, 2000; Plog and Solometo 1997; Wilcox and Haas 1994), perhaps exacerbating or deriving from concerns over access to dwindling resource supplies.

The failure of religious practices to ameliorate these stresses is a third explanation for the massive relocation of Four Corners populations. Lipe (1995) suggested that emerging religious systems to the south might have attracted immigrants from Mesa Verde, who retained few of their ritual practices once they left. Ware and Blinman (2000:401; see also Lipe 2002) argued that a "structural collapse of Pueblo culture" occurred at the end of the thirteenth century, and they postulated that the development of a diversity of ceremonial systems in the fourteenth century could be attributed to a "fundamental reshuffling of ritual beliefs and practices in postmigration Pueblo culture."

Although none of these factors alone appears to be an adequate explanation for the large-scale abandonment of the Four Corners area in the thirteenth century, taken together, "push" factors such as warfare, religious disruption, deteriorating agricultural conditions, and depleted wild resources and "pull" factors (Cameron 1995) such as new ritual systems and better agricultural and hunting conditions appear to explain the shift in the Southwestern demographic center of gravity by the 1300s.

Not surprisingly, given this scale of population movement, most destinations, whether to the west or to the east, were either unoccupied or sparsely occupied prior to immigration (see papers in Adams and Duff 2004). Unoccupied areas provided the opportunity to form new communities rather than adapt to the different cultural systems of local populations, and they probably provided access to better agricultural land than "late-comer" immigrants might have found available if they sought to join existing communities.

The largest, least occupied, but most suitable area for farming in the Southwest was the Rio Grande Valley and its vicinity. Immigrants to this area tended to avoid places occupied by indigenous populations, such as the northern Tiwa, Albuquerque, and Salinas areas. In those places, clear continuity of occupation from the pithouse period onward, continuity in ceramic styles, and lack of evidence for marked increases in population indicate the persistence of indigenous communities. Instead, immigrants selected areas that were sparsely occupied, including the Chama and Jemez areas, the Rio Abajo, the Galisteo Basin, and the upper Pecos Valley.

Interestingly, this large-scale migration took two trajectories. In much of the western Pueblo area, clusters of pueblos were occupied for only a century or so and then abandoned (Spielmann 2004). This short duration of occupation suggests that the immigrants did not develop the same ties to place or people that their ancestors had forged in their homelands. Village clusters in the western Pueblo area appear to have been culturally diverse, in that diverse ceramic technologies and styles persisted alongside one another throughout the occupation of these villages (Kaldahl, Van Keuren, and Mills 2004). As Duff (1998:33) noted, immigrants are more likely to move again, owing to their less favorable position in terms of access to resources such as productive agricultural land and prominent social roles. In many places in the western Pueblo area, migrants did not remain in one location for very long.

In contrast, immigrants to the Rio Grande area appear to have shared some commonality of cultural background and to have pursued a different strategy from those in the west, arriving in larger numbers and establishing relatively large sites from the outset. The relatively rapid construction of large villages and the development of coherent ceramic traditions within clusters of villages suggest active attempts to integrate rather than to differentiate. Aggregation might already have been a way of life in these people's homelands. In the Galisteo Basin, for example, although there is no evidence for "site-unit intrusions"—that is, immigration of populations from specific, identifiable areas—population grew dramatically (Snead, Creamer, and Van Zandt 2004; Spielmann 1996a).

In the 1200s, indigenous populations in the Sierra Blancas (Lincoln phase; Kelly 1991), the Salinas area (Early Masonry period; Caperton 1981; Spielmann 1996a), the Taos-Picuris area (Alder and Dick 1999; Fowles 2004), and the Cimarron area (Cimarron phase; Glassow 1980) also aggregated, perhaps in response to the large number of immigrants arriving from the west (e.g., Fowles 2004).

Where the Pecos Valley population came from is unclear at this point. There is a possibility that people from the immediately surrounding area moved into the valley. Alternatively, immigration from farther west is possible. Although both Kidder in 1958 and Cordell in 1998 noted the lack of Developmental period occupation in the valley, the Pecos Valley is a small area relative to the "districts" just discussed. If one looks farther afield, the Pecos Valley could be considered but one component of a low-density, adobe-pueblo-dwelling, gray-ware-making population that occupied an area from perhaps Santa Fe (early Pindi Pueblo) to the upper Pecos Valley and the eastern foothills of the Sangre de Cristo Mountains (see fig. 1.7 for the locations of these and other sites referred to in this chapter). Stuart and Gauthier (1981:305) noted a relatively high density of sites about 20 miles south of the Pecos Valley along the Pecos River, from present-day Ribera to Villanueva, New Mexico. There are also small villages around Tecolote and the Watrous Valley that date roughly to the 1100s and 1200s. These sites were probably contemporaneous with Forked Lightning, and they share ceramic inventories similar to that of Rowe Pueblo (with the exception of the local Rowe Black-on-white type; Cordell 1998:68–70). The emphasis is on Santa Fe Black-on-white, and Wiyo and Galisteo Black-on-whites are present as well. The establishment of the Rowe and later Pecos and Dick's Ruin villages might have had more to do with the emigration of people from the Ribera settlements and the Sangre de Cristo foothills, where sites appear to have been abandoned by around 1300, than with immigration from the west. Thus, aggregation in place, as in the cases of Salinas and Taos-Picuris, might be a plausible scenario.

Alternatively, this local, dispersed population might have been absorbed into communities established by immigrants from the west. The fact that utility pottery was first imported from the Pajarito Plateau by residents of Forked Lightning (Powell 2002) and only later made locally suggests the possibility of immigration from the Pajarito. Both Powell (2002) and Orcutt (2002) suggested that the marked increases in numbers of habitation sites in the Pecos Valley in the 1200–1325 and 1325–1450 periods were due to immigration, because the growth rate was too high to be accounted for by local population increase. Levine (1999:7–10), too, proposed that immigration, possibly from Arroyo Hondo Pueblo 20 kilometers to the west or from the Galisteo Basin to the south, might have contributed to the Pecos population in the early 1400s. Further research will be necessary to distinguish the relative contributions of local and immigrant populations to the late prehistoric Pecos settlements.

A second period of aggregation occurred in the Pecos Valley in the mid-1400s, after which Pecos remained the sole habitation site. Similar consolidation of population took place at the same time in the Galisteo Basin and the Salinas area (Spielmann 1996a). In these latter areas, however, people chose to live in clusters of large villages rather than in a single village. This difference in settlement organization was likely due to the size of the aggregating population and the distribution of agricultural land in the Pecos Valley. Pecos Pueblo itself is no larger than many of the individual Pueblo IV villages in clusters in other parts of the Rio Grande area, such as Gran Quivira, San Marcos, and Kuaua. Its isolation meant that any benefits that accrued to pueblos in clusters, such as the buffering of resource shortfalls, mutual defense, and the sharing of ritual personnel, were less readily available to the Pecos population.

PLAINS-PUEBLO RELATIONS

Pecos is rightly famous for its interactions with Plains Indian hunter-gatherer populations. Spanish explorers first brought this relationship to light in the documents from the Coronado expedition. Castañeda, the chronicler of Coronado's expedition to the Southwest in 1540–1541, singled out Pecos specifically, mentioning that Plains groups camped outside the village in the winter to trade (Winship 1896:274). Kidder's excavations later substantiated Pecos's relationship with Plains groups. Data from the excavations revealed a marked increase in bison bone in the Pecos middens beginning in the Glaze 4 period (late A.D. 1400s; Kidder 1932:196). During late Glaze 4, and especially during Glaze 5 (1500s), there was an influx of Plains lithic types such as snub-nosed end scrapers and beveled knives, all made of Alibates dolomite from the Texas Panhandle (Kidder 1932:44).

But Pecos was not alone in its relations with Plains Indians. All along the eastern border of the fifteenth-century Pueblo world, Pueblo and Plains populations engaged in trade and other kinds of interactions. Gran Quivira and Picuris, for example, also show marked increases in bison bone, Plains Indian stone tools, and Great Plains shell species during the late 1400s, and Spanish documents discuss Plains hunters trading at these pueblos (Spielmann 1983, 1991a:211–215; but see Harris 1999:103). Although occasional Plains artifacts are found in pueblos farther west, it is in these border pueblos that the greatest concentrations and diversity of Plains materials are found.

The development of close trading ties in the fifteenth century represented a culmination of several centuries of interaction between Southwestern and Plains populations. From the 900s to the mid-1300s, dispersed populations that practiced mixed horticulture and hunting and gathering were patchily distributed across the plains of eastern New Mexico and onto the Llano Estacado of the Texas Panhandle. Sporadic trade in obsidian, turquoise, bison, and Alibates dolomite took place among these populations (Lintz 1991). The only obvious factor distinguishing more Southwestern from more Plains-like people is ceramic technological style. Villagers in northeastern New Mexico produced coiled-and-scraped brown-ware ceramics, whereas those in Texas produced paddle-and-anvil-manufactured, cord-marked pottery (McGuire et al. 1994).

The fourteenth century witnessed a shift in settlement patterns in both the eastern Pueblo and southern Plains areas as populations in the Panhandle began to emphasize hunting over farming and those in New Mexico abandoned their settlements east of the mountains. Glassow (1980:114) proposed that environmental conditions in the Cimarron area limited the degree to which agriculture could be intensified in the face of the dramatic population growth that took place there in the thirteenth century. Jelinek (1967) and more recently Speth (2004:424–426) suggested that increasing bison herd sizes in the late prehistoric period might have made bison hunting more attractive to the peoples pursuing mixed farming, hunting, and gathering economies on the eastern border of the Pueblo world. Thus, some of the eastern mountain farmers might have become hunter-gatherers while others moved west, some to the Taos area and perhaps others into the Pecos Valley (McGuire et al. 1994; Snow 1984). Whether intercultural relations that had developed during the preceding centuries persisted through these changes is unknown.

Recent research, however, indicates that fifteenth-century Plains bands likely maintained long-term ties with specific eastern Pueblo populations (Leonard 2000), perhaps similar to the way in which more recent Jicarilla Apache families maintained inherited trade partnerships with Tewa families (Ford 1972) and intermarried with the people of Picuris (Brown 1999). To investigate whether strong ties existed between specific Plains and Pueblo populations prehistorically, Kathryn Leonard undertook a petrographic analysis of Pueblo glaze ware sherds from two groups of southern Plains sites, those from the Tierra Blanca complex in the northern Texas Panhandle along tributaries of the Red River and those from the Garza complex along the Brazos River to the south.

Her findings indicate that hunter-gatherers occupying the Tierra Blanca sites traded almost exclusively with Galisteo Basin pueblos, whereas those occupying individual Garza sites traded with people living in the Pecos, Salinas, or Galisteo Basin areas. Thus, when the residents of eastern border Pueblos, including Pecos, interacted with Plains hunter-gatherers beginning in the mid- to late 1400s, they did so with specific populations. Trade partnerships are likely to have developed between families, and social and perhaps ritual ties must have been strong between members of these populations.

Judith Habicht-Mauche's analyses of protohistoric southern Plains ceramics substantiate the sociality of ties between Plains and Pueblo populations (Habicht-Mauche 1988, 1991, 2000). Through petrographic analyses, she demonstrated that although the plain ware ceramics found on southern Plains Indian sites closely resemble Pueblo plain ware ceramics, they were not trade wares but instead were made on the plains. The adoption of a Pueblo technology by Plains people suggests close contacts for instruction in ceramic manufacture. She presents a strong argument for the adoption of Pueblo ceramic technology, specifically for processing corn, by southern Plains hunter-gatherers as corn became more important in their diet.

As I discuss later, Pecos's historical preeminence as a Plains Indian trading town focused Spanish governors' attention on the village as they sought to develop lucrative economic enterprises in the New Mexico colony.

FIGURE 3.1. Ceramic bowl in the Salado Polychrome (Gila) Pinedale style. (ASU Anthropology, no. 1959.053.00004.)

RIO GRANDE RITUAL SYSTEMS

The development of new ritual organizations and perhaps belief systems followed upon the massive migrations across the thirteenth-century Southwest. These ritual practices focused on community-wide ceremonies and might have been created to help integrate larger, possibly multiethnic communities (Adams 1991). The farthest-reaching of these ritual systems was what Crown (1994:215–217) termed the Southwestern Cult, which she suggested was particularly concerned with fertility and weather control. Plog and Solometo (1997), in contrast, argued for a greater emphasis on warfare imagery in this ritual system. Crown's data suggest that the cult originated in the Mogollon Rim area of east-central Arizona and ultimately spread through population movements and intermarriage to encompass an area from Hopi to Casas Grandes and from the mountains of Arizona to eastern New Mexico. One hallmark of the Southwestern Cult was the widespread shift from black-on-white ceramics to red- and yellow-slipped vessels and the adherence to some aspects of Pinedale-style design (fig. 3.1). Pinedale designs tend to be symmetrical, with interlocking solid and hatched motifs; motifs are repeated four or fewer times on a vessel (Crown 1994:79). This horizon style developed in the late thirteenth century.

Although the Southwestern Cult began in the western Pueblo area, as is signaled by the development of Salado Polychrome, it spread east rather rapidly. The adoption of red- and yellow-slipped ceramics in the Rio Grande coincided with the adoption of glaze paint, also developed in the west, and with the initial production of Rio Grande glaze ware (Agua Fria Glaze-on-red along with a few other types of Glaze A vessels) in the early 1300s (Spielmann 1998). These vessels, primarily bowls, were significantly larger than the local black-on-white bowls, averaging 30 rather than 20 centimeters in diameter (Spielmann 1998). Kidder (Kidder and Shepard 1936:5) noted that the larger glaze bowls at Pecos would have held at least three times the quantity of the small bowls. This increase in bowl size likely reflected an emphasis on communal feasting in the Rio Grande version of the Southwestern Cult (Spielmann 1998). As glaze-decorated replaced black-on-white bowls, both small and large glaze bowls were made.

Not all populations in the Rio Grande area adopted this new ritual system, however. Populations living north of Santa Fe (likely ancestral Tewa) developed instead a new but still black-on-white ceramic and virtually ignored glaze ware and likely all that it represented (Graves and Eckert 1998; Mera 1934, 1935). This ceramic was biscuit ware, the bowls of which exhibited the same bimodality in size as the glaze ware bowls that were being made and used farther south in the Rio Grande. In the relatively small Jemez area, yet a third ceramic, Jemez Black-on-white, also bimodal in size, was created and persisted for centuries. People living along the lower Jemez River imported some glaze ware, but glaze never represented a significant bowl type in these sites. On the northern Pajarito Plateau, at the interface between the glaze and biscuit heartlands, populations used both biscuit and glaze ware throughout the occupation (Graves and Eckert 1998).

Both of these fourteenth-century black-on-white traditions grew out of the preceding black-on-white traditions of the area, in contrast to the manufacture of glaze ware, which was both technically and stylistically an entirely new approach to ceramic production. Design analyses conducted by Graves and Eckert (1998) document that biscuit ware potters drew on a set of icons and motifs very different from those used by the glaze and Jemez potters. The most frequent biscuit ware motif in their study was the *awanyu* (Graves and Eckert 1998), which is unique to this ceramic type (Kidder 1931:90). *Awanyu* are each composed of a triangle with five pointed appendages, often connected to a "stalk." Glaze ware and Jemez Black-on-white shared a number of the same motifs, though their emphases varied.

Pecos Pueblo was situated right along the "border" between populations using biscuit ware serving bowls and those using glaze ware serving bowls. Like others in this border area, the people at Pecos used both kinds of bowls, perhaps signaling participation in both ritual systems. Kidder's excavations in fact suggest that the two ware types began to be imported into Pecos simultaneously (Kidder 1931:43). Glaze ware, however, and perhaps the ritual system with which it was associated, rapidly gained ascendancy over biscuit ware at Pecos. When first introduced, biscuit and glaze were present in about the same frequencies as the local black-on-white types. Within a few decades (that is, still in Glaze 1 times), glaze increased to 80–85% of the decorated ware, with biscuit making up only 15–20% (Kidder 1931:41). Biscuit's persistence at Pecos is interesting. Whether it represents an ethnic group, a ritual sodality, or some other social unit remains to be investigated. Fuller information on the contexts in which glaze and biscuit wares were used would be helpful. We do know that glaze ware and biscuit ware were present as grave furniture at Pecos in the same frequencies in which they were present in the middens, 80% and 15% respectively (Kidder 1958:table 6, biscuit and glaze bowls only).

SPANISH COLONIZATION

Don Juan de Oñate established the first Spanish colony in New Mexico in 1598. From that time forward, Pueblo populations were faced with Spanish demands on their land, labor, and products and with assaults on their religion. Like many other pueblos in the Rio Grande area, Pecos was missionized fairly soon after colonization. Its proximity to Santa Fe made the Pecos people and resources prime targets of Spanish governors' interests as well.

Politically and geographically the New Mexico colony lay at the very margins of the Spanish Empire, which resulted in unreliable and infrequent access to supplies from Europe and Mexico (Allen 1992; Deagan 2001; Saunders 1998). Lack of supplies was a persistent problem in the colony. Moreover, the soldiers and missionaries who colonized the area brought almost nothing in the way of capital with which to invest in their new locations (Evans 1998). Nor were there many indigenous resources that the Spaniards could extract from the Southwest that were of interest to European markets and thus from which a profit could be made. The lack of mineral wealth in the Southwest, for example, was a significant disappointment for Spanish explorers and colonists.

As a result of these factors, Spanish colonial enterprises were economically viable in the Southwest only because of the labor of indigenous people (Hayes 1974; Sheridan 1992; see also Deagan 1985; Larsen and Ruff 1994; Worth 2001). Consequently, commodities that required labor investment rather than capital investment were the focus of Spanish economic efforts (Kessell 1979).

During the seventeenth century, the products of Pueblo labor served two primary goals: support of the colony, for which they were critical (Sheridan 1992), and export, particularly to the northern Mexican mining communities of Zacatecas and later Parral. Exports served to enhance the wealth of New Mexican governors and to underwrite the supply needs of missions, which exceeded the support of the Spanish crown (Hackett 1937:192; Kessell 1979:101–170; Scholes 1936–1937). Hides were in great demand, particularly by the governors of New Mexico, for export to the mines in Parral, where the hides were made into bags that were used to haul ore (Sheridan 1992; Trigg 1999:293). Thousands of antelope and bison hides were listed among the property of New Mexican governors when they left office (Hackett 1937:188, 192, 248–249). Pecos's connections with the Great Plains made it a prime target for governors trying to obtain hides. Governor Rosas, for example, schemed to acquire additional blankets and hides from the people of Pecos in return for the freedom to practice their own religion. His actions left the Pueblos squeezed between church and state demands and expectations (Kessell 1979:157). The mechanisms by which governors accumulated such large quantities of hides are rarely discussed in the documents. Rosas, however, distributed metal knives at the missions so that they could be traded to the Pueblos for hides (Hackett 1937:130). Friars also traded corn with Pueblo Indians in return for antelope hides (Hackett 1937:192).

In addition, military personnel were awarded *encomiendas,* the right to collect tribute in goods and labor from Native people, in return for services to the crown. Officially, *encomenderos* were allowed to collect tribute in the form of one *fanega* of corn (2.57 bushels, or roughly 100 pounds; Wilson 1973) and one cotton cloth per household per year. A hide could be substituted for cloth (Hackett 1937:120) and often was in game-rich but cotton-poor areas such as the Pecos Valley (Kessell 1979:188). Pecos was a lucrative *encomienda,* providing its *encomendero* with 340 hides a year in tribute, including bison, deer, and elk hides (Kessell 1979:183–189).

More onerous, perhaps, than the demands on Pueblo products were demands on their time and labor made by governors, colonists, and friars. A Salinas Pueblo resident, indeed, attributed the cause of the Pueblo Revolt in 1680 to excessive labor demands by the Spaniards, which prevented the Pueblos from meeting their own subsistence needs (Hackett and Shelby 1942:24–25).

Each mission employed 20 to 40 Indian men and women daily for cooking, cleaning, collecting firewood, and performing religious duties (Hackett 1937:71; Kessell 1979:161; Scholes 1936–1937:158–159). Much greater numbers of Pueblos planted, tended, and harvested Spanish fields (Hackett 1937:96, 112; Scholes 1936:149; Trigg 1999:173). Pueblos also tended the large herds of cattle and sheep that were owned by missions and governors (Scholes 1936:313; Trigg 1999:48, 199).

Residents of missionized pueblos such as Pecos labored to build the enormous churches and *conventos* used by resident friars. Pecos's seventeenth-century church alone took more than 300,000 adobe bricks to build, and the massive amount of labor involved in its construction led to Pueblo complaints that they had no time to cultivate their fields (Kessell 1979:124–126). Pecos residents were also pressed into harvesting lumber and piñon nuts and then transporting them to Santa Fe, and also into knitting stockings from the sheep's wool harvested from flocks at the pueblo (Kessell 1979:171–178; Knaut 1995:103). In addition, early in the 1600s, Fray Andrés brought carpenters to Pecos to teach the men carpentry skills, which were then in great demand, particularly by friars and governors (Kessell 1979:132–133, 177–178).

Although Pueblo people were to be paid for their labor, such payments were rarely received. Lack of payment was a persistent source of tension between Spaniards and Natives in New Mexico, including the carpenters of Pecos (Kessell 1979:321–322; Scholes 1937–1941:67, 394; Trigg 1999:217–220). Indeed, a substantial loss of mission stock was blamed on a governor's decree that Natives working for friars should be recompensed, which friars refused to do (Scholes 1940:66–67).

The returns to Pueblo people for their products and labor were few. Christianity, civilization, and military protection were to be the benefits. Occasional payments for labor and products such as hides, however, do appear to have led to the transfer of some grain and domestic stock from friars to Native people (Hackett 1937:180, 191). In times of famine, the vast mission stores and herds were tapped to feed Pueblo villagers (Hackett 1937:191).

The Pueblo Revolt of 1680 set Pecos somewhat apart from other Pueblo populations in that some Pecos residents made an effort to warn Spaniards of the impending attack, although Pecos did participate in the Pueblo attack on Santa Fe (Kessell 1979:227–234). Pecos's relatively pro-Spanish attitude resurfaced with the reconquest of New Mexico by Diego de Vargas, during which Pecos people joined Spanish soldiers in retaking Santa Fe from the Tewas (Kessell 1979:262). That all did not welcome the Spaniards, however, may be documented in the renovation and expansion of the South Pueblo (also referred to as the Southern House) at Pecos. Kidder (1958:108) documented that the South Pueblo was largely a historic building, and Kessell (1979:132) proposed that a Christianized faction at Pecos was responsible for its construction.

It is unclear what Pecos gained from its association with Spaniards. It might be that certain people, especially those with more power, such as the pueblo's governor, benefited from goods provided by the Spaniards and shared these items with close associates. Moreover, between the Pueblo Revolt and the reconquest, Plains trade at Pecos appears to have waned, making some Pecos people glad of the return of Spanish goods to the trade (Spielmann 1989).

CONCLUSION

Pecos today appears to be a lonely, isolated Pueblo outpost tucked into a mountain pass at the very eastern border of the Pueblo world. Such was not the case in the centuries during which the pueblo thrived. The people of Pecos were active participants in Rio Grande–wide phenomena such as the thirteenth- and fourteenth-century pan-Southwestern immigration, the adoption of new ritual systems, and Spanish colonization. They also were one of several eastern border populations with strong connections to particular Plains Indian bands, connections that lasted for centuries. They were allies with some border people such as the Tiwas of Taos and Picuris and bitter enemies of others, such as the Galisteo Basin Tanos. The static data of the archaeological record allow us to document the energy of these regional-scale relationships.

4 A Reassessment of the Human Remains from the Upper Pecos Valley

Formerly Curated at the Peabody Museum of Archaeology and Ethnology, Harvard University

Michèle E. Morgan

Earnest Hooton's comprehensive *Indians of Pecos Pueblo,* published in 1930, presented a detailed analysis of the human remains excavated under the direction of Alfred Kidder at Pecos Pueblo from 1915 through 1924. This seminal archaeological excavation, sponsored by the Robert S. Peabody Foundation for Archaeology, Phillips Academy, Andover, Massachusetts, took place during 12 field seasons between 1915 and 1929 (Kidder 1958). Kidder set a new standard in archaeology for excavation methodology, and he developed a ceramic type collection based on the pottery excavated from Pecos Pueblo that remains salient in Southwestern archaeology (Cordell 1997; Kidder 1924, 1931; Kidder and Kidder 1917; Kidder and Shepard 1936). The human remains that Kidder and his colleagues excavated became one of the most important twentieth-century sources of information about the biology of past peoples. Not only were the remains studied to further understanding of the biology and sociology of pre- and postcontact Southwest Native Americans, but they also became a standard for interpreting biological variability and physiological adaptations in studies of humans and fossil hominins around the world (e.g., Ruff 1984, 1991, 1995; Tague 1989, 1992, 1994; Tague and Lovejoy 1986; Trinkaus 1976, 1981, 1989).

After several field seasons, Kidder invited Hooton to collaborate on the project and study the human remains (Kidder had been a graduate student at Harvard when Hooton was appointed to the faculty there in 1913). To facilitate Hooton's study, nearly all the skeletal material excavated during the first three field seasons (1915–1917) was transferred from the Robert S. Peabody Museum of Archaeology at Phillips Academy (R. S. Peabody Museum) to the Peabody Museum of Archaeology and Ethnology (PMAE) at Harvard University in 1919. Hooton spent two months in the field with Kidder in 1920, and all subsequently excavated human remains were shipped directly to the PMAE after each field season. Records at the PMAE indicate that nine separate accessions were received and catalogued between 1919 and 1929.

The Native American Graves Protection and Repatriation Act (NAGPRA) of 1990 mandated the production of a complete inventory of all human remains of Native American origin present at the PMAE. In 1994 and 1996 the PMAE and the R. S. Peabody Museum were awarded grants from the U.S. Department of the Interior to jointly accomplish this legally stipulated work on the human remains and funerary objects from Pecos Pueblo and satellite sites. Inventory work on the human remains began in 1996 and was completed in 1998. A collections improvement grant funded by the National Science Foundation in the late 1980s (see Beck, this volume) had resulted in the production of a computer database with basic information on all osteological collections at the PMAE, as well as improved storage conditions. These data provided a useful starting point for the NAGPRA inventory.

My goal in this chapter is to provide a record, with accurate biological and contextual information, for all the individuals from Pecos Pueblo and associated sites in the Pecos Valley (Forked Lightning, Loma Lothrup, Dick's Ruin, and Rowe Pueblo; see fig 2.1) that were accessioned at the PMAE, to serve as a resource for archaeologists and biological anthropologists. Other significant collections of human remains and artifacts from Pecos Pueblo are or were formerly curated by the American Museum of Natural History, the Maxwell Museum of Anthropology at the University of New Mexico, the Museum of New Mexico, the National Park Service at Pecos National Historic Park, and the R. S. Peabody Museum.

METHODS

The primary objective of the NAGPRA inventory was to provide an accurate accounting of all PMAE catalogue records from the Pecos Valley associated with human remains. This required a careful assessment of the minimum number of individuals (MNI) recovered during Kidder's fieldwork that were accessioned by the PMAE, with age and sex information, skeletal element descriptions, and determination of biological affinity. Further emphasis was placed on verifying intrasite provenience and identifying associated funerary objects. These data are presented in Appendixes A–D.

For each museum catalogue record, the corresponding field card was examined just before study of the human remains. The field cards, sorted by field number for each recorded burial, list intrasite provenience, identified ceramic stage, associated funerary objects, and some description of the skeletal material, often accompanied by a sketch. Museum catalogue numbers for the associated objects had often been annotated in pen next to the pencil-written field numbers. Discrepancies between field cards and PMAE records were noted and could often be confidently resolved by examination of the physical remains. In other cases, records were labeled as problematic and marked for additional investigation. After the PMAE completed its documentation, the PMAE and the R. S. Peabody Museum compared records and attempted to resolve remaining problems. The R. S. Peabody Museum was primarily responsible for trouble-shooting discrepancies in intrasite proveniences, associated funerary objects, and attributed ceramic stages.

Age estimates were determined primarily by evaluation of the dentition and appendicular skeleton. Degree of cranial suture fusion was also noted but was not considered to be very accurate (Ubelaker 1989, 1999). For adults, dental wear patterns were compared to Lovejoy's (1985) modal tooth-wear patterns as reproduced in White 1991. Although it is less accurate than root development (Ubelaker 1999), wear pattern evaluation was favored because it could be consistently and rapidly evaluated over the entire sample and was completely noninvasive. The pubic symphysis and auricular surface were scored following the Todd, Suchey-Brooks, and auricular surface standards described in Buikstra and Ubelaker 1994. Accurate age-at-death assessment is problematic for adults, particularly older adults (e.g., Hoppa and Vaupel 2002; Jackes 2000; Milner, Wood, and Boldsen 2000). Reporting indicator stages for several age assessment criteria per individual was strongly recommended by Jackes (2000) and Milner, Wood, and Boldsen (2000) and was done as completeness and preservation permitted. For immature individuals, stage of dental development and measurements of long-bone maximum lengths were recorded as available, and age estimates were based on charts and tables in Scheuer and Black 2000 and Ubelaker 1999. These primary aging criteria are included in Appendixes A, B, and D, along with my best overall age estimate range for each individual. Where dental and postcranial information varied appreciably, postcranial data were generally accorded greater value in arriving at an overall age estimate. A subset of individuals independently assessed by Karin Bruwelheide, Douglas Owsley, Christopher Ruff, Robert Tague, and colleagues accorded closely with my age estimates (see Bruwelheide, Owsley, and Jantz, this volume; Ruff, this volume; Tague, this volume).

Sex was determined primarily by evaluation of pelvic and cranial morphological features as described in Buikstra and Ubelaker 1994. Femoral and humeral head diameters, tibial circumference, relative robusticity, and relative size were also considered. Individuals were placed in one of five categories: male, probably male, female, probably female, and undetermined. In this analysis, sex was attributed conservatively; that is, a high degree of confidence was necessary to score an individual even as probably male or female. Few individuals aged as younger than 14 years received a scoring other than undetermined. The criteria used to evaluate sex are included in Appendixes A and B along with my determination for each individual. Assignments correlated closely with those for individuals independently assessed by Bruwelheide, Owsley, Ruff, and Tague.

Biological affinity was determined primarily by visual assessment of cranial features and relevant postcranial features (Gill and Rhine 1990). When this was insufficient to confidently determine biological affinity, cranial measurements were recorded and analyzed. A series of 83 measurements recommended by Richard Jantz and Douglas Owsley was used. These measurements are described in Howells 1973 and Buikstra and Ubelaker 1994. Richard Jantz and Joseph Powell generously analyzed these data with comparative databases available to them to assess biological affinity. (Hooton's set of craniometric data on 364 individuals was transcribed from PMAE Archives by Katherine Weisensee and is included in Appendix E. See Weisensee and Jantz, this volume, for a reanalysis of these data.)

In some cases, the available skeletal remains were too fragmentary to permit identification. The archaeological context was then considered carefully in making a determination of biological affinity with regard to the NAGPRA regulations. Eight individuals were determined to be of Caucasian biological affinity and were not repatriated under NAGPRA. All of these were examined by at least three osteologists, including me and Owsley, for confirmation. Ten additional individuals, including four identified as being of mixed Caucasian–Native American biological affinity and six identified as possibly Caucasian but which were quite fragmentary, were repatriated under NAGPRA. These 18 individuals were among 176 individuals identified as having been buried within the Pecos mission (or church) built after A.D. 1600 on the southern edge of the pueblo (see fig. 1.2).

A general assessment of pathological conditions was recorded for each individual. These were scored as categories—trauma, infection, arthritis, tumor, nutritional/physiological stress, and other—and briefly described on the basis of examination of all exposed surfaces. Time constraints did not permit detailed documentation of observed pathologies (see Beck, this volume, and Bruwelheide, Owsley, and Jantz, this volume, for additional analyses), but recorded incidences of various conditions may be usefully compared with data for other populations.

The final estimate of the minimum number of individuals (MNI) from Pecos Pueblo and the associated Pecos Valley sites of Forked Lightning, Dick's Ruin, Loma Lothrup, and Rowe Pueblo that was reported to the National Park Service under NAGPRA was 1,908. This estimate included the R. S. Peabody Museum collections that were transferred to the PMAE. It is likely a marked overestimate. A number of individuals that had been given two or more PMAE catalogue numbers and therefore were counted as two or more individuals were reassociated by me (for a total of approximately 100 individuals), but assessing the MNI of the less-well-provenienced human remains catalogued by anatomical region, element, or both was problematic. An exact enumeration of the most frequently represented skeletal element among this group was not feasible, and for NAGPRA accounting the MNI needed to be provided for each catalogue number. The actual number of individuals formerly curated at the PMAE may be closer to 1,400–1,500. Field notes suggest that approximately 2,250 graves were opened, but in many cases the skeletal elements were not removed from the site (Kidder 1958). This is often, but not always, clearly noted on the corresponding field card. As well, more than 275 PMAE catalogue records are not associated with a field card. There are 1,202 catalogue records in the PMAE database that represent single individuals, and 96 records that represent an additional 706 individuals. Of the latter 96 records, 33 describe human skeletal remains excavated from the Pecos mission. A minimum number of elements (MNE) analysis suggests that approximately 30 individuals, not 158, are represented among this group. Similarly, 25 records represent skeletal elements derived from "general excavation." An MNE approach here suggests about 30 individuals, not 289. However, given the much larger area of excavation represented in this category, 30 individuals is likely a low estimate. Another group of 29 records has no intrasite provenience data. Collectively, these 29 records suggest an MNI of 216; however, an MNE approach suggests a number closer to 30, which should perhaps be doubled or tripled given the lack of provenience information and the scale of the Pecos Pueblo site. The remaining 9 records have detailed intrasite provenience data and have a combined MNI of 43; this MNI seems reasonable given all available information.

For each catalogue number, the following were recorded where possible: field number, specimen label, site, intrasite provenience, time period (ceramic stage), MNI, age estimate (years), age category, age criteria, sex, sex criteria, pathology (category), pathology (descriptive), general remarks, cranial modification, biological affinity, recorded measurements, field card notes (regarding burial position and context), associated objects, and object description. All these data are presented in Appendixes A, B, and D. Quantified categorization of the data in Appendix A is given in

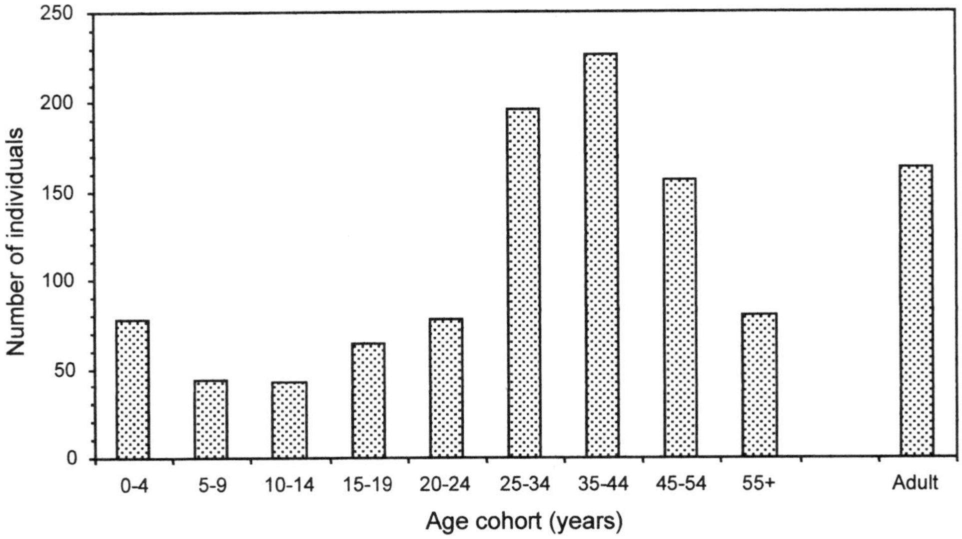

FIGURE 4.1. Mortality profile for Pecos Pueblo and Forked Lightning (n = 1,127). All adult individuals that could not be assigned to a specific age category were placed in the category "Adult."

Appendix C, along with additional data on pathologies. The accurate compilation of these data was the major objective of this work. Additional information for each catalogue record, including an element-by-element accounting of all skeletal elements, is archived at the PMAE. These contextual and observational data, in conjunction with data collected and presented by others in this volume and data amassed by A. V. Kidder, E. A. Hooton, W. W. Howells, and E. Trinkaus, among others, can perpetuate the usefulness for biological anthropologists of the upper Pecos Valley human remains excavated under the direction of A. V. Kidder.

RESULTS AND DISCUSSION

This overview focuses on the 1,017 individuals from Pecos Pueblo and the 110 from Forked Lightning that were confidently associated with a field number and were completely separated from other individuals. The analysis is not intended to be a comprehensive comparison with other Southwestern populations; rather, I compare the primary data with those reported by Hooton (1930) and indicate where strong similarities and differences exist. Some of these differences are attributed to the present inclusion of the human remains excavated after 1924. The 18 individuals identified as Caucasian (n = 8), Caucasian–Native American (n = 4), and probably Caucasian (n = 6) are included in the overall sample, but all are from the Pecos mission group, which is usually summarized separately.

Demographic Assessment

Age Distribution

Hooton reported an age profile for Pecos Pueblo based on 587 individuals, of which 19.4% were immature (< 20 years of age), 70.6% were adult (≥ 20 years of age), and 10% were unattributed (Hooton 1930:19). The mortality profile for 1,127 individuals using the age cohorts of Houdek, Buikstra, and Stojanowski (2000) is presented in figure 4.1. This profile is remarkably consistent with Hooton's: 20.2% are less than 20 years of age, and 79.8% are greater than 20 years of age. Revised adult age distributions accord with Ruff's observation (1981:148) that Hooton and Kidder tended to overestimate adult age. However, the number of infants is markedly underestimated in both these age profiles, because the majority of infant burials (probably close to 230 individuals) was not removed from the field (Kidder 1958). On the basis of field notes, Kidder reported 39.7% of 1,772 identified individuals as immature and 61.3% as adult; an additional 51 individuals were not assigned to an age category (Kidder 1958; table 4.1). Underrepresentation of infant burials and overall small sample sizes are problematic for life table reconstructions at nearly every Southwestern archaeological site; these issues have been discussed in detail by Nelson and colleagues (1994). Weisensee and

TABLE 4.1
AGE AT DEATH OF INDIVIDUALS FROM PECOS PUEBLO

Age (years)	No. individuals[a]	% of individuals of known age
0–1	322	18.17
1–3	117	6.60
3–9	120	6.77
10–20	145	8.18
20–30	107	6.04
30–50	772	43.57
50+	189	10.67
Unknown age	51	
Total	1,823	100.00

Source: Modified from Kidder 1958:284.
[a] Includes individuals that were not removed from the site.

Jantz (this volume) present demographic profiles based on a parametric model and calculate survivorship curves for males and females at Pecos Pueblo.

Sex Ratio

Across the sample of 1,127 individuals of all ages, 33.1% were identified as male or probably male, 39.1% as female or probably female, and 27.8% as undetermined (table 4.2). Non-adults accounted for almost half the unattributed individuals; nearly all the unattributed adults were fragmentary in skeletal preservation. The difference in the number of individuals identified as male or probably male and as female or probably female is statistically significant from a 1:1 ratio ($X^2 = 5.681$, $p < .025$, df = 1).

Hooton (1930) found a sex ratio skewed toward males, with male:female sex ratios ranging from 1.33:1 to 1.41:1, depending on how he subsampled the material. The overall sex ratio reported here is 0.85:1, which is nearly identical to Weisensee and Jantz's ratio of 0.86:1 (this volume). Hooton's sex determinations and methodology have been questioned by many researchers (e.g., Ruff 1981, this volume; Tague 1986, this volume; Weisensee and Jantz, this volume). Hooton's primary source of error was his misidentification of females as males, which he recognized as a potential problem (Hooton 1930:26–31). It has been suggested that Hooton was relying on a northern European model and placed more credence in cranial features than in pelvic features (Ruff 1991:150). It is true that the Pecos Pueblo females are cranially much more robust than females from some other Native North American and European populations. Nevertheless, once familiarity with the degree of intra- and intersexual variation in the collection was gained, crania could be confidently sexed in nearly all individuals. It remains puzzling why Hooton misidentified so many Southwestern U.S. female crania. The distributions of males and females at Forked Lightning and Pecos Pueblo are examined in more detail by time period later in this chapter.

Age and Sex Profiles

Integrating age and sex distribution data across the sample reveals a significant difference in the mortality profiles of males and females. Although small changes in fertility rates can have large effects on mortality profiles (e.g., Frankenberg and Konigsberg 2006; Milner, Wood, and Boldsen 2000; Wood et al. 1992), fertility rates do not directly affect observed differences in the proportional representation of males and females. The age distributions of males and females between 15 and 55+ years ($n = 758$) are significantly different ($X^2 = 15.904$, $p < .01$, df = 5). Females significantly outnumber males between the ages of 15 and 24, whereas males significantly outnumber females between 35 and 54 years, as compared with a 1:1 ratio (fig. 4.2). This may reflect variable mortality rates, with higher mortality among young females, perhaps partially associated with reproductive stresses (see Tague, this volume). However, it may also illustrate an underrepresentation of young male mortality, possibly related to differential treatment of males, higher mortality away from the Pueblo due to warfare (LeBlanc 2000), or both.

The Forked Lightning and Pecos Pueblo population can be subdivided into time periods by ceramic glaze ware stage in order to examine trends in age and sex patterns throughout the occupation. Details about the chronology and development of ceramic stages by Kidder and others are described elsewhere (Capone, this volume; table 2.2). There are a number of different ways to combine individual ceramic stages to generate meaningful sample sizes. Hooton (1930) formed four groups: Black-on-white and Glaze 1, Glazes 2 and 3, Glaze 4, and Glaze 5. Here, I use a new set of four groupings that are more evenly distributed over time according to a revised chronology of the ceramic stages (Powell and Benedict 2002; Powell, personal communication): Pecos Pueblo and Forked Lightning Black-on-white, Glazes 1 and 2, Glazes 3 and 4, and Glazes 5, 6, and "Late." Overall, the trends in differential

TABLE 4.2
DEMOGRAPHIC PROFILE OF THE PECOS PUEBLO AND FORKED LIGHTNING SAMPLE FORMERLY CURATED AT THE PMAE

	Adults				Subadults							
Ceramic phase	Total	Males	Females	Undetermined	Total	Males	Females	Undetermined	Children	Infants	Fetuses	All ages
Forked Lightning Black-on-white	65	18	19	28	11	0	2	9	20	14	0	110
Pecos Black-on-white	27	6	19	2	3	1	1	1	0	1	1	31
Pecos Black-on-white–Glaze 1	11	2	6	3	0	0	0	0	0	0	0	12
Pecos Glaze 1	54	24	25	5	3	0	0	3	4	0	0	61
Pecos Glazes 1–2	19	6	9	4	1	0	0	1	1	1	0	23
Pecos Glaze 2	35	14	16	5	1	0	1	0	3	0	0	39
Pecos Glazes 2–3	13	5	6	2	0	0	0	0	1	0	0	14
Pecos Glaze 3	104	42	51	11	13	4	3	6	6	1	0	124
Pecos Glazes 3–4	25	9	11	5	1	0	0	1	1	0	0	27
Pecos Glaze 4	67	28	33	6	6	2	4	0	4	0	0	77
Pecos Glazes 4–5	29	12	15	2	0	0	0	0	3	0	0	32
Pecos Glaze 5	46	20	23	3	6	2	3	1	7	0	0	59
Pecos Glazes 5–6	3	3	0	0	0	0	0	0	0	0	0	3
Pecos Glaze 6	26	13	12	1	2	2	0	0	2	0	0	29
Pecos Glaze 6–Late	2	1	1	0	0	0	0	0	0	0	0	3
Pecos Late	2	0	2	0	0	0	0	0	2	0	0	4
Pecos Mission	148	59	41	48	20	3	9	5	8	0	0	176
Pecos Indeterminate	233	88	119	26	25	9	10	9	27	15	4	303
Total	909	350	408	151	92	23	33	36	89	32	5	1,127

Note: Individuals were assigned to a single glaze ware stage where possible. Individuals attributed to one of two glaze ware stages were included in combined stages. Individuals with a less specific glaze ware designation were included in the Pecos indeterminate category. Age categories are based on specific dental eruption markers, discrete postcranial measurements, or both. Infants include individuals with less than a complete juvenile dentition or a postcranial age estimate of 0–2.5 years. Children include individuals with complete deciduous dentition through the emergence of the second permanent molar or a postcranial age estimate of 2.6–12 years. Subadults include individuals with second permanent molars erupted through the complete eruption of the third permanent molars, a postcranial epiphyseal closure age of 12.1–19.9 years, or both. Adults include all individuals 20 or more years of age.

TABLE 4.3
PERCENTAGES OF MALES AND FEMALES IN THE PECOS PUEBLO AND FORKED LIGHTNING SAMPLE, BY AGE AND CERAMIC STAGE

Age range (years)	Black-on-white			Glazes 1–2			Glazes 3–4			Glazes 5–Late			Summary		
	M (n = 20)	F (n = 36)	Signif.	M (n = 43)	F (n = 46)	Signif.	M (n = 82)	F (n = 99)	Signif.	M (n = 39)	F (n = 40)	Signif.	M (n = 184)	F (n = 221)	Signif.
0 to 4	0.00	0.00	NS	0.00	0.00	NS	0.00	0.00	NS	0.00	0.00	NS	0.00	0.00	NS
5 to 9	0.00	0.00	NS	0.00	0.00	NS	0.00	0.00	NS	0.00	0.00	NS	0.00	0.00	NS
10 to 14	0.00	2.78	NS	0.00	0.00	NS	3.66	1.01	NS	2.56	0.00	NS	2.17	0.90	NS
15 to 19	0.00	5.56	$p < .05$	0.00	4.35	NS	3.66	9.09	$p < .05$	7.69	7.50	NS	3.26	7.24	NS
20 to 24	3.35	13.89	$p < .01$	7.37	13.04	NS	1.43	10.11	$p < .01$	2.56	12.50	$p < .01$	3.27	11.77	$p < .02$
25 to 34	12.90	22.22	$p < .05$	9.67	36.24	$p < .01$	28.45	30.53	NS	26.92	15.83	$p < .05$	22.05	27.70	NS
35 to 44	27.50	29.17	NS	39.16	20.30	$p < .01$	40.24	29.81	NS	33.33	27.93	NS	37.14	27.38	NS
45 to 54	46.25	19.00	$p < .01$	33.33	17.74	$p < .02$	15.45	10.36	NS	18.38	22.08	NS	23.60	15.43	NS
55+	10.00	7.39	NS	10.47	8.33	NS	7.11	9.09	NS	8.54	14.18	NS	8.51	9.57	NS
Total	100.00	100.01		100.00	100.00		100.00	100.00		99.98	100.02		100.00	99.99	NS

Note: Sample includes all individuals of known sex (n = 405) attributed to the ceramic stages. Columns headed "Signif." give significant differences from the binomial distribution.
NS = not significant.

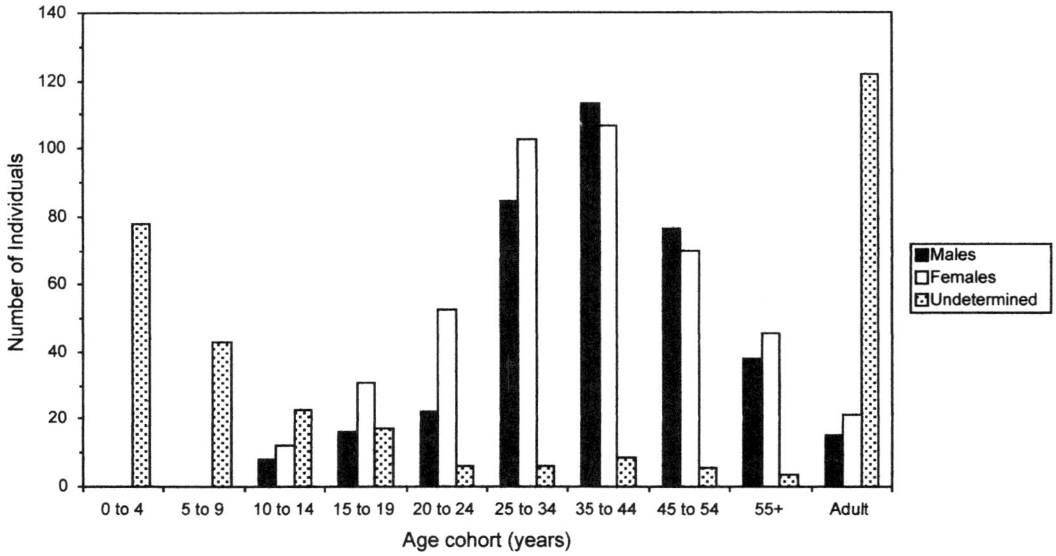

FIGURE 4.2. Mortality profile for Pecos Pueblo and Forked Lightning as numerical distribution by age and sex (n = 1,127). Note the distribution pattern of adult males and females. All adults that could not be assigned to a specific age category were placed in the category "Adult." Males include individuals identified as male and probably male; females include individuals identified as female and probably female.

mortality observed in the combined sample from Forked Lightning and Pecos Pueblo are also seen in analyses by time period.

Comparisons of the percentages of males and females of each age group in a reduced sample of well-provenienced individuals (n = 405) yield statistically significant differences from the binomial distribution for the 20–24 year age group ($p < .02$) and near significant results for the 35–44 and 45–54 age groups ($p \leq .11$) (fig. 4.3, table 4.3). The percentage of females is significantly greater than the percentage of males at the $p \leq .05$ level or higher for the 15–19 year age group (Black-on-white, Glazes 3–4), the 20–24 year age group (Black-on-white, Glazes 3–4, Glazes 5–Late), and the 25–34 year age group (Black-on-white and Glazes 1–2). Similarly, males are significantly overrepresented in the 25–34 year age group (Glazes 5–Late), the 35–44 year age group (Glazes 1–2), and the 45–54 year age group (Black-on-white, Glazes 1–2). It is interesting to speculate about whether the observed increase in male mortality in the 25–34 year age group in the later phase of Pecos Pueblo occupation (post–1600) might be attributed to changing demographic or behavioral patterns. Weisensee and Jantz (this volume) offer an elegant analysis that calculates hazard rates for males and females during different phases of Pecos's occupation.

Cranial Modification

Forty percent of the partial to complete crania from Forked Lightning and Pecos Pueblo were scored for some form of culturally induced cranial modification likely associated with infant cradleboarding practices (Cordell 1997; Hrdlička 1935; Piper 2002). Observed cranial modification was usually occipital and occasionally lambdoidal (Appendix A). The actual incidence of artificial cranial modification probably exceeded 50%. Of the 677 individuals with at least partial cranial remains, cranial modification was positively recorded for 271 individuals and was recorded as absent for 58 individuals. The remainder (n = 348) were either too fragmentary to be accurately assessed or were not scored. The observed frequency of cranial modification across glaze ware stages was remarkably consistent. In a sample of 365 individuals associated with the four glaze groupings (Black-on-white, Glazes 1–2, Glazes 3–4, Glazes 5–Late), the observed frequency ranged between 39% and 45%. This is similar to Hrdlička's late-nineteenth- and early-twentieth-century observations for the pueblos of Jemez, Isleta, and San Juan (Hrdlička 1935), but it is considerably less than Hooton's observation that more than 90% of crania were culturally modified (Hooton 1930:34, table 3.2). However, excluding individuals with only slight

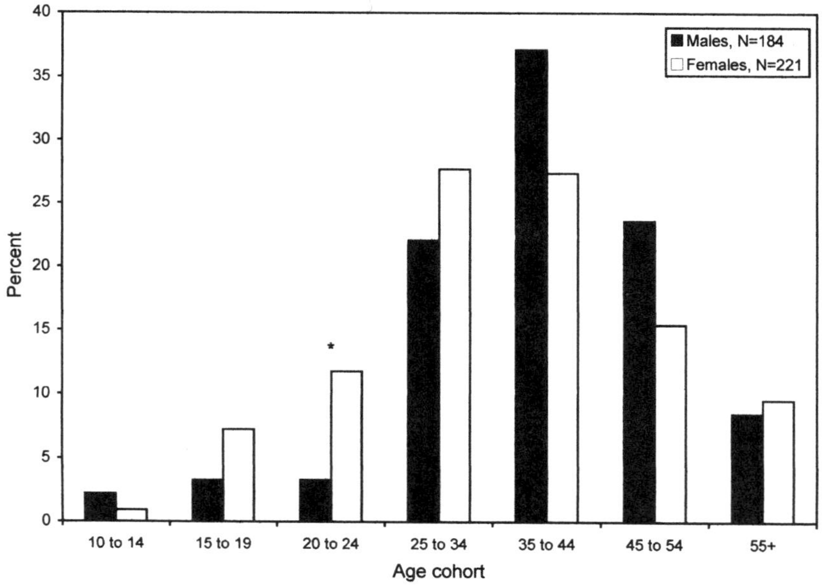

FIGURE 4.3. Mortality profile for Pecos Pueblo and Forked Lightning as percentage distribution by age and sex ($n = 405$). The asterisk indicates a significantly different distribution of males and females than would be predicted by the binomial distribution ($p < .02$). The 25–35 and 35–44 age cohorts approach significance at $p < .11$.

cranial modification, Hooton's figures decrease to 60–70% across a sample of well-provenienced individuals ($n = 291$) (Hooton 1930:33, table 3.1).

Associated Funerary Objects

Funerary objects were associated with 52% of the individuals from the combined sample of Forked Lightning and Pecos Pueblo ($n = 1,127$). They ranged from single ceramic sherds to elaborate combinations of beads, ceramics, lithics, faunal remains, and textiles (see Appendix A). The percentage of human remains from Forked Lightning with associated funerary objects in this sample was 29%, which accords with Kidder's remarks (1958) about the low incidence of imperishable mortuary offerings at Forked Lightning. For Pecos Pueblo, the percentages of well-provenienced human remains recovered with associated funerary objects ($n = 478$) from the four glaze groups (Black-on-white, Glazes 1–2, Glazes 3–4, Glazes 5–Late) averaged 64%, with a range of 60% to 68% across the ceramic stages. There were no significant age or sex differences in the frequency of associated funerary objects relative to the overall representation of human remains.

Health and Disease

In this section I summarize the recorded frequencies of specific skeletal indicators of trauma, infection, arthritis, physiological and nutritional stress, neural arch defects, tumors, and other rare pathological conditions (see Appendix C). These indicators do not represent "actual" frequencies of specific pathologies in the Pecos population, but they can serve as a useful comparative tool as long as their limitations are recognized. Comparing disease frequency data between populations assumes equal selective mortality and risk susceptibility, conditions that are rarely met, if ever (Wood et al. 1992). Biases associated with differential preservation and recovery of human remains (see, e.g., Lovell 2000) also affect assessments of disease frequency and severity. These factors, however, were fairly constant across significant portions of the excavations at Pecos Pueblo.

The majority of the human remains recovered from Pecos Pueblo and Forked Lightning were observed to have at least two types of pathology indicators (57% have two or more; 69% have at least one; $n = 1,127$). The percentages increase when only nearly complete human remains are considered (79% have two or more indicators; 88% have at least one; $n = 206$), highlighting the difficulty of working with a

TABLE 4.4
PERCENTAGES OF INDIVIDUALS FROM PECOS PUEBLO AND FORKED LIGHTNING WITH OBSERVED TRAUMA

Ceramic period or demographic subset	MNI (*n*)	Individuals with trauma	Cranial trauma	Postcranial trauma	Healed trauma	Perimortem trauma
Black-on-white	141	3.55	1.42	2.84	3.55	0.00
Glazes 1–2	122	16.39	1.64	14.75	13.11	0.00
Glazes 3–4	228	17.11	3.95	13.60	13.60	0.44
Glaze 5–Late	98	12.24	5.10	8.16	9.18	1.02
Church	176	19.32	14.20	6.25	9.09	10.23
All other individuals	362	14.09	5.25	9.39	9.94	1.93
Total	1,127	14.29	5.50	9.41	10.03	2.40
Juveniles	218	6.42	5.05	2.29	3.67	3.21
Adults	909	16.17	5.61	11.11	11.55	2.20
Males	373	21.18	8.58	13.67	15.28	3.75
Females	441	15.65	6.12	10.20	10.66	2.27

Note: See text for explanation of categories and Appendix C for data.

large assemblage that contains very fragmentary to nearly complete human skeletal remains. In what follows, I note general patterns regarding the incidence of various pathologies and compare them with Hooton's findings (1930). A more detailed contextual assessment of health and disease in the Pecos mission burials is offered by Bruwelheide, Owsley, and Jantz in chapter 11.

Trauma

Trauma, as recorded by the skeletal system, occurred quite frequently among the people of Pecos Pueblo and Forked Lightning. Nearly 15% (n = 161) of the individuals included in the sample (n = 1,127) showed evidence of healed or perimortem trauma (table 4.4). This is more than twice the percentage recorded by Hooton (4.0% cranial and 3.8% postcranial), although he scored an additional 4.8% as having lesions probably of traumatic origin (Hooton 1930:313–314, tables 10.6, 10.7, 10.9). The ratio of observed postcranial trauma to cranial trauma was nearly 2:1 (106 observations to 62 observations), and approximately two-thirds of all recorded cases were healed traumas (table 4.4). The incidence of postcranial trauma was greatest in Glazes 1–4, whereas the percentage of cranial trauma increased over time. Hooton remarked on a higher incidence of cranial trauma in the later glaze ware ceramic stages and suggested that increased warfare might have been a factor. Stodder and colleagues (2002:493) reported an overall increase in the frequency of cranial trauma in Southwestern pueblos following Spanish contact. At Pecos, the mission sample had the highest incidence of cranial trauma and trauma-caused death (10%), which is discussed in detail by Bruwelheide, Owsley, and Jantz (this volume).

Infection

Several categories of infectious responses were recorded for the cranial and postcranial skeleton (table 4.5). Dental caries and abcesses were both scored as dental infections. Infectious responses due to disease and injury were noted separately, although it was often impossible to identify the primary cause of the infection. The category "general infection" includes all observed infectious responses that could not be ascribed to one of the other categories. Thirty-one percent of the sample (n = 1,127) was scored for at least some type of infection. Twenty percent of the sample was scored for dental infection. This value increased to 26% when only individuals with at least fragmentary cranial remains were included (n = 844) and to 29% when only individuals with at least partial to complete crania were included (n = 789). Juveniles had significantly fewer dental-related infections than adults ($X^2 = 17.39$, $p < .001$, df = 1). Hooton noted a much greater incidence of dental infection (1930:120, table 4.48), with caries present in 48% of examined individuals (n = 253).

When the total sample was analyzed by ceramic stage, the Glazes 5–Late group had the highest inci-

TABLE 4.5
PERCENTAGES OF INDIVIDUALS FROM PECOS PUEBLO AND FORKED LIGHTNING WITH OBSERVED INFECTIOUS RESPONSES

Ceramic period or demographic subset	MNI (*n*)	Individuals with infection	Dental-related	Disease-related	Trauma-related	Other
Black-on-white	141	18.44	12.06	0.71	0.00	7.09
Glazes 1–2	122	39.34	22.95	3.28	1.64	13.11
Glazes 3–4	228	35.09	22.37	3.51	3.07	9.21
Glaze 5–Late	98	43.88	31.63	5.10	3.06	8.16
Church	176	21.02	11.93	3.98	1.14	7.95
All other individuals	362	31.77	22.38	1.93	2.49	8.84
Total	1,127	30.97	20.32	2.84	2.04	8.96
Juveniles	218	10.09	2.75	2.29	1.38	4.59
Adults	909	35.97	24.53	2.86	2.20	10.01
Males	373	40.21	25.47	2.95	3.22	12.06
Females	441	40.59	28.34	4.31	2.27	10.88

Note: See text for explanation of categories and Appendix C for data.

dence of dental infection, at 31.5%. The frequency of caries and abcesses was markedly higher in the Pecos mission sample, as discussed in detail by Bruwelheide, Owsley, and Jantz (this volume). The lowest frequency of dental infection occurred in the Black-on-white stage, with 12%; however, markedly fewer cranial remains were attributed to the Black-on-white period than to the other glaze ware stages. The Glazes 1–4 periods saw a 22–23% incidence of dental disease.

There were no significant differences between males and females in any category of infectious response recorded. Overall, the Pecos population seems to have had a low incidence of caries relative to other, contemporaneous Southwestern populations such as those at San Cristóbal and Hawikku, at each of which more than 50% of all individuals had caries (Stodder et al. 2002).

Arthritis

Arthritis was scored on nearly 38% of individuals (n = 1,127). The overall incidence of arthritis was lowest (19%) in the Black-on-white stage and highest (53%) in the combined sample from the latest glaze ware stages (table 4.6). Several types of arthritic responses were distinguished, including the most common degenerative joint disease (DJD) of the appendicular skeleton, spinal osteophytosis, and the rarer spondylytis deformans and septic arthritis. DJD was qualitatively subdivided into the categories mild and moderate to severe. Approximately 21% of all individuals had mild DJD, and 8% had moderate to severe DJD. Adults aged ≥ 35 years were more than twice as likely to exhibit DJD as adults aged < 35 years (n = 678). Further, 33% of adults ≥ 35 years with DJD expressed moderate to severe DJD, whereas in adults < 35 years with DJD, the incidence of moderate to severe DJD was only 9%.

The incidence of spinal osteophytosis averaged 18% across all periods, but females showed a significant increase in the incidence of spinal osteophytosis from Glaze 1 to Glaze 6 ($X^2 = 19.38$, $p < .001$, df = 3) (fig. 4.4). This might relate to changes in female activity patterns following contact with the Spaniards, such as more intensive maize processing and pottery production (Habicht-Mauche 2000; Spielmann 1996a). Other forms of arthritis, including spondylytis deformans and septic arthritis, affected approximately 5.5% of individuals. In contrast to these data, which suggest a significant health effect of arthritic disease, Hooton reported only a 4% incidence of appendicular arthritis and a 13% incidence of spinal arthritis in the sample he studied (Hooton 1930:307–308, tables 10.1 and 10.2).

Physiological and Nutritional Stress

Four types of physiological insults recorded in skeletal tissues were scored. These were enamel hypoplasia, cribra orbitalia, porotic hyperostosis, and osteoporosis. Enamel hypoplasia was scored in nearly 23% of all individuals and 27% of the well-provenienced individuals (table 4.7). Enamel hypoplasia was most frequently scored on the anterior dentition, although the

TABLE 4.6
PERCENTAGES OF INDIVIDUALS FROM PECOS PUEBLO AND FORKED LIGHTNING WITH OBSERVED ARTHRITIS

Ceramic period or demographic subset	MNI (*n*)	Individuals with arthritis	DJD mild	DJD moderate to severe	Spinal osteophytosis	Other
Black-on-white	141	19.15	12.77	2.13	8.51	2.13
Glazes 1–2	122	44.26	27.87	10.66	21.31	6.56
Glazes 3–4	228	43.42	26.32	8.77	21.49	7.02
Glaze 5–Late	98	53.06	20.41	9.18	30.61	9.18
Church	176	13.07	5.68	3.41	5.68	3.41
All other individuals	362	46.69	25.41	9.67	21.55	6.08
Total	1,127	37.98	20.76	7.63	18.19	5.68
Juveniles	218	3.67	0.46	0.00	0.46	2.75
Adults	909	45.76	25.63	9.46	22.44	6.38
Males	373	52.01	29.22	10.46	28.15	7.77
Females	441	46.71	25.85	10.20	20.86	6.58

Note: See text for explanation of categories and Appendix C for data.

posterior dentition was also affected. Cribra orbitalia, porotic hyperostosis, or both were recorded via external observation in approximately 4% of all individuals and 5% of well-provenienced individuals; these values are consistent with Hooton's reported combined incidence of 4.8% for these pathologies (Hooton 1930:319). Although cranial porosity can be misattributed to anemia, particularly when cranial modification is present (Martinson 2002), the relatively low incidence of 13% for cribra orbitalia and porotic hyperostosis among Pecos individuals < 12 years of age (n = 126), in comparison with other, contemporaneous Southwestern sites, with ranges from 18% to 87% (recorded in individuals aged < 10 years of age; Stodder et al. 2002:493), suggests that this is probably not a major factor influencing the Pecos data. Osteoporosis was recorded in approximately 13% of all individuals; the percentage of affected individuals in the Glazes 1–2 period was higher in both males and females than in subsequent stages, but not at a significant level.

Neural Arch Defects

Neural arch defects were recorded in 86 of 1,127 individuals. Because preservation of the axial skeletal elements varied markedly, it was difficult to achieve reasonable frequency estimates of neural arch defects, especially for nonsacral forms of neural arch defects, which were observed in 16 individuals. Approximately 6% of the population experienced some type of sacral neural arch defect (table 4.8). Partial to complete sacra were present in 587 individuals (52% of 1,127), suggesting that the percentages reported in table 4.8 should likely be doubled to approach realistic estimates. In support of this idea, sacral neural arch defects were noted in 11.9% of individuals with at least partial sacra. Mild sacral spina bifida (1–2 neural arches incompletely fused), also referred to as spina bifida occulta, was distinguished from moderate to severe sacral spina bifida (more than 2 sacral neural arches involved). The recorded incidence of mild sacral spina bifida was just over 4%, and the incidence of moderate to severe spina bifida was approximately 2%. In some populations the frequency of asymptomatic spina bifida occulta may range as high as 10% (Zimmerman and Kelley 1982). In the Pecos Pueblo–Forked Lightning sample, the incidence of spina bifida was significantly greater in males than in females ($X^2 = 5.113$, $p < .03$, df = 1). A similar sex bias was reported by Barnes (1994) for a small sample of sacra from the ancestral Pueblo site at Puye, New Mexico, but no such bias has been recorded for recent samples from South Africa (Buccimazza et al. 1994) and Australia (Riley, Halliday, and Lumley 1998).

Tumors

Two types of tumorous growths were recorded. The first consisted of the apparently benign button osteomas; the second included all other types of tumors.

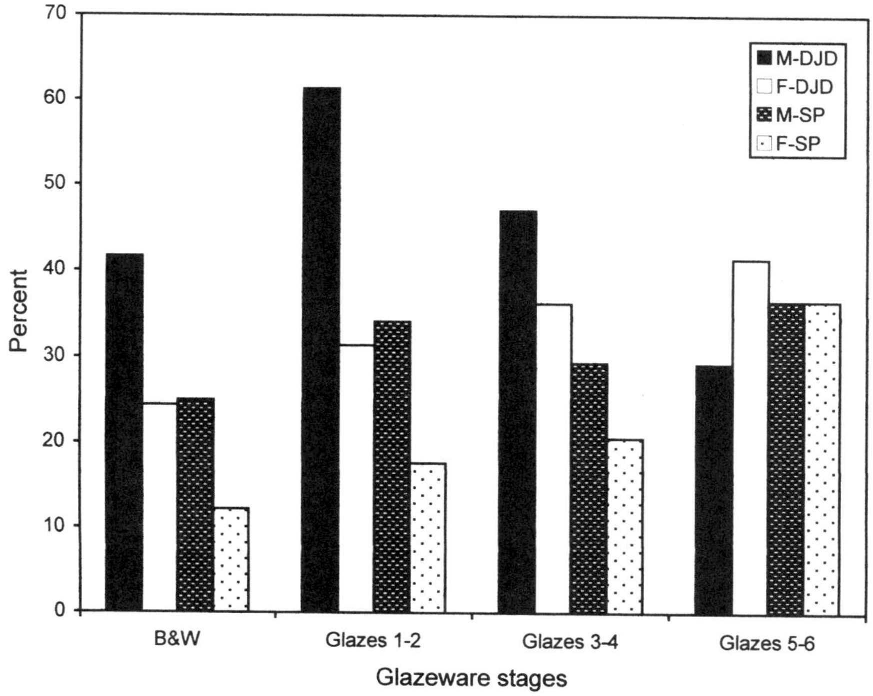

FIGURE 4.4. Percentages of males (M) and females (F) showing degenerative joint disease (DJD) and spinal osteophytosis (SP), by ceramic stage (n = 429, 194 males, 235 females). Females show a significant increase in the occurrence of spinal osteophytosis from earlier to later time periods ($X^2 = 19.38, p < .001$, df = 3). Males have a significantly higher incidence of DJD than females in the Glazes 1–2 time period ($X^2 = 8.575, p < .01$, df = 1).

TABLE 4.7
PERCENTAGES OF INDIVIDUALS FROM PECOS PUEBLO AND FORKED LIGHTNING WITH OBSERVED PHYSIOLOGICAL STRESS, NUTRITIONAL STRESS, OR BOTH

Ceramic period or demographic subset	MNI (*n*)	Enamel hypoplasia	Cribra orbitalia	Porotic hyperostosis	Osteoporosis
Black-on-white	141	24.82	4.26	0.71	6.38
Glazes 1-2	122	22.13	2.46	4.10	21.31
Glazes 3-4	228	34.21	1.75	1.32	15.35
Glaze 5-Late	98	20.41	4.08	1.02	13.27
Church	176	7.95	1.70	0.00	3.41
All other individuals	362	22.93	2.49	2.21	17.68
Total	1,127	22.80	2.57	1.60	13.58
Juveniles	218	20.18	8.26	2.75	0.46
Adults	909	23.54	1.43	1.32	16.72
Males	373	27.08	1.88	1.88	16.09
Females	441	26.76	2.04	1.59	19.73

Note: See text for explanation of categories and Appendix C for data.

TABLE 4.8
PERCENTAGES OF INDIVIDUALS FROM PECOS PUEBLO AND FORKED LIGHTNING WITH OBSERVED NEURAL ARCH DEFECTS

Ceramic period or demographic subset	MNI (*n*)	Spina bifida, slight	Spina bifida, moderate to severe	Other neural arch defects
Black-on-white	141	1.42	0.71	0.00
Glazes 1–2	122	4.10	0.82	1.64
Glazes 3–4	228	3.95	1.75	2.63
Glaze 5–Late	98	5.10	1.02	3.06
Church	176	2.27	0.00	1.14
All other individuals	362	6.08	4.42	0.83
Total	1,127	4.17	2.04	1.42
Juveniles	218	0.92	1.38	0.92
Adults	909	4.95	2.20	1.54
Males	373	7.51	3.75	2.95
Females	441	4.31	1.81	1.13

Note: See text for explanation of categories and Appendix C for data.

TABLE 4.9
PERCENTAGES OF INDIVIDUALS FROM PECOS PUEBLO AND FORKED LIGHTNING WITH OBSERVED TUMORS AND RARE PATHOLOGIES NOT DESCRIBED ELSEWHERE

Ceramic period or demographic subset	MNI (*n*)	Button osteoma	Other tumors	Other dental pathologies	Other cranial and postcranial pathologies
Black-on-white	141	1.42	0.00	3.55	2.13
Glazes 1–2	122	1.64	2.46	5.74	8.20
Glazes 3–4	228	2.63	3.95	6.58	6.14
Glaze 5–Late	98	5.10	4.08	4.08	4.08
Church	176	0.57	0.57	2.28	3.98
All other individuals	362	3.59	3.04	4.42	3.31
Total	1,127	2.57	2.48	4.52	4.43
Juveniles	218	0.00	1.38	4.59	3.21
Adults	909	3.19	2.75	4.51	4.73
Males	373	3.49	4.02	6.17	5.90
Females	441	3.63	2.72	3.85	5.44

Note: See text for explanation of categories and Appendix C for data.

Both of these classes occurred in about 2.5% of the sample (table 4.9). There were no significant differences in frequency by time period. Hooton reported a comparable incidence of 2.2% for button osteomas (1930:315, table 10.10).

Other Pathological Conditions

Last, all the other types of observed pathologies were grouped into two categories: "other dental," which included all the maleruption and developmental dental abnormalities but no infectious responses, and "other cranial/postcranial," which included all observed pathologies not readily ascribed to other categories, such as scaphocephaly and intermediately formed elements (e.g., the sacralization of the fifth lumbar vertebra). The incidence of these two groups was 5.2% and 5.3%, respectively, in the well-provenienced sample, and 4.5% and 4.4%, respectively, in the larger sample (table 4.9).

CONCLUSION

An examination of the basic data on the human remains from Pecos Pueblo and surrounding sites highlights similarities with and differences from Hooton's (1930) landmark study. Many findings, such as the overall mortality profile and the recorded incidence of specific pathologies such as cribra orbitalia and porotic hyperostosis, are similar between the two studies, but others are not. Some of the differences are attributable at least in part to the increased sample size and the refined chronological assessment presented in this chapter. However, they also reflect methodological advances in the field over the last 70 years and the increased availability of comparative studies guiding observational techniques.

One of the most striking outcomes of this study is that the incidence of debilitating trauma, disease, and nutritional stress, relative to that at other Pueblo I to Pueblo IV–historic pueblos, appears to have been low at Pecos Pueblo for most of its occupation. These data are consistent with a relatively prosperous pueblo. However, a number of biological indicators suggest a change in pattern during the later period of occupation at Pecos. For example, cranial trauma, young adult male mortality, dental caries, and arthritis, specifically spinal osteophytosis among females, increased in frequency during the Glazes 5–Late stage (after A.D. 1515). These biological indicators may reflect shifting behavioral patterns with respect to diet, activity patterns, and violence, particularly after contact with the Spaniards, and they warrant further study and comparison with other Pueblo populations. Younger males might have engaged in more violent interactions, or at least more violent interactions near the pueblo. Females might have spent more time tending crops or in other repetitive activities that increased the incidence of spinal arthritis. An increase in dental caries might reflect an increase in maize consumption, a change in maize cultivation and preparation techniques, or both.

The human remains from the upper Pecos Valley excavated under the direction of A. V. Kidder and formerly curated at the Peabody Museum of Archaeology and Ethnology were inventoried in accordance with NAGPRA regulations between 1996 and 1998. Although time constraints did not allow for extensive quantitative assessments, significant demographic and health data were obtained. As well, information about intrasite provenience and associated funerary objects was significantly improved. All these data are compiled in Appendixes A–D to serve as a comparative data set for future studies.

ACKNOWLEDGMENTS

I thank all the PMAE staff with whom I worked during the inventory phase of this project, especially Barbara Isaac, John Stubbs, Patricia Capone, and Anne-Marie Victor-Howe. I also thank the Pueblo of Jemez delegations for their contributions to and interest in the work I conducted throughout the repatriation process. The staff at the Robert S. Peabody Museum of Archaeology, Phillips Academy, particularly Leah Rosenmeier, with whom data were checked and verified, helped throughout the inventory and consultation process. Douglas Owsley, Michael Gibbons, and Catherine Smith gave generously of their time to answer my osteological questions and verify critical data. Richard Jantz and Joseph Powell separately assisted with determination of biological affinity on the basis of craniometric analyses for a number of individuals. Last, I thank Joan Merriman, Steven LeBlanc, and two anonymous reviewers for their valuable comments on this manuscript.

5 Rethinking Hooton

A Reexamination of the Pecos Cranial and Postcranial Data Using Recent Methods

Katherine E. Weisensee and Richard L. Jantz

Earnest Hooton undertook the study of the skeletal material recovered from the Pecos Pueblo site in New Mexico beginning in the early 1920s. His monograph *The Indians of Pecos Pueblo* (1930) continues to be referenced by many researchers, with more than 90 citations since 1970. The many references to this publication demonstrate the continued relevance of the Pecos sample to the field of physical anthropology. Even with the reburial of the skeletons, the Pecos sample still holds research potential, thanks in part to Hooton's thorough data collection during his original investigations. His observations, recorded on a series of data cards, provide the database used in this study to reconsider his conclusions in an updated context.

In his original investigation, Hooton addressed three main questions with regard to the Pecos population. First, he sought to determine whether the variation within the sample corresponded to changes recorded in pottery styles. Second, he hoped to determine whether the changes in pottery styles were the result of immigrants moving into the pueblo or were endogenous to the community. Third, he examined changes in the structure of the Pecos population through time in what was one of the first major paleodemographic studies of a skeletal sample. We use the original questions outlined in the Pecos study as a framework for the results of our reanalysis.

We investigate the variation within the Pecos population in relation to changes in pottery styles by using measures of biological distance between the temporal subgroups. We use Hooton's original data for both craniometrics and nonmetric traits (see Appendix E). Recently developed methods of evaluating cranial morphological variation using quantitative genetics allow for the expansion of Hooton's original work (Konigsberg and Ousley 1995; Relethford 1991; Relethford and Harpending 1994). Multivariate methods of evaluating genetic distances among populations using metric and nonmetric traits permit a broader analysis of the changes that occurred in the Pecos population over time.

The question of migration into the pueblo is addressed by analyzing the variation in the Pecos postcrania in comparison with a sample from a historic Plains Indian population known to have little variation. Variation is measured following the methods outlined in recent research (Key and Jantz 1990; Petersen 2000). The level of variation is used to give an indication of the changes in migration patterns that took place at Pecos Pueblo over time.

Finally, recent advances in paleodemographic research allow for the examination of more complex population models than did the rudimentary techniques Hooton proposed in his study of the demography of the Pecos population. We investigate the paleodemography of the sample using a parametric model to reconstruct an age-at-death distribution. This method was designed to circumvent limitations inherent in

TABLE 5.1
SEX ESTIMATIONS FOR THE PECOS SAMPLE

Observer	Total sample	Male		Female		Unknown		Possible male		Possible female	
		n	%	*n*	%	*n*	%	*n*	%	*n*	%
Hooton	374	228	60.7	125	33.2	9	2.4	9	2.4	3	0.8
Weisensee	358	179	50.0	179	50.0	—	—	—	—	—	—

applying traditional demographic techniques to prehistoric skeletal samples, which typically are too small or too poorly preserved to allow for the construction of standard life tables (Konigsberg and Frankenberg 1992, 2002; Wood et al. 2002). Parametric models for reconstructing the age-at-death distribution have recently been laid out in a series of conferences held in Rostock, Germany (Hoppa and Vaupel 2002).

In the final pages of his book, Hooton wrote, "I think it highly probable that another decade of study will modify and perhaps radically alter many of these opinions. The study of the physical anthropology of the American Indian is still in its infancy" (1930:350). Nearly eight decades have passed since Hooton made this statement, and as predicted, computational and theoretical advances have provided recent researchers with the tools to examine skeletal populations with greater sophistication. The following analysis is an attempt to apply these tools to the Pecos skeletons in order to realize Hooton's prediction.

DOES THE VARIATION WITHIN THE SAMPLE CORRESPOND TO CHANGES OBSERVED IN POTTERY STYLES?

Any discussion of the Pecos skeletal material must begin with a comment on the sex estimation used for the sample. Several researchers (Ruff 1991; Weisensee 2001; Weiss 1972) have remarked on the difficulties of sex estimation for the Pecos sample. The sex estimations used in the following analyses were based on a discriminant function analysis of the metric data. A summary of the revised sex estimates for Pecos individuals is given in table 5.1, and our sex estimates for individuals are provided in Appendix E, along with Hooton's data. For a more complete discussion of the methods used for estimating sex, see Weisensee 2008. All the following analyses of the Pecos sample use the revised sex estimations.

Hooton's first question addressed the correspondence between changing pottery styles and population variation. One major factor complicating his analysis was that the dates of the site had yet to be determined, and he believed that the occupation of the site was much longer than was later concluded. Therefore, in order to maximize the sample sizes for the different glaze ceramic stages, Hooton collapsed several of the glazes into larger groupings: Black-on-white and Glaze 1, Glazes 2 and 3, Glaze 4, and Glazes 5 and 6. All of Hooton's temporal analyses used these modified groupings. However, as his book was going to press, dendrochronological analysis provided evidence of a shorter occupation of Pecos, revising its span downward from the initial estimate of 900 years to the revised estimate of 600 years (Kidder and Shepard 1936). If this information had been available to Hooton earlier, he might have altered the way he divided the periods in his analysis.

Hooton's analysis of the cranial variation was based on standard deviations of the cranial measurements of males in each subgroup. He concluded that only the Black-on-white and Glaze 1 group varied significantly from the entire male sample and that differences decreased over time. From his analysis of the craniometrics, Hooton concluded that "the last periods become more and more typical of the group as a whole and more homogenous" (1930:74). Overall, he reported a tendency over time toward a decrease in facial height and palatal size and an increase in bizygomatic breadth. In the following sections, we present a multivariate analysis of the craniometrics and cranial nonmetric traits based on quantitative genetics in order to provide a broader perspective on the variation and change at Pecos over time.

Methods

An analysis of the biological distance represented in the Pecos sample was carried out using both Hooton's craniometric and cranial nonmetric data (Appendix E). The total sample included 154 individuals, both males and females, assigned to seven periods based generally on the style of pottery associated with the burial. The 10 metric variables used in the analysis were menton-nasion height, alveon-nasion height, maximum bizygomatic breadth, nasal height, nasal breadth, palate length, palate breadth, minimum frontal breadth, orbit height, and orbit breadth (as defined by Hooton 1954). These measurements were chosen because they mainly reflect facial morphology. This was necessary because of the presence of artificial cranial modification throughout the series, which affected the metrics of the cranial vault.

Individuals were assigned to a sex on the basis of a discriminant function analysis. The traits were first corrected for sex-related differences by centering the measurements on a mean of zero, thereby allowing the sexes to be pooled and increasing the sample size. The analyses used to assess differentiation among cranial samples assigned to seven ceramic periods—including the distance matrix, the relationship matrix, and Fst values (the total variation in the population compared with the subpopulation)—were calculated in RMET for Windows 3.0, written by John Relethford. The Fst estimate, or Wright's fixation index, is equal to the average weighted diagonal of the R matrix. Fst gives an estimate of the genetic variation distributed among populations in relation to the total and can range from 0 to 1. Fst is a useful statistic because it can be compared with results from other types of data, including allele frequencies. Comparisons of contemporary, worldwide human populations usually result in Fst estimates near 0.1 (Lewontin 1972; Nei and Roychoudhury 1974), meaning that 10% of total variation is found between populations and 90% of total variation is found within populations. The population sizes of the various ceramic periods at Pecos Pueblo were assumed to be equal for the analysis.

A principal coordinate analysis was also performed using the R matrix, which resulted in a two-dimensional canonical plot of the data, allowing for visualization of the differences among the ceramic periods. Additionally, correlation coefficients were calculated between the original variables and the principal coordinates in order to determine which variables contributed most to the variation along each axis.

Nonmetric trait data for 286 individuals were available from Hooton's data cards. These data included traits related to both the cranium and dentition. FORTRAN code written by Lyle Konigsberg was used to construct a distance matrix. The program used a tetrachoric correlation matrix for determining "an adaptation of Mahalanobis's generalized distance for polygenic threshold traits" (Konigsberg 1990:60). Nineteen of the more than 50 cranial traits that Hooton scored were used in the initial analysis. These traits were chosen because they were scored according to their presence or absence on the crania. The other traits that Hooton scored were based on the degree of expression or size of the trait, such as size of brow ridge or size of chin. The 19 traits were evaluated for age and sex dependence, and on the basis of the results, four traits were removed from the study. Nine traits were ultimately included in the analysis after some further traits were eliminated because of lack of variability or frequent missing values. The traits used in the analysis, as defined by Hooton (1954), were the pterion suture pattern and—scored as present or absent—the following: lambdoid wormian, petrous depression, myloid-hyoid ridge, dehiscences, genial tubercles, postcondyloid foramina, infraorbital suture, and parietal foramina.

A total of 201 individuals was included in the analysis. For traits that were scored bilaterally, one side was randomly selected. A distance matrix among the ceramic periods was calculated. A map of the distances between the periods based on principal coordinate analysis was also constructed. Finally, a Mantel test (Mantel 1967; Smouse and Long 1992) was used to estimate the level of correlation between the cranial metric and nonmetric distance matrices, using Mantel version 3.0, written by John Relethford. Mantel tests are used to calculate the correlation between two or more matrices.

Results

The minimum genetic distance matrix derived from the craniometric data is shown in table 5.2, with the sample sizes given in the first column. The minimum Fst for the sample is 0.037, which is considerably lower than the average Fst value of 0.070 for all of the Americas given by Cavalli-Sforza, Menozzi, and Piazza (1994). The plot of the first two eigenvectors derived from the principal components analysis using the R matrix is given in figure 5.1. The distance matrix and the principal coordinates plot both show the separation of the Black-on-white and Glaze 1 subgroups from other groups and from each other. The first axis separates Glaze 1 from the other groups. Glaze 1 is differentiated on the first axis by large orbital and nasal heights and a broad orbital breadth. The second axis separates Black-on-white from Glazes 4 and 5, with

TABLE 5.2
DISTANCE MATRIX DERIVED FROM CRANIOMETRIC DATA, MULTIPLIED BY 10

Ceramic period	*n*	Black-on-white	Glaze 1	Glaze 2	Glaze 3	Glaze 4	Glaze 5	Glaze 6
Black-on-white	13	0.0						
Glaze 1	16	16.5	0.0					
Glaze 2	19	4.1	18.3	0.0				
Glaze 3	33	4.6	8.5	4.2	0.0			
Glaze 4	38	13.0	16.1	5.4	4.7	0.0		
Glaze 5	19	10.6	12.8	5.3	3.6	3.8	0.0	
Glaze 6	16	8.5	8.6	7.7	6.8	8.6	9.1	0.0

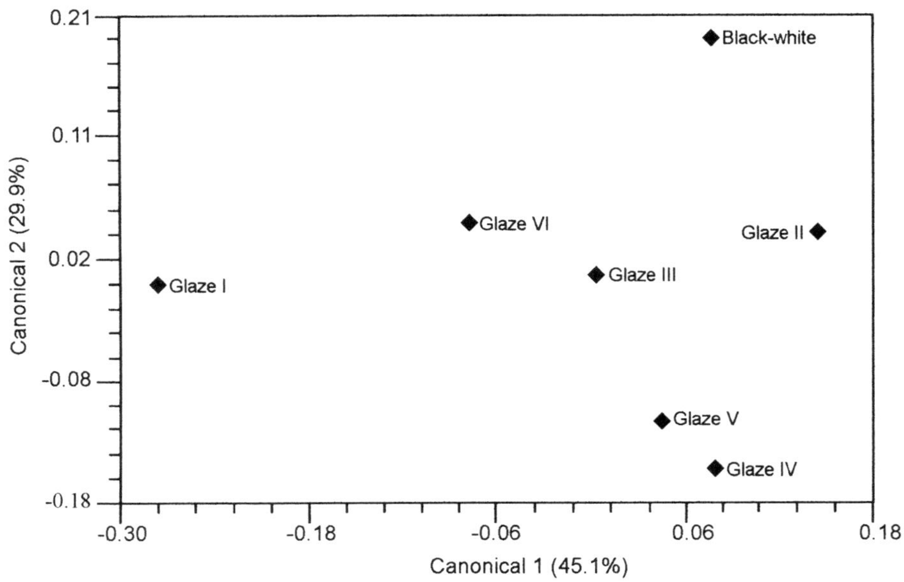

FIGURE 5.1. Principal coordinates plot of distances using craniometric data from individuals corresponding to different time periods. The first axis explains 45.1% of the variation, and the second, 29.9% of the variation.

the other groups clustering in the center. The second axis emphasizes maximum bizygomatic breadth, nasal breadth, and palatal length. The Black-on-white subgroup exhibits a narrow bizygomatic breadth, broad nasal breadth, short facial height, and decreased palatal length. Conversely, Glaze 4 exhibits a broad bizygomatic breadth, long facial height, and increased palatal length.

The distance matrix and sample sizes from the analysis of the nonmetric traits are given in table 5.3. Figure 5.2 shows the principal coordinates plot of the nonmetric data. The plot shows that the first axis separates the Glaze 1 subgroup from the others, with the Black-on-white subgroup at the opposite end of the axis. The second axis separates Glaze 3 from Glaze 5, with the other groups placed intermediately. Figure 5.3 provides a map of the differences between the craniometric and nonmetric data by superimposing the results from the metric analysis over the nonmetric results. The Mantel test demonstrates that the metric and nonmetric distance matrices are significantly correlated with one another ($r = .539$, $p = .037$), indicating that the two distance matrices show similar relationships between the subgroups.

TABLE 5.3
DISTANCE MATRIX DERIVED FROM CRANIAL NONMETRIC TRAITS, MULTIPLIED BY 10

Ceramic period	*n*	Black-on-white	Glaze 1	Glaze 2	Glaze 3	Glaze 4	Glaze 5	Glaze 6
Black-on-white	18	0						
Glaze 1	16	47.8	0					
Glaze 2	29	12.4	30.4	0				
Glaze 3	37	28.8	29.2	15.3	0			
Glaze 4	47	17.7	30.2	7.7	26.2	0		
Glaze 5	31	28.2	32.7	14.8	32.9	14.0	0	
Glaze 6	23	25.7	22.8	16.0	25.9	18.8	11.8	0

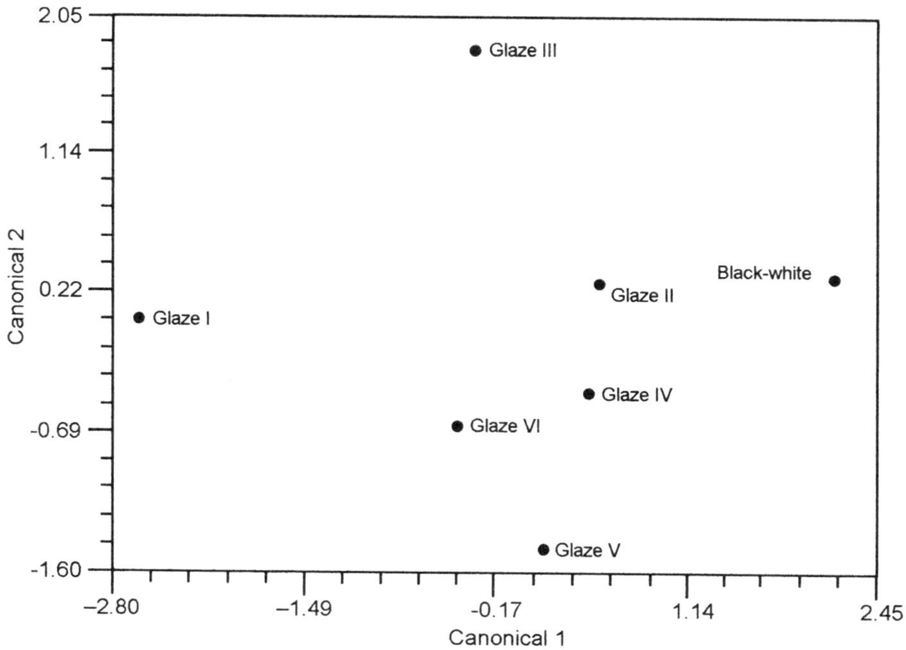

FIGURE 5.2. Principal coordinates plot of distances using nonmetric traits of the cranium and dentition.

Discussion

A comparison of the results of the variation in the craniometric and nonmetric data may be contextualized on the basis of the evolution of pottery styles at the pueblo, because that evolution is the primary source for documenting the chronology at Pecos. Kidder and Shepard (1936) gave a detailed description of the pottery styles and technologies used during the pueblo's occupation. The Black-on-white pottery was found throughout the region, from before the pueblo was established until well after its founding, and was succeeded by glaze wares. The glaze wares remained in use at Pecos for more than 300 years. According to Kidder, "earlier Glazes were almost certainly not of local or even regional manufacture. . . . Glaze I made its appearance suddenly and in fully perfected form" (Kidder and Shepard 1936:1; see Capone, this volume; Spielmann, this volume). This observation suggests that the pottery was introduced either by trade or by foreign immigrants. The abrupt change in pottery style was closely paralleled by an abrupt change in cranial morphology, as seen in the large differences between the cranial samples assigned to the Black-

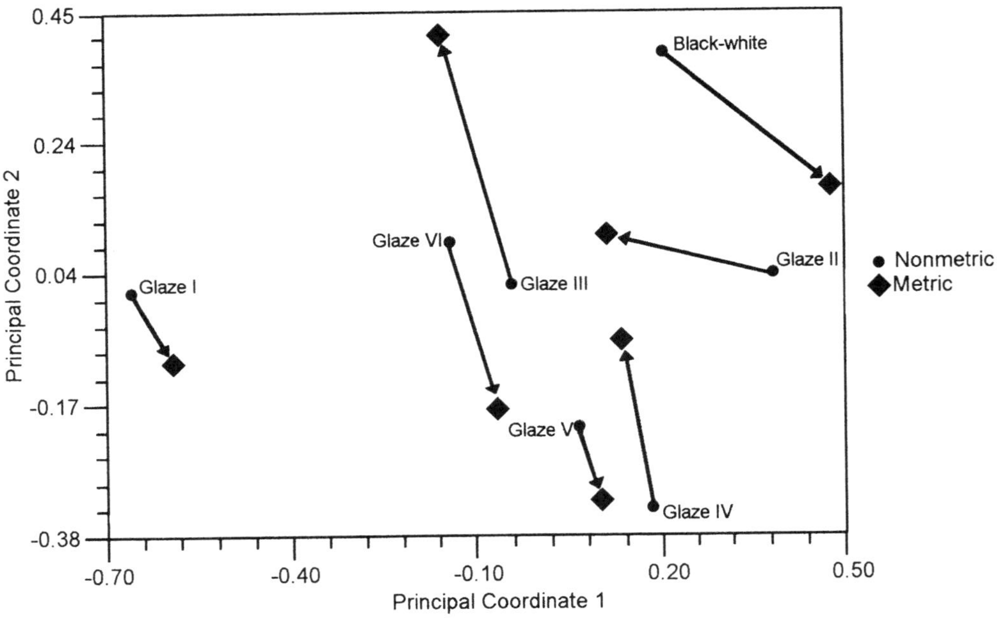

FIGURE 5.3. Procrustes plot of the first two principal coordinates, comparing the distance matrices of craniometric and cranial nonmetric data from different time periods.

on-white and Glaze 1 periods. This finding supports the idea that the Glaze 1 pottery style was introduced by immigrants rather than by local, Black-on-white-manufacturing populations trading for a new pottery style or accepting a new pottery-making technology.

According to Kidder, Glaze 2 represented a transitional ware, possibly resulting as glaze manufacturing began to take place at the pueblo. The distances between the Glaze 2 and Glaze 3 cranial samples in both the craniometric and nonmetric matrices are small, which appears to support the lack of a definitive break between these two pottery styles. The dividing line between the styles known as Glaze 3 and Glaze 4 is more easily defined. Kidder wrote, "Glaze 4 owed its presence to the taking over, lock stock and barrel, of a type developed elsewhere. And I believe that the metamorphosis must have been stimulated and hastened by strong external influences" (Kidder and Shepard 1936:161). Although there is a clear separation between the Glaze 3 and Glaze 4 cranial samples using the nonmetric data, this trend is not apparent in the craniometrics.

Glaze 5 is also distinctive from earlier glazes. It was in use at the pueblo for the longest time, it was the only glaze that originated at Pecos, and it gained only local distribution. The skeletal material classified as dating to the Glaze 5 period is undifferentiated from the Glaze 4 skeletons in both the craniometric and nonmetric analyses. Glaze 6 represents a decline in pottery skills and a degeneration of style (Kidder and Shepard 1936). It was during this phase that Pecos Pueblo was in decline, so there was probably little immigration.

The overall pattern of variability is somewhat different from the pattern that Hooton observed. Our ability to pool the sexes and therefore increase the sample size for each temporal group allows for a closer examination of the variation across all the groups, without having to collapse the phases. The analysis reveals a differentiation between the Black-on-white phase and the Glaze 1 phase that was not recorded in Hooton's results. Furthermore, by combining Glazes 2 and 3 and Glazes 5 and 6 in the initial study, Hooton could not observe differences between the samples in each pair. However, the general trend toward a more homogeneous population that Hooton recorded is observed in the decreasing distance measures in both the craniometric and nonmetric data over time. In regard to the question posed at the beginning of the section, the results suggest that the cranial variation among the subsamples does correspond to the more dramatic changes in pottery styles. The Pecos postcrania can now be used to further compare the amount of variance among the phases over time, to explore the possible causes for this variation, and to address Hooton's second question.

DO MIGRATION EVENTS CORRESPOND TO CHANGES IN POTTERY STYLES?

A second aspect of Hooton's analysis was summarized by Kidder this way: "Is there any evidence that differences in technique [pottery styles] are due to a mingling of peoples; in other words, do subtypes represent the work of a foreign element resident in the village?" (Kidder and Shepard 1936:582). Hooton used the postcranial measurements from the Pecos adult males to investigate changes in robustness and stature within the sample and to evaluate population changes at the pueblo. In reanalyzing Hooton's data, we, too, used his postcranial measurements (see Appendix E) to determine the amount of variation present at the pueblo over time. The changing level of variation within a population gives an indication of the degree to which the population is changing, possibly because of migration, and it may therefore be appropriately used to address the question posed by Kidder.

The amount of variability in a sample gives insight into the makeup of the population by demonstrating the extent to which an archaeological sample is more or less homogeneous than a known sample. Methods for determining the amount of variation within a population by comparing an archaeological hypothesis sample to a reference sample have been proposed by Key and Jantz (1990) and Petersen (2000). The earlier research by Key and Jantz (1990) used multivariate analysis to examine differences in craniometrics between two Plains Indian groups. The authors were able to distinguish between the heterogeneous Leavenworth site, known to have included several different Arikara groups, and a group of related sites in the Bad-Cheyenne region of South Dakota, the Bad River 2 Phase, by comparing the two groups to the homogeneous Larson site.

Later work by Petersen (2000) built on the earlier methods proposed by Key and Jantz. Petersen found that if alternative tests were employed, then the analysis of variability could be applied to small samples and to samples without a multivariate normal distribution, as is often the case with archaeological samples. The methods proposed by Petersen use the determinants of a covariance matrix to analyze population variation. We used Petersen's methods to determine the amount of variability in the Pecos sample.

Methods

Ten postcranial measurements were used to determine the level of variation in the Pecos sample: maximum length of humerus, maximum diameter of humerus at midshaft, maximum femur length, trochanteric oblique length, maximum subtrochanteric anteroposterior diameter, maximum subtrochanteric transverse diameter, maximum anteroposterior midshaft diameter, maximum transverse midshaft diameter, anteroposterior diameter at the nutrient foramen, and mediolateral diameter at the nutrient foramen (Hooton 1954).

Petersen (2000) presented three methods for analyzing the amount of variation within a population, and we followed these methods in our analysis. As defined by Petersen, the methods are Zhivotovsky's *F*-ratio, the Wishart bootstrap, and a nonparametric bootstrap. Petersen recommended using a nonparametric bootstrap test when raw data were available, and we employ this test to evaluate intrapopulation variability. Tests for intrapopulation variability use determinants of covariance matrices in order to compare the variability of a hypothesis sample to a reference sample. In the current study, the skeletal samples associated with the Pecos ceramic phases are used as the hypothesis samples, and the Larson site serves as the reference sample. The Larson site (39WW2), located south of Mobridge, South Dakota, consists of a protohistoric village and cemetery (Bowers 1966). The site had a short occupation, and its skeletal collection ($n = 101$) is a homogeneous population with low population variability. Therefore it provides a good baseline with which to compare the variability at Pecos.

A MATLAB script obtained from Petersen was used in order to perform the following tests. For a complete review of the tests for variability, see Petersen 2000. In general, the values of the determinants can be used to rank the amount of variability among the groups. A determinant ratio greater than 1 indicates that the amount of variability within the hypothesis sample is greater than the amount of variability present in the reference sample, and the opposite is true for ratios less than 1.

Results

The results of the analysis of variation between the Pecos sample and the Larson sample are given in table 5.4. The results include both the determinants and the determinant ratios from the covariance matrices,

TABLE 5.4
RESULTS OF NONPARAMETRIC BOOTSTRAP TEST FOR INTRAPOPULATION VARIABILITY

Sample	*n*	Determinant	Determinant ratio	*p*
Pecos	280	9.80×10^{9}	634.540	.0001
Black-on-white	21	6.01×10^{6}	0.388	.0861
Glaze 1	39	6.66×10^{8}	43.030	.0002
Glaze 2	34	1.85×10^{8}	11.970	.0004
Glaze 3	61	9.80×10^{9}	128.870	.0001
Glaze 4	65	2.09×10^{10}	1,353.000	.0001
Glaze 5	39	3.17×10^{9}	204.740	.0001
Glaze 6	21	2.20×10^{8}	14.210	.0002

Note: Reference sample is from the Larson site (39WW2) in South Dakota; *n* = 101. Number of bootstraps = 9,999.

divided by subgroup. The Black-on-white sample was found to have the smallest determinant value, and Glaze 4 the largest, which establishes the hierarchy of variability. The determinant ratios are greater than 1 for the overall Pecos sample and for every subgroup except Black-on-white. The tests of variability on the Black-on-white sample do not show a significant difference between the amount of variation present in the Larson sample and the amount present in the Black-on-white sample. For all the other subgroups, the determinant ratio is greater than 1, and the tests of variability demonstrate that the Larson sample is significantly less variable than the Pecos samples.

Discussion

An analysis of the variation within the Pecos sample reveals an interesting pattern. In the overall sample, variation is significantly larger at Pecos than in the Larson sample. However, in the earliest subgroup the variation is not significantly different from that at Larson. Variability then appears to increase, reaching a maximum during Glaze 4, and then to decrease. This pattern may be predicted from the archaeological evidence, which indicates that local people were moving into the pueblo during the earlier phases, in contrast to the later phases, when the large pueblo was being built and people may have been arriving from greater distances. It was during the middle phase that Pecos was established as a major trading center between Pueblo and Plains populations, and people likely arrived at the pueblo from farther away during that time.

In Hooton's analysis of the postcranial remains, he reported that "Glaze IV means are somewhat larger and Glaze V and VI means somewhat smaller than those of the series as a whole. It would appear that in gross dimensions the Glaze IV period marked the maximum [body size] of the pueblo population and that in the last two periods a marked or, at any rate, perceptible decline in size occurred" (Hooton 1930:181). In our analysis, the maximum variation in the sample appears during Glaze 4. This is significant in light of the changes observed in the pottery styles discussed in the previous section—namely, that Glaze 4 represents a major break with previous styles and shows strong external influence (Kidder and Shepard 1936:161). The Glaze 4–5 transition is also interesting in light of this analysis, because the determinants show a decline during Glaze 5. Specifically, the Glaze 5 pottery style is unique at Pecos because it was the only one that originated at the pueblo and the only style that had solely a local distribution (Kidder and Shepard 1936:161). The local origin of the pottery technique, along with corresponding reduction in the variation of long-bone size and robustness in adult males, shows how the changes in pottery style are correlated with changes within the population. In answer to Hooton's second question, regarding the correspondence of migration events to skeletal variation, our results seem to bear out the expected increase in variation to a maximum during Glaze 4, the time of presumed maximum immigration at Pecos Pueblo, with a subsequent decrease as the pueblo began to decline. We can now turn our attention to Hooton's final question with regard to the Pecos sample, that of paleodemography.

HOW DID THE DEMOGRAPHIC MAKEUP OF THE PUEBLO CHANGE OVER TIME?

The paleodemography of Pecos is a topic that has been taken up by several researchers (Frankenberg and Konigsberg 2006; Goldstein 1953; Hooton 1930; Mobley 1980; Palkovich 1983). However, given the latest corrections to the overall sex and age structure for the site and new methods for studying paleodemography, the topic deserves further attention. Hooton's intention in his demographic analysis was to determine the size of the population of the pueblo. Goldstein (1953) produced a more comprehensive analysis than Hooton's by applying life table methods to the original sample. In 1980, Mobley expanded on Goldstein's work by including all the burials found at Pecos and the remains uncovered at the nearby Forked Lightning site. He reported continuity between the glaze periods, whereas he had expected to see change because of the influence of immigration into the Pecos community. Mobley hypothesized that a population with a large immigration episode would have "lower subadult and old age mortality and higher overall longevity" (1980:528). Palkovich (1983) critiqued Mobley's results on the basis of Ruff's (1981) findings of the unreliability of the Pecos demographic information. Furthermore, she questioned Mobley's results on the basis of discrepancies she found in the chronology of the site. Palkovich reported a significant difference between the age distribution of the individuals from known time periods and the age distribution of undated individuals. She attributed the discrepancy to differential burials of individuals over time. According to Palkovich, the difference between dated individuals and undated individuals contradicted Mobley's finding of diachronic stability in the Pecos sample (1983:142).

As an alternative to the life table method employed for many years by anthropologists, including by those just mentioned, researchers have recently proposed innovative ways of approaching paleodemography. Using computer simulation and maximum likelihood estimation, researchers have reported success in modeling population structures using parametric models. The limitations of the life table method that the modeling analyses are able to address are fourfold. First, the specific data needed to produce life table mortality rates—data commonly available to demographers—cannot be determined from skeletal data. Second, the assumption in life table analysis that the studied population is stationary, or closed to migration, is often false for archaeological populations. Third, traditional methods are unable to account for the fact that the ages of all skeletons are not known with the same degree of certainty. Fourth, life tables are inefficient at using the limited data available in small skeletal samples (Wood et al. 2002:130).

Three principle functions are used to model mortality: hazard rate (h[a]), probability density function (PDF), and survival function. The hazard rate is defined as the risk that death will occur at a certain age, given that it has not occurred previously. The PDF is complicated but is generally an age-at-death distribution, and the survival function is the probability that an individual will survive from birth to a certain age. The hazard rate generally begins at zero and increases to one, meaning that beginning at birth the probability of dying increases. In contrast, survivorship decreases from one to zero, because the probability that the individual will survive continues to decline as he or she ages (Wood et al. 2002).

Methods

For this study, a maximum likelihood analysis was used to calculate hazard models for the population. One advantage to this type of analysis is the ability of the model to handle a wide variety of age ranges. The age estimations of the adults used for the analysis were taken from a reanalysis of the skeletal remains performed by Morgan (see chapter 4). As is typical in age estimation, Morgan assigned some individuals to five-year age ranges and others to much broader age ranges of 20 years or more, depending on the completeness of the skeleton and the presence of certain aging criteria. We used the age ranges assigned to individuals, as opposed to traditional 5- or 10-year age classes. A maximum likelihood program, MLE (Holman 2000), was used to calculate the parameters for the model. Using the age ranges assigned to individuals, two parameters were calculated for each age subgroup. The two-parameter Gompertz model was used to estimate adult mortality (Gompertz 1825; Wood et al. 2002). The two parameters were then used to calculate the hazard rate, PDF, and survivorship for the adult populations of the different subgroups. The equations for the probability density function, survivorship, and hazard rate are as follows, where *a* is a scale parameter and *b* is a shape parameter (Holman 2000):

$$f(t) = a \exp\left[bt + \frac{a}{b}(1-e^{bt})\right]$$

$$S(t) = \exp\left[\frac{\alpha}{b}(1-e^{bt})\right]$$

$$h(t) = \alpha \exp(bt)$$

In order to test the significance of the differences between the subgroups, we used a likelihood ratio statistic (Λ), which is the ratio of the log-likelihood of the reduced model to the log-likelihood of the full model and which closely follows a chi-square distribution, $2(\ln G_0 - \ln G_1)$, where G_0 and G_1 are the populations used in the comparison. In other words, the likelihood ratio statistic answers the question whether the mortality profiles of the temporal periods and the sexes are significantly different from one another. Only adult mortality was examined for the current analysis, because Hooton did not include data from subadults on his data cards. In the first comparison of mortality profiles, sex differences within the sample were explored. Second, the entire sample was divided into three temporal subgroups—early, middle, and late—and the demographic profiles for the three were compared.

Results

The mortality rates of the two sexes were compared, and the parameters and log-likelihood values are given in table 5.5. A likelihood ratio test comparing the samples indicates a significant difference between the sexes. Figure 5.4 illustrates the survivorship curves of the sexes. Figure 5.5 shows a comparison of hazard rates between males and females.

In addition to comparing the sexes, models were constructed to compare samples by ceramic time period: early (Black-on-white to Glaze 2), middle (Glazes 3 and 4), and late (Glazes 5 and 6). The parameters for each group are given in table 5.6. Likelihood ratio tests were also performed, and no significant differences among the groups were found. The survivorship curves based on the models are shown in figure 5.6. The hazard rates for the different groups are shown in figure 5.7.

TABLE 5.5
PARAMETER ESTIMATES FOR THE GOMPERTZ MODEL, BY SEX

Parameter	Males	Females
a	0.006174	0.013014
b	0.101799	0.064091
ln(L)	–369.537	–407.637

Note: *a* is the initial rate of mortality due to aging; *b* is the rate of increase in the initial rate of mortality with age (Gage 1989). Results of the likelihood ratio test for the combined samples $ln(L) = -784.4$, $\Lambda = 14.45$, $p < .0001$, df = 2. Therefore, there is a significant difference in mortality patterns between males and females.

Discussion

The male and female samples differ significantly in their demographic profiles. The mean age at death for females is 23.6 years, and for males, 24.3 years. The older age of males is expected, because of the predicted decline in survivorship of females during the childbearing years. Female survivorship surpasses that of males after about 40 years of age. The hazard model, seen in figure 5.5, indicates that males have a much higher overall hazard rate. This may be the result of the reportedly high rates of warfare and violence experienced at the pueblo.

The temporal trend in the Pecos sample shows homogeneity in the demographic profile. This indicates that no major changes corresponding to the building of the pueblo or the postcontact period can be observed in the data. Mean age at death for all three groups is about 24 years, with mean age at death for the late period being slightly younger, at 23.3 years, and that for the middle period 24.3 years. This corresponds to the mean age at death reported by Ruff (1991) of about 25 years for the Pecos sample. The hazard rate increases over time, with the earliest phase having the lowest hazard rate and the late phase having the highest.

To address Hooton's third question, then, the analysis demonstrated that although there was no significant change in age at death at Pecos over time, the hazard rate, or the risk that death would occur at a certain age, increased over time and reached a maximum during the late phase of the pueblo's occupation. Furthermore, the results suggest that a significant difference existed between males and females, with females experiencing greater risk of death earlier in adulthood and males later in adulthood.

CONCLUSION

Of three main questions posed by Hooton in his original investigation of the Pecos Pueblo skeletal sample, we addressed the first by examining the biological distances among the samples associated with different time periods. Our analysis of the internal popula-

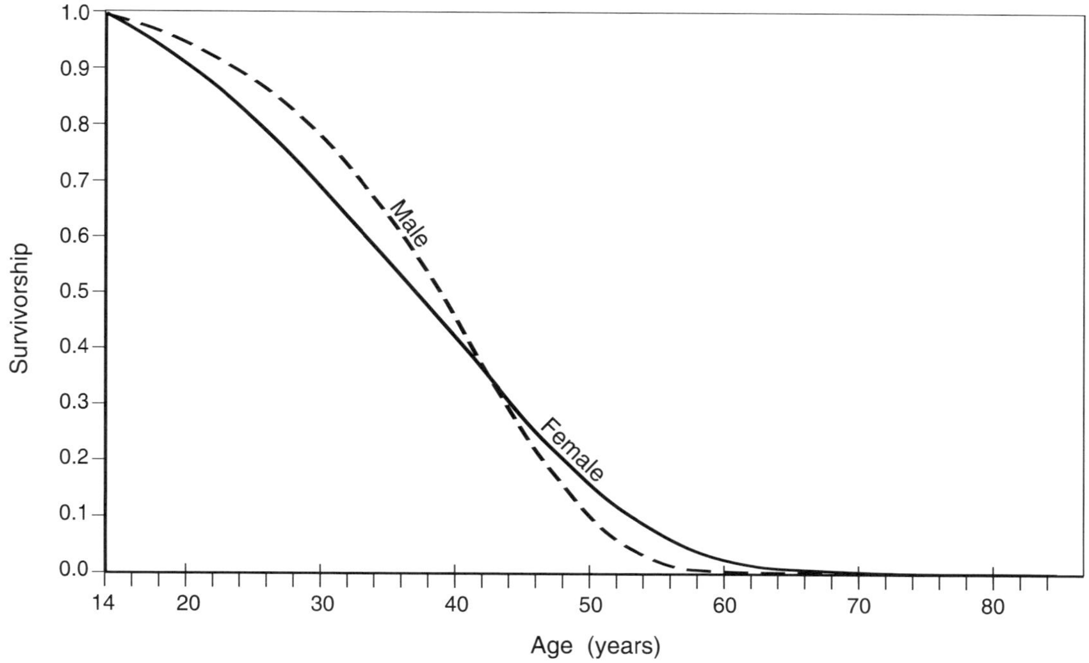

FIGURE 5.4. Survivorship by age and sex for the Pecos–Forked Lightning sample (n = 427).

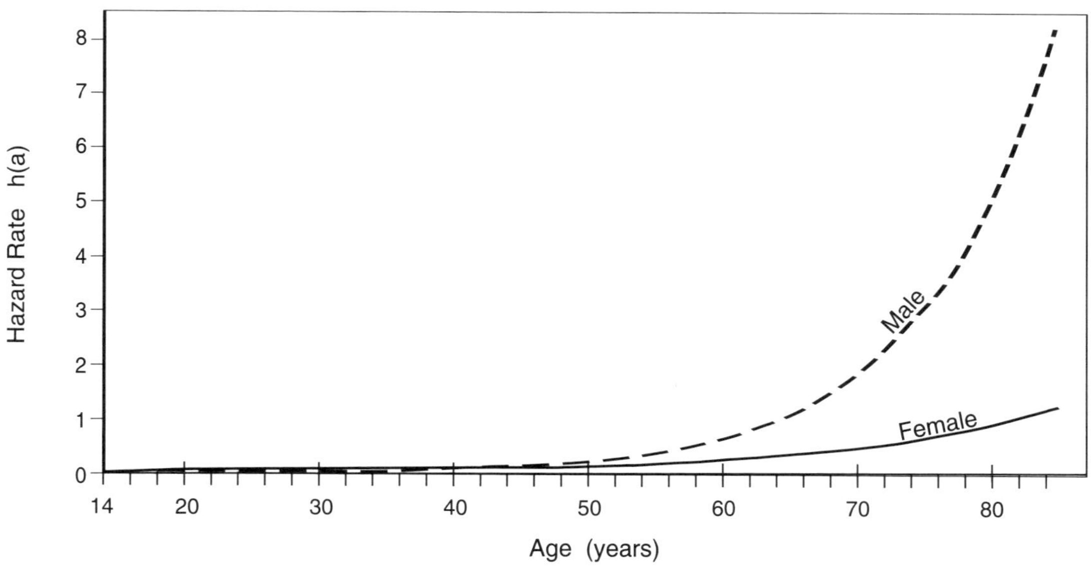

FIGURE 5.5. Hazard rates by age and sex for the Pecos–Forked Lightning sample.

tion structure agrees with the conclusions drawn by Hooton, in that the earliest occupants of the pueblo appear to be differentiated from the later groups. Hooton found that the Black-on-white and Glaze 1 cranial samples were distinct from the population as a whole. In the current study, we were able to examine the Black-on-white and Glaze 1 crania separately and found that the two groups were distinct both from the other groups and from each other. The results from the craniometric and nonmetric data agree, suggesting that the two data sets show a similar pattern of differentiation.

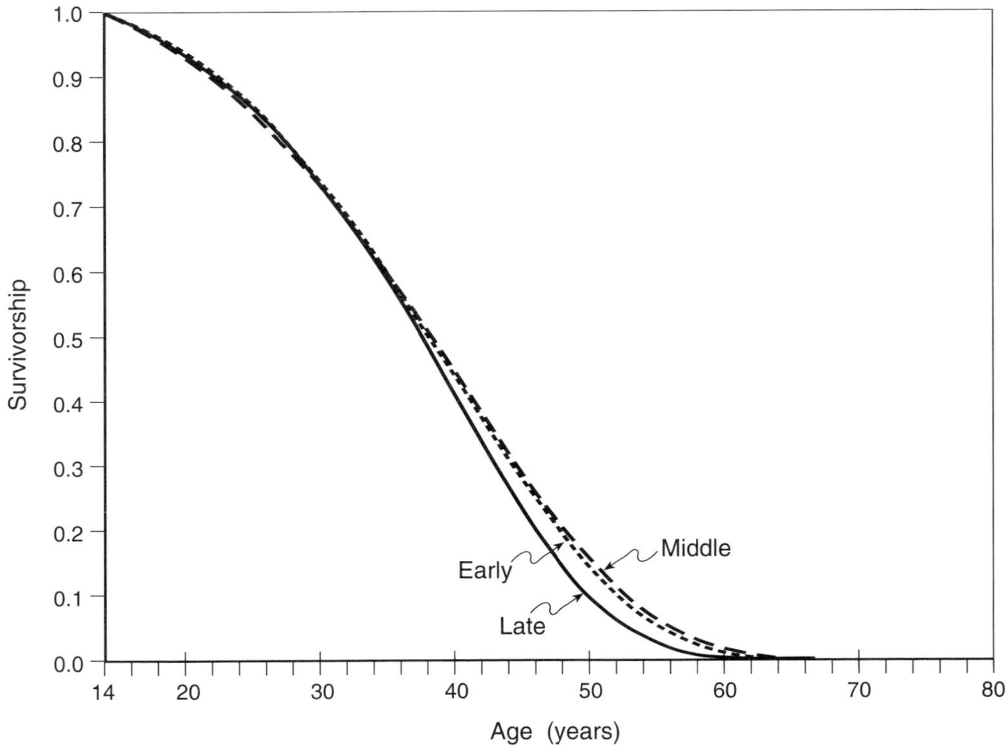

FIGURE 5.6. Survivorship by time period for the Pecos–Forked Lightning sample.

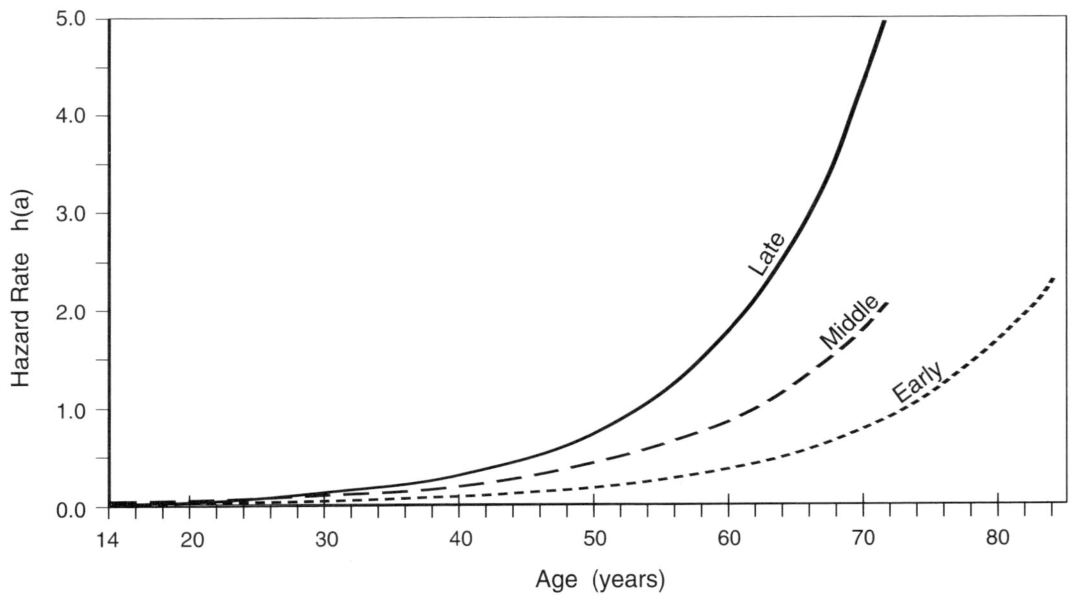

FIGURE 5.7. Hazard rates by time period for the Pecos–Forked Lightning sample.

An analysis of postcranial variation within each subsample was used to address Hooton's second question, regarding the correlation between changes in pottery styles and immigration to the pueblo. The results indicate that the Black-on-white sample contained the least variation of any of the subsamples. Variation then appears to have increased, reaching a maximum during Glaze 4. This again agrees with Hooton's finding, namely, that the postcranial data suggested an increase in heterogeneity over time at the pueblo. However, Hooton was unable to detect the increase in variation between Black-on-white and Glaze 1.

Hooton has often been criticized by those who argue that the types he identified at Pecos, such as Basketmaker, long-faced European, and Plains, were artificially superimposed on a homogeneous population (Fiedel 2004; Goodman 1997). Fiedel, for example, characterized Hooton's identification of types as "startling when one considers that [Pecos] was probably a largely endogamous community consisting of kin and affines" (2004:92). Our analysis indicates that Pecos was anything but homogeneous and instead had a long and complicated biological history. It is not necessary to defend Hooton's typology to argue that the reanalysis of Pecos supports his view of the biological heterogeneity of the Pecos skeletal series.

Finally, we were able to examine the demographic makeup of males and females and some of the demographic changes that took place at Pecos over time. We found a significant difference in mortality between males and females, with female mortality increasing during the childbearing years and falling below that of males in later adulthood. Our results show no significant differences among the ceramic time periods, with age at death remaining uniform over time. Hooton's examination of the demography of Pecos, although crude in its methods, anticipated new directions in skeletal biology. This is just one example of Hooton's forethought in his original investigation. A more important example is the database he created, which allows for continued analysis of the Pecos population as additional research questions arise, even with the reburial of the skeletal remains.

TABLE 5.6
PARAMETER ESTIMATES FOR THE GOMPERTZ MODEL, BY TIME PERIOD

Parameter	Early	Middle	Late
a	0.009937	0.010168	0.008695
b	0.077357	0.074732	0.089550
ln(L)	–271.36	–347.17	–163.41

Note: *a* is the initial rate of mortality due to aging; *b* is the rate of increase in the initial rate of mortality with age (Gage 1989). Results of the likelihood ratio test for the combined samples $ln(L) = -783.23$, $\Lambda = 2.57$, $p = .5$, df = 3. Therefore, there is no significant difference in mortality patterns between the time periods.

6 The Pecos Dentition

Discrete Morphological Traits

Christy G. Turner and Steven A. LeBlanc

The skeletal sample from Pecos Pueblo provides the opportunity to evaluate the differences and similarities between Pecos and its neighbors from an epigenetic perspective.[1] Such a comparison is interesting both because good comparative skeletal samples exist and because the cultural historical background is well known.

The essential question here is to what degree linguistic and cultural differences and similarities between social groups are reflected in their genetic makeup. It has become increasingly clear that ethnicities can be social constructs, which are strongly self-referential, rather than primarily biological groupings. To note this, however, provides little more insight than the old presumption that ethnicities were the result of social groups with long histories of relative social and biological isolation. To probe deeper into the linguistic-cultural and genetic relationship, one must understand how both genetic and social differences and similarities develop and what mechanisms lead to the degree of overlap that is observed. This is, of course, a global issue and not one that any single study can resolve; a considerable literature exists on the subject. We do not review this literature here, but simply point out that the fundamental relationship will be understood only through further specialized studies. We hope this one will contribute to that end.

Skeletal material provides a number of ways of looking at genetic differences. One of the classes of data that may be used is discrete dental traits. (Other classes of skeletal data are presented in chapters 4, 5, 7, and 9–11 of this volume.) Over the course of many years, both of us have used discrete dental traits to look at biological relationships among people. Turner has looked at very broad, worldwide patterns (Scott and Turner 1997; Turner 1984, 1990, 2002) and at more regional samples (Turner 1993, 1999), whereas LeBlanc has been interested in more local differences (Brewer-Carias, LeBlanc, and Neel 1976). Both of us believe that the utility of dental traits has been amply demonstrated at large scales. At the local level, however—the village or village cluster—the use of dental traits may require sample sizes that are not readily obtainable, an issue we return to later.

CULTURAL HISTORICAL BACKGROUND

The cultural historical setting of Pecos Pueblo has been described in chapters 2 and 3 of this volume and needs only brief summary here. Several processes led

1. The term *epigenetic* is preferred over *genetic* because it acknowledges developmental interaction between genes and the environment. Epigenetic traits usually have a strong genetic component but always incorporate some environmental effect during fetal or postnatal development. Full discussion can be found in Nichol 1990 and Scott and Turner 1997.

to the formation of the Pecos community from which the skeletal samples were drawn. The first was the long sequence of events that led to a series of presumed Tanoan-speaking farming communities in the northern Rio Grande, including the Pecos Valley. Because Pecos sat at the edge of the Puebloan world, it must also have interacted with populations down the Pecos Valley, in the mountains to the northeast, and on the Great Plains (Jelinek 1967; Spielmann 1983, 1991b; see fig. 2.2). This must have led to linguistic, biological, and social differences between this "ecotonal" population and those centered on the Rio Grande proper.

Subsequently, an influx of people moved into the area beginning in the late 1200s as the Southwestern climate changed and large parts of the ancestral Pueblo homeland were depopulated. Moreover, presumably around this time, Keres speakers initially occupied part of the Rio Grande Valley. Every bit as important, the population of the entire Southwest, including the Rio Grande and Pecos Valleys, began to go into a significant decline in size in the 1300s (Schroeder 1979a). The decline was accompanied by intensified warfare, which required social groups to maintain adequate numbers for defense (LeBlanc 1998, 1999). This in turn led populations to amalgamate as their numbers diminished. In the Pecos Valley, this can be seen directly at Rowe Pueblo, situated 5–7 kilometers down the valley from Pecos Pueblo (Cordell 1998). Rowe seems to have witnessed a large influx of newcomers, as is evidenced by the addition of a new roomblock almost as large as the original pair. It can also be seen in the subsequent abandonment of Arrowhead Ruin (J. Holden 1955), Rowe Pueblo, and several smaller pueblos in the upper Pecos area; Pecos Pueblo was the likely recipient of these people. This might also have been the case with the population of the more distant Tecolote Pueblo, to the east of the valley (W. Holden 1932).

An alliance group or polity seems to have formed in the upper Pecos Valley that was an amalgam of a number of social units, and these finally coalesced at Pecos Pueblo. Eventually, Pecos is known to have interacted intensively with Plains people in the early historic period, which most likely also resulted in gene flow between populations (see Spielmann, this volume). This process was being similarly played out all over the Rio Grande and the Southwest, so there is no reason to suspect that linguistic-cultural groups had the biological uniqueness that might otherwise be expected. Of special interest are two alliances adjacent to the Pecos group. One of these extended from the Rio Grande to Arroyo Hondo Pueblo, just south of Santa Fe. It can be termed the Tewa alliance. The other, to the southwest of Pecos, was the Galisteo alliance (figs. 1.7, 2.2). The nearest Galisteo site was about 20 kilometers from Pecos Pueblo (Schroeder 1979a; Wilcox 1981). The Galisteo alliance may have been multilingual, with the people of San Marcos possibly speaking Keres, and the remainder, Tano.

Several other alliances existed in the upper Rio Grande Valley that are relevant to this study. The site of Puye, on the Pajarito Plateau northwest of Santa Fe, participated in the "Biscuit Ware" alliance, and the site of Kwasteyu (often spelled Kwastiyukwa or Kwasteyuka) (Elliott 1982) was part of the Towa alliance in the Jemez area (see fig. 1.7). Dental samples from both these sites are used in the subsequent analysis.

As discussed by Capone in chapter 2, the terminal population of Pecos moved to the Pueblo of Jemez in 1838. There is evidence that Pecos Pueblo and the Pueblo of Jemez had been allied earlier (Kessell 1979) and that the Pecos people spoke Towa, but it is unclear how strong or how long the duration of this alliance was and how closely related linguistically the two pueblos really were (Hale and Harris 1979).[2] In this region, many linguistic and cultural differences were likely assimilated into polities that ultimately wound up speaking one language and generally shared a single cultural pattern. Here we examine the degree to which the genetic pattern seen in the dentition of individuals from Pecos Pueblo reflects the early history of the Pecos Valley and the effects of later population "shuffling" on these patterns.

THE DENTAL SAMPLE

A total of 63 individuals from Pecos was scored for 29 discrete dental traits by Turner, using definitions and standards described previously (Turner, Nichol, and Scott 1991). The use of discrete dental traits as an indirect measure of underlying genetic differences among populations was pioneered by Dahlberg (1956), among others. Differences among populations on a worldwide basis had long been recognized, but Dahlberg attempted to standardize the recording of a number of traits to facilitate comparisons of the work of different researchers. Turner added to the list of traits that could be scored and further standardized the coding meth-

2. Some linguists believe that no such link is demonstrated: "There has never been any direct evidence that the language of Pecos was indeed Towa, or for that matter, even Tanoan" (Trager 1967).

ods. The vast majority of studies of discrete dental traits now follow the procedures rigorously defined by Turner and his students (Scott and Turner 1997).

Nichol (1990) examined the degree to which these traits were under genetic control and not heavily influenced by developmental and environmental effects. He also examined the extent to which these traits were highly correlated and thus presumably influenced by the same underlying genes. From this and other studies (see Nichol 1990), a selection of 29 discrete traits was found that are under high genetic control and have low enough correlations that, to a significant degree, they represent the effects of different genes. Most of these traits are probably the result of multiple alleles, multiple loci, or, most likely, both. Some loci surely affect multiple traits. In all, these traits are not the same as a suite of proteins or actual DNA sequence differences for assessing genetic variability. However, the large number of traits and their high heritability mitigate against the lack of a simple one-to-one link between genes and traits.

Although there were approximately 2,000 skeletons from Pecos Valley sites, Turner selected only 63 from Pecos Pueblo for study, because in his experience little improvement in statistical reliability is gained beyond a sample of 50. In archaeological samples, this usually produces a minimum of 20 individuals for any given trait. The individuals were selected from the Black-on-white through Glaze 6 periods, but the great majority came from the Glaze 3 through Glaze 5 periods, or from A.D. 1425 to 1700. The sample was selected primarily on the basis of the condition of the dentition, and although it was selected without a focus on intrasite provenience, it is a reasonably random sample of the dental variability. The sample is primarily from what is termed the Pueblo IV period (A.D. 1325–1600) and as such makes a good comparison with other Pueblo IV populations. Each individual was scored for as many traits as possible, and basic data on the sample are given in table 6.1. Trait scores by individual are provided in Appendix F. Individual scoring sheets, which contain additional information on traits not used in this analysis, left-right side differences, and other features, are on file at the Peabody Museum of Archaeology and Ethnology.

The discrete morphological traits seen at Pecos fit broadly with both New World patterns and those of other Southwestern populations. These patterns include relatively high frequencies of shovel-shaped incisors, peg-shaped upper third molars, triple-rooted lower first molars, and five-cusped lower second molars, as well as a low occurrence of Carabelli's trait. No unique frequencies of any discrete trait were observed, nor were there any very unusual individuals, such as individuals with a European dental pattern, systemic pathology, or rare genetic disorder.

ANALYSIS

The traits listed in table 6.1 were scored for at least two and as many as nine levels of expression, following the protocol described in Turner et al. 1991. For analysis, the traits were combined using the method employed by Turner and colleagues (see Scott and Turner 1997), such that each trait was summarized into one of two categories, present or absent. The breakpoint given in table 6.1 is the trait number that ends the first of the categories. For example, although the trait "shovel-shaped incisors" (trait 2 in table 6.1) was initially scored as having any of eight types of expression, it was simplified as being either present or absent by combining expressions 1, 2 and 3 for the first group and the remaining expressions (4–8) for the second group. Hence its breakpoint is 3. This procedure was repeated for each of the 29 traits. Not all traits could be scored for each individual, so the 63 individuals from Pecos resulted in sample sizes per trait ranging from 26 to 62. The marked reduction in sample size for some traits is a problem in all such epigenetic analyses of archaeologically derived skeletal material. It affects the samples available for comparison with Pecos, in some cases even more dramatically than the Pecos sample. This problem is considered later in the chapter.

Traditionally, these tooth differences have been used to generate a distance matrix. This is done giving equal weight to each trait, using a standard Euclidean distance. The distance matrix has then been used to make a dendrogram, or tree diagram, showing relationships among the sample populations. For large numbers of populations this is probably the optimal method, because it is visually difficult to interpret other means of showing so many relationships. However, when the number of sample populations is less than 20, other methods have advantages.

Cluster analysis, or multidimensional scaling, has particular benefits. First, one can treat the sample differences as interval data or as ordinal data. Assuming only ordinal differences relaxes some of the assumptions made in generating the data and likely mitigates against small sample sizes. (As long as the order of the differences is preserved, the exact percentage differences between populations for the same

TABLE 6.1
DENTAL TRAIT DATA FOR PECOS PUEBLO (N = 63)

Attribute	Sample size	Breakpoint	Sum of data to breakpoint	Percentage to breakpoint	1	2	3	4	5	6	7	8	9
1. Winging I1	51	1	16	0.314	16	6	29	0					
2. Shoveling I1	51	3	7	0.137	0	0	7	29	5	9	1	0	
3. Double shoveling I1	52	2	27	0.519	27	0	0	0	0	0	25		
4. Interruptus groove I2	50	1	28	0.560	28	22							
5. Tuberculum dentale I2	51	1	32	0.627	32	0	5	0	0	0	0	9	5
6. Mesial ridge C	47	1	45	0.957	45	2	0	0					
7. Distal accessory ridge C	26	2	12	0.462	8	4	6	7	1	0			
8. Uto-Aztecan P1	53	1	53	1.000	53	0							
9. Hypocone M2	59	2	6	0.102	3	3	7	16	29	1	0		
10. Cusp 5 M1	44	1	34	0.773	34	4	6	0	0	0			
11. Carabelli cusp M1	48	2	30	0.625	15	15	6	6	3	2	1		
12. Parastyle M3	41	1	41	1.000	41	0	0	0	0	0	0	0	
13. Enamel extension M1	62	2	38	0.613	26	12	1	23	0	0	0		
14. Root number P1	48	1	47	0.979	47	1	0						
15. Root number M2	44	2	21	0.477	11	10	23	0					
16. M3 presence/absence	51	1	40	0.784	40	5	2	4					
17. Lingual cusp p2	41	2	19	0.463	0	19	19	3					
18. Y groove m2	53	2	47	0.887	11	36	6						
19. Cusp number m1	43	2	30	0.698	0	30	13						
20. Cusp number m2	50	1	2	0.040	2	32	2	14					
21. Deflecting wrinkle m1	26	3	18	0.692	10	0	8	8					
22. Trigonid crest m1	40	1	35	0.875	35	5							
23. Protostylid m1	51	1	32	0.627	32	14	3	2	0	0	0	0	
24. Cusp 7 m1	52	1	46	0.885	46	0	4	0	1	1			
25. Tomes root p1	35	4	29	0.829	29	0	0	0	0	0	0	6	
26. Root number c	30	1	30	1.000	30	0							
27. Root number m1	53	2	51	0.962	0	51	2						
28. Root number m2	45	1	15	0.333	15	30	0						
29. Odontome P1, P2, p1, p2	55	1	54	0.982	54	1							

Note: Columns 1–9 are levels of expression for each scored trait. The values given in the columns are the raw counts for each expression. The breakpoint column lists the expression number that ends the first of two categories (present or absent) into which the expressions were combined for analysis (see text). The columns headed "Sum of data to breakpoint" and "Percentage to breakpoint" give, respectively, the absolute counts and the percentages of those counts for the first ("present") category for each trait.

trait do not affect the results, so that minor differences due to sample sizes do not affect the resultant patterns.) Moreover, tree diagrams reduce the data to very hierarchical displays of relationships, resulting in the loss of some subtler aspects of the relationships. Although they do not show all aspects of the relationships perfectly, plots generated by multidimensional scaling do a better job of showing the subtleties among them. Because of all these considerations, we employed multidimensional scaling, and it is the results of these analyses that are depicted here. We also constructed dendrograms for the same comparisons, and they showed the same basic patterns that are described here.

The same multidimensional scaling procedure was used for each analysis, once with the assumption of interval data and again with the assumption of ordinal data. Although we observed minor shifts in the point plots, we noted no meaningful changes between these methods. Thus, the plots presented here seem to be a reasonable depiction of the relationships found in the data. One should keep in mind, however, the assumption that underlies all taxonomy, regardless of level—the assumption that the degree of physical similarity or dissimilarity between groups is due to their degree of relationship. For example, despite breed differences, all dogs look dentally more alike than they look like any other carnivore. All dog breeds thus are more closely related to one another than any individual dog is to any other carnivore.

COMPARISONS

We made a series of comparisons between archaeologically derived samples from the Southwest, a few from present-day Mexico, and two modern indigenous Southwestern populations. All the samples used in the comparative analysis are listed in table 6.2 (see also fig. 1.7). The original data matrix for all populations used in any part of the analysis is given in table 6.3, so other researchers can reanalyze the data by any methods they choose. The modern populations could not be scored for Tomes's root or for root numbers for canines, first premolars, and first and second molars. These traits also had very small sample sizes for several of the prehistoric samples. Because of this, these five traits were excluded from the analysis, although they are presented in table 6.3.

Overall, this selection of samples provides a fairly good data set to compare with Pecos Pueblo. It includes three other Rio Grande populations—Puye, Kwasteyu, and San Cristóbal—that were roughly contemporary. The more westerly pueblos are represented by Hawikku and Awatovi as well as by the slightly earlier Kayenta ancestral Pueblo composite sample. Pueblo Bonito represents an eastern Pueblo sample, although earlier than the Pecos sample. For the northern Southwest, additional Rio Grande pueblos and especially some Keres-speaking pueblos would have been useful, but overall, a reasonable set of comparisons is possible.

Inclusion of a number of more distant populations contextualizes the Pecos sample. Several of these population samples warrant discussion. Two central Mexican samples, those from the sites of Cuicuilco and Tlatelolco, derive from the Valley of Mexico. Whereas Tlatelolco is broadly contemporary with the Southwestern samples, Cuicuilco is Preclassic and thus 1,000 to 1,500 years older than the Southwestern samples. These two populations were run both individually (fig. 6.1) and as a composite "Mexico" group (figs. 6.2–6.4). The Mimbres sample consists of individuals from NAN Ranch Ruin and Swarts Ruin, which were fully contemporary and only about 5 kilometers apart, and from the Harris site, which dates slightly earlier (A.D. 750–1000) and lies approximately 27 kilometers up the Mimbres Valley. Individually, the NAN and Swarts samples sorted quite differently—the Harris sample was too small to plot individually—and most of the analyses were run on them as a composite "Mimbres" group. The Hohokam category is a composite sample from Classic period Hohokam sites, a grouping necessitated by small sample sizes. The Point of Pines and Grasshopper Pueblo samples, from the Mogollon area, are close in time, material culture, and location and were expected to be similar to each other.

RESULTS AND DISCUSSION

Analysis of dental traits on a worldwide basis demonstrates that these traits carry significant genetic and taxonomic information. When sample sizes are large and geographical distances between populations are great, dental traits show patterns that accord with other lines of genetic, linguistic, archaeological, and ethnographic information. When differences are slight and sample sizes are small, results can be more problematic, as is exemplified in Hawkey's (1998) analysis of South Asian populations.

The first step in our analysis was to look at samples for which we could reasonably predict the nature

TABLE 6.2
DENTAL SAMPLES USED IN THE COMPARATIVE ANALYSIS

Site or population	Maximum sample size for any trait used	Minimum sample size for any trait used	Notes	Source
Awatovi	45	14	Hopi area, primarily Pueblo IV but probably some Pueblo V	PMAE[a] and Smithsonian Museum of Natural History; Brew 1937
Hawikku	136	34	Zuni area, primarily Pueblo IV but possibly some Pueblo V	American Museum of Natural History; Smith, Woodbury, and Woodbury 1966
Kwasteyu[b]	53	9	Jemez area, presumably Towa speakers, late Pueblo IV	Reiter 1938
Puye	115	17	Pueblo IV, Tewa or at least Tanoan speakers	Morley 1910
San Cristóbal	111	27	Tano speakers; closest site to Pecos in sample	Excavated by N. C. Nelson (Nelson 1914).
Tlatelolco	173	31	Valley of Mexico, Aztec period	See Turner 1984. Combined to produce the central Mexico sample.
Cuicuilco	72	13	Valley of Mexico, Preclassic	See Turner 1984. Combined to produce the central Mexico sample.
Grasshopper Pueblo	181	63	Late Pueblo III, early Pueblo IV	Mogollon Mountains area; University of Arizona; Reid and Whittlesey 1999
Point of Pines	65	17	Late Pueblo III, early Pueblo IV	Mogollon Mountains area; University of Arizona; Bennett 1973
Pueblo Bonito	74	20	Pueblo II	Central Colorado Plateau; American Museum of Natural History
Casas Grandes	111	33	ca. 1200s–1400	Amerind Foundation; Di Peso 1974
Hohokam	99	20	Long time span, primarily Colonial and Sedentary	Composite from Classic period sites in the Phoenix area. See Turner 1987; Turner and Irish 1989.
Kayenta ancestral Pueblo	94	22	Pre–Pueblo IV	Composite from a number of sites; Museum of Northern Arizona, American Museum of Natural History, Smithsonian Museum of Natural History, PMAE
NAN Ranch Ruin	77	22	Mimbres, primarily Classic with some Late Pithouse	Mimbres Valley, combined with Harris site for Mimbres sample; Texas A&M; Shafer 2003
Swarts Ruin	32	9	Mimbres, primarily Classic with some Late Pithouse	Mimbres Valley, combined with Harris site for Mimbres sample; PMAE; Cosgrove and Cosgrove 1932
Harris Site	16	0	Mimbres, Late Pithouse	Haury 1936. Not used individually; combined with NAN and Swarts for Mimbres sample

[a] PMAE is the Peabody Museum of Archaeology and Ethnology, Harvard University.

[b] This site has been confused with Amoxiumqua, about two miles (3.5 km) away and roughly contemporary. For this analysis it does not matter which site the sample actually came from.

of the relationship between them. This gives us some idea about how robust the method and data are. The close similarity in virtually all the analytical runs between Point of Pines and the contemporary Grasshopper Pueblo, which are situated less than 100 kilometers apart, particularly demonstrates that when sample sizes are reasonably large, populations that should be similar, judging from both geography and material culture, are indeed similar (figs. 6.1–6.5). Having said this, we might, on the basis of geography, have expected the Hohokam sample to have plotted near these two villages, but it definitely did not. Similarly, the sites of Awatovi and Hawikku, parts of the Hopi and Zuni polities, respectively, were contemporaries and physically nearer to each other than to any other contemporary site in the sample. They consistently plotted very close to each other—in fact, surprisingly so, given that they probably represent individuals from completely unrelated language families. Moreover, Awatovi and Hawikku plotted very near the modern Hopi sample, as expected (fig. 6.3). Overall, then, there are several examples in which geographical proximity does correlate with genetic similarity, and in some cases this seems to be stronger than linguistic similarity.

The two Mimbres samples, from the NAN Ranch and Swarts Ruins, which are both geographically and culturally proximate to each other, did not plot closely together in the analysis of their dental traits (fig 6.1). These sites are geographically less than 7 kilometers apart and almost completely contemporary, to a greater degree than any other sites employed in the analysis. Although this might represent some unexpected biological variability in the Mimbres region and warrant further evaluation, it more likely represents a problem of sample size. The NAN sample was reasonably good, with maximal and minimal sample sizes for any trait of 77 and 22, respectively. But although the Swarts overall sample size was 55, the material was rather fragmentary; the maximal sample for any trait was 32, and the minimum, only 9. Consequently, a marked difference between the two samples is perhaps not surprising, and it certainly provides a warning about the limitations of dental traits with these sample sizes. This limitation is further accentuated by the samples from central Mexico. When plotted either individually or as a composite (figs. 6.1, 6.2), the central Mexican samples of Tlatelolco and Cuicuilco fall well within the range seen in the greater Southwest. This is a surprising outcome given the spatial distance of 1,500 kilometers between the Mexican sites and any of the nine greater Southwest samples. The plots presented here show interesting relationships between the Mexican, Mimbres, Casas Grandes, and Pueblo Bonito populations, but these fall outside the scope of this chapter and will be the subject of a future study.

A similar problem of failure to fit expectations is met with the modern Navajo sample. Because the Athabaskan-speaking Navajos derived from western Canada and arrived in the Southwest about 500 to 600 years ago, we would expect them to be distinctly different genetically from the Puebloan, Mogollon, and Hohokam Southwesterners. The Navajo dentition sample we used has been intensively analyzed by Nichol (1990) and Scott (1973), who pointed out that it appeared to be derived from a population more admixed with Pueblo people than other Navajo populations, as determined by serological analysis. Nichol hypothesized that this was because the dental sample was drawn from Navajos living near Keams Canyon in Hopi country, and this group was most likely to have had gene flow with Hopis. Moreover, some studies, such as those based on mitochondrial DNA data (Lorenz and Smith 1996) show that up to 40% of Navajo mitochondria genes derive from Puebloan populations. Yet even if the subpopulation used in this comparison is more admixed than the average Navajo subpopulation, it is still unexpected that it would plot so similarly to modern Hopi, historic Awatovi and Hawikku, and other Southwestern populations. This result further underscores that, overall, the differences in dental traits between any of these populations are not very great.

In summary, some of the expectations for the data were well met, namely, similarities between Point of Pines and Grasshopper and between Awatovi, Hawikku, and modern Hopi. Other samples did not plot as we expected, namely, Swarts and NAN Mimbres, the central Mexican samples, and modern Navajo. With these patterns in mind, we can turn to the Pecos sample itself.

The most consistent outcome of all the analyses we ran was that Pecos Pueblo was closely related to the nearby pueblo of San Cristóbal (figs. 6.1–6.5). This was expected. What is surprising is that neither Pecos nor San Cristóbal was particularly related to other pueblos in the greater Rio Grande area. Puye and Kwasteyu, both on the western side of the Rio Grande cultural area and both presumably inhabited by Tanoan speakers (although we would expect them to have spoken different languages within the Tanoan family—that is, Kwasteyu, Towa, and Puye, Tewa), were somewhat but not particularly similar to each

TABLE 6.3
DENTAL TRAITS FOR PECOS PUEBLO AND COMPARATIVE POPULATIONS

TRAITS 1–7

Site	1. Winging I1		2. Shoveling I1		3. Double shoveling I1		4. Interruption groove I2		5. Tuberculum dentale I2		6. Mesial ridge C		7. Distal accessory ridge C	
	n	%	*n*	%	*n*	%	*n*	%	*n*	%	*n*	%	*n*	%
Awatovi	34	0.412	37	0.162	34	0.647	36	0.472	37	0.622	36	1.000	14	0.214
Hawikku	100	0.380	94	0.224	74	0.554	97	0.546	105	0.800	92	1.000	34	0.265
Kwasteyu	32	0.344	24	0.167	17	0.588	26	0.500	30	0.867	30	1.000	9	0.111
Puye	88	0.261	74	0.257	47	0.766	63	0.333	91	0.868	52	0.962	17	0.235
San Cristóbal	71	0.493	68	0.103	59	0.458	74	0.419	70	0.657	56	1.000	27	0.296
Pecos Pueblo	51	0.314	51	0.137	52	0.519	50	0.560	51	0.627	47	0.957	26	0.462
Pueblo Bonito	31	0.484	41	0.220	33	0.787	46	0.457	50	0.660	47	0.979	20	0.300
Kayenta ancestral Pueblo	48	0.313	67	0.119	56	0.465	60	0.400	68	0.750	72	0.986	39	0.487
Grasshopper Pueblo	99	0.384	133	0.090	127	0.283	128	0.492	128	0.625	108	0.972	63	0.444
Point of Pines	34	0.500	33	0.030	28	0.250	44	0.545	43	0.605	42	1.000	17	0.118
NAN Ranch Ruin	41	0.390	60	0.167	52	0.519	58	0.207	57	0.526	48	0.979	22	0.182
Swarts Ruin	18	0.389	20	0.250	19	0.421	22	0.591	22	0.682	21	1.000	9	0.333
Harris Site	6	0.667	6	0.000	6	0.167	7	0.429	6	0.333	7	1.000	0	0.000
Mimbres (all)	65	0.415	86	0.175	77	0.467	87	0.322	85	0.553	76	0.987	31	0.226
Casas Grandes	33	0.455	98	0.112	98	0.613	80	0.313	83	0.446	94	0.979	64	0.250
Cuicuilco	24	0.583	29	0.069	25	0.640	28	0.571	27	0.778	29	0.966	13	0.308
Tlatelolco	31	0.645	39	0.000	36	0.250	47	0.553	47	0.617	63	1.000	34	0.294
Modern Hopi	145	0.359	151	0.245	143	0.546	143	0.476	143	0.685	147	0.980	103	0.311
Modern Navajo	121	0.289	122	0.205	120	0.584	126	0.429	131	0.603	125	0.976	92	0.293

TRAITS 8–14

Site	8. Uto-Aztecan P1		9. Hypocone M2		10. Cusp 5 M1		11. Carabelli cusp M1		12. Parastyle M3		13. Enamel extension M1		14. Root no. P1	
	n	%	*n*	%	*n*	%	*n*	%	*n*	%	*n*	%	*n*	%
Awatovi	33	0.970	33	0.212	21	0.857	20	0.700	31	1.000	44	0.523	35	0.857
Hawikku	113	0.982	89	0.180	48	0.938	55	0.564	86	0.930	135	0.489	113	0.929
Kwasteyu	38	1.000	34	0.206	18	0.833	18	0.722	26	0.923	43	0.604	46	0.935
Puye	93	1.000	70	0.129	31	0.806	22	0.591	60	0.883	115	0.548	95	0.905
San Cristóbal	73	0.986	71	0.056	34	0.647	43	0.558	55	0.964	104	0.500	90	0.889
Pecos Pueblo	53	1.000	59	0.102	44	0.773	48	0.625	41	1.000	62	0.613	48	0.979
Pueblo Bonito	51	1.000	52	0.231	48	0.875	46	0.565	33	0.939	74	0.459	67	0.836
Kayenta ancestral Pueblo	78	0.987	74	0.108	65	0.892	69	0.406	72	0.917	94	0.553	90	0.867
Grasshopper Pueblo	124	0.960	134	0.313	140	0.850	167	0.694	81	0.938	157	0.510	81	0.840
Point of Pines	44	1.000	50	0.200	36	0.778	36	0.833	43	0.930	61	0.509	55	0.855
NAN Ranch Ruin	51	1.000	61	0.213	50	0.820	51	0.588	46	0.913	54	0.556	38	0.868
Swarts Ruin	25	1.000	25	0.280	13	1.000	14	0.643	22	0.955	24	0.583	11	0.727
Harris Site	11	1.000	6	0.000	2	1.000	2	0.500	4	1.000	12	0.333	15	1.000
Mimbres (all)	87	1.000	92	0.217	65	0.862	67	0.597	72	0.931	90	0.533	64	0.875
Casas Grandes	94	0.979	82	0.135	75	0.733	76	0.224	67	0.881	32	0.625	5	0.800
Cuicuilco	50	0.980	39	0.077	29	1.000	27	0.555	36	0.972	56	0.661	45	0.889
Tlatelolco	85	1.000	102	0.147	91	0.769	99	0.545	54	0.981	122	0.492	122	0.910
Modern Hopi	153	1.000	141	0.142	121	0.967	148	0.541	47	1.000	3	0.333	0	
Modern Navajo	139	1.000	124	0.266	107	0.869	128	0.648	51	1.000	0		0	

TABLE 6.3 *(Continued)*

TRAITS 15–21

Site	15. Root no. M2		16. Peg/ reduced/ absent M3		17. Lingual cusp p2		18. Y groove m2		19. Cusp no. m1		20. Cusp no. m2		21. Deflecting wrinkle m1	
	n	%	*n*	%	*n*	%	*n*	%	*n*	%	*n*	%	*n*	%
Awatovi	21	0.666	44	0.795	36	0.945	45	0.977	27	0.556	33	0.152	20	0.750
Hawikku	97	0.474	122	0.828	100	0.840	124	0.920	82	0.659	97	0.268	48	0.792
Kwasteyu	40	0.500	49	0.857	35	0.743	44	0.886	25	0.640	35	0.286	20	0.800
Puye	100	0.550	115	0.843	72	0.820	110	0.928	40	0.750	77	0.091	33	0.879
San Cristóbal	71	0.507	81	0.864	64	0.578	96	0.938	48	0.500	85	0.094	36	0.666
Pecos Pueblo	47	0.477	51	0.784	41	0.463	53	0.887	43	0.698	50	0.040	26	0.692
Pueblo Bonito	53	0.547	64	0.891	39	0.692	63	0.937	45	0.733	53	0.151	35	0.715
Kayenta ancestral Pueblo	74	0.608	87	0.851	69	0.783	89	0.922	72	0.653	72	0.194	65	0.755
Grasshopper Pueblo	55	0.564	104	0.769	84	0.679	119	0.823	148	0.527	104	0.173	121	0.653
Point of Pines	47	0.382	58	0.879	46	0.500	57	0.912	39	0.564	54	0.130	29	0.586
NAN Ranch Ruin	37	0.486	62	0.855	55	0.691	56	0.911	62	0.597	53	0.132	41	0.635
Swarts Ruin	10	0.900	32	0.875	25	0.840	26	0.923	18	0.667	25	0.320	19	0.789
Harris Site	8	0.375	11	0.818	8	0.875	10	1.000	1	1.000	9	0.333	1	1.000
Mimbres (all)	56	0.545	105	0.857	88	0.750	92	0.924	81	0.618	87	0.207	61	0.689
Casas Grandes	2	1.000	69	0.971	83	0.699	78	0.872	75	0.667	82	0.098	49	0.633
Cuicuilco	42	0.380	53	0.849	62	0.790	69	0.927	38	0.474	59	0.017	28	0.750
Tlatelolco	88	0.420	114	0.807	90	0.667	154	0.831	132	0.348	151	0.053	96	0.739
Modern Hopi	0		67	1.000	152	0.796	28	0.928	112	0.696	114	0.158	58	0.880
Modern Navajo	0		72	0.931	133	0.759	44	0.909	93	0.645	107	0.168	58	0.758

TRAITS 22–29

Site	22. Trigonid crest m1		23. Protostylid m1		24. Cusp 7 m1		25. Tomes's root p1		26. Root no. c		27. Root no. m1		28. Root no. m2		29. Odontome (all premolars)	
	n	%	*n*	%	*n*	%	*n*	%	*n*	%	*n*	%	*n*	%	*n*	%
Awatovi	28	0.964	33	0.727	39	1.000	19	0.842	20	1.000	48	1.000	31	0.613	32	0.969
Hawikku	93	0.925	101	0.772	115	0.930	87	0.759	78	1.000	136	0.949	119	0.345	103	0.942
Kwasteyu	28	0.893	30	0.833	43	0.860	39	0.872	39	1.000	53	0.962	44	0.273	35	0.971
Puye	60	0.917	44	0.727	103	0.990	93	0.850	79	0.975	129	0.946	110	0.291	53	0.981
San Cristóbal	58	0.983	82	0.427	99	0.869	87	0.851	88	1.000	111	0.955	93	0.312	74	0.973
Pecos Pueblo	40	0.881	51	0.627	52	0.885	35	0.829	30	1.000	53	0.962	45	0.333	55	0.982
Pueblo Bonito	50	0.960	50	0.740	66	1.000	52	0.808	51	1.000	64	0.984	57	0.368	51	1.000
Kayenta ancestral Pueblo	80	1.000	83	0.542	88	0.989	70	0.800	88	0.989	88	0.932	82	0.378	69	0.957
Grasshopper Pueblo	148	0.973	150	0.387	181	0.912	54	0.833	58	1.000	135	0.933	78	0.231	108	0.981
Point of Pines	43	0.953	63	0.349	65	0.908	45	0.867	50	1.000	65	0.877	51	0.255	24	1.000
NAN Ranch Ruin	60	0.967	59	0.746	77	0.935	30	0.767	40	1.000	72	0.903	42	0.333	46	0.913
Swarts Ruin	23	1.000	28	0.929	30	0.967	17	1.000	12	1.000	29	1.000	18	0.389	23	1.000
Harris Site	2	1.000	2	1.000	6	0.833	18	0.944	16	0.938	12	1.000	10	0.100	5	0.800
Mimbres (all)	85	0.977	89	0.809	113	0.938	65	0.877	68	0.985	113	0.938	70	0.314	74	0.932
Casas Grandes	86	0.965	90	0.656	111	0.901	0		0		22	1.000	8	0.625	102	0.941
Cuicuilco	59	1.000	64	0.719	72	0.917	46	0.740	50	1.000	72	0.917	66	0.258	37	0.946
Tlatelolco	170	0.976	172	0.773	173	0.832	86	0.756	154	0.968	173	0.948	165	0.255	140	0.986
Modern Hopi	63	1.000	146	0.699	150	0.947	0		0		0		0		164	0.945
Modern Navajo	78	0.974	114	0.746	114	0.947	0		0		0		0		137	0.934

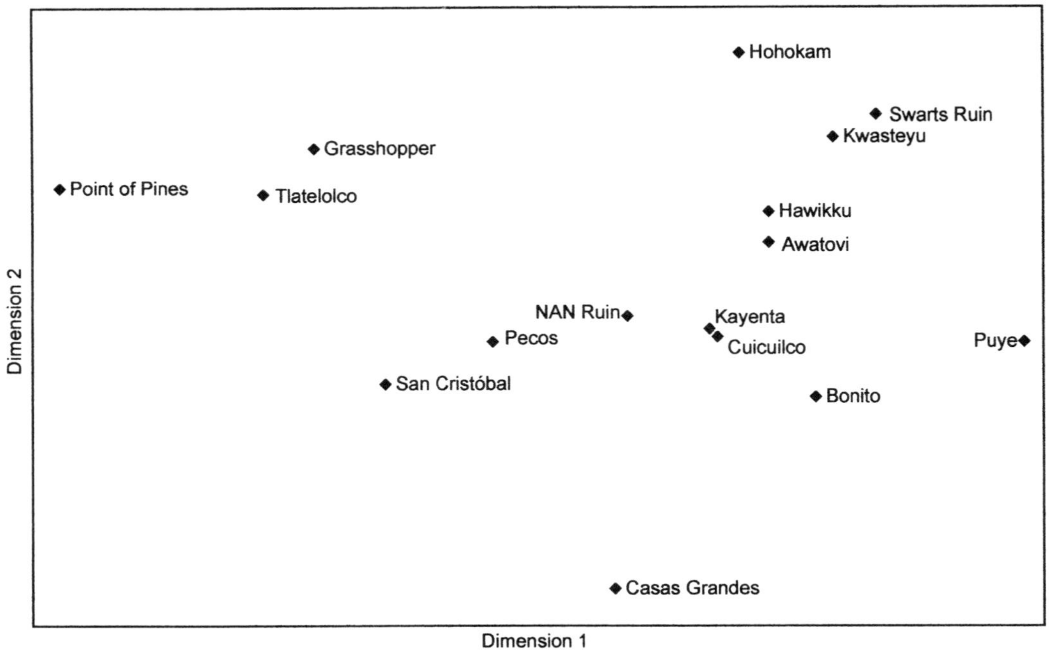

FIGURE 6.1. Two-dimensional plot of the results of multidimensional scaling of 24 dental traits showing relationships among 16 prehistoric populations, including 2 from central Mexico and 2 from the Mimbres Valley.

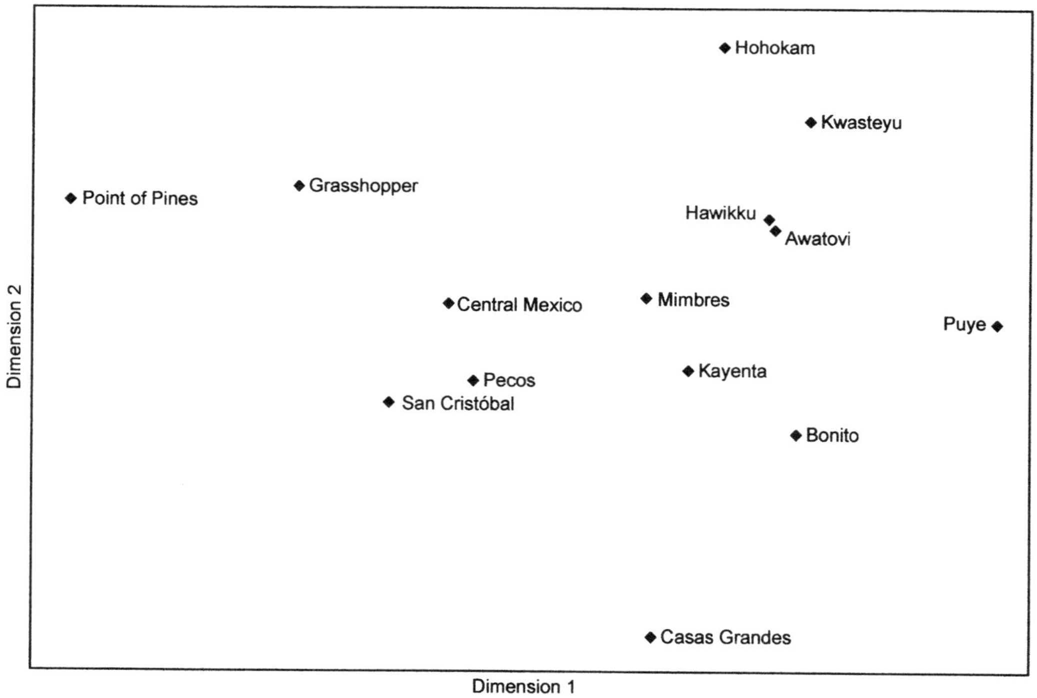

FIGURE 6.2. Two-dimensional plot of the results of multidimensional scaling of 24 dental traits showing relationships among 14 prehistoric populations. The Mimbres and central Mexican samples are combinations.

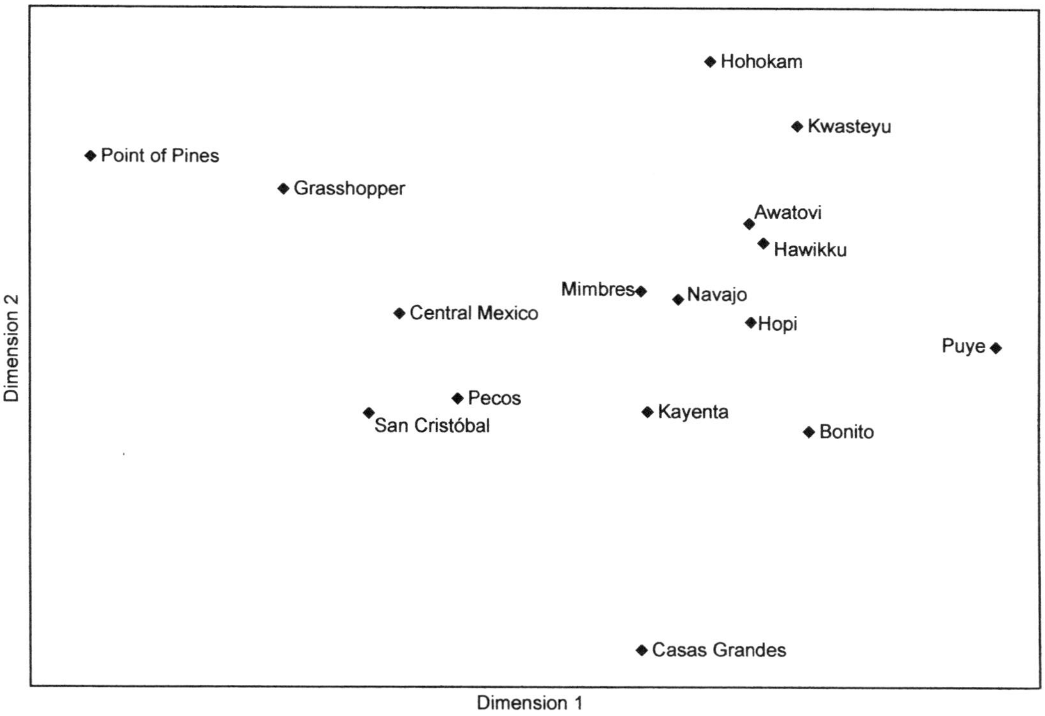

FIGURE 6.3. Two-dimensional plot of the results of multidimensional scaling of 24 dental traits showing relationships among 14 prehistoric and 2 modern (Hopi and Navajo) populations. The Mimbres and central Mexican samples are combinations.

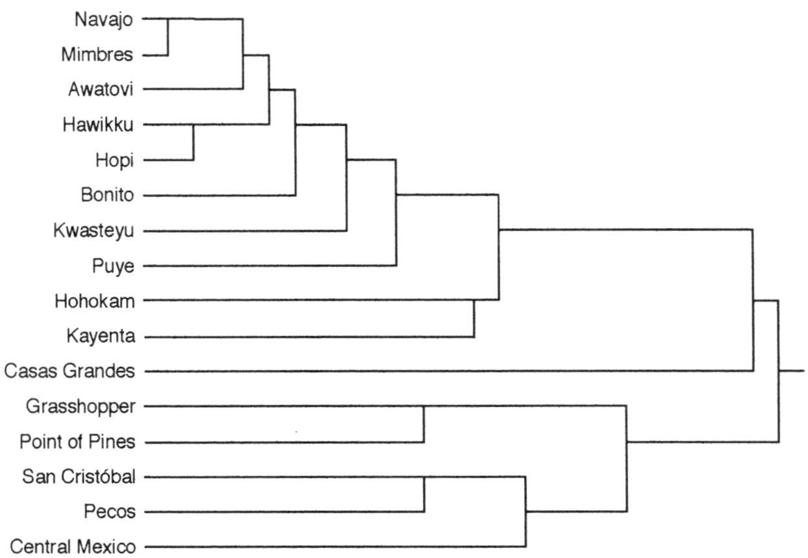

FIGURE 6.4. Dendrogram of the relationships among 14 prehistoric and 2 modern populations, using the same samples as in figure 6.3.

other and plotted closer to the western pueblos than to the geographically much closer Pecos and San Cristóbal (figs. 6.1, 6.5; see Schroeder 1979a for cultural and Hale and Harris 1979 for linguistic relationships among these pueblos). The populations of Puye and Kwasteyu should be expected to have had some admixture from ancestral Pueblo populations from farther west (see Ford, Schroeder, and Peckham 1972), so the relationships between Puye, Kwasteyu, and the more westerly Chaco, Hawikku, and Awatovi seem reasonable. It is the lack of a close relationship between Pecos and the Rio Grande pueblos of Puye and Kwasteyu that is surprising, given their geographical proximity and membership in the same language family, if they were not actually speakers of the same language.

We have no Plains Indian dental samples with which to compare Pecos, but if there were a strong Plains component in the Pecos sample, then we would expect Pecos to be more of an outlier from other Puebloan groups than it is. The apparent link to the Mogollon Mountains, the home of Point of Pines and Grasshopper Pueblo, is unexpected.

The close linkage of Pecos with San Cristóbal is of interest. Although it might have been predicted on the basis of geographical proximity, it was not a foregone conclusion. There is ample evidence of warfare among the alliance groups in the Rio Grande area. Within the Pecos group, Arrowhead Ruin seems to have been attacked and destroyed, and the Hobson-Dressler site was defensively located. Both lie in the direction of the more westerly Tewa alliance group, toward Santa Fe. In that group, Arroyo Hondo Pueblo—the member of the Tewa alliance closest to Pecos—also shows evidence of warfare. It is likely that the Pecos alliance and the Tewa alliance were competitive (Haas and Creamer 1997; LeBlanc 1999). The Spanish explorer Coronado in 1541 described passing by a large village in the Galisteo Basin, home to San Cristóbal, that had recently been destroyed by warfare (Winship 1896). The Galisteo alliance might also have been competitive with the Pecos alliance. However, the dental data suggest instead that the Galisteo and Pecos alliances were themselves allied, perhaps against the Tewa (Santa Fe area) alliance, and that the Galisteo-Pecos alliance led to considerable genetic exchange. Alternatively, both populations might have had similar founding populations, both might have interacted with Plains people more than with other Pueblos, both might have received immigrants from the same sources, or some combination of all these factors might have been the case. Nevertheless, the similarity of the Pecos and San Cristóbal populations is of interest and should be pursued by more dental studies of the Galisteo and Santa Fe alliance populations.

COMPARISONS BASED ON CRANIAL TRAITS

The dental trait study presented here is usefully compared with one based on nondental discrete traits. Twenty discrete morphological cranial traits were recorded by Mackey (1980) from 12 different Southwestern populations, including many populations that were the same or closely related to those used in the present analysis. These included Hawikku and Awatovi from the western Pueblo area. (A sample from Halona, very near Hawikku, recorded by Mackey was excluded from the present analysis because of its very small size.) The site of Giusewa, in the Towa area—the same area as Kwasteyu, which we used in the dental study—was also included. Three sites from the southern Tiwa area of the Rio Grande—Pottery Mound, Kuaua, and Tijeras Pueblo—were included in the Mackey study, but there are no comparable sites in the dental sample. The Pecos skeletons examined by Mackey included both a mission and a prehistoric sample, and there was also a sample from San Cristóbal. Mackey studied two sites from the Tewa alliance—Pindi and Arroyo Hondo Pueblos—and two from the more westerly Biscuit Ware alliance—Sapawe and Puye (the latter also included in the dental study). Thus, although the cranial sample lacks the level of geographical diversity present in the dental sample, it includes a number of the same or closely comparable sites.

The data matrix developed by Mackey was analyzed using the same multidimensional scaling algorithm used for the dental analysis (fig. 6.6). (No effort was made to determine the degree to which these traits were independent; the data were scaled and plotted just as they were presented by Mackey.) The frequencies were analyzed assuming both ordinal and interval data, with only minor differences between runs. The results were quite different from those of the dental analysis. Pecos was not particularly similar to San Cristóbal but was instead more similar to Pindi Pueblo and Pottery Mound. Moreover, the prehistoric and mission period samples from Pecos were not very similar. Although we would expect some change over time, the magnitude of the difference seems greater than one would predict. On the other hand, Awatovi was not particularly similar to Hawikku, and Pindi

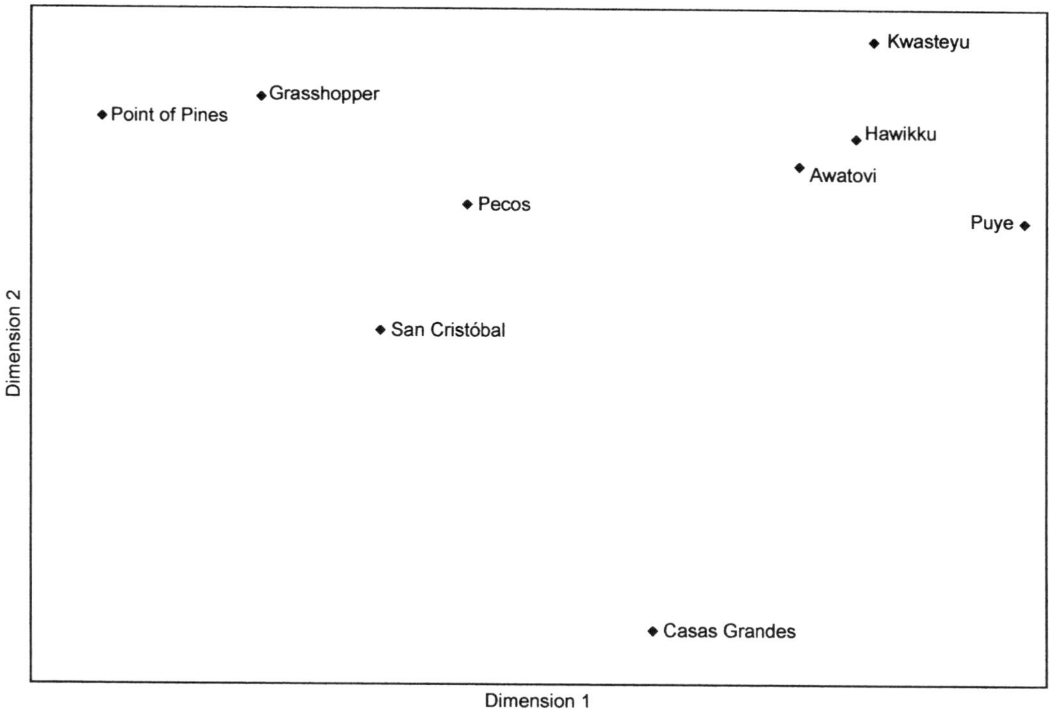

FIGURE 6.5. Two-dimensional plot of the results of multidimensional scaling of 24 dental traits showing relationships among nine Southwestern populations dating primarily to the Pueblo IV period (A.D. 1325–1600).

was only moderately similar to Arroyo Hondo, unlike what we might have expected. Overall, the results do not pattern well with geography or with presumed cultural affiliations. However, the total number of skeletons from all sites was only 368. Whereas Puye had a sample of 63 and Hawikku had 53, prehistoric and mission period Pecos had only 18 each. The Tijeras sample was small, at only 15, and Pindi was little better, with 16 skeletons. As a consequence of these small sample sizes, the results likely do not reflect heretofore unrecognized biological relationships. Again, small samples appear to play havoc with this type of analysis. Moreover, whereas age and sex have little or no influence on dental morphology, discrete osteological traits are affected, so that the age and sex composition of the osteological trait samples could also have contributed to the unexpected relationships.

CONCLUSION

All our results must be tempered by the observation made earlier, that when populations are genetically relatively similar and sample sizes are not large, analyses such as these are subject to interpretive error. The limitations inherent in dental nonmetric trait analysis of archaeological samples—incomplete data due to preservation, included populations that do not represent a single point in time, and the fact that dental nonmetric traits are an indirect and imperfect measure of underlying genetic frequencies—compound to produce a significant error component. Worldwide, intergroup differences are so great that they overcome these limitations. In other situations, such as intraregional analyses, they may not. This seems to be the same conclusion that is reached for direct genetic differences. On a worldwide level there is, for example, a relatively high correspondence between genes and languages (Cavalli-Sforza, Minch, and Mountain 1992), but this does not necessarily hold at the local level (Shields, Schmeichen, and Frazier 1993; Ward et al. 1993).

With this caveat in mind, the dental nonmetric data do suggest that Pecos Pueblo was not part of a tight epigenetic block based in the Rio Grande region or based on Tanoan speakers in particular. It is noteworthy that these data show no special epigenetic relationship between Pecos and the Jemez area. Any such

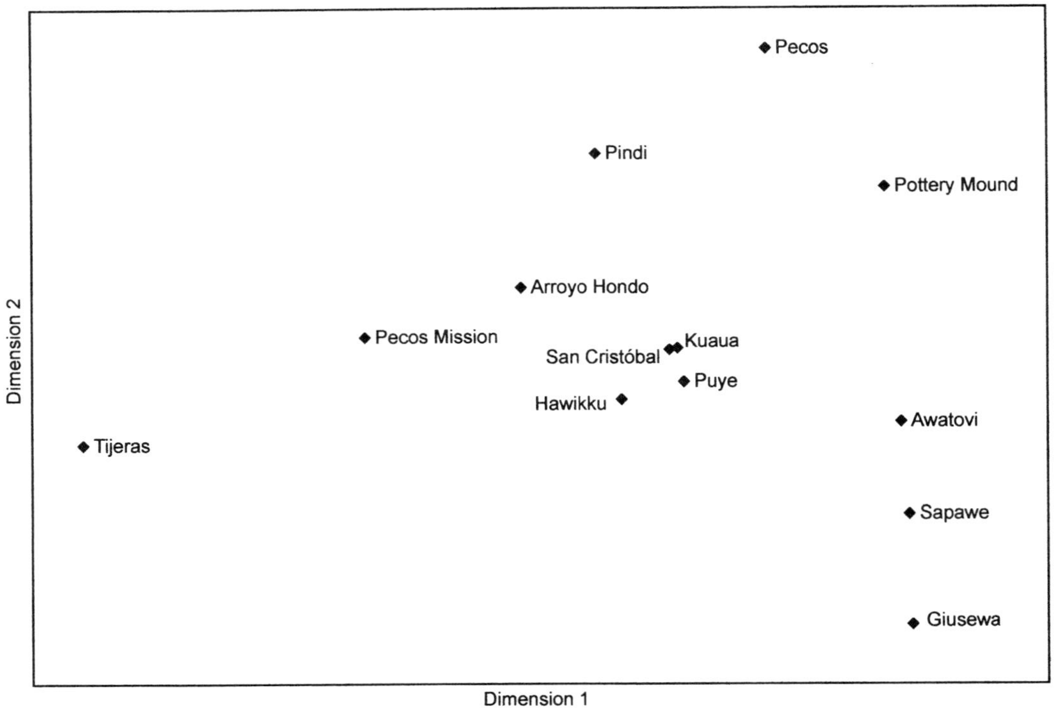

FIGURE 6.6. Two-dimensional plot of the results of multidimensional scaling of a distance matrix of discrete cranial traits for 13 Pueblo IV (A.D. 1325–1600) samples (from Mackey 1980).

relationship must have been based on cultural links and is not the result of a shared biological ancestry, except at the broadest, pan-Southwestern level. The overall conclusion must once again be that genes do not usually follow distinct microcultural or local language similarities and differences but are more likely to reflect geographical proximity, the underlying hypothesis for which is gene flow and distance.

7 The Influence of Preservation on Observability of Pathology

A Case Study from Pecos

Lane Anderson Beck

Osteologists often use pathology data as a means to compare populations across time and space. Differences in the frequencies of specific diseases are argued to indicate differences in the health of the living communities and are taken to demonstrate biocultural variation among the communities being compared. For the Southwestern Pueblo region, Pecos Pueblo is a key site for such comparisons. It has been noted that the incidence of pathology reported for Pecos Pueblo has a lower frequency in many categories of diseases than is typical of other sites (e.g., Stodder and Martin 1992). This lower frequency is often attributed to the early date of the initial analysis (Hooton 1930) and the advances made in diagnostic potential since the 1920s. From my analysis of pathology in the Pecos Valley collections (with data collection during the 1980s and 1990s), it is clear that advances in palaeopathology do not fully explain this divergence. In this chapter I review the history of the Pecos Pueblo skeletal collection with attention to potential biases that might have been created from excavation, curation, and analytical processes over the past century. This history has implications for the reported pathology at Pecos. I do not attempt to offer a thorough review of pathology at Pecos but instead use the Pecos collection to illustrate concerns that need to be explicitly considered in any analysis of pathology.

THE EXCAVATIONS

Between 1915 and 1929, Alfred Vincent Kidder spent 10 field seasons in the Pecos Valley of New Mexico excavating Pueblo ruins for the R. S. Peabody Foundation at Phillips Academy (Kidder 1958). The main site, Pecos Pueblo, sits on a low *mesilla* above Glorieta Creek, a tributary of the Pecos River (fig. 2.1). On the eastern edge of the Pueblo culture region, just southeast of Santa Fe, the Glorieta-Pecos corridor is a crossroads between the cultures of the desert Southwest and the Great Plains (fig. 2.2). Forked Lightning, a short distance away across Glorieta Creek, and Dick's Ruin, downstream on the west bank of the Pecos, are Coalition period pueblos. Loma Lothrop, north along Glorieta Creek, and Rowe Pueblo, south of Dick's Ruin on the Pecos, are early Rio Grande Classic period sites. Pecos Pueblo was occupied as early as the Coalition period and was not fully abandoned until 1838. Therefore, these five sites include population samples dating from the rise of the Pueblo tradition through European settlement of the area.

The total number of burials from these sites excavated by the Pecos expeditions exceeded 2,000 individuals. The number of individuals recovered for curation was far smaller. Kidder was particularly interested in locating burials. Indeed, he initially offered a reward for each burial discovered. That practice ceased as the number of burials mounted and threatened to consume too much of the expedition's budget (Kidder 1924). When a burial was encountered, its location was recorded on the site map for that area, and a burial number was assigned. One of the supervisory personnel then went to the burial location and filled out a field burial card, which included a sketch of the burial, a listing of any artifacts associated with it, assessments of age and sex data, and metric information about the specific three-dimensional location of the grave. When the bones were removed, the skull was marked with the field burial number (Kidder 1929).

Kidder did attempt to recover all portions of each burial but believed that complete recovery was impossible because of the fragility of the remains (Givens 1992:141). From both Kidder's notes and examination of the recovered remains it is clear that hands, feet, ribs, and vertebrae were seldom collected. As a result of these practices, the first level of excavation bias includes the hands, feet, and thoracic portion of the body. Collected remains from these parts of the body are largely restricted to those that were in exceptionally good condition or that exhibited particularly severe pathologies and thus were deemed of unusual interest.

In 1920 Kidder arranged for Earnest Hooton to participate in the excavations. He hoped that Hooton's participation would help improve the standards he had developed for field observations of the burials, and Hooton hoped his work in the field would facilitate his later laboratory analysis (Hooton 1935; Kidder 1958). To a large degree, both objectives were accomplished. Hooton provided Kidder with tables of criteria for the assessment of age and sex in order to improve the accuracy of the field observations. Unfortunately, Hooton also instructed Kidder not to bother with efforts to recover the more poorly preserved burials (Hooton 1930). As a result, for the later years of the expedition, no remains were collected from many of the Pecos burials. This second source of excavation bias is clearly seen in the demographic records for the burials. Although field burial cards report a large number of infants and children, the collection that was actually made for curation contains very few immature individuals (Kidder 1958).

CURATION PRACTICES

Curation problems with the Pecos collections began in the field. Initially, burials were stored in a field laboratory. This lab was vandalized on more than one occasion, and both collections and field records were lost (Guthe 1918; Pecos Burial Cards n.d.). Partly because of these problems, arrangements were made to ship the burials to Massachusetts for curation and analysis. The initial shipment resulted in further damage to the collection. The crates containing the skeletons were to be transported to Cambridge and placed in a warehouse. A delivery error transported them instead to the Peabody Museum of Archaeology and Ethnology (PMAE) at Harvard University. The delivery company is reported to have taken the crates off the truck and thrown them through the delivery entrance, down the steps, and into the basement of the museum. Several crates are reported to have broken open, with the skeletons falling out of their individual boxes and becoming mixed (Guthe 1918).

In the early years of the museum, collections were catalogued using the name of the expedition that made them (Hooton 1935). In some cases, an expedition involved only a single site. The Pecos expeditions involved five sites, but the cataloguing procedures listed all collections simply as "Pecos." Many of the analyses of the Pecos burials subsequent to their arrival at the Peabody thus have included a mixture of the different sites excavated. During the collection reorganization program that I directed, we were able to rectify this confusion. Each burial had been assigned a number during excavation, and many of these field numbers were recorded in the PMAE catalogue ledger. Numbers had also been written on the remains themselves in the field. During an inventory of the museum's human osteology collection in the late 1980s, we were able to locate the original field numbers for most of the burials and correctly identify the site and intrasite provenience for the majority of the collection.

During the reorganization, another curation problem became apparent. Hooton had created a large teaching collection that was stored in one of the classrooms at the PMAE. Many of the remains in the teaching collection were clearly archaeological in origin. Although some of the remains were marked with field or catalogue numbers, which allowed us to identify records of provenience, others were unmarked in any way. From the notes on the bones, we know that a large portion of this teaching collection was composed of remains from the Pecos Valley sites. It is probable that many of the unmarked bones also came from the Pecos Valley. During the reorganization, we discovered that a number of the Pecos burials recorded in the Peabody catalogue ledgers either were missing entirely or were missing elements recorded as present during the museum's original cataloguing.

HOOTON'S STUDY

Hooton's 1930 volume, *The Indians of Pecos Pueblo,* remains a landmark in the development of American physical anthropology. The preeminence of Hooton's Pecos study is due to two factors. First, he took an innovative tack for his analysis by examining a large number of individuals from a carefully controlled excavation. His treatment of the skeletons as representing a population and his analysis of this group by archaeologically determined stratigraphic periods stand in marked contrast to previous descriptive or case study approaches to human osteology (Beck 2006).

The second factor enhancing the significance of Hooton's monograph lies in the collection itself. Kidder's excavations centered on Pecos Pueblo, a site occupied continuously from roughly A.D. 1200 to the mid-nineteenth century. The excavated burials included close to 2,000 individuals representing all phases of occupation. The size of the burial sample, its outstanding preservation, and the standardization of excavation procedures made for unprecedented research potential in Hooton's day and are the major reasons the collection is still considered a valuable and unique resource.

Hooton's study used only a small subset of the remains discovered by Kidder. Fewer than 60% of the excavated burials had been collected for curation. Hooton's selection of individuals to include in his analysis was based on three criteria. First, he limited his observations to burials excavated before 1925. This eliminated the 327 individuals excavated at Pecos between 1925 and 1929, as well as the majority of individuals from Forked Lightning. He also emphasized remains that had been clearly assigned to one of the ceramic phases Kidder had defined for Pecos. Because assignment of phase was based primarily on ceramics, burials without pottery were largely excluded. In fact, according to Kidder, only 573 burials from Pecos itself were considered assignable to any specific phase (Kidder 1958). Finally, Hooton focused on the better-preserved remains. These selection procedures reduced the sample included in Hooton's analysis to approximately 25% of the excavated burials. For his pathology study, the number of individuals examined varied between 503 and 581 and included some burials from Forked Lightning as well as from Pecos Pueblo.

PEABODY NSF DATA

A National Science Foundation (NSF) inventory project that I directed for the PMAE in the late 1980s provided the first data ever assembled for the total collection of burials recovered by the Pecos expeditions. This inventory project was part of a larger effort to modernize curation of the Peabody collections. For human remains, the project had two phases, record entry and inventory. The project began by transferring catalogue records into a computer database. Field records were used to supplement the provenience information provided in the original catalogue. As I mentioned earlier, this practice enabled us to reestablish provenience data for a majority of Pecos Valley burials.

Under the NSF grant, our inventory procedures entailed a brief osteological examination of all human remains in the museum's collections. In the process, we generated data on both demography and pathology. The NSF inventory data are superficial. We spent an average of 45 minutes per skeleton to conduct a complete inventory of the remains, identify age and sex, and diagnose any pathologies. It was a cataloguing effort and not a research project. At the time I wrote the NSF grant, I was in my fourth year of graduate studies and had only limited experience with pathology. My direction of the NSF inventory project for the PMAE was a phenomenal learning experience that enabled me to observe more than 15,000 individuals from different times and places around the world. Research visitors such as Don Ortner and Jane Buikstra generously examined various individuals with me and offered their insights into various conditions. As a result, by the time I completed the project, I was a much better osteologist than I was at the beginning. Unfortunately, Pecos was the first site we observed during the NSF project. It is probable that we missed many less severe cases of disease and that the actual frequencies were higher.

On the basis of advice I received from osteologists at other museums, I opted to keep the data for the NSF project superficial in order to obtain greater standardization and accuracy throughout the examination of the more than 15,000 burials curated at the PMAE. Although I personally reviewed most individuals for pathology, I used a large number of student assistants for much of the inventory and data entry. Using superficial standards for the project did produce a high degree of consistency in the data recorded. This was simply a museum cataloging effort and was never intended to yield research-quality data.

TABLE 7.1
REPRESENTATION OF AGE CATEGORIES IN THE PECOS VALLEY COLLECTION AS ASSESSED IN THE FIELD AND IN THE LABORATORY

Age class	MNI from field data	MNI from lab data	Percentage retained to lab
Infant	439	36	8
Child	120	62	51
Subadult	145	77	53
Adult	1,068	798	75
Unknown age	51	30	60
Total	1,823	1,003	55

Note: The 1,003 individuals reported for the laboratory are those recorded as individuals during the NSF inventory project. The designation means that either an intrasite provenience was recorded or the remains were represented by more than an isolated bone.

Because of the decisions I had made in designing the NSF inventory for the Peabody, I decided to return for a final examination of the Pecos Valley collections just before their repatriation. That two-week examination, although hurried, provided "salvage pathology" notes in much greater detail than is offered by the computerized inventory records. To maximize data recovery, I tape-recorded my notes as I examined each burial, because two weeks was insufficient for the format of observations generally recommended for pathology research. The examination produced descriptive notes on pathology averaging one page per individual. Those notes, in combination with the more than 2,000 photographs I took during that visit, provide a rich resource for more detailed study of disease at Pecos. I used these notes in conjunction with the NSF data for comparison with Hooton's analysis, but the majority of the frequency data used for comparison is drawn from the NSF inventory. In many ways the superficiality of those data offers a cleaner parallel to the brief notes on pathology provided by Hooton.

DEMOGRAPHY

Turning first to demography, the field records indicate that all age and sex groups were represented among the burials uncovered during excavation in frequencies appropriate to a representative population sample. Unfortunately, this is not true for the burials that were retained following excavation. Comparison between the NSF inventory data and the field records shows that there was a serious bias by age in the recovery of skeletons. Approximately 75% of the adults discovered during excavation were retained, but only 8% of the infants were collected. Although age, sex, and burial attributes were recorded for all graves, only the better preserved skeletons were collected, following the tradition of the 1920s. The poorly preserved remains were primarily those of infants and children and were deemed to offer only limited research value. For Pecos, all subadult age groups are thus underrepresented in the museum's collections (table 7.1.) Hooton's report further incorporated this bias by focusing almost exclusively on well-preserved adults.

PATHOLOGY COMPARISON

My initial objective was to evaluate the way osteological diagnosis had changed since Hooton's original analysis. For pathology, I began by direct comparison of the NSF inventory data with Hooton's diagnoses for the subset of the burials he examined. Hooton both examined the skeletons himself and consulted with several physicians to assess disease conditions in the Pecos collection. Of the 581 skeletons included in Hooton's pathology study, he identified some category of disease in 162, for a 28% frequency of skeletal pathology. In the NSF inventory of the same individu-

TABLE 7.2
SUMMARY OF HOOTON'S REPORTED FREQUENCIES OF PATHOLOGIES

Category	Number examined	Number with pathology	Percentage with pathology
Trauma	581	42	7
Infection	503	17	3
Arthritis	503	66	13

Source: Hooton 1930.

als, we noted pathological conditions in more than 50% of the sample.

The discrepancies between the NSF observations and those reported by Hooton led me to examine his subset of the collection by category of pathology. Perhaps the increased frequencies of diseases recorded under the NSF project were the result of diagnoses of diseases not recognized in the 1920s. For example, Hooton was among the first to discuss osteoporosis symmetrica and to propose its association with nutritional deficiencies. I needed to verify that I was actually comparing the same categories of information between Hooton and the NSF data.

To avoid problems posed by changes in medical terminology and the refinement of diagnostic criteria over the preceding 70 years, I focused my examination on three broad classes of pathologies—trauma (antemortem and perimortem), infection, and arthritis. I reexamined Hooton's raw data and scored each individual for these categories. On this level the discrepancies in diagnoses became even more dramatic.

Hooton noted trauma in only 7% of the 581 individuals he examined (table 7.2). For the same individuals, I found evidence of fractures in 30% of the sample. When I reviewed specific individuals in whom Hooton had identified trauma, a pattern became clear. In many instances it was apparent that well-healed fractures without severe displacement were not recognized, whereas more serious injuries in the same individual were reported in detail. Furthermore, trauma to the hands or feet was only rarely noted. Indeed, Hooton's original observation forms do not even include information space for hands and feet.

Hooton noted infections in 17 individuals, whereas I found evidence of periosteal reactions or osteomyelitis in 69 individuals. Again, it was generally the milder cases that Hooton missed. Severe osteomyelitis was always recorded, whereas healed areas of periostitis and localized periosteal reactions were seldom noted in his analysis. Large expanses of inflammatory lesions on the cranial vault were recorded, but isolated lesions were not reported.

For arthritis, the frequencies we recorded during the NSF inventory again more than doubled those reported by Hooton. The NSF data show an increase from Hooton's 13% to our 38%. Again, Hooton's diagnoses tended to involve the more severe cases, with ankylosis or eburnation, while overlooking mild osteophytosis or marginal lipping of articular surfaces. In fact almost all his reports of spondylitis deformans are instances of ankylosis and include individuals with developmental fusion of vertebral bodies in the cervical spine, trauma (compression fractures), and advanced lytic changes produced by infectious processes such as tuberculosis.

These discrepancies should not be viewed as a criticism of Hooton but rather as evidence of the changes that have taken place in diagnostic criteria over the last 70 years. In the 1920s, when Hooton conducted his analysis, the study of dry bone pathology was in its infancy. The point to be remembered is that care must be taken to recognize how seriously underreported disease frequencies may be in the research reports of earlier decades. Having identified a pattern of underreporting of mild cases of skeletal pathology in all the categories I examined, I next returned to the total collection from Pecos Pueblo to see how the disease frequencies were expressed for the site rather than for the subset Hooton had evaluated.

PRESERVATION AND PATHOLOGY

I first compared frequencies of pathologies for Hooton's subset with those we had generated for the total population during the NSF inventory (table 7.3).

TABLE 7.3
SUMMARY OF NSF PROJECT'S REPORTED FREQUENCIES OF PATHOLOGIES IN 1,003 INDIVIDUALS

Category	Number with pathology	Percentage with pathology
Trauma	116	12
Infection	140	14
Arthritis	273	27

Note: Data are for the burials identified as individuals during the inventory project.

To my surprise, the frequencies of trauma, infection, and arthritis in the total population fell between those reported by Hooton and those I had calculated for his subset. I began some serious "data dredging" to figure out why. Given that all of our inventory data were collected during a four-month period by a single group of people and that I personally scored all the data on pathology, there was no reason to suspect any variation in scoring between the sample group and the total collection. Both groups included all time periods in comparable percentages, so a temporal bias in the sample could not explain the discrepancy. In reviewing Hooton's notes, I realized that although he had included both cranial and postcranial remains in his analysis, he had selected only well-preserved adults for his pathology study. This led me to look at the data in a slightly different manner.

Because the NSF inventory project was designed to meet collection management needs, we had created summary data fields to allow non-osteologists to answer general research inquiries. These summary fields divided the skeleton into five regions: cranial, dental, axial, appendicular, and extremities. Each region was coded for both condition and pathology. Condition coding included four levels: complete, meaning that more than 75% of the bones for that region were present and observable; partial, meaning that between 75 and 50% were present and observable; fragmentary, meaning that fewer than 50% were observable; and absent, meaning that not even fragments from that area were identified. (These standards were developed before the publication of *Standards for Data Collection from Human Skeletal Remains* [Buikstra and Ubelaker 1994] and differ from the coding defined there for the boundaries used for the partial and fragmentary categories.) Using these summary fields, I generated simple frequency data for pathologies observed on the basis of region of the skeleton and condition of preservation.

I first examined only the 74 remains for which we had coded the cranial, axial, and appendicular portions of the body as being complete (table 7.4). In each category, the frequency of disease was greater than the frequency reported by Hooton or in the NSF reevaluation of Hooton's sample. Given the excavation and curation biases reported by both Kidder and Hooton, one would expect the more fragmentary remains from Pecos to exhibit higher frequencies of pathology. Beginning during the 1920 field season, poorly preserved burials were collected only if they exhibited pathology. This means that the frequency of pathology in poorly preserved remains should be higher than the frequency reported by Hooton for the well-preserved burials. In reality, this was not the case. Rather, well-preserved remains exhibited the higher frequencies of pathology across all categories of disease (table 7.5).

Frequencies of pathology were higher when the remains were complete and lower when they were fragmentary. For example, pathological conditions were recorded for 20% of the complete crania and for only 6% of the crania scored as fragmentary. The pattern of preferential collection of pathological remains is seen most clearly in the axial area of the skeleton. In this region the frequency of pathology actually increased for the remains coded as partial before dropping for the more fragmentary remains.

TEMPORAL TRENDS IN PATHOLOGY

Among the five pueblos in the Pecos Valley, the incidence of pathologies increased over time—that is, the Coalition period population shows lower frequencies of all pathologies than the populations dated to the later horizons. A variety of explanations for this can be raised. First, the increased morbidity might be related to European diseases that entered the region in the six-

TABLE 7.4
SUMMARY OF PATHOLOGIES FOR 74 INDIVIDUALS WITH COMPLETE CRANIAL, AXIAL, AND APPENDICULAR REGIONS OF THE SKELETON

Category	Number with pathology	Percentage with pathology
Trauma	12	16
Infection	24	32
Arthritis	26	35
No pathology	19	26

Note: The categories of pathology are not mutually exclusive. Of these 74 individuals, 22% have cranial pathology scored, 49% have axial pathology scored, and 39% have appendicular pathology scored.

TABLE 7.5
FREQUENCY OF PATHOLOGY BY SKELETAL AREA FOR THE 1,003 BURIALS RECORDED AS INDIVIDUALS DURING THE NSF INVENTORY PROJECT

Skeletal area	Total MNI	MNI with pathology (% Total)	MNI C (%)	MNI C with pathology (% C)	MNI P (%)	MNI P with pathology (% P)	MNI F (%)	MNI F with pathology (% F)
Cranial	834	141	499	101	191	31	144	9
		(17)	(60)	(20)	(23)	(16)	(17)	(6)
Dental	787	187	281	90	241	60	265	37
		(24)	(36)	(32)	(31)	(25)	(33)	(14)
Axial	641	280	99	46	243	128	299	106
		(44)	(15)	(46)	(39)	(53)	(46)	(35)
Appendicular	755	275	440	181	140	52	175	42
		(36)	(58)	(41)	(19)	(37)	(23)	(24)
Extremities	543	75	103	19	140	22	300	34
		(14)	(19)	(18)	(26)	(16)	(55)	(11)

Key: C = complete (> 75% present); P = partial (50–75% present); F = fragmentary (< 50% present).

teenth and seventeenth centuries. However, the diseases themselves do not exhibit change across time; it is the incidence that alters. It might be that the disruptions of cultural patterns during this interval resulted in poorer nutrition and increased susceptibility to existing diseases. An alternative explanation that my colleagues and I are investigating centers on the decline and abandonment of the pueblo. The movement away from Pecos was gradual—a few individuals here, a family there, over the course of more than a century. Perhaps the later horizons are somewhat biased in that a healthier segment of the population moved away and left primarily the sick and elderly to live at Pecos (Kidder 1958). Although inconclusive, demographic data suggest that such selective abandonment might have been a contributing factor in the change in health over time.

CONCLUSION

The preceding is a preliminary view of the work I am conducting on the Pecos Valley collections. At the time of this writing, I am conducting a full analysis of pathology using my "salvage pathology" notes. I have transcribed the nearly 80 hours of audiotaped notes from my examination and am working with those notes and photographs to codify the data and establish more formalized diagnoses of specific conditions. My comparison with Hooton's data clearly demonstrates the need for such a reanalysis. The reestablishment of the field provenience now makes it possible to expand the study to include variations among sites at specific time periods, in order to examine regional issues. Clarification of the time periods also enlarges the dated sample to statistically appropriate sizes for all ceramic phases. I am also conducting a time-restricted survey of pathology for the 700 burials from Grasshopper Pueblo in Arizona. The burials from Grasshopper have been thoroughly documented under the NAGPRA inventory we conducted at the Arizona State Museum. My duplicate examination under conditions similar to those I experienced at the PMAE will allow me to further calibrate my two sets of observations for Pecos (the NSF and the salvage pathology) with the level of examination I routinely conduct today.

Although the Pecos collection has been reburied, its potential for analysis remains strong. Diverse data generated over the past century by different scholars can contribute to future interpretations. This chapter is meant to serve primarily as a cautionary statement regarding the need to be attentive to the excavation, curation, and analytical processes that underlie any data set.

8 Stable Isotope Data Show Temporal Stability in Diet at Pecos Pueblo and Diet Variation among Southwest Pueblos

Corina M. Kellner, Margaret Schoeninger, Katherine A. Spielmann, and Katherine Moore

Throughout the occupation of Pecos Pueblo, its subsistence economy centered on intensive maize agriculture (Kidder 1932, 1962; Spielmann 1991b). Extensive archaeological and ethnographic studies, however, reveal a set of crucial social and economic interactions that changed temporally and could have had some effect on the subsistence practices of Pecos residents. Early in the pueblo's history, its members participated in trade with other, nearby pueblos. In the mid-fifteenth century Pecos developed trading relationships with bison-hunting groups living on the southern Great Plains. Beginning in the late sixteenth century, the people of Pecos traded extensively with the Spaniards who occupied their pueblo and nearby Santa Fe (see Spielmann, this volume).

Our previous study of subsistence based on the stable isotopes of the organic fraction of bone (Spielmann, Schoeninger, and Moore 1990) failed to identify any major influence of the first two relationships on the diet of the Pecos inhabitants, the exception being the Spanish occupation. The isotope data supported the archaeological evidence for the intensive use of maize, but our limited knowledge of variation within isotope systems hampered any subtler interpretation of the data.

Since our original analysis, several lines of evidence have become available that, in combination, argue for a reevaluation of our early data set. First, isotope data have now been published for other Southwestern sites, including Grasshopper Pueblo (Ezzo 1992), Gran Quivira (Schoeninger et al. 2003), the pueblo-like community of Henderson in southeastern New Mexico (Speth 2004), and sites once used by the semisedentary groups of Antelope Creek in the Texas Panhandle (Habicht-Mauche, Levendosky, and Schoeninger 1994). We can now compare Pecos Pueblo with these other Southwestern groups to achieve a more accurate understanding of human diet over time and across space. Second, more detailed studies of health in the Southwest before and during European contact (Levine and LaBauve 1997; Ramenofsky 1996; Reff 1992, 1993; Stodder 1996; Stodder and Martin 1992; Stodder et al. 2002) clarify the extent to which disease affected diet and the subsistence economy. Finally, data from several experimental feeding studies (Ambrose and Norr 1993; DeNiro and Epstein 1978; Hare et al. 1991; Howland et al. 2002; Jim, Ambrose, and Evershed 2004; Tieszen and Fagre 1993a) serve to refine our understanding of the relationship between the stable isotope signature of diet and the tissues formed from the diet. In this chapter we consider these new data in reevaluating the stable isotope data from Pecos Pueblo.

BACKGROUND AND OUR PREVIOUS STUDY

The period preceding and including the establishment of Pecos Pueblo at approximately A.D. 1200 was a time of population mobility, aggregation, and conflict that produced widespread population increases and the first large-scale settlement of people in the Rio Grande region (Adler, Pool, and Leonard 1996; Bernardini 1998; Cordell, Doyel, and Kintigh 1993; LeBlanc 1999). Around this time, a period of drought detrimentally affected agricultural outputs in the region (Adler, Pool, and Leonard 1996; Dean, Meko, and Swetnam 1994; Larson et al. 1996). A combination of drought and conflict may account for the development of the multiple, aggregated settlements we now call pueblos (Cordell, Doyel, and Kintigh 1993; LeBlanc 1999). Pecos Pueblo became one of the largest of these communities and was important in the trading networks between pueblos, with Plains Indian groups, and with Spaniards.

Straddling the plains to the east and the rest of the Pueblo region to the west, Pecos Pueblo's location was ideal as a gateway connecting many different communities. Throughout its history Pecos served as an important node in several networks of "interethnic and interregional contacts" (Levine 1999:xix). Interpueblo trading was lively from the earliest periods (A.D. 1200–1450), although most scholars agree that, except during stressful environmental periods (Ford 1972), the exchange involved raw materials and craft items rather than food (Snow 1981, 1991). The combination of drought and conflict might also explain, at least in part, the observation that the Pueblo groups had some of the highest morbidity burdens in North and South America (Stodder et al. 2002). For this early period of habitation, archaeological evidence indicates that Pueblo subsistence was focused on maize, with variable inputs of fauna and flora local to each pueblo. Our previous stable isotope study showed heavy dependence on maize or other C_4 foods at Pecos. Local fauna, which carried a C_3 signature (e.g., antelope and mule deer), was imperceptible in the human stable isotope data, even though remains of such animals were recovered in great quantities during excavations (Kidder 1932:196).

Beginning in earnest during the mid-1400s, the inhabitants of Pecos Pueblo traded ceramics, maize, and cotton blankets for buffalo meat and hides with the semisedentary Plains hunter-gatherer-horticulturists to the east (Lintz 1984, 1991; Spielmann 1989, 1991b, 1996b). On the basis of the presence of skin preparation and leather-working tools at the pueblo, Kidder (1932) suggested that Pecos specialized in the trade of prepared buffalo hides among the Pueblos. Historic reports mention trade between the Plains and Pueblo groups with great interest, because the Spaniards prized the ubiquitous Pueblo buffalo robes (Kessell 1979:137). Such reports suggest that after 1400, Pecos inhabitants had greater access to bison meat. Bison meat should carry a C_4 signature, and we expected to see increased dependence on C_4 foods as more bison meat entered Pecos. Yet our stable isotope data showed no increased dependence on C_4 foods in the human diet during this period of specialized trade for bison (Spielmann, Schoeninger, and Moore 1990).

After initial contact in 1540, Spaniards occupied the village in 1598, and Pecos Pueblo was incorporated into Spain's economic, legal, administrative, and social networks (Levine 1999). The occupation disrupted Pecos's activity in the trading system as the Spaniards began to trade directly with semisedentary Plains groups while using horses and superior weapons to obtain Apache slaves (Kessell 1979; Snow 1981; Spielmann 1989). The Spaniards also monopolized the available agricultural land (through *encomienda*) and co-opted Pueblo labor for carpentry and to extract maize, buffalo hides, and pottery (Anderson 1999; Dozier 1970; Kessell 1979; Snow 1981; Spielmann 1989). Pecos Pueblo became the richest *encomienda* in New Mexico (Kessell 1979:188). Its contact with Spanish colonists continued until 1838, when the site was completely abandoned. The stable isotope data from the period of Spanish occupation suggested that the diet of Pecos Pueblo's inhabitants changed to include more C_3 foods. We previously argued (Spielmann, Schoeninger, and Moore 1990) that the Spaniards' demands for Pueblo labor and food forced the inhabitants of Pecos Pueblo to revert to a generalized subsistence strategy for their own consumption and that they increased their dependence on the wild floral and faunal resources surrounding the site. Since that early study, however, updated temporal attributions of the Pecos Pueblo burials require reconsideration.

NEW DATA AND CONSIDERATIONS

Stable isotope data are now available from several sites that overlapped Pecos Pueblo during some or all of its occupation (table 8.1). First, Grasshopper Pueblo (A.D. 1275–1400), in east-central Arizona, dates to the pre–Plains trade period at Pecos. Both sites were large, aggregated pueblos whose inhabitants cultivated

TABLE 8.1
HUMAN STABLE ISOTOPE VALUES FOR PECOS PUEBLO AND OTHER SOUTHWESTERN SITES

Site	n	$\delta^{13}C$‰	Standard deviation (‰)	$\delta^{15}N$‰	Standard deviation (‰)
Pre–Plains trade period (A.D. 1200–1450)					
Pecos Pueblo	9	–7.6	±0.4	9.1	±0.6
Grasshopper	53	–9.1	±1.1	7.7	±0.8
Henderson	3	–8.7	±0.3	10.4	±0.8
Antelope Creek	17	–8.3	±0.9	9.7	±0.5
Gran Quivira	3	–7.0	±0.3	9.3	±1.0
Trade period (A.D. 1450–1600)					
Pecos Pueblo	16	–7.6	±0.4	8.9	±0.6
Antelope Creek	17	–8.3	±0.9	9.7	±0.5
Gran Quivira	11	–7.0	±0.3	8.9	±0.8
Historic period (A.D. 1600–1838)					
Pecos Pueblo	15	–7.7	±0.5	9.1	±0.8
Gran Quivira	17	–7.2	±0.6	8.8	±0.6

maize. The area surrounding Grasshopper, at 1,829 meters above sea level, is marked by upland stands of ponderosa pine, piñon, and juniper (Ezzo 1991, 1992, 1993), much as Pecos is. Grasshopper was inhabited during a period of great regional migration and the consolidation of Pueblo populations into aggregated villages (Adler, Pool, and Leonard 1996; Cordell, Doyel, and Kintigh 1993). Variations in cranial modification styles suggest that the pueblo attracted migrants from other areas (Ezzo 1993:52), where they might have had different subsistence regimes.

The pueblo-like community of Henderson lies well to the south of Pecos Pueblo, near Roswell, New Mexico. It, too, dates to the pre–Plains trade period at Pecos (Speth 2004). Although Henderson was not as large as the other settled communities in the region, its sedentary people practiced maize agriculture, hunted local antelope and rabbits, and relied heavily on bison hunting and trading.

Next, the semisedentary, bison-hunting Antelope Creek groups (Habicht-Mauche, Levendosky, and Schoeninger 1994; Lintz 1984, 1986, 1991) compare temporally to Pecos Pueblo during both the pre–Plains trade period and the trade period. The people of these sites were likely trading partners with eastern pueblos such as Pecos. Trade goods such as painted ceramics, turquoise, and marine shell beads at Antelope Creek sites reinforce a Plains-Pueblo connection (Lintz 1984, 1986, 1991). These groups cultivated maize and consumed other C_4 or CAM plants such as amaranth and prickly pear cactus, but they also subsisted on the diverse wild food resources of the region (Lintz 1991). Bison was the most common animal represented at Antelope Creek archaeological sites (Habicht-Mauche, Levendosky, and Schoeninger 1994; Lintz 1991).

Finally, Gran Quivira (1315–1672), one of the Salinas pueblos, is located to the south of Pecos. It was another large, agricultural trading community inhabited during the prehistoric and historic periods (Hayes 1981; Spielmann 1991b; Stodder 1996; Stodder and Martin 1992). During the historic period, the Spanish presence was shorter-lived at Gran Quivira than at Pecos, and the site was abandoned in 1672, 166 years before Pecos was vacated. Therefore, in making comparisons with Gran Quivira, we deleted the Glaze 6 period (1675–1838) from the historic period at Pecos Pueblo. The pre–Plains trade period at Pecos Pueblo corresponds to the Early period (1315–1400) at Gran Quivira, the Pecos trade period to the Gran Quivira Middle period (1400–1550), and the Pecos historic period minus Glaze 6 to the Late period (1550–1672) at Gran Quivira.

In addition to the new data on other sites, we considered evidence for Pueblo diet during the Spanish occupation, at which time the inhabitants of Pecos Pueblo may have included several European foods in their diet. One of these was wheat, which was grown in the fields surrounding pueblos for human and presumably livestock consumption (Ford 1987; Kessell 1979). Additionally, Spanish records indicate that the Native inhabitants could have eaten sheep and goats, which would carry a C_3 signal if the animals fed on local vegetation (Ford 1987; Kessell 1979). Cattle, however, as grazers, would carry a C_4 signature. According to mission records, raiding by Plains groups and droughts in the 1700s "forced the people of Pecos Pueblo to abandon far-flung fields around the perimeter of the pueblo [and] the Spanish governor [intervened] to give them a dozen cows for their own consumption" (Kessell 1979:347). The Spanish response to such raiding might have been common as a means of ensuring the stability of the Pueblo labor force. In other parts of the greater Southwest, Jesuits gave sheep and goats to local Pueblo laborers as rewards for demonstrating "fealty" to the Christian god (Reff 1993), and it is likely that the Franciscans at Pecos Pueblo practiced such "incentives."

Societal changes, however, might not have affected all individuals in the pueblo equally. Women might have shouldered more of the workload than men, in making pottery and processing maize for trade (Habicht-Mauche 2000; Spielmann 1996b). In addition, differences in status might have developed as some people took advantage of fluctuating trade relationships in the context of societal upheaval (Vehik 2002). Therefore, in our reevaluation of the stable isotope data, we considered sex differences and group variation.

HEALTH AT PECOS PUEBLO

Several lines of evidence elucidate the nutritional status of the people who inhabited Pecos Pueblo. One is the frequency of skeletal pathologies such as cribra orbitalia and porotic hyperostosis, which in some cases are caused by nutritional deficiencies. The subadult frequency of these two markers is low at Pecos Pueblo in comparison with other Southwestern pueblos. Subadult data are commonly reported because subadults exhibit these lesions at higher frequencies and with greater severity than adults (El Najjar 1976). This is because anemic stress during the growth period spurs red blood cell production in the hemopoietic bone marrow and the concomitant expansion of bone (Stuart-Macadam 1985). Although Hooton's original study of the skeletal collection showed that 1.5% of Pecos Pueblo individuals exhibited such lesions (Hooton 1930:319), a recent reanalysis showed 2% to 4% of individuals exhibiting cribra orbitalia across time periods, including 10.5% of subadults (Morgan, this volume). In comparison, such lesions affected 15% of subadults from the pre–Plains trade period site of Grasshopper and 18% of burials from the southern trading pueblo of Gran Quivira (Stodder et al. 2002). At other Southwestern communities, the frequency of affected subadults was even higher, ranging from 45% at Pueblo Bonito in Chaco Canyon, New Mexico, to 87% at San Cristóbal Pueblo (Stodder et al. 2002). Pecos Pueblo had the lowest frequency of subadults affected with probable anemia among these Southwestern sites.

Several alternative explanations may account for the differences, some of which have dietary implications while others do not. One possible cause of the lesions is iron deficiency anemia. Maize, which is relatively low in this micronutrient, is often implicated in anemia at other Southwestern sites (Stodder et al. 2002; but see Walker 1985). Such a scenario implies that alternative sources of iron, especially meat, were in short supply. According to this line of reasoning, the relatively low frequency of cribra orbitalia and porotic hyperostosis at Pecos Pueblo would indicate that meat was available in adequate amounts to meet iron requirements.

Iron deficiency anemia, however, can be caused by factors other than dietary insufficiency, and additional biological and cultural factors deserve consideration (Holland and O'Brien 1997). Food preparation may affect iron absorption. Parasitic, viral, and bacterial diseases associated with sedentary lifestyles and careless trash disposal, resulting in contamination of water, food supplies, or both, have also been implicated (Kent 1986; Reinhard 1990; Walker 1985). At least eight species of helminthic parasites that can cause anemia have been found in human coprolites at earlier ancestral Pueblo sites (Reinhard 1988). But Pecos Pueblo had a relatively dependable water supply in the form of a perennial stream (Kessell 1979:10) and may have suffered fewer parasitic infections than did other Pueblo populations such as that of Gran Quivira, which relied on easily contaminated wells near the site (Howard 1959). Whatever the absolute cause of the lesions, the relatively low frequency at Pecos provides no support for nutritional deficiencies at the site.

Other evidence, however, suggests that the Spanish occupation was detrimental to the health of many people living at Pecos Pueblo. Although the frequency of individuals affected by skeletal infections (Morgan, this volume) stayed relatively constant over time, at approximately 10%, individual cranial traumas increased substantially during the Spanish occupation, from a range of 1.5–4.0% of the sample affected during the pre-Spanish periods to about 10% of the mission church sample (Bruwelheide, Owsley, and Jantz, this volume). Mission records also imply that epidemic diseases and raiding by Plains groups decimated the population at Pecos during the historic period (Levine and LaBauve 1997). Epidemic diseases that quickly kill the affected individual are unlikely to leave skeletal lesions. Pecos Pueblo mission burial records from 1694 to 1800 show that epidemic diseases accounted for 22% of all entries, and these deaths encompassed approximately 12 episodes of disease (Levine and LaBauve 1997). The aggregated nature of protohistoric pueblos, especially Pecos, as well as the presence of Spaniards and their livestock, might have increased disease frequencies during the historic period (Ramenofsky 1996:177).

Raiding by Plains groups, even though less detrimental to overall Pecos Pueblo health than epidemics (because it resulted in far fewer deaths), was not uncommon during the historic period and was probably linked to major episodes of drought in the region (Levine and LaBauve 1997). During the drought of the 1670s, Pecos was a major food supplier for other pueblos (Ivey 1994), but in 1676 Fray Francisco Atanasio Domínguez described terrible droughts that taxed the food stores at Pecos as well (Levine and LaBauve 1997). The loss of knowledgeable adults from the Pecos population through disease and conflict might have detrimentally affected the pueblo's ability to produce sufficient food stores for the future. It might also have weakened the pueblo's capacity to cement alliances during stressful times with other groups through marriage (Ford 1972). All these factors, in concert, must have disrupted the social system and networks that were relied on for trade relationships (Ramenofsky 1996; Reff 1992, 1993; Stodder and Martin 1992). In an analogous situation, the Guale of coastal Georgia began to eat a lower-quality diet consisting of more maize and less marine fish as the Spaniards of La Florida diverted the Guales' labor to their own profit (Schoeninger et al. 1990).

STABLE ISOTOPES AS INDICATORS OF DIET AND PROTEIN

Experimental studies on a large series of animals show that the $\delta^{13}C$ values in tissue proteins (bone collagen, shell organic, muscle, and chitin) reflect those of diet (fig. 8.1). The reported offset between diet and tissue shows a range from 1‰ to 4‰ (parts per mil, or thousand) in laboratory studies (DeNiro and Epstein 1978; Hare et al. 1991; Hobson and Clark 1992; Tieszen and Boutton 1989) and 5‰ in field studies (Lee-Thorp, Sealy, and van der Merwe 1989; Schoeninger and DeNiro 1984; Vogel 1978), which suggests that diet quality may have an effect on the size of the offset. Rats (Ambrose and Norr 1993; Jim, Ambrose, and Evershed 2004), mice (DeNiro and Epstein 1978; Tieszen and Fagre 1993a), and pigs (Hare et al. 1991; Howland et al. 2002) show an approximate mean diet-collagen offset of 3.6‰ when fed protein at levels sufficient to maintain tissue turnover without weight loss or gain. Variation exists between and within the studies (see fig. 8.1), but in general, animals on normal protein diets synthesize bone collagen that reflects total diet.

Animals probably synthesize tissue proteins from total diet when eating a low-protein diet as well, because nonessential amino acids needed for bone collagen synthesis must be synthesized de novo (Schwarcz 2000). Since nonessential amino acids make up the bulk of synthesized bone collagen, the collagen would have a $\delta^{13}C$ value reflecting total diet under such low protein conditions. The experimental data are somewhat equivocal on this point, because animals died when fed at extremely low protein levels (Ambrose and Norr 1993).

Under conditions of excess protein, bone collagen may reflect the $\delta^{13}C$ value of dietary protein rather than total diet, because it is energetically more efficient to use both essential and nonessential amino acids from diet in synthesizing tissue proteins than to synthesize them de novo (Schwarcz 2000). The experimental data for rats raised on excess amounts of protein (approximately 77% of total diet by weight; Ambrose and Norr 1993) are equivocal, but we considered it unlikely that Southwestern groups, including the people at Pecos Pueblo, had diets with excess protein in either case. Therefore, we assumed that human bone collagen $\delta^{13}C$ values reflected the $\delta^{13}C$ values of total diet at all sites.

By assuming that the total diet contributes to collagen formation, we expect that the protein and carbohydrate fractions of diet are both reflected in the $\delta^{13}C$ values of human bone collagen. This is significant because, although the main carbohydrate portion of diet at all sites was maize, the protein could have come from a C_4 plant (maize or amaranth), a C_3 plant (piñon nuts or beans), animals that primarily consumed C_4 plants (bison or cows), or animals that

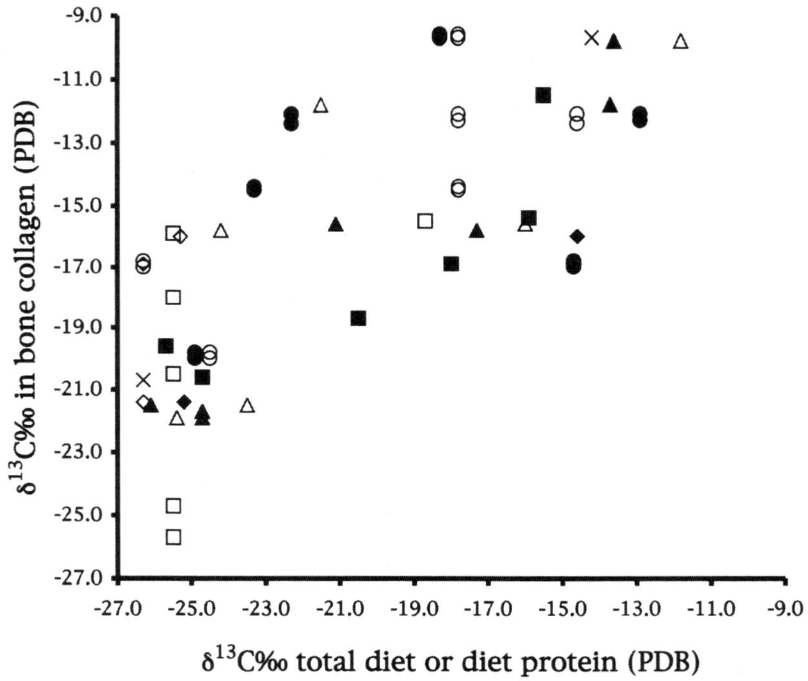

FIGURE 8.1. Correlations between $\delta^{13}C$ values in total diet or dietary protein and $\delta^{13}C$ values in bone collagen. Experimental data for rats, mice, and pigs on normal levels of dietary protein (11–27% of diet by weight or ingredients) show that $\delta^{13}C$ bone collagen values correlate positively with total dietary $\delta^{13}C$ values and with dietary protein $\delta^{13}C$ values. The $\delta^{13}C$ values for bone collagen therefore reflect total dietary $\delta^{13}C$ values. Closed symbols represent total dietary $\delta^{13}C$ values; open symbols represent dietary protein $\delta^{13}C$ values. Diamonds represent data from Ambrose and Norr 1993; triangles represent data from Tieszen and Fagre 1993; squares represent data from Howland et al. 2002; and circles represent data from Jim et al. 2004. *X*s represent animals on high-protein diets (70% of diet by weight), from Ambrose and Norr 1993. PDB stands for Peedee Belemnite carbonate, the standard for comparison for $\delta^{13}C$ values.

primarily consumed C_3 plants (antelope, mule deer, goats, sheep, or rabbits). If the protein in the diet contained a predominantly C_3 signal, then the $\delta^{13}C$ value of human bone collagen would be more strongly negative than if the protein contained a predominantly C_4 signal. This would be true even though the $\delta^{13}C$ value of human bone collagen falls within the C_4 range.

MATERIALS AND METHODS

At the time of Kidder's original excavations (1915–1929), Pecos Pueblo was the most systematically excavated site in North America, and its time sequence has been well established through dendrochronology and ceramic seriation (Kidder 1932, 1962). Sixty-four adult human skeletons from the original excavations were analyzed in our early study and are reexamined here. For the purposes of the questions examined in the present evaluation, the samples were combined into three relatively long periods on the basis of several of Kidder's ceramic phases. These periods are A.D. 1200–1450 (Black-on-white, Glazes 1 and 2), which predates the development of Plains-Pueblo interaction; 1450–1600 (Glazes 3 and 4), which comprises the Plains-Pueblo trade period; and 1600–1838 (Glazes 5 and 6), which encompasses the centuries of European contact.

To estimate the isotopic values of the foods available to humans at Pecos Pueblo, we also analyzed faunal and floral remains from Pecos and nearby sites. Excavation reports (Kidder 1932) list the animals most commonly recovered archaeologically as mule deer, rabbits, antelope, mountain sheep, and elk, with mule deer being most common. Bison bones occurred at all levels but were most numerous in deposits dating to approximately 1450 (Kidder 1932; Spielmann 1989). Remains of European domestic animals such as horses, cows, chicken, sheep, and goats were found in the Spanish-era middens at the site (Kidder 1932:196).

Turkey bones were also common; indeed, they were the most plentiful of all bird remains in Pecos Pueblo middens (Kidder 1932). Ethnohistoric data from the Southwest suggest that turkeys were kept largely for their feathers at many pueblos (Schroeder 1968; Senior and Pierce 1989; Winship 1896). In their accounts of Pecos, Spaniards specifically mentioned elaborate turkey feather robes (Kessell 1979:55). At nearby Arroyo Hondo Pueblo, however, turkey remains were recovered in sufficient abundance to suggest that the birds were a staple in the diet (Lang and Harris 1984). Some turkey remains from Gran Quivira show disarticulation (McKusick 1981), and there were cut marks on some turkey bones in refuse middens at Pecos Pueblo (Kidder 1932:197–199), although, following the custom of the time, no systematic recording and analysis of these data were carried out. Turkeys might have been important sources of protein at certain Southwestern pueblos, part of a strategy to cope with relatively limited access to other types of meat (Spielmann and Angstadt-Leto 1992).

We limited our analysis to animals commonly exploited prior to the large-scale introduction of European domesticates. Six faunal specimens from Pecos Pueblo (2 deer, 2 antelope, and 2 bison), 19 specimens from the nearby protohistoric pueblo of Gran Quivira (4 antelope, 9 hares and rabbits, 6 turkeys), and 5 bison from the nearby Texas site of Tierra Blanca and the Pecos Valley site of Henderson were analyzed isotopically. Antelope, deer, and the smaller mammals are mainly browsers that feed on the local trees and herbaceous vegetation, most of which are C_3. The predominance of C_3 plants in the area surrounding Pecos Pueblo is most likely an effect of the high altitude (more than 2,000 m above sea level) and low minimum July temperatures (see Boutton, Harrison, and Smith 1980; Cavagnaro 1988; Teeri and Stowe 1976). Bison, the meat of which was ostensibly obtained through trade with Plains groups to the east (Baugh 1984; Snow 1981; Spielmann 1991b), are grazers and feed mainly on the C_4 grasses that grow on the plains just to the east of Pecos Pueblo (see Steuter et al. 1995; Tieszen 1994). The majority of grasses on the Texas and Oklahoma plains are C_4 (Teeri and Stowe 1976).

A mixture of modern and archaeological floral materials was also analyzed for the original study. The most common floral remains from the site were maize, beans, chenopodium, piñon nuts, and amaranth (Kidder 1932), but following excavation standards at the time, archaeological samples were not collected and so were unavailable for analysis. For that reason, modern maize, chenopodium, and amaranth samples, collected from nonfertilized areas near Pecos Pueblo, were analyzed for the present study. In addition, an archaeological sample of beans from the site of Arroyo Hondo (Wetterstrom 1986) had been included in our original analysis. Piñon nuts were not available from Pecos Pueblo or Arroyo Hondo, but recently published data on modern and historic varieties from New Mexico (Pendall, Betancourt, and Leavitt 1999) and a modern sample from the Great Basin (Schoeninger 1995b) provide a comparison for the present study. All data are presented as delta (δ) values in parts per mil (‰) notation, where the standard for comparison is Peedee Belemnite carbonate (PDB) for $\delta^{13}C$ values and atmospheric nitrogen (AIR) for $\delta^{15}N$ values.

RESULTS

Forty-nine human bone collagen samples from Pecos Pueblo had acceptable carbon:nitrogen ratios (Appendix G), with an average carbon stable isotope value of –7.6‰ and an average nitrogen stable isotope value of 9.1‰. Temporal reassignment of the Pecos Pueblo burials (Appendix G) yielded 40 individuals that could be dated accurately to a specific time period. Unlike in the previous assessment, we found no temporal trend toward lighter average $\delta^{13}C$ values, and there are no significant differences in nitrogen stable isotope values between the time periods (table 8.1). There are also no differences between the sexes. The overall mean female $\delta^{13}C$ value is –7.7‰ (n = 24, sd = 0.5‰, range = –8.6 to –6.7‰), and the overall male $\delta^{13}C$ value is –7.6‰ (n = 24, sd = 0.3, range = –8.5 to –7.1‰). The overall mean female $\delta^{15}N$ value is 9.2‰ (n = 24, sd = 0.7‰, range = 8.2 to 10.1‰); the overall male $\delta^{15}N$ value is slightly lower, at 8.9‰ (n = 24, sd = 0.6, range = 8.0 to 10.4‰).

The range of variation within all periods for $\delta^{13}C$ is 1.9‰ (–8.6 to –6.7‰), which approximates the 2‰ range in $\delta^{13}C$ previously observed in animals fed monotonous diets (Schoeninger and DeNiro 1984). The range of variation for $\delta^{15}N$ is 2.4‰ (8.0 to 10.4‰), which is close to the 2‰ range in $\delta^{15}N$ reported for animals on monotonous diets. The $\delta^{15}N$ values are consistent with those reported previously for terrestrial omnivores (Schoeninger 1995a).

Most of the floral samples taken near Pecos Pueblo and at Arroyo Hondo have $\delta^{13}C$ values that are more strongly negative than that of maize (table 8.2; fig. 8.2). At the time these sites were occupied, the $\delta^{13}C$

TABLE 8.2
FLORAL STABLE ISOTOPE VALUES

Plant	*n*	$\delta^{13}C$‰	$\delta^{15}N$‰	Time period and location	Reference	Estimated $\delta^{13}C$ average value during Pueblo occupation (‰)
Chenopodium	1	–25.5	8.1	Modern; Pecos Pueblo	Spielmann et al. 1990	–24.0
Beans	1	–25.1	5.3	Archaeological; Arroyo Hondo	Wetterstrom 1986	–25.1
	1	–24.6	5.6	Archaeological; Arroyo Hondo	Wetterstrom 1986	–24.6
Piñon	8	–19.6 (avg.)	—	Modern; central New Mexico	Pendall et al. 1999	–18.5
	2	–18.2 (avg.)	—	Historic; central New Mexico	Pendall et al. 1999	–16.7
	1	–21.2	1.0	Modern; Great Basin	Schoeninger 1995b	–19.7
Amaranth	1	–11.3	9.0	Modern; Pecos Pueblo	Spielmann et al. 1990	–9.8
Maize	1	–11.2	7.6	Modern; Pecos Pueblo	Spielmann et al. 1990	–9.7
	1	–11.3	6.3	Modern; Pecos Pueblo	Spielmann et al. 1990	–9.7

value of atmospheric carbon dioxide (CO_2) was approximately 1.5‰ less negative than today's (Wahlen 1994), and we have adjusted the values in table 8.2 accordingly. The C_3 plants are chenopodium (–24‰), beans (–25‰), and piñon nuts (–20‰ to –17‰). Maize and amaranth are C_4 plants and exhibit prehistoric values averaging approximately –10‰ (Spielmann, Schoeninger, and Moore 1990). The nitrogen values for most of the flora analyzed here show that these plants use soil rather than atmospheric nitrogen, with values no lower than 5.3‰ and as high as 9.0‰ for amaranth (table 8.2). The $\delta^{15}N$ values of the piñon nuts from New Mexico were not analyzed in the study by Pendall, Betancourt, and Leavitt (1999), but Great Basin piñon has a $\delta^{15}N$ value of 1‰ (Schoeninger 1995b).

Many of the fauna also show a C_3 signal (table 8.3, fig. 8.2). The average $\delta^{13}C$ values of antelope (–18.1 ± 0.7‰, $n = 6$), deer (–18.8‰, $n = 2$), and rabbits (–17.2 ± 3.3‰, $n = 5$) all fall within the C_3 range, although two of the rabbits (–12.2‰ and –16.0‰) from Gran Quivira clearly ate C_4 plants. The hares from Gran Quivira also exhibit $\delta^{13}C$ values that indicate significant ingestion of C_4 plants (–14.1 ± 1.4‰, $n = 4$), with some eating more than others. Clearly dependent on mostly C_4 plants were the bison ($\delta^{13}C = -9.5 \pm 1.4$‰, $n = 7$) and the turkeys ($\delta^{13}C = -7.1 \pm 0.3$‰, $n = 6$). These $\delta^{13}C$ values, in combination with evidence of maize in turkey coprolites at Cedar Mesa, Utah (Matson and Chisolm 1991:455) and knowledge that Rio Grande turkeys eat seeds in the wild (Natural Resources Conservation Service 1999), lead us to conclude that most turkeys at Pecos had a bone collagen value within the C_4 range. The average $\delta^{15}N$ values of the fauna (table 8.3) range from a high of 8.5‰ for the turkeys to a low of 4.3‰ for the deer.

DISCUSSION

Overall, the skeletal remains from Pecos Pueblo have $\delta^{13}C$ values in bone collagen that indicate substantial reliance on C_4 foods. This means that the plant portion of the diet must have consisted largely of maize or amaranth, although the obvious focus of the pueblos on growing maize suggests that amaranth was har-

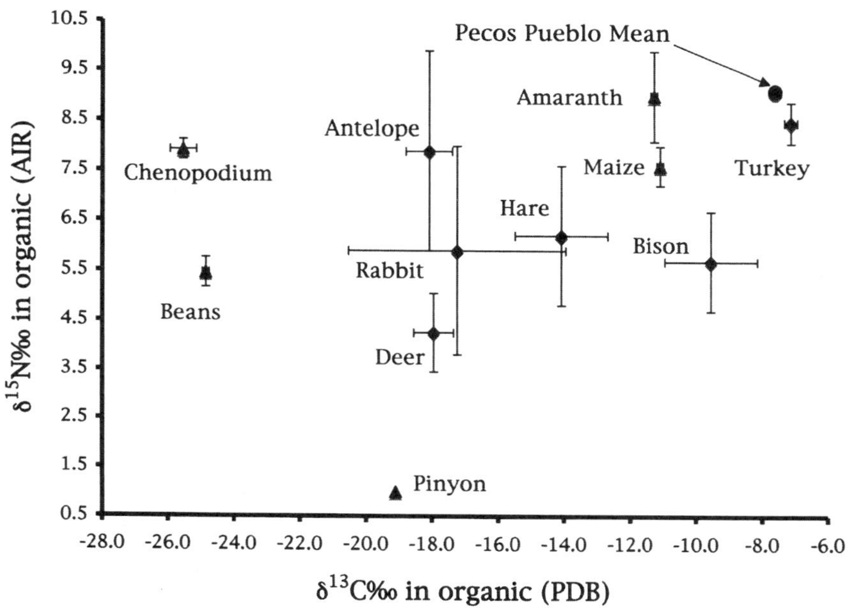

FIGURE 8.2. Mean $\delta^{13}C$ and $\delta^{15}N$ values for C_3 and C_4 plants and animals relative to mean $\delta^{13}C$ and $\delta^{15}N$ values for the Pecos Pueblo human population (large circle at upper right). The values for human bone correspond best to a diet composed predominantly of C_4 plants and animals. Lines radiating from symbols represent one standard deviation.

vested in insignificant quantities. Correspondingly, the carbon isotope data suggest that few C_3 foods such as beans, chenopodium, and piñon nuts were eaten. Because animals that consume normal or low levels of dietary protein have bone collagen $\delta^{13}C$ values that reflect the complete diet, the data indicate that the main sources of protein were also C_4 plants or animals that fed on C_4 foods. This suggests that the protein in maize was a more significant plant source of amino acids in the human diet than were beans. In terms of animal protein, the data are more consistent with the eating of animals such as bison, turkeys, and some rabbits and hares than with the eating of antelope and deer. This result is strongly at odds with the faunal information; mule deer was by far the most abundant kind of animal bone in the Pecos trash middens (Kidder 1932:196). At present, we cannot resolve the differences between the two data sets. It is impossible to reconcile the human stable isotope data with the traditional view of maize as an insignificant source of protein in the precontact Pueblo diet or with the proportion of mule deer bones in the faunal collection from Pecos.

The average human $\delta^{15}N$ value is higher than the $\delta^{15}N$ values for most of the fauna. The nitrogen stable isotope values for maize at Pecos Pueblo are high, however, and further work is needed to determine the source of the human nitrogen values. Indeed, all the modern plants grown near Pecos (chenopodium, amaranth, and maize) have $\delta^{15}N$ values that are approximately 2‰ higher than the average plant values. This indicates that the soil must have been enriched in ^{15}N. The burning of biomass could produce such enriched soil values, because up to 40% of nitrogen is lost in that process (Kuhlbusch et al. 1991). The nitrogen remaining in the soil would be enriched in ^{15}N, and plants growing in it would give higher $\delta^{15}N$ values than normal. Indigenous land management was common in many regions of precontact North America, and groups cleared fields through burning (see Hammett 1992, 1997). Alternatively, the people of Pecos might have used animal waste such as turkey dung as fertilizer. No evidence exists, however, that Southwestern agriculturalists used fire as a land management tool or turkey dung as fertilizer. We need to analyze prehistoric plants from the region in order to investigate these patterns further.

Looking at Pecos diachronically, we see no change in human $\delta^{13}C$ or $\delta^{15}N$ values during the period of increased trade with Plains hunter-gatherers or during

TABLE 8.3
FAUNAL STABLE ISOTOPE VALUES

Sample no.	Animal	Site	$\delta^{13}C$‰	$\delta^{15}N$‰	C:N ratio	Extraction yield (%)
2880	Antelope	Pecos Pueblo	–18.4	8.6	2.7	26.0
2894	Antelope	Pecos Pueblo	–18.5	6.3	2.9	13.0
2892	Deer	Pecos Pueblo	–17.5	5.0	3.0	16.4
2893	Deer	Pecos Pueblo	–18.4	3.5	2.6	21.7
2889	Bison	Pecos Pueblo	–8.6	5.8	2.8	15.2
2890	Bison	Pecos Pueblo	–11.9	7.1	2.7	16.7
2645	Bison	Tierra Blanca	–10.8	6.3	2.6	9.2
2646	Bison	Tierra Blanca	–7.8	3.8	2.7	7.3
955	Bison	Henderson	–8.9	5.1	3.0	3.8
956	Bison	Henderson	–9.6	5.9	3.6	5.5
957	Bison	Henderson	–9.2	5.8	2.6	4.2
3410	Turkey	Gran Quivira	–6.9	9.8	2.9	17.0
3411	Turkey	Gran Quivira	–6.7	8.0	2.9	14.4
3413	Turkey	Gran Quivira	–7.1	8.3	3.0	3.1
3414	Turkey	Gran Quivira	–7.4	8.8	3.0	10.7
4082	Turkey	Gran Quivira	–7.2	7.5	2.9	17.7
4084	Turkey	Gran Quivira	–7.4	8.5	3.0	18.6
4014	Rabbit	Gran Quivira	–21.1	5.4	3.0	17.7
4015	Rabbit	Gran Quivira	–12.2	7.3	3.0	16.8
4016	Rabbit	Gran Quivira	–16.0	6.3	2.9	18.9
4017	Rabbit	Gran Quivira	–18.5	3.9	2.9	18.6
4018	Rabbit	Gran Quivira	–18.4	6.6	3.4	16.0
4019	Hare	Gran Quivira	–14.3	6.2	3.0	15.5
4020	Hare	Gran Quivira	–13.5	6.3	3.0	20.3
4021	Hare	Gran Quivira	–12.6	7.9	3.0	16.9
4022	Hare	Gran Quivira	–15.9	4.4	3.3	16.4
4024	Antelope	Gran Quivira	–18.7	10.6	3.0	14.2
4025	Antelope	Gran Quivira	–16.8	7.0	2.9	12.5
4026	Antelope	Gran Quivira	–18.3	5.4	3.1	19.2
4027	Antelope	Gran Quivira	–17.8	9.5	3.3	18.2

the historic period (table 8.1). Maize remained the staple food in the diet throughout. The carbon data, combined with the high $\delta^{15}N$ values, suggest that maize, bison, turkey, rabbits or hares, and possibly cows in the historic period were the most common sources of dietary protein, whereas antelope, deer, and beans were less important in the diet. The stable nitrogen values suggest that the inhabitants of Pecos Pueblo neither increased nor decreased their meat intake, although, as mentioned previously, more work on plant nitrogen values is needed before this can be determined conclusively.

The data suggest several alternative hypotheses. The simplest interpretation is that neither the plant nor the animal portions of the Pecos diet shifted significantly over time. One possibility is that bison meat was eaten at the same levels throughout the whole occupation of Pecos, with dried bison meat coming into the pueblo during the pre–Plains trade period, when the frequency of bison bones in the deposits is low. During the trade period, when the numbers of bison bones in the deposits increase, bison meat apparently began to arrive at the pueblo on more complete carcasses. The hides were prepared for trade.

Alternatively, bison meat might have replaced another C_4 meat-protein source during the trade period. One possibility is that turkeys were eaten early in the Pecos sequence, but with the increase in available bison meat they dropped out of the diet and were kept solely for their feathers. Both hypotheses are consistent with the carbon and nitrogen isotope values but are inconsistent with the faunal remains, which are dominated by mule deer. One other possibility is that maize provided the majority of dietary protein throughout the occupation of Pecos and that the minor amounts of meat ingested (whether C_3 or C_4) had little effect on bone collagen $\delta^{13}C$ values. This last possibility seems the least likely, given the relatively good health of the inhabitants.

The stable isotope data provide no support for differential access to foodstuffs along status lines at Pecos Pueblo in any time period—something suggested by some researchers (Vehik 2002). One individual during the trade period, however, exhibited a stable isotope value that was more than two standard deviations from the mean ($\delta^{13}C$ = –8.6‰ versus the mean of –7.6‰). It is possible that this value indicates that the individual ingested more C_3 foods than the rest of the population, but we cannot conclude this given the level of precision of our data (± 0.2‰). The range of variation in $\delta^{13}C$ and $\delta^{15}N$ is nearly equivalent to the range reported in animals fed monotonous diets (Schoeninger and DeNiro 1984), which suggests that all individuals had access to similar foodstuffs. There are no differences between the sexes in either $\delta^{13}C$ or $\delta^{15}N$ values. These data, therefore, provide no evidence for a shift in gendered dietary roles as previously proposed (Habicht-Mauche 2000; Spielmann 1996b).

When the Pecos stable isotope data are compared with those from other Southwestern sites, more detailed aspects of the diet become clear. Grasshopper Pueblo, which dates to the pre–Plains trade period at Pecos and whose residents also cultivated maize, is significantly different from Pecos in its average $\delta^{13}C$ value (Ezzo 1991, 1992, 1993) (t = 3.9, p = .000). Grasshopper's mean $\delta^{13}C$ value is lower than Pecos Pueblo's by about 1.6‰ (table 8.1, fig. 8.3). Although several types of maize were used across the Southwest, it is unlikely that carbon isotopic variation in maize types is the explanation for the difference between the sites. The variation in $\delta^{13}C$ values across strains of modern corn is trivial (sd = ± 0.02–0.07; Tieszen and Fagre 1993b), and there is no association between row number and carbon stable isotope value in archaeological maize (see Schurr and Schoeninger 1995:fig. 3). Significantly, the standard deviation at Grasshopper (1.1‰) is more than twice that at Pecos Pueblo (0.4‰). This is consistent with suggestions that migrants from many areas (Ezzo 1993:52) with varied subsistence regimes populated Grasshopper Pueblo.

Some individuals who died at Grasshopper Pueblo had eaten more C_3 plants or animals that fed on C_3 foods (i.e., beans, piñon nuts, antelope, rabbits) than had the people represented at Pecos Pueblo. Others had diets similar to those of the Pecos residents. There is a significant difference in the $\delta^{15}N$ values between the two sites (t = 5.5, p = .000), with Grasshopper about 1.5‰ lower than Pecos. This may be due to the high $\delta^{15}N$ values in plants from the Pecos region, or it could indicate that many people at Grasshopper ate more piñon nuts than did the people at Pecos. The paleopathological data show a moderate level of cribra orbitalia and porotic hyperostosis at Grasshopper Pueblo relative to other Southwestern groups, which suggests that the inhabitants of the site probably ingested at least adequate dietary protein.

The three individuals from the pre–Plains trade period site of Henderson exhibit significantly lower $\delta^{13}C$ values than contemporaneous individuals from Pecos Pueblo (t = 4.3, p = .002). Henderson also has significantly higher $\delta^{15}N$ values than Pecos (t = –2.8, p = .22), with a value that is 1.4‰ higher than the mean at Pecos Pueblo. This suggests that the inhabitants of Henderson ate more C_3 foodstuffs than the inhabitants of Pecos. Even though the sample size is small, this result is intriguing, because the faunal material from Henderson suggests that the population was settled, growing maize while hunting and later trading extensively in bison products (Speth 2004:421). Even so, antelope, rabbit, and, perhaps more significantly, fish were also recovered from the site (Speth 2004), many of which would have had a C_3 signal. In addition, manos and metates from Henderson are more crudely made, larger, and unstandardized in comparison with those from other, larger pueblos (Speth 2004:421). Perhaps the people at Henderson depended to a greater extent on fauna, especially fauna with a C_3 signal, for their dietary protein than did the people at Pecos Pueblo.

The semisedentary traders from Antelope Creek, a population dating to the combined pre–Plains trade and trade periods, also show significantly lower $\delta^{13}C$ values (t = 3.0, p = .005) than the corresponding Pecos Pueblo sample. This supports suggestions that, in addition to maize, the people of Antelope Creek ate significant amounts of wild foods (Lintz 1991), which would have been C_3 in this area. The $\delta^{15}N$ values are

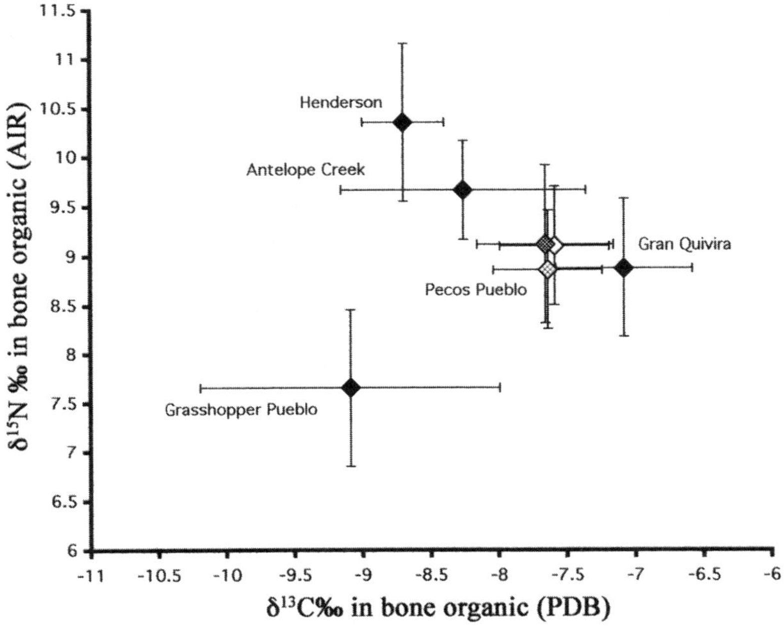

FIGURE 8.3. Values of $\delta^{13}C$ and $\delta^{13}N$ in six Southwestern skeletal samples. Pecos Pueblo shows significantly higher average $\delta^{13}C$ values (for three time periods) than Henderson, Antelope Creek, and Grasshopper Pueblo, but significantly lower values than Gran Quivira. This means that the Pecos population was consuming more C_4 foodstuffs (e.g., maize, bison, turkey) than the populations at Henderson, Antelope Creek, and Grasshopper Pueblo, but more C_3 foodstuffs (e.g., local fauna and flora, European-introduced fauna and flora, or both) than the Gran Quivira inhabitants. Pecos Pueblo means: white = pre–Plains trade period; open crosshatch = trade period; closed crosshatch = historic period. Lines radiating from the symbols represent one standard deviation.

also significantly different between the two sites (t = –3.6, p = .001), with Antelope Creek having $\delta^{15}N$ values that are about 0.5‰ higher than those at Pecos Pueblo. These data are consistent with the overall frequency of bison bones at Antelope Creek sites and with the archaeological evidence that these people focused on hunting as well as on gathering and horticulture, although more plant analyses are needed before a high degree of certainty can be achieved. As at Grasshopper Pueblo, the Antelope Creek samples have a larger range of variation than does Pecos, suggesting a more varied diet than at Pecos. The larger range among the part-time horticulturists from Antelope Creek probably reflects greater variation in diet between years for these people than was experienced by the settled agriculturists at Pecos Pueblo.

In contrast to the other two sites, Gran Quivira, the pueblo that overlaps most of the temporal period represented at Pecos, has higher $\delta^{13}C$ values than those at Pecos. The differences approach significance in the pre–Plains trade period (t = –2.2, p = .051) and are significantly different in the trade (t = –4.4, p = .000) and historic periods (t = –2.5, p = .016) (table 8.1, fig. 8.3). This indicates that although the people at both pueblos were heavily reliant on C_4 foods, those at Pecos ate more C_3 foodstuffs than did those at Gran Quivira in all time periods.

CONCLUSION

This reanalysis of stable isotope data from Pecos Pueblo and comparison with data for several other precontact Southwestern groups details significant spatial variability in diet even though the diets at all sites were mainly C_4. For the earliest period of occupation at Pecos, $\delta^{13}C$ values show that the inhabitants ingested more C_4 plants (maize), animals that fed on C_4 plants (e.g., bison, turkeys, and some rabbits and hares), or both, than did the inhabitants of distant Grasshopper Pueblo. In comparison with part-time horticulturists in the nearby Texas Panhandle (Antelope Creek), however, the people of Pecos Pueblo ate more maize. Higher maize consumption at Pecos also holds true in a comparison with the bison-rich site of Henderson, where fish remains were also recovered.

In contrast, throughout all periods (pre–Plains trade, trade, and historic), Pecos Pueblo depended on less maize, fewer animals such as bison, turkeys, and some rabbits and hares that fed on C_4 plants, or both, than did the people at Gran Quivira.

The differences in $\delta^{15}N$ values across the sites require a greater understanding of the variation in $\delta^{15}N$ values in wild and domestic plants, of fertilization, and of field management techniques before conclusions can be reached. But if supported in future studies, the data suggest that some people at Grasshopper Pueblo obtained a significant amount of dietary protein from piñon nuts, whereas those at Antelope Creek, Pecos, and Gran Quivira obtained it from bison, turkeys, and some rabbits and hares at levels that varied across the three sites.

For the temporal period when the archaeological record indicates that significant interregional trade took place, the $\delta^{13}C$ values for Pecos Pueblo suggest little change in diet relative to the earlier period, when trade was largely interpueblo and limited to nonfood items. In contrast to expectations, none of temporal subsamples exhibited an increase in foods that would have had a C_4 signal. Our data are consistent with several possibilities. First, bison meat was eaten throughout the period of occupation at Pecos, but only during the trade period was it imported on the bone. Second, during the trade period, bison meat replaced another meat that also had a C_4 signal (e.g., turkey and some rabbits or hares). Third, maize provided the main dietary protein throughout all periods of occupation at Pecos Pueblo.

Our isotope data also provide no support for dietary change during the historic period. This consistency of diet at Pecos Pueblo is remarkable for such a tumultuous time. It suggests that either the Spaniards had little to no effect on the actual food consumed by the pueblo's inhabitants or that small sample size hampered our ability to discern a dietary shift. In any case, the previous finding of a significant shift to the consumption of more C_3 foods during the historic period at the site is not supported by this reanalysis. This new conclusion highlights the importance of the correct temporal assignment of individual burials. Finally, the stable isotope data from Pecos Pueblo support no significant sex differences in any time period.

ACKNOWLEDGMENTS

We thank Michèle Morgan for inviting us to revisit these data from our earlier Pecos Pueblo study. The original study was funded by NIH Biomedical Research Support Grants to K. Spielmann (then at the University of Iowa) and to M. Schoeninger (then at Johns Hopkins University School of Medicine) and by the Wisconsin Alumni Research Foundation to M. Schoeninger. This reanalysis was funded by the University of California at San Diego to M. Schoeninger. We hope these data are relevant to the living descendants of the Pecos Pueblo population as well as to the academic community.

9 Structural Analyses of Postcranial Skeletal Remains

Christopher Ruff

Postcrania from Pecos Pueblo have figured prominently in many anthropological studies since Hooton's classic monograph (1930). These studies fall into one of three general types: those limited to the Pecos collection itself (Ruff and Hayes 1982, 1983a, 1983b), those comparing Pecos with other late Holocene samples (Burr, Ruff, and Thompson 1990; France 1988; Jurmain 1977, 1980; Ruff 1987; Tague 1989, 1992, 1994), and those using Pecos (sometimes along with other Holocene samples) as representative of "modern humans" for comparison with earlier hominins or for testing of more general theoretical relationships (Abbott, Trinkaus, and Burr 1996; Ruff 1984, 1995, 2000c; Ruff, Walker, and Trinkaus 1994; Ruff et al. 1993; Trinkaus 1976, 1981, 1989; Trinkaus and Hilton 1996; Trinkaus et al. 2002a, 2002b). Examples from all three types of applications are given later in this chapter. The preceding list is not meant to be exhaustive, but it gives some idea of the extent to which the Pecos collection has contributed to modern scientific research on the human skeleton. The importance of Pecos and its inclusion in so many different studies can be attributed to several factors: the large size of the skeletal sample and its generally excellent preservation, its well-documented provenience (e.g., Kidder 1958), and its wide accessibility to researchers through its curation at Harvard's Peabody Museum of Archaeology and Ethnology.

My own experience with the Pecos sample began in 1978, when I was a doctoral candidate at the University of Pennsylvania. I had become interested in bone structural changes with aging and how they might relate to the development of bone fragility, or osteoporosis. I decided that, as a first step, a sample from an archaeological site rather than a living human or cadaveric sample would be most feasible to study, for a variety of reasons—the difficulty of measuring the necessary parameters noninvasively or without subjecting living subjects to unjustified radiation doses, the possibility of better sampling of the entire adult age range (cadaveric specimens tend to be of advanced age), and the likelihood of greater genetic and environmental homogeneity relative to a typical modern sample. To accomplish the study objectives, the archaeological sample needed to be large, well preserved, and well provenienced. The Pecos sample fulfilled all these requirements.

To measure bone structural parameters (explained more fully later), it is necessary to obtain accurate periosteal and endosteal contours of the sections of interest. In the years since I carried out my original study, the use of computed tomographic (CT) scanning has become commonplace for this purpose (e.g., Bridges, Blitz, and Solano 2000; Ruff 1999, 2002a; Ruff and Leo 1986; Stock and Pfeiffer 2001; Sumner, Mockbee, and Morse 1985). In the late 1970s, however, CT was still a relatively new technique, limited almost exclusively to clinical use. I found radiography alone to be an inadequate substitute, and a technique using multiplane radiography and external molding, a viable alternative to CT in some circumstances

(O'Neill and Ruff 2004; Ruff et al. 1993; Trinkaus and Ruff 1989), had not yet been developed (and is cumbersome to use in large samples). Therefore, I requested permission to physically section the femora and tibiae from an adult age-representative sample from Pecos, eventually including 120 individuals divided equally between the sexes and among six age groups (five-year groups from 20 to 39 years, 40 to 49, and 50 to 59). One individual was later eliminated because of a previously undiagnosed pathology, leaving a final sample of 119.

In each case the less-well-preserved femur and tibia from individuals with both sides preserved was sectioned and analyzed, thus leaving one side intact for future study. The study included five diaphyseal sections per bone, plus a section through the femoral neck, for a total of 1,309 sections (see Ruff and Hayes 1983a for more details). The sections were photographed, manually traced on a digitizer, and input to the computer program SLICE (Nagurka and Hayes 1980), which calculated geometric section properties. As part of the study, I also re-aged and re-sexed my entire sample, and in doing so I noted some differences from Hooton and Todd's original assessments that appeared to explain some peculiarities in previous demographic estimates for Pecos (Ruff 1981). I also obtained photon absorptiometric scans of the specimens, which served as the basis for additional studies of bone density changes with age (Ruff and Hayes 1984a, 1984b).

During the following decade I revisited the Pecos Pueblo collection twice, first to collect complementary structural data on juvenile lower limb bones from 31 individuals and later to collect upper limb data for the same juveniles and for adults, the latter a subsample of 60 individuals from the adult lower limb sample with an equivalent sex and age distribution. Data were collected unilaterally from both right and left sides, with equal representation of each side within sex and age groups.[1] All data were collected using CT, either in Baltimore or in Boston. Pelvic data for a subsample of individuals with matching structural data were obtained from Robert Tague (see Tague, this volume) and have been used in several investigations of body size and shape variation in humans (e.g., Ruff 1995, 2000c). Bone cores removed from the original sample were later used in a comparative histological study of bone remodeling (Abbott, Trinkaus, and Burr 1996). Cranial capacities were retrieved from Hooton's original grave cards and used to calculate encephalization quotients for comparison with archaic humans (Ruff, Trinkaus, and Holliday 1997).

In all, data obtained from the Pecos sample have contributed to more than two dozen publications in which I have been involved over the past 20 years. The sample is still the single largest human population sample to be analyzed in terms of bone structural properties. Its size has allowed for types of comparisons that would be impossible with smaller samples, such as true "demographic" analyses. The large size of the sample in turn was made possible by the development of new technology for obtaining structural properties that became available at just about the time the original study was being carried out (Nagurka and Hayes 1980; Ruff 1989).

Bone structural data for the 91 individuals (60 adults and 31 juveniles) in the combined upper and lower limb Pecos sample, together with museum identification numbers, sexes, ages, and bone lengths, are listed in Appendix H for the femur, tibia, humerus, radius, and ulna (only 30 juvenile tibiae and humeri and 29 juvenile radii and ulnae). Data are given for the midshafts of these bones, except in the humerus, where data are for a section 40% of bone length from the distal end, to avoid the deltoid tuberosity. Note that the bone lengths used for locating sections refer to the length' measurement defined in Ruff 2002a. Maximum bone lengths, including repositioned epiphyses, are also given in Appendix H (when epiphyses were missing, maximum lengths were estimated using equations derived from individuals of similar age with preserved epiphyses). This is only a subset of the full data set collected (see the preceding and Ruff 2002a for a description of all section locations), but it is adequate for general representation of structural variation in the upper and lower limbs. This data subset is used in all subsequent analyses in this chapter. A detailed description of measurement procedures is given elsewhere (Ruff 2002a).

STRUCTURAL ANALYSES

Skeletal structural analysis is based on the premise that bone is a living tissue that will adapt to the mechanical loadings (forces) placed upon it during life, an assumption supported by many observational and experimental studies (e.g., Jarvinen et al. 2003; Jones et al. 1977; Kontulainen et al. 2002; Trinkaus,

1. Five- and 10-year age groups in the adult sample of 60 individuals with all limb bone data had 5 individuals of each sex per age group. Right and left sides were divided as equally as possible within each of these groups.

TABLE 9.1
BODY SIZE AND CROSS-SECTIONAL GEOMETRIC CHARACTERISTICS OF TWO PECOS INDIVIDUALS SHOWN IN FIGURE 9.1

							Femoral midshaft properties					
Specimen	Sex	Age (years)	Maximum femoral length (mm)	Stature (mm)	Bi-iliac breadth (mm)	Body mass (kg)	CA	%CA	J	Zp	Zp/ (BM · FL)	Ix/Iy
A	F	25–29	393	1,490	271	54.2	343	80.1	28,318	2,415	.1134	.950
B	M	25–29	444	1,642	269	65.1	341	74.5	32,143	2,556	.0884	1.183

Key to femoral midshaft properties: CA, cortical area (mm^2); %CA = (cortical area/total area) · 100; J, polar second moment of area (mm^4); Zp, polar section modulus (mm^3); Zp/(BM · FL), polar section modulus/(body mass · maximum femoral length) (mm^3/(kg · mm); Ix/Iy, ratio of second moment of area in AP plane to that in ML plane.

Churchill, and Ruff 1994; Woo et al. 1981). This adaptation, often called Wolff's Law (Wolff 1892), is likely to be complex and involve interaction with other systemic (genetic, hormonal, dietary) factors as well (Hsieh et al. 2001; Lanyon and Skerry 2001; Specker and Binkley 2003; Volkman et al. 2003). But because localized mechanical factors have such a profound effect on bone modeling and remodeling during life (Lanyon and Skerry 2001 and references therein), the preserved morphology of bones provides important clues to behavioral differences between individuals and populations living in the past (see Ruff, Holt, and Trinkaus 2006 for a recent discussion).

Long bone diaphyses can be modeled as engineering beams, which simplifies mechanical analysis (Timoshenko and Gere 1972). In a beam analysis, cross-sectional properties of the diaphysis are determined and used to calculate the strength of the bone under particular kinds of loadings. (Variation in bone material properties such as density can also occur, but this is probably much less important than geometry in determining bone strength [Erickson et al. 2002].) The following example demonstrates the general technique, as well as illustrating some other principles that are relevant to subsequent comparisons using the total sample.

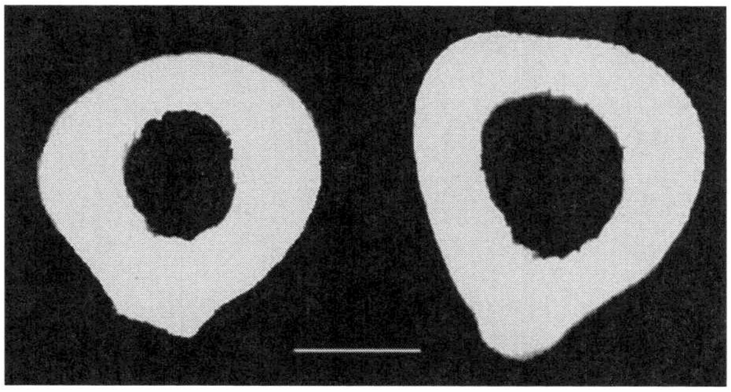

A B

FIGURE 9.1. Femoral midshaft cross sections for two individuals from the Pecos sample, a female (A) and a male (B), both 25–29 years of age. Structural properties for these sections are listed in table 9.1. Scale bar = 1 cm.

Figure 9.1 shows actual cross sections obtained from the femoral midshafts of two individuals from the Pecos Pueblo sample, a female and a male, both in their twenties. Characteristics of these individuals and sections are listed in table 9.1.

Cortical area (CA) is a measure of the amount of bone present and is proportional to pure compressional strength. Cortical areas are about equivalent in these two individuals (table 9.1). However, the distribution of CA is different, with individual A, the female, having a relatively smaller medullary canal and relatively thicker cortices. This is reflected in the higher %CA (80.1%) for this individual than for individual B (74.5%) (%CA represents the percentage of the total subperiosteal area filled by cortical bone). From a mechanical standpoint, however, neither CA nor %CA is as important as other properties referred to as second moments of area (or area moments of inertia) and section moduli. These latter properties measure bending and torsional (twisting) rigidity and strength, respectively, and are calculated as the sum of unit areas multiplied by their squared distances from a centroidal point or axis through the section (see Ruff 2000c for discussion). Because the dominant loading regimen in long bone diaphyses is bending (together with torsion in some regions), these properties are the critical ones in terms of mechanical performance. The polar second moment of area, J, can be used as both a measure of torsional rigidity and (twice) average bending rigidity. Table 9.1 shows that J is substantially (13.5%) larger in individual B, a function of his more outwardly distributed bone area. The same is true to a lesser extent of the polar section modulus, Zp, a measure of torsional and (twice) average bending strength. Thus, despite equivalent cortical areas and relatively thicker cortices in individual A, individual B has a stronger femoral midshaft, demonstrating the importance of bone distribution in determining bone strength.

Because mechanical loadings are a product of both body size and behavior, it is useful to reconstruct body size attributes as part of a structural analysis (table 9.1). Maximum femoral length differs by about 13% between the two specimens, corresponding to a difference in estimated stature (Genoves 1967) of about 10%. Both individuals are within one standard deviation of the mean stature for their sex in this sample, although the male is somewhat taller than average (Ruff 2000c). In contrast, bi-iliac (maximum pelvic) breadth is almost equivalent in the two sexes, again as in the total sample (Ruff 2000c) and in modern human males and females generally (Hiernaux 1985; Ruff 1994b). Using both stature and bi-iliac breadth, body mass can be calculated reasonably accurately (Ruff 2000b; Ruff, Trinkaus, and Holliday 1997). Furthermore, such estimates are not confounded with measures of bone strength, because they are derived from morphometric rather than mechanical principles (Ruff 2002b). The body masses calculated for the two Pecos individuals shown in figure 9.1 are both slightly above average but within one standard deviation of respective means for the sexes in the sample (Ruff 2000c).

For interpreting long bone bending and torsional strength, the most appropriate index of body size is body mass multiplied by bone length (Ruff 2000c; Selker and Carter 1989). When viewed relative to this index, femoral midshaft Zp is smaller in individual B than in individual A (table 9.1). That is, relative to the mechanical loads produced by body size alone, femur B is weaker, or less robust, than femur A. This in turn implies behavioral differences between the two individuals, whereby individual A was loading her femur more relative to body size than individual B. The ethnohistoric literature for Southwestern Native Americans documents a mechanically very rigorous lifestyle for both sexes, with women doing much of the heavy work related to agriculture (Stevenson 1901–1902). Thus, this result is not particularly surprising.

Evidence for another sex-related difference in bone structure and, by implication, behavior is well illustrated by the shapes of the two cross sections in figure 9.1. The ratio of anteroposterior to mediolateral bending rigidity, Ix/Iy, is much greater in the male femur, B, than in the female femur, A (table 9.1), as is apparent in the female's more circular cross section and in the male's more anteroposterior distribution of bone. This difference is characteristic of preindustrial populations generally and appears to be associated with sexual division of labor—specifically, greater long-distance mobility in males (Ruff 1987). As I discuss later, such a behavioral difference can be predicted to increase anteroposterior relative to mediolateral bending loads in male lower limbs. This morphological difference between the sexes was first documented in the Pecos sample (Ruff and Hayes 1983b) and has since been found in many other samples, varying systematically by subsistence strategy (Ruff 1999, 2000a).

In the sections that follow, I first present an analysis of age changes in long bone structural properties of the Pecos sample, from early childhood through late adulthood. This is followed by a review of sexual dimorphism in bone structure in Pecos and other samples, including age-sex interactions. The chapter ends with some broader evolutionary comparisons that incorporate the Pecos material.

GROWTH AND AGING

My first involvement with the Pecos sample stemmed from my interest in documenting adult patterns of bone remodeling in the femur and tibia (Ruff and Hayes 1982, 1983a, 1983b, 1984a, 1984b). This was later expanded to include growth patterns in juvenile femora, with comparisons to archaic humans (Ruff, Walker, and Trinkaus 1994). Some brief comparisons of juvenile femora, humeri, and tibiae with those of the early Upper Paleolithic Lagar Velho 1 child skeleton have also been carried out (Trinkaus et al. 2002a, 2002b). However, to date no analysis of age changes in structural properties over the entire combined Pecos sample (4 years to 50+ years) has been performed.[2] Previous studies of this kind, spanning virtually the

2. The original report on Pecos juvenile data (Ruff, Walker, and Trinkaus 1994) indicated a lower age limit of 5 years. Several individuals in the juvenile sample (under 20 years) have subsequently been re-aged, including the youngest individual, now estimated at 4 years old.

whole lifetime, are understandably rare. They have typically included only nonstructural properties such as bone mineral content or "density" (Ferretti et al. 1998; Haapasalo et al. 1996; Hui et al. 1999), or they have been limited to a single skeletal location—the second metacarpal (Garn 1970, 1972) or the midshaft femur (Feik, Thomas, and Clement 1996). Furthermore, use of modern living or cadaver-based samples increases both environmental and genetic heterogeneity and runs the risk of confounding true age trends with recent secular effects (see, e.g., Trotter, Peterson, and Wette 1968). Longitudinal sampling of living populations (Ruff 2003a; Tanner, Hughes, and Whitehouse 1981) avoids these problems but is essentially impossible to carry out over the entire life span of an individual. Thus, the Pecos sample may be the best one currently available for assessing virtually life-long trends in bone structural properties at multiple skeletal locations. Some examples are given here for the midshaft femur and mid-distal humerus (40% of bone length from the distal end).

Age trends in total subperiosteal area (TA), medullary area (MA), cortical area (CA), %CA, and polar section modulus (Zp) in the combined juvenile-adult Pecos sample are shown for the femoral midshaft in figure 9.2 and for the mid-distal humerus in figure 9.3. All data except %CA are log-transformed to preserve proportionality over different size ranges; that is, similar vertical distances in any part of the y axis range indicate equivalent percentage differences. Aging and sexing techniques have been previously described in detail (Ruff 1981; Ruff, Walker, and Trinkaus 1994). Pelvic features, especially the sciatic notch angle and ischium-pubis proportions (Krogman 1962; Washburn 1948), were primarily used to sex adults and older adolescents down to 16 years of age. Aging of adults was carried out using pubic symphyseal standards available at the time of the original adult studies (Gilbert and McKern 1973; McKern and Stewart 1957), along with functional dental wear (Miles 1963). Juveniles were aged using long bone epiphyseal closure (Krogman 1962) and stages of tooth eruption and development as determined through mandibular radiographs (Smith 1991). Ages were estimated to within 1 year among juveniles and to within 5- or 10-year periods among adults (although the oldest age group, 50–59 years, probably includes individuals greater than 59 years of age). Sex was assigned only in individuals over 15 years of age. The sexes are combined in figures 9.2 and 9.3 but are considered separately in the next section.

As in previous analyses (Ruff, Walker, and Trinkaus 1994), LOWESS curves, based on a robust, locally weighted, nonparametric curve-fitting technique (Cleveland 1979), are used to illustrate age trends for each structural parameter. A smoothing window of .4 (40% of x data range used to calculate the local curve at any point) is employed in each plot. Least squares linear regression and analysis of covariance are used to describe and compare slopes in relatively (log) linear portions of the age range. All graphing and statistics were carried out using SYSTAT 1990.

The general pattern of age changes in femoral and humeral structural properties is similar, although it also varies in slight but important ways between the two bones. In both the femur and the humerus, during childhood and early adolescence the total subperiosteal surface (TA) expands faster than the endosteal surface (MA), leading to a thickening of cortices and an increase in the percentage of cortical area (%CA) (figs. 9.2 and 9.3, plots A and B). From the third decade onward, the endosteal surface expands much faster than the subperiosteal surface, leading to a decline in %CA with aging. Total subperiosteal area increases slightly but significantly after age 20 years in both bones. However, the increase among adults is faster in the femur (slope = .0059, $r = .417$, $p < .001$) than in the humerus (slope = .0033, $r = .280$, $p < .05$). Also, the femur shows a reversal of endosteal expansion (i.e., the endosteal surface contracts) between mid-adolescence and the early twenties, whereas the humerus does not. Because of these differences in age patterns at both inner and outer bone surfaces, the femur develops a relatively thicker cortex than the humerus by early adulthood and shows a smaller decline in %CA with aging. More importantly from a biomechanical standpoint, cortical area shows a significant decline with age in the adult humerus ($r = .338$, $p < .01$) but not in the adult femur ($r = .026$, $p > .80$) (figs. 9.2 and 9.3, plot C), and the section modulus shows no age change among adults in the humerus ($r = .115$, $p > .30$) but increases in the femur ($r = .269$, $p < .05$) (plot D).

The "protective" or "compensatory" effect of continued subperiosteal expansion in adults tends to offset endosteal expansion and decline in bone material strength with aging, which are probably hormonally mediated (Garn et al. 1967; Martin and Atkinson 1977; Smith and Walker 1964). Its presence in the Pecos sample does not appear to be an artifact of differential survival or secular effects favoring larger individuals (and thus larger bones) in the older age range. Most individuals in the sample do not have bi-iliac breadths with which to carry out the body mass estimations

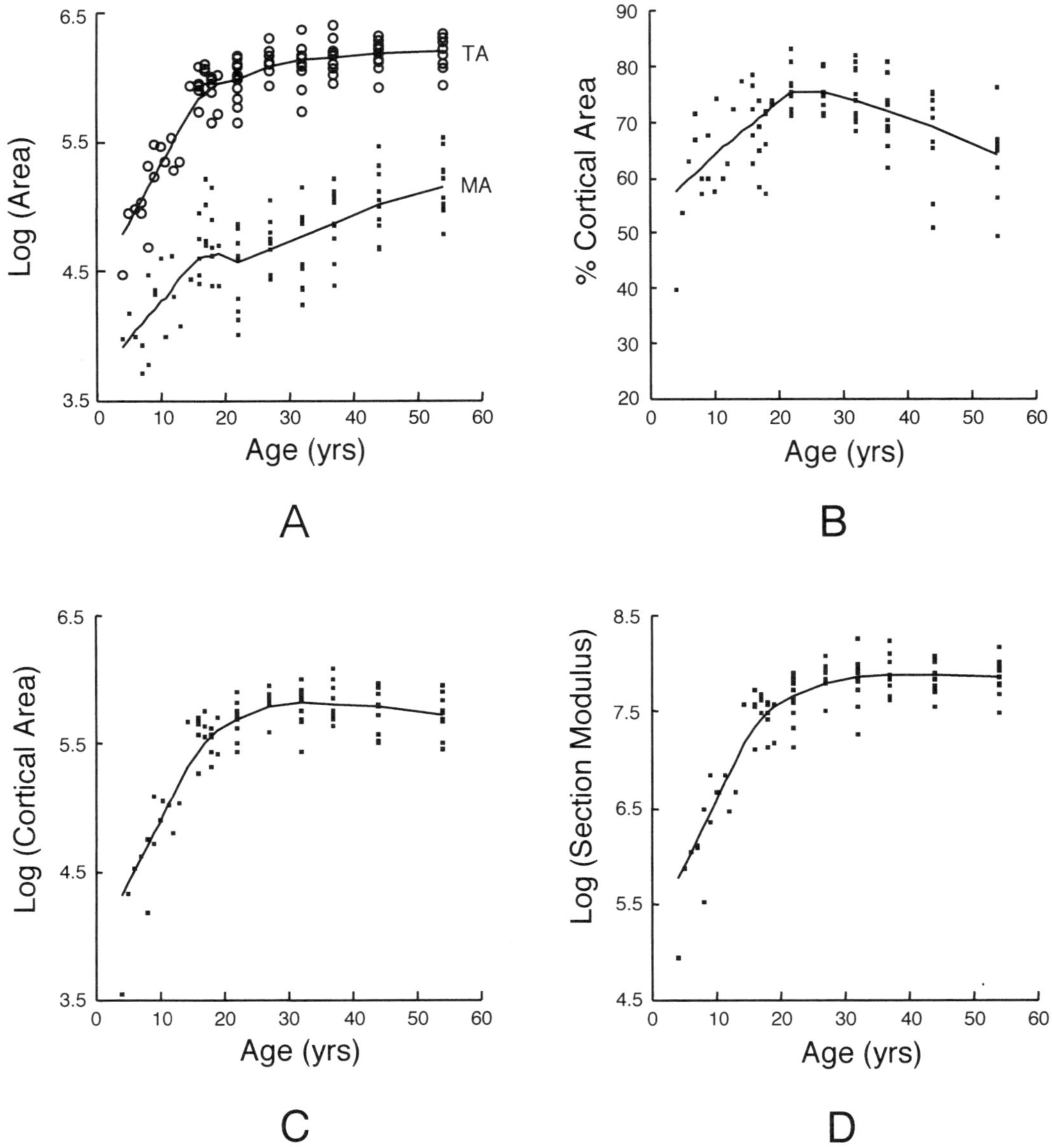

FIGURE 9.2. Age changes in the femoral midshaft in the combined juvenile and adult Pecos samples. A, total subperiosteal area (TA) and medullary area (MA); B, percentage cortical area (%CA); C, cortical area (CA); D, polar section modulus (Zp, bending and torsional strength). All data except %CA are natural log-transformed. LOWESS lines are plotted through each distribution.

demonstrated earlier. However, as is shown later in this chapter, bone length is a reasonable proxy for body mass among adults in this sample. As shown in figure 9.4, there is no significant change in either femoral or humeral length after age 20 ($p > .10$, both bones) in the sample.[3]

The contraction of the endosteal surface in late adolescence–early adulthood observed here in the femur has also been reported for the second metacarpal (Garn 1970, 1972; Johnston and Watts 1969; Smithgall et al. 1966) and in one study in the midshaft femur (Bass et al. 1999). All these studies were either of

3. More strictly, these results apply to skeletal size rather than body mass per se; it is very likely that body mass relative to stature and skeletal size increased in this sample after early adulthood (Ruff, Scott, and Liu 1991). The issue here is whether the present cross-sectional study design may be biased by (skeletally) larger individuals in the older age ranges, and the results show this not to be the case.

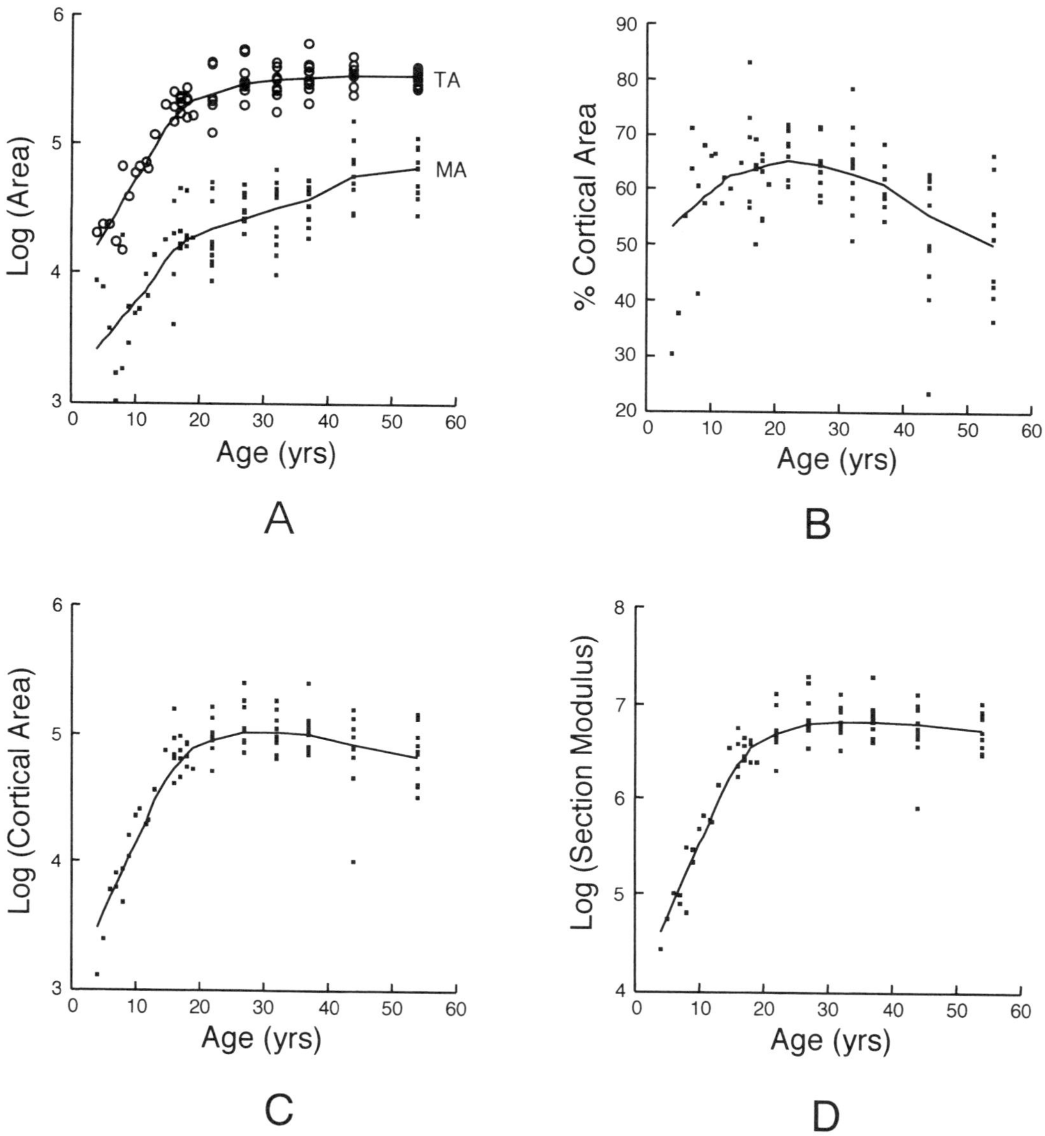

FIGURE 9.3. Age changes in the mid-distal humerus (40% of length from the distal end) in the combined Pecos sample. Data plots are as in figure 9.2.

girls or, if boys were included, showed much less endosteal contraction, suggesting a possible sex difference in this aspect of bone growth. One proposed explanation for this "mechanically inefficient" deposition of endosteal bone in girls (Schoenau et al. 2001) but not in boys is that girls need to build up greater calcium reserves to meet the later challenges of pregnancy and lactation (although see Garn 1970; Jarvinen, Kannus, and Sievanen 2003; Schoenau et al. 2001; Smithgall et al. 1966).

A sex difference in endosteal contraction in late adolescents was also observed here. If the Pecos sample is broken down by sex, females show a significant change in the slope of medullary area (MA) between 15–22 years and 22–54 years ($p < .05$), whereas males do not ($p > .10$) (fig. 9.5). The sex difference in general growth patterns at the endosteal surface is readily apparent in figure 9.5. Females not only show more endosteal contraction in late adolescence–early adulthood but also more endosteal expansion thereafter ($p \leq .05$, slope comparisons between males and females). No other cross-sectional property examined here—TA, CA, or Zp—shows a significant sex difference in slope in either age period. The greater magni-

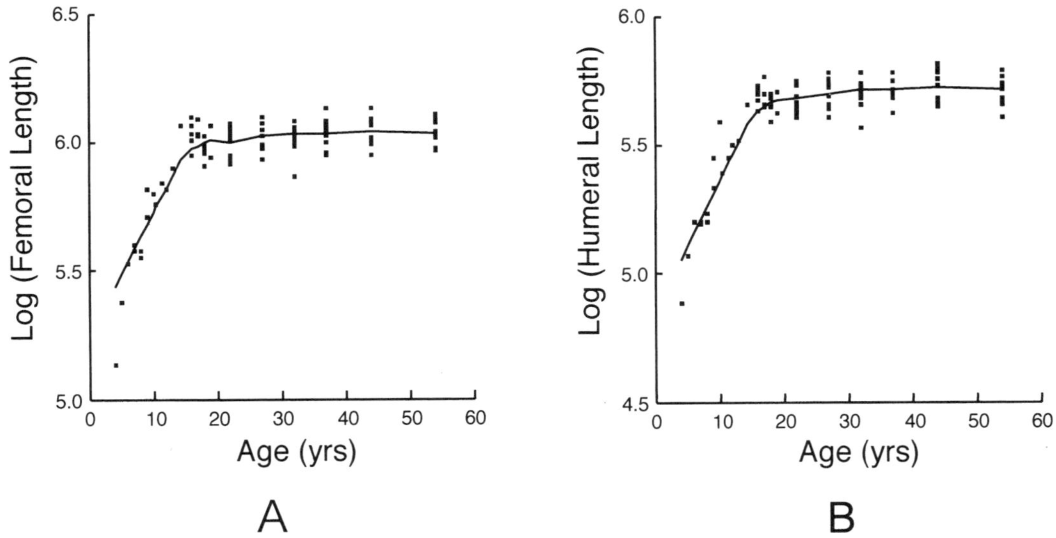

FIGURE 9.4. Age changes in femoral (A) and humeral (B) lengths in the combined Pecos sample. Data are log-transformed, with LOWESS lines plotted.

tude of change—positive and negative—at the endosteal surface throughout life in females has been discussed in particular by Garn (1970, 1972). Figure 9.5 also shows that the difference observed between the femur and humerus in the pooled sample (figs. 9.2, 9.3) is characteristic of each sex—in both males and females, there is a more "positive" change in MA (either more contraction or less expansion) in late adolescence–early adulthood in the femur than in the humerus.

Thus, a life-span analysis of femoral and humeral structural properties in the Pecos sample demonstrates both commonalities and sex- and location-specific differences in patterns of change with age. In both bones, females gain more bone endosteally during late adolescence through early adulthood, perhaps in preparation for greater endosteal bone loss in later life. In both sexes, the femur appears to benefit from both greater endosteal apposition early in adulthood and greater subperiosteal apposition later in adulthood, relative to the humerus. The more "positive" changes in femoral structure in the Pecos sample may suggest a mechanical explanation—greater life-long loading of the femur than the humerus, which could stimulate greater bone deposition at both surfaces. Indeed, there is a significant increase in the ratio of femoral to humeral strength (Zp) throughout adulthood (22–54 years) in the Pecos sample, both for the combined sexes ($r = .428$, $p < .001$) and within sex ($r = .38–.47$, $p < .05$, males and females). This is consistent with the rugged physical environment of Pecos Pueblo (see Ruff 1999), which might be expected to place more mechanical demands on the lower limb. However, as noted earlier, some of the same age-related changes have been observed in the second metacarpal in living populations, so weight-bearing per se is not the only factor involved. These other studies, though, did not directly compare changes occurring at different skeletal sites in the same individuals, which is a more rigorous test of the influence of localized mechanical loadings versus general systemic effects on bone structural changes. In any event, the results shown here demonstrate significant skeletal heterogeneity in bone modeling and remodeling during life, superimposed on a more general growth pattern that may be universal to long bones.

SEXUAL DIMORPHISM

General sex-related differences in bone structure of the femur and tibia, beyond size differences and differences due to aging, were noted in the original study of the Pecos adult material (Ruff and Hayes 1983b). Specifically, males were found to have relatively greater anteroposterior (A-P) bending rigidity, and females, relatively greater mediolateral (M-L) bending rigidity, as is reflected in their cross-sectional shapes (see fig. 9.1). Sexual dimorphism in pelvic structure, whereby females should generate larger M-L bending

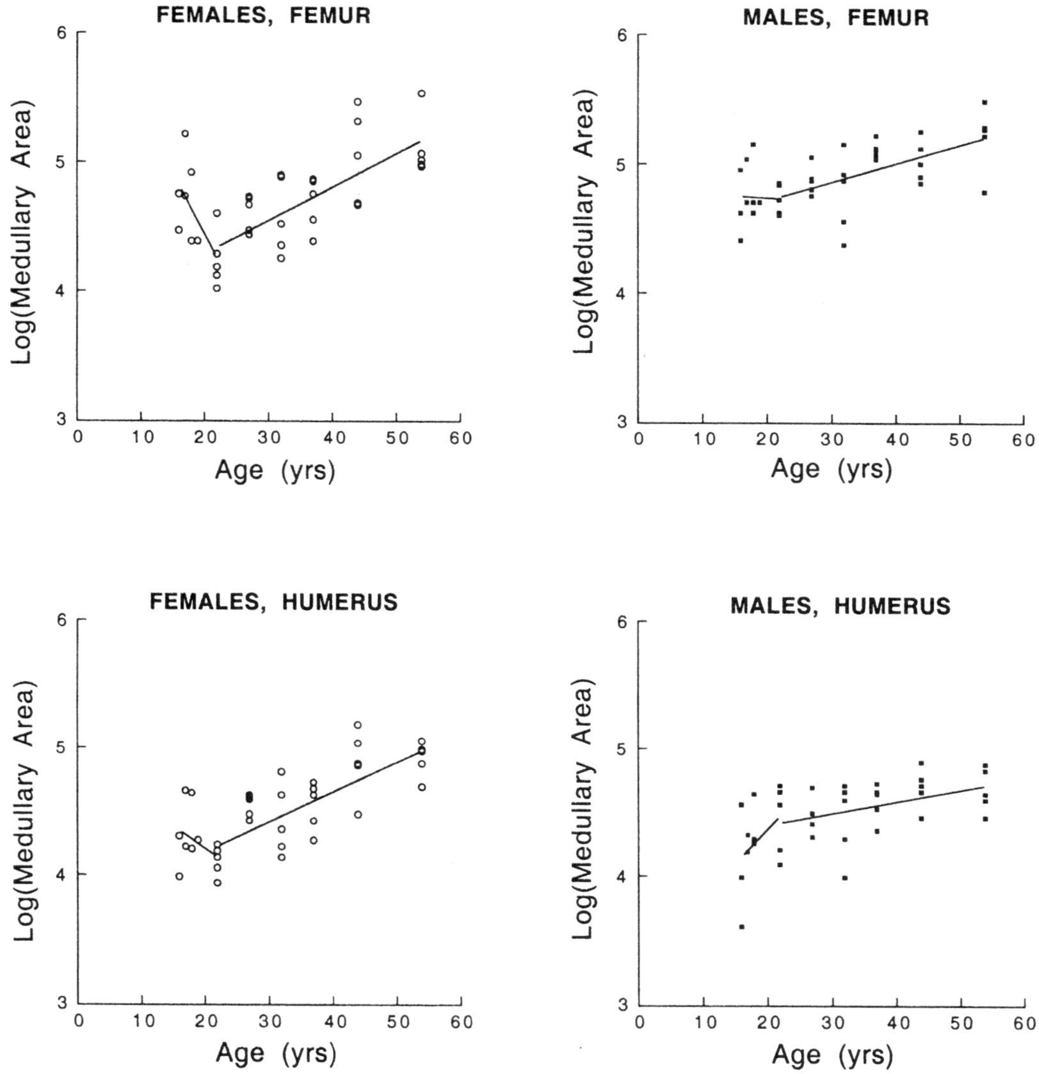

FIGURE 9.5. Age changes in medullary area in the female and male femur and humerus in Pecos older adolescents and adults. Data are log-transformed. Least squares regression lines are plotted for the 15–22 and 22–54 year age groups.

loads in their lower limbs due to their wider pelves, was considered to be a factor contributing to, but not a complete explanation for, this finding. This is because the sex-related differences in diaphyseal structure were found to be most marked in the region centered on the knee joint—that is, from the mid-tibia to the mid-femur—whereas the mechanical effects of pelvic dimorphism should be most evident in the proximal femur, nearer the hip (Ruff 1995). Therefore, a behavioral difference between the sexes that generated more A-P bending loads in the region about the knee in males was proposed to explain the observed pattern of variation. Ethnohistoric information from the American Southwest provided support for this hypothesis: males from this region, including those from the Pueblo of Jemez, which is considered to be culturally affiliated with Pecos, participated in frequent long-distance running races as well as long-distance travel over usually rough terrain (see Ruff 1991 for discussion and references; see also Ruff and Hayes 1983b). Such activities, involving powerful extension of the knee from a flexed position, should theoretically produce an increase in A-P bending loads in the lower limb, especially in the region around the knee (also see Pauwels 1980). Indeed, strain gauge studies of the tibia midshaft indicate that anteroposterior (but not

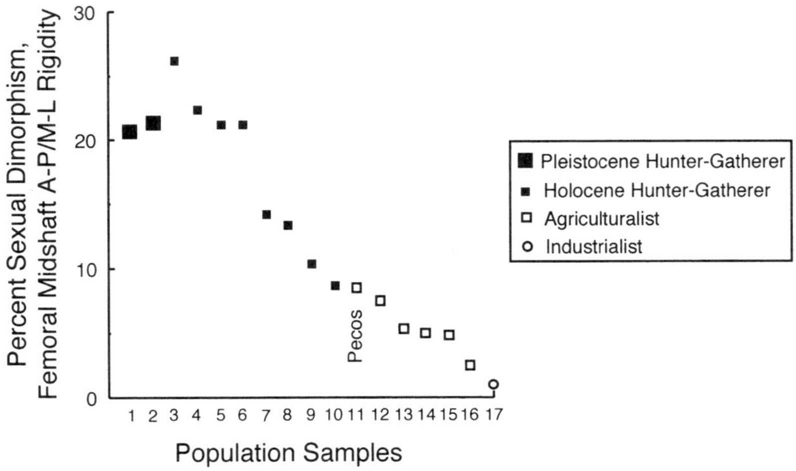

FIGURE 9.6. Percentage of sexual dimorphism, ([male–female]/female) · 100, of femoral midshaft anteroposterior to mediolateral bending rigidity in 14 prehistoric Native American population samples, two Late Pleistocene samples, and a U.S. white industrial sample. Individual sample identities and sources: 1, 2, Neandertals and early anatomically modern humans (European Upper Paleolithic) (Ruff et al. 1993; Trinkaus and Ruff 1999; Trinkaus, personal communication); 3–5, East Great Salt Lake, Stillwater, and Malheur Great Basin (Ruff 1999); 6, Delaware (Robbins, Rosenberg, and Ruff 1989); 7, 14, Archaic and Mississippian Tennessee River valley (Bridges 1985); 8, 12, 13, pre-Coalescent, Early and Late Coalescent Great Plains (Ruff 1994a); 9, 15, preagricultural and agricultural Georgia coast (Ruff and Larsen 2001); 10, 16, preagricultural and agricultural New Mexico (Brock and Ruff 1988; see Ruff 1987 for details); 11, Pecos Pueblo (Ruff 1987); 17, modern U.S. whites (Ruff 1987).

mediolateral) bending increases dramatically during running (Carter 1978).

A later analysis of modern autopsy material from U.S. whites (Ruff and Hayes 1988) found much less sexual dimorphism in cross-sectional diaphyseal shape near the knee joint (Ruff 1987). Through comparisons with several other samples, this was found to be part of a more general pattern in which hunter-gatherers had the greatest sexual dimorphism, agriculturalists intermediate levels, and industrial populations the least dimorphism in cross-sectional shape in the middle region of the lower limb (Ruff 1987). This variation in bone structure appeared to be correlated with variation in sexual division of labor, paralleling a decline in sex-related differences in mobility (and thus A-P loading of the knee region) from hunter-gatherers through agriculturalists to industrial populations. Interestingly, there was no such subsistence-related variation in cross-sectional shape of the *proximal* femur, which should be more determined by sexual dimorphism in pelvic structure, a universal distinction not associated with a particular subsistence strategy. An analysis of external metrics indicated the same patterns of sexual dimorphism in Neandertals and early anatomically modern humans, suggesting a similar sexual division of labor (Ruff 1987).

Structural data at the femoral midshaft have been obtained for several additional samples since the original analysis. Mean values of sexual dimorphism in cross-sectional shape for these samples are plotted together with those for the Pecos and U.S. white (industrial) samples in figure 9.6, keyed to subsistence strategy. Data points representing means for samples of Neandertals and early anatomically modern (early Upper Paleolithic) humans are also plotted (for details, see Ruff 2000a). The additional data confirm the earlier comparison: hunter-gatherers show systematically greater sexual dimorphism in cross-sectional shape than agriculturalists, including those from Pecos (which falls at the top end of the agriculturalists), while the industrial population shows virtually no sexual dimorphism. The Paleolithic samples again fall with modern hunter-gatherers. The comparison in figure 9.6 indicates that Pecos had fairly high values of sexual dimorphism for an agricultural population, perhaps reflecting the influence of non-subsistence-related activities (such as institutionalized racing among males) or the substantial contribution to the economy of non-agricultural food resources that required long-distance foraging by males (Ruff 1991; Ruff and Hayes 1983b). A later analysis of Pecos juveniles (Ruff 2000a) showed that sexual dimorphism in

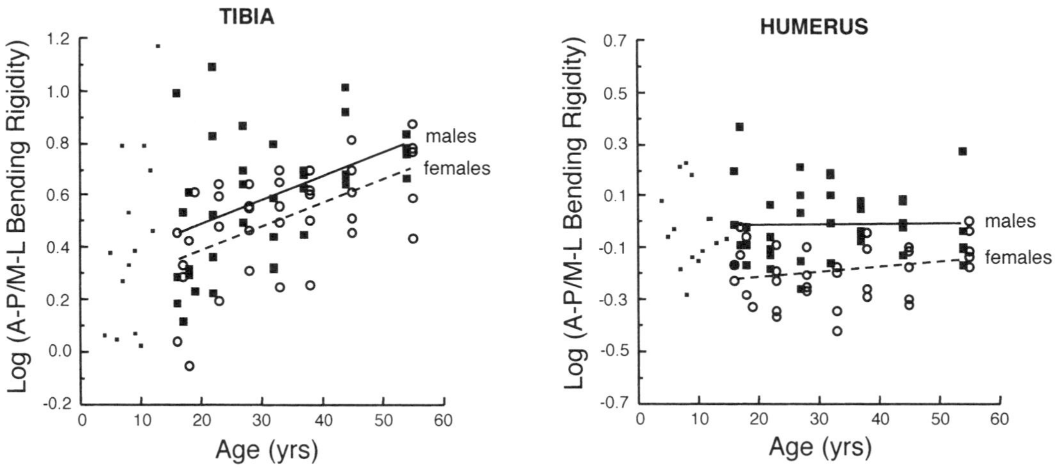

FIGURE 9.7. Age changes and sex differences in anteroposterior to mediolateral bending rigidity of the tibial midshaft and mid-distal humerus in the Pecos sample. Data are log-transformed. Key: large filled squares, males; open circles, females; small squares, (unsexed) individuals under 15 years of age. Least squares lines are plotted through males (solid) and females (dashed).

mid-distal femoral diaphyseal shape developed by late adolescence, suggesting that adult patterns of behavior were also established by this time.

It is of interest to examine this issue in the context of other skeletal regions, as well as over the entire age span represented at Pecos. Figure 9.7 shows age changes and sex differences in A-P/M-L bending rigidity of the midshaft tibia and mid-distal humerus in the combined juvenile and adult Pecos sample. For graphing and analysis of age trends, bending rigidity ratios are logged here, following arguments presented elsewhere (Ruff 2002a, 2003b) and earlier in this chapter. Both skeletal locations show significant sex differences in bone shape, with males having relatively greater A-P bending rigidity in both cases. The sex effect is relatively greater in the humerus, with an average sex difference of 20.3% ($p < .0001$, ANCOVA, effect of sex with age as a covariate), versus 9.9% for the tibia ($p < .05$, ANCOVA, effect of sex with age as a covariate). The ratio of A-P/M-L bending rigidity in the tibia increases with age in both males and females ($p < .0001$, effect of age with sex as a co-factor; also, $p < .01$, within-sex regressions), whereas there is no age effect on humeral shape ($p > .10$, all analyses). Although there is substantial overlap between individual male and female values at all ages, the average sex difference in both bones is maintained from late adolescence through adulthood (no significant sex-age interactions). Nevertheless, there is some indication of a decline in sexual dimorphism in the oldest age group (50+ years), particularly in the humerus. The humeral ratio falls to 8% in the oldest group ($p > .40$, t test between sexes), whereas it is at least 12% in every other 5- or 10-year age grouping ($p < .10$, t tests within all other groups). The tibial ratio is also low (6%) in the oldest age group but is more variable in general, ranging from –3% to 19% within age groups (only one t test between sexes, within age group, near significant at $p < .10$). Part of this variability is due to the small sample sizes within sex and age groups ($n = 5$ in adults).

Because individuals cannot be reliably sexed prior to mid-adolescence, it is impossible to extrapolate the sex-specific age trends shown in figure 9.7 backward into earlier developmental periods. However, pooled sex analyses in younger individuals indicate that adult age trends are a continuation of trends established in childhood: tibial A-P/M-L bending rigidity increases in 4- to 14-year-olds ($p < .05$, regression on age), whereas humeral shape does not change ($p > .50$). It is also impossible to detect when during childhood or adolescence sex differences in bone shape first manifest themselves, although given that the average sexual dimorphism in older adolescents (12% in both tibia and humerus) is already similar to that of young adults in their twenties (12–20% in both bones), it is likely that sexual dimorphism in bone shape begins to develop at least by early adolescence.

The sex difference in midshaft tibial shape is part of the general morphological pattern in the lower limb discussed earlier. The present analysis and a previous analysis of a mid-distal femoral section (Ruff 2000a) indicate that this sex difference is established by late adolescence. It is maintained throughout life, with some fluctuations that may be attributable to sampling error. This suggests that the behavioral factors responsible for this sex-related difference were also maintained throughout life—that is, sexual differences in long-range mobility at Pecos were present into at least the fifth and possibly the sixth decade of residents' lives. The overall increase in A-P/M-L bending rigidity throughout life in both sexes could represent a continuing adaptation to A-P bending loads in the region about the knee. Van Gerven, Armelagos, and Bartley (1969) described a similar phenomenon in the midshaft femur in a prehistoric Mississippian population sample. Interestingly, Feik and colleagues (2000) found evidence for the opposite trend in femoral midshaft shape in a modern autopsy sample: femora tended to become rounder with age in both sexes (more so in males). The difference in findings may be attributable to continued high activity levels in the prehistoric samples even in older individuals, as opposed to a decline in overall activity level with aging in most modern populations (Feik et al. 2000).

Sexual dimorphism in mid-distal humeral cross-sectional shape has not previously been documented in the Pecos sample, but it appears to be even stronger than that found for the midshaft tibia or femur (although the *distal* femur [20% of bone length from the distal end] shows dimorphism equivalent to that of the mid-distal humerus—20.8% [Ruff 1987]). These results suggest a relative increase in A-P bending loads on the male humerus, possibly caused by greater use of the flexors and extensors of the arm and forearm (biceps, triceps, etc.). This sex effect appears to be limited to the humerus, however: sex differences in radial and ulnar midshaft shape are nonsignificant ($p > .30$, t tests between sexes, both bones). Various biomechanical factors could explain this difference, including the addition of pronation and supination movements in the forearm that could create different loading regimens during flexion and extension motions, or possibly even sex-specific activities such as corn grinding that preferentially loaded the forearm in females (Bridges 1989). The partial independence of arm and forearm mechanical loadings, and thus of bone robusticity and shape, was also suggested by a temporal analysis of Illinois Native American skeletons (Bridges, Blitz, and Solano 2000).

On the basis of results for the lower limb, it might be predicted that sexual dimorphism in upper limb bone cross-sectional shape, when present, would also show some relationship to subsistence strategy, declining from hunter-gatherers through agriculturalists to industrial populations. Structural data for upper limb bones are less abundant than those for the lower limb, but two studies of regional temporal series provide some support for this prediction. Bridges, Blitz, and Solano (2000), in their study of Illinois samples, presented data that indicate a reduction in sexual dimorphism in midshaft humeral shape (maximum to minimum bending rigidity) in a Mississippian agricultural sample relative to earlier, mixed-economy samples from the same region (their table 4, my comparison between males and females). Humeri from Georgia coast samples show a similar decline in sexual dimorphism of bone shape (A-P/M-L bending rigidity), measured at 35% of bone length from the distal end, from earlier hunter-gatherers to later horticulturalists (Ruff and Larsen 2001). However, in this case females had higher A-P/M-L ratios than males, a result that appeared to be caused by the greater development of flexor and extensor ridges in male humeri, which would preferentially increase M-L dimensions (Ruff and Larsen 2001:fig. 5.13). The use of a slightly more proximal section in the present study—at 40% rather than 35% of length from the distal end—probably helped to avoid this complication and might explain the difference in results.

The use of maximum to minimum bending rigidity at humeral midshaft in the Bridges, Blitz, and Solano (2000) study introduces a number of further issues, including possible differences in orientation of these properties and variation in development of the deltoid tuberosity, which overlaps the midshaft. Thus, it is important when carrying out such analyses to compare only similar properties collected at the same skeletal locations. In general, upper limb structural properties are difficult to interpret because their mechanical loadings are more complex and less stereotyped than those of the lower limb. However, greater similarity in upper limb bone properties between males and females should, on balance, indicate greater similarity in activity patterns. For example, using the lower limb model, it would be predicted that modern industrial populations, with the least systematic difference between the sexes in activity patterns, would show the least sexual dimorphism in upper limb bone cross-sectional shape. A twentieth-century East African sample (probably intermediate between "industrial" and "agricultural"; see Ruff 2000a) in fact has no sexual dimorphism in A-P/M-L bending rigidity of the humeral 40% section (identical mean values

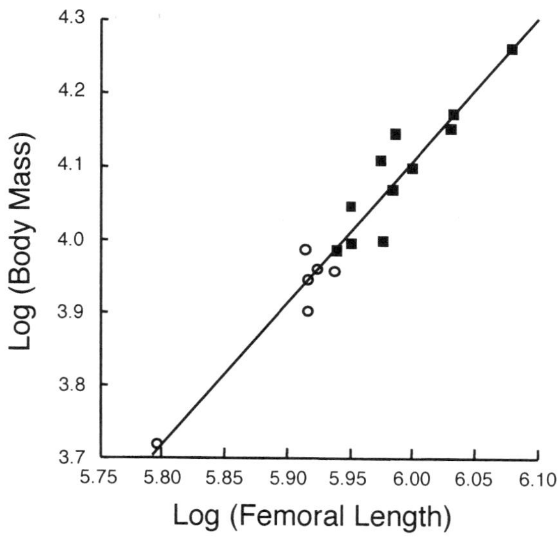

FIGURE 9.8. Prediction of body mass from femoral length (least squares) in a subset of the adult Pecos sample. Body mass was calculated from reconstructed stature and bi-iliac breadth (Ruff, Trinkaus, and Holliday 1997). Data are log-transformed. Large filled squares represent males, and open circles, females.

for both sexes). The same index in the femoral midshaft also shows minimal sexual dimorphism (< 4%) in this sample.

HOMININ EVOLUTION

The previous two sections illustrated the value of the Pecos sample for both within- and between-population comparisons of postcranial structure. The third and final way in which the Pecos sample has contributed to skeletal structural research is as a general representative of modern humans for both general theoretical studies and comparison with earlier, pre-Holocene humans. Given the rapid changes in behavior and morphology that have accompanied recent industrialization and mechanization in many parts of the world, the use of a preindustrial, relatively robust population as a baseline representative of Holocene humans, rather than a cadaver-based or living sample, has much in its favor.

As discussed earlier in the chapter, to properly interpret measures of bone strength requires scaling of strengths to an appropriate body size parameter, which for bending and torsional strength is body mass multiplied by bone length. When estimating body mass for such analyses, it is preferable to avoid mechanically based methods—that is, methods that assume a particular mechanical relationship between a skeletal structure and body size. Thus, bi-iliac breadth and stature were used in a "morphometric" model (Ruff 2000c; Ruff, Trinkaus, and Holliday 1997) to estimate body mass in the two Pecos individuals represented in figure 9.1 (see table 9.1). But although many individuals in archaeological and paleontological contexts have relatively well-preserved long bones from which stature can be estimated, few have well-preserved pelves on which to measure bi-iliac breadth. In such cases, bi-iliac breadth can be estimated using known (and strong) geographic clines in body breadth (Ruff 1994b), known body proportions from closely related populations, or both (Ruff, Trinkaus, and Holliday 1997; Trinkaus et al. 2003).

Alternatively, body mass may be estimated directly from long bone lengths, if it can be shown that the relationship between the two is close to constant—that is, that body proportions do not vary. Figure 9.8 is a plot of body mass, as estimated from bi-iliac breadth and stature in 17 Pecos adults, against femoral length for the same individuals (all data log-transformed). The correlation between the two is excellent (r = .96), and the standard error of estimate of body mass from femoral length is quite low (3.5%). There is no apparent sex difference in proportions (p > .80, ANCOVA, effect of sex with femoral length as a covariate). Therefore, it is justifiable to estimate body mass in the Pecos sample using this sample-specific relationship, and in the following analysis this is done for the other 43 individuals in the adult Pecos long bone structural sample (n = 60). The success of this approach here suggests that this may be a viable alternative technique for estimating body mass in other samples with sufficient numbers of individuals of "known" (i.e., bi-iliac- and stature-determined) body mass to develop a similar sample-specific equation. It illustrates again the value of the Pecos sample for investigating general body size and shape relationships in humans (for other examples see Ruff 1995, 2000c).

Figure 9.9 shows the bending and torsional strength of the femoral midshaft (Zp), divided by the product of body mass and femoral length (y axis), plotted against dates for 110 Pleistocene and early Holocene (Mesolithic) individuals and 60 Pecos adults. Data for the Pleistocene and early Holocene specimens were derived from previous (Ruff et al. 1993) and ongoing collaborations (Ruff, Trinkaus, and Holt 2000). There is a significant, log-linear decline in femoral strength relative to body size—what I have referred to previously as "robusticity" or "robustness"—throughout the Pleistocene ($r = .546$, $p < .0001$). The Pecos sample is significantly less robust than the earlier specimens ($p < .0001$, t test), even the Mesolithic sample ($p < .0001$). The behavioral implications of this temporal trend have been explored in more detail elsewhere (Ruff et al. 1993) but are probably explained largely by an inverse relationship between technological level and mechanical loading of the skeleton. This trend within human ancestors extends back into the Pliocene, when Australopithecines had even stronger long bones relative to body size, more similar to those of nonhuman primates (Ruff 1998; Ruff, McHenry, and Thackeray 1999). Figure 9.9 also shows that there is a great deal of variation in femoral robusticity at any one time, including within the Pecos sample. This is to be expected in a morphological characteristic as developmentally plastic and mechanically sensitive as long bone cross-sectional geometry (Ruff 2003a; Ruff, Walker, and Trinkaus 1994; Trinkaus, Churchill, and Ruff 1994). In any population or temporal period, there will be individuals who are more or less active than others and thus develop more or less robust skeletons. General temporal trends, however, are still clearly apparent within this environmental "noise."

Because cranial capacities for many of the more complete Pecos crania were determined by Hooton, it is also possible to use these along with body mass estimates from associated postcrania to investigate temporal trends in relative brain size—that is, encephalization. This has been done in figure 9.10 for 29 individuals in the Pecos sample (cranial capacities were obtained from Hooton's original grave cards), along with similar data for 43 earlier human specimens: 32 "early anatomically modern" (EAM) humans (10,000–90,000 years B.P.), 9 Neandertals (45,000–150,000 years B.P.), 1 late *H. erectus*/early *H. sapiens*, the Jinnuishan specimen (280,000 B.P.), and 1 early *Homo erectus/ergaster*, KNM-WT 15000, or the Nariokotome boy (1.5 million years B.P.) (Rosenberg, Lü, and Ruff 2006; Ruff, Trinkaus, and Holliday 1997). Regression lines are drawn through the Pecos and EAM samples, which are very similar to each other in relative brain size. Neandertals, however, while overlapping in relative brain size with both the Pecos and EAM samples, are clearly lower on average, with most individuals near the bottom edge of the Pecos-EAM distribution. The late *H. erectus*/early *H. sapiens* specimen (Jinnuishan) falls below but not too far from the more recent specimens, and the early *Homo* specimen (KNM-WT 15000) is far below all the others.

These results are consistent with those based on a broader sampling of individually unmatched brain and body sizes for modern and earlier human temporal groups, indicating that Neandertals were slightly less encephalized than modern humans but that the major increase in encephalization during the evolution of *Homo* occurred earlier, within *H. erectus* (Ruff, Trinkaus, and Holliday 1997). Figure 9.10 also illustrates the importance of factoring body size into such analyses: all the pre-Holocene specimens are large in body size relative to the Pecos sample, so that comparisons of raw brain size (which is correlated with body size) would be misleading.

CONCLUSION

The Pecos Pueblo skeletal sample, because of its size, preservation, and provenience, has been an invaluable resource for studies of postcranial skeletal structure, from population-specific analyses of growth and aging to comparative analyses with broadly contemporary and earlier human populations. Results from these studies have been instrumental in demonstrating the dynamic nature of limb bones throughout life. Long bone cortical structure is sensitive to general systemic factors such as variation in body size and shape and hormonal changes with aging, as well as to specific mechanical factors such as sex differences in mobility patterns and evolutionary changes in activity levels. These studies have also shown that these two aspects of bone growth and remodeling—systemic and specific—are themselves interrelated; for example, skeletal structural adaptation to hormonally mediated bone loss with aging may vary according to local mechanical loadings. Thus, both endosteal contraction in late adolescence and continuing subperiosteal expansion throughout adulthood are greater in the femur than in the humerus, possibly in response to relatively greater lower limb loadings in the rugged Pecos Pueblo environment. Females show more endosteal contraction in late adolescence than males, perhaps to partially offset greater endosteal expansion later in life. To my knowledge, these are the first structural comparisons of this kind—between the sexes and between weight-bearing and non-weight-bearing major long bones, over the majority of the life span—in the clinical or anthropological literature.

These kinds of observations also highlight the importance of large, well-documented archaeological collections like that from Pecos in clarifying some aspects of skeletal biology that are more difficult to

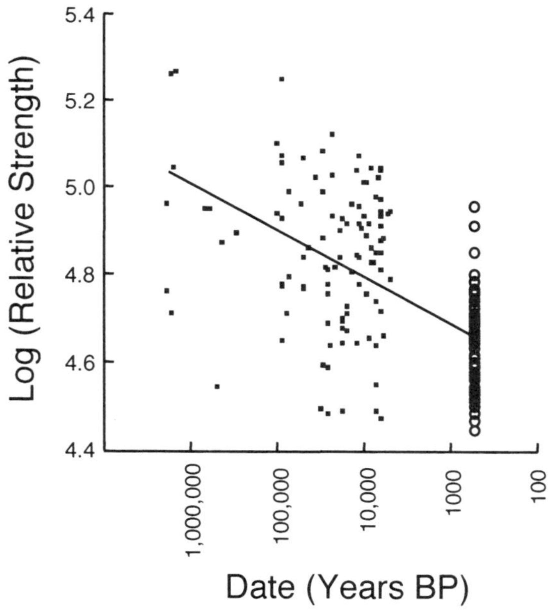

FIGURE 9.9. Temporal changes in log-transformed femoral midshaft polar section modulus (Zp), divided by the product of body mass and femoral length, in 110 paleontological specimens (small squares) and 60 Pecos adult specimens (circles). A least squares regression line is plotted through all points. The temporal *(x)* axis is also logarithmic. Data from Ruff, Trinkaus, and Holt 2000.

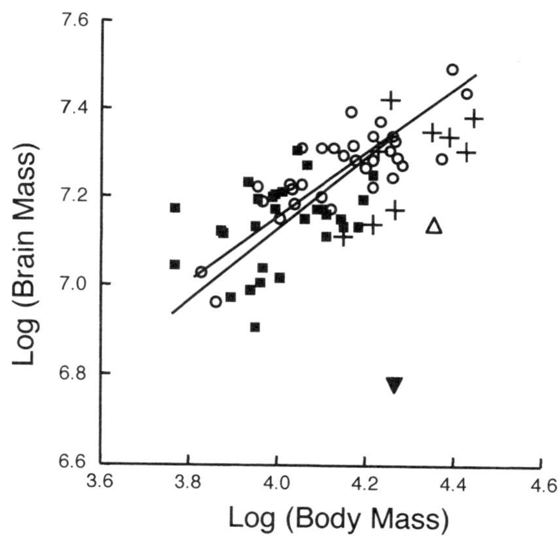

FIGURE 9.10. Encephalization in modern and earlier humans, data log-transformed. Key: filled squares, Pecos; open circles, early anatomically modern humans (Upper Paleolithic and Skhul-Qafzeh); crosses, Neandertals; open triangle, Jinnuishan; filled triangle, KNM-WT 15000. Least squares regression lines are plotted through the Pecos and EAM samples. Data from Ruff, Trinkaus, and Holliday 1997 and Rosenberg, Lü, and Ruff 2006.

address in in-vivo studies. Osteoporosis is a serious medical and social issue today. Although the general importance of bone structural changes in osteoporosis has long been recognized, it is difficult to actually demonstrate many of these changes noninvasively—that is, in a living population (but see Beck et al. 1996, 2000). The age changes in geometry documented for the Pecos sample shed new light on the mechanisms underlying bone fragility (or the avoidance of fragility) in old age. The relative genetic homogeneity of the Pecos sample in comparison with most clinical samples also helps to better identify environmental influences on bone form. Finally, physically active preindustrial samples such as the Pecos Pueblo collection may provide a suitable "baseline" against which to compare more sedentary recent populations (Ruff 2000a; Ruff and Hayes 1988).

A general sex difference in lower limb bone structure related to subsistence activities and differences in mobility patterns, now known for many populations, was also first demonstrated in the Pecos sample. Males in preindustrial societies have lower limb bone shapes that are more adapted for long-distance travel than those of females, corresponding to documented behavioral differences in recent populations. It is shown here that similar sex-related differences in upper limb bone structure exist in the Pecos sample, although these are more difficult to interpret in terms of specific behaviors because of the multifunctionality of the upper limb. Cross-sectional shape of the tibia changes systematically throughout adulthood in both sexes but remains constant in the humerus, again demonstrating the heterogeneity of bone modeling and remodeling in different areas of the skeleton. Sex-related differences in bone shape are established by late adolescence and maintained throughout adulthood, with a possible slight decline in the oldest age group, suggesting that behavioral differences between the sexes were also maintained through most of adulthood. There is also evidence that activity levels in general remained high throughout adulthood at Pecos.

When expressed relative to body size, however, the strength of the femur in the Pecos sample is reduced relative to that of much earlier humans, part of a general temporal trend toward reduced skeletal robusticity characteristic of human evolution. Body size estimates for Pecos individuals, based on postcranial dimensions, are also useful for evaluating other long-term temporal trends, including an increase in encephalization (brain size relative to body size) in modern humans relative to archaic humans.

These examples give some indication of how valuable the Pecos skeletal collection has been for addressing a variety of structural and functional questions. Only with a large, well-documented sample of this kind can such interrelationships be effectively explored. The Pecos sample is now unavailable for new studies, but its legacy in skeletal structural research will continue.

ACKNOWLEDGMENTS

I would like to thank Erik Trinkaus and Brigitte Holt for providing femoral robusticity and cranial capacity data for Pleistocene–early Holocene specimens, Erik Trinkaus and Trent Holliday for collaboration in Pleistocene *Homo* body size and encephalization studies, Tom Berger for obtaining individual cranial capacities from Hooton's original Pecos grave cards, Karen Rosenberg for her collaboration on the Jinnuishan specimen, Alan Walker for his collaboration on KNM-WT 15000, and, most of all, the Harvard Peabody Museum of Archaeology and Ethnology for allowing access to and study of the Pecos Pueblo skeletal collection over the past 30 years. The work reported here was supported in part by the National Science Foundation and the Wenner-Gren Foundation for Anthropological Research.

10 The Obstetric Pelvis of Pecos Pueblo

Robert G. Tague

The Pecos Pueblo skeletal population informs us of adaptations of indigenous Americans and extends our understanding of variation in the anatomy and biology of premodern and modern people. I studied Pecos Pueblo as part of my dissertation on obstetrical adaptations of the human pelvis (Tague 1986). I selected Pecos because of the large number of individuals in the collection and because it provided geographical and temporal diversity with three other Native American populations in my study. I have used my data on Pecos Pueblo in seven projects beyond the dissertation (Tague 1988, 1989, 1991, 1992, 1994, 2000; Tague and Lovejoy 1986), and I continue to find new applications for these data.

In my dissertation and previous publications, I analyzed Pecos Pueblo as part of a suite of samples. My interest was in identifying commonalities among different groups of people in order to make broad inferences about evolutionary and developmental determinants of human pelvic anatomy. Some of the commonalities I found were the following. First, the female pelvis is larger than that of the male principally in the posterior aspect of each pelvic plane and the breadth of the lower pelvic planes. The implication is that the female pelvis is enlarged primarily by separation of bony elements and to a lesser extent by differential elongation of bony elements. Second, the most sexually dimorphic aspects of the pelvis, with the pelves of females being larger than those of males, are also the most variable in size. This result is opposite to theoretical expectation. The reason is that the hormone relaxin, which is secreted during pregnancy, causes increased mobility of the hipbones and sacrum at the pelvic joints, thereby temporarily increasing pelvic capacity at delivery. This hormonally mediated modulation of pelvic size attenuates the theoretically expected relationship between dimorphism and variability. Third, the sexes are not significantly different from one another in pelvic variability, a result that is again the opposite of theoretical expectation. The reason is that female pelvic anatomy is the default anatomy; male anatomy results from modulation of growth of the default anatomy by testicular androgens. The intrinsic variability of the default, female anatomy, is thereby shifted en masse to the male anatomy.

In the present study, I depart from my earlier methodological approach of identifying interpopulational commonalities and concentrate my analysis on Pecos Pueblo. I demonstrate that the pelves in the Pecos Pueblo skeletal collection are distinctive in the size and shape of the pelvic inlet. I then discuss the correlates and obstetrical implications of this distinctive anatomy.

The pelvis is subdivided anatomically into false and true pelvic cavities (fig. 10.1). The false pelvis is above the linea terminalis and the promontory and alae of the sacrum (see fig. 10.2 for definitions), and the true pelvis is below them. Only the true pelvis is

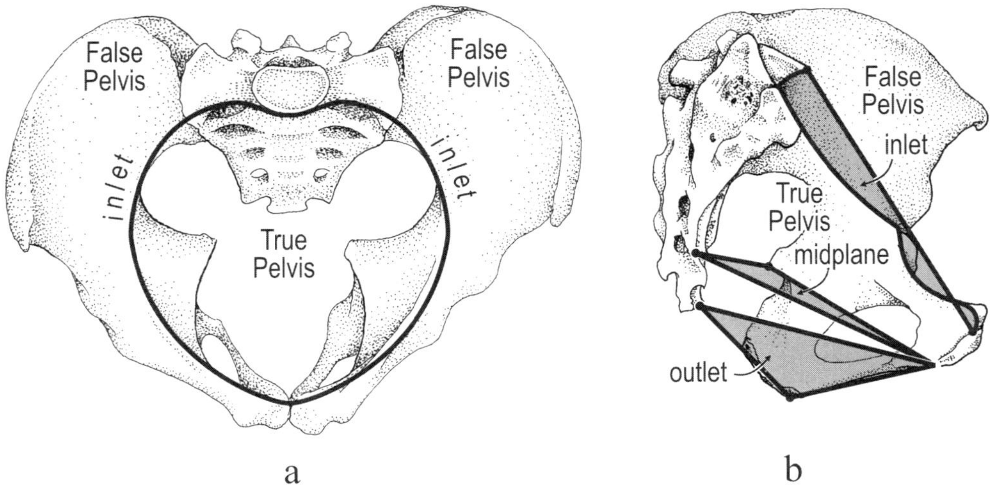

FIGURE 10.1. Anatomical subdivisions of the pelvis into the false and true pelvis as shown in a superior view (a) of the whole pelvis and a medial view (b) of the left hipbone and sacrum. The drawings show three planes of obstetrical interest in the true pelvis: the inlet (a, b), the midplane (b), and the outlet (b).

obstetrically relevant; it is the bony birth canal. Because some obstetrically important pelvic dimensions are highly variable and of low correlation with one another (Tague 1989), there is no singularly optimal size or shape of the true pelvis. Rather, pelvic anatomy should be categorized simply as obstetrically "sufficient" or "insufficient," depending on whether or not there can be successful vaginal delivery with survival of mother and newborn. A number of studies have characterized pelvic size and shape, most of them based on European and North American females. A leading textbook on obstetrics and gynecology, *Williams Obstetrics* (Cunningham et al. 2001), uses Caldwell and Moloy's (1933) classification of the female pelvis. Caldwell and Moloy (1938) reported that the "gynecoid" pelvic type was modal, being observed in approximately 42% of females. In the gynecoid pelvis,

> the posterior sagittal diameter of the inlet is only slightly shorter than the anterior sagittal. . . . Because the transverse diameter of the inlet is either slightly greater than or about the same as the anteroposterior diameter, the inlet is either slightly oval or round. The sidewalls of the pelvis are straight, the spines are not prominent, the pubic arch is wide, and the transverse diameter at the ischial spines is 10 cm or more. The sacrum is inclined neither anteriorly nor posteriorly. The sacrosciatic notch is well rounded and never narrow. (Cunningham et al. 2001:57)

Although the Pecos Pueblo skeletal collection has been repatriated and reburied, it can continue to contribute to scientific research. I hope that other researchers will find use for the data I present in Appendix K.

MATERIALS AND METHODS

I compare Pecos Pueblo with two other data sets: (1) five skeletal samples—Haida, Indian Knoll, Libben, "blacks," and "whites"—that I measured, and (2) samples analyzed by others and reported in the published literature. Haida, Indian Knoll, and Libben are samples of prehistoric and historic Native American populations; they and Pecos Pueblo are temporally and geographically diverse from each other (Hooton 1930; Lovejoy et al. 1977; Webb 1974). Haida skeletal remains were collected from the Queen Charlotte Islands, British Columbia, by the Field Columbian Museum Expedition in 1897. The skeletal material was likely historic, though it was not dated. Traditionally the Haidas have lived in permanent communities, with fishing and collecting as their principal economy (Stearns 1981). The Indian Knoll skeletal material came from a Late Archaic site in Ohio County, Kentucky. Recalibrated radiocarbon dates suggest an initial occupation of the site sometime between 4160 ± 315 B.C. and 2558 ± 365 B.C. (Webb 1974). The site was occupied for several hundred years, although the terminal date is unknown. The people of Indian Knoll were hunter-gatherers whose settlement pattern was one of prolonged yet seasonal residency. The skeletal remains from the Libben site, in Ottawa County, Ohio, date to the Late Woodland period, radiocarbon dated between A.D. 800 and 1100. The people of Libben were probably sedentary hunter-gatherers (Lovejoy et al. 1977). The blacks and whites were sampled from the Hamann-Todd collection, which was assembled during the late nineteenth and early twentieth centuries. The skeletons in this collection came principally from unclaimed bodies at hospitals in Cleveland, Ohio, and at the Cuyahoga, Ohio, county morgue (Kern 2006). These individuals were born primarily in North America and Europe. Categorization of the ancestry of the individuals in the Hamann-Todd collection—"black" for African American and "white" for European American—was based on records at the Cleveland Museum of Natural History.

The sexes of the blacks and whites were known from records at the Cleveland Museum of Natural History. I assigned sex to the Native Americans using standard traits of the pelvis and skull (Tague 1989). I made an initial assignment of sex using the hipbones, seriating them from the most "male-like" to the most "female-like" using both Phenice's (1969) technique (i.e., ventral arc of pubis, subpubic concavity, and medial border of inferior pubic ramus) and the greater sciatic notch. I subsequently evaluated the skulls on a case-by-case basis, because seriation was impractical, and I measured the lengths of the femur and clavicle and the head diameter of the femur. If the anatomy of the skull or size of the nonpelvic bones was inconsistent with the sex assignment from the hipbones, I reevaluated the sex of the individual.

Although the accuracy of these sex assignments was unknowable, my assignments were the same as those of Webb (1974) for 208 of 212 individuals from Indian Knoll and of Lovejoy (C. Owen Lovejoy, personal communication) for all 161 individuals from Libben. Among studies of individuals of known sex, Meindl and colleagues (1985) reported a 3% error rate in sexing skeletons using both the pelvis and skull, Phenice (1969) missexed 4% of skeletons using the three traits of the hipbone mentioned earlier, and Sutherland and Suchey (1991) missexed 4% of skeletons using only the ventral arc. Lovell (1989) and MacLaughlin and Bruce (1990) reported lower levels of accuracy in sex identification using the three traits of the hipbone in Phenice 1969. Meindl and colleagues (1985) argued that a researcher would more likely missex males than females.

Only adults were used in the present study. The distinction between adult and subadult was based on epiphyseal fusion. I classified an individual as adult if the epiphyses of the distal femur and proximal tibia had fused (for females) or had begun to fuse (for males) to the diaphyses and if the epiphyses of the iliac crest and ischial ramus had begun to fuse (for both sexes). Judging from the Indian Knoll sample, individuals with these stages of epiphyseal fusion were approximately 18 years of age and older (Johnston 1961). All the individuals in this study from the Hamann-Todd collection had fused femoral and tibial epiphyses, and all but two had completely fused epiphyses of the iliac crest and ischial ramus. Records of the Cleveland Museum of Natural History show that all blacks and whites in this study were 18 years of age and older.

I used only adults for which I could take at least one pelvic measurement. On the basis of this stipulation, sample sizes for the Native Americans were as follows: 160 females and 133 males for Pecos Pueblo (13 additional individuals with pelvic data were not included in this study or previous studies by me; see Appendix K, note 2), 22 females and 30 males for Haida, 88 females and 124 males for Indian Knoll, and 68 females and 93 males for Libben. I randomly selected 100 blacks and 100 whites, with 50 of each sex. Within each group, sample sizes for some pelvic and nonpelvic measurements differed because of the variable preservation of bones.

I articulated the hipbones and sacrum to take some measurements of the pelvis. The sacroiliac joints were held together with thin strips of adhesive tape, and the pelvis was encircled with rubber bands. The two pubes touched in the midline; I did not make compensation for an interpubic disk. I took 25 measurements of the hipbone, sacrum, and pelvis, selected principally to characterize the size and shape of the true pelvis among the three planes typically recognized by obstetricians—the inlet, midplane, and outlet (figs. 10.1, 10.2). The inlet is bounded posteriorly by the promontory and alae of the sacrum, laterally by the linea terminalis, and anteriorly by the superior border of the pubic symphysis. The midplane is bounded posteriorly by the junction of the fourth and fifth sacral vertebrae, laterally by the ischial spine to the pubic corpus, and anteriorly by the inferior border of the pubic symphysis. The outlet is bounded posteriorly by the inferior margin of the fifth sacral vertebra, laterally by the ischial tuberosity and ischiopubic ramus, and anteriorly by the inferior border of the pubic symphysis.

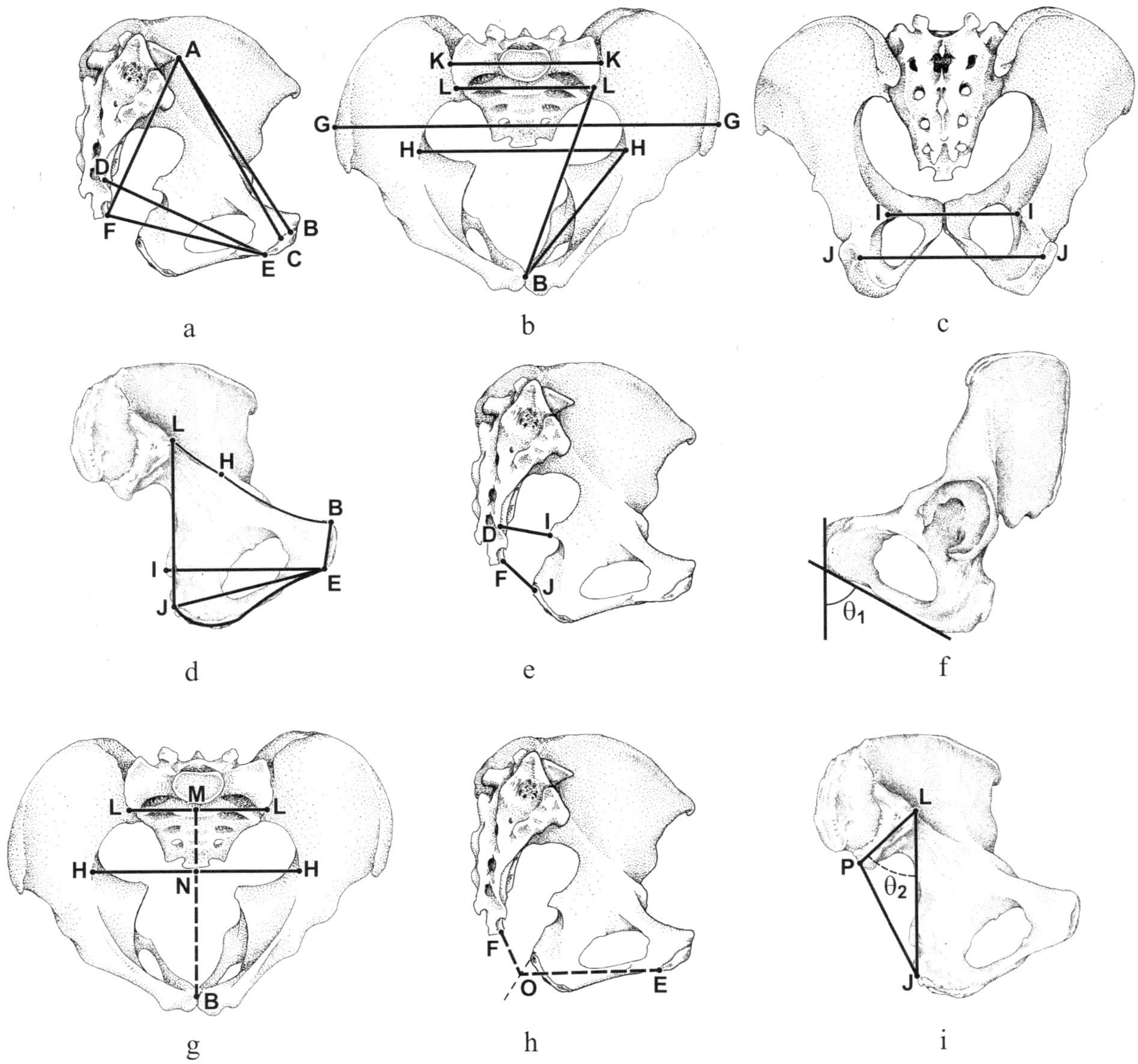

Except for the bi-iliac diameter, all pelvic variables are obstetrically relevant, although some variables are considered to be more important determinants of obstetric success than others (Tague 2000). Some pelvic measurements were either duplicated (e.g., anteroposterior and transverse diameters) or modified (e.g., depth) from those reported in the literature. Other measurements (e.g., curved length of linea terminalis) were developed for this study. The study did not include measurements based on lumbar or coccygeal vertebrae fused to the sacrum. From the suite of 25 measurements, I computed eight other pelvic variables (fig. 10.2). I also took three measurements of nonpelvic bones: maximum lengths of femur and clavicle and maximum diameter of femoral head. I used sliding calipers, osteometric board, curvometer, and protractor to take linear, curvilinear, and angular measurements. Bi-iliac diameter and maximum lengths of femur and clavicle were measured to the nearest millimeter. All other linear measures were taken to the nearest 0.1 millimeter. Curvilinear measures were taken to the nearest 0.64 millimeter.[1] With

1. The curvometer is calibrated in units of 0.05 inch (1.27 mm). Data were recorded to the nearest half of a calibrated unit (i.e., to the nearest 0.64 mm).

FIGURE 10.2. Osteometric points of measurement and computed measurements of the hipbone, sacrum, and pelvis as shown on medial views of the left hipbone and sacrum articulated (a, e, h), superior views of the articulated pelvis (b, g), a dorsal view of the articulated pelvis (c), medial views of the left hipbone (d, i), and a ventral view of the left hipbone (f). The measurements and the equations used to calculate some of them are as follows:

Direct Measurements (n = 25), Shown with Solid Lines

Anteroposterior Diameters (n = 4)

Inlet, anatomical conjugate: promontory of sacrum (i.e., most ventrally projecting aspect on superior border of body of first sacral vertebra) to dorsomedial aspect of superior border of pubis (a, A–B)
Obstetrical conjugate: minimum distance between sacral promontory and dorsomedial aspect of pubis (a, A–C)
Midplane: junction of fourth and fifth sacral vertebrae to dorsomedial aspect of inferior border of pubis (a, D–E)
Outlet: apex of fifth sacral vertebra to dorsomedial aspect of inferior border of pubis (a, E–F)

Transverse Diameters (n = 4)

Bi-iliac diameter: maximum distance between tubercles of iliac crest (b, G–G)
Inlet: maximum distance between linea terminales, with the diameter visually aligned to be perpendicular to the anteroposterior diameter (b, H–H)
Midplane: minimum distance between ischial spines (c, I–I)
Outlet: distance between ischial tuberosities, with the point on the ischial tuberosity being the inner margin of the transverse ridge (Williams 1995:668, fig. 6.269), which demarcates the attachment of the common tendon of biceps femoris and semitendinosus from that of adductor magnus (c, J–J)

Sacrum (n = 4)

Maximum breadth (b, K–K)
Breadth at inlet: straight distance across ventral surface of sacrum where sacrum met linea terminalis of ilium when pelvis was articulated (b, L–L)
Straight and curved lengths (two measurements): sacral promontory to apex of fifth sacral vertebra (a, A–F)

Anterior Space (n = 4)

Inlet: dorsomedial aspect of superior border of pubis to point of transverse diameter of inlet on linea terminalis (b, B–H)
Midplane: dorsomedial aspect of inferior border of pubis to ischial spine (d, E–I)
Outlet, straight and curved lengths (two measurements): dorsomedial aspect of inferior border of pubis to ischial tuberosity (d, E–J)

Posterior Space (n = 4)

Inlet, straight and curved lengths (two measurements): point of transverse diameter of inlet to auricular surface of ilium on linea terminalis (d, H–L)
Midplane: junction of fourth and fifth sacral vertebrae to ischial spine (e, D–I)
Outlet: apex of fifth sacral vertebra to ischial tuberosity (e, F–J)

Other Direct Measures (n = 5)

Linea terminalis, straight and curved lengths (2 measurements): dorsomedial aspect of superior border of pubis to auricular surface of ilium (b and d, B–L)
Depth: ischial tuberosity to junction of linea terminalis and auricular surface of ilium (d, J–L)

Continued on next page

FIGURE 10.2. *Continued*

Height of pubic symphysis: superior border of pubis to inferior border of pubis at dorsomedial aspect (d, B–E)

Subpubic angle (f, θ_1): Hipbone was aligned with the symphyseal face of the pubis, parallel with a vertical upright. Hipbone was illuminated from behind, thereby creating a sharp–shadow image of the pubis and ischiopubic ramus. A tangential line along the ramus was drawn to intersect with a vertical line along the symphyseal face of the pubis. The angle was measured. If there were two tangents along the ramus, the average of the angles was recorded. An individual's total subpubic angle is the sum of the angles for the left and right hipbones. If only one hipbone was available, the total subpubic angle is double the value of that single hipbone.

Computed Measurements (n = 8), Shown with Dotted Lines

Anteroposterior diameter at plane of linea terminalis: dorsomedial aspect of superior border of pubis to midpoint of sacral breadth at inlet (g, B–M, eq. 1 [below])

Anterior sagittal diameter of inlet: dorsomedial aspect of superior border of pubis to midpoint of transverse diameter of inlet (g, B–N, eq. 2)

Posterior sagittal diameter of inlet: midpoint of sacral breadth at inlet to midpoint of transverse diameter of inlet (g, M–N, eq. 3)

Anterior sagittal diameter of outlet: dorsomedial aspect of inferior border of pubis to midpoint of transverse diameter of outlet (h, E–O, eq. 4)

Posterior sagittal diameter of outlet: apex of fifth sacral vertebra to midpoint of transverse diameter of outlet (h, F–O, eq. 5)

Angulation of sacrum (i, θ_2): angle between sides 1 and 2 of the following triangle: side 1, ischial tuberosity to junction of linea terminalis and auricular surface of ilium (i, J–L); side 2, junction of linea terminalis and auricular surface of ilium to posterior inferior iliac spine (i, L–P); side 3, ischial tuberosity to posterior inferior iliac spine (i, J–P)

Area of inlet (a and b, eq. 6)

Area of outlet (c, d, and e, eq. 11)

Equations for the Computed Measurements

(1) Anteroposterior diameter at plane of linea terminalis = ([straight length of linea terminalis]^2 – [breadth of sacrum at inlet/2]^2)$^{0.5}$

(2) Anterior sagittal diameter of inlet = ([anterior space of inlet]^2 – [transverse diameter of inlet/2]^2)$^{0.5}$

(3) Posterior sagittal diameter of inlet = anteroposterior diameter at plane of linea terminalis – anterior sagittal diameter of inlet

(4) Anterior sagittal diameter of outlet = ([straight length of anterior space of outlet]^2 – [transverse diameter of outlet/2]^2)$^{0.5}$

(5) Posterior sagittal diameter of outlet = ([posterior space of outlet]^2 – [transverse diameter of outlet/2]^2)$^{0.5}$

(6) Area of inlet = π(obstetrical conjugate diameter of inlet/2)(transverse diameter of inlet/2)

The area of outlet is based on Heron's formula, representing summation of areas of anterior (ANT) and posterior (PST) triangles, with TRV representing the transverse diameter of outlet and s_1 and s_2 representing half the perimeter lengths of the ANT and PST triangles, respectively, where

(7) s_1 = ([2(straight length of ANT space)] + TRV)/2

(8) s_2 = ([2(PST space)] + TRV)/2

(9) area of ANT triangle = (s_1[(s_1 – straight length of ANT space)^2][s_1 – TRV])$^{0.5}$

(10) area of PST triangle = (s_2[(s_2 – PST space)^2][s_2 – TRV])$^{0.5}$

(11) area of outlet = area of ANT triangle + area of PST triangle

TABLE 10.1
COMPARISON OF SEX ASSIGNMENTS BY HOOTON AND TAGUE

Sex assignment	n
Male by Tague; male by Hooton	99
Female by Tague; female by Hooton	99
Female by Tague; male by Hooton	17
Female by Tague; sex indeterminate by Hooton	4
Total	219

the exception of subpubic angle and height of pubic symphysis, the measurements were repeated on 26 individuals from the Hamann-Todd collection, more than a decade after the original measurements were taken. The medians for measurement precision were greater than 97% for all 23 pelvic and 3 nonpelvic variables. The medians for precision were also greater than 97% for all eight variables computed from the measurements:

> measurement precision = (1 – [| original measurement – repeat measurement | / original measurement])100.

Statistical analyses using SPSS software (© SPSS, Chicago, Illinois) included ANOVA, chi-square test, Kruskal-Wallis test, multiple comparisons test, partition of contingency table, Pearson's correlation coefficient, sign test, and Wilcoxon-Mann-Whitney test. Statistical significance was set at $p \leq .05$.

RESULTS AND DISCUSSION

My assignments of sex for some individuals from Pecos Pueblo differ from those made by Earnest Hooton, which were provided to me by Michèle Morgan (personal communication). Of the 293 individuals that I sexed and include in this study, Hooton reported either a sex or "sex indeterminate" for 219. For these 219 individuals, as can be seen in table 10.1, Hooton and I agreed fully on the sample of males, and we agreed on 82.5% of those that I sexed as female (99 out of 120 individuals). These percentages of agreement are similar to those reported by Ruff (1981).

Hooton remarked that "not infrequently in the Pecos series skeletons with exaggeratedly masculine crania and long bones have features which one expects to see only in the female pelvis" (1930:27). In table 10.2, I use ANOVA, the Kruskal-Wallis test, and the multiple comparisons test to compare my sex assignments with Hooton's with respect to three nonpelvic and four pelvic variables: maximum length of femur and clavicle, maximum diameter of femoral head, curved length of linea terminalis, posterior space of outlet, angulation of sacrum, and subpubic angle. In each of the six samples that I studied, males were significantly larger than females for the three nonpelvic variables, and females were significantly larger than males for the four pelvic variables (Tague 1992). The three groups compared here are as follows: Group 1, individuals that Hooton and I agree are male (n = 99); Group 2, individuals that I sexed as female but Hooton sexed as male (n = 17); and Group 3, individuals that Hooton and I agree are female (n = 99). Results show that all ANOVA and Kruskal-Wallis tests are significantly different from zero. Multiple comparisons tests show that (1) for femoral length, clavicular length, and femoral head diameter, Group 1 is significantly larger than Groups 2 and 3, and Groups 2 and 3 are nonsignificantly different from each other; and (2) for linea terminalis length, posterior space of outlet, angulation of sacrum, and subpubic angle, Groups 2 and 3 are nonsignificantly different from one another, and both groups are significantly larger than Group 1. Therefore, the 17 individuals that I sexed as female but Hooton sexed as male (Group 2) do not, as a group, have "masculine long bones." Moreover, these 17 individuals are nonsignificantly different in size for four pelvic variables in comparison with individuals that both Hooton and I agree are female. Collectively, the individuals I sexed as female but Hooton sexed as male are female-like in both nonpelvic and pelvic size.

Table 10.3 presents summary statistics from my study and from Hooton 1930 and Reynolds 1930 for females and males of Pecos Pueblo for four pelvic and three nonpelvic measurements. Edward Reynolds was the director of Harvard University's Peabody Museum at the time of the study of the Pecos collection and the publication of *The Indians of Pecos Pueblo: A Study of Their Skeletal Remains*. In his Appendix 2 to that book, he reported summary statistics on eight measurements and five indices of the inlet of the Pecos Pueblo pelves. I selected the seven aforementioned variables for comparison because they were duplicated between my study and those of Hooton, Reynolds, or both. Sample sizes in my study are typically smaller than those of Hooton and Reynolds, for several reasons. First, Hooton provided a complete profile of the Pecos Pueblo collection, including analyses of demography, osteometry, and pathology. Correspondingly, he exam-

TABLE 10.2
ANOVA, KRUSKAL-WALLIS, AND MULTIPLE COMPARISONS TESTS COMPARING MEANS AND MEAN RANKS OF PELVIC AND NONPELVIC VARIABLES FOR PECOS PUEBLO, BY HOOTON AND TAGUE SEX ASSIGNMENTS

Variable	Group 1, Mean ± sd (*n*)	Group 2, Mean ± sd (*n*)	Group 3, Mean ± sd (*n*)	ANOVA (A) or Kruskal-Wallis (K-W) (*p*)	Multiple comparisons (direction of significant difference between groups in parentheses)		
					p, Groups 1:2	*p*, Groups 1:3	*p*, Groups 2:3
Femoral length (mm)	429.2 ± 19.1 (80)	403.7 ± 12.6 (13)	390.5 ± 17.4 (76)	<.001 (K-W)	<.01 (1>2)	<.001 (1>3)	>.10 (NS)
Femoral head diameter (mm)	43.8 ± 2.1 (82)	39.9 ± 1.8 (16)	38.8 ± 1.7 (73)	<.001 (A)	<.001 (1>2)	<.001 (1>3)	>.20 (NS)
Clavicular length (mm)	150.4 ± 6.9 (51)	138.6 ± 5.2 (8)	133.1 ± 6.8 (38)	<.001 (K-W)	<.01 (1>2)	<.001 (1>3)	>.50 (NS)
Linea terminalis length, curved (mm)	124.9 ± 7.7 (81)	134.0 ± 6.9 (14)	129.9 ± 7.3 (85)	<.001 (K-W)	<.001 (2>1)	<.001 (3>1)	>.10 (NS)
Posterior space of outlet (mm)	71.4 ± 6.8 (21)	81.4 ± 6.6 (7)	81.4 ± 8.1 (22)	<.001 (K-W)	<.05 (2>1)	<.001 (3>1)	>.50 (NS)
Angulation of sacrum (°)	61.5 ± 5.3 (55)	70.5 ± 5.1 (12)	69.2 ± 5.7 (58)	<.001 (K-W)	<.001 (2>1)	<.001 (3>1)	>.50 (NS)
Subpubic angle (°)	62.3 ± 7.6 (76)	83.0 ± 5.3 (12)	87.2 ± 10.1 (68)	<.001 (K-W)	<.001 (2>1)	<.001 (3>1)	>.50 (NS)

Note: Group 1 = Tague: male, Hooton: male; Group 2 = Tague: female, Hooton: male; Group 3 = Tague: female, Hooton: female. The Kruskal-Wallis test is used if the groups differ significantly in their variances or if sample size is < 15 for any one of the groups. NS = nonsignificant.

TABLE 10.3
COMPARISON OF MEANS FOR PECOS PUEBLO VARIABLES REPORTED BY TAGUE, HOOTON (1930), AND REYNOLDS (1930)

Variable (mm)	Females, mean (*n*)			Males, mean (*n*)		
	Tague	Hooton	Reynolds	Tague	Hooton	Reynolds
Obstetrical conjugate	91.0 (42)	93.46 (56)	96.73 (62)	90.6 (35)	95.56 (69)	95.00 (53)
Transverse diameter of inlet	133.5 (54)	130.12 (59)	129.64 (63)	126.0 (41)	125.58 (74)	120.49 (55)
Bi-iliac diameter	261.8 (38)	261.52 (56)	—	267.2 (35)	270.42 (72)	—
Straight length of sacrum	96.2 (71)	100.78 (46)	—	100.9 (64)	104.73 (66)	—
Maximum length of femur, left	394.3 (83)	396.05 (105)	—	428.1 (65)	426.71 (140)	—
Maximum length of femur, right	392.3 (75)	394.26 (102)	—	429.2 (63)	427.22 (142)	—
Maximum diameter of femoral head, left	39.2 (78)	38.35 (109)	—	44.1 (70)	42.95 (154)	—
Maximum diameter of femoral head, right	38.8 (82)	38.35 (109)	—	43.6 (61)	42.99 (151)	—
Maximum length of clavicle, left	134.9 (44)	134.89 (47)	—	151.4 (50)	150.52 (73)	—
Maximum length of clavicle, right	132.2 (42)	131.63 (41)	—	150.3 (41)	149.33 (72)	—

ined the "whole collection of innominates and sacra" (Reynolds 1930:375). In contrast, Reynolds was interested in sexual dimorphism of the pelvis as it related to obstetrical adaptation. Toward this end, he used only a subset of the Pecos Pueblo collection, "118 pelves which are reasonably complete and unbroken" (Reynolds 1930:375). I, too, was interested in understanding pelvic anatomy as an obstetrical adaptation, and I used only individuals for which I could take at least one pelvic measurement. Therefore, my sample for nonpelvic variables does not include all adults of Pecos Pueblo. Second, Hooton and Reynolds may have used individuals of younger ages than I did. Third, some individuals used by Hooton and Reynolds may not have been available for my study. Fourth, some bones may have been damaged subsequent to Hooton's and Reynolds's studies and therefore are not included in mine. Sample sizes also differed between Hooton's and Reynolds's studies. The larger sample of males in Hooton's study than in Reynolds's is explainable by their different research agendas. The unexpectedly larger sample of females in Reynolds's study than in Hooton's may have been due to either a more inclusive use of fragmentary female specimens by Reynolds or his independent sexing of the skeletal material, although the latter was mentioned by neither Hooton (1930) nor Reynolds (1930).

Means differ trivially between my study and Hooton's for lengths of femur and clavicle, head diameter of femur, and bi-iliac diameter. However, means from my study differ from those of Reynolds by more than 5 millimeters for the obstetrical conjugate diameter of inlet of females and transverse diameter of inlet of males. Hooton and Reynolds also differ by 5 millimeters for the transverse diameter of inlet of males. The reason Reynolds and I differ for the transverse diameter of inlet may be that we chose different points of measurement. Reynolds (1930:375) measured the "transverse diameter of the brim . . . [at] the level . . . where this diameter is the least, usually a little below the ilio-pectineal line." In contrast, I measured the transverse diameter at the level of the ilio-pectineal line (i.e., linea terminalis; fig. 10.2b, H–H). A reason for the difference in the obstetrical conjugate of inlet may again be methodological. Obstetrical conjugate diameter is the "shortest distance between the promontory of the sacrum and the symphysis pubis" (Cunningham et al. 2001:55). Reynolds (1930) predefined the point on the pubis prior to taking the measurement. In contrast, I measured the obstetrical conjugate by placing one tip of the caliper on the sacral promontory and then finding the point on the inner border of the pubis that yielded the minimum diameter. Although this methodological difference may explain part of the discrepancy between the means in my study and Reynolds's, it does not likely explain the entire 5 millimeter difference. The remaining difference must be due to sample size, method of pelvic articulation, or difference in sex assignment.

Summary Statistics

Table 10.4 presents summary statistics for 15 pelvic and 3 nonpelvic variables for females and males in the samples referred to as Pecos Pueblo, Haida, Indian Knoll, Libben, blacks, and whites. I do not include all the variables shown in figure 10.2, because of small sample sizes for some (e.g., transverse diameter of midplane) and redundancy for others (e.g., straight and curved lengths of sacrum). Results of ANOVA and the Kruskal-Wallis test show that the six samples of females are nonsignificantly different in their mean ranks for anterior sagittal diameter of outlet and means for sacral length. They differ significantly in their means or mean ranks for the other 13 pelvic and 3 nonpelvic variables. For males, the six samples differ significantly in their means or mean ranks for all 18 variables. For variables in which the samples differ significantly, I use the multiple comparisons test to identify which samples differ from the others (Appendix I).

On the basis of results given in Appendix I, I use the sign test for pairwise comparisons among the six samples of females and separately among the six samples of males, in order to determine whether one sample is significantly different from another in overall pelvic size when considering the 15 pelvic variables simultaneously (table 10.5). For females, results show that the pelvis in the Haida sample is significantly larger than that in the Pecos Pueblo and black samples. All other pairwise comparisons for females are nonsignificant. For males, the pelvis in Haida and whites is significantly larger than that in Pecos Pueblo and blacks. The pelvis in Libben also is significantly larger than that in Pecos Pueblo. All other pairwise comparisons are nonsignificant.

Appendix J presents summary statistics from the literature and from my research for six pelvic variables for which I believe researchers are comparable in their measurements. There are 88 samples for females and 33 for males. Some samples, however, may be redundant. For example, there are nine samples of white females from the United States. Results show that for females, Pecos Pueblo has the shortest anteroposterior diameter of inlet; its obstetrical conjugate diameter is the shortest among 46 samples (fig. 10.3). Some researchers are not explicit about whether they are reporting the obstetrical or the anatomical conjugate (fig. 10.2a, A–C and A–B, respectively) as the anteroposterior diameter of the inlet. If I use the anatomical conjugate (which is longer than the obstetrical conjugate) for Pecos Pueblo and compare it with whichever anteroposterior diameter is reported in other samples (when both the obstetrical and anatomical conjugate diameters are reported, I choose the shorter obstetrical conjugate), Pecos Pueblo females still have the shortest anteroposterior diameter of inlet among 77 samples. Pecos Pueblo females also have a relatively short sacrum and short anteroposterior diameter of outlet, ranking eighth among 33 samples and tied for fifteenth among 42 samples, respectively. In contrast, Pecos females have a broad transverse diameter of inlet, ranking fifty-ninth among 71 samples. At the pelvic inlet for Pecos Pueblo, the combination of short anteroposterior diameter and broad transverse diameter results in a markedly transversely oval inlet. The index of inlet shape, computed as (obstetrical conjugate/transverse diameter)100, is 68.5, the lowest index among 42 samples (fig. 10.3).

Results for males from Pecos Pueblo are similar to those for females for the pelvic inlet: a short anteroposterior diameter (ranking first among 15 samples for the obstetrical conjugate and third among 26 samples in comparisons of the anatomical conjugate with whichever anteroposterior diameter of inlet is reported in other samples) and a broad transverse diameter (ranking twentieth among 26 samples). The index of inlet shape for Pecos Pueblo males is 71.6, the lowest among 15 samples (fig. 10.3).

Results in tables 10.4 and 10.5 and Appendix J show that overall, females and males from Pecos Pueblo have small pelves. Pecos Pueblo is distinctive in its short anteroposterior diameter of inlet. Indeed, 85.7% (36 of 42) of Pecos Pueblo females have a contracted anteroposterior diameter of inlet; Cunningham and colleagues (2001) defined an obstetrical conjugate of less than 100 millimeters as being contracted. I discuss the obstetrical implications of pelvic contraction later. The prevalence of contraction of the anteroposterior diameter of inlet among females in the other samples is as follows: Haida, 20% (3 of 15); Indian Knoll, 14.3% (4 of 28); Libben, 63.6% (7 of 11); blacks, 18.8% (9 of 48); and whites, 10.6% (5 of 47). One female in a sample of 54 from Pecos Pueblo has a contracted transverse diameter of inlet; Cunningham and colleagues (2001) defined a transverse diameter of inlet of less than 120 millimeters as being contracted.

Among Pecos Pueblo females, the markedly transversely oval pelvic inlet is characterized as platypelloid. Thoms (1941, 1946) defined the inlet as platypelloid when the transverse diameter is at least 30 millimeters longer than the obstetrical conjugate diameter. Approximately 90% (38 of 42) of Pecos Pueblo females have a platypelloid inlet. The prevalence of platypelloidy among females in the other five samples is as follows: Haida, 26.7% (4 of 15); Indian Knoll, 28.6% (8 of 28); Libben, 81.8% (9 of 11); blacks, 12.5% (6 of 48); and whites, 31.9% (15 of 47). The prevalence of platypelloidy in other populations ranges from approximately 2% to 23% (Greulich, Thoms, and Twaddle 1939; Manahan, Marquez, and Mangay 1954; Ross 1975; Sibley, Armelagos, and Van Gerven 1992; Sibthorpe and Allbrook 1958; Thoms 1939, 1941, 1946; Wei, Chen, and Chen 1956). Clinically, platypelloidy is associated with obstetrical difficulty and operative intervention (Thoms, Foote, and Friedman 1939). Therefore, I concentrate the remaining analysis and discussion on the anatomical correlates and obstetrical implications of the short anteroposterior diameter of inlet and platypelloid pelvis in Pecos Pueblo females.

Anatomical Correlates of Anteroposterior Diameter of Inlet and Platypelloid Inlet

Pecos Pueblo females have a short anteroposterior diameter of pelvic inlet because of their small body size, short linea terminalis, and protruding sacral promontory. Anteroposterior diameter of inlet is positively correlated with stature (Bernard 1952; Holland et al. 1982; Ince and Young 1940; Jordaan 1976; Thoms 1947) and femoral length (Tague 2000); femoral length is often used to estimate stature (Feldesman and Fountain 1996; Feldesman, Kleckner, and Lundy 1990; Krogman and İşcan 1986). Pecos Pueblo females have short femurs (table 10.4) and, by inference, short stature. However, a short femur alone cannot explain the short anteroposterior diameter of inlet in Pecos Pueblo. Haida females, for example, are nonsignificantly different from Pecos Pueblo females in femoral length, yet Haida females have a significantly longer anteroposterior diameter of inlet than Pecos Pueblo females (table 10.4, Appendix I).

Anteroposterior diameter of inlet is positively correlated with curved length of linea terminalis (Tague 2000). Pecos Pueblo females have a significantly shorter curved length of linea terminalis (132 ± 8, n = 135) than females in the groups labeled Haida (144 ± 9, n = 22), Indian Knoll (139 ± 9, n = 65), Libben (141 ± 9, n = 29), black (137 ± 9, n = 49), and white (144 ± 8, n = 49) (t-test results not presented; Tague 1992). Linea terminalis length, in turn, is positively correlated with a measure of body size—diameter of femoral head (Tague 2000). Pecos Pueblo females have a significantly smaller mean femoral head diameter than females in the Haida, Libben, black, and white groups, but they are nonsignificantly different from females in the Indian Knoll sample (table 10.4, Appendix I).

Pecos Pueblo females also have sacral promontories that protrude anteriorly. I classify a sacral promontory as protruding when the anatomical conjugate diameter is shorter than the anteroposterior diameter at the plane of the linea terminalis (fig. 10.2a, A–B, and fig. 10.2g, B–M, respectively).[2] Approximately 76% of Pecos Pueblo females have a protruding sacral promontory, a larger percentage than in any of the other five samples (table 10.6). The six samples of males, with a range of 0% to 30.3%, do not differ significantly among themselves in the prevalence of protruding sacral promontory (chi-square test, X^2 = 10.47, df = 5, p = .063). The six samples of females, however, do differ in this prevalence (chi-square test, X^2 = 73.27, df = 5, p < .001). Partition analysis among the samples of females shows that Pecos Pueblo and Libben do not differ significantly in their prevalences of protruding sacral promontory but have significantly higher prevalences than Haida, Indian Knoll, blacks, and whites. The last four samples do not differ significantly among themselves (results not presented). Interestingly, only in the Pecos Pueblo and Libben samples do females not have a significantly longer anteroposterior diameter of inlet than males (Tague 1992). The high prevalence of sacral protrusion among females in Pecos Pueblo and Libben may be the reason for the absence of sexual dimorphism in the anteroposterior diameter.

The reason for the protruding sacral promontory in Pecos Pueblo females is abstruse, not being due to angulation of sacrum, curvature of sacrum, or curvature of linea terminalis (table 10.7). With Pecos Pueblo skeletal material now reburied, the opportunity is lost

2. If the promontory of the sacrum (fig. 10.2a, A) is in the same coronal plane and directly above the midpoint of sacral breadth at the inlet (fig. 10.2g, M), then the anatomical conjugate diameter and anteroposterior diameter at the plane of the linea terminalis would be two sides of a right triangle. The anatomical conjugate diameter would be the hypotenuse and thereby longer than the anteroposterior diameter at the plane of the linea terminalis. The sacral promontory "protrudes" anteriorly when the former diameter is shorter than the latter.

TABLE 10.4
SUMMARY STATISTICS (MEAN ± SD [*N*]) AND RESULTS OF ANOVA OR KRUSKAL-WALLIS TEST WITHIN SEX FOR PELVIC AND NONPELVIC VARIABLES IN SIX POPULATION SAMPLES

Variable	Pecos Pueblo	Haida	Indian Knoll	Libben	Blacks	Whites	*p*
Females							
Anteroposterior diameter of inlet (mm)	91.0 ± 7.7 (42)	112.0 ± 10.9 (15)	108.2 ± 7.1 (28)	97.9 ± 8.0 (11)	110.4 ± 12.5 (48)	110.8 ± 11.0 (47)	<.0001 (K-W)
Anteroposterior diameter of outlet (mm)	114.0 ± 9.0 (43)	134.8 ± 9.9 (15)	114.5 ± 9.3 (30)	121.2 ± 5.6 (8)	122.9 ± 10.4 (46)	118.7 ± 8.6 (47)	<.0001 (K-W)
Transverse diameter of inlet (mm)	133.5 ± 5.9 (54)	134.7 ± 8.1 (19)	134.0 ± 6.7 (33)	134.3 ± 7.3 (12)	122.9 ± 8.3 (50)	134.0 ± 8.5 (50)	<.0001 (K-W)
Transverse diameter of outlet (mm)	115.5 ± 9.6 (29)	123.6 ± 8.8 (11)	127.4 ± 8.5 (23)	127.4 ± 12.6 (6)	112.6 ± 11.3 (50)	119.7 ± 9.1 (49)	<.0001 (K-W)
Posterior sagittal diameter of inlet (mm)	27.4 ± 3.3 (42)	32.8 ± 3.4 (19)	30.3 ± 3.3 (21)	28.1 ± 5.9 (10)	31.1 ± 5.3 (49)	33.1 ± 5.3 (46)	<.0001 (K-W)
Posterior sagittal diameter of outlet (mm)	59.2 ± 9.3 (25)	71.2 ± 7.1 (10)	61.2 ± 9.6 (22)	66.6 ± 11.0 (5)	69.4 ± 9.4 (46)	63.7 ± 10.7 (46)	.0001 (K-W)
Anterior sagittal diameter of inlet (mm)	68.8 ± 5.9 (53)	79.6 ± 6.8 (19)	72.9 ± 7.4 (32)	71.3 ± 4.4 (12)	73.8 ± 6.4 (49)	74.1 ± 6.0 (49)	<.0001 (K-W)
Anterior sagittal diameter of outlet (mm)	67.1 ± 5.4 (29)	69.0 ± 10.1 (11)	64.9 ± 4.5 (23)	66.9 ± 3.4 (6)	65.2 ± 5.9 (50)	66.4 ± 5.8 (48)	.53 (K-W)
Breadth of sacrum at inlet (mm)	109.2 ± 5.0 (61)	110.0 ± 6.8 (19)	105.6 ± 5.5 (42)	111.2 ± 5.9 (29)	102.4 ± 7.9 (50)	109.3 ± 6.9 (47)	<.0001 (A)
Straight length of sacrum (mm)	96.2 ± 9.7 (71)	100.2 ± 7.7 (17)	99.1 ± 10.1 (55)	100.0 ± 7.5 (31)	96.0 ± 10.4 (45)	99.6 ± 12.1 (47)	.17 (A)
Angulation of sacrum (°)	69.3 ± 5.7 (98)	73.3 ± 6.6 (16)	67.9 ± 5.7 (72)	72.5 ± 5.1 (42)	71.0 ± 5.6 (50)	71.0 ± 7.0 (50)	.0001 (A)
Depth of pelvis (mm)	98.6 ± 5.1 (113)	104.5 ± 6.7 (16)	102.2 ± 5.3 (79)	103.3 ± 5.1 (50)	104.1 ± 6.4 (50)	105.6 ± 6.2 (50)	<.0001 (A)
Subpubic angle (°)	85.9 ± 10.0 (114)	93.0 ± 12.3 (19)	98.1 ± 8.4 (58)	95.3 ± 10.7 (21)	85.2 ± 10.4 (49)	88.4 ± 8.5 (50)	<.0001 (A)
Area of inlet (mm^2)	9,504.2 ± 942.4 (42)	11,899.8 ± 1,311.1 (15)	11,349.1 ± 1,048.0 (28)	10,394.3 ± 1,183.9 (11)	10,645.1 ± 1,477.0 (48)	11,663.7 ± 1,342.6 (47)	<.0001 (A)
Area of outlet (mm^2)	7,308.6 ± 942.8 (25)	8,647.5 ± 539.4 (10)	8,034.0 ± 962.1 (22)	8,673.2 ± 1,402.6 (5)	7,557.2 ± 950.9 (46)	7,743.9 ± 745.3 (45)	.0003 (K-W)
Femoral length (mm)	392.8 ± 18.5 (119)	391.5 ± 12.4 (20)	410.0 ± 18.1 (77)	426.2 ± 19.6 (39)	443.0 ± 24.2 (50)	417.8 ± 23.4 (49)	<.0001 (K-W)
Femoral head diameter (mm)	39.2 ± 1.8 (121)	41.9 ± 2.0 (20)	39.2 ± 1.7 (84)	41.7 ± 1.8 (48)	42.9 ± 2.4 (49)	43.2 ± 2.1 (49)	<.0001 (A)
Clavicular length (mm)	133.5 ± 6.9 (62)	139.1 ± 4.8 (17)	131.1 ± 5.6 (66)	141.3 ± 6.7 (27)	141.0 ± 9.1 (48)	135.3 ± 7.8 (48)	<.0001 (K-W)

TABLE 10.4 (*Continued*)

Variable	Pecos Pueblo	Haida	Indian Knoll	Libben	Blacks	Whites	*p*
Males							
Anteroposterior diameter of inlet (mm)	90.6 ± 8.2 (35)	102.7 ± 6.4 (19)	103.1 ± 8.0 (36)	101.8 ± 10.7 (18)	96.5 ± 8.4 (47)	101.9 ± 9.1 (47)	<.0001 (A)
Anteroposterior diameter of outlet (mm)	107.5 ± 7.5 (37)	117.0 ± 5.8 (20)	108.6 ± 7.3 (32)	116.0 ± 5.4 (21)	113.2 ± 6.7 (48)	111.2 ± 7.8 (46)	<.0001 (A)
Transverse diameter of inlet (mm)	126.0 ± 7.4 (41)	131.6 ± 7.0 (23)	125.1 ± 6.2 (46)	132.4 ± 7.7 (27)	117.7 ± 8.1 (50)	129.7 ± 8.3 (50)	<.0001 (A)
Transverse diameter of outlet (mm)	95.6 ± 8.6 (33)	102.1 ± 8.0 (17)	99.7 ± 7.6 (33)	103.4 ± 7.9 (19)	94.7 ± 10.2 (50)	101.6 ± 9.5 (48)	.0001 (A)
Posterior sagittal diameter of inlet (mm)	23.7 ± 3.4 (36)	28.8 ± 3.6 (22)	25.5 ± 2.9 (29)	27.6 ± 4.2 (24)	23.7 ± 3.5 (47)	27.4 ± 4.5 (45)	<.0001 (A)
Posterior sagittal diameter of outlet (mm)	51.9 ± 7.2 (32)	56.6 ± 5.7 (16)	52.4 ± 6.5 (25)	56.2 ± 8.0 (17)	59.6 ± 8.5 (48)	56.8 ± 8.5 (43)	.0002 (A)
Anterior sagittal diameter of inlet (mm)	69.0 ± 5.6 (40)	75.0 ± 4.7 (23)	72.3 ± 5.9 (45)	75.7 ± 7.0 (25)	73.2 ± 6.0 (50)	75.9 ± 6.1 (50)	<.0001 (A)
Anterior sagittal diameter of outlet (mm)	71.7 ± 5.9 (33)	74.8 ± 4.6 (17)	71.1 ± 6.8 (32)	75.6 ± 4.6 (19)	68.5 ± 5.3 (50)	71.8 ± 5.8 (48)	<.0001 (A)
Breadth of sacrum at inlet (mm)	105.4 ± 6.2 (53)	110.6 ± 4.8 (23)	106.5 ± 5.5 (43)	110.2 ± 5.5 (43)	99.4 ± 6.7 (47)	109.4 ± 6.8 (45)	<.0001 (A)
Straight length of sacrum (mm)	100.9 ± 7.8 (64)	111.9 ± 7.5 (22)	104.4 ± 8.0 (59)	104.7 ± 8.7 (39)	99.8 ± 7.9 (44)	102.7 ± 9.2 (42)	<.0001 (A)
Angulation of sacrum (°)	61.5 ± 5.0 (73)	60.3 ± 4.7 (19)	61.6 ± 4.4 (100)	60.9 ± 4.0 (54)	64.5 ± 5.3 (50)	65.2 ± 4.7 (49)	<.0001 (A)
Depth of pelvis (mm)	101.6 ± 5.6 (103)	112.2 ± 5.0 (21)	103.6 ± 5.9 (112)	105.7 ± 5.5 (75)	107.5 ± 6.0 (50)	112.3 ± 6.1 (50)	<.0001 (A)
Subpubic angle (°)	61.5 ± 8.2 (104)	65.4 ± 8.2 (29)	73.8 ± 8.4 (74)	68.8 ± 7.8 (46)	65.8 ± 8.7 (50)	63.7 ± 7.8 (50)	<.0001 (A)
Area of inlet (mm^2)	9,026.9 ± 1,038.6 (35)	10,583.8 ± 844.5 (19)	10,160.2 ± 954.2 (36)	10,540.3 ± 1,465.9 (18)	8,961.5 ± 1,089.8 (47)	10,376.5 ± 1,256.3 (47)	<.0001 (A)
Area of outlet (mm^2)	5,913.9 ± 680.4 (32)	6,667.9 ± 585.0 (16)	6,179.3 ± 620.2 (24)	6,778.1 ± 621.6 (17)	6,068.9 ± 754.1 (48)	6,485.5 ± 855.2 (43)	.0001 (A)
Femoral length (mm)	428.3 ± 18.9 (98)	428.2 ± 16.9 (27)	437.1 ± 22.1 (109)	459.6 ± 18.8 (65)	477.6 ± 29.2 (49)	458.5 ± 22.2 (49)	<.0001 (K-W)
Femoral head diameter (mm)	43.9 ± 2.0 (101)	46.6 ± 1.7 (27)	43.2 ± 2.1 (109)	45.9 ± 2.1 (73)	48.7 ± 2.7 (49)	49.3 ± 2.2 (48)	<.0001 (A)
Clavicular length (mm)	150.4 ± 6.7 (64)	156.5 ± 8.7 (23)	144.3 ± 7.9 (103)	158.8 ± 8.3 (47)	154.7 ± 7.1 (50)	151.7 ± 8.2 (47)	<.0001 (A)

Note: Anteroposterior diameter of inlet is the obstetrical conjugate diameter. The Kruskal-Wallis test is used if the samples differ significantly in their variances or if sample size is < 15 for any one of the samples. A = ANOVA; K-W = Kruskal-Wallis.

TABLE 10.5
PAIRWISE COMPARISONS FOR 15 PELVIC MEASUREMENTS AMONG SIX POPULATION SAMPLES USING SIGN TEST

	Pairwise comparison		Number of significant and nonsignificant differences in means or mean ranks between samples A and B			
Sex	Sample A	Sample B	A > B	B > A	A = B	p
Female	Pecos	Haida	0	9	6	.004
	Pecos	Indian Knoll	0	5	10	.063
	Pecos	Libben	0	3	12	.250
	Pecos	Blacks	2	7	6	.180
	Pecos	Whites	0	5	10	.063
	Haida	Indian Knoll	2	0	13	.500
	Haida	Libben	1	0	14	NS
	Haida	Blacks	6	0	9	.031
	Haida	Whites	1	0	14	NS
	Indian Knoll	Libben	0	2	13	.500
	Indian Knoll	Blacks	3	2	10	1.000
	Indian Knoll	Whites	1	1	13	1.000
	Libben	Blacks	3	1	11	.625
	Libben	Whites	0	1	14	NS
	Blacks	Whites	0	4	11	.125
Male	Pecos	Haida	0	10	5	.002
	Pecos	Indian Knoll	0	3	12	.250
	Pecos	Libben	0	11	4	.001
	Pecos	Blacks	2	7	6	.180
	Pecos	Whites	0	9	6	.004
	Haida	Indian Knoll	5	1	9	.219
	Haida	Libben	2	0	13	.500
	Haida	Blacks	8	1	6	.039
	Haida	Whites	2	1	12	1.000
	Indian Knoll	Libben	1	2	12	1.000
	Indian Knoll	Blacks	5	4	6	1.000
	Indian Knoll	Whites	1	4	10	.375
	Libben	Blacks	7	1	7	.070
	Libben	Whites	1	2	12	1.000
	Blacks	Whites	0	8	7	.008

Key: > and = represent statistically significant and nonsignificant differences, respectively, in means or mean ranks; see Appendix I. NS = nonsignificant, for cases in which a probability could not be computed because there was only one significant difference between the samples.

to assess their protruding sacral promontory. Further research on this interesting anatomy is warranted in other populations because it explains, in part, the distinctively short and clinically contracted anteroposterior diameter of inlet in the Pecos Pueblo pelves. The importance of the Pecos collection with respect to this anatomy is evident from the observation that 28 of these females show sacral protrusion, whereas only 17 do so in the five other skeletal collections combined (table 10.6).

Collectively, Pecos Pueblo females have a short anteroposterior diameter of pelvic inlet because of their small body size and protruding sacral promontory. Body size has a genetic component but is also responsive to diet and health. The agriculturally based diet and lifestyle of Pecos Pueblo may have con-

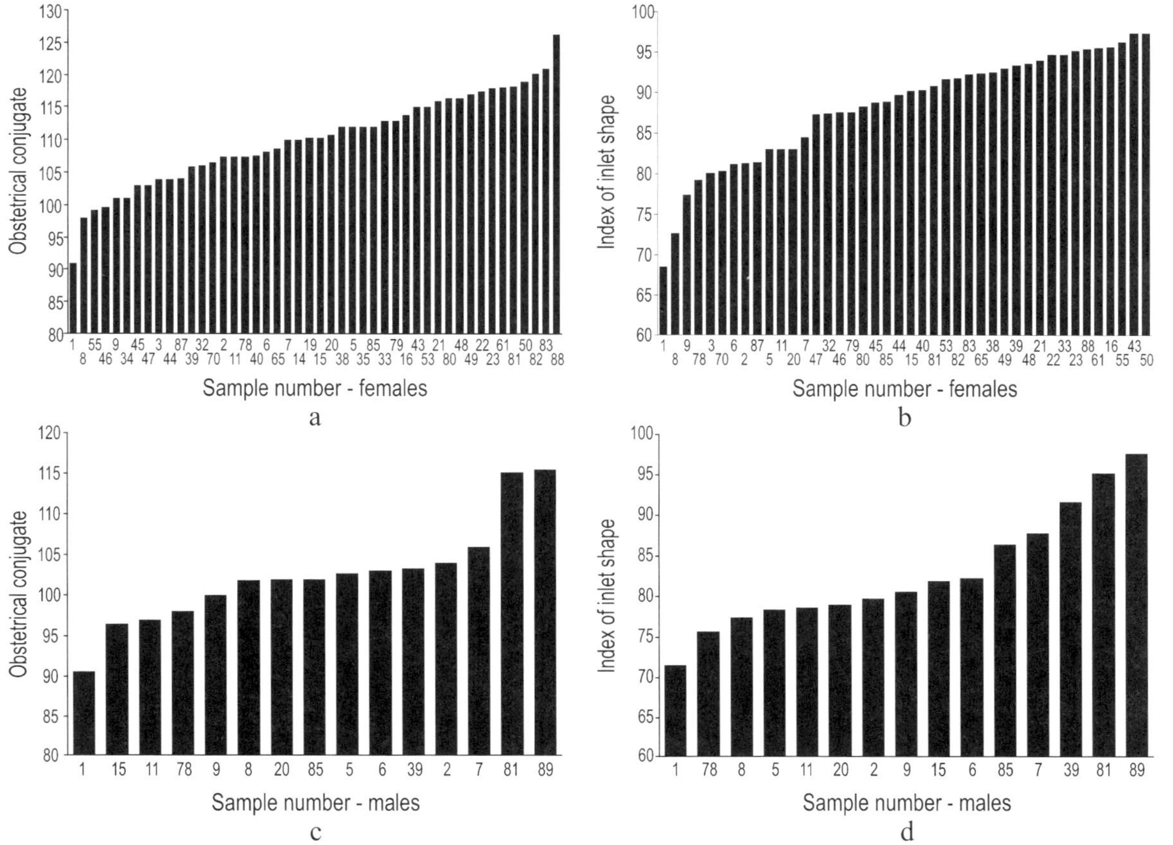

FIGURE 10.3. Bar graphs of obstetrical conjugate diameter of pelvic inlet (a, c) and index of inlet shape (b, d) for females and males. The populations associated with the sample numbers are given in Appendix J; Pecos Pueblo is sample number 1.

tributed to the small body size among females (cf. Larsen 1981; but see Bridges 1989). Because I did not analyze the pathologies of the individuals in this study, I cannot ascertain the relationship between health and pelvic size or body size at Pecos Pueblo. Finally, my awareness of the contribution of sacral protrusion to the anteroposterior diameter of inlet came subsequent to the reburial of the Pecos Pueblo skeletal collection. I must defer to a future study to explain the genetic or developmental basis of this anatomy.

In contrast with the inlet, the pelvic outlet among Pecos Pueblo females cannot be considered contracted. The index of outlet capacity (computed as transverse diameter of outlet + posterior sagittal diameter of outlet) can be used to assess the obstetrical adequacy of the outlet.[3] I can compare the index of outlet

TABLE 10.6
PREVALENCE OF PROTRUDING SACRAL PROMONTORY AMONG FEMALES AND MALES IN SIX POPULATION SAMPLES

	Females		Males	
Sample	%	*n*	%	*n*
Pecos Pueblo	75.7	28 of 37	30.3	10 of 33
Haida	26.3	5 of 19	14.3	3 of 21
Indian Knoll	0.0	0 of 20	0.0	0 of 28
Libben	44.4	4 of 9	17.4	4 of 23
Blacks	10.2	5 of 49	14.9	7 of 47
Whites	6.5	3 of 46	19.0	8 of 42

3. Floberg and colleagues (1986) considered summation of the lengths of the transverse diameters of midplane and outlet and posterior sagittal diameter of outlet to be an informative summary statistic for the obstetrical capacity of the lower aspect of the pelvis. I could not measure the transverse diameter of midplane for any female in Pecos Pueblo because of damage to the bones, so I used only the transverse and posterior sagittal diameters of the outlet to assess pelvic outlet capacity.

TABLE 10.7
RESULTS OF WILCOXON-MANN-WHITNEY TEST FOR RELATIONSHIP BETWEEN PRESENCE OR ABSENCE OF PROTRUDING SACRAL PROMONTORY AND OTHER PELVIC VARIABLES IN PECOS PUEBLO FEMALES

	Protruding sacral promontory, mean rank (mean ± sd, *n*)		
Other pelvic variable	**No**	**Yes**	***p***
Angulation of sacrum (°)	17.50 (69.6 ± 5.0, 8)	15.48 (68.4 ± 5.2, 23)	.59
Index of sacral curvature	16.13 (91.1 ± 3.7, 8)	14.57 (90.3 ± 4.0, 21)	.66
Index of linea terminalis curvature	19.00 (84.3 ± 2.2, 8)	18.36 (84.2 ± 2.8, 28)	.88

Note: Index of sacral curvature = (straight length of sacrum/curved length of sacrum)100; index of linea terminalis curvature = (straight length of linea terminalis/curved length of linea terminalis)100.

capacity in the Pecos sample with indexes for the other five samples in my study but not with those for other samples presented in the literature, because researchers differ in their measurement of the transverse diameter of outlet (Emmons 1913; Heyns 1944; Thoms 1941; Thoms and Greulich 1940). The index of outlet capacity for Pecos Pueblo females is nonsignificantly different from the indexes for females in the Indian Knoll, Libben, black, and white groups, but it is significantly smaller than that for Haida females (table 10.8; results of Kruskal-Wallis and multiple comparisons tests not presented). Notably, of the 154 females designated Pecos Pueblo, Haida, Indian Knoll, Libben, blacks, and whites for which the index of outlet capacity can be computed, the five with the smallest indices are all from Pecos Pueblo. These five constitute 20% of the Pecos sample. Consequently, 20% of Pecos Pueblo females have a confined pelvic outlet. Although these five females also have a contracted pelvic inlet, the correlation between obstetrical conjugate diameter of inlet and index of outlet capacity is nonsignificantly different from zero among Pecos Pueblo females ($r = .33$, $n = 22$, $p = .14$, two-tailed test).

The pelvic inlet among Pecos Pueblo females is platypelloid. Platypelloidy is due primarily to a short anteroposterior diameter. For Pecos Pueblo females, the correlations for pelvic inlet shape with obstetrical conjugate diameter and transverse diameter of inlet are 0.87 ($p < .001$) and –0.42 ($p = .005$), respectively ($n = 42$, two-tailed test). Heyns (1947) and Jordaan (1976) also reported that inlet shape is associated more with the anteroposterior diameter than with the transverse diameter. The reason for this association is that the anteroposterior diameter is more variable than the transverse diameter (coefficients of variation can be computed from data in table 10.4).

Among Pecos Pueblo females, shape of the pelvic inlet is significantly correlated with shape of the outlet (index of outlet shape computed as [anteroposterior diameter/transverse diameter]100): $r = .43$ ($n = 22$, $p = .045$, two-tailed test). Despite this positive association, the shape of the outlet is approximately circular rather than platypelloid (100.6 ± 10.5, $n = 25$).

Platypelloidy is considered an abnormal pelvic shape, often associated with rickets (Thoms 1939; Thoms, Foote, and Friedman 1939). In a richetic pelvis, "shortening of the conjugate diameter [is] associated with a lengthening of the transverse" (Nicholson 1945:132). Alternatively stated, the linea terminalis is markedly curved in a richetic pelvis. A comparison of the index of linea terminalis curvature between Pecos Pueblo females and females in the other five samples shows that this index in the Pecos sample is (1) nonsignificantly different from that in Indian Knoll and blacks, (2) significantly higher than that in Libben and whites, and (3) significantly lower than that in Haida (table 10.9; results of ANOVA and multiple comparisons test not presented). Therefore, rickets is not the likely explanation for the platypelloid pelvis at Pecos Pueblo.

Obstetrical Implications of Pelvis at Pecos Pueblo

The mechanism of birth is the manner in which the fetus passes through the birth canal. Modally, the fetus is oriented such that its head is the leading, or "presenting," part during entry into the true pelvis (fig.

TABLE 10.8
INDEX OF OUTLET CAPACITY AMONG FEMALES IN SIX POPULATION SAMPLES

Sample	Index of outlet capacity (mm), mean ± sd (*n*)
Pecos Pueblo	175.1 ± 16.0 (25)
Haida	195.2 ± 11.1 (10)
Indian Knoll	188.6 ± 13.8 (22)
Libben	195.1 ± 23.9 (5)
Blacks	181.8 ± 14.0 (46)
Whites	183.3 ± 12.4 (46)

Note: Index of outlet capacity = transverse diameter of outlet + posterior sagittal diameter of outlet.

TABLE 10.9
INDEX OF LINEA TERMINALIS CURVATURE AMONG FEMALES IN SIX POPULATION SAMPLES

Sample	Index, mean ± sd (*n*)
Pecos Pueblo	84.5 ± 2.4 (135)
Haida	86.5 ± 1.9 (22)
Indian Knoll	84.3 ± 2.5 (65)
Libben	83.5 ± 3.2 (25)
Blacks	85.2 ± 1.6 (49)
Whites	83.3 ± 1.9 (49)

Note: Index of linea terminalis curvature = (straight length of linea terminalis/curved length of linea terminalis)100.

10.4). During either entrance into or descent through the birth canal, the fetus flexes its head so that the back of its head (the occiput) is the presenting part. Moreover, the fetus becomes oriented so that the long diameter of its head (the anteroposterior diameter) is aligned with the long diameter of the maternal pelvis (Tague and Lovejoy 1986). This alignment is optimal to ensure that one oval structure, the fetal head, fits within another oval structure, the pelvic plane of the maternal pelvis. The adult female pelvis is transversely oval at the inlet, anteroposteriorly oval at the midplane, and approximately circular at the outlet (Tague and Lovejoy 1986). At the inlet, the anteroposterior diameter of the fetal head aligns with the transverse diameter of the maternal pelvis (fig. 10.4a). As the shape of the female pelvis changes from inlet to midplane, the fetus reorients its head to maintain optimal alignment. This turning of the fetal head is called "internal rotation," and it occurs at or just below the midplane. With internal rotation, the anteroposterior diameter of the fetal head aligns with the anteroposterior diameter of the maternal pelvis, and the fetal occiput is situated in the anterior aspect of the pelvis (fig. 10.4b). Delivery then occurs through the subpubic space, with the anteroposterior diameters of fetal head and maternal pelvis remaining aligned (fig. 10.4c). The fetal occiput is delivered first, with the suboccipital area of the fetal head pressed against the inferior aspect of the maternal pubic symphysis. The remainder of the fetal head is delivered by extension of the head, using the symphysis as a pivot. (See Walrath 2003 for a discussion of variation in the mechanism of birth.)

The mechanism of birth at Pecos Pueblo likely corresponded with that just described. The pelvic inlet among Pecos Pueblo females is platypelloid. The obligate orientation at the inlet would have been for the anteroposterior diameter of the fetal head to be aligned with the transverse diameter of the maternal pelvis. Less predictable for Pecos Pueblo is at which pelvic plane internal rotation of the fetal head occurred, because I could not measure the transverse diameter of the midplane (nor, correspondingly, compute the index of pelvic midplane shape) for any female because of damage to the ischial spines. Because the pelvic outlet is circular for Pecos Pueblo females, the fetus would likely have rotated its head either before or at the outlet.

Despite the foregoing description as the modal mechanism of birth at Pecos Pueblo, this study shows that females had a markedly short anteroposterior diameter of pelvic inlet and a correspondingly high frequency of inlet contraction. Inlet contraction is associated with anomalies in fetal presentation. "In women with contracted pelves, face and shoulder presentations are encountered three times more frequently, and cord prolapse occurs four to six times more frequently" (Cunningham et al. 2001:437). With a face presentation, the fetus's chin is the leading part because the fetus extends its head as it enters the true pelvis (fig. 10.4d). Despite this anomaly, most face presentations result in spontaneous, successful deliveries. The fetus rotates its head in the lower part of the pelvic cavity, so that its chin is positioned in the anterior aspect of the maternal pelvis, underneath the pubic symphysis (fig. 10.4e). Delivery is achieved when the

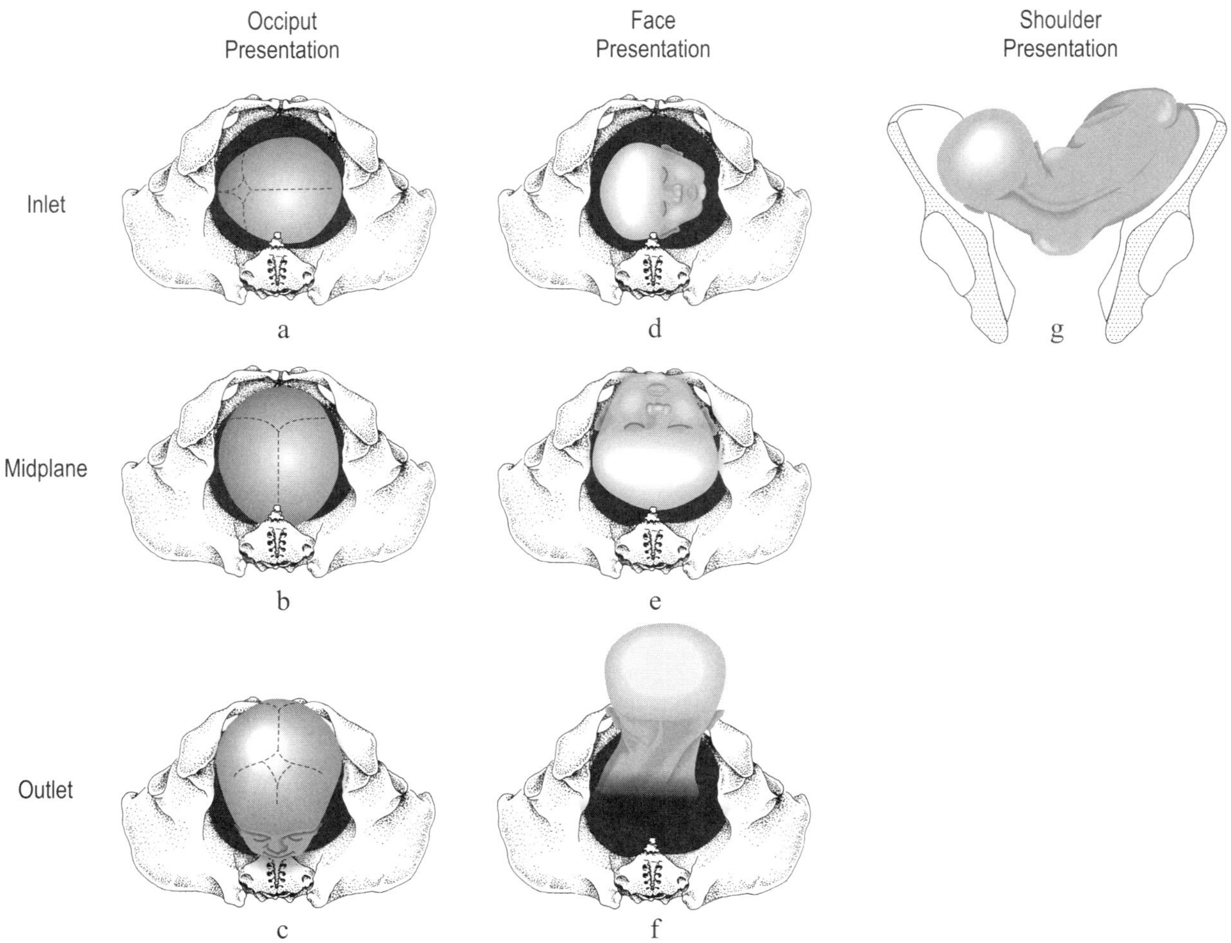

FIGURE 10.4. Positions of the fetal head in the maternal pelvis during delivery with occiput and face presentations, as shown at the pelvic inlet, midplane, and outlet. The position of the fetus with a shoulder presentation is also shown at the pelvic inlet. Dotted lines on the fetal head represent the vault sutures; the diamond shape (a, c) is the anterior fontanelle. The fontanelle is not visible at the midplane because of flexion of the head. The maternal pelvis is shown from inferior (a–f) and coronal (g) views. Redrawn from Oxorn 1980: 137, 175, 248.

fetus flexes its head, with the maternal pubic symphysis serving as a pivot (fig. 10.4f). Labor is longer and more difficult with a face presentation than an occipital presentation, and the mother's perineum may be partly torn during delivery. Importantly, if the fetus rotates its head posteriorly rather than anteriorly in a face presentation, its chin will face the mother's sacrum, and delivery is impossible. Without surgical intervention, an option unavailable at Pecos Pueblo, both fetus and mother will die.

Shoulder presentation is a serious malposition and "if not properly managed, . . . the mother and fetus die" (Cunningham et al. 2001:456). The fetus lies transversely at the pelvic inlet and cannot descend into the true pelvis (fig. 10.4g). "Proper management" in modern obstetrics is typically cesarean section, although there may be an attempt to reorient the fetus by external manipulation of the mother's abdomen prior to labor. At Pecos Pueblo, only the latter would have been an option.

Finally, prolapse of the umbilical cord means that part or all of the cord lies beside or below the presenting part of the fetus during delivery. The cord may be compressed, thereby reducing or cutting off the blood supply to the fetus. Depending on the speed of delivery, prolapse of the umbilical cord could lead to fetal death.

Despite the high prevalence of inlet contraction among Pecos Pueblo females, anomalies in fetal presentation and delivery may have been ameliorated by small fetal size. The transverse diameter of the fetal head (i.e., biparietal diameter) should be at least 5 millimeters smaller than the obstetrical conjugate diameter of the maternal pelvis to allow fetal entrance into the birth canal (Cunningham et al. 2001). With a mean of 91 millimeters for the obstetrical conjugate diameter in the Pecos sample (table 10.4), the fetal biparietal diameter needed to have been ≤ 86 millimeters. This mean biparietal diameter is comparable to that in other populations: 85 millimeters and 86.8 millimeters in China (Garner 1918; King and Teh 1937) and 86.6 millimeters in Ghana (Adadevoh, Hobbs, and Elkins 1989). Small mothers also give birth to small babies, and Pecos Pueblo females are small with respect to the three nonpelvic measures used in this study (table 10.4, Appendix I).

Of notable interest are the five females mentioned earlier that have the smallest indices of pelvic outlet capacity. The catalogue numbers of these females are 60006, 60061, 60084, 60143, and 60736. All five also have a contracted anteroposterior diameter of inlet. Catalogue number 60736 seems to have a generally contracted pelvis, because she also has a contracted transverse diameter of inlet. As noted earlier, this female is singular in the Pecos Pueblo sample in having a contracted transverse diameter. Although catalogue number 60006 has a short clavicle and 60736 has a small femoral head diameter, these five females are not distinctively small in their nonpelvic dimensions (Appendix K). Therefore, birth for these females might have been "laborious." Because these five females constitute 20% of the sample for pelvic outlet capacity, and because most females have a contracted inlet, the implication is that, in general, labor might have been prolonged and birth difficult for Pecos Pueblo females. Thoms, Foote, and Friedman (1939) reported a high prevalence of prolonged labor in females with a small pelvic inlet or outlet.

I cannot draw an inference about the prevalence of maternal and neonatal mortality at Pecos Pueblo, although some prehistoric populations show high prevalences of such mortality (Arriaza, Allison, and Gerszten 1988; Sibley, Armelagos, and Van Gerven 1992). Of the five females from Pecos Pueblo just mentioned, I aged three as younger than age 30 (60084, 60143, and 60736) and two as older than 40 (60006 and 60061). Three of these females (60006, 60061, and 60084) also had areas of bone resorption ("pitting") on either the dorsal aspect of the pubic symphysis or the preauricular area of the ilium more pronounced than that of any male in the Pecos sample. Such areas of resorption may indicate that these females experienced pregnancy, parturition, or both (Tague 1988). The two females with no resorption of the pubis and with "male-like" resorption of the preauricular area (60143 and 60736) were under the age of 30; 60143 was 19 or 20. Overall, however, only a minority of adult females in the Pecos sample show resorption more severe than that of males—36% for the pubis and 47% for the preauricular area (Tague 1988). The implication of this result is consistent with that of previous research (Suchey et al. 1979): some females who have been pregnant or given birth do not develop these pelvic pittings. Correspondingly, it is unlikely that only a minority of adult females at Pecos Pueblo had been pregnant or given birth by the time of their death. Although some studies of females of known parity show a positive relationship between resorption at the pubis or preauricular area and number of births, defining parity status from pitting of the pubis and preauricular area is unreliable (Tague 1988).

CONCLUSION

Pecos Pueblo females have a contracted pelvic inlet that is platypelloid in shape. At least 20% of females also have a confined pelvic outlet. These anatomies are not obstetrically favorable. The platypelloid inlet made almost obligate a fetal pelvic orientation in which the anteroposterior diameter of the fetus's head was aligned with the transverse diameter of the pelvic inlet. Although most females likely had successful, uneventful deliveries, a substantial minority may have had difficulty in delivery because of a small pelvis involving both contraction of the inlet and confinement of the outlet. Moreover, the high prevalence of contracted pelvic inlet might have induced some fetuses to assume obstetrically unfavorable presentations. These presentations might have resulted in maternal morbidity, maternal and fetal mortality, or both.

The high prevalence of contraction of the pelvic inlet and platypelloid inlet shape distinguishes the Pecos Pueblo sample from other populations. The contracted inlet was likely due to the combination of small adult body size and protrusion of the sacral promontory. The latter anatomy warrants additional research to determine its developmental basis and correlates with other anatomies.

Appendix K presents the original metric and angular measurements I took for the individuals of Pecos Pueblo. I hope these data enable other researchers to continue to study the Pecos Pueblo pelvis.

ACKNOWLEDGMENTS

I thank the following for the opportunity to study skeletal material in their care: Glen Cole, Department of Anthropology, Field Museum of Natural History; Lathel Duffield and Nancy O'Malley, W. S. Webb Museum of Anthropology, University of Kentucky; Bruce Latimer and Lyman Jellema, Department of Physical Anthropology, Cleveland Museum of Natural History; C. Owen Lovejoy, Department of Anthropology, Kent State University; and David Pilbeam, Department of Anthropology and Peabody Museum of Archaeology and Ethnology, Harvard University. George Milner, when he was director of the W. S. Webb Museum of Anthropology at the University of Kentucky, provided me with Charles Snow's data on femoral head diameter for Indian Knoll. Michèle Morgan, of the Peabody Museum of Archaeology and Ethnology, Harvard University, provided me with E. A. Hooton's sex assignments for the Pecos Pueblo collection. Mary Lee Eggart, of the Department of Geography and Anthropology, Louisiana State University, drew the illustrations.

11 Burials from the Fourth Mission Church at Pecos

Karin S. Bruwelheide, Douglas W. Owsley, and Richard L. Jantz

In this chapter we report on the human skeletal remains excavated from the fourth and last Spanish mission church at Pecos Pueblo under the direction of Alfred Kidder during his 1915 and 1925 field seasons. The demographic composition of the skeletal series and the dental and bone health of the individuals are assessed. These results are discussed relative to the history of the pueblo and its decline during the eighteenth century.

Fray Andrés Juárez (Suárez) completed the construction of the great second church at Pecos, dedicated to Nuestra Señora de los Ángeles de Porciúncula in the 1620s (Hayes 1974; Kessell 1979). Little is known of the first church, located northeast of the pueblo and believed to have been constructed by one of the early missionaries at Pecos some years before 1622. It was small and used for only a short time. In sharp contrast, the second church was an architectural anomaly and achievement in both time and place, a bold testament to Spanish presence in the Southwest and to the steadfastness of the early friars who brought the structure into being.

The Pueblo Revolt of 1680 brought a temporary retreat of the Spaniards and a brief hiatus in their occupation of the Southwest. The revolt also resulted in the near complete destruction of the church at Pecos. Two years after the Spanish reconquest of 1692, a small chapel was in use at Pecos. It reportedly was built next to the rubble of the previous, grandiose structure and was being enlarged when a second revolt was suppressed in 1696. This chapel was replaced 10 years later by the fourth and final church at Pecos, which was erected on the mound formed by the pre-rebellion second church, although it was smaller and positioned with a slightly different orientation (Hayes 1974). Throughout the eighteenth century the church at Pecos suffered from sporadic religious leadership, but it continued to function as an institution until 1829, when the church center moved to the nearby village of San Miguel del Vado in response to the pueblo's declining population and the expansion of Hispanic settlements in the Pecos Valley (Hayes 1974; Kessell 1979).

The foundation and walls of the final mission church remained standing long after the abandonment of the pueblo in 1838. An 1846 account by Colonel William H. Emory, a member of the American Army of Occupation during the Mexican-American War, mentioned the church and monastery and provided a drawing of a still relatively complete but deteriorating structure (Kidder 1962). In 1881, archaeologist Adolph Bandelier described the ruins of Pecos, including the church, in his report to the Archaeological Institute of America (Bandelier 1881). By this time, the roof beams of the church had been removed and the adobe walls were crumbling.

In 1915, archaeologist Alfred Kidder began excavations at Pecos. Repair work on the fourth church during the first field season involved reinforcement of its foundation and walls. This work, under the direction of Jesse Nusbaum, revealed numerous bones that Kidder described as "a great mass of skeletal material of the historic period . . . in and near the mission" (Kidder 1962:98). Some of the remains were in relatively intact wooden coffins (fig. 11.1), and two of these reportedly contained the remains of priests (Kidder 1958). Kidder assumed that the skeletons from the church were demographically distinct from those of the pueblo at large, likely belonging to Indians and non-Indians alike and temporally representing the final stage of the pueblo.

With the exception of a few selected remains, skeletons from the church uncovered during the 1915 field season were reburied and not analyzed. During the next several years Kidder and his crew removed numerous burials from the pueblo, avoiding the context of the church. Looting at the site by artifact collectors, however, disturbed burials within the church, and in 1925 a single feature located "midway between both sides and ends of the nave" (Kidder 1958:304) was excavated to mitigate some of this disturbance and to recover information and skeletal remains relating to the later inhabitants of the pueblo. Twenty-five complete skeletons and 31 skulls and other bones of disturbed burials were reportedly recovered. Of the 25 skeletons, 21 were identified as adults, 2 as adolescents, and 2 as children of about 8 years. Eighteen individuals were classified as male, three as female, and four as of unknown sex. Of the 31 unassociated skulls, 23 were identified as male, 5 as probably male, and 3 as children of indeterminate sex, ages 7, 11, and 12 years (Kidder 1958).

These remains were transferred to Harvard University's Peabody Museum of Archaeology and Ethnology, as was the case for the previously excavated bones from the church and the larger skeletal series recovered from the habitation and refuse middens of the pueblo. Because of the unique temporal and contextual identity of the fourth church series and their late transfer to the museum, this set of skeletons, with the exception of a few remains retained from the 1915 excavation, was not included in Hooton's (1930) osteological analysis of Pecos Pueblo, nor have they been reported elsewhere. As Hooton (1930:vii) wrote in the preface to *The Indians of Pecos Pueblo*, "The considerable number of skeletons found in the nave of the church at Pecos, all of them of Historical date and some of them undoubtedly the result of recent Mexican burials, have not been analyzed as a group. I have left this series open in the hope that subsequent additions to it may come from future excavations."

FIGURE 11.1. Exposed coffin burials at the foot of the reconstructed sanctuary steps of the fourth Pecos mission church, 1915 field season. (Courtesy Palace of the Governors Photo Archives [NMHM/DCA], 41034.)

In 1939, additional excavations in the second and fourth churches did take place, resulting in the collection of approximately 51 sets of remains. This work, conducted as a joint research expedition of the University of New Mexico and the Museum of New Mexico, was led by archeologist William Witkind. These remains stayed in New Mexico until their repatriation to the Pueblo of Jemez on May 14, 1999. Information on this series is not part of this study. A partial report on the burials excavated by Witkind is on file at the Museum of New Mexico (Heather Edgar, personal communication, 2004).

Hooton's plan for a comprehensive study of the burials excavated from the churches at Pecos Pueblo never materialized. Classification of the church burials as separate from the larger and earlier Pueblo sample by both Kidder and Hooton, along with the commingled proveniences of some individual remains, left the fourth church skeletal series at the Peabody Museum unstudied for the better part of the twentieth century.

ANALYSIS OF SKELETONS FROM THE FOURTH MISSION CHURCH

In 1998 the skeletons recovered during the 1915 and 1925 field seasons were examined in order to document the remains prior to reburial. Initial inspection indicated that a majority of bones from the church had previously been separated into subsets by type—

cranium, humerus, radius, ulna, femur, tibia, and fibula—resulting in the loss of bone associations for many individuals. This sorting was likely implemented in response to the disturbed and commingled recovery state of many bones.

Extensive re-sorting was conducted in order to regroup long bones and mandibles with represented crania. Reassociation of elements was based on consistencies in size, color, preservation (i.e., taphonomic characteristics), age and sex determinations, articulation of corresponding joints, and pathological features. Although only partial reassociation of the assemblage was possible, given the cost-related time constraints for data recovery and the likelihood that some skeletons were only partially recovered in the field, approximately one-half of the crania were nevertheless matched with a mandible, some postcranial elements, or both. Unassigned, isolated bones were recorded individually using a standardized system for coding commingled ossuary remains described by Jantz and colleagues (2002).

Good preservation was the norm for this series and facilitated the reassociation. One individual displayed remnants of desiccated tissue. Green, copper oxide staining as a result of artifact association was not observed, but dark mottling from contact with coffin wood was widespread, and discoloration from fabric was apparent on the bones of two individuals. The fabric-related discoloration was characterized by dark brown, woven pattern impressions on the surfaces of some bones. A few crania and mandibles were coated with a thin layer of wax, presumably an early means of stabilization to help ensure preservation. Most of the bones, however, were complete and free of residue.

The documentation used standard methodology in skeletal biology (Buikstra and Ubelaker 1994) and followed data collection procedures described by Owsley and Jantz (1989), including compilation of bone inventories and determinations of age, sex, and ancestry. Bone and dental pathological conditions were identified and recorded using a detailed scoring format that facilitated frequency tabulation of injuries and pathological conditions in bones and teeth by age and sex.

DEMOGRAPHIC COMPOSITION

Total bone counts for major elements in the church series are presented in tables 11.1 and 11.2. In an effort to avoid inflating the count of the minimum number of individuals (MNI) present, minimally represented, partial postcranial bones were not included in the documentation because of an inability to associate these pieces with more intact elements. Their numbers are given in the note to table 11.2.

The MNI for this series, determined on the basis of complete or nearly complete crania, is 68 individuals, which are listed in table 11.3. The number of identified remains is higher than Kidder's preliminary count from the church, cited earlier, in part because it includes the few remains recovered in 1915. The left femur provides the highest postcranial bone count, 42, along with pieces of 10 others. Cranial bones provide the greatest representation for both children and adults, with a count of 11 frontals, right and left parietals, occipitals, left temporals, and right and left maxillae for children, and 57 frontals and left and right parietals for adults. Few postcranial bones of children are present, the highest counts being three each for the left radius and femur and left and right ulnae. The left femur provides the largest adult postcranial bone count, 39.

Of the 68 individuals represented by a cranium and associated mandible, by postcranial bones, or both, 11 are infants and children aged 14 years or younger (table 11.4). No bones of infants under the age of one year are present, and only a single individual is represented in the one- to four-year age range. For purposes of analysis, the remains of individuals aged greater than 14 years were divided into the general categories of younger (15–34 years) and older (35+ years) adult males and females. The bones of at least 57 adults are present, with the sample unevenly divided by sex: 37 males (65%) and 20 females (35%). Young adult males make up a majority of the church series ($n = 22$, 32.4%). Older adult males are the second largest group, followed by older adult females. Young adult females and adolescents aged 10 to 14 years each compose 10% of the total series.

Although males outnumbered females in Kidder's report (1958) and in this assessment, a marked discrepancy exists in the number of females initially identified by Kidder. Kidder recognized only three females. A high male-to-female ratio remains, however, and is consistent with the pattern noted for all other archaeological periods at Pecos, especially the postcontact period (Hooton 1930). Ages assigned to juveniles were also adjusted during this examination. Advances in osteological methodology involving new identification techniques and access to comparative skeletal databases contributed greatly to our assessment of the Pecos church skeletons.

TABLE 11.1
PECOS PUEBLO CHURCH SERIES CRANIAL AND MANDIBULAR BONE INVENTORY

	Frontal		Parietal				Occipital		Temporal				Zygomatic				Maxilla				Mandible	
			Left		Right				Left		Right		Left		Right		Left		Right			
	C	P	C	P	C	P	C	P	C	P	C	P	C	P	C	P	C	P	C	P	C	P
Number	68	0	66	2	68	0	66	1	64	0	65	1	57	5	59	3	63	1	64	0	35	2
Total	68		68		68		67		64		66		62		62		64		64		37	

Key: C = complete; P = partial.

TABLE 11.2
PECOS PUEBLO CHURCH SERIES POSTCRANIAL BONE INVENTORY

	Manubrium		Sternum		Scapula				Clavicle				Innominate				Acetabulum				Sacrum	
					Left		Right		Left		Right		Left		Right		Left		Right			
	C	P	C	P	C	P	C	P	C	P	C	P	C	P	C	P	C	P	C	P	C	P
Number	11	0	8	1	15	2	14	4	17	0	15	0	21	1	19	3	21	0	21	0	19	1
Total	11		9		17		18		17		15		22		22		21		21		20	

	Patella				Talus				Calcaneus				Humerus				Radius			
	Left		Right		Left		Right		Left		Right		Left		Right		Left		Right	
	C	P	C	P	C	P	C	P	C	P	C	P	C	P	C	P	C	P	C	P
Number	3	0	5	0	7	0	7	0	7	0	6	0	30	1	31	0	25	0	26	0
Total	3		5		7		7		7		6		31		31		25		26	

	Ulna				Femur				Tibia				Fibula			
	Left		Right		Left		Right		Left		Right		Left		Right	
	C	P	C	P	C	P	C	P	C	P	C	P	C	P	C	P
Number	26	0	26	0	41	1	38	1	39	0	36	0	14	0	16	0
Total	26		26		42		39		39		36		14		16	

Key: C = complete; P = partial.

Note: Additional long bone segments were present but were not included in the base counts or documented for pathological conditions because of incompleteness: 15 left and 15 right partial humeri; 6 left and 3 right partial radii; 5 left and 13 right partial ulnae; 10 left and 14 right partial femora; 6 left and 6 right partial tibiae.

TABLE 11.3
THE 68 SETS OF SKELETAL REMAINS FROM THE PECOS CHURCH

ID number	Sex	Age (yrs.)	Ancestry[a]	Preservation[b]	Anomalies and/or pathological conditions
59888	F	11.5–12.5	I	G (cranium)	—
59889	F	60+	I	G (cranium, sort)[c]	Osteoporosis, degenerative joint disease
59891	M	30–34	E	G (skeleton)	Heavy calculus, enamel hypoplasia defects, tooth staining, 5th lumbar spondylolysis, multiple Schmorl's nodes
59892	M	50–54	I	G (cranium, mandible)	Healed nasal bones and left maxilla fractures, porosity on mandibular condyles
59894	F	11.5–12.5	E	G (cranium, sort)	Mild ectocranial porosity
59895	F	35–39	E	G (cranium, sort)	Slight degenerative changes, tooth staining
59896	I	1.5–2.5	I	G (cranium)	Healed occipital fracture, perimortem bladed fracture in left parietal
59897	F	45–49	I	G (cranium, mandible, sort)	Palatine tori, slight degenerative changes
59898	F	18.5–19.5	E/I	G (cranium)	—
59899	F	60+	I	F (cranium)	Cranium dipped in wax
59900	I	7.5–8.5	I	G (cranium)	Slight cribra, slight ectocranial porosity
59901	M	20–24	E	G (cranium, mandible, sort)	Trace joint porosity, multiple Schmorl's nodes
59902	M	25–29	I	G (cranium, mandible)	Perimortem bladed cranial vault and facial fractures
59903	F	20–24	I	G (cranium)	Dipped in wax
59904	M	60+	I	F (cranium)	Perimortem bladed trauma
59906	M	25–29	I	G (cranium)	Perimortem fractures, cuts indicative of scalping
60633	M	30–34	I	G (partial skeleton)	Remodeled lytic lesions widespread on cranium and postcranial skeleton
60634	M	40–44	I	G (skeleton)	Slight to moderate degenerative changes
60648	M	45–49	I	F (cranium, mandible)	Healed depression fractures on cranium
60649	M	60+	E	G (cranium)	Lytic lesions on cranium
60651	M	25–29	I	G (cranium, mandible)	Healed right malar fracture, perimortem vault and facial fractures, mastoiditis (left temporal)
60652	F	20–24	I	G (cranium)	—
60653	F	35–39	I	G (cranium)	Temporomandibular joint erosion and porosity
60654	M	50–54	I	G (cranium, mandible)	Temporomandibular joint porosity
60663	M	55–59	I	G (skeleton)	Partial 5th lumbar spondylolysis, degenerative joint changes (weaver?), healed nasal fractures
60667	M	20–24	I	F (cranium, mandible)	—
60680	M	11.5–12.5	I	G (skeleton, sort)	Coronal synostosis
60681	F	45-49	I	G (partial skeleton)	Degenerative joint changes, enthesophyte development on femora, healed depression fracture on frontal
60685	F	25–29	I	G (partial skeleton)	Vertebral degeneration (12th thoracic has anterior wedging)

Continued on next page

TABLE 11.3 *(Continued)*

ID number	Sex	Age (yrs.)	Ancestry[a]	Preservation[b]	Anomalies and/or pathological conditions
60691	M	50–54	I	F (skeleton)	Vertebral degeneration, 5th lumbar spondylolysis, enthesophyte on humerus, antemortem fracture of T1/T2, severe antemortem facial trauma
60696	F	60+	I	G (partial skeleton)	Osteoporosis, degenerative joint changes
60704	F	20–24	I	G (skeleton)	Multiple Harris lines noted in radiograph images of long bones
60706	F	50–54	E/I	G (skeleton)	Temporomandibular joint erosion, degenerative joint changes, possible perimortem cut on 2nd lumbar
60708	F	13.5–14.5	I	F (skeleton)	Widespread abnormal, woven bone formation (periostitis)
60711	M	25–29	I	G (skeleton)	5th lumbar spondylolysis, left scapula periostitis, healed right radius and humerus fractures, fused right sacroiliac joint, perimortem cut on left rib 9 and right femur, cuts indicative of scalping
60712	M	5.5–6.5	E	G (cranium)	Deciduous carious lesions
60714	M	70+	I	F (partial skeleton)	Osteoporosis, degenerative joint changes, healed right ulna fracture, tibiae periostitis, enthesophyte formations on several bones
60717	M	30–34	E/I	F (partial skeleton)	Perimortem fracture of right temporal, cuts indicative of scalping
60718	M	35–39	I	G (cranium, mandible)	Antemortem frontal fractures
60719	F	12.5–13.5	I	G (cranium)	Slight cribra orbitalia, healed right nasal fracture
60720	M	60+	I	G (cranium, mandible)	Temporomandibular joint erosion, healed left orbit fracture, maxillary sinusitis, tooth staining
60721	M	20–24	I	G (cranium)	Synostosis of left lambdiodal suture, temporomandibular joint erosion
60724	M	20–24	I	F (cranium)	Slight ectocranial porosis
60734/35	M	14.5–15.5	E	G (skeleton)	5th lumbar spondylolysis, multiple Schmorl's nodes
60737	M	27–32	I	G (partial skeleton, sort)	Slight ectocranial porosis, right 6th rib periostitis
60747	M	27–34	I	G (cranium)	Slight ectocranial porosis
60748	M	30–34	E	G (cranium)	Cuts indicative of scalping, antemortem cranial fractures, slight ectocranial porosis
60749	M	23–27	I	G (cranium)	Slight ectocranial porosis, antemortem left parietal and right maxilla and nasal fractures
60750	F	22–26	E/I	G (cranium, mandible, sort)	Perimortem frontal fracture
60751	M	25–29	I	G (cranium, mandible)	Perimortem chops and cuts on skull
60753	M	30–39	I	F (cranium, mandible)	Three small lytic lesions in left palate
60757	M	20–24	I	G (cranium, mandible)	Perimortem chops on skull
60758	F?	5.5–6.5	I	G (cranium, mandible)	Perimortem chops on skull

Continued on next page

TABLE 11.3 *(Continued)*

ID number	Sex	Age (yrs.)	Ancestry[a]	Preservation[b]	Anomalies and/or pathological conditions
60759	F	40–49	I	G (cranium)	Temporomandibular joint erosion and porosity
60760	F	10.5–11.5	I	F (cranium)	Perimortem skull fractures, cuts indicative of scalping
60761	M	50–59	I	F (cranium, mandible, sort)	Perimortem bladed, frontal fracture
60762	F	40–49	I	F (cranium, sort)	Perimortem bladed vault fractures, temporomandibular joint erosion
60764	M	22–27	E	G (cranium, mandible)	—
60766	M	35–44	I	F (cranium)	Antemortem right parietal depression fracture
60767	F	40–49	I	G (cranium, mandible)	Mild ectocranial porosis
60768	M	30–34	I	G (cranium)	Mild ectocranial porosis
60769	M	27–34	E	G (cranium, mandible, sort)	Frontal, right parietal, and mandible periostitis, possible pipe facet
60770	M	40–44	E/I	G (cranium)	Mild ectocranial porosis, temporomandibular joint erosion, antemortem frontal fracture
60779	F	30–34	I	G (skeleton)	Osteoporosis, lytic rib lesion, vertebral ankylosis and depressions, antemortem compression fracture of 12th thoracic (possible tuberculosis)
60905	F	11.5–12.5	E/I	G (cranium)	Perimortem chops on cranium
61122	F	30–39	I	G (cranium)	Temporomandibular joint erosion, perimortem chops on cranium
61155	F	50–59	I	G (cranium)	Temporomandibular joint erosion
61289	M	35–44	I	F (cranium)	Perimortem bladed fractures to right parietal, left and right temporals, small tumor in frontal sinus

[a] I = Indian; E = European or Hispanic; E/I = European/Indian or Hispanic/Indian mixed ancestry.

[b] G = good; F = fair; P = poor.

[c] The term *sort* identifies commingled postcranial elements that were matched with specific crania but constituted less than half the skeleton.

The development of new methods and the effects of advancing analytical capability are especially apparent in the determination of ancestry. The ancestry of the individuals buried in the church was an unresolved question for Kidder (1958), who believed that most of the remains were those of Indians, because of the burial positioning of the arms. On the basis of morphology, Hooton (1930:212, figs. VI-20, VI-21) identified at least two skulls from the church as "almost certainly Mexican or of Mexican admixture." These skulls were assigned to his "Long-faced European" morphological type.

Techniques have developed beyond the methods applied by Kidder and Hooton for the determination of ancestry. At present, ethnic identity can be more precisely documented through craniometric analysis. Craniometric analyses were used to determine the ancestry of 37 older subadults and adults with complete or nearly complete crania. Infants, young children, and adults with broken or warped crania that could not be measured were evaluated visually using nonmetric morphological criteria (Gill and Rhine 1990), including the presence or absence of cranial modification. Occipital and, less frequently, lambdoidal

TABLE 11.4
PECOS PUEBLO CHURCH SERIES BY AGE, SEX, AND ANCESTRY

Age (yrs.)/Sex	Native American		European		Native American/ European		Total	
	n	%	*n*	%	*n*	%	*n*	%
0–0.5	0	0	0	0	0	0	0	0
1–4	1	1.9	0	0	0	0	1	1.5
5–9	2	3.8	1	10	0	0	3	4.4
10–14	5	9.6	1	10	1	16.7	7	10.3
Males 15–34	15	28.8	6	60	1	16.7	22	32.4
Males 35+	13	25.0	1	10	1	16.7	15	22.1
Females 15–34	5	9.6	0	0	2	33.3	7	10.3
Females 35+	11	21.2	1	10	1	16.7	13	19.1
Total	52	76.5	10	14.7	6	8.8	68	100.0

flattening were common in the Indian component of the Pecos church series. Cranial modification was usually asymmetrical and relatively minor.

Judging from the history of the site, we could expect the church sample to be a mixture of Native Americans, Europeans of Spanish ancestry, and possibly hybrid offspring of Indians and whites. Two approaches were used to evaluate the church sample. In the first method, each measurable skull was classified against reference samples of whites and Native Americans, and its "posterior probability" was calculated—that is, the probability that the skull belonged to the ancestral group to which it was most similar, on the assumption that it must belong to one group or the other. Our interpretation was that skulls with posterior probabilities of 0.8 or greater for a group belonged to that group, whereas those with posterior probabilities of less than 0.8 could belong to either group. Accordingly, we designated skulls with posterior probabilities of 0.8 or greater either white or Indian. Crania with posterior probabilities more evenly allocated between whites and Native Americans display a combination of features, likely a result of admixture. Following this metric classification system, 22 crania showed features strongly characteristic of Indians, eight were identified as European, and six suggested probable admixture.

The second analysis was developed from the expectation that a sample composed of a mixture of ancestries would be morphologically more heterogeneous and metrically more variable than a single-ancestry group. This assertion was evaluated by calculating Mahalanobis distances between all pairs of Pecos church crania using the pooled within-group covariance matrix of the single-ancestry groups. The equation is

$$D^2_{(i,j)} = (x_i - x_j)\, W^{-1} (x_i - x_j),$$

where W^{-1} is the pooled within-group phenotypic covariance matrix of the single-ancestry samples and x_i and x_j are the mean vectors for skulls *i* and *j*, respectively (see Jantz and Owsley 2001 for additional details). Two crania drawn at random from a population with covariance matrix W have an expected distance (D) of $\sqrt{(2p-1)}$, with variance equal to 1.0, where *p* is the number of variables (DeFrise-Gussenhoven 1967).

The mean pair-wise distance of all 630 pairs of Pecos crania is 6.249 ± 0.0476, sd = 1.19587, as against an expected value of 5.916. The expected value falls outside the 95% confidence interval of the observed mean, indicating that the observed value is significantly larger than the expected, as would be the case in a heterogeneous sample.

The interskull distances can be displayed on linear axes by extracting the principal coordinates of the Mahalanobis distances (Gower 1972). The crania are displayed on the first two axes in figure 11.2. The first axis was chosen to exhibit maximum interskull variance; the second accounts for the maximum residual variance. Together the two axes explain 33% of the interskull variance. Since the algorithm has no information regarding the ancestry of the crania, their distribution provides a test of the hypothesis that ancestry, as assessed previously in the discriminant analysis, is important in explaining variation among the crania. As shown in figure 11.2, individuals that

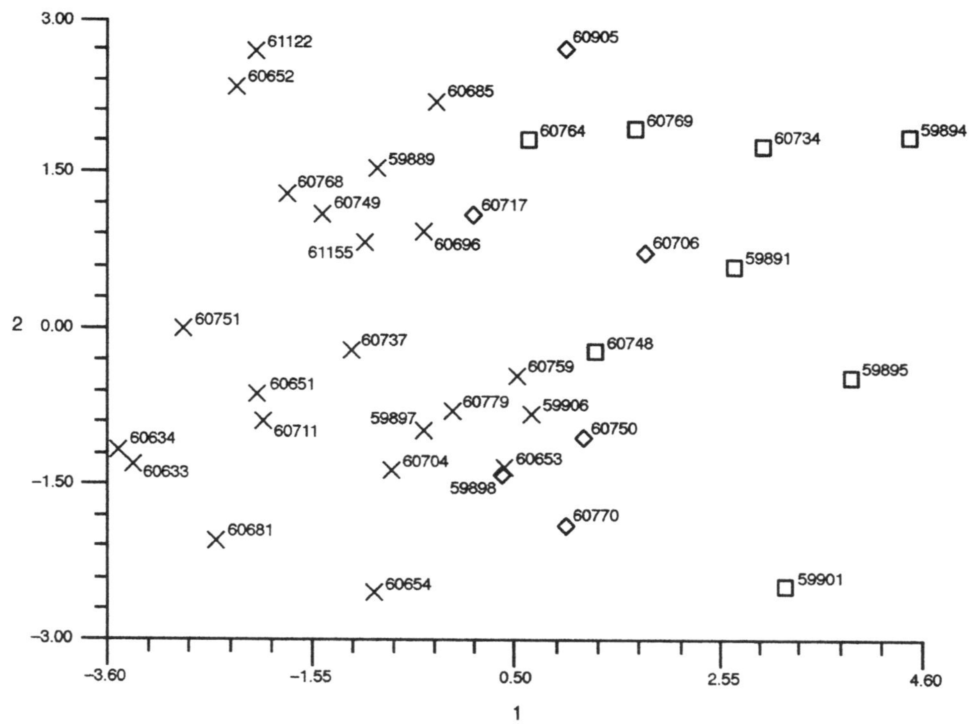

FIGURE 11.2. Linear axes showing interskull distances. The first axis (1, shown on the *x*-axis in the diagram) exhibits maximum interskull variance; the second (2, shown on the *y*-axis) accounts for the maximum residual variance. *Key:* squares, white or European/Hispanic; diamonds, mixed European–Native American ancestry; Xs, Native American.

were classified as white are positive on axis 1 and, with the exception of catalogue number 59901, on axis 2 as well. Crania designated as Native American have negative scores on axis 1, whereas those designated white-Indian hybrids have intermediate scores, close to the midpoint between the two parent populations. Table 11.5 presents a post hoc test for the mean scores of the three groups for the first two principal coordinates. The first principal coordinate shows highly significant variation among the designated groups, supporting the hypothesis that it is detecting variation among ancestries. The second principal coordinate is not significant.

Structure coefficients indicate that individuals with positive scores on the first principal coordinate can be described as having narrow, small faces with high facial projections and narrow vault breadths. Individuals with negative scores display the reverse (fig. 11.3). These descriptions are the ones normally applied to differences between whites and Native Americans.

Both craniometric and nonmetric analyses indicate that the majority of individuals in this series are Native American (n = 52, 76%). Ten (15%) are identified as white, or having European ancestry, and six (9%) are classified craniometrically as having mixed ancestry, indicated by a combination of Native American and European features.

The three ancestry subgroups differ not only in the number of individuals represented but also by age and sex (table 11.4). Individuals with Native American ancestry are represented by nearly all age groups, from 1.5 to 60+ years. Young adult males are a slight majority (29%), but older adult males (25%) and females (21%) are also well represented. The 10 individuals identified as having European ancestry are predominately young adult males (60%). One female aged 35–39 years and one male aged 60+ years are also represented, as are two children aged 11.5–12.5 years and 5.5–6.5 years. The six individuals with mixed ancestry have nearly equal representation in each of the age and sex groups, from children aged 10 years and above through adulthood. This subgroup includes one child aged 11.5–12.5 years, one young adult male, one older adult male, two young adult females, and one older adult female.

TABLE 11.5
POST HOC TESTS OF DESIGNATED GROUP SCORES ON FIRST TWO PRINCIPAL COORDINATES

Statistic	First principal coordinate[a]			Second principal coordinate[b]		
	Indian	E/I	European	Indian	E/I	European
n	22	6	8	22	6	8
Mean	–1.334	0.865	2.569	–0.088	0.001	0.568
sd	1.209	0.607	1.316	1.464	1.772	1.587

[a] $F_{2,33} = 35.47, p < .001$
[b] $F_{2,33} = 0.540, p > .05$

DENTAL HEALTH

In order to assess individual well-being and the health of the identified subgroups and skeletal series as a whole, the remains were examined for the presence of multiple pathological conditions in teeth and bones. Teeth and alveolar sockets were examined for the presence of carious lesions, abscesses, and antemortem loss. The degree of occlusal surface attrition and the presence of task activity wear (e.g., pipe facets) were also recorded.

Table 11.6 summarizes the dental inventory and pathology of adults from the Pecos church. Caries affected 10.8% of all represented adult teeth. Maxillary and mandibular molars have the largest percentages of lesions, whereas mandibular incisors and canines show no evidence of decay. Of 1,274 observable tooth sockets, 15.7% show antemortem loss. The maxillary molars have the highest percentage of antemortem loss, followed by maxillary incisors and premolars and mandibular molars, which display similar loss frequencies. Mandibular premolars and canines have the

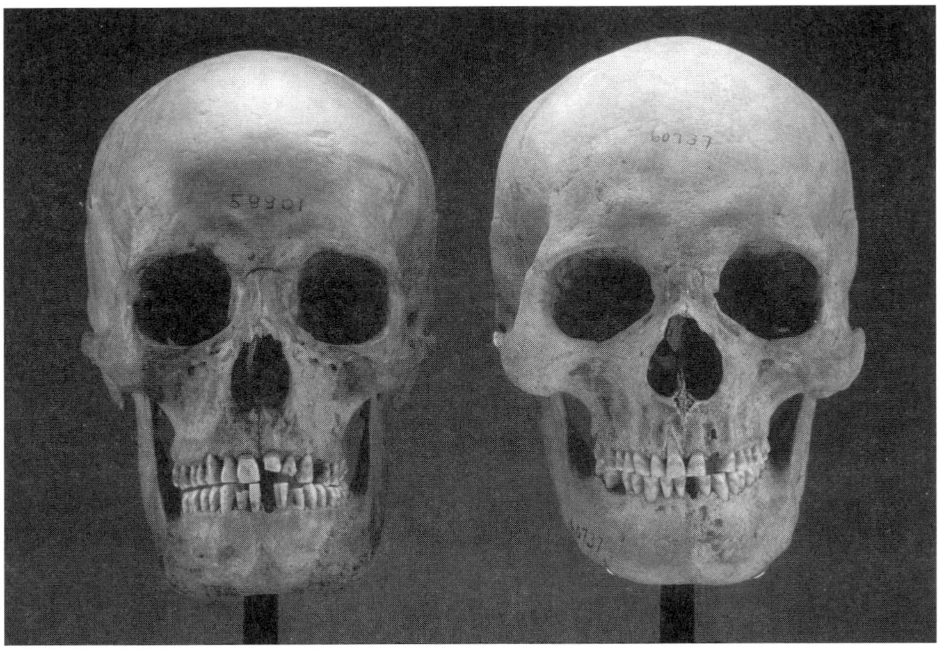

FIGURE 11.3. Morphology reflected in two skulls from the Pecos church: European (left, cat. no. 59901) and Native American (right, cat. no. 60737). Left: PMAE 19-6-10/59901.0; right: PMAE 25-3-10/60737.0.

TABLE 11.6
DECAYED, MISSING, AND ABSCESSED TEETH IN CHURCH ADULTS, BY TOOTH TYPE

Teeth and arcade	Teeth (*n*)	Carious (*n*)	% Carious	Sockets (*n*)	Lost antemortem (*n*)	% Lost antemortem	Alveolar abscesses (*n*)	% Alveolar abscesses
Incisors								
Maxillae	53	1	1.9	200	31	15.5	20	10.0
Mandible	42	0	0.0	121	16	13.2	5	4.1
Canines								
Maxillae	36	1	2.8	101	9	8.9	14	13.9
Mandible	33	0	0.0	61	4	6.6	4	6.6
Premolars								
Maxillae	96	5	5.2	202	31	15.3	16	7.9
Mandible	73	3	4.1	122	7	5.7	6	4.9
Molars								
Maxillae	153	28	18.3	298	77	25.8	37	12.4
Mandible	116	27	23.3	169	25	14.8	20	11.8
Total	602	65	10.8	1,274	200	15.7	122	9.6

lowest percentages of antemortem loss. Alveolar abscesses were noted in 9.6% of the represented sockets. Although abscesses are present for all tooth types, maxillary canines and molars exhibit the largest percentages while the mandibular incisors and premolars are least affected.

Examined by age and sex (table 11.7), older adults have larger percentages of carious lesions, antemortem loss, and alveolar abscesses, and females have poorer dental health overall. The percentage of carious lesions in teeth of adult males is 8.8%; the percentage in females is nearly double that value (16.6%). Older males show a slightly higher rate of tooth decay than females, 16.5% versus 13.4%, whereas the percentage of cavities in young adult female teeth (18.9%) is nearly three times higher than that of young males (6.9%).

A similar pattern is noted for antemortem tooth loss. Tooth loss in older adults affected 35.7% of observable sockets. Young adults had a low percentage of antemortem loss (1.0%). Statistically, females have a larger percentage of antemortem loss than males (19.9% and 13.9%, respectively). However, the frequency of antemortem loss in sockets of older males (36.3%) is just slightly higher than that of older females (34.8%). Young females, on the other hand, have a 1% difference in loss, 1.8% versus 0.7% in young males.

Alveolar abscesses are evident in 16.9% of older adult tooth sockets, versus 4.2% of young adult sockets. For males, 8.2% of sockets show the condition, whereas 12.7% of female sockets are affected. Larger percentages of abscesses are noted in both younger and older females (6.5% of tooth sockets in the former, 17.9% in the latter) relative to younger and older males (3.5% and 16.2%, respectively).

Differences in male and female diets could have caused the sex-related differences in dental pathology. However, dietary analysis of approximately 100 skeletons from the pre- and postcontact periods of Pecos Pueblo (Spielmann, Schoeninger, and Moore 1990) showed no significant differences in bone strontium concentrations of males and females from any represented time period, indicating no noticeable difference in the proportion of meat versus plant foods in male and female diets. Likewise, stable carbon and nitrogen isotope values were consistent within and across the represented time periods (Spielmann, Schoeninger, and Moore 1990; Kellner et al., this volume). Poorer female dental health in the church series apparently signals a difference between the sexes that was not detected through chemical analyses of the Indian population. This difference might be linked to the historical origins of these demographic units, with no young adult European females but several young males of European ancestry represented. The possibility that the dental health of the mission population was influenced by factors related to ethnic composition is difficult to evaluate, because of the small sample sizes available. Never-

TABLE 11.7
DECAYED, MISSING, AND ABSCESSED TEETH IN CHURCH ADULTS, BY AGE AND SEX

CARIOUS LESIONS

	Male			Female			Total		
Age (yrs.)	Teeth (*n*)	Carious (*n*)	%	Teeth (*n*)	Carious (*n*)	%	Teeth (*n*)	Carious (*n*)	%
15–34	360	25	6.9	90	17	18.9	450	42	9.3
35+	85	14	16.5	67	9	13.4	152	23	15.1
Total	445	39	8.8	157	26	16.6	602	65	10.8

ANTEMORTEM LOSS

	Male			Female			Total		
Age (yrs.)	Sockets (*n*)	Lost antemortem (*n*)	%	Sockets (*n*)	Lost antemortem (*n*)	%	Sockets (*n*)	Lost antemortem (*n*)	%
15–34	564	4	0.7	170	3	1.8	734	7	1.0
35+	333	121	36.3	207	72	34.8	540	193	35.7
Total	897	125	13.9	377	75	19.9	1,274	200	15.7

ALVEOLAR ABSCESSES

	Male			Female			Total		
Age (yrs.)	Sockets (*n*)	Alveolar abscesses (*n*)	%	Sockets (*n*)	Alveolar abscesses (*n*)	%	Sockets (*n*)	Alveolar abscesses (*n*)	%
15–34	564	20	3.5	170	11	6.5	734	31	4.2
35+	333	54	16.2	207	37	17.9	540	91	16.9
Total	897	74	8.2	377	48	12.7	1,274	122	9.6

theless, a comparison of the dental health of young adult Indian males with males of European ancestry shows a significant difference in the occurrence of tooth decay ($X^2 = 7.42$, df = 1, $p < .01$) (table 11.8). Of the young adult Indian male teeth, 9.4% were affected by caries, versus 1.8% of the teeth of young adult European males. Antemortem tooth loss and alveolar abscesses were scarce for both groups, with no significant differences noted. The difference in percentages of caries indicates that the dental pathology of the church series is not wholly reflective of Native American health but is instead a composite, influenced by the addition of European and Mexican immigrants with potentially differing diets and associated oral pathology.

Ethnic differences are also suggested by particular activities reflected in the dentition. Tobacco use, inferred by distinctive dark staining on the lingual surfaces of the teeth (fig. 11.4) and pipe wear facets are present mostly in individuals of European ancestry. One young adult European male has a probable pipe stem wear facet in his anterior teeth. Dark staining was present in a second young adult European male, an older European female, and one older Indian male. The priest's 1731 mission supply inventory for Pecos lists fine-cut tobacco (Kessell 1979:330), confirming its presence at the site.

Despite heterogeneity in the dental data, the dental pathology profile of this series generally corre-

TABLE 11.8
PATHOLOGY IN THE PERMANENT DENTITION OF YOUNG ADULT MALES, BY ANCESTRY

Teeth/Arcade	Teeth (*n*)		Carious (*n*)		% Carious		Sockets (*n*)		Lost antemortem (*n*)		% Lost antemortem		Alveolar abscesses (*n*)		% Alveolar abscesses	
	I	E	I	E	I	E	I	E	I	E	I	E	I	E	I	E
Incisors																
Maxillae	19	15	0	0	0.0	0.0	54	24	0	1	0.0	4.2	2	0	3.7	0.0
Mandible	12	8	0	0	0.0	0.0	38	20	0	0	0.0	0.0	0	0	0.0	0.0
Canines																
Maxillae	16	6	0	0	0.0	0.0	28	12	0	0	0.0	0.0	1	1	3.6	8.3
Mandible	7	7	0	0	0.0	0.0	18	10	0	0	0.0	0.0	0	0	0.0	0.0
Premolars																
Maxillae	40	15	0	0	0.0	0.0	56	24	0	0	0.0	0.0	1	0	1.8	0.0
Mandible	20	12	0	0	0.0	0.0	36	20	0	0	0.0	0.0	0	1	0.0	5.0
Molars																
Maxillae	67	27	10	1	14.9	3.7	83	34	2	0	2.4	0.0	4	0	4.8	0.0
Mandible	41	20	11	1	26.8	5.0	50	25	1	0	2.0	0.0	2	4	4.0	16.0
Total	222	110	21	2	9.4	1.8	363	169	3	1	0.8	0.6	10	6	2.7	3.5

Key: I = Indian; E = European.

sponds to those of other groups in the Southwest. Direct comparison with previous reports is difficult, however, especially in regard to age, sex, or both, because of differences in methodology and reporting of data. Data are typically presented as numbers or percentages of individuals affected, rather than as numbers and percentages of affected teeth or tooth sockets by type (Berry 1985; Hooton 1930; Stodder 1989; Stodder and Martin 1992). Because many individuals have incomplete representation of the dentition, general percentages can be misleading. Age and sex criteria are also rarely figured into the presented results but potentially can influence a study's outcome, as is evidenced in the Pecos data by the larger percentages of dental pathologies noted in older church adults and by the differences between males and females.

At the broadest level, when tabulated as the number of individuals affected by one or more carious lesions or one or more abscesses, Pecos church adults show 51% and 63% involvement, respectively. These percentages place the Pecos church series within the mid-range for the prevalence of caries and in the higher range for the prevalence of abscesses relative to Southwestern samples summarized by Stodder and Martin (1992), which ranged in date from the fifth to the nineteenth century. Percentages of individuals with caries reported for these prehistoric and historic Southwest groups range from a low of 9% for the prehistoric series of Turkey Creek (A.D. 1000–1285) to a high of 85% reported for both Chaco Canyon (A.D. 700–1300) and the postcontact pueblo of Gran Quivira (A.D. 1550–1672). Percentages most similar to those reported for the Pecos church sample are from the prehistoric series (A.D. 1275–1400) of Grasshopper Pueblo (52%) and the pre- to postcontact period (A.D. 1300–historic) skeletons of San Cristóbal (57%) and Hawikku (53%). The percentages of individuals with abscesses range from a reported 11% for the precontact series of Swarts Ruin (A.D. 600–1000) to 66% recorded for the postcontact skeletal series of Gran Quivira. These studies generally show increases in dental pathology over time and stress a correlation between poorer dental health and intensification of maize horticulture.

Dental pathology data reported by Hooton (1930) for the pre- and postcontact occupations at Pecos, which include general information on antemortem tooth loss and pyorrhea, reflect a downward trend in dental health during the Spanish colonial period. The church remains show a larger percentage of individuals with caries than in any of the precontact ceramic periods except for Glaze 4, or A.D. 1350–1500 (61%), and the largest percentage of individuals with abscesses for any of the reported periods of Pecos.

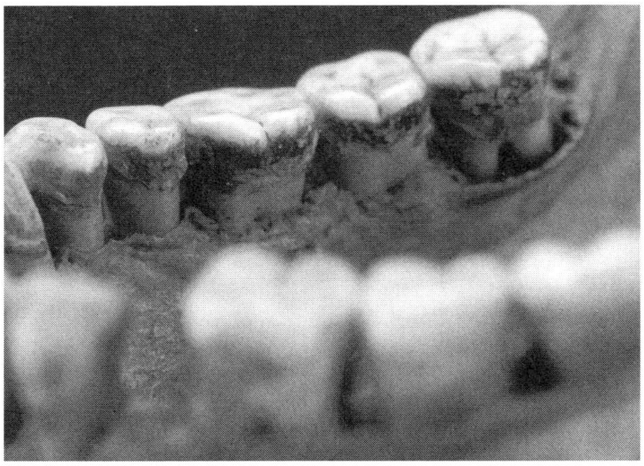

FIGURE 11.4. Dark staining indicative of tobacco use on the lingual surfaces of mandibular premolars and molars of an adult male (cat. no. 59891). PMAE 19-6-10/59891.0.

Direct comparison of Pecos church dental pathology with other Spanish mission samples is possible for the roughly contemporaneous Texas missions of Nuestra Señora del Refugio (ca. 1795–1830) and San Juan Capistrano, Room 26 (ca. 1764–1790) and Room 17 (ca. 1794–1850), because strictly comparable data are available (Jantz et al. 2002; Owsley et al. 1999). Karankawa Indians composed most of the Refugio series, although some European admixture was identified. The aboriginal group is believed to have periodically continued its traditional subsistence pattern of hunting and gathering while occupying the mission (Ricklis 1996). Coahuiltecan Indians were the primary group represented by the two San Juan Capistrano samples. Although this group also originally practiced a hunting and gathering subsistence strategy, over time it became more dependent on mission foods, primarily maize, and by the late eighteenth century had culturally and socially adapted to mission life (Francis 1999; Humphreys 1971).

The dental health of the Pecos mission series falls within the general pattern observed at these two Texas missions (table 11.9). Pecos church teeth exhibit a slightly larger percentage of carious lesions than that documented for Refugio but a smaller percentage than that for remains from Rooms 26 and 17 of San Juan Capistrano. Pecos tooth sockets have a slightly larger percentage of abscesses than the other mission samples and a frequency of antemortem tooth loss as high as the later occupation of San Juan Capistrano.

TABLE 11.9
DECAYED, MISSING, AND ABSCESSED TEETH IN ADULTS FROM THREE SPANISH MISSION SKELETAL SERIES

Mission series	Teeth (*n*)	Carious (*n*)	% Carious	Sockets (*n*)	Lost ante-mortem (*n*)	% Lost ante-mortem	Alveolar abscesses (*n*)	% Alveolar abscesses
Pecos church	602	65	10.8	1,274	200	15.7	122	9.6
Capistrano Room 26	432	58	13.4	674	83	12.3	53	7.9
Capistrano Room 17	281	55	19.6	581	92	15.8	15	5.2
Refugio	1,427	124	8.7	1,200	95	7.9	99	8.3

In general, the dental pathology profile for the Pecos church series is fairly consistent with that of San Juan Capistrano. This pattern might be anticipated given the heavy reliance on maize documented for all occupations at Pecos (Spielmann, Schoeninger, and Moore 1990). One aspect of dental health evident at Pecos and Refugio is a tendency toward increased dental pathology in young women relative to men. Tooth decay in young adult females from Refugio (Jantz et al. 2002) was 8.9%, and at Pecos, 18.9%, versus 4.5% and 6.9%, respectively, in young males. Further dietary studies are needed to reveal the source of this pattern of poorer female dental health in mission populations.

BONE HEALTH

The bones from the Pecos church were examined for the presence of lytic lesions, abnormal subperiosteal bone formation (caused by periostitis), osteomyelitis, arthritic changes, evidence of muscle pulls and tears (enthesopathies), anomalies, markers of anemic conditions and nutritional status (i.e., cribra orbitalia, porotic hyperostosis, and osteopenia), and traumatic injuries (antemortem and perimortem).

An osteolytic response was noted on the bones of five adults (7.3% of the total series), including four males and one female. An infectious process was responsible for lytic bone loss and external porosity in the left mastoid process of a young adult Indian male (fig. 11.5), which is enlarged relative to the right side. Postmortem damage has broken the tip, but internally the left mastoid process is mostly hollow, with only a thin cortical shell. Endocranially, the wall of the sigmoid sulcus is perforated, indicating spread of the infection into the brain cavity (fig. 11.6). These changes are consistent with mastoiditis, a severe, chronic osteomyelitis that can develop following a middle ear infection.

Treponemal disease is indicated in at least one and possibly as many as three individuals from the church. A young male exhibits multiple lesions, both active and well healed, in the cranium and postcranium (fig. 11.7). At least seven remodeled lesions were noted in the frontal bone, four in the left parietal, and one in the right parietal. Partial representation of the postcranial skeleton revealed localized areas of lytic resorption in the clavicles, scapulae, humeri, ribs, ilia, and sacrum. Although the cranial lesions are consistent with healed syphilitic lesions (remodeled caries sicca), the pattern represented by the postcranial skeleton does not fully support this diagnosis and is hindered by missing elements. In general, the postcranial infectious process followed a bilateral distribution that involved lytic resorption of localized areas of bone, with some predilection for metaphyseal and epiphyseal areas.

Two additional cases showing focal areas of lytic bone loss are present in older adult males. An Indian male represented only by a skull had three lytic lesions in the maxillary palate (fig. 11.8), two of which had penetrated into the nasal chamber. A European male had multifocal lytic destruction in the frontal and both parietal bones. In some locations the condition had penetrated the inner and outer tables (fig. 11.9). Because no postcranial elements were associated with this individual, a probable diagnosis was not assigned, but the condition seems most suggestive of carcinoma or mycotic infection.

Treponemal infections were identified in three crania from precontact occupations at Pecos (Hooton 1930). In these cases the frontal bone showed remodeled caries sicca, and in at least two individuals, the nasal cavity was involved and lytic lesions had perforated the palate. These changes were noted as being suggestive of syphilis.

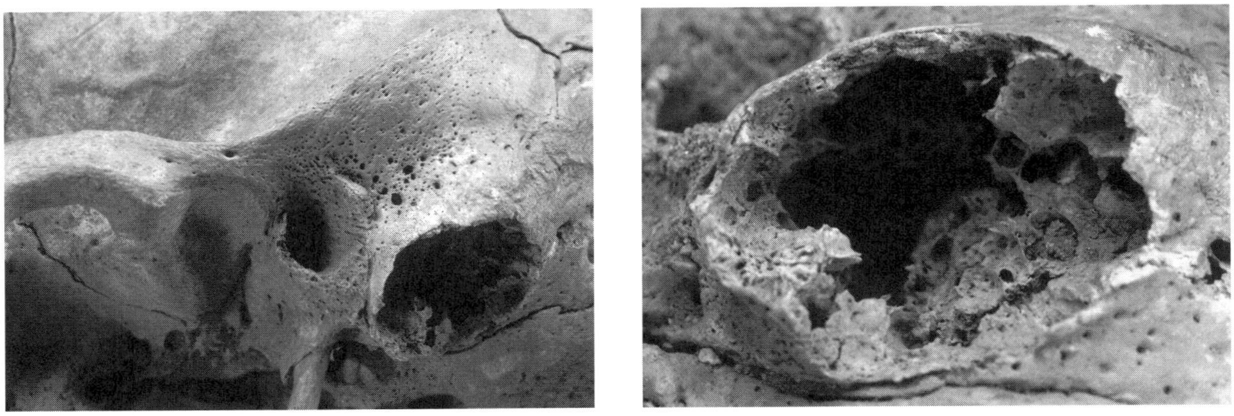

FIGURE 11.5. Enlarged left mastoid process (left) of an adult male (cat. no. 60651) showing marked porosity due to mastoiditis. Postmortem breakage of the tip has revealed a hollow internal cavity (right) attributable to lytic destruction of the sinus chamber. PMAE 25-3-10/60651.0.

Possible tubercular changes are evident in the skeleton of a young adult Indian female from the church (cat. no. 60779). Osteopenia, lytic rib and vertebral lesions, and vertebral fracture and ankylosis characterize this debilitating condition. Lytic lesions are present on the left ninth rib and the inferior end-plate of the ninth thoracic vertebra. The condition was accompanied by extreme osteopenia (loss of bone density) throughout the postcranial skeleton, marked flaring of the sternal rib ends, ankylosis of thoracic vertebrae 8, 9, and 10 with depression of the vertebral end-plates, and anterior wedging and fracturing of a partially represented twelfth thoracic vertebra. The condition clearly reflects a serious nutritional problem in addition to possible tuberculosis.

Tuberculosis was endemic in the Southwest, and high rates of the disease have been reported for pre- and postcontact populations, such as Chavez Pass and San Cristóbal, which had four adults and five adults, respectively, showing evidence of the disease (Stodder and Martin 1992). Hooton (1930) identified two skeletons in the larger Pecos series with Pott's disease, or tuberculosis of the spine. One was a middle-aged male from the Spanish colonial period. The other was a middle-aged female from the Glaze 4 period. Although changes consistent with tuberculosis in the church population were found only in the one young female, previous documentation regarding the condition (Ortner and Putschar 1981) has shown that skeletal manifestations of the disease are present in only a small percentage of the infected population.

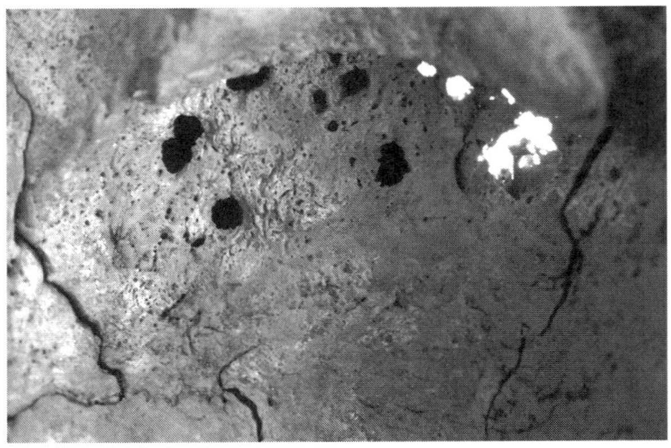

FIGURE 11.6. Perforated internal wall of the sigmoid sulcus posterior to the left petrous temporal of individual 60651. PMAE 25-3-10/60651.0.

Proliferative lesions, characterized by abnormal subperiosteal bone formation (periostitis), either woven (indicating an active process) or sclerotic (surface remodeling indicating a healing or healed condition), were evident on the bones of five individuals (7.3% of the total series): two young adult Indian males, one young adult European male, one older adult Indian male, and one subadult Indian female. The most severe case involved the adolescent, aged about 14 years. Her condition was characterized by widespread cortical surface formation of woven, subperiosteal, plaque-like bone on the left humerus, right radius, and the left and right ulnae, femora, tibiae, and fibulae. The tibiae displayed the most severe changes, with periostitis affecting the anterior, medial, and lateral surfaces of the entire diaphysis. Radiographs show no medullary involvement, but multiple Harris lines (lines of arrested growth) are present in the femora.

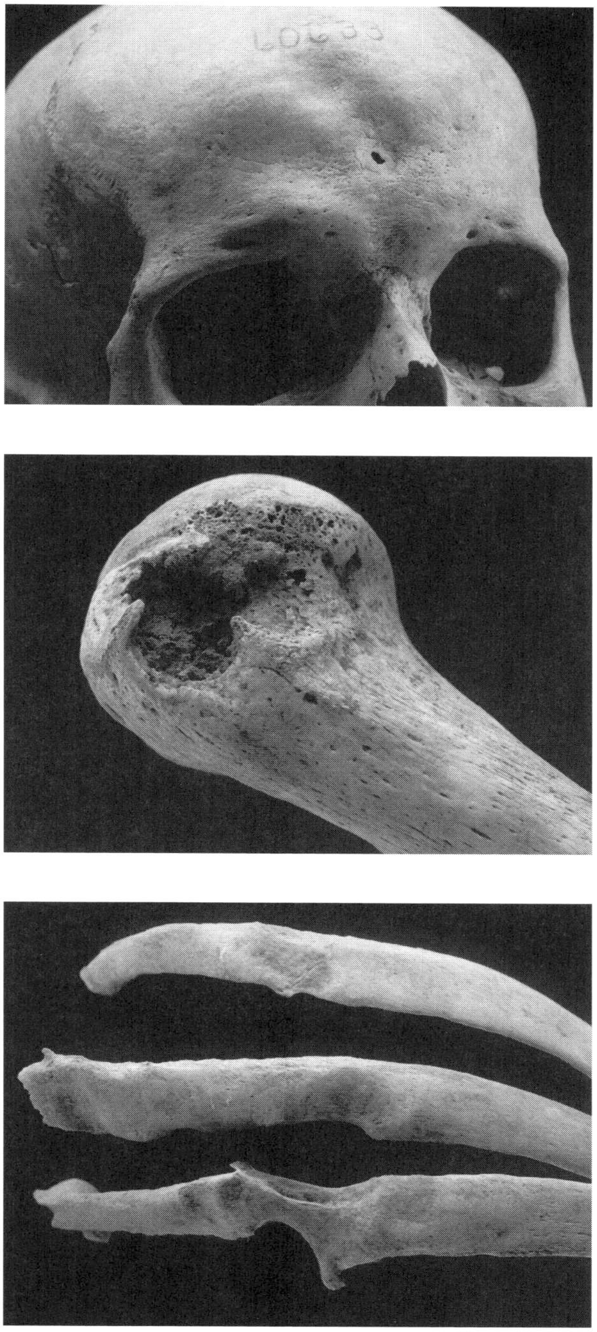

FIGURE 11.7. Remodeled lytic lesions in the cranial and postcranial bones of an adult male (cat. no. 60633). A lesion located above glabella (top) has perforated the frontal sinus. The greater tuberosity of the left humerus (middle) and right ribs 8, 9, and 10 (bottom) show localized areas of lytic resorption. PMAE 25-3-10/60633.0.

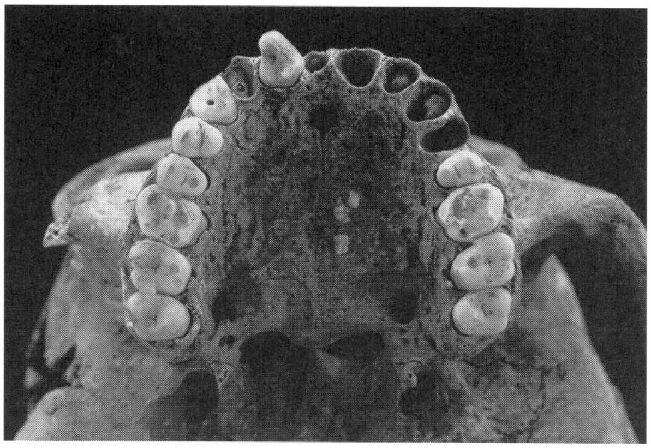

FIGURE 11.8. Lytic lesions in the hard palate of an adult male (cat. no. 60753). Lighter areas in the palate represent perforations into the nasal cavity. PMAE 25-3-10/60753.0.

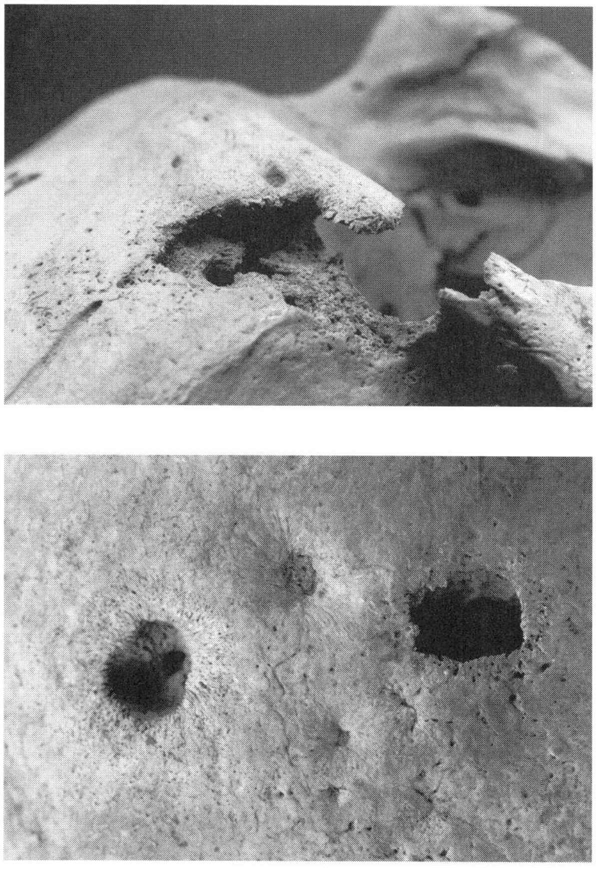

FIGURE 11.9. Frontal (top) and parietal bones (bottom) of an adult European male (cat. no. 60649) with lytic destruction. Remodeled cavitations and active lytic lesions are present. PMAE 19-6-10/60649.0.

Five long bones that were not associated with a specific cranium also displayed abnormal bone formation caused by periostitis. Four of these, left and right humeri and tibiae, exhibited severe forms of the condition and likely represented a single individual, because of consistency in relative size and bone preservation. Their diaphyses were enlarged because of extensive cortical thickening, and their surfaces were roughened with marked vascularity. Treponemal infection is a likely diagnosis for this set of remains. Their association with a specific cranium could not be established with certainty, so they are included in the following tabulation for the total series as individual bones.

New bone formation due to periostitis was noted on 2 of 31 left (6.5%) and 1 of 31 right (3.2%) humeri, 1 of 26 right radii (3.8%), 1 of 42 left (2.4%) and 1 of 39 right (2.6%) femora, 3 of 39 left (7.7%) and 3 of 36 right (8.3%) tibiae, and 1 of 14 left (7.1%) and 2 of 16 right (12.5%) fibulae. Therefore, percentages of long bones exhibiting abnormal, subperiosteal bone by element and side range from zero for ulnae and the left radius to 12.5 for the right fibula.

For purposes of comparison, abnormal bone formation and/or bone loss attributed to nonspecific infections (in this series characterized by mastoiditis and periostitis) is present in 8.8% (6 of 68) of the Pecos church individuals. This percentage is an underestimate because it is based on the total count of individuals, many of which were represented by a cranium only. This minimum assessment of bone infections in sets of church remains identified as individuals (excluding unmatched isolated elements) is smaller than that reported for the postcontact New Mexico skeletal series of San Cristóbal (23%) and Hawikku (36%) (Stodder 1994) but larger than the percentages of skeletal infection presented for other Southwestern prehistoric and early contact period sites (Stodder and Martin 1992). Hooton (1930) reported bone infection in a small percentage of 503 Pecos skeletons from the Black-on-white through Glaze 6 periods. Thirteen skeletons were documented as showing evidence of periostitis (2.6%), and there were four cases of osteomyelitis (0.8%). The later occupations at Pecos, represented by Glaze 5 and 6 skeletons, had six cases of periostitis out of 93 sets of remains (6.5%), but no examples of osteomyelitis. Relative to Hooton's data, our study indicates an increase in the incidence of skeletal infection during the postcontact period. Morgan (this volume), reassessing the entire Pecos series, reports larger percentages of skeletons with infection than were identified by Hooton (1930). In Morgan's study, evidence of infection (not ascribed to specific diseases or related to trauma) varies from 7% in the Black-on-white period to 13% in the Glaze 1 and 2 periods, 9.2% in Glazes 3 and 4, and 8.2% in Glaze 5 and later. Morgan additionally reports an 8% incidence of skeletal infection in the church series, a slightly lower estimate than that presented in this study.

Arthritic changes, characterized by bony lipping, porosity of a joint surface or margin, or both, were recorded in the bones of 31 adults (45.6% of the total series). Most changes were slight in severity, with older adults displaying the most advanced degeneration. Because several individuals were represented by a cranium only, it is again important to calculate the prevalence of arthritic changes with regard to specific long bones. Degenerative changes were noted in 9 adult left humeri (29%), 6 right humeri (19.4%), 7 left radii (28%), 7 right radii (27%), 10 left ulnae (38.5%), 8 right ulnae (30.8%), 7 left femora (16.7%), 6 right femora (15.4%), 6 left tibiae (15.4%), and 8 right tibiae (22.2%). Forty percent of adult crania showed erosion and/or porosity of one or both temporomandibular joints. Of the 17 adults with associated vertebrae, 12 individuals, or 70.6%, had some degree of arthritic degeneration. Of the 12, eight were aged older than 40 years.

Arthritis was the most common affliction noted by Hooton (1930) for the Pecos bones. For the 503 skeletons (adults and children) from the Black-on-white through Glaze 6 periods, Hooton reported that 13% had spondylitis deformans (spinal arthritis). He scored 4% of long and short bones as having degenerative changes and noted that these conditions increased over time, with later periods showing generally larger percentages of arthritic changes. Morgan (this volume) similarly identifies an increase in spinal column arthritis in the Pecos remains, from an incidence of 8.5% in the earliest series to a 30.6% occurrence in remains from the final stages of occupation. The percentages we present for the church skeletons follow this pattern but reflect a difference in scoring criteria for this condition.

Schmorl's depressions (or nodes) and spondylolysis were tabulated as a means of evaluating the severity of spinal column strain and trauma. Schmorl's depressions in intervertebral end-plates occur when axial loading and compression of the spine result in herniation of the nucleus pulposus of the vertebral disk into the centrum (Resnick and Niwayama 1978). Spondylolysis is defined as partial or complete separation of the neural arch from the vertebral body, usually in the region of the pars interarticularis (Mann and Murphy 1990). Although a genetic predisposition for this condition has been noted (Cyron, Hutton, and Troup 1976; Roche and Rowe 1952; Stewart 1956), its formation is generally a response to sustained stress on or trauma to the lower spine.

Multiple Schmorl's depressions are present in 5 of the 17 church individuals with associated vertebrae. Only males exhibited this condition, the youngest aged about 15 years. This male and two other young adults with Schmorl's depressions were identified as having European ancestry. One young adult and one older adult Indian male were also affected. The number of Schmorl's depressions in the vertebrae of the European males ranged from 8 to as many as 17. The indigenous males each had three affected vertebrae.

Spondylolysis was present in the fifth lumbar vertebra of five adult males: two young adults with European ancestry and one young and two older adult Native Americans. The two males identified as European also exhibited multiple Schmorl's depressions. Spondylolysis was not concurrent with Schmorl's depressions in the spines of Indian males.

The presence of Schmorl's depressions and spondylolysis only in males suggests that these conditions were caused by activities more limited to this segment of the population. Variation between European and Indian males in the severity and number of Schmorl's depressions also suggests differences in activities. Horseback riding has been cited as a factor contributing to the large number of Schmorl's depressions in Native American males from Nebraska (Sandness and Reinhard 1992). The occurrence of multiple Schmorl's depressions has also been noted in skeletal remains of Spanish noblemen dating to the early nineteenth century, which show additional markers attributable to horseback riding (Owsley et al. 2002). The high incidence of the condition, combined with the co-occurrence of spondylolysis in young adult European males, suggests that these men were equestrians. Although some pueblo natives were also experienced equestrians, the availability of horses to most Pecos inhabitants was limited, especially during the last century of the pueblo's occupation (Kessell 1979). Schmorl's depressions in the two Indian males likely resulted from labor such as packing heavy loads and lifting large bundles. Spondylolysis is also a reflection of strenuous activity and could have resulted from the bent posture associated with planting and harvesting. A similar posture has been associated with the development of spondylolysis in other Native American groups (Merbs 1983; Sandness and Reinhard 1992). Weaving, which was considered a male activity in Pueblo society (Kidder 1962), might have required the legs to be extended in a sitting position for long periods of time. The skeleton of an older Indian male (cat. no. 60663) has degenerative changes indicative of the repetitive motions of weaving, in addition to spondylolysis of the fifth lumbar vertebra.

Enthesophyte formations were noted on nine bones of four individuals. Four of the bones belonged to an elderly Indian male, aged 70+ years. Two other older adult Indian males and an older Indian female exhibited the condition.

Cranial synostosis, or premature closure of cranial sutures, was present in two individuals with Indian ancestry. A child aged about 11.5–12.5 years had premature closure of the left half of the coronal suture with distortion of the cranium characterized by enlargement of the right side of the vault. The second, a male aged 20–24 years, had early closure of the left lambdoid suture. This union occurred after skull growth had essentially ceased; the vault shape was normal. The few cases of premature suture closure noted for the larger Pecos Pueblo series were more severe forms of the condition. Hooton (1930) documented three adults and one child with marked scaphocephaly, reflecting early closure of the sagittal sutures. Morgan (this volume) identifies three additional crania from the larger Pecos series with scaphocephaly.

Cranial lesions characteristic of cribra orbitalia were observed in three juvenile crania, representing 4.4% of the total sample crania ($n = 68$) and 27.3% of individuals under 15 years of age ($n = 11$). Percentages of cribra orbitalia in the early to later occupations at Pecos reported by Morgan (this volume) vary little from that documented for the total church series; they range from 4.3% in individuals from the Black-on-white period to 4.1% in remains from Glaze 5 through later periods at Pecos. No cases of porotic hyperostosis were observed, although 15 of the 68 church crania (22%) showed some degree of microporosity in the outer table of the cranial vault. The highest incidence of porotic hyperostosis in the larger Pecos series reported by Morgan (this volume) is 4.1% in remains from the Glaze 1 and 2 periods. The incidence of cribra orbitalia and porotic hyperostosis in the Pecos church juveniles is low relative to the incidence in several other Southwestern groups, which have reported values as high as 60–80% for infants and children with anemic conditions, as inferred by the presence of this type of pathology (Stodder and Martin 1992). However, some sites in the Southwest, including Navajo Reservoir (16%), Gran Quivira (18%), Grasshopper

Pueblo (15%), and Arroyo Hondo (22%), show a lower prevalence of these conditions in children 0–10 years old (Stodder and Martin 1992). Although a correlation between conditions of anemia and other skeletal indicators of infection has been reported, none of the three adolescents from the Pecos church with cribra orbitalia was scored for evidence of skeletal infection. The small sample and incompleteness of the children's skeletons, however, undoubtedly affects the results for the series.

The etiology of cribra orbitalia and porotic hyperostosis was beginning to be recognized when Hooton published his analysis of the Pecos skeletons. He referred to these conditions jointly as "osteoporosis symmetrica" and discussed its presence in what he referred to as "a considerable number of crania of infants and children from the Pecos collection" (Hooton 1930:318). Adults from the pueblo were described as having healed forms of the condition, characterized by patches of pitted and thickened compact bone tissue. Of the 581 skeletons Hooton analyzed for the condition, 19 (3.2%) had definite evidence of the disorder, including 9 children and young adults (Hooton 1930:319). This result is not so different from that recorded for the church, and the percentage is much smaller than percentages reported for most other groups in the Southwest.

Senile osteoporosis was documented in two elderly Indian females and one elderly Indian male. Their condition was characterized by marked loss of cortical and trabecular bone. The young Indian female scored for tuberculosis also displayed widespread bone loss secondary to the chronic disease process.

Bone injuries were evaluated for features that distinguished them from postmortem deterioration and were identified as healed or perimortem and by location (i.e., on which bone) by two experienced observers. Traumatic fractures in the cranial vault (frontal, parietals, occipital, and temporals) were also measured and scored for the type of force that produced each defect (i.e., blunt, crushing, edged/bladed, or projectile). If an injury was caused by blunt force trauma, characterized by depressed bone, its shape was indicated (e.g., circular or oval). Crushing injuries represent more extreme examples of blunt force trauma, with increased severity and greater bone involvement. Edged, or bladed, fractures are characterized by sharp-force, linear penetrations in bone that often define the size and shape of the weapon used to inflict the trauma. Injuries identified as "projectile" represent trauma caused by arrows or bullets.

Traumatic fractures were common in the skeletons from the fourth Pecos church and were often extreme in severity. Although the most severe trauma was recorded for the bones of the skull, postcranial injuries were also noted. Seven healed and three perimortem injuries were present in adult long bones (table 11.10). No side preference was indicated by the limb fractures, and no postcranial injuries were observed in children. The occurrence of long bone fractures is higher than that reported by Hooton (1930) for 503 Pecos Pueblo skeletons examined for pathological conditions. In that study, 3.7% had long bone or pelvic bone fractures, with the highest percentage (5.8) noted in the Glaze 2 and 6 periods. Limb fractures in the church remains were noted in 2 of the 68 recognized individuals and in 5 sets of unassociated bones. On the assumption that all the bones exhibiting fractures belonged to one of the 68 sets of remains identified from the church, the frequency of long bone trauma by individual is estimated as 10%, more than double the percentage reported by Hooton. Morgan (this volume) reports a postcranial trauma percentage as high as 14.8 for Pecos remains from the Glaze 1 and 2 periods. Considering the partial representation of several individuals from the church and lower counts for long bones and other postcranial elements relative to skulls, the percentage of postcranial trauma is probably a marked underestimate for the church sample.

TABLE 11.10
PECOS CHURCH SERIES
LONG BONE FRACTURES

	Left		Right	
Bone	–/+	%	–/+	%
Humerus	29/2	6.4	30/1	3.2
Radius	24/1	4.0	25/1	3.8
Ulna	25/1	3.8	25/1	3.8
Femur	41/1	2.4	38/1	2.6
Tibia	38/1	2.6	36/0	0.0
Fibula	14/0	0.0	16/0	0.0

Key: –/+ = number of bones without fracture/number bones with fracture.

Support for this assertion was present in two relatively complete skeletons from the church with multiple injuries. A young adult male had trauma-induced ankylosis of the right sacroiliac joint and healed fractures in his right humerus and radius. He also had a perimortem bladed fracture in his ninth left rib and a perimortem conical puncture wound in his proximal right femur (fig. 11.10). Horizontal cuts on his frontal

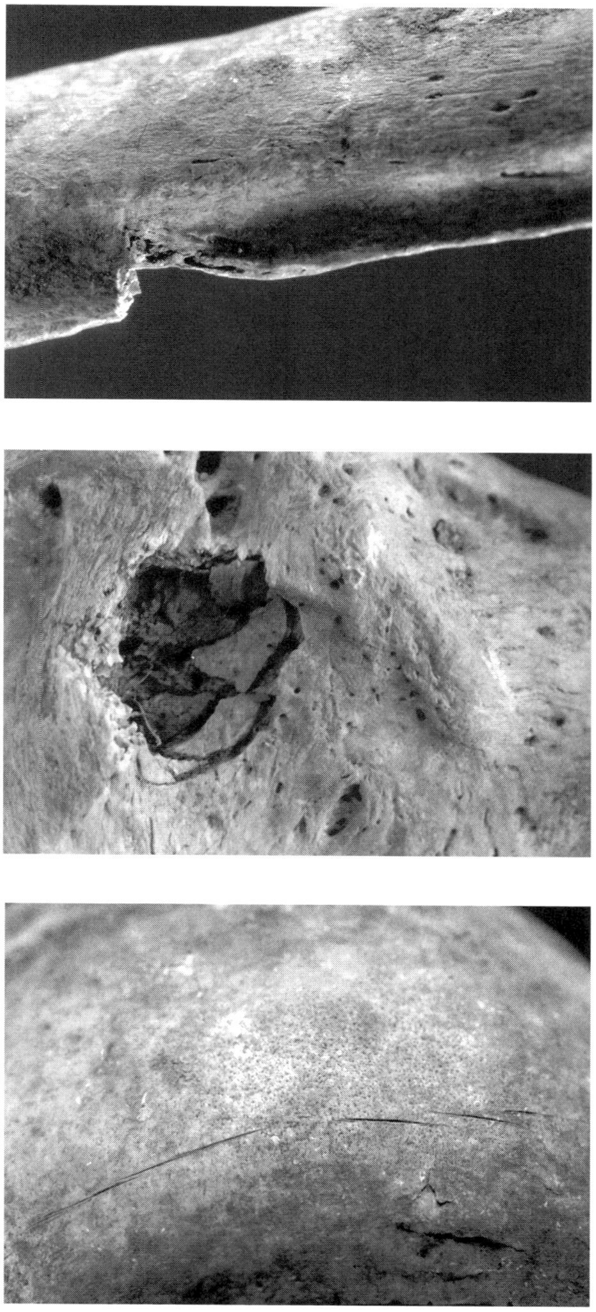

FIGURE 11.10. Perimortem trauma in the cranial and postcranial bones of an adult male (cat. no. 60711). A cut is present in the inferior border of a ninth left rib (top). The defect's left margin shows recent postmortem breakage. A conical puncture wound is present in the greater trochanter of the right femur (middle), and sharply defined transverse cuts extend across the frontal bone, indicating scalping (bottom). PMAE 25-3-10/60711.0.

bone revealed that he was also scalped. An older male exhibited first and second thoracic vertebrae with spinous processes that were short and irregular because of antemortem fracturing. This individual also had three depressed areas on his frontal bone, and the left half of his face was dramatically distorted by a blow that removed and displaced bones on that side of the head (fig. 11.11). This injury, caused by a blow from a celt-like, edged weapon, showed a remarkable degree of healing.

These examples were not unusual in the church series. A pattern of extreme violence and traumatic death is evident for children and adults of both sexes, as reflected by an exceedingly high rate of cranial trauma. These injuries were severe, not only in the type of fracturing that occurred but also in the number of injuries sustained per individual.

Of the 11 individuals aged less than 15 years, four (36%) had mortal cranial injuries. The youngest victim, a child aged 1.5–2.5 years, had a healed depression fracture on the occipital and a perimortem blade fracture in the left parietal. The next youngest victim, aged 5.5–6.5 years, had multiple bladed, perimortem injuries in the frontal, left maxilla, and left parietal. Two other children with mortal injuries were aged 10.5–12.5 years. One had a large fracture from a blow to the right side of the head (fig. 11.12) along with coarse cuts indicative of scalping. The other child exhibited numerous chops made by a bladed stone weapon, fragments of which were embedded in the left parietal (fig. 11.13).

Twenty-four of the 57 adult skulls from the church (42.1%) had skull fractures (table 11.11). A total of 63 healed and perimortem injuries were documented in the 685 represented cranial bones and mandibles (table 11.12). Thirty-three vault fractures (52.4% of the total adult cranial injuries recorded) and 30 facial fractures (47.6%) were present. The frontal bone had the highest incidence of injury (22.8%) for the adult sample, followed by the parietals (12.3%), nasals (11.3%), and maxillae (10.4%). The mandible showed the fifth highest trauma frequency (6.1%), followed by the zygomatic (4.8%), temporal (4.6%), and occipital (1.8%) bones. The frontal bone had the largest percentage of injury for both sexes, with males and females nearly equally affected (males = 24.3%, females = 20.0%). Whereas the parietals showed the second largest percentage of injury for males, the nasal bones had the second largest percentage of injury for females.

The majority of injuries were classified as perimortem, indicating a fatality. Twenty-six fractures (41.3%) showed evidence of healing. The number of

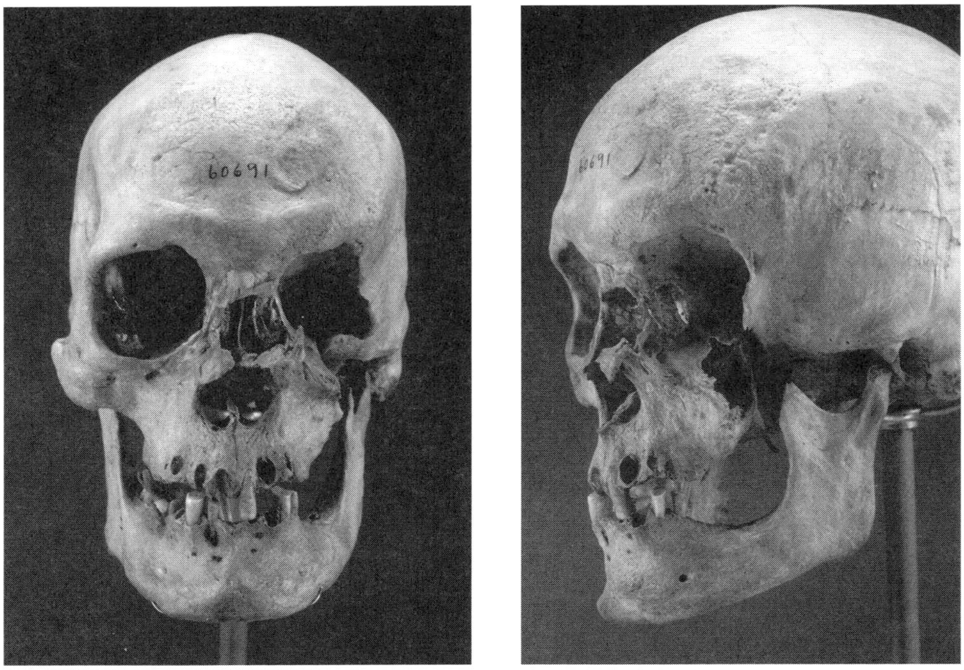

FIGURE 11.11. Frontal (left) and left lateral (right) views of the skull of an adult male (cat. no. 60691) with extensive antemortem trauma to the face. A severe blow crushed and displaced the nasal bones and the medial, lateral, and inferior borders of the left orbit. The lateral aspect of the supraorbital torus of the frontal bone, the left malar, and the zygomatic arch were detached, followed by extensive remodeling. PMAE 25-3-10/60691.0.

perimortem and healed defects for bones of the vault and those of the face differed only slightly. In the vault, 20 fractures, or 60.6%, were identified as unhealed, and 13 (39.4%) as healed. For the bones of the face, including the mandible, 17 (56.7%) fractures were perimortem and 13 (43.3%) were healed.

Sharp force trauma produced the majority of male and female vault injuries and accounts for most of the trauma observed in the parietal, occipital, and temporal bones (figs. 11.14, 11.15; table 11.13). Blunt force (fig. 11.16) and bladed trauma were equally represented in the frontal bones (46%). Projectile injuries were not observed in the skeletons from the church but were evident in postcranial remains from the surrounding pueblo in the form of stone points embedded in bone and one possible bullet hole in the innominate of a historic period male (Hooton 1930).

The percentage of males showing trauma to the head was more than two and a half times larger than the percentage of females showing trauma to the head (table 11.11), a statistically significant difference (X^2 = 5.09, df = 1, $p < .05$). Most of the individuals from the church exhibiting cranial trauma were older adult

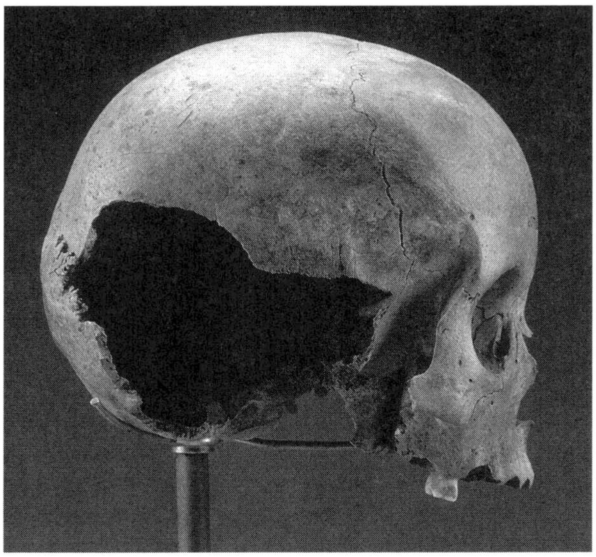

FIGURE 11.12. Perimortem fracture from a blow to the right posterior parietal of a juvenile female (cat. no. 60760). PMAE 25-3-10/60760.0.

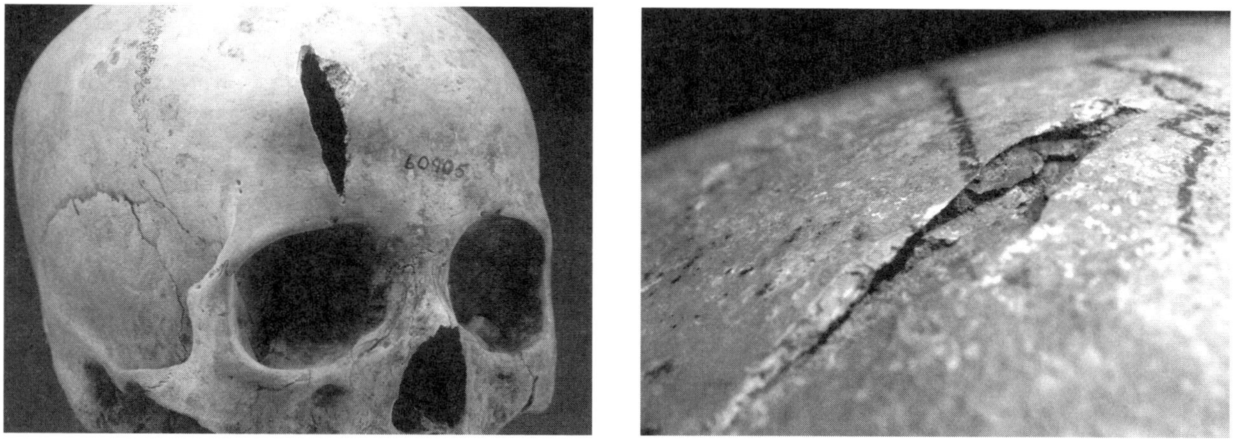

FIGURE 11.13. Perimortem bladed trauma in the frontal (left) and left parietal (right) bones of a juvenile female (cat. no. 60905). The size and shape of the weapon is defined by the defect in the frontal bone, which measures 43 millimeters in length, with a maximum breadth of 8 millimeters. A 10-millimeter-long fragment of chipped stone remains embedded in the bladed fracture of the left parietal. PMAE 27-6-10/60905.0.

males (45.8%), followed by young adult males (37.5%), older adult females (12.5%), and young adult females (4.2%). In addition to having a larger percentage of individuals with cranial trauma, males exhibited more cranial injuries per individual. Fifty-five of the 63 recorded defects (87.3%; table 11.12) were present in 20 males, for an average of 2.8 defects per injured male. Eight skull injuries were noted in four females, for an average of two defects per injured female. Males also showed nearly equal proportions of remodeled and perimortem injuries (table 11.12). In contrast, only one of the eight injuries recorded for females was healed, this being a depressed fracture on the frontal bone.

When examined by ancestry, the great majority of skull injuries (92%, n = 58) were present in Native Americans. Three fractures were recorded for individuals of mixed ancestry. Individuals of European ancestry exhibited two healed injuries (table 11.11). Relative to their numbers, individuals identified as having mixed ancestry were the most affected group, with

TABLE 11.11
ADULTS WITH SKULL INJURIES, BY ANCESTRY, AGE, AND SEX

	Total adult trauma		Female trauma		Male trauma	
Ancestry/Age (yrs.)	–/+	% with injuries	–/+	% with injuries	–/+	% with injuries
Indians	25/19	43.2	13/3	18.8	12/16	57.1
15–34	14/6	30.0	5/0	0.0	9/6	40.0
35+	11/13	54.2	8/3	27.3	3/10	76.9
Europeans	6/2	25.0	1/0	0.0	5/2	28.6
15–34	5/2	28.6	0/0	—	5/2	28.6
35+	2/0	0.0	1/0	0.0	1/0	0.0
Indian/Europeans	2/3	60.0	2/1	33.3	0/2	100.0
15–34	1/2	66.7	1/1	50.0	0/1	100.0
35+	1/1	50.0	1/0	0.0	0/1	100.0
Total sample	33/24	42.1	16/4	20.0	17/20	54.1

Key: –/+ = number of individuals without skull injuries/number of individuals with skull injuries.

TABLE 11.12
SKULL INJURIES BY BONE AND SEX FOR CHURCH ADULTS

	Females						Males					
	Trauma		Healed		Perimortem		Trauma		Healed		Perimortem	
Bone	−/+	% [b]	*n*	% [c]	*n*	% [c]	−/+	% [b]	*n*	% [c]	*n*	% [c]
Vault												
Frontal	16/4	20.0	1	25.0	3	75.0	28/9	24.3	6	66.7	3	33.3
Parietal [a]	39/1	2.5	0	0.0	1	100.0	61/13	17.6	5	41.7	8	66.7
Occipital	19/0	0.0	0	0.0	0	0.0	36/1	2.7	0	0.0	1	100.0
Temporal	38/0	0.0	0	0.0	0	0.0	66/5	7.0	1	20.0	4	80.0
Total vault	112/5	4.3	1	20.0	4	80.0	191/28	12.8	12	42.9	16	57.1
Face												
Zygomatic	36/0	0.0	0	0.0	0	0.0	63/5	7.4	1	20.0	4	80.0
Maxilla	35/1	2.8	0	0.0	1	100.0	60/10	14.3	5	50.0	5	50.0
Nasal	34/2	5.6	0	0.0	2	100.0	60/10	14.3	7	70.0	3	30.0
Mandible	8/0	0.0	0	0.0	0	0.0	23/2	8.0	0	0.0	2	100.0
Total face	113/3	2.6	0	0.0	3	100.0	206/27	11.6	13	48.1	14	51.9
Total	225/8	3.4	1	12.5	7	87.5	397/55	12.2	25	45.5	30	54.5

Key: −/+ = number of bones without skull injuries/number of bones with skull injuries.

[a] Paired bones (parietal, temporal, zygomatic, maxilla, and nasal) are combined for total counts.

[b] Percentage of bones with skull injuries.

[c] Percentage of fractures per bone by sex.

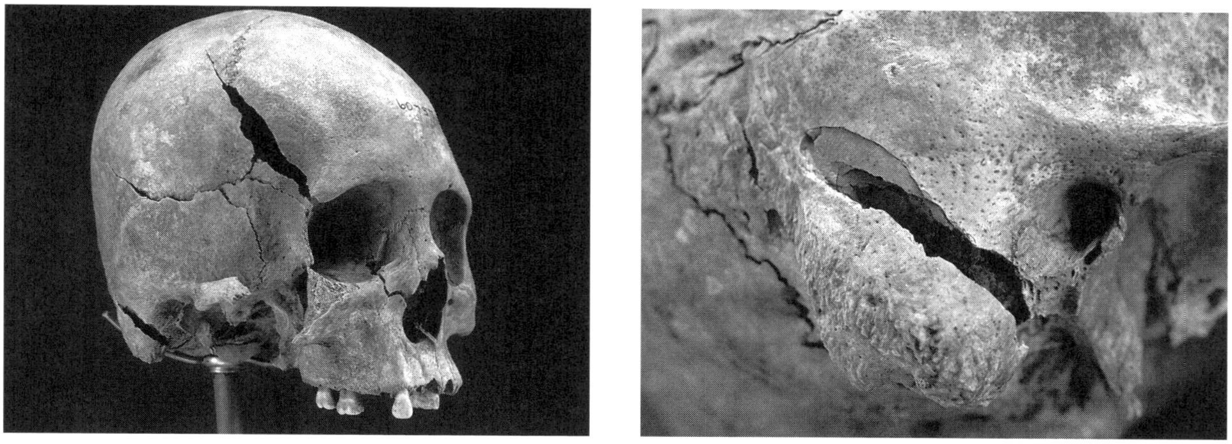

FIGURE 11.14. Similarly oriented sharp-force injuries in the right frontal (left) and mastoid process (right) of an adult male (cat. no. 60757). The penetrating defect in the frontal measures 72 millimeters in length by 10 millimeters in maximum breadth. The mastoid fracture measures 33 millimeters in length and 4 millimeters in breadth. PMAE 25-3-10/60757.0.

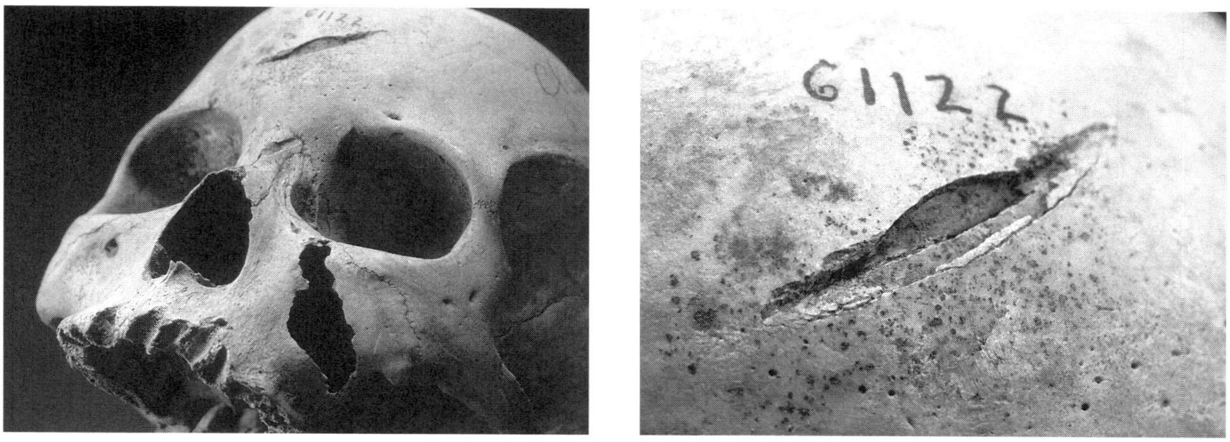

FIGURE 11.15. Adult female (cat. no. 61122) with multiple sharp-force fractures (left) in the forehead, nasal, and left mid-face. A close-up of the frontal injury (right) shows inward crushing. Below this primary fracture is a smaller defect produced by a secondary impact when the weapon skipped following the initial blow. PMAE 25-3-10/61122.0.

three of the five scored for cranial trauma. Individuals of European ancestry were the least affected group.

Cuts on the cranial vault also are indicative of violence, and if unhealed, they signify traumatic death followed by scalping. Two young adult Native American males, one young adult European male (fig. 11.17), and one young adult male of mixed ancestry displayed fine cuts caused by scalping. With the exception of the European, the young adult males had additional evidence of perimortem injuries and traumatic death. In one (cat. no. 60751), cuts on the posterior margin of the foramen magnum indicated decapitation. One juvenile female (cat. no. 60760) also had numerous coarse cuts on the superior cranial vault, indicative of scalping.

Taking into account individuals with evidence of scalping, injuries scored as perimortem, or both, at least 4 children and 14 adults in the Pecos church series (26.5%) died of traumatic injuries. Among the adults (table 11.14), these individuals included 11 Native Americans (79% of adult trauma deaths), 2 persons of mixed ancestry (14%), and 1 European (7%). Two Native American females and one female of mixed ancestry died from mortal injuries, resulting in

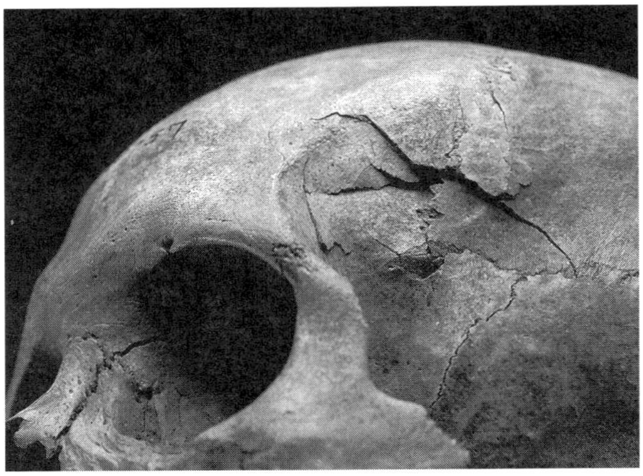

FIGURE 11.16. Perimortem depression fracture caused by blunt force trauma in the left anterior cranium of an adult male (cat. no. 60757). PMAE 25-3-10/60757.0.

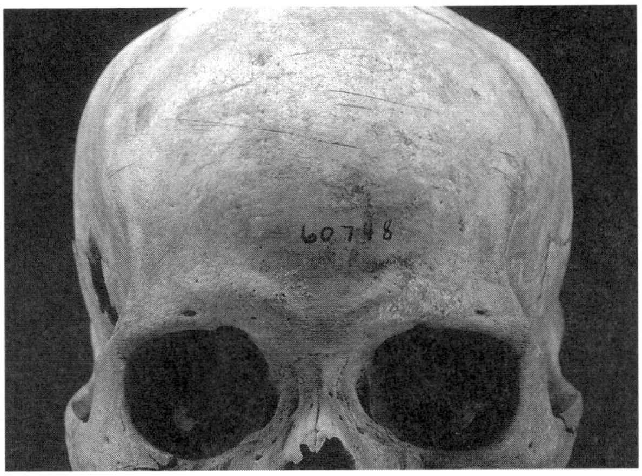

FIGURE 11.17. Fine, transverse and diagonally oriented cuts across the frontal bone of a young adult European male (cat. no. 60748), caused by scalping. PMAE 25-3-10/60748.0.

a 15% incidence of adult female traumatic death in the church sample. Nine Native American males, one European male, and one male of mixed ancestry reflect an incidence of male traumatic death of 29.7%. Of the 29 young adults represented in the study, 9 (31%) had perimortem injuries, and 5 of the 28 older adults (18%) showed evidence of perimortem trauma. Young adult males were the most likely to have experienced perimortem trauma.

The occurrence of skull injuries in the church series is high relative to the skeletal sample from Pecos's earlier occupations and to most other North American groups. Hooton (1930) reported 9%, 51 out of 581 Pecos Pueblo skeletons, as having skull fractures or probable traumatic skull lesions. The largest percentages were noted for the earliest and latest periods, represented by the Black-on-white and Glaze 1 remains (12%) and the skeletons from Glazes 5 and 6 (16%), respectively. A lower incidence of cranial trauma was reported for Glazes 2–3 and Glaze 4, 3.5% and 8%, respectively. Although Hooton offered warfare as an explanation for the cranial injuries of males, he attributed the five female cases of skull fracture to accidental falls, collapsed dwellings, or marital disputes. Morgan (this volume) documents smaller numbers of cranial traumas for the Pecos series. The smallest percentage of cranial trauma was noted in remains from the Black-on-white period, with percentages increasing to 5.1 in skeletons from the Glaze 5 period and later.

TABLE 11.13
ADULT CRANIAL VAULT TRAUMA, BY BONE AND TYPE OF FORCE

			Blunt force		Crushing		Edged		Projectile	
Bone	*n*	Total fractures	*n*	%	*n*	%	*n*	%	*n*	%
Frontal	57	13	6	46.2	1	7.7	6	46.1	0	0
Parietal[a]	114	14	5	35.7	2	14.3	7	50.0	0	0
Occipital	56	1	0	0.0	0	0.0	1	100.0	0	0
Temporal	109	5	0	0.0	1	20.0	4	80.0	0	0
Total	336	33	11	33.3	4	12.1	18	54.5	0	0

Note: % = percentage of total fractures.

[a] Paired bones (parietal and temporal) were combined for a total count.

TABLE 11.14
ADULT TRAUMA DEATH, BY ANCESTRY, AGE, AND SEX

Ancestry/Age (yrs.)	Total adult trauma death		Female trauma death		Male trauma death	
	–/+[a]	% trauma death	–/+	% trauma death	–/+	% trauma death
Indians	33/11	25.0	14/2	12.5	19/9	32.1
15–34	14/6	30.0	5/0	0.0	9/6	40.0
35+	19/5	20.8	9/2	18.2	10/3	23.1
Europeans	7/1	12.5	1/0	0.0	6/1	14.3
15–34	5/1	16.7	0/0	0.0	5/1	16.7
35+	2/0	0.0	1/0	0.0	1/0	0.0
Indian/Europeans	3/2	40.0	2/1	33.3	1/1	50.0
15–34	1/2	66.7	1/1	50.0	0/1	100.0
35+	2/0	0.0	1/0	0.0	1/0	0.0
Total sample	43/14	24.6	17/3	15.0	26/11	29.7

[a] –/+ = number of adults not scored for trauma death/number of adults scored for trauma death.

Stodder (1994) listed the percentages of cranial trauma from other Southwestern sites in her analysis of two New Mexican protohistoric villages. The incidence ranged from an absence of injury at Arroyo Hondo Pueblo to a high of 15% for San Cristóbal. Forty-one percent of Gallina skeletal remains were reported as having trauma (Stodder and Martin 1992), but it is unclear what percentage of this trauma was cranial versus postcranial. Percentages of cranial trauma documented for central California range from 2.1% (Newman 1957) to 4.4% (Jurmain 2001). Coastal and island skeletal series from southern California and Baja California show larger percentages of cranial trauma: 19.3% (Walker 1989) and 30% (Tyson 1977), respectively. Whereas Tyson related the large percentage of healed cranial injuries in the Baja sample to self-inflicted wounds, Walker attributed the high prevalence of cranial trauma in his sample to interpersonal violence. Violence has also been offered as the explanation for the high incidence of skull injuries (23.4%) in historic period Native Americans from the Great Plains (Owsley and Bruwelheide 2008). Equestrian tribes from this region had an especially high prevalence of skull fractures (33%). A majority of these defects were attributed to conflict, as evidenced by high frequencies of injuries, particularly perimortem trauma, the nearly equal distribution of fractures in the vault and face, standardization in shape or appearance of the defects, and more frequent occurrence in young adult males, particularly equestrian males, for which 41% of the sample was affected.

A similar pattern of skull fractures is evident in the series from the fourth church at Pecos. The number of individuals with skull injuries is high, and several individuals exhibit multiple defects. When tabulated by bone, the injuries are almost evenly divided between the vault (frontal, parietals, occipital, and temporals) and the face (maxillae, nasals, zygomatics, and mandible), and there is a clear predilection for frontal, parietal, maxillary, and nasal fractures. The high incidence of perimortem trauma on the anterior-superior vault and especially the face is highly suggestive of injuries sustained in hand-to-hand combat. Stewart (1973) noted a similar correlation between facial injuries and the strategy of deadly combat practiced by the Mohave Indians of California.

Violence, not accidental injury, is also supported by the type of trauma observed. Stone and metal blade injuries accounted for the majority of fractures in the cranial vault and face. Typically, these were penetrating defects with the shape of the weapon bit often delineated by the wound. These types of injuries, the context of the remains, and the absence of gunshot wounds argue against conflict with Spaniards and are indicative of tribal raiding and warfare. All segments of the mission population were victims. The greater number of adult males with injuries, particularly males of Native American and mixed ancestry, reflects greater exposure to violence. Occasionally these men survived the assaults, as is indicated by the relatively frequent occurrence of healed traumas in males. For women and children who were the unfortunate victims of such violence, survival was less likely.

DISCUSSION

Life was precarious for the people of Pecos Pueblo during the eighteenth and early nineteenth centuries. This period in Southwest history was highly dynamic and characterized by changing demographic profiles, epidemic diseases, and warfare. For Pecos, which had long served as the gateway between the pueblos and the southern plains, the events of the eighteenth century were especially traumatic, as is reflected in its severe population decline over time. The total population numbered around 2,000 at the time of contact in the sixteenth century, judging from census data compiled by Spanish authorities, sacramental records, and personal writings of the missionaries (Kessell 1979; Simmons 1979). By the beginning of the eighteenth century, this number had decreased by half. In the 1790s only about 150 Indians remained. During the same decade, individuals identified as Hispanic, or Hispanos, appeared on the census rolls for the Pecos River valley, numbering approximately 150. In 1821, 54 Indians were listed in the census, and Hispanos numbered 738.

Multiple factors were responsible for the abandonment of Pecos Pueblo, but the specific causes and the process of decline are subject to debate. Historic documents (Kessell 1979) and sacramental records (Levine 1999; Levine and LaBauve 1997) have been studied in order to address this question, but until now no comprehensive examination of the physical remains of the mission people who lived and died during the terminal century of the pueblo has been completed. Yet some factors related to health, such as evidence of specific diseases and nutrition, "can be directly evaluated only through the study of human biological remains" (Stodder and Martin 1992:55). In this regard, the 68 individuals from the final church offer revealing data about the people and events of Pecos, in terms of both demography and the degree to which the major factors used to define its eighteenth-century occupation—Spanish encroachment, disease, and violence—contributed to its abandonment.

Spanish Encroachment

The demographic structure of the church series confirms a Spanish presence at the pueblo, but it is limited and does not reflect the large number of non-Native immigrants who settled in the Pecos Valley during the last years of the eighteenth century and the early nineteenth century. Only 10 individuals (15% of the total series) were identified as having European ancestry, and individuals with mixed European and Native American ancestry represent a smaller percentage of the burial sample (9%). These percentages do not correspond to reported census data for the area, at least for the last 50 years of the pueblo's occupation, and are seemingly inconsistent with Pecos's sacramental records. According to Levine's (1999) summary of records from 1695 to 1829, Pecos natives made up 57% of the burial entries. Residents of the outlying Pecos Valley Hispanic communities, referred to in the records as *vecinos,* or neighbors, represent the remainder (43%), even though they first began appearing in the Pecos books only in 1795. If the non-Native population of the valley used the church interior for burial with regularity, then the European and mixed-ancestry components of the study sample should be proportionately larger. Even including individuals identified as having mixed ancestry, the non-Native component of the church series made up only 24% of the total sample.

Several variables might have affected the observed proportion of Indian to non-Indian remains in the church series. The objective of the archaeological investigation and the location of burial likely influenced the number of non-Indian remains excavated. For example, Kidder (1958) noted the presence of two coffins near the altar steps, presumably containing the remains of priests, that were not removed during the 1915 field season. Also for that field season, he mentioned the presence of three, likely non-Native adults that were buried "just outside the sanctuary" with several Mexican infants (Kidder 1958:305). He made no further reference to Mexican infants or any other non-Indian series, so these remains presumably were left in place. This area within the church, close to the altar, traditionally was the preferred burial location for the more devout and economically advantaged, given that higher fees were typically charged for burial there. Areas closer to the door were less costly (Di Peso 1974, vol. 3). The Hispanic community not only represented the more Christian segment of society but also generally had greater economic advantages than the Indians of Pecos and therefore greater desire for and access to preferred burial locations in the church. From this perspective, it would have been advantageous to this study if all remains near the sanctuary had been recovered in 1915, but only a select few were removed. Kidder's second excavation in the church was centrally located, possibly alleviating some of the bias toward a particular socioeconomic group. In the later excavation there is no reference to reburial of remains as was the case in 1915.

Despite these inherent biases, an alternative and more likely explanation for the lower-than-expected number of non-Indian remains in the sample is reduced use of the church structure by the late eighteenth century. Church and governmental data indicate that the functions of the mission greatly deteriorated during the eighteenth and early nineteenth centuries (Kessell 1979). By the latter half of the eighteenth century Pecos no longer had a resident priest but was a *visita* of Santa Fe. The difficulty of keeping a priest at the mission, which was considered remote and dangerous, is evident in the number of friars who served it. From 1704 through 1794, at least 58 missionaries served at Pecos (Kessell 1979), and as many as 70 missionaries served from 1694 to 1829 (Levine 1999). During the eighteenth century, most were temporary priests based in Santa Fe, where they conducted the majority of their business (Kessell 1979). The frequent and extended absence of a priest and the long distance to the pueblo prompted the rapidly expanding Spanish and mixed communities to establish their own places of worship in the early nineteenth century. The settlement of San Miguel del Vado, established in 1794, lay 18 miles east of Pecos Pueblo. As early as 1804 its residents petitioned for a chapel, because of the "long and perilous journey to Pecos" (Hayes 1974:17). In 1812 a resident priest was officially assigned to San Miguel and visited Pecos only intermittently. In 1827 the Church of San Antonio was established in the modern village of Pecos. This information suggests that by 1812 the sacraments offered to the Hispanic communities in the upper Pecos Valley were mostly performed outside of Pecos Pueblo, even though they were recorded in the mission's sacramental records until 1829, at which time the church center was moved to San Miguel del Vado.

Historical trends explain why so few non-Indian remains were found in the eighteenth-century burial series from the fourth Pecos church and the disproportionate number of males in the European group of the series. Despite the eighteenth century's having been called the "era of accommodation" (Sando 1992:77–78), few non-Natives ventured into the Pecos Valley before the 1800s because of the threat posed by raiding southern Plains tribes. A peace treaty with the Comanches in 1786 and the offering of land grants in 1794 opened the area to Hispanic settlement, as is reflected in the sudden appearance of non-Natives in the census and sacramental records. Until this time the non-Native population at Pecos Pueblo consisted mostly of men, including a succession of priests servicing the mission, the occasional Spanish government official, or *alcalde,* assigned the task of monitoring trade and Indian relations, and a small number of presidio soldiers stationed for defense. Historical documents first mention the presence of a 15-man garrison at Pecos in the 1740s (Kessell 1979). Although a detachment camped at the pueblo as late as 1808 (Schroeder 1979b), it likely fluctuated in numbers and rotated along the New Mexico frontier. In 1760 only four Spanish soldiers were documented as living in the garrison (Adams 1954). Sixteen years later, the detailed report on Pecos by the visiting Fray Francisco Atanasio Domínguez (Adams and Chavez 1956) named only one Spaniard, *alcalde* José Herrera. Census data from 1790 name two, a man and his 13-year-old son. No Spaniards were listed four years later (Kessell 1979). Although no reference was made to soldiers being accompanied by families, the practice was not uncommon, and archaeological evidence at the site testifies to the presence of living quarters and structures attributed to a Spanish secular community (Hayes 1974). However, even if families were present, the Spaniards were few and their presence was evidently sporadic throughout the eighteenth century.

The small number of individuals of Spanish descent at Pecos and the societal roles they filled during most of the eighteenth century are consistent with the demographic profile of the European component of the church series, which was composed mostly of young adult males (6 out of 10). These results are in contrast to the mixed-ancestry component of the series, which had nearly equal representation of age and sex groups from adolescence to adulthood. Skeletal features indicative of horseback riding in three of the young adult European males provide corroborating evidence that these men were likely members of the Spanish garrison. Although the church series does not reflect the heavy influx of Hispanic settlers into the Pecos Valley, in many respects the demography of the church sample seems entirely consistent with the events and people of Pecos, especially during the eighteenth century.

Disease

Postcontact epidemic diseases, both before and during the eighteenth century, have been identified as the major factor responsible for the decline of most Native American populations, including those of the Southwest (Dobyns 1989, 1992; Reff 1992; Schroeder 1979b; Simmons 1979; Ubelaker 1988; Upham 1992). Although the timing of epidemics and their effects on the Pueblo Indians have been questioned (Palkovich 1994), multiple outbreaks were recorded for many of these communities during the eighteenth and nineteenth centuries, and they often had devastating effects (Stodder and Martin 1992).

In the case of Pecos, the demise of the population has been attributed largely to epidemic disease (Kessel 1979; Levine 1999). This argument is based on the Pecos sacramental records, which, in addition to listing numbers of persons who died during specific periods, sometimes listed causes of death. Although disease was not always cited, Kessell (1979:378) interpreted the "bunching of deaths" in the burial books as indicators of epidemic disease and suggested that 10 major epidemics struck Pecos from 1696 to 1826. Following this assumption, Levine (1999:64–65) estimated that epidemics accounted for about 22% of all entries in the burial records from 1694 to 1800 and 58% of deaths during the "epidemic years."

Mortality caused by acute infectious diseases is not clearly evident in the church series. Failure to identify the severe effects of epidemic disease suggested by the burial records is partly attributable to the fact that most of these infections were lethal without becoming manifest in bone. Attempting to detect changes in mortality patterns caused by epidemic disease, such as increased occurrence of infant or adolescent deaths, is also problematic for the church series, given the sample size and the clear bias in favor of adults. It is possible to detect a slight increase in the percentage of individuals with evidence of chronic conditions that affected the skeleton relative to Hooton's (1930) assessment for earlier periods at Pecos. The increase in bone infections noted for the church series may be even more significant in light of its lack of complete sets of remains and the likelihood of underrepresentation of pathological conditions that this implies. Nevertheless, infectious diseases and epidemics were just two of the factors that disrupted the demographic stability and health of the population.

Violence

Warfare also undoubtedly had a major effect on the health of the residents of Pecos during the eighteenth century. Just as the pueblo's role as a trade center along the eastern border of the Southwest and southern plains heightened people's exposure to epidemic diseases (Dobyns 1992), it also increased the risk of violence. The people of Pecos were sporadically engaged in warfare both before and after Spanish contact (Jones 1966; Schroeder 1979b; Simmons 1979). The church sample shows that violence during the final century of the pueblo was especially common and more deadly than at any other time in Pecos's history. The large proportions of children and adults with cranial trauma (36% and 42%, respectively) represent a more than twofold increase over the percentage of individuals with cranial injuries identified in Hooton's (1930) analysis of the Glazes 5 and 6 remains. Most skull injuries in the church series were made by celt-like stone weapons or hatchets, and many were unhealed, indicating the traumatic death of at least one in every four individuals. The pattern of trauma and consistency in the types of defects argue for conflict with indigenous enemies, notably Apaches and Comanches.

The Apaches and Comanches alternated as trading partners with and enemies of Pecos during the eighteenth and early nineteenth centuries. As enemies, they conducted numerous raids against the pueblo. Sacramental records list 36 raids resulting in 60 Pecos deaths between 1697 and 1829 (Levine 1999). Although the identity of the raiding party was not always given, Apaches were cited as responsible for the attacks from 1697 to 1728 and from 1790 to 1829. Comanche raids were listed for the years 1739 to 1786. Attacks ranged from small-scale raids resulting in scattered deaths to large-scale attacks on the entire pueblo. In both cases, consistencies can be found between the written records and the skeletal remains from the church. The 1697 decapitation death of a man by Apaches (Kessell 1979) is consistent with evidence of decapitation in an adult male from the church (cat. no. 60751). A major attack by Comanches in 1746 resulted in the deaths of more than a dozen people, including women and children. A boy killed during this attack was described as dying from "arrows and hatchet blows" (Kessell 1979:372). Kidder (1958) described the remains of an adolescent male excavated from the church in 1925 that had a chipped stone projectile point in his rib cage. Bladed trauma, such as that sustained from hatchet blows, characterized the majority of the church individuals with injuries and all of the juveniles with injuries. A second major attack by Comanches, in 1748, caused the deaths of 13 men, including 11 from Pecos, 1 civilian from Santa Fe, and 1 Apache living at the pueblo. All 13 were reportedly buried in the church that same day (Kessell 1979). Although these commonalities do not prove that the skeletal remains of the church series are those of the persons described in historical accounts, the osteological evidence from the church offers a striking parallel to documented eighteenth-century events at Pecos.

These recorded assaults were undoubtedly only some of the more notable attacks against the pueblo. More frequent and deadly attacks would have been required to justify the actions of the Pecos people and the Spanish government, who, in the first part of the eighteenth century, joined forces to combat the Apaches and later waged war on the Comanches. The threat was sufficient that by the mid-eighteenth century Pecos was fortified with defensive structures, including towers at the gates, and soldiers were stationed at the pueblo for its protection (Kessell 1979). Only in this context are statements like those made by Fray Domínguez during his visit to Pecos in 1776 logical. Domínguez reported that the irrigated land was of no use to the people of Pecos, because they were "so very much besieged by the enemy" (Adams and Chavez 1956:213).

Some scholars (Kessell 1979; Levine 1999; Levine and LaBauve 1997) have argued that raiding and warfare played only a minor role in Pecos's demise relative to other factors, particularly epidemic disease. Kessell (1979) believed that the assaults against Pecos were more newsworthy than they were deleterious in terms of numbers of deaths. Kessell (1979) and Levine (1999) cited sacramental records as the basis for emphasizing the secondary role of warfare and the importance of epidemic disease. They attributed epidemic and episodic deaths to persons listed as having died from disease, as well as to any clustering of an "unusually high number of deaths within a short period of time" (Levine 1999:60). In contrast, they recognized deaths from violence only when deaths were specifically identified as such in the sacramental records. By this tabulation, raiding accounted for only 4% of deaths from 1697 to 1829, versus 22% attributed to disease. Levine (1999) found additional support for the relative effects of disease versus violence in Hooton's (1930) analysis of skeletons from the pueblo, in which he documented an increase in disease frequency over time and little evidence for warfare.

The church series, which was excluded from Hooton's study, presents compelling evidence for the major effects of raiding on the population. The percentage of church individuals who died from trauma was much larger than the 4% recorded in the burial entries. This evidence argues that some clusters of deaths in the records may represent victims of raids. Furthermore, many casualties, particularly those occurring during more dangerous periods, likely went unrecorded, because sacramental records were kept only sporadically, and a friar might not have visited the pueblo during periods of heightened conflict. Also, as the aforementioned statement by Domínguez implies, raiding extended beyond the number dead; it had a profound influence on the well-being of the living. In this respect, Kessell (1979:371) was likely correct in stating that the disruption of trade caused by the persistent raids of the Comanches "probably figured more heavily in the decline of Pecos than all the notorious Comanche assaults put together."

Whether the attacks or the threat of violence was as significant as epidemic disease in the health and lives of the Pecos people remains uncertain. Just as the sacramental record may be biased in favor of epidemic disease, so the skeletal record may be biased toward violence and warfare, since traumatic injuries are readily identifiable in bone. Burial within the church itself may be biased in favor of trauma deaths, for there are accounts of victims of raids being buried in the structure. Burial within the church might have been a service offered to victims. Nevertheless, the skeletal pathology is dramatic and shows that warfare played an undeniable role in the decline of Pecos. The threat of violent death was significant for all members of the population, especially males. Not only did this violence affect those whose lives were lost and their immediate families, but the fear of attack undoubtedly threatened food supplies and the activities of daily life.

CONCLUSION

As this and other studies reveal, the multiple factors related to the decline of the population of Pecos Pueblo are expressed with varying degrees of clarity depending upon the resource used for the investigation. Biological, sacramental, and historical records offer illuminating data on the events and people of Pecos. Combined, these resources contribute greatly to our understanding of the pueblo's history.

In this final chapter on Pecos's history, the Spanish presence is confirmed by human remains from the church. The heavy influx of Hispanic settlers into the Pecos Valley was not detected, likely because most remains predate their arrival. Death from infectious disease is also written in the biological record of the fourth church. A pattern of declining health is supported by the prevalence of infections in the burial sample. Unfortunately, the demographic composition of the church series, the partial representation of a majority of skeletons, and the inability to detect many acute viral infections in bone do not allow the full story of disease prevalence to be told. In this regard, sacramental records from the mission are informative. Violence, more than any other factor, is reflected in the biological record of the fourth church and underscores the final chapter and decades of Pecos history. The high rate of traumatic death in the series shows that violence exacted a heavy toll on all segments of the population.

Although Kidder eventually promoted the theory that interrelated factors resulted in the abandonment of Pecos, at one point he cited the arrival of the Comanches in the southwestern Great Plains as the chief cause of the pueblo's demise (Kidder 1962). His theory of the nomadic band adopting a semiparasitic existence, the host being the pueblo towns possessing stores of corn, extended back to the early establishment of the Pueblo life-way. As Kidder postulated, no great incursions were necessary to destroy these communities. A consistent raiding force would have had far-reaching consequences, because "the parasite ultimately destroys its host" (Kidder 1962:336). The skeletal evidence from the fourth church at Pecos, representing a population already reduced in number and weakened by disease and famine, provides an example of this process.

Few skeletal collections from mission churches of the Southwest have been studied. Such samples have tremendous potential for elucidating the lives of the peoples and cultures that once sustained them and the effects of Spanish colonialism on indigenous populations. The Pecos church series contributes significantly to the record of mission populations and of this pueblo in particular. The church series augments the already well-documented skeletal record of the preceding periods of occupation and presents the final chapter in the Pecos biological record.

ACKNOWLEDGMENTS

This research was conducted with the cooperation and assistance of Michèle Morgan, Patricia Capone, and Steven LeBlanc of the Peabody Museum of Archaeology and Ethnology, Harvard University. The Peabody Museum graciously allowed access to and use of its facilities during the documentation. Chip Clark, Smithsonian science photographer, provided skeletal images for this chapter. Thanks are also extended to the many people who helped collect portions of the data, including Shannon Novak, Dana Kollman, and Parvene Hamzavi.

12 Conclusion

Michèle E. Morgan

The Native American Graves Protection and Repatriation Act (NAGPRA) has had a profound effect on recent studies of Pecos Pueblo and will affect those of the future. The May 1999 repatriation remains the single largest physical repatriation that has taken place to date under NAGPRA. It required coordination and cooperation among many institutions and Native American tribes. The several-year consultation preceding the repatriation provided opportunities for members of these institutions and Native American tribes to discuss many issues related to the history of Pecos Pueblo, the significance of the preserved cultural and physical remains of the people who lived there, and interactions between the people of Pecos Pueblo and many other Indian nations. These rewarding discussions, although sometimes difficult, aired many viewpoints and perspectives and provided a strong foundation for future collaborations and consultations among and between participating institutions and Native American nations.

The Peabody Museum of Archaeology and Ethnology (PMAE) and the Robert S. Peabody Museum of Andover, Massachusetts, were fortunate to receive grants from the U.S. Department of the Interior that funded the NAGPRA inventory and consultation. These grants supported NAGPRA-related research that resulted in improved intellectual control of the field notes, objects, and human remains and stimulated visits by Karin Bruwelheide, Douglas Owsley, Richard Jantz, Lane Beck, and their colleagues to pursue or augment previous research prior to repatriation. The Pueblo of Jemez was supportive of this work and, together with the PMAE and Robert S. Peabody Museum, desired to see the process accomplished in a careful and thorough manner before physical repatriation took place.

This volume is a product of the NAGPRA-initiated final examination of the human remains and artifacts excavated by A. V. Kidder and his colleagues. Because of the enduring importance of Kidder's and E. A. Hooton's landmark studies, the idea originated to publish the data gathered during the NAGPRA inventory, in order to update and augment previously published data on these collections. In addition, this volume includes perspectives from scholars who have studied many different facets of Pecos Pueblo, although the majority of the chapters report on research conducted on the human remains formerly curated at the PMAE.

The primary descriptive work on Pecos Pueblo and Forked Lightning was completed in the first half of the twentieth century by Kidder, Hooton, and their colleagues. Their research paradigm focused on typology. On the basis of glaze wares found at Pecos, Kidder developed a ceramic typology used to seriate ceramics chronologically over a 500-year period; it became a keystone for interpreting Southwestern ceramics (Elliott 1995; Kidder 1931, 1958; Willey and

Sabloff 1993). Hooton used morphometrics to study biological variation and sought to organize the observed variation into types (Hooton 1930). Both scholars recognized that Pecos Pueblo developed rapidly into a focal point in a complex trade network. Both attempted to assess the degree and complexity of this exchange and the implications of external influences on Pecos Pueblo—Kidder, culturally, through the study of objects found at Pecos, and Hooton, biologically, through the study of variation in cranial morphology.

Present understanding of Pecos Pueblo and the history of the upper Pecos Valley builds on the pioneering work of Kidder and Hooton. The development of new research tools and techniques such as stable carbon isotope analysis, cranial and dental nonmetric analyses, CT scans, and novel statistical approaches to analyzing biological distance—in conjunction with the availability of comparative data sets from the Southwest and other parts of the world—have yielded a body of multidisciplinary research based on an integrative holistic approach rather than a typological one.

The objectives of this volume were twofold: first, to review significant findings and interpret them within current scholarly contexts, and second, to present new data on and interpretations of the behavior and biology of the people who lived at and visited Pecos Pueblo. Scholars are no longer able to apply new research techniques to the human remains and associated funerary objects that have been reinterred; in future they will apply new methods of analysis only to existing data. Toward that end we have provided, from the PMAE records, considerable primary data about the human remains and the associated funerary objects excavated by Kidder and colleagues (Appendixes A–D). The complete osteological and NAGPRA inventory data are archived at the PMAE. Contributing authors generously supplied primary data so that it could be made available to interested scholars (Appendixes E–K).

BIOLOGICAL VARIATION AND SOCIAL NETWORKS

The origin of the Puebloan people who first inhabited the Pecos Valley in the twelfth century remains unknown. A number of fairly large Pueblo sites, including Rowe, Dick's Ruin, Loma Lothrup, Arrowhead Ruin, LA 267, LA 680, and Forked Lightning, developed in the upper Pecos Valley beginning in the late twelfth century (Cordell 1998; Spielmann, this volume; see fig. 2.1). Inhabitants of many of these sites may have moved to Pecos Pueblo shortly after its inception in the early fourteenth century. By the middle of the fifteenth century, Pecos Pueblo was the only occupied site in the upper Pecos Valley (Capone, this volume; Spielmann, this volume).

Interpueblo trade for raw materials and craft items began at Pecos Pueblo as soon as the village was established. Insights into the ritual customs of the region inform our understanding of the social networks in which Pecos engaged. Pecos Pueblo incorporated elements of both the "northern" ritual system, associated with black-on-white biscuit ceramic ware, and the more "southern" ritual system, termed the Southwestern Cult (Crown 1994), which was associated with red- and yellow-slipped glaze ware vessels (Spielmann, this volume). These two ceramic styles, biscuit ware and glaze ware, largely replaced the earlier black-on-white ceramic ware at Pecos. Both types of serving bowls persisted for centuries at Pecos Pueblo, although glaze ware rapidly became dominant over biscuit ware (Kidder 1958; Spielmann, this volume). The transition from the initial black-on-white ceramic ware to Glaze 1 ware was accompanied by a significant change in the composition of the population at Pecos Pueblo. Weisensee and Jantz (this volume), in a reanalysis of Hooton's cranial metric and nonmetric data, found measurable morphological change as well as an increase in intrapueblo biological variation. (Hooton [1930] concluded that the earliest occupants were morphologically different from the later occupants, but he did not discern that variability also increased.) Weisensee and Jantz suggest that the ceramic transition from black-on-white to Glaze 1 ware was accompanied by a period of assimilation of people. They offer the biological evidence to support Spielmann's observations, based on a substantial body of archaeological evidence, that initially the Rio Grande immigrants actively aggregated in large numbers.

For two centuries, beginning in the late 1400s, Pecos Pueblo developed special trading relationships with specific groups of Plains Indian hunter-gatherers. Pecos was not alone in this eastern trade; Gran Quivira and Picuris also developed trade relationships with Plains groups (Spielmann, this volume). Pecos Pueblo traded with pueblos to the west as well. This facilitated the exchange of goods along both east-west and west-east gradients. Cotton from the western pueblos was traded along with maize and ceramics for buffalo meat from the plains (Kellner et al., this volume; Lintz 1991; Spielmann 1991b, this volume).

The late-fifteenth- to early-sixteenth-century period associated with Glaze 4 ware is the time for which Pecos shows the greatest biological diversity as assessed via morphometrics. Weisensee and Jantz's analyses suggest that significant numbers of people were joining the pueblo from greater distances than had done so previously. A subsequent decrease in variability was accompanied by population decline and the development of the only in situ glaze ware change, from Glaze 4 to Glaze 5.

Surprisingly, Pecos Pueblo appears not to have shared close biological ties with other Rio Grande populations. Attempts to identify genetic links with nearby sites in the Rio Grande Valley using dental nonmetric trait analyses yielded no straightforward relationship between genetic, geographic, and linguistic proximity (Turner and LeBlanc, this volume). Although Pecos Pueblo does exhibit biological ties with San Cristóbal, it does not do so with other Rio Grande pueblos such as Puye and Kwasteyu (Kwastiyukwa). It is possible that Pecos Pueblo and San Cristóbal had similar founding populations, and there might have been an influx of people into the Pecos Valley from the Galisteo Basin in the early 1400s, as well as from Arroyo Hondo Pueblo (Levine 1999; Spielmann, this volume; Turner and LeBlanc, this volume).

The arrival of Spaniards in the region in the sixteenth century disturbed the social fabric among the Pueblo and Plains tribes. Hostile interactions with Plains Comanches and Apaches increased over the following century, and interpueblo relationships changed. Uneasy alliances developed between former enemies such as Pecos and Taos, whereas other networks, such as the long-standing Pecos and Tano alliance, became less stable following the Pueblo Revolt of 1680 (Kessell 1987). The Spaniards were resented for their imposition of Christianity as well as their substantial labor and tithing demands. Despite this, the people of Pecos increasingly relied on the Spaniards for protection against their enemies, particularly groups from the Comanche and Apache tribes, beginning in the late seventeenth century, following the return of the Spaniards after the Pueblo Revolt. The biological affinities of the individuals buried in the fourth mission church suggest that considerable admixture between Puebloans and Europeans took place during this period (Bruwelheide, Owsley, and Jantz, this volume).

DEMOGRAPHY

Population size at Pecos Pueblo reached a maximum of approximately 2,000 persons in the sixteenth century. By the early eighteenth century the population had declined to about 1,000 persons, and it fell to near 100 residents by 1800 (Kessell 1987:489ff.; Levine and LaBauve 1997). Estimates of population structure made on the basis of skeletal remains must factor in preservation and collection biases (Beck, this volume; Bello et al. 2006; Nelson et al. 1994). Beck (this volume) reports that nearly 10 times as many adults as children that were identified in the field were collected, and therefore all subadult age groups are underrepresented. However, the available dietary and health data are not in conflict with the depiction of a relatively well-nourished and healthy population, in comparison with other Southwestern Pueblo I through Pueblo IV and historic sites (Martin et al. 2001; Nelson et al. 1994; Stodder et al. 2002).

The mortality profiles of adult females and adult males from Pecos differ. Analyses by both Morgan (this volume) and Weisensee and Jantz (this volume) indicate that females tended to encounter greater risks of death at younger ages than males. This may be related to child-bearing costs for females. Morphological support for the observation of high female mortality during child-bearing years is provided by Tague (this volume). Tague's detailed study of the pelvis reveals that the average Pecos Pueblo female pelvis was quite platypelloid in inlet shape relative to the pelves of other Native American and non-Native populations. Pecos Pueblo females thus might have faced greater than average risk during childbirth. The relatively narrow anterior-posterior pelvic inlet shape observed in Pecos pelves would have created more difficult deliveries than a nonplatypelloid pelvis, in the absence of other confounding factors. Tague's findings support the interpretation that the higher mortality of adult females under 40 years of age might have been due, at least in part, to increased morbidity associated with parturition. In at least six cases, field notes indicated that females had been buried with newborns (Morgan, this volume).

Higher mortality rates among males 40+ years of age (Weisensee and Jantz, this volume), as well as among younger males after A.D. 1515 (the Glaze 5–Late group analyzed by Morgan), suggest that trauma resulting from conflict may have been a major factor in male mortality, especially after 1515. High perimortem trauma is readily evident in the mission church sample analyzed by Bruwelheide, Owsley, and Jantz. The demographic composition of individuals buried in the church was strongly biased toward males. Although a notable number of Caucasian and mixed Caucasian–Native American individuals were buried in the church, there were no Caucasians identified as female. This finding supports ethnographic records indicating that few Spanish women were present in the Pecos Valley before 1800, because of instability in the region that was generated primarily by recurring hostile interactions with Plains tribes (Bruwelheide, Owsley, and Jantz, this volume).

DIET

Maize agriculture provided the staple diet at Pecos. This was not unusual among Southwestern Pueblo populations of the time. Pecos Pueblo, however, sits at a high elevation, near the limits for maize agriculture, which is constrained by the length of the growing season (Capone, this volume). Despite these marginal conditions, the people of Pecos were able to grow enough maize to develop surpluses and maintain a considerable population for more than 500 years.

The maize diet was supplemented by meat, derived primarily from bison and turkey rather than from deer, antelope, and rabbits, as was typical of the Pueblo Indians (Kellner et al., this volume). Beans appear not to have been a substantial part of the diet. Stable isotope data presented by Kellner and colleagues indicate a remarkably consistent dietary signal at Pecos Pueblo over several hundred years, with no statistically significant differences between males and females. Developments and disruptions in trade relationships with Plains tribes, beginning in the mid-fifteenth century, and in trade and labor relationships with Spaniards after the early seventeenth century, had little effect on diet as assessed through bone chemistry.

Comparisons with other Southwestern Pueblo sites indicate that Pecos Pueblo relied on maize more than Grasshopper Pueblo did but less than Gran Quivira. Like Pecos, Gran Quivira exhibits isotopic stability in dietary composition from the pre–Plains trade period through the historic period.

ACTIVITY PATTERNS

Activity levels among Pecos Pueblo inhabitants were likely high from childhood throughout adulthood. Sex differences in bone structure were seen in both adolescents and adults, in both upper and lower limb skeletal elements (Ruff, this volume). The nature of the sex differences observed in adolescents, particularly in the deposition of endosteal bone, suggests that hormonal influences were contributing factors, along with activity levels. Pecos Pueblo exhibited greater sexual dimorphism in aspects of the structure of the femur than other North American agricultural populations. These biological differences support ethnographic observations of long-distance travel and running events among Pecos Pueblo males, which induced high anterior-posterior bending loads. Females apparently engaged in activities favoring greater medial-lateral bending loads (Ruff 1991, this volume; Ruff and Hayes 1983b).

HEALTH

Observed skeletal indicators of disease and trauma suggest that the population at Pecos Pueblo was relatively healthy in comparison with contemporaneous Southwestern Pueblo populations. Hooton's (1930) assessments are widely cited in the literature; in this volume, observations by Beck, Morgan, and Bruwelheide and colleagues supplement and refine Hooton's data. Overall, Pecos inhabitants appear to have sustained lower rates of nutritional insufficiencies and chronic diseases than residents of many other contemporaneous sites. Access to a clean water supply might have played a significant role in reducing the incidence of some types of diseases, such as infection-induced anemia, as suggested by Kellner and colleagues (this volume).

Although diet appears to have remained relatively constant throughout the occupation of Pecos Pueblo, both ethnohistoric records and study of the Pecos skeletal remains suggest that rates of contagious diseases and other maladies might have increased following contact with Spaniards. Available written records indicate that 12 epidemics of infectious diseases, including measles and smallpox, were responsible for 22% of all recorded deaths at Pecos Pueblo between 1694 and 1829 (Levine and LaBauve 1997). Beck (this volume) reports an overall increase in the incidence of trauma, infection, and arthritis in the Pecos skeletal remains. She suggests that the observed population decline of the late eighteenth century took place partly because of the emigration of healthy people.

Morgan also notes changes in the patterns of some biological indicators after A.D. 1515 that are further exemplified in the study of the church sample by Bruwelheide, Owsley, and Jantz. These patterns include increases in cranial trauma, dental caries, and mortality of young adult males. In addition, the incidence of spinal osteophytosis increases, especially among females. This may reflect a shift in labor patterns and greater frequency of heavy or repetitive tasks that exert marked stress on the vertebral column (Morgan, this volume). Females from the mission church, although a relatively small sample, have greater numbers of caries than males. During this time period Pecos was most similar to Grasshopper Pueblo, San Cristóbal, and Hawikku in its rate of dental disease, and it falls near the median of a comparative sample of Southwestern sites (Bruwelheide, Owsley, and Jantz, this volume). A change associated with food cultivation practices, preparation techniques, ingestion behaviors, or a combination thereof, rather than in the basic dietary regime, likely accounted for the increased incidence of dental disease.

CONFLICT

Conflict has been proposed as a significant factor influencing the decline of Pecos Pueblo (Bruwelheide, Owsley, and Jantz, this volume). Pecos had a considerable fighting force, as Spaniards recorded multiple times between 1541 and the early eighteenth century (Kessell 1987:489). Trauma death was more evident in the fourth mission church burials than during any other period of Pecos occupation; nearly a quarter of the adult crania examined had perimortem fractures (Bruwelheide, Owsley, and Jantz, this volume). The majority of injuries were delivered by stone and metal blades, which are more indicative of conflict with Comanches and Apaches than with Spaniards, even though tension with the Spaniards existed, especially over payments for services rendered (Spielmann, this volume). An increase in young male mortality was observed in Pecos Glaze 5–Late burials that might have been related to increased defensive conflict (Morgan, this volume). Kessell (1987) noted that the high incidence of raiding during the 1700s resulted in the Spaniards' building fortifications at Pecos Pueblo and in a loss of agricultural yields. Although Puebloan burials should record trauma deaths occurring at or near Pecos Pueblo, they might not reflect the results of offensive warfare activities away from the pueblo.

SYNTHESIS

Pecos Pueblo appears to have begun as a place of consolidation for people dwelling in the upper Pecos River valley. The arrival of glaze ware, replacing black-on-white ceramics, in the early fifteenth century apparently was accompanied by an influx of people who differed physically, as determined through metric analyses. By the mid-fifteenth century, Pecos Pueblo was the only occupied site in the upper Pecos Valley. Well situated for defense and living near a reliable water source, the population of Pecos Pueblo increased markedly, likely reaching a maximum of several thousand persons in the late fifteenth or early sixteenth century. Trade networks expanded beyond nearby Pueblo settlements to include Plains Indians. The greatest variability within the Pecos Pueblo population seems to have existed during the Glaze 4 ceramic stage, between approximately A.D. 1475 and 1515. From the Black-on-white ceramic stage through Glaze 4, despite the changes noted previously, certain aspects of Pecos Pueblo lifestyles appear to have remained fairly stable. These included diet, overall health, the mortality profiles of males and females, and activity patterns. Maize was a staple of the diet, and buffalo and turkey were primary sources of meat. Young adult females appear to have had higher rates of mortality than young adult males; it appears that childbirth was more difficult for Pecos Pueblo females than for females from other populations, judging from detailed studies of the pelvis. Activity levels were high throughout life.

Disruptions of these patterns took place after the Spaniards arrived, and particularly during the seventeenth century. Studies of those buried in the fourth mission church indicate poorer overall health as well as a greater incidence of death by trauma, particularly among young adult males. Most of the conflict appears to have been against other Native Americans, particularly Comanches and Apaches. Historical records indicate that alliance networks underwent a marked shift in the years following the Pueblo Revolt of 1680, and the residents of Pecos increasingly relied on Spaniards to protect them. Eventually, the cumulative effects of intense periods of disease, conflict, and drought reduced the number of inhabitants at Pecos Pueblo below sustainability. In a well-documented event, the final group of Pecos Indians left their pueblo to join the Pueblo of Jemez in 1838. Despite this merging of communities, Pecos traditions and tribal governmental officers continue.

Nearly three-quarters of a century later, academic scholars, led by Kidder, began an intensive study of Pecos Pueblo in order to gain understanding of the lifestyles of the people who had lived there over many centuries. Kidder and Hooton published synthetic volumes on their research that continue to be widely used (e.g., Kidder 1924, 1931, 1958; Hooton 1930). The study of Pecos Pueblo has advanced the field of American archaeology, particularly Southwestern archaeology, in fundamental and enduring ways, exemplified by the establishment of the annual Pecos Conference. Before another century had passed, the human remains, along with many associated belongings, were reinterred at Pecos National Historical Park.

REPATRIATION AND THE FUTURE OF PECOS PUEBLO

Pecos Pueblo endures as a living historical monument—for descendants who continue to practice traditions developed many centuries ago, for scientists studying the biological, social, and environmental factors that shaped the development and practices of the people of Pecos Pueblo (the Cicuye), and for those interested in understanding a part of the history of the American Southwest.

This volume presents research culminating from many years of study of the human remains and material culture of the people of Pecos Pueblo; the appendixes will allow future researchers to further analyze the primary data. The careful analysis of all the available primary sources of data relating to the human remains formerly curated at the PMAE between 1996 and 1998 would not have been accomplished without the NAGPRA mandate. The participating institutions and the Pueblo of Jemez desired to see this work done in full compliance with NAGPRA prior to repatriation. The physical repatriation was accompanied by ceremonial events attended by both Native Americans and invited members from museums and institutions. The spirit in which the entire repatriation took place marks a positive shift toward increasing cooperation between Native peoples and American academic establishments in assessing Native history through analysis of material culture and biological data. The lessons learned from Pecos Pueblo will positively inform current study and serve as a model for future collaborative work between academic researchers and Native communities.

Appendixes A–K

Data presented in these appendixes are also available for download at www.peabody.harvard.edu/pecos.

As a further resource, the Pecos expedition field cards and Hooton's Pecos data cards have been scanned by Peabody Museum of Archaeology and Ethnology Archives and may be requested in electronic format.

Appendix A

Information on Human Remains from the Upper Pecos Valley That Were Used in the Chapter 4 Analysis

Michèle E. Morgan

This appendix covers 1,127 individuals—1,017 from Pecos Pueblo and 110 from Forked Lightning—that were confidently associated with field numbers and were completely separated from other individuals. Additional data on pathologies for these 1,127 individuals are presented in Appendix C. Appendix B includes all other human remains from the upper Pecos Valley. The data for each Peabody Number continue from the left page to the right page.

Key to Column Heads	
Peabody number	Unique number assigned by the Peabody Museum of Archaeology and Ethnology
Object description	Brief, standardized description of human remains
FN	Field number
Specimen label	All numbers, words, etc., recorded on human remains. Notations in the form of "(MEM 11/96)" indicate that the author labeled the specimens on the given date.
Site	DR (Dick's Ruin), FL (Forked Lightning), LL (Loma Lothrup), PP (Pecos Pueblo), R (Rowe)
Intrasite provenience	Information quoted directly from field cards and other primary sources
Ceramic stage	Abbreviated, e.g., "B&W" (Black-on-white), "Pres. Gl. 2 or 3" (Presumably Glaze 2 or 3).
MNI	Minimum number of individuals
Age (yrs.)	Age estimate in years, or more generally as fetus, infant, child, subadult, adult
Ad	Adult. Describes developmental age on the basis of physical maturation of the remains determined by full skeletal maturation, which could include eruption of third molars and fusion of epiphyses and the basilar suture.

Su	Subadult. Describes developmental age on the basis of physical maturation of the remains determined by the eruption of the second permanent premolar or a postcranial age estimate of ≥ 12 years.
Ch	Child. Describes developmental age on the basis of physical maturation of the remains determined by the presence of complete deciduous dentition or a postcranial age estimate of ≥ 2.5 years.
In	Infant. Describes developmental age on the basis of physical maturation of the remains determined by long bone measurements and tooth maturation.
Fe	Fetus. Describes developmental age on the basis of physical maturation of the remains determined by long bone measurements and tooth maturation.
Unk. age	Unknown age. An insufficient number of features were present to allow determination of the physical maturation of the remains.
M	Identified as male.
M?	Identified as probably male.
F	Identified as female.
F?	Identified as probably female.
Unk. sex	Undetermined sex. An insufficient number of features were present to allow a confident determination of the sex of the individual.
Age criteria	Criteria used to assess age: D (dental stage as reproduced in White 1991), T (Todd), S (Suchey-Brooks), AS (auricular surface), PC (postcranial age based on epiphyseal fusion and/or metric data)
Sex criteria	Female (indicates all or nearly all criteria support female); F all features (all criteria indicate female); ?Female (majority of criteria support female); Male (indicates all or nearly all criteria support male); M all features (all criteria indicate male); ?Male (majority of criteria support male); Indet. (indicates that sex could not be confidently determined). Pelvic criteria abbreviated, "VA" (ventral arc), "SPC" (subpubic cavity), "IPR" (ischiopubic ramus), "SN" (sciatic notch), "PS" (preauricular sulcus); cranial criteria abbreviated, "NC" (nuchal crest), "MP" (mastoid process), "SM" (supraorbital margin), "SR" (supraorbital ridge), "ME" (mental eminence)
Pathology type	Presence of any of six categories of pathology: trauma, infection, arthritis, tumor, nutritional/physiological stress ("Nutr."), other (includes neurological and all other pathologies)
Pathology description	Brief notes describing pathology, abbreviated, e.g., "DJD" (degenerative joint disease), "TMJ" (temporal mandibular joint)
Cranial mod.	Cranial modification, abbreviated, e.g., "Sl. occ. def." (slight occipital deformation), "Lamb." (lambdoidal), "Unobs." (unobservable), "Not obs." (not observed).
Associated funerary objects	Objects listed on NAGPRA inventory, quoting directly from field cards (FCs). ID numbers preceded by "no." are field numbers; those in parentheses are Peabody Andover numbers.
Field context	Notes recorded on field card that pertain to context. "Agua Fria," "Pajaritan," and "Frijolito" refer to categories of pottery.
Notes	Additional descriptive notes

Peabody number	Object description	FN	Specimen label	Site	Intrasite provenience	Ceramic stage	MNI	Age (yrs.)	Ad	Su	Ch	In	Fe	Unk. age	M	M?	F	F?	Unk. sex	Age criteria	Sex criteria
00-0-10/N9468.101	Fragmentary cranial human remains of a child of unknown sex.	Bb17	Bb17	FL	Sta. 1, room bb4, depth 6′, height 1′	Probably B&W	1	9–13	0	0	1	0	0	0	0	0	0	0	1	D 8 dms present, 8 molars	Indet.
00-0-10/N9468.38	Fragmentary cranial human remains of an adult of unknown sex.	1992	1992	PP	1550E100, depth 4′6″, height 1′	Probably Gl. 2 or 3	1	Adult	1	0	0	0	0	0	0	0	0	0	1		Indet.
00-0-10/N9468.40	Very fragmentary cranial human remains of an adult of unknown sex.	Bb	9468-40 (MEM 11/96)	FL		Probably B&W	1	Adult	1	0	0	0	0	0	0	0	0	0	1		Indet.
00-0-10/N9468.42	Very fragmentary cranial human remains of a subadult of unknown sex.	Bb	9468-42 (MEM 11/96)	FL		Probably B&W	1	12–18	0	1	0	0	0	0	0	0	0	0	1		Indet.
00-0-10/N9468.43	Fragmentary cranial human remains of an adult of unknown sex.	Bb26	Bb26	FL	Room 15, depth 2′	Probably B&W	1	Adult	1	0	0	0	0	0	0	0	0	0	1		Indet.
00-0-10/N9468.44	Fragmentary cranial human remains of an infant of unknown sex.	Bb84	Bb84	FL	Trench j, depth 5′, on bottom	Probably B&W	1	1–2	0	0	0	1	0	0	0	0	0	0	1	D 4 dms not all erupted 1–2 yrs	Indet.
00-0-10/N9468.45	Fragmentary cranial human remains of a child of unknown sex.	Bb102	Bb102	FL	Room 79, nw corner, depth 72″, height–in grave cut into	Probably B&W	1	2–5	0	0	1	0	0	0	0	0	0	0	1	D 8 dms present	Indet.
00-0-10/N9468.46	Very fragmentary cranial human remains of an infant of unknown sex.	Bb	N9468-46 (MEM 11/96)	FL		Probably B&W	1	1.5–2.5	0	0	0	1	0	0	0	0	0	0	1	D 1.5–2.5 yrs	Indet.
00-0-10/N9468.47	Fragmentary cranial human remains of an adult of unknown sex.	Bb91	N9468-47 (bag with teeth MEM 11/96)	FL	Trench k, depth 14″, height ?	Probably B&W	1	Adult	1	0	0	0	0	0	0	0	0	0	1		Indet.
00-0-10/N9468.48	Fragmentary cranial human remains of a subadult of unknown sex.	Bb22	Bb22	FL	Trench c, depth 3′, height 2′	Probably B&W	1	Sub-adult	0	1	0	0	0	0	0	0	0	0	1		Indet.
00-0-10/N9468.49	Fragmentary cranial human remains of an adult of unknown sex.	Bb20	Bb20	FL	Trench c, depth ?, on bottom	Probably B&W	1	Adult	1	0	0	0	0	0	0	0	0	0	1		Indet.
00-0-10/N9468.50	Fragmentary cranial human remains of an adult of unknown sex.	Bb28	Bb28	FL	Trench c, subsoil, depth 4′8″, in scoop of bottom	Probably B&W	1	Adult	1	0	0	0	0	0	0	0	0	0	1		Indet.
00-0-10/N9468.52	Fragmentary cranial human remains of an adult of unknown sex.	Bb47	Bb47	FL	S4E13, depth 4′6″, on bottom	Probably B&W	1	Adult	1	0	0	0	0	0	0	0	0	0	1		Indet.
00-0-10/N9468.53	Fragmentary cranial human remains of an adult of unknown sex.	Bb50	Bb50	FL	Room 15, depth 4′, in scoop of bottom	Probably B&W	1	Adult	1	0	0	0	0	0	0	0	0	0	1		Indet.
00-0-10/N9468.54	Fragmentary cranial human remains of an adult of unknown sex.	Bb122	N9468-54 (MEM 11/96)	FL	N60E18, head at sw corner, depth 52″, height 8″	Probably B&W	1	Adult	1	0	0	0	0	0	0	0	0	0	1		Indet.
00-0-10/N9468.55	Very fragmentary cranial human remains of a child of unknown sex.	Bb31	Bb37 and bb31	FL	Trench c, bb31: dpth 2′, hght 3′ ± ?, bb37: dpth 18″, hght 2′	Probably B&W	1	Child	0	0	1	0	0	0	0	0	0	0	1		Indet.
00-0-10/N9468.56	Fragmentary cranial human remains of a subadult of unknown sex.	Bb46	Bb46	FL	Room 40, depth 4′6″, height ?	Probably B&W	1	10–15	0	1	0	0	0	0	0	0	0	0	1	D M2s erupted	Indet.
00-0-10/N9468.57	Fragmentary cranial human remains of a child of unknown sex.	Bb89	N9468-57 (labeled bag, MEM 11/96)	FL	Room 11, under base of n wall, depth 4″, scoop in bottom	Probably B&W	1	Child	0	0	1	0	0	0	0	0	0	0	1	D 4 dms	Indet.
00-0-10/N9468.58	Fragmentary cranial human remains of a subadult of unknown sex.	Bb97	N9468-58 (labeled, MEM 11/96)	FL	Trench l-k, depth 21″, height ?	Probably B&W	1	12–18	0	1	0	0	0	0	0	0	0	0	1		Indet.
00-0-10/N9468.59	Fragmentary cranial human remains of an adult of unknown sex.	Bb113	Bb113	FL	N56E10, depth 7′6″, height–grave dug into bottom	Probably B&W	1	Adult	1	0	0	0	0	0	0	0	0	0	1		Indet.

Peabody number	Pathology type	Pathology description	Cranial mod.	Associated funerary objects	Field context	Notes
00-0-10/N9468.101			Unobs.			
00-0-10/N9468.38	Nutr.	Enamel hypoplasia	Unobs.	B&W and Gl.1 red sherds, no numbers		
00-0-10/N9468.40			Unobs.			
00-0-10/N9468.42			Unobs.			
00-0-10/N9468.43			Unobs.			Field card suggests male
00-0-10/N9468.44			Unobs.			
00-0-10/N9468.45			Unobs.			Field card suggests male
00-0-10/N9468.46			Unobs.			
00-0-10/N9468.47			Unobs.	B&W bowl no. 209 (76649), 1000+ small white shell beads listed, no number		Field number problematic
00-0-10/N9468.48			Unobs.			
00-0-10/N9468.49			Unobs.			Field card suggests male, possible mixing with N9468-50 and N9468-51
00-0-10/N9468.50			Unobs.	Bone awl no. 71 (76550)		Possible mixing with N9468-49 and N9468-51 (added 2 teeth from N-9468-51, MEM 11/96)
00-0-10/N9468.52	Nutr.	Enamel hypoplasia	Unobs.			
00-0-10/N9468.53	Nutr.	Enamel hypoplasia	Unobs.			
00-0-10/N9468.54	Nutr.	Enamel hypoplasia	Unobs.			
00-0-10/N9468.55			Unobs.			Possible mixing with N-9468-56
00-0-10/N9468.56	Nutr.	Enamel hypoplasia	Unobs.			Possible mixing with N-9468-55
00-0-10/N9468.57			Unobs.			
00-0-10/N9468.58	Nutr.	Enamel hypoplasia	Unobs.			Field card suggests female
00-0-10/N9468.59			Unobs.	2 B&W bowls nos. 326 (76650), 327 (76651)		Field card suggests male

Peabody number	Object description	FN	Specimen label	Site	Intrasite provenience	Ceramic stage	MNI	Age (yrs.)	Ad	Su	Ch	In	Fe	Unk. age	M	M?	F	F?	Unk. sex	Age criteria	Sex criteria
00-0-10/N9468.60	Fragmentary cranial human remains of an adult of unknown sex.	Bb44	Bb44	FL	Room 37, se corner, depth 4′, on scoop of subsoil	Probably B&W	1	Adult	1	0	0	0	0	0	0	0	0	0	1		Indet.
00-0-10/N9468.61	Fragmentary cranial human remains of an adult of unknown sex.	Bb18	Bb18	FL	Trench c, depth head and knees 2′10″, height 1′6″	Probably B&W	1	Adult	1	0	0	0	0	0	0	0	0	0	1		Indet.
00-0-10/N9468.63	Fragmentary cranial human remains of an infant of unknown sex.	516	516	PP	1400E100, test xii, base of cut 6	Indet.	1	1–1.5	0	0	0	1	0	0	0	0	0	0	1	D 1.5 ± .5 yrs	Indet.
00-0-10/N9468a.12	Fragmentary cranial human remains of an adult of unknown sex.	1710	1710	PP	1250E0, trench e of kiva xii, depth 3′, height 2′+	Indet.	1	Adult	1	0	0	0	0	0	0	0	0	0	1		Indet.
19-6-10/59801.0	Nearly complete cranial and postcranial human remains of an adult male.	318	58901/318	PP	1250E100, depth 6′	Indet.	1	35–44	1	0	0	0	0	0	1	0	0	0	0	D f-g/g, T 9, AS 5	Male
19-6-10/59802.0	Nearly complete cranial and postcranial human remains of an adult female.	558	59802/558	PP	1550W50, depth 40″, height 6″ scoop in red clay	Probably B&W	1	35–44	1	0	0	0	0	0	0	0	1	0	0	D g/g, AS 5	Female
19-6-10/59803.0	Nearly complete cranial and postcranial human remains of an adult female.	490	59803/490	PP	1000E150, depth 71″	Indet.	1	35–49	1	0	0	0	0	0	0	0	1	0	0	D g/g, T 10, AS 5	Female
19-6-10/59804.0	Nearly complete cranial and postcranial human remains of an adult female.	541	59804/541	PP	1550W50, depth 16″	Probably Gl. 1 or 2	1	30–39	1	0	0	0	0	0	0	0	1	0	0	D f-g/f-g, AS 3	Female
19-6-10/59805.0	Nearly complete cranial and postcranial human remains of an adult male.	377	59805/377	PP	1550E100, depth 35″, height–above red clay 35″	Probably Gl. 5	1	35–45	1	0	0	0	0	0	1	0	0	0	0	D h/h, AS 4	Male
19-6-10/59806.0	Nearly complete cranial and postcranial human remains of an adult male.	374	59806/374	PP	1550E100, depth 77″, height–below red clay 6″	Probably Gl. 2 or 3	1	45–55+	1	0	0	0	0	0	1	0	0	0	0	D g-h/g-i, AS 7	Male
19-6-10/59807.0	Nearly complete cranial and postcranial human remains of a subadult that is probably male.	510	59807/510	PP	1400E100, in SE corner of test XII, cut 6, height 33″	Very probably Gl. 3	1	15–18	0	1	0	0	0	0	0	1	0	0	0	D b2-c, T 1, AS < 1, PC 15–18 yrs	?Male
19-6-10/59808.0	Nearly complete cranial and postcranial human remains of a subadult that is probably male.	619	59808	PP	1550W0, depth 73″, below red clay 13″	Indet.	1	14–16	0	1	0	0	0	0	0	1	0	0	0	D b1-b2, T 1, AS 1, PC 13–16 yrs	?Male
19-6-10/59809.0	Nearly complete cranial and postcranial human remains of an adult female.	583	59809	PP	1550W0, depth 6–23″, above red clay 48″	Indet.	1	18–24	1	0	0	0	0	0	0	0	1	0	0	D c/b2, T 1, AS 2	Female
19-6-10/59810.0	Nearly complete cranial and postcranial human remains of an adult female.	614	59810/614	PP	1550W0, north terrace, depth 11–19″, above red clay 33–49″	Pres. Gl. 6	1	50–59	1	0	0	0	0	0	0	0	1	0	0	D g/?, AS 7	Female
19-6-10/59811.0	Nearly complete cranial and postcranial human remains of an adult female.	489	59811/489	PP	1000E150, depth 50″, height–above red clay 16″	Probably Gl. 4	1	35–44	1	0	0	0	0	0	0	0	1	0	0	D g, AS 5	Female
19-6-10/59812.0	Nearly complete cranial and fragmentary postcranial human remains of an adult male.	556	59812/556	PP	1550W50, just below surface, above red clay 27″	Probably Gl. 1 or 2	1	35–49	1	0	0	0	0	0	1	0	0	0	0	D right side f, left side i	Male
19-6-10/59813.0	Nearly complete cranial and postcranial human remains of an adult female.	564	59813/564	PP	1550W50, depth 5–10″, height above red clay 37″	Probably Gl. 6	1	40–44	1	0	0	0	0	0	0	0	1	0	0	D g, AS 5–6	Female

Peabody number	Pathology type	Pathology description	Cranial mod.	Associated funerary objects	Field context	Notes
00-0-10/N9468.60			Unobs.	Frag. bone awl no. 143 (76551), stone ax no. 134 (76614), stone tubular large pipe no. 136 (76618), shell bead necklace no. 106 (76628), medicine sack imprint containing: bone tube no. 138 (76566), bone whistle no. 137 (76574), mooreheadite piece no. 139 (76617), paint stone no. 140 (76622), 3 medicine stones nos. 141 (76624), 35 (76625), greenish slip or lithic 144 (76634), medicine sack 2 contained 12 shrine stones no. 142 (76626)		
00-0-10/N9468.61	Nutr.	Enamel hypoplasia	Unobs.	Stone ax no. 61 (76613)		
00-0-10/N9468.63			Unobs.	Miniature pot no. 1681 (67315), rubbing stone no. 1682 (68824), small black pot no. 1729 (67073), Gl. 2 bowl no. 1680 (67072)		
00-0-10/N9468a.12	Nutr.		Unobs.	Gl. 3 bowl no. 696 (74930)		
19-6-10/59801.0			Not obs.	Gl. 4 bowl no. 879 (68271)		
19-6-10/59802.0	Infection, other	Possible tuberculosis–L elbow, vt/t and vt/l, ribs, spina bifida sacrum	Not obs.	Fur and skin under pelvis and at head, FC		
19-6-10/59803.0	Trauma, infection, arthritis, nutr.	Compression fracture L parietal, periodontal disease, spinal osteophytosis, osteoporosis	Occ. def.	Bowl Gl. 2 no. 1834 (67058)		
19-6-10/59804.0			Not obs.	Black pot no. 1802 (67084), bone tube no. 1824 (68021), shaft straightener no. 1823 (68873)	Infant no. 542 under left arm	
19-6-10/59805.0	Infection, arthritis, other	Swelling at bregma, spinal osteophytosis retained milk incisor	Not obs.	Stone ax no. 1178 (68732), projectile point no. 1180 (68424)		
19-6-10/59806.0	Arthritis, nutr.	Spinal osteophytosis, DJD, slight osteoporosis	Not obs.			
19-6-10/59807.0	Infection	Periostitis of L tibia	Not obs.	3 stone pendants no. 1884 (69132), 3 chipped stones no. 1885 (68937), broken bone disk no. 1886 (68128)		
19-6-10/59808.0			Not obs.	Large B&W bowl no. 3831 (67148)		
19-6-10/59809.0			Not obs.	Bone awl no. 1914 (67846), bone tube in the making(?) no. 1915 (68027)		
19-6-10/59810.0	Arthritis	Spinal osteophytosis	Not obs.	Projectile point no. 2008 (68446), clay object no. 2007 (67630), broken bone implement no. 2006 (68100)		
19-6-10/59811.0	Arthritis, nutr.	Spinal osteophytosis, osteoporosis	Not obs.	Sherd tool no. 1613 (67228), copper ore no. 1616 (69204), antler tool frag. no. 1612 (69263), polishing stone no. 1614 (68826), stone scraper no. 1615 (68835)		
19-6-10/59812.0			Not obs.	Pottery, black, no number	Flexed on right side, head north	
19-6-10/59813.0	Arthritis	Spinal osteophytosis	Not obs.	Worked sherd no. 1881 (67230), projectile point no. 1882 (68442)	On back, head east	

Peabody number	Object description	FN	Specimen label	Site	Intrasite provenience	Ceramic stage	MNI	Age (yrs.)	Ad	Su	Ch	In	Fe	Unk. age	M	M?	F	F?	Unk. sex	Age criteria	Sex criteria
19-6-10/59814.0	Nearly complete cranial and postcranial human remains of an adult female.	59	59814/59	PP	1600E/150, SW corner, depth 5′, 6″ above red clay	Gl. 3	1	35–39	1	0	0	0	0	0	0	0	1	0	0	D d, AS 4	Female
19-6-10/59814.1	Fragmentary cranial human remains of an adult of unknown sex.	59	59814-1	PP	1600E150, SW corner	Gl. 3	1	Adult	1	0	0	0	0	0	0	0	0	0	1		Indet.
19-6-10/59815.0	Nearly complete cranial and postcranial human remains of an adult female.	329	59815/329	PP		Gl. 4 or 5	1	30–39	1	0	0	0	0	0	0	0	1	0	0	D d-e/d-e, AS 4	Female
19-6-10/59816.0	Nearly complete cranial and postcranial human remains of an adult male.	197	59816	PP	North terrace, depth 63″	Probably Gl. 5	1	25–34	1	0	0	0	0	0	1	0	0	0	0	D e, AS 3	Male
19-6-10/59817.0	Nearly complete cranial and postcranial human remains of an adult female.	525	59817/525	PP	North terrace, depth 56″, height–on red clay	Indet.	1	35–44	1	0	0	0	0	0	0	0	1	0	0	D f/g, AS 5	Female
19-6-10/59818.0	Fragmentary cranial and partial postcranial human remains of an adult female.	262	59818	PP	North terrace, depth 4′	Indet.	1	25–29	1	0	0	0	0	0	0	0	1	0	0	D e/d-e, AS 2, T 6, SB 4	Female
19-6-10/59818.1	Very fragmentary cranial and postcranial human remains of a subadult of unknown sex.	262		PP	North terrace, depth 4′	Indet.	1	16–20	0	1	0	0	0	0	0	0	0	0	1	D b1	Indet.
19-6-10/59819.0	Fragmentary cranial and postcranial human remains of an adult female.	391	59819	PP	1550E100, depth 71″, height–on red clay	Indet., poss. Gl. 2	1	50–59	1	0	0	0	0	0	0	0	1	0	0	D essentially edentulous; AS 7, cranial sutures quite fused	Female
19-6-10/59820.0	Nearly complete cranial and postcranial human remains of an adult female.	233	59820	PP	1450E100	Indet.	1	25–34	1	0	0	0	0	0	0	0	1	0	0	D d-e/d; AS 3–4	Female
19-6-10/59821.0	Nearly complete cranial and postcranial human remains of a subadult of unknown sex.	381	59821/381	PP	1550E100, above red clay 3′6″	Indet.	1	13–16	0	1	0	0	0	0	0	0	0	0	1	D M3s beginning to erupt b1/b2, PC 13–18 yrs	Indet.
19-6-10/59822.0	Nearly complete cranial and postcranial human remains of a subadult male.	417	59822/417	PP	1400E100, depth 46″	Probably Gl. 6	1	18–19	0	1	0	0	0	0	1	0	0	0	0	D b1-b2, M3s not in, T 1, AS 1	Male
19-6-10/59823.0	Nearly complete cranial and postcranial human remains of an adult male.	426	59823	PP	1400E100, test xi, cut 5, depth 83″	Probably Gl. 4	1	20–29	1	0	0	0	0	0	1	0	0	0	0	D c/c-d, AS 3, spheno-occipital closed, clavicle just fused	Male
19-6-10/59824.0	Nearly complete cranial and postcranial human remains of an adult female.	196	59824	PP	North terrace, depth 60″, 6″ above red	Probably Gl. 5	1	20–29	1	0	0	0	0	0	0	0	1	0	0	D c, T 5, AS 3–4?	Female
19-6-10/59825.0	Partial cranial and nearly complete postcranial human remains of an adult male.	386	59825/386	PP	1550E100, depth 50″, height–above red clay 21″	Probably Gl. 4	1	35–39	1	0	0	0	0	0	1	0	0	0	0	D g-h, AS 4	Male
19-6-10/59825.1	Fragmentary postcranial human remains of an adult of unknown sex.	386		PP	1550E100, depth 50″, height–above red clay 21″	Probably Gl. 4	1	Adult	1	0	0	0	0	0	0	0	0	0	1		Indet.
19-6-10/59826.0	Nearly complete cranial and postcranial human remains of an adult that is probably female.	306	59826	PP	1250 E100, depth 6″, on red clay	Gl. 4	1	30–34	1	0	0	0	0	0	0	0	0	1	0	D f/e, AS 3	Pelvis F, Robust skull-M
19-6-10/59827.0	Nearly complete cranial and postcranial human remains of an adult male.	319	59827	PP	1200E150, depth 26″, on red clay	Indet.	1	25–29	1	0	0	0	0	0	1	0	0	0	0	D e-f/f, AS 2	Male

Peabody number	Pathology type	Pathology description	Cranial mod.	Associated funerary objects	Field context	Notes
19-6-10/59814.0	Infection, arthritis, tumor	Syphilis–nasal region, DJD, exostosis on L fibula	Not obs.	Gl. 3 bowl no. 247 (65061), broken ax head no. 243 (69631)	Flexed, head north, face down	
19-6-10/59814.1			Unobs.	Gl. 3 bowl no. 247 (65061), broken ax head no. 243 (69631)		Field number and provenience problematic
19-6-10/59815.0	Arthritis	DJD	Not obs.	Gl. 5 Pecos white bowl no. 978 (67010), Bisc. B bowl no. 979 (67011), Gl. 5 olla bowl no. 983 (67013), Tularosa B/W sherd (67668), 2 bone awls no. 995 (67829), 2 stone scrapers no. 996 (68519)		
19-6-10/59816.0	Trauma, nutr., other	Healed compression fracture on frontal, osteoporosis, malerupted upper L canine	Not obs.	Fine scraper no. 1061 (68504), 2 bone awls no. 1062 (67818), turquoise bead no. 1071 (69088)		
19-6-10/59817.0	Arthritis	Mild spinal osteophytosis, ankylosed vt/l to sacrum	Not obs.	Gl. 3 broken bowl no. 1810 (67076)		
19-6-10/59818.0			Unobs.	Stone tool no. 483 (68856), 2 stone scrapers no. 484 (68508)	Pajaritan	
19-6-10/59818.1			Unobs.	Stone tool no. 483 (68856), 2 stone scrapers no. 484 (68508)	Pajaritan	Field number and provenience problematic
19-6-10/59819.0	Infection, nutr.	Pathological endocranial surface, osteoporosis	Unobs.		Under no. 381	
19-6-10/59820.0	Trauma, infection, arthritis, other	Dislocation L TMJ, slight spinal osteophytosis l vt/l and vt/s, DJD elbows, spina bifida vt/t 11, impacted M3	Not obs.	Gl. 5 olla bowl no. 313 (66982), Gl. 3 bowl no. 358 (66984)	Impression of yucca fiber? Fabric in lumbar region	
19-6-10/59821.0	Infection	Impacted M3	Not obs.	Gl. 5 bowl no. 1194 (67027), shell pendant no. 1184 (69114)		
19-6-10/59822.0	Infection	Lytic defects on R Mt3 and navicular	Not obs.	Worked obsidian no. 1337 (68688), lump of paint no. 1338 (69233), antler tool no. 1350 (68258), head of effigy no. 1336 (67559)		
19-6-10/59823.0	Infection, arthritis, tumor	Lytic defects on R Mt3, ankylosis and spondylitis vt/l and vt/s, exotosis on 1st rib	Not obs.	Bone awl no. 1375 (67833), stone scraper no. 1376 (68526), bone awl no. 1378 (67834), stone knife no. 1379 (68598), animal effigy no. 1380 (67560), clay pipe no. 1387 (67440)		
19-6-10/59824.0	Trauma, arthritis, tumor	Healed fracture R 2nd rib, DJD sacroiliac jt, exotosis on L clavicle	Not obs.		Print of fine cloth at head, Pajaritan pottery	
19-6-10/59825.0	Infection, arthritis, nutr., other	Spinal osteophytosis vt/t and vt/l, enamel hypoplasia, impacted M3, spondylolysis of vt/l5	Not obs.	Pottery, Frijolito, no number	Frijolito	
19-6-10/59825.1			No cran.		Frijolito	
19-6-10/59826.0			Not obs.	Bisc. bowl no. 880 (67002)		
19-6-10/59827.0	Arthritis	Slight spinal osteophytosis sacrum	Not obs.	Gl. 4 red pot no. 871 (67004)		

Peabody number	Object description	FN	Specimen label	Site	Intrasite provenience	Ceramic stage	MNI	Age (yrs.)	Ad	Su	Ch	In	Fe	Unk. age	M	M?	F	F?	Unk. sex	Age criteria	Sex criteria
19-6-10/59828.0	Nearly complete cranial and postcranial human remains of an adult female.	93	59828	PP	1.45E+103	Probably Gl. 5	1	18–22	1	0	0	0	0	0	0	0	1	0	0	D b2/c, AS 1, prox. clavicle not fused, spheno-occipital fused	Female
19-6-10/59829.0	Nearly complete cranial and postcranial human remains of an adult female.	81	59829	PP	1.55E+103	Gl. 1	1	18–24	1	0	0	0	0	0	0	0	1	0	0	D c, AS 1	Female
19-6-10/59830.0	Nearly complete cranial and partial postcranial human remains of an adult male.	15	59830	PP	1600E150, head 8″ below ground, feet 18″, 3′ above red	Probably Gl. 6	1	25–34	1	0	0	0	0	0	1	0	0	0	0	D d/e, AS 3	Male
19-6-10/59830.1	Very fragmentary cranial and postcranial human remains of an adult of unknown sex.	17	60350-8/17	PP	1600E150, sw corner	Indet.	1	Adult	1	0	0	0	0	0	0	0	0	0	1		Indet.
19-6-10/59832.0	Nearly complete cranial and fragmentary postcranial human remains of an adult male.	298	59832	PP	1250E100, depth 6′6″, 12″ above red clay	Indet.	1	35–45	1	0	0	0	0	0	1	0	0	0	0	D g/g-h	Male
19-6-10/59833.0	Fragmentary cranial and postcranial human remains of an adult male.	148	59833	PP	1500E100, depth 5 1/2′, 2′ above red clay	Indet.	1	35–44	1	0	0	0	0	0	1	0	0	0	0	D g, T 8, SB 5, AS 5	Male
19-6-10/59834.0	Partial cranial and postcranial human remains of an adult female.	266	59834	PP	1450E100	Probably Gl. 3	1	22–29	1	0	0	0	0	0	0	0	1	0	0	D d, T 3, AS 2, SB 2	Female
19-6-10/59835.0	Nearly complete cranial and postcranial human remains of an adult female.	242	59835	PP	1450E100, depth about 10′	Probably Gl. 3	1	18–24	1	0	0	0	0	0	0	0	1	0	0	D b2/b1, AS 2 (iliac crests fusing)	Female
19-6-10/59836.0	Partial cranial and fragmentary postcranial human remains of an adult female.	313	59836	PP	1200E150, depth 2′8″, on red soil	Very probably Gl. 4	1	30–39	1	0	0	0	0	0	0	0	1	0	0	D e-h uneven, AS 4	Female
19-6-10/59837.0	Partial cranial and fragmentary postcranial human remains of an adult male.	369	59837	PP	1550E100, depth 52″, height–above red clay 25″	Pres. Gl. 5	1	25–29	1	0	0	0	0	0	1	0	0	0	0	D e	Male
19-6-10/59838.0	Nearly complete cranial and partial postcranial human remains of an adult male.	360	59838	PP	1400E150, depth 10′6″, on red clay	Probably Gl. 2 or 3	1	45–55+	1	0	0	0	0	0	1	0	0	0	0	D h+/i+	Male
19-6-10/59839.0	Nearly complete cranial and postcranial human remains of an adult female.	478	59839	PP	1000E150, depth 30″, height–above red clay 44″	Probably Gl. 4 or 5	1	40–44	1	0	0	0	0	0	0	0	1	0	0	D h, T 8, AS 5, SB 4	Female
19-6-10/59841.0	Nearly complete cranial and postcranial human remains of an adult female.	504	59841/504	PP	North terrace, depth 64″, height–on red clay	Indet.	1	27–34	1	0	0	0	0	0	0	0	1	0	0	D e/e-f, AS 3, T 5	Female
19-6-10/59842.0	Nearly complete cranial and partial postcranial human remains of an adult male.	536	59842	PP	1600W50, depth 13″, height–on shale	Probably Gl. 1 or 2	1	40–49	1	0	0	0	0	0	1	0	0	0	0	D h+/i, AS 5	Male
19-6-10/59843.0	Nearly complete cranial and partial postcranial human remains of an adult female.	250	59843/250	PP	North terrace depth 3′, on red	Indet.	1	30–39	1	0	0	0	0	0	0	0	1	0	0	D g, PC 30-34 yrs	Female
19-6-10/59844.0	Very fragmentary postcranial human remains of an adult of unknown sex.	551	59844a	PP	1550W0, depth 3″, height–above red clay 28″	Probably B&W or Gl. 1	1	Adult	1	0	0	0	0	0	0	0	0	0	1		Indet.
19-6-10/59845.0	Nearly complete cranial and postcranial human remains of an adult female.	552	59845/552	PP	1550W50, depth 14″, height–above red clay 28″	Probably B&W or Gl. 1	1	30–39	1	0	0	0	0	0	0	0	1	0	0	D g/f, AS 3	Female
19-6-10/59846.0	Nearly complete cranial and postcranial human remains of an adult female.	385	59846	PP	1550E100, depth 42″, height–above red clay 32″	Indet.	1	40–49	1	0	0	0	0	0	0	0	1	0	0	D g/g-h, T 10, SB 6, AS 5-6	F all features
19-6-10/59847.0	Nearly complete cranial and postcranial human remains of a subadult that is probably female.	335	59847	PP	1250E100, depth 6′	Probably Gl. 3	1	15–17	0	1	0	0	0	0	0	0	1	0	0	D b2, PC 15-17 yrs	?Female

Peabody number	Pathology type	Pathology description	Cranial mod.	Associated funerary objects	Field context	Notes
19-6-10/59828.0	Infection	Impacted M3	Not obs.	Small effigy no. 483 (65588), sherds Pecos white		
19-6-10/59829.0			Not obs.	Gl. 1 bowl no. 288 (65017), projectile point no. 289 (65976)	Print of twilled rush matting under body	
19-6-10/59830.0	Infection	Periodontal disease mandible	Not obs.	Black sherd of unusual shape (65763)		
19-6-10/59830.1	Arthritis, nutr.	Severe DJD at TMJ, enamel hypoplasia	Unobs.			
19-6-10/59832.0	Infection	Periodontal disease	Not obs.	Gl. 4 bowl no. 980 (66999)		Unusual face shape, narrow palate, teeth not very shoveled
19-6-10/59833.0			Unobs.	Gl. 2 (?) red bowl no. 1048 (65021)		
19-6-10/59834.0	Trauma, infection, tumor	Healed fracture R femur, L parietal and frontal–infection, possible cancer–R elbow and spine.	Not obs.	Frags. of 2 miniature pots no. 977 (67312)		Copper stains on facial bones
19-6-10/59835.0	Other	Incomplete fusion neural arch vt/t11	Not obs.	Miniature ladle no. 353 (67313)	Agua Fria and Frijolito	
19-6-10/59836.0	Infection, arthritis, nutr.	Periostitis L and R femur and tibia, severe DJD appendicular and spinal, osteoporosis	Not obs.			
19-6-10/59837.0	Infection, tumor	Abscess, exotosis on 5th Mt	Not obs.	Pajaritan, no number	Pajaritan pottery	
19-6-10/59838.0	Infection, arthritis, tumor	Periodontal infection, spinal osteophytosis with collapsed disks, DJD right shoulder and feet, exotoses on L and R tibia	Not obs.			
19-6-10/59839.0	Infection, arthritis, nutr.	Periodontal disease, spinal osteophytosis, osteoporosis	Not obs.	Bone awl no. 1563 (67842)	Pottery distinctly Pajaritan	
19-6-10/59841.0	Trauma, arthritis	Healed fracture L radius, mild spinal osteophytosis vt/l and vt/s	Not obs.	Small polishing stone no. 1726 (68827)–lost, piece of obsidian no. 1725 (68690), Gl. 3 bowl no. 1668 (67068)		
19-6-10/59842.0	Trauma, infection, arthritis, nutr.	Healed fracture pedal phalange and R fibula, mandifular infection, DJD, osteoporosis	Not obs.	Sherd no. 1759 (67164)		
19-6-10/59843.0	Infection, arthritis	Dental abscess, slight DJD elbows	Not obs.	Agua Fria sherds, no number.	Fetus 7–8 months in pelvis appears to have descended just prior to death	
19-6-10/59844.0	Trauma	Healed fracture L glenoid fossa	No cran.	Worked potsherd no. 1850 (67229)		
19-6-10/59845.0	Trauma, infection, arthritis, nutr., tumor	Scalping, healed periostitis R femur, slight spinal osteophytosis, enamel hypoplasia, frontal button osteoma	Not obs.	Obsidian projectile point no. 1849 (68441)		
19-6-10/59846.0	Trauma, infection, arthritis, other	Dislocation L TMJ, periodontal infection, moderate DJD, spinal osteophytosis, spina bifida vt/s4–5	Not obs.	Gl. 5 olla bowl no. 1196 (67030), bone bead no. 1179 (68016)		Clay on long bones
19-6-10/59847.0			Not obs.		Remains of hide marking found over whole body	

Peabody number	Object description	FN	Specimen label	Site	Intrasite provenience	Ceramic stage	MNI	Age (yrs.)	Ad	Su	Ch	In	Fe	Unk. age	M	M?	F	F?	Unk. sex	Age criteria	Sex criteria
19-6-10/59848.0	Nearly complete cranial and postcranial human remains of a child of unknown sex.	172	59848/172	PP	1500E100, terrace, depth 7′, 1′ above red clay	Indet.	1	8–10	0	0	1	0	0	0	0	0	0	0	1	D 5 dm present, 10 ± 2.5 years	Indet.
19-6-10/59849.0	Nearly complete cranial and postcranial human remains of an adult female.	626	59849/626	PP	1550W0, depth 68″, below red clay 11″	Indet.	1	20–29	1	0	0	0	0	0	0	0	1	0	0	D d-e, T 4-5, SB 2-3, AS 1	Female
19-6-10/59850.0	Nearly complete cranial and partial postcranial human remains of an adult female.	191	59850	PP	North terrace, depth about 6′	Indet.	1	20–24	1	0	0	0	0	0	0	0	1	0	0	D c/d, AS 2, iliac crest partially fused	Female
19-6-10/59851.0	Nearly complete cranial and partial postcranial human remains of an adult male.	633	59851	PP	1550W0, depth 46″, above red clay 22″	Probably Gl. 1	1	50–59+	1	0	0	0	0	0	1	0	0	0	0	D H+/i+, AS 8	Male
19-6-10/59852.0	Nearly complete cranial and fragmentary postcranial human remains of an adult male.	607	59852/607	PP	North terrace, depth 9–11″, height–above rock 1.5″	Indet.	1	35–39	1	0	0	0	0	0	1	0	0	0	0	D f-h uneven, AS 4	Male
19-6-10/59853.0	Partial cranial and nearly complete postcranial human remains of an adult male.	484	59853	PP	1400E100, test xii, base of cut 5	Probably Gl. 4	1	50–59	1	0	0	0	0	0	1	0	0	0	0	D old, nearly edentulous; AS 7	M all features
19-6-10/59854.0	Partial cranial and postcranial human remains of an adult female.	530	59854	PP	1600W50, depth 12″, height–15″ above shale	Probably Gl. 2	1	50–59	1	0	0	0	0	0	0	0	1	0	0	D old, nearly edentulous, AS 7	F all features
19-6-10/59855.0	Very fragmentary postcranial human remains of an adult of unknown sex.	410	59855	PP	1550E100, depth 79″, height–above red clay 8″	Probably Gl. 6	1	Adult	1	0	0	0	0	0	0	0	0	0	1		Indet.
19-6-10/59856.0	Nearly complete cranial and postcranial human remains of an adult female.	431	59856	PP	1100E150, depth 22″, height–above red clay 35″	Indet.	1	35–45	1	0	0	0	0	0	0	0	1	0	0	D h/h-i uneven, T 7, SB 3, AS 4	Female
19-6-10/59857.0	Nearly complete cranial and postcranial human remains of an adult female.	468	59857/f28	PP	1400E100, test xii, near top of cut 4	Probably Gl. 5	1	40–49	1	0	0	0	0	0	0	0	1	0	0	D h? (almost edentulous), T 9, SB 4, AS 5	F all features
19-6-10/59859.0	Nearly complete cranial and partial postcranial human remains of an adult female.	171	59859/171	PP	1500E100, depth 7′6″, on red clay	Probably Gl. 1	1	30–39	1	0	0	0	0	0	0	0	1	0	0	D c-f uneven, AS 5	F all features
19-6-10/59861.0	Partial cranial and nearly complete postcranial human remains of an adult female.	497	59861	PP	1400E100 (test xii, cut 6)	Indet.	1	45–55+	1	0	0	0	0	0	0	0	1	0	0	D g (nearly edentulous), AS 7	Female
19-6-10/59862.0	Nearly complete cranial and postcranial human remains of an adult male.	512	59862/512	PP	1000E100, depth 89″, height–21″ below red clay	Indet.	1	30–34	1	0	0	0	0	0	1	0	0	0	0	D f-h uneven, T 6, SB 3, AS 3	M all features
19-6-10/59863.0	Nearly complete cranial and postcranial human remains of an adult male.	670	59863	PP	Test ix, cut 7, north terrace, depth 62″, above rocks 24″	Probably Gl. 3	1	35–39	1	0	0	0	0	0	1	0	0	0	0	D f-g/f, AS 4?	M all features
19-6-10/59864.0	Nearly complete cranial and fragmentary postcranial human remains of an adult female.	601	59864	PP	North terrace	Indet.	1	25–29	1	0	0	0	0	0	0	0	1	0	0	D d-e/e. AS 2	F all features
19-6-10/59865.0	Partial cranial and postcranial human remains of an adult male.	494	59865	PP	1400E100, test xii, top of cut 6	Probably Gl. 3	1	40–49	1	0	0	0	0	0	1	0	0	0	0	D h+/i+, T 8, SB 4-5, AS 5	M all features
19-6-10/59866.0	Nearly complete cranial and postcranial human remains of an adult male.	94	59866	PP	1450E150, depth 4′, above red 7′6″	Probably Gl. 6	1	27–34	1	0	0	0	0	0	1	0	0	0	0	D g uneven, T 5, SB 3, AS 3	Male
19-6-10/59867.0	Very fragmentary cranial and nearly complete postcranial human remains of a subadult female.	28	59867	PP	1550E150, 2′6″ below surface, 2′ above red	Probably Gl. 4	1	16–20	0	1	0	0	0	0	0	0	1	0	0	D c M3s unerupted, AS 1, clavicle not fused, iliac crest stage 1	F all features

Peabody number	Pathology type	Pathology description	Cranial mod.	Associated funerary objects	Field context	Notes
19-6-10/59848.0			Not obs.	Large Bisc. A bowl no. 1216 (67074)		
19-6-10/59849.0			Not obs.	B&W bowl no. 2038 (67124), bone awl no. 2054 (67848), frag. of effigy no. 2055 (67564), lump of clay no. 2056 (67641)		Reassociated previously missing extremities that were located in 59845-1; adult (no. 626) & child (no. 627) buried together (on same field card); there is no field number or field card for the child.
19-6-10/59850.0			Not obs.	Gl. 4 pot no. 876 (66963)		
19-6-10/59851.0	Arthritis, nutr.	Marked DJD, spinal osteophytosis, slight osteoporosis	Not obs.	Stone drill no. 2093 (69664)	Pottery, red and yellow Agua Fria	
19-6-10/59852.0	Nutr.	Enamel hypoplasia	No cran. def.	Gl. 3 pots nos. 2011 (67117), 2012 (67118)	Original grave (of 607) is in Agua Fria and B&W rubbish	
19-6-10/59853.0	Infection, arthritis, nutr.	Periodontal infection, DJD, slight osteoporosis	Not obs.			
19-6-10/59854.0	Trauma, infection, arthritis, nutr.	Healed compression fracture R parietal, thick diploic expansion, spinal osteophytosis, DJD especially knee, osteoporosis	Occ. def.			
19-6-10/59855.0			No cran.			
19-6-10/59856.0	Arthritis, nutr.	Spinal osteophytosis, DJD at elbows, enamel hypoplasia	Not obs.	Gl. 3 bowl no. 1440 (67043)		
19-6-10/59857.0	Trauma, infection, arthritis	Healed fracture R humerus, L fibula, periodontal disease, marked antemortem tooth loss, periostitis L tibia, spinal osteophytosis	Not obs.		Pecos white	
19-6-10/59859.0	Arthritis, nutr., other	DJD, slight osteoporosis, vt/s5 not fused	Occ. def.	Chipped implement no. 1226 (65997)		
19-6-10/59861.0	Arthritis, nutr.	Marked DJD, spinal osteophytosis, osteoporosis	Not obs.	Large Gl. 2 bowl no. 1628 (67063)		
19-6-10/59862.0	Infection, other	Healed periostitis tibia, supernumerary incisor	Occ. def.	Gl. 3 bowl no. 1731 (67071), 6 bone tubes no. 1712 (68024), 10 eagle claws no. 1713 (69147), lumps of paint no. 1718 (69235), stone ball no. 1717, 2 stone scrapers no. 1715 (68536), clay pipe no. 1714 (68944), stone drill no. 1716 (68662)		
19-6-10/59863.0	Arthritis, other	Mild DJD, spinal osteophytosis, rotated premolar	Lamb. def.	Agua Fria sherds, no number.	Agua Fria sherds	
19-6-10/59864.0	Infection	Syphilitic lesions on face	Not obs.	Broken Bisc. B bowl no. 2236 (67113) or Gl. 1 red bowl no. 1990 (67116)		
19-6-10/59865.0	Arthritis, nutr.	DJD, spinal osteophytosis, enamel hypoplasia	Not obs.	Black jar no. 1624 (67060), projectile point no. 1611 (68436)		
19-6-10/59866.0	Arthritis	Slight DJD	No cran. def.	Human effigy no. 534 (65587), also Pecos white sherds no. 538	Pecos white sherds	
19-6-10/59867.0			Unobs.	Projectile point (65977), 3 bone implements found nearby (65387)	Gl 4? Because at some level and close to no. 79	

Peabody number	Object description	FN	Specimen label	Site	Intrasite provenience	Ceramic stage	MNI	Age (yrs.)	Ad	Su	Ch	In	Fe	Unk. age	M	M?	F	F?	Unk. sex	Age criteria	Sex criteria
19-6-10/59868.0	Nearly complete cranial and postcranial human remains of an adult male.	159	59868/159	PP	1400E100, depth 12′, 5′ above red clay	Probably Gl. 3 or 4	1	45–55+	1	0	0	0	0	0	1	0	0	0	0	D h/i, AS 8	M all features
19-6-10/59869.0	Very fragmentary cranial human remains of an adult of unknown sex.	36	60350-26/3 6	PP	1600E150, 2′ below surface, 3′ above red	Very probably Gl. 4	1	Adult	1	0	0	0	0	0	0	0	0	0	1		Indet.
19-6-10/59870.0	Nearly complete cranial and postcranial human remains of an adult male.	20	59870/m25	PP	1550E100, nw corner, 3′ under ground, 1 1/2′ above red clay	Pres. Gl. 4	1	35–49	1	0	0	0	0	0	1	0	0	0	0	D f-g/f, T 9, SB 5, AS 6	M all features
19-6-10/59871.0	Nearly complete cranial and postcranial human remains of an adult female.	91	59871	PP	1400E100, depth 4′6″, 6′6″ above red clay	Probably Gl. 6	1	25–29	1	0	0	0	0	0	0	0	1	0	0	D e, AS 1-2, T 4, SB 2	F all features
19-6-10/59871.1	Nearly complete cranial human remains of an adult male.	89	89/59871-1	PP	1500E100, depth 9′, 1′ above red clay	Probably Gl. 2	1	30–39	1	0	0	0	0	0	1	0	0	0	0	D f-g	Male
19-6-10/59873.0	Fragmentary cranial and partial postcranial human remains of an adult male.	471	59873	PP	1000E150, depth 48″, height–above red clay 14″	Probably Gl. 5	1	15–17	0	1	0	0	0	0	1	0	0	0	0	D b1, PC < 17 yrs	Male
19-6-10/59874.0	Fragmentary cranial and partial postcranial human remains of an adult male.	225	59874/225	PP	North terrace, depth 2′	Indet.	1	50–59	1	0	0	0	0	0	1	0	0	0	0	D edentulous, fused cranial sutures, T 10, SB 6, AS 7	Male
19-6-10/59874.1	Very fragmentary cranial human remains of an adult of unknown sex.	226	226	PP	North terrace, depth 3′	Indet.	1	Adult	1	0	0	0	0	0	0	0	0	0	1		Indet.
19-6-10/59875.0	Nearly complete cranial and partial postcranial human remains of an adult male.	199	59875	PP	1600E150, depth 2″, above red about 4′	Probably Gl. 3	1	35–39	1	0	0	0	0	0	1	0	0	0	0	D h, AS 4	Male
19-6-10/59876.0	Nearly complete cranial and postcranial human remains of an adult male.	435	59876	PP	1100E150, depth 19″, height–above red clay 28″	Probably Gl. 5 or 6	1	45–54	1	0	0	0	0	0	1	0	0	0	0	D g-h/h, T 10, SB 6, AS 7	Male
19-6-10/59877.0	Fragmentary cranial and partial postcranial human remains of an adult female.	416	59877	PP	1400E100, southern end cut 3, test XI, depth 53″	Probably Gl. 6	1	30–39	1	0	0	0	0	0	0	0	1	0	0	D f, AS 4	Female
19-6-10/59877.1	Very fragmentary postcranial human remains of an adult male.	416		PP	1400E100, southern end cut 3, test XI, depth 53″	Probably Gl. 6	1	Adult	1	0	0	0	0	0	1	0	0	0	0	Very large elements	Male- very large
19-6-10/59878.1	Very fragmentary postcranial human remains of an adult male.	C	59878, 59879	PP	Church	Indet.	1	>35	1	0	0	0	0	0	1	0	0	0	0	PC >35 yrs	Male
19-6-10/59878.3	Very fragmentary postcranial human remains of an adult of unknown sex.	C	59878-17, 59879-12	PP	Church	Indet.	1	<35	1	0	0	0	0	0	0	0	0	0	1	PC < 35 yrs	Indet.
19-6-10/59878.4	Very fragmentary postcranial human remains of an adult of unknown sex.	C	59878-8, 59878-18	PP	Church	Indet.	1	Adult	1	0	0	0	0	0	0	0	0	0	1		Indet.
19-6-10/59879.1	Very fragmentary postcranial human remains of an adult of unknown sex.	C	59879-1	PP	Church	Indet.	1		1	0	0	0	0	0	0	0	0	0	1		Indet.
19-6-10/59879.2	Very fragmentary postcranial human remains of an adult male.	C	59878-5, -14; 59879-2, -6.	PP	Church	Indet.	1	>35	1	0	0	0	0	0	1	0	0	0	0	PC >35 yrs	Male
19-6-10/59879.3	Very fragmentary postcranial human remains of a subadult male.	C	59879-3, 59878-12	PP	Church	Indet.	1	16–19	0	1	0	0	0	0	1	0	0	0	0	PC 16–19 yrs	Male
19-6-10/59879.4	Very fragmentary postcranial human remains of an adult of unknown sex.	C	59879-4, 59878-13	PP	Church	Indet.	1		1	0	0	0	0	0	0	0	0	0	1		Indet.
19-6-10/59879.5	Very fragmentary postcranial human remains of an adult of unknown sex.	C	59879	PP	Church	Indet.	1		1	0	0	0	0	0	0	0	0	0	1		Indet.
19-6-10/59879.6	Very fragmentary postcranial human remains of an adult of unknown sex.	C	59879	PP	Church	Indet.	1		1	0	0	0	0	0	0	0	0	0	1		Indet.
19-6-10/59879.7	Very fragmentary postcranial human remains of an adult that is probably female.	C	59879-10	PP	Church	Indet.	1		1	0	0	0	0	0	0	0	0	1	0		?Female
19-6-10/59881.0	Very fragmentary cranial human remains of an adult male.	C	59881-1	PP	Church	Indet.	1	40–49	1	0	0	0	0	0	1	0	0	0	0	D h	Male

Peabody number	Pathology type	Pathology description	Cranial mod.	Associated funerary objects	Field context	Notes
19-6-10/59868.0	Arthritis	Marked DJD, spinal osteophytosis	Occ. def.	Agua Fria sherd no. 16659 (65761)		
19-6-10/59869.0	Nutr.	Enamel hypoplasia	Unobs.	Projectile point (75912)		Field card indicates more of individual present
19-6-10/59870.0	Arthritis, nutr., other	DJD, slight osteoporosis, spina bifida vt/s4–5,	Occ. def.			
19-6-10/59871.0		Slight DJD	Occ. def.	Shell beads no. 210 (77067)	Burial at length	
19-6-10/59871.1	Nutr.	Enamel hypoplasia	Not obs.	Part of turquoise pendant no. 445 (66269)		
19-6-10/59873.0			Unobs.	Bone awl no. 1540 (67841), 2 bone tubes no. 1539 (68023), hair and wrappings (67654)	Pecos white sherds	Traces of hair and skin on skull
19-6-10/59874.0	Arthritis, nutr.	Spinal osteophytosis and Schmorl's nodes, slight osteoporosis	Unobs.	Bowl Gl. 4 red Pajaritan no. 200 (66980), projectile point no. 199 (68407)		
19-6-10/59874.1			Unobs.	Strand of yellowish hair no. 361 (69246)		
19-6-10/59875.0	Arthritis, nutr.	Slight DJD, Schmorl's nodes–vt/l, osteoporosis	Not obs.			
19-6-10/59876.0	Infection, arthritis, other	Healed lesion to frontal, ankylosis vt/c3–4, Schmorl's nodes vt/t, DJD, spinal osteophytosis, vt/s4–5 not fused, very different facial shape	No cran. def.	Broken obsidian projectile point, no number		
19-6-10/59877.0	Arthritis	DJD	Unobs.	2 pieces shell no. 1335 (69046)		
19-6-10/59877.1	Arthritis	Arthritic tarsals	No cran.	2 pieces shell no. 1335 (69046)		Field number and provenience problematic
19-6-10/59878.1	Other	Enthesophyte on linea aspera of L femur at midshaft	No cran.			Reassociated by Owsley team as F (1/98)
19-6-10/59878.3			No cran.			Reassociated by Owsley team as Q (1/98)
19-6-10/59878.4			No cran.			Reassociated by Owsley team as N (1/98)
19-6-10/59879.1			No cran.			
19-6-10/59879.2	Trauma	Healed fracture R prox. tibia (condyle)	No cran.			Reassociated by Owsley team. Designated C.
19-6-10/59879.3			No cran.			Reassociated by Owsley team and designated D (16–19 yrs.)
19-6-10/59879.4			No cran.			Reassociated by Owsley team. No designation labeled.
19-6-10/59879.5			No cran.			Reassociated by Owsley team as Z (1/98)
19-6-10/59879.6			No cran.			Tibia
19-6-10/59879.7			No cran.			Reassociated by Owsley team as Q (1/98). They indicate not Native American.
19-6-10/59881.0			Unobs.			

Peabody number	Object description	FN	Specimen label	Site	Intrasite provenience	Ceramic stage	MNI	Age (yrs.)	Ad	Su	Ch	In	Fe	Unk. age	M	M?	F	F?	Unk. sex	Age criteria	Sex criteria
19-6-10/59881.1	Very fragmentary cranial human remains of an adult male.	C	59881-3	PP	Church	Indet.	1	20–24	1	0	0	0	0	0	1	0	0	0	0	D d	Male
19-6-10/59881.2	Very fragmentary cranial human remains of an adult female.	C	59881-4	PP	Church	Indet.	1	40–49	1	0	0	0	0	0	0	0	1	0	0	D h	Female
19-6-10/59882.1	Fragmentary postcranial human remains of an adult male.	C	59882-5, 59878-6, 59878-19, 60110-10	PP	Church	Indet.	1	20–29	1	0	0	0	0	0	1	0	0	0	0	PC 20–29 yrs	Male
19-6-10/59883.0	Very fragmentary postcranial human remains of an adult of unknown sex.	C	59883	PP	Church	Indet.	1		1	0	0	0	0	0	0	0	0	0	1		Indet.
19-6-10/59885.1	Very fragmentary postcranial human remains of an adult of unknown sex.	C	59885-1, 59886-8	PP	Church	Indet.	1		1	0	0	0	0	0	0	0	0	0	1		Indet.
19-6-10/59885.3	Very fragmentary postcranial human remains of an adult of unknown sex.	C	59885-3, 59886-3, 59886-9	PP	Church	Indet.	1		1	0	0	0	0	0	0	0	0	0	1		Indet.
19-6-10/59885.4	Very fragmentary postcranial human remains of an adult of unknown sex.	C	59885-4, 59886-6	PP	Church	Indet.	1		1	0	0	0	0	0	0	0	0	0	1		Indet.
19-6-10/59885.5	Very fragmentary postcranial human remains of an adult of unknown sex.	C	59885-5, 59886-4	PP	Church	Indet.	1		1	0	0	0	0	0	0	0	0	0	1		Indet.
19-6-10/59888.0	Nearly complete cranial human remains of a subadult that is probably female.	C	59888	PP	Church	Indet.	1	12–15	0	1	0	0	0	0	0	0	0	1	0	D b2/d-e 15 ± 3 yrs, spheno-occipital not fused, (Owsley aged 11-12 yrs)	Female
19-6-10/59889.0	Nearly complete cranial and fragmentary postcranial human remains of an adult female.	C	59889, 59878-16, 60977	PP	Church	Indet.	1	50–59	1	0	0	0	0	0	0	0	1	0	0	D edentulous maxilla; very old	Female
19-6-10/59890.0	Nearly complete cranial human remains of an adult female. Individual is of mixed Native American/Caucasian ancestry.	C/mex 3	59890, mex 3	PP	Church	Indet.	1		1	0	0	0	0	0	0	0	1	0	0		Female
19-6-10/59891.0	Nearly complete cranial and postcranial human remains of an adult male. Individual is of Caucasian ancestry.	C/mex 1	59891	PP	Church	Indet.	1	27–34	1	0	0	0	0	0	1	0	0	0	0	D e, T 5, SB 3, AS 3	Male
19-6-10/59892.0	Nearly complete cranial human remains of an adult that is probably male. Individual is of mixed Native American/Caucasian ancestry.	C/mex 1	59892, 59881-2	PP	Church	Indet.	1	40–49	1	0	0	0	0	0	0	1	0	0	0	D h, 40–50 yrs	?Male
19-6-10/59893.0	Nearly complete cranial human remains of an adult male.	C/mex 1	59893	PP	Church	Indet.	1	>35	1	0	0	0	0	0	1	0	0	0	0	D edentulous, >40 yrs	Male
19-6-10/59894.0	Nearly complete cranial and fragmentary postcranial human remains of a subadult female. Individual is of Caucasian ancestry.	C/mex 1	59894 (f91 on innominate)	PP	Church	Indet.	1	16–20	0	1	0	0	0	0	0	0	1	0	0	B1, spheno-occipital just fused PC 11.5–12.5 yrs	Female
19-6-10/59895.0	Nearly complete cranial and fragmentary postcranial human remains of an adult female. Individual is of Caucasian ancestry.	C/mex 1	59895	PP	Church	Indet.	1	30–39	1	0	0	0	0	0	0	0	1	0	0	D f-g	Female
19-6-10/59896.0	Nearly complete cranial human remains of a child of unknown sex.	C/mex 1	59896	PP	Church	Indet.	1	2–3	0	0	1	0	0	0	0	0	0	0	1	D 3-4 ± 1yrs, Owsley aged 1.5–2.5 yrs	Indet.
19-6-10/59897.0	Nearly complete cranial and fragmentary postcranial human remains of an adult female.	C/mex 1	59897	PP	Church	Indet.	1	40–49	1	0	0	0	0	0	0	0	1	0	0	D g, T 10, SB 6, AS 6	Female
19-6-10/59898.0	Nearly complete cranial human remains of an adult female.	C/mex 1	59898	PP	Church	Indet.	1	25–29	1	0	0	0	0	0	0	0	1	0	0	D d-e, spheno-occipital fused	Female
19-6-10/59899.0	Nearly complete cranial human remains of an adult that is probably male.	C/mex 1	59899	PP	Church	Indet.	1	>35	1	0	0	0	0	0	0	1	0	0	0	Heavy antemortem tooth loss	?Male

Peabody number	Pathology type	Pathology description	Cranial mod.	Associated funerary objects	Field context	Notes
19-6-10/59881.1			Unobs.			
19-6-10/59881.2			Unobs.			
19-6-10/59882.1			No cran.			Reassociated by Owsley team as G (1/98)
19-6-10/59883.0			No cran.			
19-6-10/59885.1			No cran.			Includes radius 59885-1 from 59885 and ulna 59886-8, renumbered as 59885-1 from 59886 (MEM 12/96)
19-6-10/59885.3			No cran.			Includes radius 59885-3 from 59885 and ulna 59886-3 renumbered as 59885-3 from 59886 (MEM 12/96). Owsley team added radius and labeled XX.
19-6-10/59885.4			No cran.			Includes radius 59885-4 from 59885 and ulna 59886-6, renumbered as 59885-4 from 59886 (MEM 12/96)
19-6-10/59885.5			No cran.			Includes radius 59885-5 from 59885 and ulna 59886-4 renumbered as 59885-5 from 59886 (MEM 12/96)
19-6-10/59888.0			No cran.			Owsley identified as 11–12 yrs female.
19-6-10/59889.0	Infection, arthritis, nutr.	Infection with severe antemortem tooth loss, mild DJD, marked osteoporosis,	Not obs.			
19-6-10/59890.0	Other	Atlas fused to skull	Not obs.			
19-6-10/59891.0	Arthritis	Slight DJD, spinal osteophytosis	Not obs.			
19-6-10/59892.0	Trauma, infection, other	Healed nasal fracture, abscesses, basicranial porosity, sagittal keel	Not obs.			
19-6-10/59893.0	Other	Marked sagittal keel	Not obs.			
19-6-10/59894.0			Not obs.			
19-6-10/59895.0	Other	Microcephalic	Not obs.			
19-6-10/59896.0	Trauma, nutr., other	Cranial perimortem trauma, healed trauma on occipital, cribra orbitalia, trace metopic suture	Not obs.			
19-6-10/59897.0	Arthritis	TMJ has DJD	Not obs.			May not be Native American, very narrow skull
19-6-10/59898.0	Trauma, infection	Trauma to frontal, healed lesion on parietals,	Not obs.			
19-6-10/59899.0			Not obs.			

Peabody number	Object description	FN	Specimen label	Site	Intrasite provenience	Ceramic stage	MNI	Age (yrs.)	Ad	Su	Ch	In	Fe	Unk. age	M	M?	F	F?	Unk. sex	Age criteria	Sex criteria
19-6-10/59900.0	Nearly complete cranial human remains of a child of unknown sex.	C/mex 1	59900	PP	Church	Indet.	1	8–10	0	0	1	0	0	0	0	0	0	0	1	D 10 ± 2.5 yrs	Indet.
19-6-10/59901.0	Nearly complete cranial and postcranial human remains of an adult male. Individual is of Caucasian ancestry.	C	59901, 61171, 30119, 60116, 60113, 60110, 60120	PP	Church	Indet.	1	20–24	1	0	0	0	0	0	1	0	0	0	0	D c-d	Male
19-6-10/59902.0	Nearly complete cranial human remains of an adult male.	C	59902	PP	Church	Indet.	1	20–24	1	0	0	0	0	0	1	0	0	0	0	D c/b2, basi-occipital just fused	Male
19-6-10/59903.0	Fragmentary cranial human remains of an adult female.	C	59903	PP	Church	Indet.	1		1	0	0	0	0	0	0	0	1	0	0		Female
19-6-10/59904.0	Partial cranial human remains of an adult male.	C	59904/church	PP	Church	Indet.	1	>35	1	0	0	0	0	0	1	0	0	0	0	Edentulous maxilla	Male
19-6-10/59905.0	Nearly complete cranial human remains of an adult male. Individual is of mixed Native American/Caucasian ancestry.	C	59905	PP	Church	Indet.	1	>35	1	0	0	0	0	0	1	0	0	0	0	Edentulous premortem, cranial sutures quite fused	Male
19-6-10/59906.0	Nearly complete cranial human remains of an adult male.	C	59906	PP	Church	Indet.	1	25–29	1	0	0	0	0	0	1	0	0	0	0	D e-f	Male
19-6-10/59907.0	Nearly complete cranial and postcranial human remains of an adult male.	379	59907	PP	1550E100, depth 59″, height–above red clay 19″	Probably Gl. 5	1	25–29	1	0	0	0	0	0	1	0	0	0	0	D e, AS 2	Male
19-6-10/59908.0	Nearly complete cranial and postcranial human remains of an adult female.	260	59908/f42	PP	1450E100, trench b, depth 2′8″	Indet.	1	25–29	1	0	0	0	0	0	0	0	1	0	0	D f-g, AS 2	F all features
19-6-10/59909.0	Nearly complete cranial and postcranial human remains of a child of unknown sex.	10	59909	PP	1600E150, 1550E150, 3′ below surface, 18″ above red soil	Indet., poss. Gl. 3 or 4	1	8.5–9.5	0	0	1	0	0	0	0	0	0	0	1	D 10–11 yrs, PC 8 yrs	Indet.
19-6-10/59910.0	Nearly complete cranial and postcranial human remains of an adult female.	562	59910	PP	North terrace, depth 42″, height above red clay 18″	Indet.	1	25–29	1	0	0	0	0	0	0	0	1	0	0	D d/e, AS 2	F all features
19-6-10/59911.0	Nearly complete cranial and partial postcranial human remains of an adult female.	547	59911	PP	1550W50, depth 15″, height–9″ above red subsoil	Indet.	1	35–45	1	0	0	0	0	0	0	0	1	0	0	D h/i, AS 4	Female
19-6-10/59912.0	Nearly complete cranial and fragmentary postcranial human remains of an adult female.	157	59912/157/f33	PP	1400E100, depth 12′, 5′ above red clay	Indet.	1	45–49	1	0	0	0	0	0	0	0	1	0	0	D h/i old, T 10, SB 6, AS 6	F all features
19-6-10/59913.0	Fragmentary cranial and partial postcranial human remains of an adult female.	553	59913	PP	1550W50, depth 23″, height above red clay 15″	Indet.	1	40–44	1	0	0	0	0	0	0	0	1	0	0	D worn, T 8, SB 5, AS 5	Female
19-6-10/59914.0	Partial cranial and postcranial human remains of an adult female.	365	59914/365	PP	North terrace, depth 48″, 1′6″ above red soil	Pres. Gl. 3	1	20–29	1	0	0	0	0	0	0	0	1	0	0	D d, AS 2	Female
19-6-10/59915.0	Nearly complete cranial and partial postcranial human remains of an adult male.	125	59915/125/m17	PP	1450E100, depth 7′, 6′ above red	Probably Gl. 5	1	20–24	1	0	0	0	0	0	1	0	0	0	0	D c-d, AS 1, iliac crest 1	Male- most features, but SN 3
19-6-10/59916.0	Nearly complete cranial and fragmentary postcranial human remains of an adult that is probably female.	187	59916	PP	1500E50, about 4′ below surface, 4′ above red	Pres. Gl. 5	1	>40	1	0	0	0	0	0	0	0	0	1	0	D edentulous	?Female small MP
19-6-10/59917.0	Nearly complete cranial and postcranial human remains of an adult female.	467	59917/467	PP	1450E100,	Probably Gl. 2	1	30–34	1	0	0	0	0	0	0	0	1	0	0	D h/h, AS 3	Female

Peabody number	Pathology type	Pathology description	Cranial mod.	Associated funerary objects	Field context	Notes
19-6-10/59900.0	Nutr.	Slight cribra orbitalia	Occ. def.			
19-6-10/59901.0	Infection	Small lesion on frontal	Not obs.			
19-6-10/59902.0	Trauma	Trauma death, blows to frontal	Not obs.			Unusual dentition on left side.
19-6-10/59903.0			Unobs.			
19-6-10/59904.0	Trauma, infection	Trauma death, blow to parietals, infected L mastoid	Not obs.			
19-6-10/59905.0			Not obs.			
19-6-10/59906.0	Trauma, infection, other	Trauma–scalping, healed fracture nasals, severe dental abscess lower L M2, dental maleruption	Not obs.			
19-6-10/59907.0	Arthritis, other	DJD, spondylolysis of vt/l6, dental –L upper C is located distal to upper P1	Not obs.	Sherds, no number	Probably Pajaritan pottery	High slanting forehead
19-6-10/59908.0	Trauma, infection	Trauma to R temporal and R parietal, dental abscesses	Not obs.	Broken red Gl. 5 bowl no. 460 (66991)		
19-6-10/59909.0	Nutr.	Enamel hypoplasia	Not obs.			Field card and ledger list different age
19-6-10/59910.0	Infection, tumor	Dental abscess, large tumor on R frontal	Not obs.	Gl. 3 pot no. 1988 (67091), obsidian knife no. 1879 (68601), textile (67656)		
19-6-10/59911.0	Infection, arthritis, nutr.	Healed abscess, some DJD, spinal osteophytosis, osteoporosis	Occ. def.	2 large B&W bowls nos. 1762 (67086), 1843 (67087)		
19-6-10/59912.0	Infection, arthritis, nutr.	Abscesses, DJD, osteoporosis	Occ. def.	Bisc. B bowl no. 1013 (65033), black olla nos. 1050 and 1229		
19-6-10/59913.0	Arthritis, nutr.	DJD, osteoporosis	Occ. def.	Red bowl pieces no. 1808 (67088), B&W bowl frags. no. 1809 (67089)		
19-6-10/59914.0	Infection, arthritis	Healed periostitis L tibia, slight DJD	Not obs.			
19-6-10/59915.0	Trauma, arthritis, nutr., tumor, other	Healed fracture pedal phalanges, vt/l spinal osteophytosis, developing spondylytis deformans, enamel hypoplasia, button osteoma, spina bifida vt/s3–5 with incomplete fusion vt/s1–2	Occ. def.	Pecos white sherds (no number assigned)	Pecos white sherds	
19-6-10/59916.0	Infection, arthritis, other	Marked antemortem tooth loss, spinal osteophytosis, occipital bun, extra bone between parietals and frontal	Not obs.	Bone awl no. 1233 (65388)		
19-6-10/59917.0	Arthritis	Mild DJD	Occ. def.	Black olla no. 1545 (67054)		

Peabody number	Object description	FN	Specimen label	Site	Intrasite provenience	Ceramic stage	MNI	Age (yrs.)	Ad	Su	Ch	In	Fe	Unk. age	M	M?	F	F?	Unk. sex	Age criteria	Sex criteria
19-6-10/59918.0	Nearly complete cranial and postcranial human remains of an adult male.	175	59918	PP	1400E100, depth 12′, prob 6′ above red clay	Probably Gl. 5	1	25–34	1	0	0	0	0	0	1	0	0	0	0	D e, AS 30–35	Male
19-6-10/59919.0	Nearly complete cranial and postcranial human remains of an adult male.	49	59919	PP	1650W50, on red soil–depth 18″	Indet., early	1	30–39	1	0	0	0	0	0	1	0	0	0	0	D g-h, T 6, SB 2, AS 4	Male
19-6-10/59920.0	Nearly complete cranial and postcranial human remains of an adult female.	57	59920/003	PP	1600E150, depth 2′8″, 12″ above red	Pres. Gl. 3 or 4	1	40–49	1	0	0	0	0	0	0	0	1	0	0	D h/i, T 8, AS 5	Female
19-6-10/59921.0	Partial cranial and nearly complete postcranial human remains of an adult male.	101	59921/101/m18	PP	1400E100, depth 6′, 6′ above red clay	Probably Gl. 4 or 5	1	30–34	1	0	0	0	0	0	1	0	0	0	0	D f, AS 3	Male
19-6-10/59922.0	Partial cranial and fragmentary postcranial human remains of an adult female.	430	59922	PP	1400E100, cut 5, test xi	Probably Gl. 4	1	30–39	1	0	0	0	0	0	0	0	1	0	0	D h/i, AS 3	Female
19-6-10/59923.0	Partial cranial and nearly complete postcranial human remains of an adult female.	195	59923/195/f35	PP	North terrace, depth 62″, on red	Indet. - Gl. 1 to 4	1	40–44	1	0	0	0	0	0	0	0	1	0	0	D g/h, T 8, SB 5, AS 5	Female
19-6-10/59924.0	Partial cranial and postcranial human remains of an adult male.	341	59924/341	PP	1250E100, depth 3′2″	Probably Gl. 6	1	30–39	1	0	0	0	0	0	1	0	0	0	0	D h/i, T 7, AS 3	M all features
19-6-10/59925.0	Nearly complete cranial and partial postcranial human remains of an adult female.	44	59925/44	PP	1550E0, north terrace	Indet.	1	25–29	1	0	0	0	0	0	0	0	1	0	0	D e, T 4, SB 2, AS 2	Female, robust
19-6-10/59951.0	Nearly complete cranial and partial postcranial human remains of an adult male.	68	59951/6035 0-11/68	PP	1600E100, sw corner, depth 2′6″, 2′6″ above red clay	Very probably Gl. 3	1	40–49	1	0	0	0	0	0	1	0	0	0	0	D h+ almost edentulous	Male
19-6-10/59952.0	Nearly complete cranial and partial postcranial human remains of an adult female.	623	59952/623	PP	1550W0, depth 48″, above red clay 17″	Indet.	1	40–49	1	0	0	0	0	0	0	0	1	0	0	D h+ severe tooth loss, AS 5–6	Female
19-6-10/59953.0	Nearly complete cranial and partial postcranial human remains of an adult male.	317	59953/317	PP	North terrace, depth 3′6″	Probably Gl. 4	1	25–29	1	0	0	0	0	0	1	0	0	0	0	D d, AS 2	Male
19-6-10/59954.0	Nearly complete cranial and fragmentary postcranial human remains of an adult male.	546	59954/546	PP	1550W50, depth 18″	Probably Gl. 2	1	45–49	1	0	0	0	0	0	1	0	0	0	0	D g-h uneven and tooth loss, AS 6	Male
19-6-10/59955.0	Nearly complete cranial human remains of an adult male.	290	59995/290	PP	1200E150, depth 10″	Pres. Gl. 3 or 4	1	25–29	1	0	0	0	0	0	1	0	0	0	0	D e	Male
19-6-10/59956.0	Partial cranial and nearly complete postcranial human remains of an adult male.	58	59956	PP	1600E150, about 4′6″ below surface	Probably Gl. 3	1	25–34	1	0	0	0	0	0	1	0	0	0	0	D e, AS 3	M all features
19-6-10/59957.0	Nearly complete cranial and postcranial human remains of an adult male.	53	59957	PP	1550W	Indet., poss. late Gl. 5	1	50–59	1	0	0	0	0	0	1	0	0	0	0	D h, AS 7	Male
19-6-10/59958.0	Nearly complete cranial and postcranial human remains of an adult male.	128	59958/128	PP	1450E100, depth 9′ from surface	Probably Gl. 4 or 5	1	40–49	1	0	0	0	0	0	1	0	0	0	0	D h/i, T 9, SB 5, AS 5	M all features
19-6-10/59959.0	Partial cranial and postcranial human remains of an adult female.	156	59959	PP	1400E100, depth 14′, 3′ above red clay	Indet.	1	40–44	1	0	0	0	0	0	0	0	1	0	0	D h tooth loss, AS 5	F all features
19-6-10/59960.0	Nearly complete cranial and fragmentary postcranial human remains of an adult female.	375	59960/375	PP	1450E100	Indet.	1	20–34	1	0	0	0	0	0	0	0	1	0	0	D d-f	Female
19-6-10/59961.0	Nearly complete cranial human remains of a child of unknown sex.	348	59961/348	PP	North terrace, depth 54″, 2″ above bedrock	Probably Gl. 2 or 3	1	6–8	0	0	1	0	0	0	0	0	0	0	1	D 8 ± 2 yrs	Indet.
19-6-10/59962.0	Nearly complete cranial and postcranial human remains of an adult male.	258	59962	PP	North terrace, depth 5′	Indet.	1	25–34	1	0	0	0	0	0	1	0	0	0	0	D f, AS 2, T 6, SB 3	M all features

Peabody number	Pathology type	Pathology description	Cranial mod.	Associated funerary objects	Field context	Notes
19-6-10/59918.0	Arthritis, other	Slight DJD, metopic suture	Occ. def.	Antler or armlet (?) no numbers		
19-6-10/59919.0	Arthritis	Slight DJD	Occ. def.	Projectile point no. 118		
19-6-10/59920.0	Arthritis, nutr.	DJD, spinal osteophytosis, enamel hypoplasia, osteoporosis	Occ. def.			
19-6-10/59921.0	Arthritis, nutr., other	Spinal osteophytosis, enamel hypoplasia, incomplete fusion cervical and sacral vertebrae	Occ. def.			
19-6-10/59922.0	Trauma, arthritis	Trauma–1st L Mt, DJD	Not obs.	Broken projectile point no. 1388 (68430)		Field number and provenience problematic
19-6-10/59923.0	Arthritis	DJD	Occ. def.		Agua Fria sherds	
19-6-10/59924.0	Arthritis	Slight spinal osteophytosis	Not obs.	Projectile point and chipped knife no. 1065 (68421), coyote tooth no. 1066 (69146)	Pecos white	
19-6-10/59925.0			Occ. def.	Gl. 3 bowl no. 102 (65076), Gl. 3 bowl no. 101 (65006) with something inside		
19-6-10/59951.0	Arthritis, other	Spinal osteophytosis, 2 unerupted canines	Not obs.	Chipped implement no. 1246 (65999), polishing stone and worked stone no. 1244 (66210)		
19-6-10/59952.0	Arthritis, nutr., tumor	Moderate DJD, osteoporosis, button osteoma frontal	Occ. def.	Partial B&W pot no. 2028 (67121), B&W or early Bisc. bowl no. 2029 (67122), small B&W bowl no. 2033 (67123)		
19-6-10/59953.0	Infection, arthritis, nutr.	Healed periostitis L tibia, slight DJD, enamel hypoplasia	Occ. def.	Bone awl no. 822 (67826), corncob and wrappings no. 822 (67826)		Unusual cranial morphology
19-6-10/59954.0	Infection, arthritis, nutr., other	Dental abscesses, DJD, porotic hyperostosis, supernumerary tooth visible in mandible below P1	Not obs.			Two puncture marks on L parietal
19-6-10/59955.0	Other	Upper R M3 is peg tooth, upper L M3 congenitally absent	Not obs.			Very prognathic
19-6-10/59956.0	Arthritis	Slight DJD	Occ. def.	B&W, Agua Fria, and corrugated sherds, no number.	B&W, Agua Fria, corrugated	
19-6-10/59957.0	Arthritis, other	Moderate DJD, ribs fused to sternum, lower P rotated 180° in mandible	No cr. def.			
19-6-10/59958.0	Arthritis	Marked DJD, especially knees and feet, and marked spinal osteophytosis	Occ. def.			Pick mark L parietal
19-6-10/59959.0	Arthritis	Slight DJD	Occ. def.	Bisc. B bowl no. 1013 (65033), Gl. 2 red bowl no. 1014 (65053), projectile point no. 1047 (65993), unusual Bisc. sherd no. 1047 (65760), worked sherd no. 1047 (65697)	Red Glaze 2	
19-6-10/59960.0			Not obs.			
19-6-10/59961.0			Not obs.			
19-6-10/59962.0	Arthritis	DJD, spinal osteophytosis	Not obs.	Gl. 4 bowl frags. no. 681 (66990)		

Peabody number	Object description	FN	Specimen label	Site	Intrasite provenience	Ceramic stage	MNI	Age (yrs.)	Ad	Su	Ch	In	Fe	Unk. age	M	M?	F	F?	Unk. sex	Age criteria	Sex criteria
19-6-10/59963.0	Nearly complete cranial and postcranial human remains of an adult male.	325	59963	PP	Indeterminate	Poss. Gl. 4	1	40–49	1	0	0	0	0	0	1	0	0	0	0	D h/i, AS 5	Male
19-6-10/59963.1	Very fragmentary cranial and postcranial human remains of an adult of unknown sex.	325		PP	Indeterminate	Poss. Gl. 4	1		1	0	0	0	0	0	0	0	0	0	1		Indet.
19-6-10/59964.0	Partial cranial and nearly complete postcranial human remains of an adult female.	334	59964	PP	1250E100, depth 5′8″	Probably Gl. 3 or 4	1	20–21	1	0	0	0	0	0	0	0	1	0	0	D c-d, T 2, prox. clavicle not fused, iliac crest 1	Female
19-6-10/59965.0	Nearly complete cranial and postcranial human remains of an adult female.	563	59965/f47	PP	1550W50, depth 44″, height in red clay	Probably B&W	1	35–39	1	0	0	0	0	0	0	0	1	0	0	D g/h, AS 4, T 7, SB 4	Female
19-6-10/59966.0	Nearly complete cranial and postcranial human remains of an adult male.	495	59966	PP	1000E150, depth 21″, height–above red clay 53″	Indet.	1	40–44	1	0	0	0	0	0	1	0	0	0	0	D g/h, AS 5	Male
19-6-10/59967.0	Nearly complete cranial and postcranial human remains of an adult male.	499	59967/6034 7-10	PP	1400E100, test xii, cut 6	Indet.	1	40–49	1	0	0	0	0	0	1	0	0	0	0	D h/i, AS 5	Male
19-6-10/59968.0	Nearly complete cranial and partial postcranial human remains of a subadult of unknown sex.	428	59968	PP	1000E150, depth 51″	Indet.	1	18–19	0	1	0	0	0	0	0	0	0	0	1	D b2, T 1, iliac crest 1	Indet.
19-6-10/59969.0	Nearly complete cranial and postcranial human remains of an adult female.	378	59969/378	PP	1550E100, depth 63″, height–above red clay 17″	Probably Gl. 4	1	30–34	1	0	0	0	0	0	0	0	1	0	0	D f-g, AS 3	Female
19-6-10/59970.0	Nearly complete cranial human remains of an adult male.	555	59970/555	PP	1550W50, depth 11″, height above red clay 14″	Probably B&W or Gl. 1	1	45–55+	1	0	0	0	0	0	1	0	0	0	0	D h+/i antemortem tooth loss	Male
19-6-10/59971.0	Partial cranial and postcranial human remains of an adult female.	169	59971	PP	1450E100, depth 6′6″, about 8–11′ above red	Indet., poss. Gl. 5	1	25–29	1	0	0	0	0	0	0	0	1	0	0	D e-f, AS 2	F all features
19-6-10/59972.0	Partial cranial and postcranial human remains of an adult male.	349	59972	PP	1250E100, depth 5′6″	Probably Gl. 4 or 5	1	40–49	1	0	0	0	0	0	1	0	0	0	0	D g, AS 6	M all features
19-6-10/59973.0	Nearly complete cranial and partial postcranial human remains of a subadult of unknown sex.	613	59973/613	PP	1550W0, depth 13″, above red clay 35″, 4″ above stone layer	Indet.	1	10–12	0	1	0	0	0	0	0	0	0	0	1	D b1, 15 ± 3 yrs, PC 10–11.5 yrs	Indet.
19-6-10/59974.0	Nearly complete cranial and partial postcranial human remains of a subadult female.	35	59974	PP	1550E150, 3′ below surface, 8″–1′ above red	Very probably Gl. 4	1	18–24	0	1	0	0	0	0	0	0	1	0	0	D c, PC young adult	Female
19-6-10/59975.0	Partial cranial and nearly complete postcranial human remains of an adult male.	457	59975	PP	1100E100, depth 62″	Indet.	1	45–50+	1	0	0	0	0	0	1	0	0	0	0	D h+/i+, AS 6	Male
19-6-10/59976.0	Nearly complete cranial and postcranial human remains of an adult female.	433	59976	PP	1400E100, cut 5, test xi	Indet.	1	30–34	1	0	0	0	0	0	0	0	1	0	0	D f/f, AS 3	Female
19-6-10/59977.0	Nearly complete cranial and postcranial human remains of an adult female.	84	59977/84	PP	1500E100, depth 6′, height 18″ above red	Pres. Gl. 5	1	50–59	1	0	0	0	0	0	0	0	1	0	0	D edentulous/i+, AS 7, T 10	Female
19-6-10/59978.0	Nearly complete cranial and postcranial human remains of an adult female.	131	59978/F36/F43	PP	1450E100, depth 7′, 6′ above red clay	Probably Gl. 4 or 5	1	50–59	1	0	0	0	0	0	0	0	1	0	0	Almost edentulous, AS 7	F all features
19-6-10/59979.0	Nearly complete cranial and partial postcranial human remains of an adult female.	132	59979	PP	1450E100, depth 8′, 5′ above red clay	Probably Gl. 4	1	35–39	1	0	0	0	0	0	0	0	1	0	0	D h/i, AS 4	Female
19-6-10/59980.0	Nearly complete cranial and postcranial human remains of an adult female.	462	59980/462	PP	1400E100, test xii, cut 2	Probably Gl. 6	1	40–44	1	0	0	0	0	0	0	0	1	0	0	D h/i, almost edentulous, AS 5, T 8, SB 4	F all features

Peabody number	Pathology type	Pathology description	Cranial mod.	Associated funerary objects	Field context	Notes
19-6-10/59963.0	Arthritis, nutr.	Moderate DJD, osteoporosis	Slight occ. def.	Early Gl. 4 bowl no. 878 (68272)		
19-6-10/59963.1			Unobs.	Early Gl. 4 bowl no. 878 (68272)		
19-6-10/59964.0			Not obs.			
19-6-10/59965.0	Infection, arthritis, other	Healed periostitis R tibia, slight DJD, partial spinal bifida–offset neural arch–in sacrum	Not obs.			
19-6-10/59966.0	Trauma, infection, arthritis, nutr.	Healed fracture L radius and fibula, dental abscesses, DJD, spinal osteophytosis, enamel hypoplasia	Occ. def.	Gl. 4 parts of olla no. 1626 (67061), bone implement in abdominal region, no number	With infant	
19-6-10/59967.0	Trauma, arthritis, nutr., other	Trauma L acetabulum fracture–major dislocation with healing, spinal osteophytosis, marked DJD, osteoporosis, spina bifida	Not obs.	Gl. 3–4 bowl no. 1630 (67065), projectile point no. 1667 (68438)		
19-6-10/59968.0	Infection, arthritis, nutr.	Vt/t spine fused and compressed– lytic, porotic hyperostosis	Not obs.			
19-6-10/59969.0	Arthritis, nutr.	Ankylosis of vt/t2 and vt/t3, enamel hypoplasia	Occ. def.	Obsidian knife no. 1185 (68597)		
19-6-10/59970.0	Infection	Mild periodontal disease	Not obs.			
19-6-10/59971.0	Arthritis, nutr.	Moderate DJD–arms, enamel hypoplasia	Not obs.			
19-6-10/59972.0	Trauma, arthritis, nutr.	Healed fracture Mt, spinal osteophytosis, enamel hypoplasia, osteoporosis	Not obs.	Projectile point no. 1102 (68423)		
19-6-10/59973.0	Other	Maleruption of canine	Not obs.	Gl. 1 yellow bowl no. 1983 (67120)		
19-6-10/59974.0	Nutr.	Enamel hypoplasia	Occ. def.	Perforated shell (66334) and bone awl (65394)		
19-6-10/59975.0	Arthritis, nutr.	Vt/t–marked spinal osteophytosis, osteoporosis	Not obs.	Gl. 3 bowl no. 1476 (67052)		
19-6-10/59976.0	Trauma, arthritis, nutr.	Trauma to R parietal?, ankylosed vt/l5 and osteophytosis, DJD, osteoporosis	Occ. def.	Bone bead no. 1410 (68019), shaft straightener no. 1408 (68872), Gl. 5 bowl no. 1839 (67045), pipe frag. no. 1409 (67441)		
19-6-10/59977.0	Infection, arthritis, nutr.	Essentially edentulous antemortem, marked DJD and spinal osteophytosis, osteoporosis	Not obs.	Worked sherd no. 329, bone tube no. 330		
19-6-10/59978.0	Infection, arthritis	Essentially edentulous antemortem, DJD, spinal osteophytosis	Occ. def.	Chipped flint point no. 894 (65983)–missing		
19-6-10/59979.0	Arthritis, nutr.	Moderate DJD, spinal osteophytosis, enamel hypoplasia	Not obs.			
19-6-10/59980.0	Infection, arthritis	Severe antemortem tooth loss, moderate DJD, spinal osteophytosis,	Occ. def.	Stone scraper no. 1510 (68530)		

Peabody number	Object description	FN	Specimen label	Site	Intrasite provenience	Ceramic stage	MNI	Age (yrs.)	Ad	Su	Ch	In	Fe	Unk. age	M	M?	F	F?	Unk. sex	Age criteria	Sex criteria
19-6-10/59981.0	Nearly complete cranial and postcranial human remains of an adult male.	198	59981	PP	North terrace, below surface 48″, above red about 3′	Indet.	1	20–29	1	0	0	0	0	0	1	0	0	0	0	D d/d, T 4, AS 2	M all features
19-6-10/59982.0	Nearly complete cranial and postcranial human remains of an adult male.	415	59982/m24	PP	1400E100, test xi, cut 3, depth 51″	Probably Gl. 6	1	35–45	1	0	0	0	0	0	1	0	0	0	0	D g/f-g, AS 5, T 8	M all features
19-6-10/59983.0	Nearly complete cranial and partial postcranial human remains of a subadult that is probably male.	466	59983	PP	1400E100, test xii, base of cut 3	Probably Gl. 5	1	14–16	0	1	0	0	0	0	0	1	0	0	0	D b1-b2 15 ± 3 yrs, PC 14–17 yrs	?Male
19-6-10/59984.0	Nearly complete cranial and postcranial human remains of an adult male.	483	59984	PP	1400E100, test xii, cut 5	Probably Gl. 3 or 4	1	40–49	1	0	0	0	0	0	1	0	0	0	0	D h/i+ almost edentulous., AS 5	M all features
19-6-10/59985.0	Nearly complete cranial and postcranial human remains of an adult that is probably male.	211	59985	PP	1250E100, near terrace, depth 5′, just above red	Indet.	1	20–29	1	0	0	0	0	0	0	1	0	0	0	D d, T 5	?Male
19-6-10/59986.0	Nearly complete cranial and postcranial human remains of an adult female.	264	59986	PP	North terrace, depth 4′	Pres. Gl. 3	1	35–39	1	0	0	0	0	0	0	0	1	0	0	D h+/h-I, T 7, AS 4	F all features
19-6-10/59986.1	Fragmentary postcranial human remains of an adult that is probably male.	264		PP	North terrace, depth 4′	Pres. Gl. 3	1		1	0	0	0	0	0	0	1	0	0	0		?Male-very large extremities
19-6-10/59987.0	Nearly complete cranial and postcranial human remains of an adult male.	265	59987/265	PP	1450E100, depth 13′6″	Probably Gl. 2 or 3	1	22–29	1	0	0	0	0	0	1	0	0	0	0	D c-d/d, AS 2, T 3	M all features
19-6-10/59988.0	Fragmentary cranial and partial postcranial human remains of an adult female.	423	59988	PP	1000E150, depth 48″, height–above red clay 13″	Indet.	1	25–29	1	0	0	0	0	0	0	0	1	0	0	D f?, AS 2, T 5	Female
19-6-10/59989.0	Fragmentary cranial and nearly complete postcranial human remains of an adult female.	593	59989	PP	North terrace, depth 55″, height 3–4″ above rock	Indet.	1	40–49	1	0	0	0	0	0	0	0	1	0	0	D h/i, AS 5	Female
19-6-10/59990.0	Partial cranial and nearly complete postcranial human remains of an adult male.	503	59990/503	PP	North terrace, depth 60″	Probably Gl. 3	1	30–34	1	0	0	0	0	0	1	0	0	0	0	D f, AS 3	M all features
19-6-10/59991.0	Nearly complete cranial and postcranial human remains of an adult female.	514	59991	PP	1000E150, depth 40″, height–above red clay 34″	Probably Gl. 4	1	40–44	1	0	0	0	0	0	0	0	1	0	0	D h/h, T 8, AS 5	Female
19-6-10/59992.0	Nearly complete cranial and partial postcranial human remains of a child of unknown sex.	447	59992/447	PP	1000E150, depth 23″, height–above red clay 39″	Probably Gl. 5	1	8–9	0	0	1	0	0	0	0	0	0	0	1	D 11–12 ± 2.5 yrs, PC 8 yrs	Indet.
19-6-10/59993.0	Nearly complete cranial and postcranial human remains of an adult female.	446	59993/446	PP	1400E100, test xi, top of cut 7	Indet.	1	40–44	1	0	0	0	0	0	0	0	1	0	0	D g-h/h, AS 5	F all features
19-6-10/59994.0	Partial cranial and nearly complete postcranial human remains of an adult male.	208	59994	PP	1100E150, depth 8″, 2′6″ above red	Indet.	1	30–34	1	0	0	0	0	0	1	0	0	0	0	D g/g-h, T 6, SB 3-4, AS 3	Male
19-6-10/59995.0	Nearly complete cranial and postcranial human remains of an adult female.	403	59995	PP	1550E100, depth 38″, height–above red clay 33″	Probably Gl. 4	1	30–39	1	0	0	0	0	0	0	0	1	0	0	D f-g, T 6, AS 4	F all features
19-6-10/59996.0	Nearly complete cranial and postcranial human remains of an adult female.	650	59996/f49	PP	North terrace, depth 18″, above red clay 12″	Probably Gl. 3	1	18–19	1	0	0	0	0	0	0	0	1	0	0	D b2, iliac crest stage 1, T 1	Female
19-6-10/59997.0	Partial cranial and nearly complete postcranial human remains of an adult male.	398	59997	PP	1550E100, depth 74″. height–above red clay 8″	Indet.	1	25–34	1	0	0	0	0	0	1	0	0	0	0	D d, AS 3, T 5	Male
19-6-10/59998.0	Nearly complete cranial and postcranial human remains of an adult male.	420	59998/m12/m16	PP	1400E100, test xi, cut 5, depth 83″	Indet.	1	20–24	1	0	0	0	0	0	1	0	0	0	0	D c-d, T 2, AS 1, prox. clavicle not fused	Male

Peabody number	Pathology type	Pathology description	Cranial mod.	Associated funerary objects	Field context	Notes
19-6-10/59981.0	Nutr.	Enamel hypoplasia	Occ. def.	Bone awl no. 1030 (67819), Bisc. B bowl no. 1851 (66965), turtle carapace ornament no. 1104 (69134)		
19-6-10/59982.0	Arthritis, nutr.	Slight DJD, enamel hypoplasia,	Occ. def.	Stone projectile point (no number listed)		
19-6-10/59983.0	Infection	R parietal healed periosteal expansion	Not obs.			
19-6-10/59984.0	Infection, arthritis, nutr.	Extensive antemortem tooth loss, marked DJD, spinal osteophytosis with ankylosis, enamel hypoplasia	Occ. def.	Fossil shell no. 1575 (69193)		
19-6-10/59985.0	Arthritis, other	Slight DJD, scaphocephaly–intensive	Not obs.	Flat bone, perforated (no number given)	Frijolito grave?	
19-6-10/59986.0	Trauma, infection, arthritis, nutr.	Pedal phalange fracture, some antemortem tooth loss spinal osteophytosis, moderate DJD, slight osteoporosis, enamel hypoplasia	Occ. def.	Sherds, no number		
19-6-10/59986.1	Trauma	Fracture of hallux	No cran.	Sherds, no number		
19-6-10/59987.0	Infection, nutr., other	L calcaneus–lytic infection, enamel hypoplasia, vt/l5 neural arch defect	Occ. def.	Agua Fria sherds, no number	Distinct traces of twilled rush matting under body (FC), Agua Fria sherds	
19-6-10/59988.0	Arthritis, nutr.	Slight DJD, enamel hypoplasia	Unobs.	Gl. 6 bowl no. 1842 (67041)		
19-6-10/59989.0	Infection, arthritis, nutr.	Caries, moderate DJD, osteoporosis	Occ. def.	Bisc. A bowl no. 1993 (67105), bone awl no. 1994 (67847)		
19-6-10/59990.0	Trauma, arthritis, other	Fracture right hand Mt and ph, slight DJD, spinal osteophytosis, vt/c incomplete fusion	Not obs.	Agua Fria red and yellow sherds, no number		
19-6-10/59991.0	Trauma, infection, arthritis, nutr.	R clavicle–healed fracture, some tooth loss, DJD, marked spinal osteophytosis, osteoporosis	Lamb. def.	Fragment of effigy no. 1709 (67561), polishing stone no. 1710 (68828)	Frijolito pottery	
19-6-10/59992.0			Not obs.	Miniature bowl no. 1457 (67316), lump red paint no. 1458 (69234)	Pajaritan pottery	
19-6-10/59993.0	Arthritis, nutr.	Moderate DJD, spinal osteophytosis, osteoporosis, enamel hypoplasia	Occ. def.	Bisc. B bowl no. 1443 (67049)		
19-6-10/59994.0	Arthritis, other	Spinal osteophytosis vt/l, spina bifida/incomplete fusion	Occ. def.			
19-6-10/59995.0	Trauma, arthritis, tumor	Healed fracture nasals, slight DJD, spinal osteophytosis, frontal–button osteoma	Occ. def.	Scraper no. 1292 (68524), obsidian piece no. 1293 (68687)		
19-6-10/59996.0	Other	Minor spina bifida in sacral region	Occ. def.	Broken projectile point no. 2258 (68450)		Time period problematic according to notes on field card
19-6-10/59997.0	Arthritis	Schmorl's node vt/l4	Not obs.	Gl. 2 bowl no. 1837 (67035)		
19-6-10/59998.0	Nutr.	Enamel hypoplasia	Occ. def.	Gl. 4 bowl no. 1385 (67040), bone awl, no number		

Peabody number	Object description	FN	Specimen label	Site	Intrasite provenience	Ceramic stage	MNI	Age (yrs.)	Ad	Su	Ch	In	Fe	Unk. age	M	M?	F	F?	Unk. sex	Age criteria	Sex criteria
19-6-10/59999.0	Nearly complete cranial and partial postcranial human remains of an adult that is probably female.	252	59999	PP	North terrace, depth 4′6″, 8″ above rock bottom	Indet.	1	35–39	1	0	0	0	0	0	0	0	1	0	0	D h+/i+, AS 4	?Female (pelvis-F, SM,SR-M)
19-6-10/60000.0	Nearly complete cranial and partial postcranial human remains of an adult male.	248	60000/m11	PP	1400E150, depth 10′, 1′ above red subsoil	Pres. Gl. 4	1	35–39	1	0	0	0	0	0	1	0	0	0	0	D g, AS 4	Male
19-6-10/60001.0	Nearly complete cranial and postcranial human remains of an adult female.	249	60001/252	PP	1450E100, depth 10′ ±, about 1′ above red	Indet.	1	30–34	1	0	0	0	0	0	0	0	1	0	0	D f-g, AS 3	Female
19-6-10/60002.0	Nearly complete cranial and partial postcranial human remains of an adult male.	421	60002/421	PP	1000E150, depth 37″, height–above red clay 26″	Indet.	1	45–49	1	0	0	0	0	0	1	0	0	0	0	D h/i, AS 6	Male
19-6-10/60003.0	Nearly complete cranial and postcranial human remains of an adult female.	554	60003/554	PP	1550W0 top sq. 5, depth 27″, height above red clay 22″	Probably Gl. 1 or 2	1	30–34	1	0	0	0	0	0	0	0	1	0	0	D g/e, AS 3	Female
19-6-10/60004.0	Nearly complete cranial and postcranial human remains of an adult female.	597	60004/f33	PP	1550W0, depth 41–58″, above red clay 9–27″	Indet.	1	40–44	1	0	0	0	0	0	0	0	1	0	0	D g/g-h, T 8, AS 5	Female
19-6-10/60005.0	Nearly complete cranial and postcranial human remains of an adult female.	672	60005/672	PP	1550E100, 57 inches south of 1550 line, depth 30″	Probably Gl. 4 or 5	1	40–49	1	0	0	0	0	0	0	0	1	0	0	D h+/i+, AS 6, T 8	Female
19-6-10/60006.0	Nearly complete cranial and postcranial human remains of an adult female.	261	60006/261	PP	North terrace, depth 3′10″, above red 3′	Pres. Gl. 4 or 5	1	40–44	1	0	0	0	0	0	0	0	1	0	0	D h/i, AS 5, T 8	Female
19-6-10/60007.0	Partial cranial and fragmentary postcranial human remains of an adult female.	397	60007/397	PP	North terrace, depth 5′8″, height–above red clay 2–3″	Indet.	1	35–39	1	0	0	0	0	0	0	0	1	0	0	D h/i, AS 4	Female
19-6-10/60008.0	Nearly complete cranial and partial postcranial human remains of an adult female.	73	60008	PP	1600E150, depth 2′6″, 2′ above clay	Probably Gl. 4	1	30–49	1	0	0	0	0	0	0	0	1	0	0	D h+/i+, AS 6, T 6	Female
19-6-10/60009.0	Nearly complete cranial and postcranial human remains of an adult female.	241	60009/241	PP	1450E100, depth about 3′	Very probably Gl. 6	1	50–59+	1	0	0	0	0	0	0	0	1	0	0	Very old	Female
19-6-10/60010.0	Nearly complete cranial and postcranial human remains of an adult female.	263	60010	PP	North terrace, depth 4′	Indet.	1	40–49	1	0	0	0	0	0	0	0	1	0	0	D h+/i+, AS 5, T 9	Female
19-6-10/60011.0	Nearly complete cranial and postcranial human remains of an adult female.	332	60011/332/f32	PP	1250E100, depth 6′1″	Probably Gl. 3 or 4	1	30–34	1	0	0	0	0	0	0	0	1	0	0	D f/f, AS 3, T 6	Female
19-6-10/60012.0	Partial cranial and postcranial human remains of an adult female.	366	60012/366	PP	North terrace, depth 56″, 1′ above red soil	Indet.	1	35–39	1	0	0	0	0	0	0	0	1	0	0	D g (uneven wear)	?Female
19-6-10/60013.0	Nearly complete cranial and postcranial human remains of an adult female.	620	60013	PP	1550W0, depth 56″, above red clay 5″	Very probably B&W	1	20–24	1	0	0	0	0	0	0	0	1	0	0	D b2-c/c, AS 1, iliac crest stage 1, prox. clavicle not fused	?Male (cranium M, pelvis indet.)
19-6-10/60014.0	Nearly complete cranial and postcranial human remains of a child of unknown sex.	2	60014	PP	1600E150, depth 4′	Pres. Gl. 4 or 5	1	5–6	0	0	1	0	0	0	0	0	0	0	1	D 7 ± 2 yrs, PC 4.5–5.5 yrs	Indet.
19-6-10/60016.0	Nearly complete cranial and fragmentary postcranial human remains of an adult that is probably male.	560	60016/560	PP	1550W50, depth 14″, height above red clay 20″	Probably Gl. 2	1	20–29	1	0	0	0	0	0	0	1	0	0	0	D e/d	?Male
19-6-10/60017.0	Nearly complete cranial and partial postcranial human remains of an adult female.	611	60017/611	PP	1550W0, depth 66″, below top of red clay 13″	Very probably B&W	1	50–59	1	0	0	0	0	0	0	0	1	0	0	D h+/i+, AS 7	Female
19-6-10/60018.0	Nearly complete cranial and postcranial human remains of an adult male.	528	60018/m7/m18	PP	1600W50, depth 18″	Probably Gl. 2	1	40–44	1	0	0	0	0	0	1	0	0	0	0	D g+/I, AS 5, T8	M all features

Peabody number	Pathology type	Pathology description	Cranial mod.	Associated funerary objects	Field context	Notes
19-6-10/59999.0	Infection, arthritis, nutr.	Antemortem tooth loss, DJD, osteoporosis	Occ. def.	Sherds, no number	B&W, Agua Fria	
19-6-10/60000.0	Infection, other	L tibia–healed periostitis, 3rd molars congenitally absent	Occ. def.	Pajaritan sherds, no number	Pajaritan pottery	
19-6-10/60001.0	Arthritis, nutr.	Spinal osteophytosis vt/l, slight DJD, osteoporosis	Occ. def.	Gl. 2 bowl no. 443 (66986)		
19-6-10/60002.0	Arthritis, nutr., other	DJD, enamel hypoplasia, mild osteoporosis, sacral spinal bifida, crooked teeth–affected wear	Occ. def.	Stone scraper no. 1356 (68526), bone tubes no. 1352 (68018), antler tool no. 1355 (68259)		
19-6-10/60003.0	Nutr., other	Enamel hypoplasia, small spina bifida vt/s2–5	No cran. def.			Skull looks different–very wide orbital spacing and square orbits, shoveled incisors, metopism, ancestry?
19-6-10/60004.0	Arthritis, nutr., other	DJD, enamel hypoplasia, spinal osteophytosis, minor spina bifida in sacral region	Occ. def.	Bisc. A bowl no. 2032 (67108), Agua Fria red scoop no. 1956 (67107), projectile point no. 2027 (68445)		
19-6-10/60005.0	Arthritis , nutr.	Marked DJD and osteoporosis, especially at L sacroiliac	Occ. def.	Frag. of miniature pot no. 2283 (67319), bone bead no. 2284 (68030), 2 bone awls no. 2282 (67849), polishing stone no. 2285 (68832)	Pajaritan pottery	
19-6-10/60006.0	Infection, arthritis, nutr.	Molar tooth loss, extensive DJD and spinal osteophytosis, osteoporosis, slight enamel hypoplasia	Not obs.	Bone awl no. 515 (67824)		
19-6-10/60007.0	Arthritis	DJD, spinal osteophytosis	Occ. def.	Gl. 4 bowl no. 1311 (67034), projectile point no. 1310 (68427), fragments of hide? wrappings (67649)	Body wrapped in hide, fragments saved, no number	
19-6-10/60008.0	Infection, arthritis, nutr., other	Some tooth loss, abscesses, spinal osteophytosis, slight DJD, enamel hypoplasia, nasal septum defect	Occ. def.			
19-6-10/60009.0	Infection, arthritis, nutr., other	Antemortem tooth loss–edentulous, severe DJD, TMJ, osteoporosis, incomplete fusion vt/l5	Occ. def.	Sherds, no number	Burial at length, Pajaritan sherds, 1 polished black olla sherd	
19-6-10/60010.0	Infection, arthritis	A few caries, abscesses, antemortem tooth loss, lytic lesion in feet, spinal osteophytosis and spondylitis on v/t with ankylosis	Not obs.	Gl. 5 olla bowl no. 1838 (66992), 2 projectile points no. 523 (68410), bone awl no. 524 (67825)		
19-6-10/60011.0	Arthritis, nutr.	Slight DJD, enamel hypoplasia	Occ. def.			
19-6-10/60012.0	Trauma, arthritis, nutr.	Trauma L humerus, DJD, osteoporosis	Occ. def.	Gl. 4 red bowl no. 1170 (67024), hair and feather cloth (67647)		
19-6-10/60013.0	Nutr.	Enamel hypoplasia, mild cribra orbitalia	Not obs.	Piece B&W pottery no. 2026 (67231)		
19-6-10/60014.0			Not obs.			
19-6-10/60016.0	Arthritis	Slight DJD right arm	Occ. def.	4 pcs bone tubes no. 1904 (68025), worked antler no. 1903 (68265), Gl. 1 Agua Fria bowl no. 1905 (67090)		
19-6-10/60017.0	Arthritis, nutr.	DJD, osteoporosis	?Occ. def.		Many stones over skeleton, as if grave had been filled with stones	
19-6-10/60018.0	Trauma, arthritis	Healed fracture R 12th rib and v/t12, healed fracture R femur, healed fracture vt/t3 spinous process, marked spinal osteophytosis, especially lumbar, slight DJD	No cran. def.	Animal tooth and projectile point left in trench		

Peabody number	Object description	FN	Specimen label	Site	Intrasite provenience	Ceramic stage	MNI	Age (yrs.)	Ad	Su	Ch	In	Fe	Unk. age	M	M?	F	F?	Unk. sex	Age criteria	Sex criteria
19-6-10/60018.1	Very fragmentary cranial and postcranial human remains of an adult female.	527	527/60339-3/60349-8	PP	1660W50, depth 11″, height–2″ above rock	Probably Gl. 2	1		1	0	0	0	0	0	0	0	1	0	0	Adult, no teeth	Female size
19-6-10/60019.0	Nearly complete cranial and postcranial human remains of an adult male.	568	60019	PP	North terrace, depth 36″, height 3″ above bedrock	Indet.	1	25–29	1	0	0	0	0	0	1	0	0	0	0	D e/e, AS 2, T 5, SB 3	Male
19-6-10/60020.0	Fragmentary cranial and nearly complete postcranial human remains of an adult female.	316	60020/316	PP	1250E100, depth 7′, on red clay	Indet.	1	18–24	1	0	0	0	0	0	0	0	1	0	0	D e-f, AS 1	Female
19-6-10/60020.1	Fragmentary cranial human remains of an adult that is probably female.	380 ?	60020	PP	1550E100, depth 16″, height–above red clay 57″	Probably Gl. 6	1		1	0	0	0	0	0	0	0	0	1	0		?Female
19-6-10/60021.0	Nearly complete cranial and postcranial human remains of an adult female.	311	60021/311/f31	PP	1250E100, depth 5′, 1′6″ above red soil	Probably Gl. 4	1	25–29	1	0	0	0	0	0	0	0	1	0	0	D d-e/e, AS 2, T 5	Female
19-6-10/60022.0	Partial cranial and nearly complete postcranial human remains of an adult male.	316 ?	60022/m20	PP	1250E100, depth 7′, on red clay	Indet.	1	25–34	1	0	0	0	0	0	1	0	0	0	0	D d-e, T 6, SB 4, AS 3	Male
19-6-10/60022.1	Fragmentary cranial human remains of an adult of unknown sex.	316 ?	60022-1	PP	1250E100, depth 7′, on red clay	Indet.	1	35–39	1	0	0	0	0	0	0	0	0	0	1	D g? (very fragmen tary)	Indet.
19-6-10/60023.0	Nearly complete cranial and postcranial human remains of a subadult female.	309	60023	PP	1250E100, depth 5′	Probably Gl. 5	1	16–22	0	1	0	0	0	0	0	0	1	0	0	D b2-c, AS 1, iliac crest1, ischial crest 1	Female
19-6-10/60025.0	Nearly complete cranial and postcranial human remains of an adult female.	376	60025/376	PP	1550E100, depth 77″, height–above red clay 18″	Probably Gl. 2	1	50–59	1	0	0	0	0	0	0	0	1	0	0	D h+/i+, AS 7	Female
19-6-10/60026.0	Nearly complete cranial and postcranial human remains of an adult male.	314	60026/314	PP	1400E150, depth 3′4″	Probably Gl. 6 or mod.	1	35–44	1	0	0	0	0	0	1	0	0	0	0	D h+/i+, T8, AS 4	Male
19-6-10/60027.0	Nearly complete cranial and partial postcranial human remains of an adult male.	206	60027/m9	PP	950E150, head + feet depth 1′, middle of body 2′	Probably Gl. 6	1	35–39	1	0	0	0	0	0	1	0	0	0	0	D f-g, AS 4	Male
19-6-10/60028.0	Fragmentary cranial and partial postcranial human remains of an adult male.	6	60028/m22	PP	1600E150, depth 1′	Probably Gl. 6	1	50–59	1	0	0	0	0	0	1	0	0	0	0	D edentulous, AS 7, T 10	Male
19-6-10/60029.0	Nearly complete cranial and partial postcranial human remains of an adult female.	382	60029/382/f37	PP	1550E100, depth 32″ height–above red clay 34″	Probably Gl. 3 or 4	1	25–34	1	0	0	0	0	0	0	0	1	0	0	D d-e, AS 3	Female
19-6-10/60030.0	Nearly complete cranial and partial postcranial human remains of an adult male.	432	60030	PP	1100E150, depth 25″, height–above red clay 35″	Probably Gl. 4	1	30–34 (?old er 40–50)	1	0	0	0	0	0	1	0	0	0	0	AS 3	Female
19-6-10/60031.0	Nearly complete cranial human remains of an adult female.	393	60031	PP	West dump trench, depth 60″	Probably Gl. 1 or 2	1	45–55+	1	0	0	0	0	0	0	0	1	0	0	D h+/i+	Female
19-6-10/60032.0	Nearly complete cranial human remains of an adult female.	609	60032	PP	North terrace, west of room Aa	Probably B&W or Gl. 1	1	25–29	1	0	0	0	0	0	0	0	1	0	0	D e	Female
19-6-10/60033.0	Nearly complete cranial human remains of an adult that is probably female.	407	60033/407	PP	1550E100, depth 87″, height–above red clay 19″	Pres. Gl. 2 or 3	1	45–54	1	0	0	0	0	0	0	0	0	1	0	D h/i severe antemortem loss	?Female
19-6-10/60034.0	Nearly complete cranial human remains of an adult that is probably male.	430	60034/430	PP	1400E100, test xi, cut 5	Probably Gl. 4	1	30–39	1	0	0	0	0	0	0	1	0	0	0	D g/d-very uneven	?Male
19-6-10/60035.0	Nearly complete cranial and partial postcranial human remains of an adult female.	76	60035	PP	1600E150, depth 4′, 6″ above red	Pres. Gl. 3	1	60+	1	0	0	0	0	0	0	0	1	0	0	D edentulous, AS 8	Female
19-6-10/60036.0	Nearly complete cranial and partial postcranial human remains of an adult female.	54	60036/f38	PP	1650W50, depth 20″	Indet., early	1	30–34	1	0	0	0	0	0	0	0	1	0	0	D g, AS 3	Female

Peabody number	Pathology type	Pathology description	Cranial mod.	Associated funerary objects	Field context	Notes
19-6-10/60018.1	Nutr.	Osteoporosis	Occ. def.	Broken bowl awl (no number)–lost in trenches		
19-6-10/60019.0	Arthritis, nutr.	Spinal osteophytosis, cribra orbitalia, slight enamel hypoplasia	?Occ. def.	Broken Frijolito bowl no. 1909 (67094)		
19-6-10/60020.0	Other	Spina bifida vt/s4–5	Unobs.	Stone ax no. 802 (68731), scoop of broken Bisc. B bowl no. 824 (67224)	Pecos white	
19-6-10/60020.1	Infection	Caries, few teeth remaining, thick diploe	Poss. cran. def.	Pecos white sherds		Field number problematic
19-6-10/60021.0	Infection	Caries	Occ. def.			
19-6-10/60022.0	Trauma, arthritis, nutr.	Trauma L elbow joint–humerus, radius and ulna affected, spinal osteophytosis especially vt/l, some osteoporosis	Not obs.	Stone ax no. 802 (68731), scoop of broken Bisc. B bowl no. 824 (67224)		Field number and provenience problematic
19-6-10/60022.1	Infection	Dental abscess	Unobs.	Stone ax no. 802 (68731), scoop of broken Bisc. B bowl no. 824 (67224)		Field number and provenience problematic
19-6-10/60023.0	Arthritis, other	Frontal–lytic lesion, peg incisors	Occ. def.	Turquoise no. 868 (69081), projectile point no. 869 (68414)		
19-6-10/60025.0	Trauma, infection, arthritis, nutr.	Trauma Mt1, periostitis–R tibia, spinal osteophytosis, osteoporosis, enamel hypoplasia	Occ. def.	B&W, no number		
19-6-10/60026.0	Infection, arthritis	Dental abscesses, DJD–moderate, spinal osteophytosis	No cran. def.	Stone ax no. 821 (68730)		
19-6-10/60027.0	Arthritis, nutr.	DJD–elbow, enamel hypoplasia	Lamb. def.		Burial at length	
19-6-10/60028.0	Infection, arthritis, nutr.	Extensive periostitis–appendicular skeleton, lesion on L parietal, spinal osteophytosis, extensive DJD, osteoporosis	Occ./ Lamb. def.			
19-6-10/60029.0	Other	Maleruption maxillary premolar, supernumerary tooth	Occ. def.	Projectile point no. 1183 (68425)	Frijolito pottery	
19-6-10/60030.0	Arthritis, nutr.	Slight DJD, spinal osteophytosis, osteoporosis	?Lamb. def.	Black olla no. 1441 (67044), antler tool frag. no. 1404 (68260)		
19-6-10/60031.0			No cran. def.		Probably 1 B&W sherd	
19-6-10/60032.0	Trauma, infection	Trauma to right side face, extensive infection, lost eye?	Lamb. def.			
19-6-10/60033.0			Not obs.			
19-6-10/60034.0	Infection	Dental abscess lower L P2	Occ. def.	Projectile point no. 1388 (68430)		Field number and provenience problematic
19-6-10/60035.0	Infection, arthritis, nutr.	Healed periostitis in L tibia, spinal osteophytosis, DJD, severe osteoporosis	Occ. def.			
19-6-10/60036.0			Occ. def.			

Peabody number	Object description	FN	Specimen label	Site	Intrasite provenience	Ceramic stage	MNI	Age (yrs.)	Ad	Su	Ch	In	Fe	Unk. age	M	M?	F	F?	Unk. sex	Age criteria	Sex criteria
19-6-10/60038.0	Nearly complete cranial and postcranial human remains of an adult male.	271	60038	PP	1450E100, depth 18′	Probably Gl. 3	1	35–44	1	0	0	0	0	0	1	0	0	0	0	D f, AS 5	M all features
19-6-10/60039.0	Nearly complete cranial and partial postcranial human remains of an adult male.	4	60039	PP	1600E150	Pres. Gl. 4 or 5	1	50–59	1	0	0	0	0	0	1	0	0	0	0	D h/i, AS 7, T 10	M all features
19-6-10/60040.0	Partial cranial and postcranial human remains of an adult female.	310	60040/310	PP	1200E150, depth 1′	Probably Gl. 5	1	45–49	1	0	0	0	0	0	0	0	1	0	0	D h/i, AS 6	Female
19-6-10/60041.0	Nearly complete cranial and postcranial human remains of an adult female.	419	60041	PP	1400E100, test xi, base of cut 4, depth 50″	Probably Gl. 4	1	25–34	1	0	0	0	0	0	0	0	1	0	0	D d/e, T 6	Female
19-6-10/60042.0	Nearly complete cranial and postcranial human remains of an adult male.	402	60042/402	PP	1550E100, depth 13″, height–above red clay 57″	Pres. Gl. 3	1	50–59	1	0	0	0	0	0	1	0	0	0	0	D h/i, AS 7	Male (pelvis M, cranium indet.)
19-6-10/60043.0	Nearly complete cranial and postcranial human remains of a subadult that is probably female.	324	60043/324	PP	1400E150, depth 8′	Indet.	1	10–14	0	1	0	0	0	0	0	0	0	1	0	D b2 but M3s unerupted, T < 1, AS < 1	?Female
19-6-10/60044.0	Nearly complete cranial and postcranial human remains of an adult male.	394	60044/394/m8	PP	1550E100, depth 52″, height–above red clay 1″	Probably Gl. 2	1	18–26	1	0	0	0	0	0	1	0	0	0	0	D b2, T 5, AS 2	Male
19-6-10/60045.0	Nearly complete cranial and postcranial human remains of a subadult female.	388	60045/388	PP	1550E100, depth 36″, height–above red clay 32″	Indet.	1	11–15	0	1	0	0	0	0	0	0	1	0	0	D b2, AS 1	Female
19-6-10/60046.0	Nearly complete cranial and postcranial human remains of an adult male.	205	60046/205/m1	PP	950E150, depth 3′10″	Probably Gl. 5 or 6	1	30–39	1	0	0	0	0	0	1	0	0	0	0	D g-h/g. AS 3–4	M all features
19-6-10/60047.0	Nearly complete cranial and postcranial human remains of an adult that is probably female.	557	60047/557	PP	1550W50, depth 19″, height above red clay 24″	Probably Gl. 1 or 2	1	25–29	1	0	0	0	0	0	0	0	0	1	0	D d-e/c-d, AS 2	?Female, (pelvis F, cranium M)
19-6-10/60048.0	Nearly complete cranial and partial postcranial human remains of an adult male.	8	60048	PP	1600E150, 3′ below surface, 2′ above red clay	Probably Gl. 5 or 6	1	45–49	1	0	0	0	0	0	1	0	0	0	0	D h/i, AS 6	M all features
19-6-10/60049.0	Nearly complete cranial and postcranial human remains of an adult female.	115	60049/f13/f34	PP	1450E100, depth 5′, 6′ above bottom	Very probably Gl. 4	1	45–59	1	0	0	0	0	0	0	0	1	0	0	D h/i, T 10, AS 6–7	Female
19-6-10/60050.0	Nearly complete cranial and postcranial human remains of an adult male.	630	60050/630	PP	1550W0, depth 67″, above red clay 12″	Probably B&W or Gl. 1	1	50–59	1	0	0	0	0	0	1	0	0	0	0	D nearly edentulous, T10, AS 7	Male
19-6-10/60051.0	Nearly complete cranial and partial postcranial human remains of an adult male.	124	60051/124	PP	1400E100, depth 11–12′, in scoop of red clay	Indet.	1	40–49	1	0	0	0	0	0	1	0	0	0	0	D g/f-g	Male
19-6-10/60052.0	Nearly complete cranial and postcranial human remains of an adult female.	141	60052/141/f44	PP	1550E100, depth 6′6″, 6″ above red soil	Very probably Gl. 2	1	25–29	1	0	0	0	0	0	0	0	1	0	0	D c-d, AS 2	Female
19-6-10/60053.0	Nearly complete cranial human remains of an adult male.	55	60053	PP	East slope below north end of south building, depth 2′	Indet., late	1	35–45	1	0	0	0	0	0	1	0	0	0	0	D g-h	Male
19-6-10/60054.0	Nearly complete cranial and fragmentary postcranial human remains of an adult female.	7	60054	PP	1600E150, depth 2′	Pres. Gl. 4 or 5	1	40–55+	1	0	0	0	0	0	0	0	1	0	0	D g (uneven), old 50-59 yrs	Female
19-6-10/60055.0	Nearly complete cranial and postcranial human remains of an adult female.	520	60055/520/f46	PP	North terrace, depth 42″, height–above red clay 18″	Probably Gl. 3 or 4	1	25–34	1	0	0	0	0	0	0	0	1	0	0	D d-e/d, T8, AS 3	Female

Peabody number	Pathology type	Pathology description	Cranial mod.	Associated funerary objects	Field context	Notes
19-6-10/60038.0	Trauma, infection, arthritis, nutr., other	Depression fracture on L and R parietals, tibia–saber shaped, spinal osteophytosis, osteoporosis	Occ. def.	2 projectile points no. 514 (68411)	Probably Agua Fria	
19-6-10/60039.0	Infection, arthritis, nutr., tumor	Dental infection, mandible–bony abscesses lingual surface, spinal osteophytosis, moderate osteoporosis, button osteoma frontal	Not obs.			
19-6-10/60040.0	Trauma, arthritis	Trauma cranium, DJD vt, spinal osteophytosis	Occ. def.	Turquoise bead no. 718 (69090), stone palette no. 719 (68875), bone bead no. 720 (68012)		
19-6-10/60041.0	Nutr.	Enamel hypoplasia	No cran. def.	Projectile point no. 1351 (68429)		Facial features non-Pueblo?
19-6-10/60042.0	Infection, arthritis, nutr.	Severe dental abscesses, spinal osteophytosis, vt–ankylosis, osteoporosis	Occ. def.	Projectile point (no number listed)		
19-6-10/60043.0			Not obs.	Scoops, Gl. 5 no. 883 (67008)		
19-6-10/60044.0	Nutr., other	Porotic hyperostosis, maleruption of L incisor	No cran. def.			
19-6-10/60045.0	Tumor	Mt 3 and 4 exotoses	Lamb. def.	Gl. 4 bowl no. 1172 (67031), antler tool no. 1182 (68257)		
19-6-10/60046.0	Infection, arthritis	Abscesses mandibular molars, spinal osteophytosis	Not obs.			
19-6-10/60047.0			No cran. def.	Pottery, Agua Fria red, no number	Probably Agua Fria	
19-6-10/60048.0	Infection, nutr.	Dental abscesses with molar tooth loss, marked osteoporosis	Not obs.	Potsherds (no cat. no.), 2 large stones over femora	Burial at length	
19-6-10/60049.0	Infection, arthritis, nutr.	Many dental abscesses, DJD especially elbow, spinal osteophytosis, osteoporosis	No cran. def.	Catalogued as no. 957: miniature pot (65838), bone bead (65395), projectile point (65987), piece of turquoise (66267)		
19-6-10/60050.0	Arthritis, nutr., other	Severe DJD, severe osteoporosis, spina bifida–sacrum	Occ. def.	Biscuit sherd, no number	One undecorated biscuit sherd	
19-6-10/60051.0	Infection, arthritis, nutr., tumor	Healed lesion on frontal, spinal osteophytosis, enamel hypoplasia, exotosis on 2nd Mt	Occ. def.	Gl. 2 bowls nos. 925 (65027), 1028 (65024)		
19-6-10/60052.0	Nutr.	Enamel hypoplasia	No cran. def.	Bone points no. 946 (65389), bone bead no. 947 (65390), blunt bone implement (65391), frag. animal effigy no. 865 (65545), thumb-marked sherd no. 865 (65756), miniature pot (65835), ceramic pot no. 565 (65836), obsidian projectile points no. 923 (65986), black crystals no. 924 (66205)		
19-6-10/60053.0	Infection	Dental abscesses with tooth loss	No cran. def.			
19-6-10/60054.0	Arthritis, nutr.	Spinal osteophytosis, feet–arthritic, DJD, osteoporosis, enamel hypoplasia	Occ. def.	Piece of cloth (66353)– missing	May have been mixed with 7a	
19-6-10/60055.0	Arthritis, nutr., tumor	Slight DJD elbows, enamel hypoplasia, exotosis on L 2nd Mt	Occ. def.	Stone drill no. 1766 (68663), obsidian knife no. 1767 (68599), bone implement no. 1768 (68098)	Infant no. 521 under right arm against breast	

Peabody number	Object description	FN	Specimen label	Site	Intrasite provenience	Ceramic stage	MNI	Age (yrs.)	Ad	Su	Ch	In	Fe	Unk. age	M	M?	F	F?	Unk. sex	Age criteria	Sex criteria
19-6-10/60056.0	Nearly complete cranial and partial postcranial human remains of a child of unknown sex.	621	60056/621	PP	1550W0, depth 27″, above red clay 34″	Probably Gl. 1 or 2	1	4.5–5.5	0	0	1	0	0	0	0	0	0	0	1	D 7 ± 2 yrs, PC 4.5–5.5 yrs	Indet.
19-6-10/60057.0	Nearly complete cranial and postcranial human remains of an adult female.	47	60057	PP	1650W50, 1′ under ground, 6″ above red	Indet., early	1	50–59	1	0	0	0	0	0	0	0	1	0	0	D g-h, T10, AS7	Female
19-6-10/60058.0	Nearly complete cranial and postcranial human remains of an adult male.	108	60058	PP	1400E100, depth 11′, 1′ above red clay	Very probably Gl. 2	1	50–59	1	0	0	0	0	0	1	0	0	0	0	D h, AS7	M all features
19-6-10/60059.0	Partial cranial and postcranial human remains of an adult female.	133	60059/133	PP	1450E100, depth 8′, 4′ above red clay	Probably Gl. 3	1	25–34	1	0	0	0	0	0	0	0	1	0	0	D e, AS3	F all features
19-6-10/60060.0	Nearly complete cranial and partial postcranial human remains of an adult female.	363	60060/363	PP	1550E100, depth 48″, 28″ above red clay	Probably Gl. 5	1	35–45	1	0	0	0	0	0	0	0	1	0	0	D g uneven, T8, SB 5, AS4	F all features
19-6-10/60061.0	Nearly complete cranial and postcranial human remains of an adult female.	639	60061/639	PP	North terrace, depth 2–8″, above rock	Probably Gl. 6	1	50–59	1	0	0	0	0	0	0	0	1	0	0	D h/h, T10, AS 8	Female
19-6-10/60062.0	Nearly complete cranial and postcranial human remains of an adult female.	321	60062/321	PP	North terrace, depth 4′	Indet., Gl. 1 to 4	1	40–49	1	0	0	0	0	0	0	0	1	0	0	D h+/i+, AS 4–5	Female
19-6-10/60063.0	Nearly complete cranial and postcranial human remains of an adult female.	537	60063/537	PP	1550W50, depth 17″	Indet.	1	45–54	1	0	0	0	0	0	0	0	1	0	0	D g-h, T10, AS7	F all features
19-6-10/60064.0	Nearly complete cranial and postcranial human remains of an adult male.	475	60064/475	PP	1000E100, depth 40″, height–above red clay 30″	Probably Gl. 4 or 5	1	40–44	1	0	0	0	0	0	1	0	0	0	0	D g-h/h, T8, SB 4, AS5	M all features
19-6-10/60065.0	Nearly complete cranial and postcranial human remains of an adult male.	237	60065/237	PP	1450E100, depth 3–4′	Pres. Gl. 5	1	40–44	1	0	0	0	0	0	1	0	0	0	0	D f-h uneven, T 8, AS5	M all features
19-6-10/60066.0	Fragmentary cranial and partial postcranial human remains of an adult female.	323	60066	PP	1400E150, depth 9′	Indet.	1	60+	1	0	0	0	0	0	0	0	1	0	0	AS 8, cranial sutures 50+ yrs	F all features
19-6-10/60067.0	Nearly complete cranial and partial postcranial human remains of a subadult female.	177	60067/117	PP	1400E100, depth 14′	Indet.	1	15–19	0	1	0	0	0	0	0	0	1	0	0	D b1, T2, AS2	Female
19-6-10/60068.0	Nearly complete cranial human remains of an adult female.	7a	60068/7a	PP	1600E150, depth 2′8″	Pres. Gl. 4 or 5	1	45–55+	1	0	0	0	0	0	0	0	1	0	0	D h+/i+	Female
19-6-10/60070.0	Partial cranial and postcranial human remains of an adult female.	114	60070/114	PP	1400E100, depth 10′, 2 1/2′ above red	Pres. Gl. 3	1	40–49	1	0	0	0	0	0	0	0	1	0	0	D h+/i+ (severe ante-mortem molar loss), AS5	Female
19-6-10/60071.0	Nearly complete cranial and postcranial human remains of an adult male.	359	60071	PP	1350E150, depth 9′7″, 17″ above red clay	Indet.	1	35–39	1	0	0	0	0	0	1	0	0	0	0	D h/h-i, T7, AS 4	M all features
19-6-10/60072.0	Partial cranial and nearly complete postcranial human remains of an adult female.	144	60072/144	PP	1400E100, depth 12′, 5′ above red clay	Pres. Gl. 4 or 5	1	40–49	1	0	0	0	0	0	0	0	1	0	0	D i+, AS5	Female
19-6-10/60073.0	Partial cranial and nearly complete postcranial human remains of an adult female.	353	60073/353	PP	1550E100,depth 30″, 48″ above red clay	Probably Gl. 6	1	30–34	1	0	0	0	0	0	0	0	1	0	0	D g, AS3	Female
19-6-10/60074.0	Nearly complete cranial and postcranial human remains of an adult male.	150	60074/150	PP	1400E100	Probably Gl. 3	1	25–29	1	0	0	0	0	0	1	0	0	0	0	D e-f/e, T4, AS 2	M all features
19-6-10/60075.0	Nearly complete cranial and postcranial human remains of an adult male.	655	60075	PP	1550W0, depth 35″, above red clay 25″	Probably Gl. 1	1	35–45	1	0	0	0	0	0	1	0	0	0	0	D h/i (all molars lost antemortem), AS 4–5	M all features

Peabody number	Pathology type	Pathology description	Cranial mod.	Associated funerary objects	Field context	Notes
19-6-10/60056.0			Not obs.	Antler tool no. 2063 (68266), bone tool no. 2064 (68101)		
19-6-10/60057.0	Trauma, infection, arthritis, nutr., tumor	Healed fracture to R Mc, abscesses, spinal osteophytosis, moderate DJD, large button osteoma on frontal	Not obs.	2 scrapers (65979), shell pendant (66290), bone awl (65393), 2 sherds no. 114.		
19-6-10/60058.0	Arthritis, nutr.	Severe DJD, severe spinal osteophytosis, osteoporosis	Occ. def.			
19-6-10/60059.0	Infection, arthritis, nutr., other	Impacted M3, spondylolysis vt/l5, slight DJD, enamel hypoplasia,	No cran. def.			
19-6-10/60060.0	Infection, arthritis	Several molar abscesses, slight DJD	Occ. def.	Pajaritan sherds, no number	Pajaritan pottery	
19-6-10/60061.0	Trauma, infection, arthritis	Trauma R Mc3, osteomyelitis–L humerus, clavicle, ulna, tibia, abscessing molars, marked DJD, spinal osteophytosis, ankylosis vt/c3–4, vt/t3–4	Occ. def.	Gl. 3 bowl no. 2073 (67132)	Burial at length	
19-6-10/60062.0	Infection, arthritis, nutr.	Infection L mastoid, upper R M2 abscess slight DJD, spinal osteophytosis, slight enamel hypoplasia	Not obs.			
19-6-10/60063.0	Trauma, infection, arthritis, tumor	Healed fracture R clavicle, abscesses with severe antemortem maxillary premolar and molar tooth loss, moderate DJD, spinal osteophytosis, 10 button osteomas frontal and parietals	Occ. def.	2 eagle talons no. 1782 (69149), stone ball no. 1784 (68905), Gl. 2 bowl frag. no. 1760 (67081), small black pot no. 1761 (67082), wolf tooth no. 1783 (69148), obsidian projectile point no. 1785 (68440), flint knife no. 1786 (68600)		
19-6-10/60064.0	Arthritis, nutr., tumor	DJD–arms, spinal osteophytosis, enamel hypoplasia, diffuse button osteomas L parietal	Occ. def.	Pajaritan sherds, no number	Pajaritan pottery	
19-6-10/60065.0	Infection, arthritis, nutr., tumor	Abscesses–upper R and L M1s, moderate DJD, enamel hypoplasia, slight osteoporosis, small osteoma–frontal	No cran. def.			
19-6-10/60066.0	Arthritis, nutr.	Moderate DJD, spinal osteophytosis, moderate osteoporosis	No cran. def.	Gl. 4 bowl no. 1836 (67007), no. 881 (67005), small black pot no. 882 (67006), 2 bone awls no. 858 (67827), antler tool no. 859 (68252)		
19-6-10/60067.0	Trauma, infection, arthritis, tumor	Septic infection right hip fracture, infection L mastoid, healed fracture nasals, L Mc2 – exotosis	Occ. def.	Bowl no. 1142, scraper no. 1169 (65994)		
19-6-10/60068.0	Nutr.	Mild cribra orbitalia	Occ. def.			
19-6-10/60070.0	Trauma, arthritis, nutr., tumor	Trauma–R dist. radius, v/c 2–3 ankylosed, severe DJD–especially right elbow, severe spinal osteophytosis, moderate osteoporosis, tumor–L patella	Occ. def.	Bone bead no. 1082 (65392)		
19-6-10/60071.0	Infection, arthritis, nutr.	Abscess lower R M2, spinal osteophytosis, moderate DJD, moderate osteoporosis	No cran. def.	Gl. 3 bowl no. 1150 (67020)		
19-6-10/60072.0	Infection, arthritis, nutr.	Mandibular dental abscesses and tooth loss, moderate DJD, marked spinal osteophytosis, slight osteoporosis	No cran. def.	Bisc. and Pecos-Frijolito sherds, no numbers	Sherds, biscuit and Pecos, Frijolito	
19-6-10/60073.0	Arthritis	Slight DJD, spinal osteophytosis	Unobs.		Burial at length	
19-6-10/60074.0	Nutr.	Enamel hypoplasia	Occ. def.			
19-6-10/60075.0	Trauma, arthritis	Ankylosis L calcaneus / talus and v/c 2–3, spinal osteophytosis, spondylytis deformans, moderate DJD–especially L tibia, fibula, and L and R clavicles	No cran. def.	Black pot no. 2117 (67136), bone tube no. 2264 (68029), bone flute no. 2196 (68118), shell bead no. 2263 (69072), 3 pcs crystal no. 2262 (68932) 4 crystals no. 2261 (68031)		

Peabody number	Object description	FN	Specimen label	Site	Intrasite provenience	Ceramic stage	MNI	Age (yrs.)	Ad	Su	Ch	In	Fe	Unk. age	M	M?	F	F?	Unk. sex	Age criteria	Sex criteria
19-6-10/60076.0	Nearly complete cranial and postcranial human remains of an adult female.	612	60076	PP	1550W50, depth 40″, below red clay 15″	Indet.	1	40–49	1	0	0	0	0	0	0	0	1	0	0	D h/i, AS 5–6	Female
19-6-10/60077.0	Partial cranial and postcranial human remains of a subadult female.	577	60077	PP	1550W50, depth 45″, below upper edge of green clay 17″	Probably B&W	1	18–20	0	1	0	0	0	0	0	0	1	0	0	D b1, AS 1, iliac crest incomplete fusion	Female
19-6-10/60078.0	Nearly complete cranial and partial postcranial human remains of a child of unknown sex.	463	60078/463	PP	1400E100, test xii, top of cut 3, depth 53″	Probably Gl. 4 or 5	1	2–3	0	0	1	0	0	0	0	0	0	0	1	D 4 ± 1 yrs, PC 1.5–2.5 yrs	Indet.
19-6-10/60079.0	Nearly complete cranial and partial postcranial human remains of a child of unknown sex.	24	60079	PP	1550E150, center west side, 4′ under surface, 1′ above red clay	Indet.	1	4.5–5.5	0	0	1	0	0	0	0	0	0	0	1	D 6.5 ± 2 yrs, PC 4.5–5.5 yrs	Indet.
19-6-10/60080.0	Nearly complete cranial and postcranial human remains of an adult male.	625	60080/625	PP	1550W0, depth 62″, above red clay 14″	Probably Gl. 1	1	30–39	1	0	0	0	0	0	1	0	0	0	0	D e, AS 4	Male
19-6-10/60081.0	Nearly complete cranial and postcranial human remains of an adult female.	571	60081/571 (see remarks)	PP	1550W50, depth 34″, above red clay 17″	Indet.	1	30–34	1	0	0	0	0	0	0	0	1	0	0	D g, AS 3	Female
19-6-10/60082.0	Nearly complete cranial and partial postcranial human remains of an adult female.	43	60082	PP	1550E0, depth 58″	Probably Bisc. B	1	40–44	1	0	0	0	0	0	0	0	1	0	0	D g, T 8, AS 5	F all features
19-6-10/60083.0	Nearly complete cranial and partial postcranial human remains of an adult female.	21	60083/21/f sub2	PP	1550E100, nw corner, 3′ under ground, 1 1/2′ above clay	Pres. Gl. 4 or 5	1	60+	1	0	0	0	0	0	0	0	1	0	0	D h/i (severe antemortem tooth loss), AS 8	Female
19-6-10/60084.0	Nearly complete cranial and postcranial human remains of an adult female.	452	60084/Fsub 1	PP	1400E100, test xii, base of cut 1, depth 18″	Probably Gl. 6	1	35–45	1	0	0	0	0	0	0	0	1	0	0	T 8, AS 4	Female
19-6-10/60085.0	Nearly complete cranial and partial postcranial human remains of an adult male.	429	60085/429	PP	1000E150, depth 52″, height–above red clay 13″	Probably Gl. 5	1	25–34	1	0	0	0	0	0	1	0	0	0	0	D f/e	Male
19-6-10/60086.0	Nearly complete cranial and postcranial human remains of an adult male.	174	60086/174	PP	1450E100depth 2′6″	Probably Gl. 5	1	35–39	1	0	0	0	0	0	1	0	0	0	0	D h severe antemortem tooth loss, AS 4, T 7	Male
19-6-10/60088.0	Nearly complete cranial and postcranial human remains of an adult male.	453	60088/453	PP	1400E100, test xii, cut 2, depth 28″	Probably Gl. 6	1	35–45	1	0	0	0	0	0	1	0	0	0	0	D h+/i+, AS 4, T 8	Male
19-6-10/60089.0	Nearly complete cranial and postcranial human remains of an adult female.	170	60089, 170	PP	1450E100, depth about 4′6″	Probably Gl. 4 or 5	1	40–49	1	0	0	0	0	0	0	0	1	0	0	D g-h/h, T9, AS 6	F all features
19-6-10/60091.0	Partial cranial and postcranial human remains of an adult female.	165	60091, 165	PP	1500E100, depth 5′6″, 2′ above red	Pres. Gl. 3 or 4	1	35–39	1	0	0	0	0	0	0	0	1	0	0	D g, AS 4	Female
19-6-10/60092.0	Nearly complete cranial and partial postcranial human remains of an adult male.	622	60092/622	PP	1550W0, depth 46″, above red clay 35″	Probably Gl. 1	1	35–39	1	0	0	0	0	0	1	0	0	0	0	D f-g	Male
19-6-10/60093.0	Nearly complete cranial and partial postcranial human remains of an adult female.	575	60093/575	PP	1550W50, depth 30″, above red clay 22″	Probably Gl. 1	1	27–34	1	0	0	0	0	0	0	0	1	0	0	D e-f/e, T 5, AS 3	F all features
19-6-10/60094.0	Nearly complete cranial and postcranial human remains of an adult female.	11	60094/11	PP	1550E150, red clay 4′ from surface	Pres. Gl. 3 or 4	1	25–29	1	0	0	0	0	0	0	0	1	0	0	D f, T 4, AS 1	Female
19-6-10/60095.0	Nearly complete cranial and partial postcranial human remains of an adult female.	181	60095/181	PP	1500E50, terrace, depth 7′	Indet.	1	50–59	1	0	0	0	0	0	0	0	1	0	0	D h+/i+, AS 7	Female

Peabody number	Pathology type	Pathology description	Cranial mod.	Associated funerary objects	Field context	Notes
19-6-10/60076.0	Arthritis, nutr.	Moderate DJD, ankylosis v/t, moderate osteoporosis, enamel hypoplasia	Lamb. def.	Part of B&W bowl no. 1986 (67119)		Different cranial features
19-6-10/60077.0	Infection	Periostitis on Ti, Fe	Slight lamb. def.			
19-6-10/60078.0			Not obs.			
19-6-10/60079.0			Not obs.			
19-6-10/60080.0	Arthritis	Vt/l Schmorl's nodes, slight DJD	Lamb. def.	6 shell beads no. 2071 (69071), 8 shell pendants no. 2070 (69118), 1 turquoise stone no. 2072 (69128), 5 projectile points no. 2069 (68447)		
19-6-10/60081.0	Infection, arthritis	Dental abscesses, bowed tibia, slight DJD	Occ. def.	Broken pot, Gl. 2 no. 1987 (67096)		
19-6-10/60082.0	Trauma, infection, arthritis	Healed trauma–R clavicle, severe infection right mastoid, abscess upper LM1, slight DJD	Occ. def.	Large Bisc. B bowl (66899), broken red ware bowl (modern?) (65010)		
19-6-10/60083.0	Infection, arthritis	Impacted molar, L tibia–healed periostitis, DJD–elbow, vt spondylytis deformans	No cran. def.			
19-6-10/60084.0	Infection, arthritis, nutr.	Molars lost antemortem, marked DJD–especially elbows, spinal osteophytosis and spondylitis, osteoporosis	Occ. def.		Burial at length	
19-6-10/60085.0	Nutr.	Enamel hypoplasia	Not obs.	Leather bag fragment (67651)	Probably Frijolito sherds, remains of skin bag, pottery smoother	
19-6-10/60086.0	Infection, arthritis, nutr.	Molar loss antemortem, spinal osteophytosis, slight DJD, enamel hypoplasia	Occ. def.	3 bone beads no. 1139 (66351)	Late Pecos white sherds	
19-6-10/60088.0	Infection, arthritis	Severe caries lower molars, several abscesses upper teeth, moderate spinal osteophytosis, slight DJD	Occ. def.	2 projectile points no. 1506 (68433), pipe (no number), bone needle no. 1504 (68131), antler tool no. 1505 (68262)	Burial at length	
19-6-10/60089.0	Arthritis, nutr.	Moderate DJD, marked spinal osteophytosis, slight enamel hypoplasia, moderate osteoporosis	Occ. def.	Antler no. 1187 (65385), 2 projectile points no. 1187 (65995)		
19-6-10/60091.0	Infection, arthritis, nutr.	Marked antemortem tooth loss, moderate DJD–especially elbow, knee, vt, marked spinal osteophytosis–Schmorl's nodes, marked osteoporosis	Occ. def.			
19-6-10/60092.0	Arthritis, nutr., other	Moderate DJD, slight osteoporosis, fusion of lower R I2–C, extra upper right incisor, pronounced mandibular torus	Slight occ. def.	Bone implement no. 2059b (68102)	See no. 619, which notes that leg may have been lost before death (L tibia and fibula not present)	
19-6-10/60093.0	Arthritis	Slight DJD	Slight occ. def.	Bone awl no. 1913 (68099)		
19-6-10/60094.0	Infection, nutr.	Abscess lower L M2, enamel hypoplasia	Occ. def.	Sherds		
19-6-10/60095.0	Arthritis, nutr.	Slight DJD, slight osteoporosis (femora have very flat necks)	Occ. def.	Black olla no. 1184, red olla fragments no. 1186 (65056) Gl. 1–2 red bowl no. 1182 (65048), Gl. 2? red bowl no. 1185 (65015)		

Peabody number	Object description	FN	Specimen label	Site	Intrasite provenience	Ceramic stage	MNI	Age (yrs.)	Ad	Su	Ch	In	Fe	Unk. age	M	M?	F	F?	Unk. sex	Age criteria	Sex criteria
19-6-10/60096.0	Nearly complete cranial and postcranial human remains of an adult male.	532	60096/532/m18	PP	1000W50, depth 24″, height–on shale	Indet.	1	50–59	1	0	0	0	0	0	1	0	0	0	0	D h+/i+, T 10, SB 6, AS7	M all features
19-6-10/60097.0	Nearly complete cranial and partial postcranial human remains of an adult female.	658	60097	PP	1000E150, depth 39″, above red clay 43″	Indet.	1	19–21	1	0	0	0	0	0	0	0	1	0	0	D b2, AS 1	F all features
19-6-10/60098.0	Nearly complete cranial and partial postcranial human remains of an adult female.	244	60098/244	PP	1450E100, depth about 6–7′	Indet.	1	18–24	1	0	0	0	0	0	0	0	1	0	0	D b2, AS B	F all features
19-6-10/60099.0	Nearly complete cranial and partial postcranial human remains of an adult female.	464	60099	PP	1400E100, test xii, base of cut 2	Probably Gl. 5	1	25–29	1	0	0	0	0	0	0	0	1	0	0	D f, T 5, SB 3, AS d	F all features
19-6-10/60100.0	Nearly complete cranial and partial postcranial human remains of an adult female.	465	60100/465	PP	1400E100, test xii, top of cut 3	Probably Gl. 6	1	20–29	1	0	0	0	0	0	0	0	1	0	0	D c/d, T 5, SB 3, AS 4	Female
19-6-10/60101.0	Nearly complete cranial and partial postcranial human remains of an adult female.	117	60101	PP	1450E100, depth 4′, 4′ above red	Pres. Gl. 5	1	40–50+	1	0	0	0	0	0	0	0	1	0	0	D g (uneven due to ante-mortem loss), AS 7	Female
19-6-10/60102.0	Nearly complete cranial and postcranial human remains of an adult male.	450	60102	PP	1400E100, test xi, bottom of cut 7, ne corner	Indet.	1	30–39	1	0	0	0	0	0	1	0	0	0	0	D f, T 7, SB 4, AS 3	Male
19-6-10/60103.0	Nearly complete cranial and postcranial human remains of an adult male.	444	60103	PP	1100E150, on red clay, covered with brown earth	Indet.	1	25–34	1	0	0	0	0	0	1	0	0	0	0	D e-f	Male (pelvis M, cranium indet.)
19-6-10/60104.0	Nearly complete cranial human remains of an adult male.	Bb1	60104	FL	East and north arroyos	Probably B&W	1		1	0	0	0	0	0	1	0	0	0	0		Male
19-6-10/60106.0	Nearly complete cranial and postcranial human remains of an adult male.	46	60106	PP	1550E0, depth 6′	Pres. Gl. 3 or 4	1	27–34	1	0	0	0	0	0	1	0	0	0	0	D e, T 5, SB 3-4, AS 3	Male
19-6-10/60107.0	Nearly complete cranial and postcranial human remains of a subadult male.	455	60107	PP	1400E100, test xii, cut 2, depth 33″	Probably Gl. 6	1	17–20	0	1	0	0	0	0	1	0	0	0	0	D b2, T 1?, SB 1, AS < 1 (epiphyseal lines still prominent, prox. clavicle still unfused, right iliac crest unfused)	Male (all features except pelvis IPR)
19-6-10/60108.0	Nearly complete cranial and postcranial human remains of an adult female.	254	60108, f44	PP	1450E100, depth 8′, ±10′ above red	Very probably Gl. 4	1	50–59	1	0	0	0	0	0	0	0	1	0	0	D h/i, T 10, SB 6, AS 6	F all features
19-6-10/60109.0	Nearly complete cranial and postcranial human remains of an adult male.	270	60109	PP	1450E100, test x, base of cut 6, depth 13′	Probably Gl. 3	1	30–39	1	0	0	0	0	0	1	0	0	0	0	D f, T 6, SB 4, AS 5	M all features
19-6-10/60110.1	Fragmentary postcranial human remains of an adult of unknown sex.	C	60110-1	PP	Church	Indet.	1		1	0	0	0	0	0	0	0	0	0	1		Indet.
19-6-10/60110.2	Very fragmentary postcranial human remains of an adult of unknown sex.	C	60110	PP	Church	Indet.	1		1	0	0	0	0	0	0	0	0	0	1		Indet.
19-6-10/60111.1	Very fragmentary postcranial human remains of an adult female.	C	60111-5, 60111-7	PP	Church	Indet.	1	20–34	1	0	0	0	0	0	0	0	1	0	0	PC < 35 yrs	Female
19-6-10/60112.0	Fragmentary postcranial human remains of an adult of unknown sex.	C	60112-2, 60112-4	PP	Church	Indet.	1		1	0	0	0	0	0	0	0	0	0	1		Indet.

Peabody number	Pathology type	Pathology description	Cranial mod.	Associated funerary objects	Field context	Notes
19-6-10/60096.0	Trauma, infection, arthritis, nutr.	Fibula–healed fracture?, extensive remodeling of limb bones especially tibia, slight DJD, spinal osteophytosis, enamel hypoplasia	Not obs.	Gl. 2 bowl no. 1756 (67079), stone scraper no. 1781 (68537)		
19-6-10/60097.0	Infection	2 cranial lesions–R parietal, os inca, dental abscess, thickening upper tibias, R clavicle–periostitis	Lamb. def.	Gl. 4 bowl no. 2127 (67137)		
19-6-10/60098.0			Lamb. def.	Early Gl. 4 bowl no. 354 (66985), 2 used flakes nos. 13631, 13632 (68604)		
19-6-10/60099.0			Lamb. def.	27 turquoise beads no. 1527 (69082)	Pajaritan pottery	
19-6-10/60100.0	Infection	Possible tuberculosis? vt/t have severe resorption, sacrum also affected, some type of severe infection	Not obs.		Burial at length	
19-6-10/60101.0	Trauma, infection, arthritis	Healed trauma–L clavicle, abscess and molar tooth loss, vt/c ankylosis 2–3 with arthritis	Not obs.	Pajaritan and a few doubtful Pecos white sherds (no numbers)	Glaze 2	
19-6-10/60102.0	Nutr., other	Enamel hypoplasia, peg upper M3s, unusual canine morphology	Not obs.	Gl. 2 bowl no. 1446 (67050)	Glaze 2	
19-6-10/60103.0	Arthritis	Ankylosis of vt/c 3–4	Not obs.	Gl. 2 bowl no. 1445 (67048), projectile point no. 1459 (68432) scraper no. 1460 (68528)	Glaze 2	
19-6-10/60104.0	Infection	Marked antemortem tooth loss	Not obs.			Square orbits, different facial features, mastoids not that large, no field card
19-6-10/60106.0			Not obs.	Mano found near (no. 118/no cat. no.)		
19-6-10/60107.0	Other	Spondylolysis vt/l5 neural arch	Not obs.	Flint and obsidian projectile points no. 1507 (68434), stone hammer no. 1508 (68733)	Burial at length	Marked torsion L femur (was sectioned)
19-6-10/60108.0	Nutr., other	Enamel hypoplasia, developmentally only 4 of lower 6 anterior teeth present, lateral incisors and canines are a morphologically intermediate tooth	Occ. def.			R clavicle sampled for isotopes
19-6-10/60109.0	Arthritis, nutr.	DJD, enamel hypoplasia	Occ. def.	Bone awl (no number listed)		
19-6-10/60110.1			No cran.			Reassociated dispersed elements 60112-1, -3, 60111-8, -3, 60110-1, -3, and recatalogued as 60110-1.
19-6-10/60110.2			No cran.			Femur
19-6-10/60111.1			No cran.			Tibia, reassociated by Owsley team as GG (1/98)
19-6-10/60112.0	Trauma	Healed fracture on L fibula	No cran.			Removed and recatalogued 2 fibulae as 60110-1

Peabody number	Object description	FN	Specimen label	Site	Intrasite provenience	Ceramic stage	MNI	Age (yrs.)	Ad	Su	Ch	In	Fe	Unk. age	M	M?	F	F?	Unk. sex	Age criteria	Sex criteria
19-6-10/60113.0	Fragmentary postcranial human remains of an adult male.	C	60113-1 and 60113-3, 59884, 59885, 59886, 60110, 60111, 59880, 59887	PP	Church	Indet.	1		1	0	0	0	0	0	1	0	0	0	0		Male
19-6-10/60113.1	Fragmentary postcranial human remains of an adult female.	C	60113-4/60113-6	PP	Church	Indet.	1		1	0	0	0	0	0	0	0	1	0	0		Female
19-6-10/60113.2	Fragmentary postcranial human remains of an adult male.	C	60113-2/60113-5	PP	Church	Indet.	1		1	0	0	0	0	0	1	0	0	0	0		Male
19-6-10/60116.1	Fragmentary postcranial human remains of an adult of unknown sex.	C	60116-1	PP	Church	Indet.	1		1	0	0	0	0	0	0	0	0	0	1		Indet.
19-6-10/60116.2	Fragmentary postcranial human remains of an adult of unknown sex.	C	60116-2	PP	Church	Indet.	1		1	0	0	0	0	0	0	0	0	0	1		Indet.
19-6-10/60117.0	Fragmentary postcranial human remains of an adult of unknown sex.	C	60117-4	PP	Church	Indet.	1		1	0	0	0	0	0	0	0	0	0	1		Indet.
19-6-10/60118.0	Fragmentary postcranial human remains of an adult male.	C	60116, 59885, 60118, 59882, 60110, 60111	PP	Church	Indet.	1	35–44	1	0	0	0	0	0	1	0	0	0	0	PC 38–43 yrs	Male
19-6-10/60119.0	Very fragmentary postcranial human remains of an adult of unknown sex.	C	60119-1	PP	Church	Indet.	1		1	0	0	0	0	0	0	0	0	0	1		Indet.
19-6-10/60133.0	Fragmentary cranial and nearly complete postcranial human remains of an adult male.	320	60133/m6	PP	1100E150, depth 8″, 2′ above red soil	Pres. Gl. 3 or 4	1	25–34	1	0	0	0	0	0	1	0	0	0	0	T 6, SB 3, AS -, D e	Male
19-6-10/60134.0	Fragmentary postcranial human remains of an adult female.	243	60134	PP	1450E100,	Indet., poss. Gl. 4	1	20–24	1	0	0	0	0	0	0	0	1	0	0	T 1-3, SB 1-2, AS 1	F all features
19-6-10/60135.0	Nearly complete cranial and postcranial human remains of an adult male.	140	60135/140	PP	1450E100depth 11′, 1′ above red clay	Indet.	1	25–34	1	0	0	0	0	0	1	0	0	0	0	D e, AS 3	M all features
19-6-10/60136.0	Fragmentary cranial and partial postcranial human remains of an adult male.	Bb2	60136	FL	East and north arroyos	Probably B&W	1	45–55+	1	0	0	0	0	0	1	0	0	0	0	D h+/i+	Male
19-6-10/60137.0	Nearly complete postcranial human remains of an adult female.	327a	60137	PP		Probably Gl. 4	1	40–44	1	0	0	0	0	0	0	0	1	0	0	AS 5	F all features
19-6-10/60138.0	Fragmentary cranial and partial postcranial human remains of an adult female.	305	60138	PP	1200E150, depth 1′, 1′ above red soil	Indet.	1	50–59	1	0	0	0	0	0	0	0	1	0	0	D f, T 10, SB 6, AS 7	F all features
19-6-10/60140.0	Nearly complete cranial and fragmentary postcranial human remains of a child of unknown sex.	232	60140	PP	1450E100, depth about 8′	Very probably Gl. 3	1	5–7	0	0	1	0	0	0	0	0	0	0	1	D 9 ± 2 yrs, PC 4.5–4.5 yrs	Indet.
19-6-10/60142.0	Nearly complete cranial and postcranial human remains of a subadult of unknown sex..	595	60142	PP	North terrace, depth 38″, on bedrock	Indet.	1	18–24	0	1	0	0	0	0	0	0	0	0	1	D b2-d, T 1, SB 1, AS 1, M3s still erupting	Indet.
19-6-10/60143.0	Nearly complete cranial and postcranial human remains of an adult female.	358	60143	PP	1400E150, depth 7′6″, 39″ above red clay	Indet.	1	20–24	1	0	0	0	0	0	0	0	1	0	0	D c, T 2, SB 1 AS 1 (prox. clavicle not fused)	F all features
19-6-10/60144.0	Nearly complete cranial and partial postcranial human remains of an adult female.	273	60144	PP	1450E100, depth 8′	Indet.	1	20–29	1	0	0	0	0	0	0	0	1	0	0	D c, AS 2	F all features
19-6-10/60145.0	Partial cranial and postcranial human remains of an adult that is probably female.	652	60145	PP	1550W0, depth 12″, above red clay 37″	Probably Gl. 1	1	60+	1	0	0	0	0	0	0	0	0	1	0	D h+/i+, AS 8	?Female (Pelvis-F, skull M except SM, SR)

Peabody number	Pathology type	Pathology description	Cranial mod.	Associated funerary objects	Field context	Notes
19-6-10/60113.0			No cran.			Owsley team reassociated as T. They suggest not Native American.
19-6-10/60113.1			No cran.			
19-6-10/60113.2			No cran.			
19-6-10/60116.1			No cran.			Reassociated by Owsley team as WW; reassociated dispersed elements 60116-1, 60117-3, 60118-2 and recatalogued as 60116-1, MEM 2/97.
19-6-10/60116.2			No cran.			Reassociated dispersed elements 60118-4, -1, 60117-2, -1, 60116-3, -4. recatalogued as 60116-2, MEM 2/97.
19-6-10/60117.0			No cran.			Removed and recatalogued 3 radii as 60116-1, 60116-2
19-6-10/60118.0	Trauma	Healed fractures dist. L tibia and dist. L radius	No cran.			Owsley team reassociated and designated M (male 38-43).
19-6-10/60119.0			No cran.			
19-6-10/60133.0	Nutr.	Enamel hypoplasia	Unobs.	4 scrapers no. 962 (68509), 2 hammerstones no. 961 (68857), fossil shell no. 963 (69191)		
19-6-10/60134.0			No cran.			
19-6-10/60135.0	Arthritis, other	DJD, spina bifida vt/s1–2 and vt/s 4–5	Not obs.	Bowl Gl. 3–4 no. 1022 (65044), Gl. 3 bowl no. 1023 (65055), projectile point no. 962 (65990)		
19-6-10/60136.0	Other	Bone very brittle, perforated on right acromion	Unobs.			No field card
19-6-10/60137.0	Arthritis, other	Spinal osteophytosis, generalized DJD, 5th Mcs very foreshortened	Not obs.			
19-6-10/60138.0	Arthritis	DJD	Unobs.	Gl. 3–4 bowl no. 688 (67001), drilled stone no. 733 (68885)		
19-6-10/60140.0			Not obs.			
19-6-10/60142.0	Nutr., other	Enamel hypoplasia, M3s unusual shape	Occ. def.			
19-6-10/60143.0	Infection, other	Caries upper L M1, vt/s4–5 not fused	Not obs.	Gl. 4 Pajaritan bowl no. 1148 (67019)		
19-6-10/60144.0	Trauma	Healed fractures–R radius, R ulna, R humerus	Not obs.	Gl. 5 Pajaritan bowl no. 507 (66993), frag. notched deer scapula no. 516 (68138)		
19-6-10/60145.0	Infection, arthritis, other	R tibia–infection, healed, periostitis; extensive DJD, sacral spina bifida	Occ. def.	Red and yellow Agua Fria sherds, no number	Agua Fria red and yellow, probably Glaze 1	

Peabody number	Object description	FN	Specimen label	Site	Intrasite provenience	Ceramic stage	MNI	Age (yrs.)	Ad	Su	Ch	In	Fe	Unk. age	M	M?	F	F?	Unk. sex	Age criteria	Sex criteria
19-6-10/60146.0	Fragmentary postcranial human remains of an adult female.	653	60146	PP	1550W0, depth 19″, above red clay 43″	Indet.	1	35–39	1	0	0	0	0	0	0	0	1	0	0	AS 4	Female
19-6-10/60146a	Fragmentary cranial and postcranial human remains of an adult of unknown sex.	653		PP	1550W0, depth 19″, above red clay 43″	Indet.	1	Fetus	0	0	0	0	1	0	0	0	0	0	1		Indet.
19-6-10/60147.0	Fragmentary postcranial human remains of an adult female.	126	60147	PP	1400E100, depth about 12 feet down, 1′ above red clay	Pres. Gl. 2 or 3	1	45–59	1	0	0	0	0	0	0	0	1	0	0	T 9, SB 5, AS 7	F all features
19-6-10/60148.0	Partial cranial and nearly complete postcranial human remains of an adult male.	161	60148	PP	1450E100, depth 14 1/2′, 1′ above red	Indet., early	1	35–39	1	0	0	0	0	0	1	0	0	0	0	D g, AS 4	M all features
19-6-10/60149.0	Nearly complete cranial and partial postcranial human remains of a child of unknown sex.	240	60149/240	PP	1450E100, below surface 4′	Probably Gl. 6	1	7–8	0	0	1	0	0	0	0	0	0	0	1	D 9 ± 2 yrs, PC 7.5–8.5 yrs	Indet.
19-6-10/60150.0	Nearly complete cranial and partial postcranial human remains of a subadult of unknown sex.	228	60150/228	PP	1450E100, 8 feet from surface	Very probably Gl. 3	1	9–11	0	1	0	0	0	0	0	0	0	0	1	D 11 ± 2.3 yrs, PC 8.5–9.5 yrs	Indet.
19-6-10/60151.0	Partial cranial and nearly complete postcranial human remains of a child of unknown sex.	576	60150	PP	1550W50, depth 39″, above red clay 9″	Indet.	1	5–6	0	0	1	0	0	0	0	0	0	0	1	D 6–7 yrs, PC 4.5–6.5 yrs	Indet.
19-6-10/60152.0	Fragmentary cranial and postcranial human remains of an adult female.	168	60152	PP	1500E100, terrace, depth 9′, on red clay	Indet.	1	18–24	1	0	0	0	0	0	0	0	1	0	0	D b2/c, T 1, SB 1, AS 1	Female
19-6-10/60154.0	Fragmentary cranial and partial postcranial human remains of an adult male.	458	601545	PP	1100E100, depth 27″, height–above red clay 44″	Indet.	1	30–39	1	0	0	0	0	0	1	0	0	0	0	D g, T 7, SB 4, AS 3	Male
19-6-10/60155.0	Fragmentary cranial and partial postcranial human remains of a child of unknown sex.	404	60155	PP	1550E100, depth 49″, height–above red clay 22″	Pres. Gl. 3	1	6–7	0	0	1	0	0	0	0	0	0	0	1	D 9 ± 2 yrs, PC 5.5–6.5 yrs	Indet.
19-6-10/60156.0	Fragmentary cranial and postcranial human remains of an adult male.	474	60156	PP	North terrace, depth 3′	Indet.	1	50–59	1	0	0	0	0	0	1	0	0	0	0	D i+	Male
19-6-10/60157.0	Fragmentary cranial and nearly complete postcranial human remains of a subadult male.	127	60157	PP	1450E100, depth 8′, 4′ above red clay	Probably Gl. 4	1	14–16	0	1	0	0	0	0	1	0	0	0	0	D 16–22 yrs, PC 13.5–14.5 yrs	M all features
19-6-10/60158.0	Fragmentary cranial and postcranial human remains of an adult male.	456	60158	PP	1000E150, depth 28″, height–above red clay 24″	Probably Gl. 5	1	60+	1	0	0	0	0	0	1	0	0	0	0	D i+, T 10, SB 6, AS 8	Male
19-6-10/60159.0	Fragmentary cranial and partial postcranial human remains of an adult female.	673	60159	PP	1500W0, depth 30″, on red clay	Probably B&W	1	50–59	1	0	0	0	0	0	0	0	1	0	0	D h+, AS 7	F all features
19-6-10/60161.0	Fragmentary postcranial human remains of an adult male.	654	60161	PP	1550W0, depth 30″, above red clay 24″	Indet.	1	45–49	1	0	0	0	0	0	1	0	0	0	0	T 9, SB 5, AS 6	M all features
19-6-10/60162.0	Nearly complete cranial and partial postcranial human remains of a child of unknown sex.	486	60162	PP	1400E100, test xii, base of cut 5	Probably Gl. 3	1	5.5–6.5	0	0	1	0	0	0	0	0	0	0	1	D 8 ± 2 yrs, PC 5.5–6.5 yrs	Indet.
19-6-10/60163.0	Fragmentary cranial and partial postcranial human remains of an adult male.	383	60163	PP	1550E100, depth 66″, height–above red clay 12″	Indet.	1	50–59	1	0	0	0	0	0	1	0	0	0	0	D edentulous, AS 8	M all features
19-6-10/60164.0	Nearly complete cranial and postcranial human remains of an adult female.	307	60164	PP	1450E100, test x, base of cut 3	Probably Gl. 4 or 5	1	35–39	1	0	0	0	0	0	0	0	1	0	0	D g, AS 4	F all features
19-6-10/60165.0	Nearly complete cranial and partial postcranial human remains of an adult female.	277	60165	PP	North terrace, depth 6′	Indet.	1	35–49	1	0	0	0	0	0	0	0	1	0	0	D e, AS 7	F all features
19-6-10/60166.0	Fragmentary cranial human remains of an adult male.	287	60166/287	PP	1200E150, eroded, depth 4″	Pres. Gl. 3 or 4	1	35–45	1	0	0	0	0	0	1	0	0	0	0	D g-h	M all features
19-6-10/60167.0	Fragmentary postcranial human remains of an adult of unknown sex.	288a	288, 60358, 60343-15	PP	1200E150, depth 6′, on red clay	Pres. Gl. 3 or 4	1	Adult	1	0	0	0	0	0	0	0	0	0	1		Indet.

Peabody number	Pathology type	Pathology description	Cranial mod.	Associated funerary objects	Field context	Notes
19-6-10/60146.0			No cran.	Frag. Bisc. A bowl no. 2100 (67135), piece red paint no. 2103 (69237), stone knife no. 2104, fossil no. 2102 (69195)	Agua Fria	
19-6-10/60146a			Unobs.	Frag. Bisc. A bowl no. 2100 (67135), piece red paint no. 2103 (69237), stone knife no. 2104, fossil no. 2102 (69195)	Agua Fria	
19-6-10/60147.0	Arthritis, nutr., other	DJD, 2 v/t, and 4 vt/l ankylosed, osteoporosis, spina bifida vt/s1–5, v/c 2–3	No cran.	Stone knife no. 922 (65985)		
19-6-10/60148.0	Trauma, arthritis, nutr.	Healed fracture R humerus, spinal osteophytosis, enamel hypoplasia	Not obs.	Projectile point no. 1188 (65996)		Field card says early time period
19-6-10/60149.0	Nutr.	Enamel hypoplasia	Not obs.		Burial at length, Pecos white and 2 pcs polished black sherds	
19-6-10/60150.0	Infection, nutr.	Abscess upper R M1, enamel hypoplasia	Not obs.			
19-6-10/60151.0			Not obs.	B&W handle frag. no. 1965 (67097), bone awl no. 1964 (67845)		
19-6-10/60152.0	Nutr., other	Enamel hypoplasia, scaphocephaly	Unobs.	Gl. 1 red bowl no. 1116 (65016)	Glaze 1 red	
19-6-10/60154.0			Unobs.	Gl. 4 bowl no. 1477 (67053), polishing stone no. 1512 (68825), frag. miniature pot no. 1511 (67318), bone awl no. 1513 (67839)	Glaze 4	
19-6-10/60155.0	Nutr., other	Enamel hypoplasia, scaphocephaly	Unobs.			
19-6-10/60156.0	Infection, arthritis, nutr.	L and R tibia–infection, spinal osteophytosis, DJD, osteoporosis	Unobs.	Broken bowl, Gl. 3 no. 3869 (67055), stone scraper no. 1666 (68532), turquoise pendant no. 1665 (69121)		
19-6-10/60157.0	Arthritis, nutr.	Vt/c 2–3 ankylosed, basicranium–vt/c1 fused, enamel hypoplasia	Unobs.	Pecos white, Gl. 5 sherd no. 1055 (65040)		
19-6-10/60158.0	Infection, arthritis, nutr.	Dental abscesses, spinal osteophytosis, osteoporosis	Unobs.	Incised pebble no. 1456 (68934), wrappings (67653)	Pecos white	
19-6-10/60159.0	Infection, arthritis, nutr.	Occipital lesion, spinal osteophytosis, osteoporosis	Unobs.			
19-6-10/60161.0			No cran.	Agua Fria sherds, no number	Agua Fria?	
19-6-10/60162.0			Not obs.	Drilled piece of pottery no. 1610 (67227)	Glaze 3, probably	
19-6-10/60163.0	Infection, arthritis, nutr.	Infection of L femur, extensive DJD vertebrae and limbs, remodeling hip joints, osteoporosis	Unobs.	Gl. 3 white bowl no. 1316 (67028)	B& and Agua Fria sherds	
19-6-10/60164.0	Nutr.	Enamel hypoplasia	Not obs.	2 projectile points no. 729 (68413), antler tool no. 731 (68251)		Femoral neck similar to chimpanzee
19-6-10/60165.0	Arthritis, nutr.	DJD, enamel hypoplasia, osteoporosis	Not obs.	Gl. 2 bowl no. 536 (66995)		Crania and postcrania may not be associated
19-6-10/60166.0			Unobs.			
19-6-10/60167.0	Trauma, arthritis, nutr.	Compression fracture vt/l2, spinal osteophytosis, osteoporosis	No cran.			Possibly should be 20-6-10/60358.5

Peabody number	Object description	FN	Specimen label	Site	Intrasite provenience	Ceramic stage	MNI	Age (yrs.)	Ad	Su	Ch	In	Fe	Unk. age	M	M?	F	F?	Unk. sex	Age criteria	Sex criteria
19-6-10/60168.0	Fragmentary cranial and partial postcranial human remains of an adult female.	276	60168	PP	North terrace, depth 5′6″	Pres. Gl. 2	1	35–49	1	0	0	0	0	0	0	0	1	0	0	D e, T 9, SB 5, AS 7	Female
19-6-10/60169.0	Partial cranial and postcranial human remains of an adult female.	207	60169/207	PP	1100E150, depth 2′, on red soil	Indet.	1	40–50+	1	0	0	0	0	0	0	0	1	0	0	T 10, SB 5, AS 5	F all features
19-6-10/60171.0	Fragmentary cranial and nearly complete postcranial human remains of an adult female.	112	60171	PP	1500E100, depth 9′, on red clay	Probably Gl. 2	1	40–44	1	0	0	0	0	0	0	0	1	0	0	D h+, T 4, B 3, AS 5	F all features
19-6-10/60172.0	Fragmentary postcranial human remains of an adult male.	461	60172	PP	1400E100, test xii, cut 2, depth 24″	Probably Gl. 6	1	45–54+	1	0	0	0	0	0	1	0	0	0	0	T 10, SB 6	Male
19-6-10/60173.0	Fragmentary cranial and partial postcranial human remains of an adult female.	368	60173	PP	1550E100, depth 60″, 14″ above red clay	Indet.	1	60+	1	0	0	0	0	0	0	0	1	0	0	h+/i+, T10, SB 6, AS 8	Female
19-6-10/60174.0	Nearly complete cranial and postcranial human remains of an adult female.	449	60174/f46	PP	1100E100, depth 32″, height–above red clay 27″	Probably Gl. 4 or 5	1	45–49	1	0	0	0	0	0	0	0	1	0	0	D h+/i+, T9, SB 5-6, AS 6	F all features
19-6-10/60175.0	Partial cranial and fragmentary postcranial human remains of an adult female.	85	60175	PP	1550E100, depth 7′6″	Probably Gl. 2	1	18–22	1	0	0	0	0	0	0	0	1	0	0	D c	Female
19-6-10/60176.0	Nearly complete cranial and partial postcranial human remains of an adult male.	5	60176	PP	1600E150, depth 3′	Pres. Gl. 4 or 5	1	50–59	1	0	0	0	0	0	1	0	0	0	0	D h+, T 9–10, SB 5, AS 7	M all features
19-6-10/60177.0	Partial cranial and fragmentary postcranial human remains of an adult male.	637	60177	PP	North terrace, sw corner of room x	Indet.	1	16–20	1	0	0	0	0	0	1	0	0	0	0	D b2	Male
19-6-10/60178.0	Fragmentary postcranial human remains of an adult male.	548	60178	PP	1550W50	Probably Gl. 6	1		1	0	0	0	0	0	1	0	0	0	0		M all features
19-6-10/60179.0	Fragmentary cranial and postcranial human remains of an adult female.	220	60179	PP	1200E100	Indet.	1	25–39	1	0	0	0	0	0	0	0	1	0	0	D d/e, AS 4	F all features
19-6-10/60180.0	Fragmentary cranial and nearly complete postcranial human remains of an adult male.	338	60180/338/d	PP	1250E100, depth 5′9″	Probably Gl. 4 or 5	1	50–59	1	0	0	0	0	0	1	0	0	0	0	D h+/h, T 1-, SB 6, AS 7	M all features
19-6-10/60181.0	Partial cranial and nearly complete postcranial human remains of an adult male.	584	60181	PP	North terrace, depth 17–24″, height–4″ above rock	Indet.	1	40–44	1	0	0	0	0	0	1	0	0	0	0	D g, T 8, SB 4, AS 5	Male
19-6-10/60182.0	Fragmentary postcranial human remains of an adult male.	145	60182	PP	1450E100	Bisc. B or Gl. 1 to 2?	1	50–59	1	0	0	0	0	0	1	0	0	0	0	AS 7	Male
19-6-10/60183.0	Nearly complete cranial and postcranial human remains of a child of unknown sex.	493	60183	PP	1000E150, depth 65″, height–above red clay 5″	Probably Gl. 5	1	4.5–5.5	0	0	1	0	0	0	0	0	0	0	1	D 7 ± 2 yrs, PC 4.5–5.5 yrs	Indet.
19-6-10/60183.1	Nearly complete cranial human remains of an adult female.	493?	60183	PP	1000E150, depth 65″, height–above red clay 5″	Probably Gl. 5	1	20–24	1	0	0	0	0	0	0	0	1	0	0	D d	F all features
19-6-10/60184.0	Nearly complete cranial and fragmentary postcranial human remains of a child of unknown sex.	344	60184	PP	1250E100, depth 4′	Probably Gl. 5	1	1–2	0	0	1	0	0	0	0	0	0	0	1	D 3 ± 1 yrs, PC 0.5–1.5 yrs	Indet.
19-6-10/60185.0	Very fragmentary cranial and postcranial human remains of a child of unknown sex.	194		PP	North terrace, 55″ below surface, 18″ above red	Indet.	1		0	0	1	0	0	0	0	0	0	0	1		Indet.
19-6-10/60185a	Fragmentary cranial and partial postcranial human remains of an infant of unknown sex.	194	60185-a	PP	North terrace	Indet.	1	0–0.5	0	0	0	1	0	0	0	0	0	0	1	PC NB 0.5 yrs	Indet.
19-6-10/60186.0	Fragmentary cranial and postcranial human remains of a fetus of unknown sex.	275	60186	PP	North terrace, depth 2′	Indet., probably early	1	Fetus	0	0	0	0	1	0	0	0	0	0	1	PC fetus	Indet.
19-6-10/60187.1	Very fragmentary cranial human remains of a child of unknown sex.	129	60187-1/129	PP	1500E100, depth 5.5′, 6′ above red clay	Pres. Gl. 5	1	3–4	0	0	1	0	0	0	0	0	0	0	1	D 4 ± 1 yrs	Indet.

Peabody number	Pathology type	Pathology description	Cranial mod.	Associated funerary objects	Field context	Notes
19-6-10/60168.0	Arthritis, nutr.	DJD, spinal osteophytosis, Schmorl's nodes, osteoporosis	Unobs.	Obsidian projectile point no. 551 (68412)		Crania and postcrania may not be associated
19-6-10/60169.0	Arthritis, nutr.	Spinal osteophytosis, osteoporosis	Not obs.	Projectile point no. 243 (68405)		Field card says provenience doubtful
19-6-10/60171.0	Trauma, nutr., other	Vt/c 2–3 ankylosis due to trauma, porotic hyperostosis	Unobs.			
19-6-10/60172.0	Arthritis, nutr.	Spinal osteophytosis, osteoporosis	No cran.		Burial at length	
19-6-10/60173.0	Arthritis	Spinal osteophytosis	Unobs.	B&W sherd, no number	1 B&W sherd	
19-6-10/60174.0	Infection, arthritis, nutr.	Infected R tibia and frontal, DJD, osteoporosis	Not obs.	Stone scraper no. 1455 (68529), bone tool frag. no. 1454 (68097)		
19-6-10/60175.0			Not obs.		Agua Fria and B&W	Humeri–unusually flat on ventro-medial surface
19-6-10/60176.0	Arthritis, nutr.	DJD, osteoporosis	Not obs.			
19-6-10/60177.0			Not obs.	Gl. 1 red bowl no. 2037 (67131)		
19-6-10/60178.0			No cran.		Burial at length	
19-6-10/60179.0	Arthritis, nutr.	DJD, slight enamel hypoplasia	Unobs.	Broken black jar no. 144 (66974), Gl. 2 yellow bowl no. 142 (66972), broken "duck pot" no. 143 (66973)		
19-6-10/60180.0	Arthritis, nutr., other	Patellae slight lipping, enamel hypoplasia, vt/c show pronounced split spinous process, maxilla–upper L C and P1 reversed, P1 deformed	Unobs.	Projectile point no. 1028 (68420)		
19-6-10/60181.0	Arthritis, nutr.	DJD, enamel hypoplasia	Not obs.	Stone scraper no. 1959 (68538), projectile point no. 1958 (68444), Gl. 3? bowl no. 1957 (67099)		
19-6-10/60182.0	Arthritis, nutr.	DJD, osteoporosis	No cran.	Fibrolite ax no. 446 (68729), Gl. 2 red bowl no. 1025 (65071), Bisc. B bowl no. 1021 (65002)		
19-6-10/60183.0	Nutr.	Cribra orbitalia	Occ. def.	Pecos white sherds, no number		
19-6-10/60183.1			Occ. def.	Pecos white sherds, no number	Pecos white sherds	Field number and provenience problematic
19-6-10/60184.0	Nutr.	Cribra orbitalia, porotic hyperostosis	Not obs.			May be some mixing with 60185, 60185a & 60186
19-6-10/60185.0			Unobs.	Traces of red ochre-looking material at feet (FC)		May be some mixing with 60185, 60185a & 60186, field number and provenience problematic
19-6-10/60185a	Nutr.	Porotic hyperostosis	Unobs.	Traces of red ochre-looking material at feet (FC)		May be some mixing with 60184, 60185 & 60186
19-6-10/60186.0	Infection	Cranium–infection, cloaca in frag. cranial bone	Unobs.	Large black olla, damaged, no. 537 (66994)		May be some mixing with 60184, 60185, & 60185a; fetus was found in pot
19-6-10/60187.1			Unobs.			

Peabody number	Object description	FN	Specimen label	Site	Intrasite provenience	Ceramic stage	MNI	Age (yrs.)	Ad	Su	Ch	In	Fe	Unk. age	M	M?	F	F?	Unk. sex	Age criteria	Sex criteria
19-6-10/60187.10	Very fragmentary cranial human remains of a child of unknown sex.	102	60187-10/102	PP	1400E150, depth 3′, 6′ above red	Pres. Gl. 5	1	6–8	0	0	1	0	0	0	0	0	0	0	1	8 ± 2 yrs	Indet.
19-6-10/60187.2	Very fragmentary cranial human remains of a child of unknown sex.	79	79	PP	1550E150, nw corner, depth 2′, height 2′ above red clay	Probably Gl. 4	1	4–6	0	0	1	0	0	0	0	0	0	0	1	D 6 ± 2 yrs	Indet.
19-6-10/60187.3	Very fragmentary cranial human remains of a child of unknown sex.	160	60187-3/160?	PP	1400E100, depth 11′6″, 6′6″ above red	Probably Gl. 4	1	2–3	0	0	1	0	0	0	0	0	0	0	1	D 3 ± 1 yrs	Indet.
19-6-10/60187.4	Very fragmentary cranial human remains of a child of unknown sex.	95	95	PP	1500E100, depth 9′, 1′ above red clay	Probably Gl. 2	1	4–6	0	0	1	0	0	0	0	0	0	0	1	D 6 ± 2 yrs	Indet.
19-6-10/60187.5	Very fragmentary cranial human remains of a child of unknown sex.	149	60187-5, 149	PP	1450E100, depth 8′, 4′ above red clay	Indet., poss. Gl. 3	1	1–1.5	0	0	0	1	0	0	0	0	0	0	1	1.5 ± 0.5 yrs	Indet.
19-6-10/60187.6	Very fragmentary cranial human remains of a child of unknown sex.	111	60187-6/111	PP	1500E100, depth 8′, 4′ above red clay	Indet., poss. Gl. 4	1	3–4	0	0	1	0	0	0	0	0	0	0	1	D 4 ± 1 yrs	Indet.
19-6-10/60187.7	Very fragmentary cranial human remains of a child of unknown sex.	130	60187-7, 130	PP	1450E100, depth 11–12′, on red clay	Pres. Gl. 1 or 2	1	1–1.5	0	0	0	1	0	0	0	0	0	0	1	D 1.5 ± 0.5 yrs	Indet.
19-6-10/60187.8	Very fragmentary cranial human remains of a child of unknown sex.	104	60187-8, 104	PP	1500E100, depth 3′	Indet.	1	2–3	0	0	1	0	0	0	0	0	0	0	1	D 3 ± 1 yrs	Indet.
19-6-10/60187.9	Very fragmentary cranial human remains of a child of unknown sex.	110	60187-9, 110	PP	1500E100, depth 8.5′, 4′ above red clay	Indet.	1	0.5–0.75	0	0	0	1	0	0	0	0	0	0	1	0.75 ± 0.25 yrs	Indet.
19-6-10/60189.0	Partial postcranial human remains of an adult male.	565	60189	PP	1550W50, depth 13–17″, above red clay 23–29″	Probably Gl. 2	1	50–59	1	0	0	0	0	0	1	0	0	0	0	T 9-10, SB 6, AS 7	Male
19-6-10/60190.0	Partial cranial and postcranial human remains of a subadult of unknown sex.	151	60190	PP	1450E100, depth 12′, on red clay	Pres. Gl. 1 or 2	1	9–11	0	1	0	0	0	0	0	0	0	0	1	D 10–12 yrs, PC 8.5–10.5 yrs	Indet.
19-6-10/60191.0	Partial cranial and fragmentary postcranial human remains of an adult female.	92	60191	PP	1550E100, depth 4′6″	Pres. Gl. 4 or 5	1	50–59	1	0	0	0	0	0	0	0	1	0	0	D >45 yrs, AS 7	F all features
19-6-10/60192.0	Fragmentary cranial and nearly complete postcranial human remains of an adult female.	83	60192	PP	1550E100, depth 5′6″	B&W	1	35–39	1	0	0	0	0	0	0	0	1	0	0	D g, T 7, SB 4	Female
19-6-10/60192.1	Very fragmentary cranial and postcranial human remains of a fetus of unknown sex.	83		PP	1550E100, depth 5′6″	B&W	1	Fetal	0	0	0	0	1	0	0	0	0	0	1	Fetus	Indet.
19-6-10/60193.0	Partial cranial and nearly complete postcranial human remains of an adult female.	70	60193/5/16	PP	1550E150, in scoop in red clay 4′ below surface	Very probably Gl. 3	1	45–55+	1	0	0	0	0	0	0	0	1	0	0	D g-h, AS 7	Female
19-6-10/60624.0	Very fragmentary postcranial human remains of an adult of unknown sex.	1270	60624	PP	1300W100, depth 3′6″, height 2′ above bottom	Probably Gl. 1r	1		1	0	0	0	0	0	0	0	0	0	1		Indet.
19-6-10/60646.0	Partial cranial and very fragmentary postcranial human remains of an adult female.	1863	60646	PP	800W100, depth 4′, height–1′ into clay	Indet.	1	20–29	1	0	0	0	0	0	0	0	1	0	0	D d-e	Female
19-6-10/60647.0	Nearly complete cranial and postcranial human remains of an adult female.	1859	60647	PP	800W100, depth 2′6″, height 3′	Indet.	1	17–20	0	1	0	0	0	0	0	0	1	0	0	D b2, PC young adult	Female
19-6-10/60648.0	Nearly complete cranial human remains of an adult male.	C24	60648	PP	Church nave, depth 3′, height 1′	Indet.	1	30–39	1	0	0	0	0	0	1	0	0	0	0	D f-g	Male
19-6-10/60648.1	Very fragmentary cranial human remains of an adult of unknown sex.	C24		PP	Church nave, depth 3′, height 1′	Indet.	1	45–54	1	0	0	0	0	0	0	0	0	0	1	D i	Indet.
19-6-10/60649.0	Nearly complete cranial human remains of an adult male. Individual is of Caucasian ancestry.	C43	60649	PP	Church nave, depth 3′, height 1′	Indet.	1	50–59	1	0	0	0	0	0	1	0	0	0	0	D h++	Male
19-6-10/60650.0	Nearly complete cranial human remains of an adult male.	1637	60650/1637	PP	Trench n, room 335, depth 3′, height 3′	Indet.	1	40–49	1	0	0	0	0	0	1	0	0	0	0	D h+	Male

Peabody number	Pathology type	Pathology description	Cranial mod.	Associated funerary objects	Field context	Notes
19-6-10/60187.10			Unobs.			
19-6-10/60187.2			Unobs.	Gl. 4 Pajaritan bowl no. 279 (65057)		
19-6-10/60187.3			Unobs.	Parts of Gl. 4 bowl no. 279 (65057)		
19-6-10/60187.4			Unobs.			
19-6-10/60187.5			Unobs.			
19-6-10/60187.6			Unobs.			
19-6-10/60187.7			Unobs.			
19-6-10/60187.8			Unobs.			
19-6-10/60187.9			Unobs.	Large Bisc. A sherd (65035), broken worked sandstone slab no. 9345 (66212)		
19-6-10/60189.0	Trauma, arthritis, nutr., tumor, other	Healed fracture R clavicle, vertebrae exhibit spondylolysis and ankylosis, moderate DJD appendicular, osteoporosis, tumorous growth	No cran.	Copper piece no. 1880 (69294)	Agua Fria sherds	
19-6-10/60190.0	Nutr.	Enamel hypoplasia	Occ. def.	Polishing stone no. 1033 (66211)		
19-6-10/60191.0	Infection, arthritis	Dental abcesses and severe antemortem tooth loss, DJD	Not obs.	Smoothing stone (no number)		
19-6-10/60192.0	Infection, arthritis, other	Periostitis of R femur, L and R tibia, L humerus, ankylosis vt/c2–3, vt/s5 not fused	Unobs.	Bisc. A "scoop" no. 238 (65064), bone bead no. 328 (65396)		
19-6-10/60192.1			Unobs.	Bisc. A "scoop" no. 238 (65064), bone bead no. 328 (65396)	With mother	Field card notes fetal skeleton in pelvis of female
19-6-10/60193.0	Infection, arthritis	Dental abscess upper L P1, DJD	Cran. def.			Sternum twice perforated
19-6-10/60624.0	Trauma, infection	Healed fracture L fibula–Pott's (?)	No cran.			L fibula only present, but field notes say cranium and long bones present
19-6-10/60646.0	Arthritis, nutr.	DJD, enamel hypoplasia	Prob. occ. def.	Gl. 2 bowl no. 1531 (74977), carved stone projectile straightener no. 1527 (76104), 2 bone beads, red paint, no. 1551 (76333), metallic powder (77117)		
19-6-10/60647.0	Infection, other	M3s impacted, partial spina bifida vt/s4, disjunction of other sacral elements, increased porosity frontal	Not obs.			
19-6-10/60648.0	Infection	Frontal lesion, infection of mandible?	Not obs.	Shell beads, no number	Burial at length	
19-6-10/60648.1			Unobs.	Shell beads, no number		
19-6-10/60649.0	Infection	Severe cranial vault infection affecting right orbit and surrounding areas, 5 partially healed lesions in parietals	No cran. def.			
19-6-10/60650.0	Infection	Extensive antemortem tooth loss	Prob. cran. def.	Gl. 2 bowl no. 200 (74893)		

Peabody number	Object description	FN	Specimen label	Site	Intrasite provenience	Ceramic stage	MNI	Age (yrs.)	Ad	Su	Ch	In	Fe	Unk. age	M	M?	F	F?	Unk. sex	Age criteria	Sex criteria
19-6-10/60958.0	Very fragmentary cranial and fragmentary postcranial human remains of an adult that is probably female.	C	60958	PP	Church	Indet.	1		1	0	0	0	0	0	0	0	0	1	0		?Female
19-6-10/60959.1	Very fragmentary postcranial human remains of an adult male.	C		PP	Church	Indet.	1	<35	1	0	0	0	0	0	1	0	0	0	0	< 35 yrs	Male
19-6-10/60959.2	Very fragmentary postcranial human remains of an adult of unknown sex.	C	60977-7, 60959	PP	Church	Indet.	1		1	0	0	0	0	0	0	0	0	0	1		Indet.
19-6-10/60959.3	Very fragmentary postcranial human remains of an adult of unknown sex.	C		PP	Church	Indet.	1		1	0	0	0	0	0	0	0	0	0	1		Indet.
19-6-10/60959.4	Fragmentary postcranial human remains of an adult of unknown sex.	C		PP	Church	Indet.	1		1	0	0	0	0	0	0	0	0	0	1		Indet.
19-6-10/60959.5	Fragmentary postcranial human remains of an adult of unknown sex.	C	60759, 60970, 60762, 60763	PP	Church	Indet.	1	20–29	1	0	0	0	0	0	0	0	0	0	1	PC young adult	Indet.
19-6-10/60962.1	Very fragmentary postcranial human remains of an adult of unknown sex.	C		PP	Church	Indet.	1		1	0	0	0	0	0	0	0	0	0	1		Indet.
19-6-10/60962.2	Very fragmentary postcranial human remains of an adult of unknown sex.	C	60962 (pencil)	PP	Church	Indet.	1		1	0	0	0	0	0	0	0	0	0	1		Indet.
19-6-10/60962.3	Very fragmentary postcranial human remains of an adult of unknown sex.	C		PP	Church	Indet.	1		1	0	0	0	0	0	0	0	0	0	1		Indet.
19-6-10/60962.4	Very fragmentary postcranial human remains of a child of unknown sex.	C		PP	Church	Indet.	1		0	0	1	0	0	0	0	0	0	0	1		Indet.
19-6-10/60963.1	Very fragmentary postcranial human remains of an adult of unknown sex.	C	60963	PP	Church	Indet.	1		1	0	0	0	0	0	0	0	0	0	1		Indet.
19-6-10/60963.2	Very fragmentary postcranial human remains of an adult of unknown sex.	C	60963	PP	Church	Indet.	1		1	0	0	0	0	0	0	0	0	0	1		Indet.
19-6-10/60965.0	Very fragmentary postcranial human remains of a subadult female.	C	60965	PP	Church	Indet.	1	18–19	0	1	0	0	0	0	0	0	1	0	0	T 1, SB 1, AS 1	Female
19-6-10/60966.0	Very fragmentary cranial human remains of an adult that is probably female.	C		PP	Church	Indet.	1	16–20	1	0	0	0	0	0	0	0	0	1	0	D b2	?Female (ME and tooth size)
19-6-10/60967.0	Very fragmentary postcranial human remains of an adult male.	C	60967	PP	Church	Indet.	1	30–44	1	0	0	0	0	0	1	0	0	0	0	T 6, SB 4, AS 5	Male
19-6-10/60970.4	Very fragmentary postcranial human remains of an adult of unknown sex.	C	60970	PP	Church	Indet.	1		1	0	0	0	0	0	0	0	0	0	1		Indet.
19-6-10/60970.5	Very fragmentary postcranial human remains of an adult of unknown sex.	C	60970	PP	Church	Indet.	1		1	0	0	0	0	0	0	0	0	0	1		Indet.
19-6-10/60971.1	Very fragmentary postcranial human remains of an adult female.	C		PP	Church	Indet.	1	20–34	1	0	0	0	0	0	0	0	1	0	0	PC < 35 yrs	Female
19-6-10/60971.10	Very fragmentary postcranial human remains of an adult of unknown sex.	C		PP	Church	Indet.	1		1	0	0	0	0	0	0	0	0	0	1		Indet.
19-6-10/60971.11	Very fragmentary postcranial human remains of an adult of unknown sex.	C		PP	Church	Indet.	1		1	0	0	0	0	0	0	0	0	0	1		Indet.
19-6-10/60971.12	Very fragmentary postcranial human remains of an adult of unknown sex.	C		PP	Church	Indet.	1		1	0	0	0	0	0	0	0	0	0	1		Indet.
19-6-10/60971.2	Very fragmentary postcranial human remains of an adult female.	C		PP	Church	Indet.	1	20–34	1	0	0	0	0	0	0	0	1	0	0	PC < 35 yrs	Female
19-6-10/60971.3	Very fragmentary postcranial human remains of an adult female.	C		PP	Church	Indet.	1	17–22	1	0	0	0	0	0	0	0	1	0	0	PC 17–22 yrs	Female
19-6-10/60971.4	Very fragmentary postcranial human remains of an adult female.	C		PP	Church	Indet.	1	20–34	1	0	0	0	0	0	0	0	1	0	0	PC < 35 yrs	Female
19-6-10/60971.5	Very fragmentary postcranial human remains of an adult male.	C		PP	Church	Indet.	1	>35	1	0	0	0	0	0	1	0	0	0	0	PC >35 yrs	Male
19-6-10/60971.6	Very fragmentary postcranial human remains of an adult male.	C		PP	Church	Indet.	1	25–34	1	0	0	0	0	0	1	0	0	0	0	PC 25–34 yrs	Male
19-6-10/60971.7	Very fragmentary postcranial human remains of a subadult of unknown sex.	C		PP	Church	Indet.	1	12–18	0	1	0	0	0	0	0	0	0	0	1		Indet.

Peabody number	Pathology type	Pathology description	Cranial mod.	Associated funerary objects	Field context	Notes
19-6-10/60958.0			Unobs.			
19-6-10/60959.1	Trauma	Perimortem trauma to distal anterior humerus with blade	No cran.			L humerus only present; Owsley team labeled this OO.
19-6-10/60959.2			No cran.			Humeri only; Owsley team reassociated this. No designation recorded.
19-6-10/60959.3			No cran.			Humeri and R ulna; Owsley team reassociated this. No designation recorded.
19-6-10/60959.4			No cran.			Reassociated by Owsley team as S. (1/98)
19-6-10/60959.5			No cran.			
19-6-10/60962.1			No cran.			Reassociated by Owsley team. No designation recorded (1/98).
19-6-10/60962.2			No cran.			Reassociated by Owsley team. No designation recorded (1/98).
19-6-10/60962.3			No cran.			Radii, reassociated by Owsley team. No designation recorded (1/98).
19-6-10/60962.4			No cran.			Reassociated by Owsley team. No designation recorded (1/98).
19-6-10/60963.1			No cran.			Reassociated by Owsley team. No designation recorded (1/98).
19-6-10/60963.2			No cran.			Reassociated by Owsley team. No designation recorded (1/98).
19-6-10/60965.0			No cran.			Pelvis only
19-6-10/60966.0	Nutr.	Enamel hypoplasia	Unobs.			
19-6-10/60967.0			No cran.			2 innominates
19-6-10/60970.4			No cran.			Reassociated by Owsley team. No designation recorded (1/98).
19-6-10/60970.5			No cran.			Reassociated by Owsley team. No designation recorded (1/98).
19-6-10/60971.1			No cran.			Femora, reassociated by Owsley team as H (1/98)
19-6-10/60971.10			No cran.			Reassociated by Owsley team as O (1/98)
19-6-10/60971.11			No cran.			Reassociated by Owsley team as X (1/98)
19-6-10/60971.12			No cran.			Reassociated by Owsley team as L (1/98)
19-6-10/60971.2	Nutr.	Osteoporosis	No cran.			Femora, reassociated by Owsley team as I (1/98)
19-6-10/60971.3			No cran.			Reassociated by Owsley team as J (1/98)
19-6-10/60971.4			No cran.			Reassociated by Owsley team as K (1/98). They indicate not Native American. No squatting facet on tibia.
19-6-10/60971.5	Arthritis	DJD on distal femur	No cran.			Reassociated by Owsley team as E (1/98)
19-6-10/60971.6			No cran.			Tibia, reassociated by Owsley team as HH (1/98)
19-6-10/60971.7			No cran.			Reassociated by Owsley team as R (1/98)

Peabody number	Object description	FN	Specimen label	Site	Intrasite provenience	Ceramic stage	MNI	Age (yrs.)	Ad	Su	Ch	In	Fe	Unk. age	M	M?	F	F?	Unk. sex	Age criteria	Sex criteria
19-6-10/60971.8	Very fragmentary postcranial human remains of a subadult male.	C		PP	Church	Indet.	1	15–19	0	1	0	0	0	0	1	0	0	0	0	PC 16–18 yrs	Male
19-6-10/60971.9	Very fragmentary postcranial human remains of an adult of unknown sex.	C		PP	Church	Indet.	1		1	0	0	0	0	0	0	0	0	0	1		Indet.
19-6-10/60972.1	Very fragmentary postcranial human remains of an adult of unknown sex.	C		PP	Church	Indet.	1		1	0	0	0	0	0	0	0	0	0	1		Indet.
19-6-10/60972.2	Very fragmentary postcranial human remains of an adult of unknown sex.	C		PP	Church	Indet.	1		1	0	0	0	0	0	0	0	0	0	1		Indet.
19-6-10/60972.3	Very fragmentary postcranial human remains of an adult female.	C		PP	Church	Indet.	1		1	0	0	0	0	0	0	0	1	0	0		Female
19-6-10/60972.5	Very fragmentary postcranial human remains of an adult that is probably male.	C		PP	Church	Indet.	1		1	0	0	0	0	0	0	1	0	0	0		?Male
19-6-10/60972.6	Very fragmentary postcranial human remains of a subadult of unknown sex.	C		PP	Church	Indet.	1	12–18	0	1	0	0	0	0	0	0	0	0	1		Indet.
19-6-10/60975.2	Very fragmentary postcranial human remains of an adult of unknown sex.	C		PP	Church	Indet.	1		1	0	0	0	0	0	0	0	0	0	1		Indet.
19-6-10/60977.10	Very fragmentary postcranial human remains of an adult of unknown sex.	C	60977-19	PP	Church	Indet.	1		1	0	0	0	0	0	0	0	0	0	1		Indet.
19-6-10/60977.11	Fragmentary postcranial human remains of an adult that is probably female.	C	60645, 60977-5	PP	Church	Indet.	1	>35	1	0	0	0	0	0	0	0	0	1	0	>35 yrs	?Female
19-6-10/60977.12	Very fragmentary cranial and postcranial human remains of an adult female.	C		PP	Church	Indet.	1	30–39	1	0	0	0	0	0	0	0	1	0	0	D f-g	Female
19-6-10/60977.3	Very fragmentary postcranial human remains of an adult of unknown sex.	1340	60977-22/1 340	PP	Room 308, under e wall, depth 7′, height 8–12″	Probably late Gl. 1R, Gl. 1 or Gl. 2	1		1	0	0	0	0	0	0	0	0	0	1		Indet.
19-6-10/60977.4	Fragmentary cranial and very fragmentary postcranial human remains of an adult female.	291	60977-17, 291	PP	1200E150, depth 6″, 1′ above red clay	Pres. Gl. 3 or 4	1		1	0	0	0	0	0	0	0	1	0	0		Female
19-6-10/60977.5	Fragmentary postcranial human remains of an adult of unknown sex.	C	60977-4, -2; 59886-2, -7	PP	Church	Indet.	1		1	0	0	0	0	0	0	0	0	0	1		Indet.
19-6-10/60977.6	Very fragmentary postcranial human remains of an adult of unknown sex.	C	60977-26, 60959	PP	Church	Indet.	1		1	0	0	0	0	0	0	0	0	0	1		Indet.
19-6-10/60977.7	Very fragmentary postcranial human remains of an adult of unknown sex.	C	60977-6, 60977-18; 60970	PP	Church	Indet.	1		1	0	0	0	0	0	0	0	0	0	1		Indet.
19-6-10/60977.8	Very fragmentary postcranial human remains of an adult of unknown sex.	C	60977, 60977-27, 60970, 60962	PP	Church	Indet.	1		1	0	0	0	0	0	0	0	0	0	1		Indet.
19-6-10/60977.9	Very fragmentary postcranial human remains of a child of unknown sex.	C	60977-1, 60963	PP	Church	Indet.	1	Child	0	0	1	0	0	0	0	0	0	0	1		Indet.
19-6-10/61111.0	Nearly complete cranial human remains of a child of unknown sex.	Bb54	61111/bb54	FL	Rm 43, sw corner, depth pelvis 12″ skull 14″, face 20″ abv flr	B&W	1	10–12	0	0	1	0	0	0	0	0	0	0	1	D 10–12 yrs	Indet.
19-6-10/61112.0	Nearly complete cranial and postcranial human remains of an adult male.	Bb82	61112/bb82	FL	Room 74, depth 2′, height about 1′6″ above room floor	Probably B&W	1	40–49	1	0	0	0	0	0	1	0	0	0	0	D g/h-i, T 9, SB 5	M all features
19-6-10/61113.0	Nearly complete cranial human remains of a subadult of unknown sex.	965	61113	PP	1400W100, depth 5′5″, height 1′10″	Probably Gl. 1	1	10–12	0	1	0	0	0	0	0	0	0	0	1	D 12 ± 2.5 yrs	Indet.
19-6-10/61114.0	Partial cranial and fragmentary postcranial human remains of an adult female.	Bb92	61114/BB9 2	FL	Trench k, depth 3′, height ?	Probably B&W	1	20–24	1	0	0	0	0	0	0	0	1	0	0	D c-d	Female
19-6-10/61120.0	Partial cranial human remains of an adult that is probably female.	Bb64	61120, bb64	FL	Room 57, se corner, depth 2′3″, on floor	Probably B&W	1	25–29	1	0	0	0	0	0	0	0	0	1	0	D e	?Female

Peabody number	Pathology type	Pathology description	Cranial mod.	Associated funerary objects	Field context	Notes
19-6-10/60971.8			No cran.			Reassociated by Owsley team as A (1/98), probably not Native American
19-6-10/60971.9			No cran.			Reassociated by Owsley team as W (1/98)
19-6-10/60972.1			No cran.			Reassociated by Owsley team as R (1/98)
19-6-10/60972.2			No cran.			Reassociated by Owsley team as U (1/98), unusual cut marks on prox. femur
19-6-10/60972.3			No cran.			Reassociated by Owsley team as DD (1/98). Note: possible match with 60762
19-6-10/60972.5			No cran.			1 tibia; Owsley team indicates element is European (1/98)
19-6-10/60972.6			No cran.			Reassociated by Owsley team as D (1/98). MEM thinks this element is not part of D.
19-6-10/60975.2			No cran.			Tarsals
19-6-10/60977.10			No cran.			Reassociated by Owsley team as QQ (1/98)
19-6-10/60977.11	Infection	Severe lytic infection–tibiae, humeri	No cran.			
19-6-10/60977.12			Unobs.			
19-6-10/60977.3	Arthritis	DJD	No cran.	Gl. 1 yellow sherd, no number		1 humerus
19-6-10/60977.4			Unobs.			
19-6-10/60977.5			No cran.			Reassociated by Owsley team. No designation recorded (1/98).
19-6-10/60977.6			No cran.			2 humeri; Owsley team reassociated this. No designation recorded (1/98).
19-6-10/60977.7			No cran.			Reassociated by Owsley team. No designation recorded (1/98).
19-6-10/60977.8			No cran.			Reassociated by Owsley team. No designation recorded (1/98).
19-6-10/60977.9			No cran.			Reassociated by Owsley team. No designation recorded (1/98).
19-6-10/61111.0	Infection, other	Healed lesion frontal, peg-shaped lateral incisors	Lamb. def.	2 pcs broken manos, no number		
19-6-10/61112.0	Infection, arthritis, nutr.	Abscess lower M2, porotic thickening DJD, spinal osteophytosis, enamel hypoplasia	Lamb. def.			
19-6-10/61113.0			Not obs.			
19-6-10/61114.0	Infection	Dental abscesses– upper L M2, lower L M1, large cavity lower L M2	Lamb. def.	Seed bead no. 219 (77055)		
19-6-10/61120.0			Lamb. def.			Large brow ridge, small mastoids, field number and provenience problematic

Peabody number	Object description	FN	Specimen label	Site	Intrasite provenience	Ceramic stage	MNI	Age (yrs.)	Ad	Su	Ch	In	Fe	Unk. age	M	M?	F	F?	Unk. sex	Age criteria	Sex criteria
19-6-10/N687.0	Fragmentary cranial human remains of an adult of unknown sex.	Bb93	Bb93	FL	W173N121, depth 2′6″, on bottom	Probably B&W	1	40–44	1	0	0	0	0	0	0	0	0	0	1	D h	Indet.
19-6-10/N690.0	Fragmentary cranial human remains of an adult male.	Bb120	Bb120	FL	Room 81, trench p, depth 40″, height–on floor	Probably B&W	1	45–54	1	0	0	0	0	0	1	0	0	0	0	D i	Male
20-6-10/60194.0	Nearly complete cranial and partial postcranial human remains of an adult male.	871	60194	PP	1290E120, depth 6′4″, height 3′	Gl. 5	1	50–59	1	0	0	0	0	0	1	0	0	0	0	D i+, T 10, SB 5–6, AS 7	M all features
20-6-10/60195.0	Partial cranial and postcranial human remains of a subadult that is probably male.	806	60195	PP	1300E140	Probably Gl. 3	1	17–20	0	1	0	0	0	0	0	1	0	0	0	D d, PC 17–20 yrs (iliac crest not fused, ischial tuberosity stage 1 fusion)	
20-6-10/60196.0	Partial cranial and fragmentary postcranial human remains of an adult female.	796	60196	PP	1350E100, depth 3′4″, height 1′±	Indet. late (post-Co-lumbian)	1	40–49	1	0	0	0	0	0	0	0	1	0	0	D h+/i+, AS 5–6	Female
20-6-10/60197.0	Partial cranial and postcranial human remains of an adult male.	926	60197	PP	590W70, depth 2′2″, height 2′6″	Probably Gl. 5	1	30–34	1	0	0	0	0	0	1	0	0	0	0	D e, T 6, SB 3, AS 3	Male
20-6-10/60198.0	Partial cranial and postcranial human remains of an adult male.	883	60198	PP	1310E110, depth 5′1″, height 5′	Pres. Gl. 4 or 5	1	27–39	1	0	0	0	0	0	1	0	0	0	0	D h/i, T 5, SB 3, AS 3	M all features
20-6-10/60199.0	Nearly complete cranial and partial postcranial human remains of an adult female.	732	60199	PP	1450W50, against cross wall, depth 8′, height–on bottom	B&W	1	25–29	1	0	0	0	0	0	0	0	1	0	0	D e-f, AS 2	Female
20-6-10/60200.0	Nearly complete cranial and postcranial human remains of an adult male.	709	60200	PP	1450W50, depth 5′, height 3′	Pres. Gl. 5	1	45–49	1	0	0	0	0	0	1	0	0	0	0	D g-h, AS 6	Male
20-6-10/60201.0	Partial cranial and postcranial human remains of an adult male.	864	60201/x4	PP	1290E120, depth 4′4″, height 4′	Gl. 3	1	30–39	1	0	0	0	0	0	1	0	0	0	0	D f, T 6-7, SB 3	M all features
20-6-10/60202.0	Fragmentary postcranial human remains of an adult male.	703	60202/f27	PP	1450W100, west trench, depth 4′	Gl. 2	1	40–44	1	0	0	0	0	0	1	0	0	0	0	T 8, SB 4, AS 5	M all features
20-6-10/60203.0	Fragmentary postcranial human remains of an adult male.	810	60203	PP	1330E140, depth 5′5″	Very probably Gl. 4	1	35–39	1	0	0	0	0	0	1	0	0	0	0	T 7, SB 4, AS 4	M all features
20-6-10/60204.0	Fragmentary cranial and postcranial human remains of a child of unknown sex.	913	60204	PP	570W80, depth 3′, height 6′	Probably Gl. 3	1	1–3	0	0	1	0	0	0	0	0	0	0	1	D 4 ± 1 yrs, PC 1.0–1.5 yrs	Indet.
20-6-10/60205.0	Fragmentary cranial and partial postcranial human remains of an adult female.	897	60205	PP	750W130, depth 4′6″, height 2′	Gl. 4	1	16–20	0	1	0	0	0	0	0	0	1	0	0	D b2, PC 16–20 yrs	Female
20-6-10/60206.0	Nearly complete cranial and partial postcranial human remains of a child of unknown sex.	928	60206, 60208	PP	750E120, depth 8′3″, height–on bottom	Probably Gl. 3	1	4.5–5.5	0	0	1	0	0	0	0	0	0	0	1	D 5 ± 1 yrs, PC 4.5–5.5 yrs	Indet.
20-6-10/60207.0	Partial cranial and postcranial human remains of an adult female.	916	60207	PP	580W70, depth 6″, height 4′	Late (post-Col umbian)	1	45–55+	1	0	0	0	0	0	0	0	1	0	0	D h+/i+	Female
20-6-10/60208.0	Nearly complete cranial and partial postcranial human remains of a child of unknown sex.	893	60208	PP	740W150, depth 1′, height 1′6″	Gl. 5	1	6.5–7.5	0	0	1	0	0	0	0	0	0	0	1	D 9-10 ± 2 yrs, PC 6.5–7.5 yrs	Indet.
20-6-10/60209.0	Fragmentary cranial and postcranial human remains of a child of unknown sex.	831	60209	PP	1220E130, depth 4′1″, height 3′6″	Indet.	1	4.5–5.5	0	0	1	0	0	0	0	0	0	0	1	PC 4.5–5.5 yrs	Indet.
20-6-10/60210.0	Fragmentary cranial and partial postcranial human remains of an adult male.	948	60210	PP	Room 33, depth 10′10″, height ?, very near mesa	Gl. 1	1	25–29	1	0	0	0	0	0	1	0	0	0	0	D G, T 5, SB 4?, AS 2	Male

Peabody number	Pathology type	Pathology description	Cranial mod.	Associated funerary objects	Field context	Notes
19-6-10/N687.0			Unobs.			
19-6-10/N690.0	Nutr.	Enamel hypoplasia	Unobs.			Drawing on field card suggests skeleton was complete although poorly preserved.
20-6-10/60194.0	Infection, arthritis, tumor	Multiple dental abscesses, severe molar antemortem tooth loss, DJD, button osteoma on frontal	Not obs.	Gl. 5 olla bowl no. 993 (69836)		
20-6-10/60195.0	Trauma, infection	Infection with joint deformation right acetabulum and femoral head–trauma related	Not obs.			Rodent gnawing
20-6-10/60196.0	Infection, arthritis, nutr.	Pott's disease? (2 vt/l fused, infection in frontal bone above L orbit), severe antemortem tooth loss, DJD, osteoporosis	Not obs.		Burial at length?	
20-6-10/60197.0			Not obs.			
20-6-10/60198.0	Infection, arthritis	Severe antemortem molar loss, long bones–arthritic lipping, especially on arm	Occipital def.	Gl. 3 and 4 sherds, no numbers		
20-6-10/60199.0	Infection, nutr., other	Infection L tibia, small infection R tibia, enamel hypoplasia, unusual shaped upper L M3	Occ. def.	Large B&W bowl no. 1013 (69789), small B&W bowl no. 1012 (69781), rubbing stone no. 454 (70600)		
20-6-10/60200.0	Infection, nutr., other	Dental abscesses, enamel hypoplasia, spina bifida vt/s5–6,	Not obs.			
20-6-10/60201.0	Arthritis, nutr.	DJD, enamel hypoplasia	Not obs.	Gl. 3 bowl no. 941 (69834)		
20-6-10/60202.0	Arthritis	DJD vertebrae and ulnae	No cran.	Gl. 2 bowl no. 983 (69768)		
20-6-10/60203.0			No cran.			
20-6-10/60204.0	Other	Lower R I1 and I2 are fused	Unobs.			
20-6-10/60205.0	Nutr.	Enamel hypoplasia	Occ. def.	Gl. 4 bowl no. 9506 (69847), stone knife no. 752 (70499)		
20-6-10/60206.0			Not obs.	Gl. 3 and Bisc. B sherds in grave, no number		
20-6-10/60207.0	Infection, arthritis, tumor	Dental abscesse , DJD–L knee and elsewhere, button osteoma on frontal	Not obs.		Burial at length, probably post-Columbian	
20-6-10/60208.0			Not obs.	Gl. 5 bowl no. 868 (69842), small bone tube no. 766 (70261), bone tool frag. no. 766b (70273), antler tip frag. no. 766a (70364), broken projectile point no. 766a (70421), broken knife no. 766d (70498), obsidian flake no. 766c (70538), small lump ochre no. 766b (70871)	Glaze 5	
20-6-10/60209.0			Unobs.			
20-6-10/60210.0			Unobs.	Bisc. A bowl no. 1154 (69870), 24 pierced shells no. 1175 (70810)		

Peabody number	Object description	FN	Specimen label	Site	Intrasite provenience	Ceramic stage	MNI	Age (yrs.)	Ad	Su	Ch	In	Fe	Unk. age	M	M?	F	F?	Unk. sex	Age criteria	Sex criteria
20-6-10/60211.0	Nearly complete cranial and postcranial human remains of an adult female.	911	60211	PP	750W120, under defense wall, depth 8′2″, height 1′	Gl. 3	1	50–59	1	0	0	0	0	0	0	0	1	0	0	T 10, SB 6, AS 7	F all features
20-6-10/60212.0	Fragmentary cranial and nearly complete postcranial human remains of an adult male.	834	60212/m4	PP	1330E13`0, depth 6′9″, height indeterminate	Very probably Gl. 4	1	50–59	1	0	0	0	0	0	1	0	0	0	0	D h+/i+, AS 7	M all features
20-6-10/60213.0	Nearly complete cranial and partial postcranial human remains of an adult female.	811	60213	PP	1300E140, depth 6′3″, on bottom	Gl. 2	1	20–24	1	0	0	0	0	0	0	0	1	0	0	D d-e/e, AS 1	F all features
20-6-10/60214.0	Fragmentary cranial and nearly complete postcranial human remains of a subadult that is probably female.	827	60214	PP	1270E130, depth 5′5″, height –on bottom	Pres. Gl. 3	1	12.5–13.5	0	1	0	0	0	0	0	0	0	1	0	D 15 ± 3 yrs, PC 12.5–13.5 yrs	?Female
20-6-10/60215.0	Partial cranial and postcranial human remains of an adult male.	850	60215	PP	1310E130, depth 4′8″, height 3′	Gl. 3	1	25–34	1	0	0	0	0	0	1	0	0	0	0	D d-e, AS 3	M all features
20-6-10/60216.0	Nearly complete cranial and postcranial human remains of an adult female.	846	60216	PP	1310E130, depth 6′, height 2′	Probably Gl. 3 or 4	1	60+	1	0	0	0	0	0	0	0	1	0	0	D h/i, T 10, SB 6, AS 8	Female, all features except SR (3)
20-6-10/60217.0	Nearly complete cranial and postcranial human remains of a child of unknown sex.	851	60217	PP	1290E120, depth 4′4″, height 1′8″	Probably Gl. 4 (or 5?)	1	8–9	0	0	1	0	0	0	0	0	0	0	1	D 9 ± 2 yrs, PC 8–9 yrs	Indet.
20-6-10/60218.0	Nearly complete cranial and postcranial human remains of an adult male.	815	60218	PP	1310E130, depth 4′4″, 2′ above bottom	Gl. 5	1	40–49	1	0	0	0	0	0	1	0	0	0	0	D h/h-i. T 8, SB 6, AS 6	M all features
20-6-10/60219.0	Nearly complete cranial and postcranial human remains of an adult male.	822	60219	PP	1260E130, depth 6′, height–on bottom	Gl. 3	1	50–59	1	0	0	0	0	0	1	0	0	0	0	D h+/i, T 10, SB 6, AS 7	M all features
20-6-10/60220.0	Nearly complete cranial and partial postcranial human remains of an adult that is probably male.	814	60220	PP	1280E130, depth 5′, 1′ above bottom	Gl. 3	1	30–39	1	0	0	0	0	0	0	1	0	0	0	D f-g, AS 4	?Male (Skull MP 2, SM 3, SR 2, NC 5, ME 5, Pelvis SN M)
20-6-10/60221.0	Very fragmentary cranial human remains of an adult of unknown sex.	75a	60221 (crossed out) 60258-29	PP		Indet.	1		1	0	0	0	0	0	0	0	0	0	1		Indet.
20-6-10/60222.0	Nearly complete cranial and partial postcranial human remains of an adult female.	860	60222/f29	PP	1310E130, depth 6′, height 2′	Gl. 1	1	18–22	1	0	0	0	0	0	0	0	1	0	0	D c, AS 1, iliac crest incom-plete fusion	F all features
20-6-10/60223.0	Very fragmentary postcranial human remains of an adult of unknown sex.	890	60223	PP	750W160, depth 24″, height–on bottom	Probably Gl. 3 or 4	1		1	0	0	0	0	0	0	0	0	0	1		Indet.
20-6-10/60224.0	Partial cranial and postcranial human remains of an adult male.	918	60224	PP	750W120, depth 5′11″, height 2′	Probably Gl. 4	1	35–39	1	0	0	0	0	0	1	0	0	0	0	D g/g-h uneven, T 7, SB 4, AS 4	M all features
20-6-10/60225.0	Nearly complete cranial and partial postcranial human remains of an adult female.	833	60225	PP	1300E130, depth 4′2″, height 2′	Very probably Gl. 4	1	50–59	1	0	0	0	0	0	0	0	1	0	0	AS 7	Female
20-6-10/60226.0	Nearly complete cranial and postcranial human remains of an adult male.	756	60226/x3	PP	1450W100 feet in, n. wall of cistern, depth 2′6″, height 12′±	Indet., Late (post-Col umbian)	1	45–54	1	0	0	0	0	0	1	0	0	0	0	D h+/i+, T 10, SB 5-6, AS 5	M all features
20-6-10/60227.0	Nearly complete cranial and partial postcranial human remains of an adult female.	733	60227/f26	PP	1450W50, west trench, depth 6′	Probably Gl. 1	1	30–39	1	0	0	0	0	0	0	0	1	0	0	D e, AS 5	Female
20-6-10/60228.0	Nearly complete cranial and postcranial human remains of an adult female.	740	60228/f30	PP	1555E149, depth 4′6″, height 6″	Gl. 4	1	35–44	1	0	0	0	0	0	0	0	1	0	0	D F, T 7, SB 4, AS 5	F all features

Peabody number	Pathology type	Pathology description	Cranial mod.	Associated funerary objects	Field context	Notes
20-6-10/60211.0	Arthritis	DJD, spinal osteophytosis	Not obs.	Gl. 3 bowl no. 1105 (69851)		
20-6-10/60212.0	Trauma, infection, arthritis, other	Pseudoarthrosis of L clavicle, osteomyelitis R ulna, periodontal disease, DJD, spinal osteophytosis, extra vt/c, v/l5 sacralized	Unobs.	Pipe and awl, no number		Teeth are very worn
20-6-10/60213.0			Not obs.	Gl. 2 scoop no. 719 (69818)		
20-6-10/60214.0			Unobs.	Thin olla frags. no. 1037 (69825)		
20-6-10/60215.0	Nutr.	Enamel hypoplasia	Not obs.	Gl. 3 bowl no. 997 (69829)		Rodent gnawing
20-6-10/60216.0	Infection, arthritis, nutr.	Periodontal disease, spinal osteophytosis, osteoporosis	Not obs.	Gl. 3 sherds, no number		Reassociated human remains from 60323 (MEM 5/98).
20-6-10/60217.0	Other	Retention deciduous incisors	Occ. def.	Gl. 4 sherds, no number		
20-6-10/60218.0	Infection, nutr., tumor, other	Dental abscesses, healed lesion on frontal, enamel hypoplasia, button osteoma on frontal, spina bifida vt/s4–5	Occ. def.	Late Gl. 5 bowl no. 851 (69821)		
20-6-10/60219.0			Occ. def.	Bisc. B bowl no. 850 (69824)		
20-6-10/60220.0	Arthritis, nutr.	DJD, enamel hypoplasia	Occ. def.	Gl. 3 bowl no. 863 (69820)		
20-6-10/60221.0	Other	Upper L M3 is peg tooth	Unobs.			
20-6-10/60222.0	Infection, nutr.	Healed periostitis on L tibia, enamel hypoplasia	Occ. def.	Large Gl. 1 sherd no. 1187 (69833)		
20-6-10/60223.0			No cran.			
20-6-10/60224.0	Infection, arthritis, nutr.	Several abscessing teeth, spinal osteophytosis, enamel hypoplasia	Not obs.	Snub-nosed scraper no. 782 (70449)		
20-6-10/60225.0	Arthritis, tumor, other	DJD, button osteoma–frontal, interesting exotoses on lingual surface mandible, scaphocephaly, 2 upper 1st premolars not present (no space)	Not obs.			
20-6-10/60226.0	Infection, arthritis, other	Infection of right ilium, spinal osteophytosis, DJD, thickening sagital suture, spina bifida	Not obs.		Burial at length	
20-6-10/60227.0			Occ. def.			
20-6-10/60228.0	Trauma, infection, arthritis, nutr.	Healed fracture L humerus, healed lesion on R parietal, spinal osteophytosis, enamel hypoplasia	Not obs.	Large Gl. 4 bowl no. 857 (69787)		

Peabody number	Object description	FN	Specimen label	Site	Intrasite provenience	Ceramic stage	MNI	Age (yrs.)	Ad	Su	Ch	In	Fe	Unk. age	M	M?	F	F?	Unk. sex	Age criteria	Sex criteria
20-6-10/60229.0	Partial cranial and postcranial human remains of an adult female.	736	60229	PP	1450W50, depth 5′6″, height 3′	Probably Gl. 5	1	45–59	1	0	0	0	0	0	0	0	1	0	0	D h/i, T 9, AS 7	
20-6-10/60230.0	Nearly complete cranial and postcranial human remains of an adult male.	820	60230	PP	1270E130, depth 6′, height–on bottom	Gl. 4	1	40–44	1	0	0	0	0	0	1	0	0	0	0	D g-h, AS 5	Male
20-6-10/60231.0	Nearly complete cranial and partial postcranial human remains of an adult female.	806a	60231/60230	PP	1300E140	Probably Gl. 3	1	60+	1	0	0	0	0	0	0	0	1	0	0	D edentulous, old, AS 8	Female
20-6-10/60232.0	Partial cranial and postcranial human remains of an adult male.	784	60232	PP	Room 29, depth 7′, height–on rocks	Very probably Gl. 1	1	50–59	1	0	0	0	0	0	1	0	0	0	0	AS 7	
20-6-10/60233.0	Fragmentary cranial and nearly complete postcranial human remains of an adult female.	844	60233	PP	1320E130, depth 8′2″	Probably Gl. 1	1	45–59	1	0	0	0	0	0	0	0	1	0	0	T 9, AS 7, molar tooth loss	Female
20-6-10/60234.0	Partial cranial and postcranial human remains of an adult female.	872	60234	PP	1300E120, depth 9.5′, height 1′	Probably Gl. 2	1	40–49	1	0	0	0	0	0	0	0	1	0	0	D h/h, AS 5, tooth loss	F all features
20-6-10/60235.0	Nearly complete cranial and partial postcranial human remains of an adult male.	876	60235	PP	1280E120, depth 5′8″, height 2′	Probably Gl. 1	1	20–29	1	0	0	0	0	0	1	0	0	0	0	D e-f/f (no M3s), AS 1-2	M all features
20-6-10/60236.0	Nearly complete cranial and fragmentary postcranial human remains of an adult male.	707	60236	PP	1450W100, west trench, depth 4′, on bottom	Probably Gl. 2	1	45–54	1	0	0	0	0	0	1	0	0	0	0	D I	Male
20-6-10/60237.0	Nearly complete cranial and fragmentary postcranial human remains of an adult male.	932	60237	PP	750W120, depth 4′, height 4′6″	Probably Gl. 5	1	35–45	1	0	0	0	0	0	1	0	0	0	0	D g/h	Male
20-6-10/60238.0	Nearly complete cranial and partial postcranial human remains of an adult male.	914	60238/e/m 25	PP	590W80, depth 1′1″, height 2′	Probably Gl. 3	1	30–39	1	0	0	0	0	0	1	0	0	0	0	D g, T 6, SB 3–4, AS 4	M all features
20-6-10/60239.0	Nearly complete cranial and postcranial human remains of an adult female.	707a	60239/f28/f47	PP	1450W100, west trench, depth 4′, on bottom	Probably Gl. 2	1	18–25	1	0	0	0	0	0	0	0	1	0	0	D c/c-d, T 2–3, AS 1–2	Female
20-6-10/60240.0	Fragmentary cranial and postcranial human remains of an adult female.	803	60240	PP	Kiva 6	Gl. 1	1	>35	1	0	0	0	0	0	0	0	1	0	0	>35	F all features
20-6-10/60241.0	Nearly complete cranial and postcranial human remains of an adult female.	929	60241	PP	750W120, depth 8′3″, height–on bottom	Probably Gl. 3	1	45–49	1	0	0	0	0	0	0	0	1	0	0	D g, AS 6	
20-6-10/60242.0	Nearly complete cranial and postcranial human remains of an adult male.	798	60242	PP	1290E140, depth 4′4″, height 1′6″	Very probably Gl. 4	1	45–55+	1	0	0	0	0	0	1	0	0	0	0	D h+/i+	Male
20-6-10/60243.0	Fragmentary cranial and partial postcranial human remains of an adult male.	812	60243	PP	1260E130, depth 3′10″	Probably Gl. 4	1	30–34	1	0	0	0	0	0	1	0	0	0	0	D h, AS 3	Male
20-6-10/60244.0	Nearly complete cranial and postcranial human remains of an adult male.	769	60244/m3	PP	1550E100, depth 4′2″, height 3′	Gl. 3	1	35–44	1	0	0	0	0	0	1	0	0	0	0	D f-g, T 7, SB 5, AS 5	M all features
20-6-10/60245.0	Nearly complete cranial and fragmentary postcranial human remains of a subadult of unknown sex.	877	60245	PP	1300E120, depth 2′6″, height 5′6″	Probably Gl. 5	1	9–10	0	1	0	0	0	0	0	0	0	0	1	D 12 ± 2.5 yrs (M2s just occluding), PC 8.5–9.5 yrs	Indet.
20-6-10/60246.0	Nearly complete cranial and partial postcranial human remains of an adult female.	758	60246	PP	Kiva 6, depth 6′6″, height 6″?	Gl. 3 (poss. Gl. 2)	1	50–59	1	0	0	0	0	0	0	0	1	0	0	D h+, AS 7	F all features

Peabody number	Pathology type	Pathology description	Cranial mod.	Associated funerary objects	Field context	Notes
20-6-10/60229.0	Infection, arthritis, nutr.	Ankylosis vt/l5–vt/s1 with associated infection, reabsorption anterior surface of vt/s1, DJD, spinal osteophytosis, osteoporosis	Not obs.	Frags. early Bisc. A bowl no. 1010 (69783)		
20-6-10/60230.0	Arthritis, nutr.	Spinal osteophytosis, DJD, osteoporosis	Not obs.	Gl. 4 bowl no. 857 (69823)		
20-6-10/60231.0	Infection, arthritis, nutr., other	Edentulous, spondylytis deformans of vt/l and vt/t11–12, osteoporosis, sacralization vt/l5	Not obs.			Specimen labels problematic
20-6-10/60232.0	Arthritis, nutr.	DJD, osteoporosis	Not obs.			Incorrect field number
20-6-10/60233.0	Infection, arthritis, nutr.	Marked antemortem tooth loss, marked DJD, mild osteoporosis	Unobs.	Portions of B&W? (no number listed)		
20-6-10/60234.0	Infection, nutr.	Dental abscesses, osteoporosis, enamel hypoplasia	Occ. def.	Black olla no. 1019 (69387), 2 bone awls no. 673 (70175), projectile point no. 674 (70420), Gl. 1 sherds, no number		
20-6-10/60235.0	Infection, nutr., other	Periostitis on tibia, porotic hyperostosis? (unusual striations on parietals, small lesions on frontal and parietals), vt/l5 spina bifida (perforated body), upper L P1 rotated	No cran. def.	Worked Gl. 1 sherd no. 677 (69928), Gl. 1 sherds, no number		
20-6-10/60236.0	Infection, arthritis	Extensive molar tooth loss, DJD	Occ. def.			
20-6-10/60237.0	Infection, arthritis	Impacted upper L canine protruding through maxilla, spinal osteophytosis	Occ. def.	Frags. of wrappings (70958), turquoise beads no. 870 (70801), stone scraper no. 845 (70450), obsidian projectile point no. 845 (70422)		
20-6-10/60238.0	Nutr., other	Enamel hypoplasia, exotoses on lingual side mandible	Not obs.	Bone tube no. 781 (70262)		
20-6-10/60239.0	Infection, nutr., other	Dental abscess, enamel hypoplasia, porotic hyperostosis in occipital, wormian bone at bregma	Not obs.			Field number and provenience problematic
20-6-10/60240.0	Trauma, infection, arthritis, nutr.	Trauma prox. R humerus (atrophy), antemortem edentulous, DJD, marked spinal osteophytosis, osteoporosis	Unobs.	Gl. 1 yellow bowls nos. 8116 (69814), 847 (69815)		
20-6-10/60241.0	Trauma, infection, arthritis, other	Healed fracture R dist. humerus, dental abscesses, DJD assoc. with fracture, spina bifida vt/s4–5	Occ. def.			
20-6-10/60242.0	Trauma, infection, arthritis	Fracture both humeral heads, healed fracture right innominate and healed hole in ilium (bullet? arrow wound?) with fusion of sacrum to innominates, thickening along sagittal suture, severe arthritis throughout	Occ. def.	Bisc. B bowl no. 1170 (69811)		
20-6-10/60243.0	Infection, arthritis, other	Antemortem tooth loss, dental abscesses, spinal osteophytosis, spina bifida vt/s4–5	Lamb. def.			
20-6-10/60244.0	Trauma, arthritis, nutr., other	Healed fracture L femur, DJD, enamel hypoplasia, ossification defect vt/c6–7 transverse process, spina bifida vt/s4–5	Not obs.	Gl. 3 bowl no. 869, scraper, no number		
20-6-10/60245.0	Nutr.	Enamel hypoplasia	Not obs.			
20-6-10/60246.0	Infection, arthritis, nutr.	Extensive antemortem tooth loss, teeth very worn, DJD, ankylosis 3 vt/t, osteoporosis	Occ. def.	Bisc. B bowl, Awanyu, no. 1172 (69792)		

Peabody number	Object description	FN	Specimen label	Site	Intrasite provenience	Ceramic stage	MNI	Age (yrs.)	Ad	Su	Ch	In	Fe	Unk. age	M	M?	F	F?	Unk. sex	Age criteria	Sex criteria
20-6-10/60247.0	Partial cranial and postcranial human remains of an adult female.	754	60247	PP	1572E154, depth 4′10″, height–on red clay	Very probably Gl. 2	1	25–29	1	0	0	0	0	0	0	0	1	0	0	D d, AS 2	F all features
20-6-10/60248.0	Partial cranial and postcranial human remains of an adult female.	702	60248	PP	900E150, depth 5′, height 6′	Probably Gl. 2	1	40–49	1	0	0	0	0	0	0	0	1	0	0	D f-g, AS 6	
20-6-10/60249.0	Partial cranial and fragmentary postcranial human remains of an adult female.	683	60249	PP	1500W50, depth 1′6″, height 6″	Very probably Gl. 1	1	45–54	1	0	0	0	0	0	0	0	1	0	0	D h/i,	Female
20-6-10/60250.0	Very fragmentary cranial and postcranial human remains of an adult of unknown sex.	892	60250/8	PP	750W160, depth 5″, height 2′4″	Pres. Gl. 5	1		1	0	0	0	0	0	0	0	0	0	1		Indet.
20-6-10/60251.0	Fragmentary cranial and partial postcranial human remains of an adult female.	767	60251	PP	1540E120, depth 5′, height 3′	Indet.	1	27–34	1	0	0	0	0	0	0	0	1	0	0	D g, T 5, AS 3	Female
20-6-10/60252.0	Nearly complete cranial and postcranial human remains of an adult female.	720	60252/f48	PP	900E150, depth 5′8″, height 8″ in red clay	Gl. 3	1	30–34	1	0	0	0	0	0	0	0	1	0	0	D f/f-g, T 6, AS 3	F all features
20-6-10/60253.0	Very fragmentary cranial and fragmentary postcranial human remains of an adult male.	711	60253	PP	850E150, depth 5′, height 6″	Pres. Gl. 3	1	25–29	1	0	0	0	0	0	1	0	0	0	0	AS 2	Male
20-6-10/60254.0	Very fragmentary cranial and partial postcranial human remains of an adult male.	772	60254/772	PP	1540E120, depth 5′, height 3′	Gl. 1	1	60+	1	0	0	0	0	0	1	0	0	0	0	T 10+, AS 8	M all features
20-6-10/60255.0	Nearly complete cranial and postcranial human remains of an adult female.	715	60255	PP	850E150, depth 4′, height 1′	Gl. 3	1	50–59	1	0	0	0	0	0	0	0	1	0	0	D h+/i+, AS 7	
20-6-10/60256.0	Nearly complete cranial and fragmentary postcranial human remains of an adult male.	941	60256	PP	Room 27, depth 6′3″, height ?	Probably Gl. 1	1	18–22	1	0	0	0	0	0	1	0	0	0	0	D c	M all features
20-6-10/60257.0	Nearly complete cranial and postcranial human remains of an adult female.	717	60257	PP	1450W100, trench 2 feet east of old wall, depth 4′, on rock	Probably Gl. 3	1	35–45	1	0	0	0	0	0	0	0	1	0	0	D h+/i+, T 6, AS 4–5	F all features
20-6-10/60258.0	Nearly complete cranial and postcranial human remains of an adult female.	752	60258, 752	PP	1546E138, depth 4′6″, height 1′3″	Probably Gl. 3	1	30–39	1	0	0	0	0	0	0	0	1	0	0	D f, T 7, SB 4, AS 4	Pelvis F, skull mixed ME,NC,SR-M, MP,SM-F)
20-6-10/60259.0	Nearly complete cranial and postcranial human remains of an adult female.	708	60259	PP	1450W100 w. trench by s. wall, depth 3′6″, height 1′	Gl. 1	1	25–29	1	0	0	0	0	0	0	0	1	0	0	D e, SB 2, T 4, AS 7 (pathological)	F all features
20-6-10/60260.0	Nearly complete cranial and postcranial human remains of an adult male.	750	60260	PP	1549E139, depth 4′, height 1′9″	Gl. 4	1	30–39	1	0	0	0	0	0	1	0	0	0	0	D f, T 7, SB 4, AS 5	M all features
20-6-10/60261.0	Nearly complete cranial and postcranial human remains of an adult male.	718	60261	PP	1450W100, west trench by north wall, depth 2′6″	Probably Gl. 2	1	35–45	1	0	0	0	0	0	1	0	0	0	0	D g-h/h, T 7, SB 4	M all features
20-6-10/60262.0	Nearly complete cranial and postcranial human remains of an adult female.	753	60262/ 60261	PP	1570E153, depth 4′2″, height 9″	Very probably Gl. 2	1	18–24	1	0	0	0	0	0	0	0	1	0	0	D c, PC < 25 yrs	F all features
20-6-10/60263.0	Nearly complete cranial and partial postcranial human remains of a child of unknown sex.	821	60263	PP	1270E130, depth 5′5″, height–on bottom	Pres. Gl. 3 or 4	1	3–4	0	0	1	0	0	0	0	0	0	0	1	D 6 ± 2 yrs, PC 3–4 yrs	Indet.
20-6-10/60264.0	Fragmentary cranial and postcranial human remains of an adult female.	858	60264	PP	1260E120, depth 3′5″, height 4′	Gl. 5	1	18–22	1	0	0	0	0	0	0	0	1	0	0	D c, AS and T N/A	F all features
20-6-10/60264.1	Very fragmentary postcranial human remains of a child of unknown sex.	858		PP	1260E120, depth 3′5″, height 4′	Gl. 5	1		0	0	1	0	0	0	0	0	0	0	1		
20-6-10/60265.0	Very fragmentary cranial and fragmentary postcranial human remains of an adult female.	712	60265	PP	1450W100, west trench, depth 2′3″, height 2′	Pres. Gl. 3	1	18–24	1	0	0	0	0	0	0	0	1	0	0	AS 2	Female

Peabody number	Pathology type	Pathology description	Cranial mod.	Associated funerary objects	Field context	Notes
20-6-10/60247.0	Infection, other	Upper R M3 impacted, peg lower R P2	Not obs.	Polishing stone no. 1199 (70601)		
20-6-10/60248.0	Infection, arthritis, nutr.	Extensive antemortem molar loss, slight DJD, osteoporosis	Not obs.	Gl. 3 sherds, no number		
20-6-10/60249.0	Trauma, infection, nutr.	Trauma L ulna and radius, L femur, dental abscesses, slight osteoporosis, severe enamel hypoplasia	Occ. def.			
20-6-10/60250.0			Unobs.			Field card says middle-aged male
20-6-10/60251.0	Trauma, infection, nutr., other	Healed fracture–R humerus, periostitis R tibia, extensive molar loss, enamel hypoplasia, U R C and P1 are reversed	Unobs.			
20-6-10/60252.0	Nutr., other	Enamel hypoplasia, spina bifida vt/s4–5	Not obs.			Field card description does not match human remains
20-6-10/60253.0	Infection	Dental abscess	Unobs.	Trefoil flint point, flakes (no number listed)		
20-6-10/60254.0	Arthritis, nutr.	Severe DJD and osteoporosis, spinal osteophytosis and ankylosis vt/l, spondylytis deformans	Unobs.	Gl. 1 yellow bowl no. 961 (69803)		
20-6-10/60255.0	Trauma, infection, arthritis, nutr.	R femur–severe lytic infection related to trauma, DJD, severe osteoporosis	Occ. def.	Gl. 3 bowl no. 991 (69775)		
20-6-10/60256.0	Nutr.	Mild porotic hyperostosis	Not obs.	Reused olla base, Gl. 1 no. 1004 (69916)		
20-6-10/60257.0	Trauma, infection, arthritis	Healed fracture–L radius, impacted upper I1s emerging in palate, significant antemortem tooth loss, some DJD	Occ. def.			
20-6-10/60258.0	Infection, nutr.	Dental abscess–molar, lytic lesions on parietal, enamel hypoplasia	Not obs.	2 antler implements no. 244 (70363), 2 bone flakes, B&W sherds and a few Gl. 1 sherds (no numbers)		
20-6-10/60259.0	Infection, arthritis, other	Periostitis at sacroiliac and associated DJD, vt/s4–5 not fused	Occ. def.	Gl. 1 bowl no. 987 (69771), black olla no. 998 (69772)		
20-6-10/60260.0	Infection, arthritis, nutr.	Lytic infection calcanei, costal frag. ossified to sternum, sacral ankylosis to ilia, spinal osteophytosis, DJD, enamel hypoplasia	Not obs.	Bisc. B scoop (no number listed)		Very large brow ridges, unusual skull shape
20-6-10/60261.0	Infection, arthritis, nutr., other	Dental abscess, osteophytes on tibia, enamel hypoplasia, dolicocephalic	Not obs.			
20-6-10/60262.0	Nutr.	Enamel hypoplasia	Not obs.			
20-6-10/60263.0			Not obs.	Shell pendant, no number		
20-6-10/60264.0	Infection, arthritis, nutr.	Lesion on dist. articular surface of pollux proximal phalange, DJD, enamel hypoplasia	Unobs.	Gl. 5 bowl no. 944 (69832)		
20-6-10/60264.1			No cran.	Gl. 5 bowl no. 944 (69832)		Field card description does not match human remains
20-6-10/60265.0			Unobs.			

Peabody number	Object description	FN	Specimen label	Site	Intrasite provenience	Ceramic stage	MNI	Age (yrs.)	Ad	Su	Ch	In	Fe	Unk. age	M	M?	F	F?	Unk. sex	Age criteria	Sex criteria
20-6-10/60266.0	Fragmentary cranial and nearly complete postcranial human remains of an adult male.	894	60266	PP	Kiva 6, depth 12′, height in scoop 6″ deep into clay	B&W	1	30–39	1	0	0	0	0	0	1	0	0	0	0	D f, AS 4	M all features
20-6-10/60267.0	Nearly complete cranial and postcranial human remains of an adult male.	894a	60267	PP	750W130, depth 6′, height–on bottom	Gl. 3	1	35–39	1	0	0	0	0	0	1	0	0	0	0	D i+, T 7, AS 4	M all features
20-6-10/60268.0	Fragmentary cranial and partial postcranial human remains of an adult male.	748	60268	PP	1540E134, depth 4′4″, height 9″	Probably Gl. 3	1	20–34	1	0	0	0	0	0	1	0	0	0	0	D d, AS 3–4	M all features
20-6-10/60269.0	Fragmentary cranial and partial postcranial human remains of a child of unknown sex.	837	60269/ 60239	PP	1310E130, depth 4′, height 3′6″	Late (post-Col umbian)	1	2.5–3.5	0	0	1	0	0	0	0	0	0	0	1	D 4 ± 1 yrs, PC 2.5–3.5 yrs	Indet.
20-6-10/60270.0	Nearly complete cranial and partial postcranial human remains of a child of unknown sex.	825	60270	PP	1310E140, depth 3′, height 4′6″ above bottom	Late (post-Col umbian)	1	6–9	0	0	1	0	0	0	0	0	0	0	1	M1s erupted, all dec molars present	Indet.
20-6-10/60271.0	Nearly complete cranial and partial postcranial human remains of an adult female.	842	60271	PP	1290E130, depth 3′7″, height 2′	Probably Gl. 4 or 5	1	18–24	1	0	0	0	0	0	0	0	1	0	0	D c, AS 1	F all features
20-6-10/60272.0	Nearly complete cranial and partial postcranial human remains of an adult female.	915	60272	PP	750W120, depth 7′, height–on bottom	Gl. 3	1	18–24	1	0	0	0	0	0	0	0	1	0	0	D c-d, T 2, SB 1, AS 1	F all features
20-6-10/60273.0	Fragmentary postcranial human remains of an adult male.	707b	60273	PP	1450W100, west trench, depth 4′, on bottom	Probably Gl. 2	1	30–44	1	0	0	0	0	0	1	0	0	0	0	AS 3–5	Male
20-6-10/60274.0	Nearly complete cranial and postcranial human remains of an adult male.	685	60274	PP	1500W50, depth 2′6″, height 8″	Very probably Gl. 1	1	40–54	1	0	0	0	0	0	1	0	0	0	0	D h/i	M all features
20-6-10/60275.0	Fragmentary cranial and partial postcranial human remains of an adult male.	698	60275/m2	PP	1400W50, room 25 along west wall, depth 8′2″, on rock	Probably Gl. 1	1	35–45	1	0	0	0	0	0	1	0	0	0	0	D f-g, T 7–8, AS 5	Male
20-6-10/60276.0	Nearly complete cranial and partial postcranial human remains of an adult male.	773	60276	PP	1550E120, depth 3′9″, height 3′	Pres. Gl. 2 or 3	1	45–54+	1	0	0	0	0	0	1	0	0	0	0	D h+/i+, AS 6	Male
20-6-10/60277.0	Nearly complete cranial and postcranial human remains of an adult male.	790	60277	PP	1550E150, depth 4′2″, height 1′6″	Gl. 3	1	30–34	1	0	0	0	0	0	1	0	0	0	0	D e (uneven), T 7, SB 3, AS 3	Male
20-6-10/60278.0	Partial cranial and postcranial human remains of an adult male.	793	60278	PP	1600E150, depth 2′7″, height 2′9″	Gl. 3	1	40–44	1	0	0	0	0	0	1	0	0	0	0	D h/i, T 8, AS 5	M all features
20-6-10/60279.0	Partial cranial and nearly complete postcranial human remains of an adult male.	836	60279	PP	1270E130, depth 5′3″, height 1′	Probably Gl. 3	1	50–59	1	0	0	0	0	0	1	0	0	0	0	h+/i+, AS 7	M all features
20-6-10/60280.0	Nearly complete cranial and partial postcranial human remains of an adult female.	840	60280	PP	1220E130?, depth 5′	Very probably Gl. 4	1	30–39	1	0	0	0	0	0	0	0	1	0	0	D g-h, AS 3	F all features
20-6-10/60281.0	Partial cranial and postcranial human remains of an adult female.	856	60281/n/93	PP	1290E120, depth 6′2″	Gl. 2	1	30–34	1	0	0	0	0	0	0	0	1	0	0	D g, AS 3	F all features
20-6-10/60282.0	Nearly complete cranial and partial postcranial human remains of an adult female.	863	60282	PP	1300E130, depth 3′, height 5′4″	Very probably Gl. 5	1	35–45	1	0	0	0	0	0	0	0	1	0	0	D h/i, AS 3–4	F all features
20-6-10/60283.0	Nearly complete cranial and partial postcranial human remains of an adult male.	904	60283	PP	750W120, depth 4′, height 2′6″	Pres. Gl. 4 or 5	1	35–39	1	0	0	0	0	0	1	0	0	0	0	D g/h, AS 4	Male
20-6-10/60284.0	Fragmentary cranial and partial postcranial human remains of an adult female.	906	60284	PP	580W90, depth 2′6″, height–on bottom	Probably Gl. 3	1	30–34	1	0	0	0	0	0	0	0	1	0	0	D g, T 6, AS 3	F all features

Peabody number	Pathology type	Pathology description	Cranial mod.	Associated funerary objects	Field context	Notes
20-6-10/60266.0	Trauma, arthritis, nutr.	Fusion of pedal phalanges assoc. with trauma, some spinal osteophytosis, DJD, enamel hypoplasia	Occ. def.	B&W bowl no. 848 (69843), frags. black olla no. 1039 (69844), Gl. 3 bowl frags. no. 990 (69845), tubular pipe no. 775 (70025)		
20-6-10/60267.0	Infection, arthritis, other	Dental abscess, severe spinal osteophytosis, DJD, spina bifida–sacrum	Not obs.	Gl. 3 bowl no. 990 (69845), pipe under chest		
20-6-10/60268.0	Trauma, infection, nutr.	Healed fracture L radius, lytic infection L tibia, enamel hypoplasia	Unobs.	6 stone implements no. 243 (70537)		
20-6-10/60269.0			Unobs.	Shell fragment, Olivella shell beads, and crinoid stem, no. 572 (70796)	Burial at length	
20-6-10/60270.0	Nutr., other	Cribra orbitalia, ankylosis 2 vt/c	Not obs.			
20-6-10/60271.0	Infection	Dental abscesses, cavities in molars	Poss. cran. def.			
20-6-10/60272.0	Trauma	Trauma–2 broken ribs ankylosed	Occ. def.	Gl. 3 bowl no. 1104 (69852), stone knife no. 780 (70501)		
20-6-10/60273.0	Trauma, arthritis	Trauma R radius–healed with displacement Coll's fracture, spinal osteophytosis, some DJD	No cran.			Field card description does not match human remains
20-6-10/60274.0	Infection, arthritis, nutr.	Molar tooth loss antemortem, spondylytis deformans, spinal osteophytosis, DJD, osteoporosis	Not obs.	Gl. 1 potsherds		Perforated sternum
20-6-10/60275.0	Other	Scaphocephalic shape but sutures visible	Unobs.	Clay cylinders no. 442 (70080), 7 stones found in leather pouch no. 1161 (70660)		
20-6-10/60276.0	Infection, arthritis, nutr., other	Lytic lesion R parietal, spinal osteophytosis, osteoporosis, nonfusion of vt/s4–5	Cran. def.?			
20-6-10/60277.0	Arthritis, nutr., other	Ankylosis R capitate and trapezoid fused, calcanei and tali fused, DJD assoc with ankylosis, enamel hypoplasia, interesting pillar on upper R I2, rotation upper L P2, massive brow ridge development	Not obs.			
20-6-10/60278.0	Trauma, arthritis, nutr., other	Trauma–R fibula healed fracture, vt/t and l severe spondylytis deformans, DJD, osteoporosis, vt/s4–5 not fused	Not obs.	Early Gl. 3 bowl no. 854 (69808)		
20-6-10/60279.0	Infection, arthritis, nutr.	Extensive antemortem tooth loss, DJD, enamel hypoplasia	Occ. def.			
20-6-10/60280.0	Infection, arthritis, nutr.	Dental abscesses, extensive antemortem tooth loss, slight–moderate DJD, enamel hypoplasia	Lamb. def.			
20-6-10/60281.0	Arthritis	DJD, arthritis at pubic symp., vt/l5 fused to sacrum	Occ. def.	Part of Gl. 2 bowl no. 949 (69831) with traces of meal		
20-6-10/60282.0	Arthritis, nutr.	DJD, enamel hypoplasia	Occ. def.	Gl. 3 sherds, no number		
20-6-10/60283.0	Infection, arthritis, nutr., other	Lesion on frontal, infection right side mandible, minor DJD, enamel hypoplasia, spina bifida–sacrum	Unobs.	Gl. 4 sherds, no number		Shape of cranium "different"
20-6-10/60284.0			Unobs.			

Peabody number	Object description	FN	Specimen label	Site	Intrasite provenience	Ceramic stage	MNI	Age (yrs.)	Ad	Su	Ch	In	Fe	Unk. age	M	M?	F	F?	Unk. sex	Age criteria	Sex criteria
20-6-10/60285.0	Fragmentary cranial and partial postcranial human remains of an adult female.	922	60285	PP	750W120, depth 7′4″	Gl. 3	1	35–39	1	0	0	0	0	0	0	0	1	0	0	D g, T 7, AS 4	Female
20-6-10/60286.0	Nearly complete cranial and fragmentary postcranial human remains of an adult male.	819	60286/m5	PP	1290E130, depth 2′7″	Very probably Gl. 4 (in Gl. 3 sherds)	1	35–39	1	0	0	0	0	0	1	0	0	0	0	D h, AS 4	Male
20-6-10/60287.0	Partial cranial and fragmentary postcranial human remains of an adult female.	826	60287	PP	1310E140, depth 7′6″, height–on bottom	Pres. Gl. 3 or 4	1	50–59	1	0	0	0	0	0	0	0	1	0	0	h++/i++, PC 50–59 yrs	F all features
20-6-10/60288.0	Partial cranial and postcranial human remains of an adult female.	755	60288	PP	1546E138, depth 4′10″	Probably Gl. 3	1	20–29	1	0	0	0	0	0	0	0	1	0	0	D c, AS 2	F all features
20-6-10/60289.0	Fragmentary cranial and postcranial human remains of an adult female.	849	60289	PP	1290E130, depth 3′6″, height 2′	Probably Gl. 3	1	35+	1	0	0	0	0	0	0	0	1	0	0	old >35 yrs	Female
20-6-10/60290.0	Fragmentary cranial and nearly complete postcranial human remains of an adult female.	896a	60290	PP	750W150, depth 3′4″, height 6″ above bottom	Gl. 5	1	45–54	1	0	0	0	0	0	0	0	1	0	0	D h+, T 10, AS 6	F all features
20-6-10/60291.0	Fragmentary cranial and partial postcranial human remains of an adult female.	838	60291	PP	1290E130, depth 5′3″, height 1′	Pres. Gl. 3	1	25–34	1	0	0	0	0	0	0	0	1	0	0	D e, AS 3	Female
20-6-10/60292.0	Fragmentary cranial and partial postcranial human remains of an adult female.	697	60292	PP	850E150, 2 feet e of defense wall, depth 4′, height 3″	Indet., Late (post-Columbian)	1	18–24	1	0	0	0	0	0	0	0	1	0	0	D b2, AS 1	Female
20-6-10/60293.0	Nearly complete cranial and partial postcranial human remains of an adult male.	947	60293	PP	Room 43, depth 54″, height ?	Gl. 3	1	30–34	1	0	0	0	0	0	1	0	0	0	0	D e-f, AS 3	M all features
20-6-10/60294.0	Partial cranial and postcranial human remains of an adult male.	776	60294	PP	1550E120, depth 3′4″, height 4′6″	Pres. Gl. 3	1	30–39	1	0	0	0	0	0	1	0	0	0	0	D h+/i+, AS 3	M all features
20-6-10/60295.0	Fragmentary cranial and partial postcranial human remains of an adult that is probably female.	789	60295	PP	1590E150, depth 5′, height–on bottom	Probably Gl. 3	1	25–29	1	0	0	0	0	0	0	0	0	1	0	D e	?Female
20-6-10/60296.0	Partial cranial and nearly complete postcranial human remains of an adult male.	949	60296	PP	Room 42, sw corner, depth 7′6″, height–on mesa	B&W	1	30–39	1	0	0	0	0	0	1	0	0	0	0	D f, AS 4, T 6	M all features
20-6-10/60297.0	Nearly complete cranial and partial postcranial human remains of an adult female.	862	60297/f49	PP	1330E130, depth 3′10″, height 4′	Probably Gl. 5	1	35–49	1	0	0	0	0	0	0	0	1	0	0	D h/i, AS 4, T 10	F all features
20-6-10/60298.0	Nearly complete cranial and partial postcranial human remains of an adult female.	808	60298	PP	1260E130	Probably Gl. 4	1	25–34	1	0	0	0	0	0	0	0	1	0	0	D e-f, T 4–5, AS 3	Pelvis clearly F
20-6-10/60299.0	Partial cranial and postcranial human remains of an adult male.	910	60299	PP	750W120, depth 5′5″, height 2′	Probably Gl. 4	1	40–44	1	0	0	0	0	0	1	0	0	0	0	D ?/h, AS 5	M all features
20-6-10/60300.0	Nearly complete cranial and partial postcranial human remains of an adult female.	818	60300	PP	1320E130, depth 4′6″	Pres. Gl. 4	1	25–34	1	0	0	0	0	0	0	0	1	0	0	D f, PC 25–35 yrs	Female
20-6-10/60301.0	Nearly complete cranial and partial postcranial human remains of an adult female.	828	60301	PP	1280E130, depth 4′10″	Probably Gl. 5	1	35–44	1	0	0	0	0	0	0	0	1	0	0	D h uneven, AS 4–5	Female
20-6-10/60302.0	Nearly complete cranial and postcranial human remains of an adult female.	848	60302	PP	1290E130, depth 4′, height 2′	Probably Gl. 3	1	30–39	1	0	0	0	0	0	0	0	1	0	0	D g/g, T 7, AS 3	F all features
20-6-10/60303.0	Nearly complete cranial and postcranial human remains of an adult female.	830	60303	PP	1310E140, depth 3′8″, height 3′6″ above bottom	Late (post-Columbian)	1	20–29	1	0	0	0	0	0	0	0	1	0	0	D c-d, T 5, AS 2	F all features

Peabody number	Pathology type	Pathology description	Cranial mod.	Associated funerary objects	Field context	Notes
20-6-10/60285.0	Arthritis, nutr.	Slight DJD, enamel hypoplasia	Unobs.	Gl. 3 bowl no. 1103 (69855)		
20-6-10/60286.0	Trauma, infection, arthritis, other	L parietal–small lesion with healed crack due to trauma, spinal osteophytosis, ankylosis vt/c6–7, spondylytis deformans, 1st costals marked arthritis, maleruption rt max teeth–premolars	Slight cran. def.	Gl. 3 sherds, no numbers		
20-6-10/60287.0	Infection, arthritis, nutr.	Extreme antemortem tooth loss, severe DJD, osteoporosis	Not obs.			
20-6-10/60288.0	Infection	Lytic infection R tibia, dental abscess 2 large cavities	Not obs.			
20-6-10/60289.0	Infection	Lesions on parietals	Unobs.			
20-6-10/60290.0	Arthritis, nutr.	DJD, osteoporosis	Unobs.	Gl. 5 bowl no. 929 (69846)		
20-6-10/60291.0	Infection, nutr.	Cavities and tooth loss, enamel hypoplasia	Unobs.			
20-6-10/60292.0	Trauma, infection, nutr., other	Periostitis R radius assoc. with trauma, periostitis R tibia, infection of L sacroiliac joint, extensive porotic hyperostosis, thick diploe, vt/l5 partially fused to sacrum	?Occ. def.	Early Gl. 5 sherds, no number	Burial at length, post-Columbian	
20-6-10/60293.0	Trauma, nutr.	Trauma–blow to R parietal extends over to left side, enamel hypoplasia	Not obs.	Gl. 3 bowl frag. no. 1159 (69869)		
20-6-10/60294.0	Trauma, infection, arthritis	Healed trauma–R humerus, extreme antemortem tooth loss, slight DJD, spinal osteophytosis	Not obs.			
20-6-10/60295.0	Infection, arthritis	Vt–ankylosis, vt/l lesions–tuberculosis?, lower 3rd molars impacted and infected, slight DJD	Unobs.			
20-6-10/60296.0	Arthritis, nutr.	Slight DJD, enamel hypoplasia	Not obs.	B&W bowl no. 1165 (69871)		Perforated sternum
20-6-10/60297.0	Nutr.	Enamel hypoplasia	Occ. def.	Human effigy no. 650 (70064), Gl. 4–5 potsherds, no numbers		
20-6-10/60298.0	Infection, arthritis, nutr., other	Frontal lesion, slight DJD, enamel hypoplasia, vt/s5 not fused	Not obs.	Infant skull frag. assoc., no number		Field card description does not match human remains
20-6-10/60299.0	Arthritis, other	DJD, extra pillar on lower R M3	Not obs.			
20-6-10/60300.0	Infection, arthritis	Tuberculosis, lumbar vts–absorption bodies, ankylosis	Not obs.			
20-6-10/60301.0	Trauma, infection, arthritis, nutr.	L humerus–healed fracture, extensive antemortem tooth loss, systemic infection affecting cranium and appendicular skeleton (treponemal-like), R femur infection, slight on L femur, extensive lesions left side cranium up to temporal line, DJD, osteoporosis	Not obs.			
20-6-10/60302.0	Infection, arthritis, nutr.	Antemortem loss mandibular molars, spinal osteophytosis, slight DJD, enamel hypoplasia canine	Not obs.	Gl. 1 sherds, no number		
20-6-10/60303.0	Arthritis, nutr.	DJD, enamel hypoplasia	Occ. def.		Burial at length	

Peabody number	Object description	FN	Specimen label	Site	Intrasite provenience	Ceramic stage	MNI	Age (yrs.)	Ad	Su	Ch	In	Fe	Unk. age	M	M?	F	F?	Unk. sex	Age criteria	Sex criteria
20-6-10/60304.0	Partial cranial and postcranial human remains of an adult female.	886	60304	PP	750W160, depth 1′3″, height–on bottom	Probably Gl. 5	1	16–20	0	1	0	0	0	0	0	0	1	0	0	D b2-c, T 1	Female
20-6-10/60305.0	Partial cranial and postcranial human remains of a subadult male.	899	60305	PP	750W130, depth 5′6″	Probably Gl. 3	1	14.5–15.5	0	1	0	0	0	0	1	0	0	0	0	D 15 ± 3 yrs, AS 1, PC 14–15 yrs	
20-6-10/60306.0	Partial cranial and nearly complete postcranial human remains of an adult male.	865	60306	PP	1260E120, depth 4′6″, height 3′9″	Probably Gl. 4	1	30–39	1	0	0	0	0	0	1	0	0	0	0	D f, T 7, AS 4	M all features
20-6-10/60307.0	Nearly complete cranial and postcranial human remains of an adult female.	843	60307	PP	Kiva 6, depth 3′6″, height 5′	Gl. 1	1	45–49	1	0	0	0	0	0	0	0	1	0	0	D h/i, AS 6	
20-6-10/60308.0	Partial cranial and nearly complete postcranial human remains of an adult female.	898	60308	PP	760W130, depth 5′	Gl. 3	1	25–34	1	0	0	0	0	0	0	0	1	0	0	D e, AS 3	Female
20-6-10/60309.0	Partial cranial and postcranial human remains of an adult female.	935	60309	PP	1310E110, depth 6′, height 4′	Probably Gl. 5	1	40–49	1	0	0	0	0	0	0	0	1	0	0	D ?/i, T 8, AS 5–6	F all features
20-6-10/60310.0	Fragmentary cranial human remains of an adult female.	885	60310	PP	740W170, sw trench 26 ft e of beginning, depth 12″, height 3″	Probably Gl. 5	1	50–59	1	0	0	0	0	0	0	0	1	0	0	>50 yrs	Female
20-6-10/60311.0	Partial cranial and postcranial human remains of an adult female.	713	60311	PP	900E150, depth 3′, height 1′	Gl. 3	1	25–34	1	0	0	0	0	0	0	0	1	0	0	D f, T 7?, AS 2	Female
20-6-10/60312.0	Partial cranial and postcranial human remains of an adult female.	852	60312	PP	1300E130, depth 7′, height 6″	Probably Gl. 2 or 3	1	18–24	1	0	0	0	0	0	0	0	1	0	0	D c (few teeth present), T 1, AS 1, young adult prox. clavicles not fused, iliac crest stage 1	F all features
20-6-10/60313.0	Nearly complete cranial and postcranial human remains of an adult male.	795	60313	PP	1570E140, depth 3′4″, height 2′6″	Gl. 3	1	40–49	1	0	0	0	0	0	1	0	0	0	0	D h+/i+, T 8, AS 6	Male (all except SR, ME)
20-6-10/60314.0	Nearly complete cranial and partial postcranial human remains of an adult male.	804	60314	PP	1290E140	Very probably Gl. 3	1	30–34	1	0	0	0	0	0	1	0	0	0	0	D f, T 6, AS 3	M all features
20-6-10/60315.0	Nearly complete cranial and postcranial human remains of an adult male.	704	60315	PP	1450W100, west trench, depth 2″, height 3′8″	Gl. 3	1	35–45	1	0	0	0	0	0	1	0	0	0	0	D g/g, T 8, SB 4, AS 5	M all features
20-6-10/60316.0	Nearly complete cranial and partial postcranial human remains of an adult male.	896	60316	PP	750W150, depth 3′4″, height 6″	Gl. 5	1	30–34	1	0	0	0	0	0	1	0	0	0	0	D h, AS 3	M all features
20-6-10/60317.0	Partial cranial and postcranial human remains of an adult male.	817	60317,l,eo9	PP	1310E140, depth 4′8″	Gl. 3	1	50–59	1	0	0	0	0	0	1	0	0	0	0	AS 7 (but TB makes look older, probably in 40s)	Male
20-6-10/60318.0	Nearly complete cranial and postcranial human remains of an adult male.	680	60318	PP	1500W50, depth 8″, height 1′6″	Indet., Late (post-Columbian)	1	40–44	1	0	0	0	0	0	1	0	0	0	0	D g-h/h, T 8, AS 5	M all features
20-6-10/60319.0	Nearly complete cranial and partial postcranial human remains of an adult that is probably male.	782	60319	PP	1550E120, depth 3′4″, height 4′8″	Gl. 3	1	30–34	1	0	0	0	0	0	0	1	0	0	0	D h/i, AS 3	? Male
20-6-10/60320.0	Nearly complete cranial and partial postcranial human remains of an adult of unknown sex.	777	60320	PP	Room 29, depth 6′6″, height 6″	Gl. 2	1		1	0	0	0	0	0	0	0	0	0	1	D c, PC looks older than 35 yrs	Indet.
20-6-10/60321.0	Fragmentary cranial human remains of an adult female.	765	60321	PP	Kiva 6, depth 4′2″, height 4′8″	Gl. 2	1		1	0	0	0	0	0	0	0	1	0	0		Female

Peabody number	Pathology type	Pathology description	Cranial mod.	Associated funerary objects	Field context	Notes
20-6-10/60304.0	Infection, nutr.	Lesions on vt/ls and L parietal–tuberculosis?, enamel hypoplasia on C	Poss. occ. def.	Gl. 5 sherds, no number		
20-6-10/60305.0	Nutr.	Enamel hypoplasia	Not obs.	2 small knives no. 778 (70500)		
20-6-10/60306.0	Arthritis, nutr.	Vt/l5 fused to sacrum, arthritic lipping on carpals, enamel hypoplasia	Occ. def.	Gl. 3 sherds, no number		
20-6-10/60307.0	Infection, arthritis, nutr.	Dental abscess, arthritis–limb bones, DJD, enamel hypoplasia	Occ. def.	Gl. 1 yellow bowl no. 849 (69828), frag. olla base no. 718 (69926)		
20-6-10/60308.0	Infection, arthritis, nutr.	Infection of L sacroiliac, slight DJD, enamel hypoplasia	Not obs.	Gl. 3 bowl no. 975 (69848)		Perforated sternum
20-6-10/60309.0	Trauma, arthritis, nutr.	Healed fracture–distal L radius, spondylytis lumbar vt, marked osteoporosis	Occ. def.	Gl. 5 sherds, no number		
20-6-10/60310.0			Unobs.	Piece of iron under head, Gl. 5 sherds (no numbers listed)		
20-6-10/60311.0	Trauma, infection, other	4 perimortem blows to L parietal, healed fracture L elbow–extensive remodeling dist. humerus and prox. ulna, marked bowing L tibia and fibula, scaphocephaly	Not obs.	Gl. 3 bowl no. 995 (69773)		
20-6-10/60312.0			Occ. def.	Gl. 1 sherds, no number		
20-6-10/60313.0	Trauma, infection, arthritis, nutr., other	Healed fracture–dist. L radius, extensive antemortem tooth loss, DJD, spondylytis deformans, osteoporosis, vt/s3–5 not fused	Occ. def.	Gl. 3 bowl no. 955 (69819)		
20-6-10/60314.0	Trauma, infection, arthritis, nutr.	Healed fracture pedal phalange (bone spur), healed fracture prox. L fibula, dental abscess, spondylytis deformans–lumbar vt, DJD, enamel hypoplasia, increased porosity frontal and parietals	Occ. def.			
20-6-10/60315.0	Infection, nutr.	Impacted upper L M3, osteoporosis	No cran. def.	Gl. 3 bowl no. 947 (69770), small black bowl no. 946 (69769), jade (?) flake no. 134 (70841)		
20-6-10/60316.0	Arthritis, nutr.	Spinal osteophytosis, osteoporosis	Not obs.	Gl. 5 bowl no. 929 (69846)		
20-6-10/60317.0	Trauma, infection, nutr., tumor	Trauma L fibula, slight porosity on parietal and frontal, lytic lesions on vertebrae - possible tuberculosis? extensive antemortem tooth loss, extensive osteoporosis, exotosis on lingual side mandible	Not obs.	Frag. Gl. 3 bowl no. 865 (69822)		
20-6-10/60318.0	Infection, nutr.	Lesions on parietals, frontal, pronounced occipital bun, infection L tibia and femur, R and L clavicles, slight infection in L and R fibula, R tibia–possible syphilis, dental abscess, enamel hypoplasia	Not obs.	Bisc. A bowls no. 720 (69758) and no. 981 (69759)	Burial at length	
20-6-10/60319.0	Arthritis, nutr.	Slight DJD, enamel hypoplasia	Not obs.	Gl. 3 sherds, no number		Cranium looks different, unusual incisor polish
20-6-10/60320.0	Arthritis, nutr.	Spinal osteophytosis, enamel hypoplasia	Occ. def.			Postcrania large adult male yet cranium gracile, clearly an immigrant, cranial deformation is very pronounced
20-6-10/60321.0			Unobs.			Associated package of hair

Peabody number	Object description	FN	Specimen label	Site	Intrasite provenience	Ceramic stage	MNI	Age (yrs.)	Ad	Su	Ch	In	Fe	Unk. age	M	M?	F	F?	Unk. sex	Age criteria	Sex criteria
20-6-10/60322.0	Fragmentary cranial and postcranial human remains of an adult of unknown sex.	854	60322	PP	1280E130, depth 3′, height 3′	Indet.	1		1	0	0	0	0	0	0	0	0	0	1		Indet.
20-6-10/60324.1	Very fragmentary postcranial human remains of an adult of unknown sex.	179	179, 60324-7	PP	1400E100, depth 13′	Indet.	1		1	0	0	0	0	0	0	0	0	0	1		Indet.
20-6-10/60327.1	Very fragmentary postcranial human remains of an adult of unknown sex.	136	136, 60327-6	PP	1500E100, depth 9′, 2′ above red clay	Indet.	1		1	0	0	0	0	0	0	0	0	0	1		Indet.
20-6-10/60332.1	Fragmentary postcranial human remains of an adult of unknown sex.	531	60332-2/60333-3/60326-1/60338-1/531	PP	1600W50, depth 14″, height–on shale	Indet.	1		1	0	0	0	0	0	0	0	0	0	1		Indet.
20-6-10/60333.1	Fragmentary postcranial human remains of an adult of unknown sex.	185	60333-1, 60346-20, 185	PP	1500E50, terrace, depth 3′	Indet.	1		1	0	0	0	0	0	0	0	0	0	1		Indet.
20-6-10/60336.1	Fragmentary postcranial human remains of an adult of unknown sex.	285	60336-4, 285	PP	Trench b, depth 12′, on red soil	Indet.	1		1	0	0	0	0	0	0	0	0	0	1		Indet.
20-6-10/60337.1	Fragmentary postcranial human remains of an adult of unknown sex.	501	60337-14, 501 (581 or 531?)	PP	North terrace, depth 3′6″	Indet.	1		1	0	0	0	0	0	0	0	0	0	1		Indet.
20-6-10/60337.2	Fragmentary postcranial human remains of an adult of unknown sex.	538	60337-11, 538	PP	1600W50, depth 4″, height–on shale	Probably Gl. 2	1		1	0	0	0	0	0	0	0	0	0	1		Indet.
20-6-10/60337.3	Fragmentary postcranial human remains of an adult of unknown sex.	192	60337-4, 192	PP	1450E100, depth 12′8″	Very probably Gl. 1	1		1	0	0	0	0	0	0	0	0	0	1		Indet.
20-6-10/60339.1	Fragmentary postcranial human remains of an adult of unknown sex.	9	60338-9/9	PP	1550E150	Indet., poss. Gl. 4	1		1	0	0	0	0	0	0	0	0	0	1		Indet.
20-6-10/60343.1	Fragmentary postcranial human remains of an adult of unknown sex.	308	60343-23, 308	PP	1450E100, test x, at base of cut 3	Probably Gl. 4 or 5	1		1	0	0	0	0	0	0	0	0	0	1	Adult	Indet.
20-6-10/60344.1	Fragmentary postcranial human remains of an adult male.	253	60344-7, 31, 60343-2, 253	PP	1450E100, kiva 2, depth 3′ in ash	Indet.	1	Adult	1	0	0	0	0	0	1	0	0	0	0	Adult	Male
20-6-10/60350.1	Fragmentary cranial human remains of a subadult of unknown sex.	107	60350, 41, 107	PP	1450E100, depth 3′, 9′ above red clay	Indet.	1	12–15	0	1	0	0	0	0	0	0	0	0	1	M3s unerupted	Indet.
20-6-10/60350.10	Fragmentary cranial human remains of a subadult of unknown sex.	336	60350-19	PP	1250E100depth 3′8″	Probably Gl. 3 or poss. 4	1	10–12	0	1	0	0	0	0	0	0	0	0	1	1 M2 unerupted, 11–12 yrs	Indet.
20-6-10/60350.11	Fragmentary cranial human remains of a child of unknown sex.	414	60350-16/414	PP	1400E100, depth 2′6″, in cut 2, test XI	Probably Gl. 6	1	2–5	0	0	1	0	0	0	0	0	0	0	1	2–6 yrs (M1s unerupted)	Indet.
20-6-10/60350.4	Fragmentary cranial human remains of an adult of unknown sex.	67	60350-2/67	PP	1600E150	Gl. 3	1		1	0	0	0	0	0	0	0	0	0	1		Indet.
20-6-10/60350.5	Fragmentary cranial human remains of an adult of unknown sex.	77	60350-22/77 (do not see 77 on specimen, MEM 7/97)	PP	1600E150	Pres. Gl. 3	1		1	0	0	0	0	0	0	0	0	0	1		Indet.
20-6-10/60350.6	Fragmentary cranial human remains of an adult of unknown sex.	90	60350-10/90 (do not see 90 on specimen, MEM 7/97)	PP	1450E150, depth 7′, 4′ above red clay	Pres. Gl. 4 or 5	1		1	0	0	0	0	0	0	0	0	0	1		Indet.
20-6-10/60350.7	Fragmentary cranial human remains of an adult of unknown sex.	96	60350-4/96	PP	1500E100, depth 9′, height 1′ above red clay	Gl. 2	1		1	0	0	0	0	0	0	0	0	0	1		Indet.
20-6-10/60350.8	Fragmentary cranial human remains of a child of unknown sex.	119	60350-38, 119	PP	1450E100, depth 4′, 5′ above red clay	Probably Gl. 4	1	3–5	0	0	1	0	0	0	0	0	0	0	1	3–6 yrs (M1s unerupted)	Indet.
20-6-10/60350.9	Fragmentary cranial human remains of an adult of unknown sex.	296	60350-15	PP	1250E100, depth 4′6″	Indet.	1		1	0	0	0	0	0	0	0	0	0	1		Indet.

Peabody number	Pathology type	Pathology description	Cranial mod.	Associated funerary objects	Field context	Notes
20-6-10/60322.0			Unobs.	Pipe, deer horn, 2 cut bones, no numbers listed		Number split into 60322-0, 60322-1 (MEM 6/97)
20-6-10/60324.1			No cran.			NSF inventory catalogued this as 0-59949 (MEM 4/98)
20-6-10/60327.1			No cran.	Frag. Gl. 4 bowl no. 1026 (65067), broken projectile points nos. 1027 (65992), 895 (65984)		NSF inventory catalogued this as 0-59950 (MEM 4/98)
20-6-10/60332.1	Arthritis	DJD	No cran.	Gl. 2 bowl no. 1755 (67077), Gl. 2 red bowl no. 1765 (67078), antler tip no. 1775 (68264), frag. of effigy no. 1776 (67562)		
20-6-10/60333.1	Arthritis	DJD	No cran.	Gl. 4 bowl no. 1217 (65007)		
20-6-10/60336.1			No cran.	Gl. 1 large red bowl no. 685 (66997), frags. small clay vessel no. 689 (66998)		
20-6-10/60337.1	Arthritis	DJD	No cran.	Gl. 1–2 Frijolito (early) bowl over ribs no. 1580 (67067)		
20-6-10/60337.2			No cran.			
20-6-10/60337.3	Arthritis	DJD	No cran.			
20-6-10/60339.1			No cran.	Pieces of worked hematite, obsidian projectile point, Olivella shell, shell object, quartz pebble, all catalogued as no. 49 (72266)		Field card number assigned on basis of specimen label and description match
20-6-10/60343.1			No cran.	2 projectile points no. 729 (68413), antler tool over ribs no. 731 (68251)		
20-6-10/60344.1	Arthritis	DJD	No cran.	Olla bowl Gl. 5 no. 1835 (66988), bone tool no. 475 (67822)		
20-6-10/60350.1	Nutr.	Enamel hypoplasia	Unobs.	Objects no. 760	Burial at length	
20-6-10/60350.10	Nutr.	Enamel hypoplasia	Unobs.	Projectile point no. 1015 (68419), Gl. 4 bowl no. 981 (67012)		Field card indicates more complete individual
20-6-10/60350.11			Unobs.	Pecos white pottery, projectile points no. 1324 (68428), bone bead no. 1325 (68017), stone implement no. 1327 (68660)		Field card indicates more complete individual
20-6-10/60350.4			Unobs.	Gl. 3 bowl no. 256 (65049)		Field card indicates more complete individual
20-6-10/60350.5			Unobs.			Field card indicates more complete individual
20-6-10/60350.6			Unobs.			Field card indicates more complete individual
20-6-10/60350.7	Other	Mandibular incisor fused to canine (counted as 2 teeth)	Unobs.	Gl. 2 yellow bowl no. 400 (65031) (contained flour)		Field card indicates more complete individual
20-6-10/60350.8	Nutr.	Enamel hypoplasia	Unobs.	Frag. black olla no. 1018, bone awl no. 834 (65258)		
20-6-10/60350.9			Unobs.			Field card indicates more complete individual

Peabody number	Object description	FN	Specimen label	Site	Intrasite provenience	Ceramic stage	MNI	Age (yrs.)	Ad	Su	Ch	In	Fe	Unk. age	M	M?	F	F?	Unk. sex	Age criteria	Sex criteria
20-6-10/60358.1	Fragmentary postcranial human remains of an adult of unknown sex.	134	60358-7, Pecos 134	PP	1450E100, depth 11′, 1′ above red	Indet.	1		1	0	0	0	0	0	0	0	0	0	1		Indet.
20-6-10/60358.2	Fragmentary postcranial human remains of a subadult of unknown sex.	638	60358-11,12,13,15/638	PP	North terrace, depth 25″, 8″ above bedrock	Pres. Gl. 1, 2, or 3	1		0	1	0	0	0	0	0	0	0	0	1		Indet.
20-6-10/60358.3	Fragmentary postcranial human remains of an adult of unknown sex.	572	60358-10/572	PP	1550W50	Probably B&W or Gl. 1	1		1	0	0	0	0	0	0	0	0	0	1		Indet.
20-6-10/60358.4	Fragmentary cranial human remains of an adult of unknown sex.	401	60358/401	PP	1550E100, depth 70″, height–on red clay	Indet.	1		1	0	0	0	0	0	0	0	0	0	1		Indet.
24-4-10/60426.0	Nearly complete cranial and fragmentary postcranial human remains of an adult male.	1107	60426	PP	Room 71, depth 8″, height 1′	Probably Gl. 4	1	40–54	1	0	0	0	0	0	1	0	0	0	0	D h/i	Male
24-4-10/60427.0	Fragmentary postcranial human remains of an adult of unknown sex.	1104	60427	PP	1350W100, depth 2′10″, height 2′	Poss. Gl. 3	1		1	0	0	0	0	0	0	0	0	0	1		Indet.
24-4-10/60427a	Partial cranial and fragmentary postcranial human remains of a subadult of unknown sex.	1104	60427	PP	1350W100, depth 2′10″, height 2′	Poss. Gl. 3	1	18–22	0	1	0	0	0	0	0	0	0	0	1	D c	Indet.
24-4-10/60428.0	Nearly complete cranial and fragmentary postcranial human remains of an adult female.	1115	60428	PP	Room 46, depth 82″, height 20″	Indet.	1	30–34	1	0	0	0	0	0	0	0	1	0	0	D f, T 6, AS 3	F all features
24-4-10/60429.0	Nearly complete cranial human remains of an adult male.	1132	60429	PP	Just west of room 53, depth 5′10″, height 2′	Probably Gl. 4	1	45–55+	1	0	0	0	0	0	1	0	0	0	0	D h+/i+	Male
24-4-10/60430.0	Nearly complete cranial and fragmentary postcranial human remains of an adult female.	1548	60430	PP	1250W150, depth 8″, height 2′±	Gl. 1 or 2	1	30–34	1	0	0	0	0	0	0	0	1	0	0	D f	Female
24-4-10/60431.0	Nearly complete cranial and fragmentary postcranial human remains of a child of unknown sex.	1394	60431/STR	PP	1200W150b, depth 2′6″, height–on bottom	Probably Gl. 1 red	1	9.5–10.5	0	1	0	0	0	0	0	0	0	0	1	D 10 ± 2.5 yrs, PC 9.5–10.5 yrs	Indet.
24-4-10/60432.0	Fragmentary cranial and postcranial human remains of an adult male.	1301	60432	PP	1300W100, depth 3′, height 2′6″	Probably Gl. 3	1		1	0	0	0	0	0	1	0	0	0	0		Male
24-4-10/60433.0	Partial cranial and fragmentary postcranial human remains of an adult female.	1286	60433	PP	1350W100, south wall, depth 2′6″, height 6′± above bottom	Probably Gl. 5	1	40–49	1	0	0	0	0	0	0	0	1	0	0	D h+/i+, AS 5	Female
24-4-10/60434.0	Partial cranial and fragmentary postcranial human remains of an adult female.	1291	60434	PP	1350W100, south wall, depth 5′, height 1′	Probably Gl. 2 or 3	1	35+	1	0	0	0	0	0	0	0	1	0	0	D and PC >35 yrs	Female
24-4-10/60435.0	Fragmentary cranial and postcranial human remains of an adult female.	1273	60435	PP	1350W100, south wall, depth 4′, height 3′	Late Gl. 3 or early Gl. 4	1	35–45	1	0	0	0	0	0	0	0	1	0	0	D g-h	Female
24-4-10/60436.0	Fragmentary cranial and postcranial human remains of an adult female.	1517	60436/1517	PP	Room c, depth 5′6″, height ?	Gl. 3, 4, or 5	1	50–59	1	0	0	0	0	0	0	0	1	0	0	AS 7, T 10	Female
24-4-10/60437.0	Partial cranial and postcranial human remains of an adult female.	1102	60437	PP	1450W100, depth 2′2″, height 3′1″	Gl. 4	1	50–59	1	0	0	0	0	0	0	0	1	0	0	D i+, AS 7	Female
24-4-10/60438.0	Fragmentary cranial and postcranial human remains of an adult female.	1385	60438	PP	1100W150	Probably Gl. 5	1	25–34	1	0	0	0	0	0	0	0	1	0	0	D e-f	Female
24-4-10/60439.0	Nearly complete cranial human remains of an adult female.	1594	60439	PP	Room 328, through south wall, depth 6′, height 1′	Probably Gl. 1 or 2	1	30–34	1	0	0	0	0	0	0	0	1	0	0	D f/f	
24-4-10/60440.0	Nearly complete cranial and fragmentary postcranial human remains of an adult female.	1276	60440	PP	1350W100, west wall, depth 2′6″, height 4′ above bottom	Probably Gl. 4	1	35–39	1	0	0	0	0	0	0	0	1	0	0	D f/g-h, very uneven, AS 4	Female
24-4-10/60441.0	Fragmentary cranial and postcranial human remains of an adult of unknown sex.	1600	60441	PP	Room 328, sub, depth 8′, height–on bottom	Probably B&W	1		1	0	0	0	0	0	0	0	0	0	1		Indet.

Peabody number	Pathology type	Pathology description	Cranial mod.	Associated funerary objects	Field context	Notes
20-6-10/60358.1	Trauma	Healed fracture–R ulna	No cran.	Gl. 2 bowl no. 1017 (65047)		
20-6-10/60358.2	Trauma, infection	L femur–trauma and infection	No cran.			
20-6-10/60358.3	Infection	Infection in L tibia	No cran.	Projectile point no. 1911 (68443)		
20-6-10/60358.4			Unobs.	Gl. 1 yellow no. 1314 (67037), fossil shell no. 1323 (69192)		
24-4-10/60426.0	Trauma	Healed fracture–L radius and L ulna	Not obs.	Hopi pot no. 759 (71135), 4 bone tubes no. 804 (71567), bone flute frag. no. 803 (71659), bone tube no. 1777 (73849)		Skull very different shape, not part of population; very narrow face and mandible
24-4-10/60427.0	Trauma	Healed fracture–L femur	No cran.			
24-4-10/60427a	Arthritis	Slight DJD	Not obs.			Field number and provenience problematic
24-4-10/60428.0	Arthritis, nutr.	Slight DJD, enamel hypoplasia	Occ. def.	Gl. 1 yellow bowl no. 934 (71139)		
24-4-10/60429.0	Infection	Dental abscesses, extensive antemortem tooth loss	No cran. def.	Bone awl frag. no. 886 (71522), Gl. 4 bowl no. 937 (71144)		
24-4-10/60430.0	Infection, nutr.	Extensive antemortem molar tooth loss, enamel hypoplasia	Not obs.	Gl. 1 yellow bowl no. 1999 (72746)		
24-4-10/60431.0	Nutr.	Cribra orbitalia	Occ. def.			
24-4-10/60432.0	Nutr.	Enamel hypoplasia	Unobs.	Obsidian frag, no number listed		Interesting dental wear
24-4-10/60433.0	Infection, arthritis	Occipital lesion and pronounced torus, DJD, spinal osteophytosis	Not obs.			
24-4-10/60434.0	Infection, arthritis, nutr.	Antemortem tooth loss and abscessing, slight DJD, ankylosed 5vt/l5, slight osteoporosis	Not obs.	Poss. lump of copper ore		
24-4-10/60435.0			Unobs.			
24-4-10/60436.0	Trauma, infection, arthritis, nutr.	Healed fracture L radius that affected ulna, extensive antemortem molar loss, severe DJD, spinal osteoporosis, osteoporosis	Unobs.			Field number and provenience problematic
24-4-10/60437.0	Arthritis, nutr.	Spinal osteophytosis, slight DJD, slight osteoporosis	Not obs.			
24-4-10/60438.0			Unobs.	Gl. 4 sherds, no numbers		
24-4-10/60439.0			Not obs.	Bisc. B bowl no. 2193 (72772), bone awl no. 2195 (73634), feather scraper no. 2192 (74817)		
24-4-10/60440.0	Trauma, infection, nutr., other	Healed fracture–R zygomatic, dental abscess upper M1, enamel hypoplasia, malocclusion due to misaligned L upper premolars, remodeling of sacroiliac joint	Not obs.	Projectile point no. 49 (74257)		
24-4-10/60441.0	Infection, nutr.	Dental abscesses, osteoporosis	Unobs.			

Peabody number	Object description	FN	Specimen label	Site	Intrasite provenience	Ceramic stage	MNI	Age (yrs.)	Ad	Su	Ch	In	Fe	Unk. age	M	M?	F	F?	Unk. sex	Age criteria	Sex criteria
24-4-10/60442.0	Nearly complete cranial and fragmentary postcranial human remains of an adult female.	1268	60442	PP	1350W100, west wall, depth 3′6″, height 1′6″ above bottom	Probably Gl. 1R	1	30–39	1	0	0	0	0	0	0	0	1	0	0	D g-h/i, AS 3	Female
24-4-10/60443.0	Partial cranial and fragmentary postcranial human remains of an adult female.	1305	60443	PP	1300W100, depth 3′, height 3′	Probably Gl. 3	1	35–49	1	0	0	0	0	0	0	0	1	0	0	D f-g/g, AS 6	Female
24-4-10/60444.0	Fragmentary cranial and postcranial human remains of an adult male.	1290	60444	PP	1350W100, s wall, 4′ E of NS wall of room 92, d. 4′6″, ht 1′6″	Gl. 1 or 2	1	40–44	1	0	0	0	0	0	1	0	0	0	0	D g-h/g-I, AS 5	Male
24-4-10/60445.0	Nearly complete cranial and fragmentary postcranial human remains of an adult male.	1019	60445	PP	Room 51, depth 5′6″	Probably Gl. 1r	1	45–55+	1	0	0	0	0	0	1	0	0	0	0	D h+/i+	Male
24-4-10/60446.0	Nearly complete cranial and partial postcranial human remains of an adult female.	1013	60446	PP	Room 46, depth 5′6″	Gl. 1	1	45–49	1	0	0	0	0	0	0	0	1	0	0	D h/i+, AS 6	Female
24-4-10/60447.0	Nearly complete cranial and fragmentary postcranial human remains of a subadult of unknown sex.	1518	60447	PP	Kiva 10, depth 4′, height 3′	Gl. 1	1	15–19	0	1	0	0	0	0	0	0	0	0	1	D b1-b2/b1-b2, PC 15–20 yrs, epiphyses	Indet.
24-4-10/60448.0	Fragmentary postcranial human remains of an adult of unknown sex.	1365	60448	PP	1250W150, depth 3′4″, height ?	Probably Gl. 3	1	25–34	1	0	0	0	0	0	0	0	0	0	1	young adult	Indet.
24-4-10/60449.0	Fragmentary cranial and postcranial human remains of an adult female.	1519	60449	PP	1350W150, depth 3′, height 3′6″	Probably Gl. 4	1	40–49	1	0	0	0	0	0	0	0	1	0	0	D h-i, AS 5	Female
24-4-10/60450.0	Fragmentary cranial and partial postcranial human remains of an adult female.	1537	60450	PP	1200W200, west extension, depth 3′6″, height–on bottom	Indet.	1	20–29	1	0	0	0	0	0	0	0	1	0	0	D e	Female
24-4-10/60450.1	Very fragmentary cranial human remains of a child of unknown sex.	1537		PP	1200W200, w extension	Indet.	1	2–5	0	0	1	0	0	0	0	0	0	0	1	Young child due to wear on teeth	Indet.
24-4-10/60451.0	Nearly complete cranial and partial postcranial human remains of an adult female.	1586	60451	PP	Sub room e, depth 4′6″, height 2′6″	Probably Gl. 1	1	50–59	1	0	0	0	0	0	0	0	1	0	0	D edentulous/i+, AS 7	Female
24-4-10/60452.0	Nearly complete cranial and partial postcranial human remains of an adult female.	1585	60452	PP	Room a sub, depth 5′, height 3′6″	Gl. 3	1	25–29	1	0	0	0	0	0	0	0	1	0	0	D d, AS 2	Female
24-4-10/60453.0	Nearly complete cranial and fragmentary postcranial human remains of an adult male.	1010	60453	PP	Room 46, depth 5.2′	Gl. 3	1	35+	1	0	0	0	0	0	1	0	0	0	0	Cranial sutures	Male
24-4-10/60454.0	Nearly complete cranial and postcranial human remains of an adult male.	1015	60454	PP	Room 50, nw corner, depth 63″, height ?	Probably Gl. 2	1	45–49	1	0	0	0	0	0	1	0	0	0	0	D g-h? (teeth poor), cranial sutures not markedly fused, AS 6	Male
24-4-10/60455.0	Nearly complete cranial and postcranial human remains of an adult female.	1358	60455	PP	Room 44, depth 8′, height 1′	Probably Gl. 1	1	35–39	1	0	0	0	0	0	0	0	1	0	0	D g/f-g, T 7, AS 4, cranial sutures not well fused	
24-4-10/60456.0	Nearly complete cranial and fragmentary postcranial human remains of an adult male.	1068	60456	PP	Room 19, ne corner, depth 7′6″, height–on rock	Indet.	1	30–34	1	0	0	0	0	0	1	0	0	0	0	D f/f, cranial sutures not well fused, AS 3	Male
24-4-10/60457.0	Nearly complete cranial and fragmentary postcranial human remains of an adult female.	1041	60457/1041	PP	Room 52, depth 6′6″	Indet.	1	30–34	1	0	0	0	0	0	0	0	1	0	0	D uneven wear on ant. dentition e, post. dentition h, AS 3	F all features
24-4-10/60458.0	Nearly complete cranial and postcranial human remains of an adult male.	1426	60458	PP	1200W200d, depth 3′, height–on bottom	Indet.	1	30–34	1	0	0	0	0	0	1	0	0	0	0	D f/f, T 6	M all features

Peabody number	Pathology type	Pathology description	Cranial mod.	Associated funerary objects	Field context	Notes
24-4-10/60442.0	Nutr.	Enamel hypoplasia	Not obs.	Bead frag., no number		
24-4-10/60443.0	Nutr., other	Osteoporosis, upper L M3 emerging partially buccally	Not obs.	Black bowl (no number listed)		
24-4-10/60444.0	Trauma, infection, arthritis	Healed fracture with displacement L Mt II, upper L M2 and lower R M1–dental abscess, DJD	Unobs.	Gl. 2 bowl no. 151 (72645), shell no. 224 (74709)		
24-4-10/60445.0	Trauma, arthritis	Healed fracture R dist. radius affecting ulna, DJD, bone spur at dist. R tibia	Not obs.	Crushed black duck pot no. 1063 (71084)		
24-4-10/60446.0	Arthritis	Extensive spinal osteophytosis	Occ. def.			
24-4-10/60447.0			?Occ. def.			
24-4-10/60448.0			No cran.			
24-4-10/60449.0	Arthritis	Slight DJD	Unobs.			
24-4-10/60450.0	Other	Neural arch defect in 1 vt/1	Unobs.	Frag. Gl. 2 red bowl no. 1925 (72742)		Field card and ledger descriptions do not match
24-4-10/60450.1	Other	2 fused incisors	Unobs.	Frag. Gl. 2 red bowl no. 1925 (72742)		Field card and ledger descriptions do not match
24-4-10/60451.0	Arthritis, nutr.	Spinal osteophytosis especially lumbar region, DJD especially elbows, spondylitis of vt/l5, arthritic occipital condyle and axis with atlas ankylosed to occipital, right TMJ pathological, moderate osteoporosis	Occ. def.	Bisc. A bowl no. 2162 (72764)		
24-4-10/60452.0	Infection, nutr.	Dental abscesses, enamel hypoplasia	Occ. def.	Small black jar no. 2161 (72763), textile no. 2192 (74816)		
24-4-10/60453.0	Arthritis, nutr., tumor	Vt/l 1&2–ankylosed, spondylytis deformans, spinal osteophytosis, osteoporosis, button osteoma–frontal, prox. tibias and patellae–bone spurs	No cran. def.	2 Bisc. B bowls nos. 1058 (71081), 715 (71080), 1 Gl. 2 bowl no. 701 (71082)		
24-4-10/60454.0	Infection, arthritis	Dental abscesses, DJD, spondylytis deformans	Poss. occ. def.			
24-4-10/60455.0	Infection, arthritis	Syphilis–extensive cranial lesions, frontal, parietals and nasals, R femur–infection, spinal osteophytosis	Occ. def.			
24-4-10/60456.0	Infection	Extensive postcranial infection–L ulna, tibia, femur, fibula; R clavicle, tibia, fibula, femur	Not obs.	2 Gl. 1 red bowls nos. 820 (71123), 604 (71124)		
24-4-10/60457.0	Nutr.	Enamel hypoplasia	Occ. def.	Gl. 1 red bowl no. 535 (71115)		
24-4-10/60458.0	Arthritis, other	Slight DJD, arthritic sacroiliac joint, spina bifida–vt/s4–5	Not obs.	Gl. 4 bowls nos. 1086 (72698), 1087 (72699), 2 worked stones, no numbers		

Peabody number	Object description	FN	Specimen label	Site	Intrasite provenience	Ceramic stage	MNI	Age (yrs.)	Ad	Su	Ch	In	Fe	Unk. age	M	M?	F	F?	Unk. sex	Age criteria	Sex criteria
24-4-10/60459.0	Very fragmentary postcranial human remains of an adult of unknown sex.	1111 & 1112	60459	PP	1500W50, near room, depth 2′4″, height–on bottom	Probably Gl. 4	1		1	0	0	0	0	0	0	0	0	0	1		Indet.
24-4-10/60460.0	Fragmentary postcranial human remains of an adult female.	1141	60460	PP	Sw room 53, depth 3′6″, height 4′6″	Probably Gl. 4	1	35+	1	0	0	0	0	0	0	0	1	0	0	T 9	Female
24-4-10/60461.0	Very fragmentary postcranial human remains of an adult of unknown sex.	1549	60461	PP	1200W150b, north extension, depth 3′, height–on bottom	Probably Gl. 3 or 4	1		1	0	0	0	0	0	0	0	0	0	1		Indet.
24-4-10/60462.0	Very fragmentary postcranial human remains of an adult of unknown sex.	1195	60462	PP	Plaza room 2, s. wall, depth 3′10″, height–on bottom	B&W or Gl. 1	1		1	0	0	0	0	0	0	0	0	0	1		Indet.
24-4-10/60463.0	Nearly complete cranial human remains of an adult female.	1029	60463	PP	Room 39, depth 5′ so. roof beams	Indet.	1	16–20	1	0	0	0	0	0	0	0	1	0	0	D b2?/b1, M3s erupting, spheno-occipi-tal fused, teeth chipped	Female (gracile but SR marked)
24-4-10/60464.0	Fragmentary cranial human remains of an adult male.	1437	60464/1437	PP	1150W200, north wall, depth 3′, height–on bottom	Pres. Gl. 4	1		1	0	0	0	0	0	1	0	0	0	0		Male
24-4-10/60465.0	Partial cranial and fragmentary postcranial human remains of an adult female.	1023	60465	PP	Room 46, depth 5′2″, height 24″	Indet., not later than Gl. 3	1	45–49	1	0	0	0	0	0	0	0	1	0	0	D h/i, AS 6	Female
24-4-10/60466.0	Nearly complete cranial and fragmentary postcranial human remains of an adult male.	970	60466	PP	1400W100, depth 4′6″, height 3′	Gl. 4	1	35–45	1	0	0	0	0	0	1	0	0	0	0	D g/h, AS 5	Male
24-4-10/60467.0	Fragmentary cranial and postcranial human remains of an adult male.	1064	60467	PP	1450W100	Gl. 1 (or 2 or 3)	1	30–34	1	0	0	0	0	0	1	0	0	0	0	D ?/f. AS 3	M all features
24-4-10/60468.0	Fragmentary postcranial human remains of an adult of unknown sex.	1475	60468	PP	1200W200c, room 260, depth 5′, height–on bottom	Gl. 1	1		1	0	0	0	0	0	0	0	0	0	1		Indet.
24-4-10/60469.0	Fragmentary cranial and postcranial human remains of an adult female.	1188	60469	PP	Room 56, nw corner, depth 4′	Probably Gl. 4	1	16–20	1	0	0	0	0	0	0	0	1	0	0	D b2	Female
24-4-10/60470.0	Fragmentary postcranial human remains of an adult male.	1457	60470	PP	1250W200, depth 1′, height 4′6″	Probably Gl. 4 or 5	1	20–24	1	0	0	0	0	0	1	0	0	0	0	Epiphyseal lines visible; young adult	Male
24-4-10/60471.0	Fragmentary cranial and postcranial human remains of an adult female.	1294	60471	PP	1350W150, west wall, depth 1′6″, height 5″	Probably Gl. 5	1	30–39	1	0	0	0	0	0	0	0	1	0	0	D f-g	Female
24-4-10/60472.0	Nearly complete cranial and fragmentary postcranial human remains of an adult female.	1094	60472	PP	1500W50, depth 1′8″, height 6″	Indet.	1	25–29	1	0	0	0	0	0	0	0	1	0	0	D e	Female
24-4-10/60473.0	Nearly complete cranial and fragmentary postcranial human remains of an adult male.	1065	60473	PP	Room 46, depth 4′6″, height 2′6″	Probably Gl. 1	1	40–49	1	0	0	0	0	0	1	0	0	0	0	D edentulous, AS 5	Male
24-4-10/60474.0	Fragmentary cranial and postcranial human remains of an adult female.	1459	60474	PP	Room 214, through floor, depth 8′, height 1′+	Gl. 3	1		1	0	0	0	0	0	0	0	1	0	0		Female
24-4-10/60475.0	Fragmentary postcranial human remains of an adult female.	1062	60475	PP	1450W100, depth 4′2″, height 1′	Probably Gl. 2 or 3	1	20–24	1	0	0	0	0	0	0	0	1	0	0	AS 1, incom-plete fusion of iliac crest and ischial tuber-osity epiphyses	Female
24-4-10/60476.0	Nearly complete cranial and fragmentary postcranial human remains of an adult female.	1196	60476	PP	Plaza room 2, depth 3′10″, height–on bottom	B&W or Gl. 1	1	35–45	1	0	0	0	0	0	0	0	1	0	0	D g-h	Female
24-4-10/60477.0	Partial cranial human remains of a subadult female.	1400	60477	PP	1200W150c, depth 1′6″, height 1′6″	Indet.	1	15–19	0	1	0	0	0	0	0	0	1	0	0	D c, M3s not fully erupted (15–20 yrs), spheno-occipit al not fused.	

Peabody number	Pathology type	Pathology description	Cranial mod.	Associated funerary objects	Field context	Notes
24-4-10/60459.0			No cran.			Large individual, field number problematic
24-4-10/60460.0	Trauma, arthritis	Healed fracture L fibula, slight DJD	No cran.			Field card indicates 18–20 years
24-4-10/60461.0	Arthritis, nutr.	Slight DJD, osteoporosis	No cran.			
24-4-10/60462.0			No cran.			Field card says old, field number problematic
24-4-10/60463.0	Infection	Small lesion L parietal	Not obs.	Gl. 1 bowl no. 348 (71094)		
24-4-10/60464.0			Unobs.			
24-4-10/60465.0	Arthritis	DJD	Not obs.			
24-4-10/60466.0	Nutr.	Enamel hypoplasia	Occ. def.	Gl. 4 bowl frag. no. 126 (71055)		
24-4-10/60467.0	Nutr., other	Enamel hypoplasia, spina bifida vt/s1–5	Unobs.			Unusual obdurator foramina on R innominate
24-4-10/60468.0			No cran.	Gl. 1 red bowl no. 1365 (72716)		
24-4-10/60469.0	Nutr., other	Enamel hypoplasia, vt/s4–5 not fused	Unobs.			
24-4-10/60470.0			No cran.			
24-4-10/60471.0			Unobs.	Bone bead no. 178 (73802)		
24-4-10/60472.0	Other	Metopic suture	Not obs.	Sherds of Gl. 1 scoop, no number	Glaze1 ? Early	
24-4-10/60473.0	Infection, arthritis, tumor	Dental abscesses and extensive tooth loss, arthritis at occipital condyle, button osteoma on frontal	Occ. def.	String of turquoise beads no. 827 (72192), string of shell beads no. 828 (72237)		
24-4-10/60474.0	Arthritis, nutr., other	Unusual mandibular exotoses and DJD at TMJ, enamel hypoplasia	Unobs.	Gl. 3 bowl no. 1366 (72714)		
24-4-10/60475.0	Other	Pathological robusticity especially on arms?, spondylolysis vt/l5	No cran.	Part of Gl. 1 yellow bowl no. 603 (71122)		
24-4-10/60476.0	Arthritis, nutr., other	Slight DJD at dist. femur, diploic explansion, unusual upper R canine	Not obs.			
24-4-10/60477.0	Nutr.	Enamel hypoplasia	Not obs.			

Peabody number	Object description	FN	Specimen label	Site	Intrasite provenience	Ceramic stage	MNI	Age (yrs.)	Ad	Su	Ch	In	Fe	Unk. age	M	M?	F	F?	Unk. sex	Age criteria	Sex criteria
24-4-10/60477.1	Fragmentary cranial human remains of an adult of unknown sex.	1400		PP	1200W150c, depth 1′6″, height 1′6″	Indet.	1	>35	1	0	0	0	0	0	0	0	0	0	1	D anth+, post e/f?	Indet.
24-4-10/60479.0	Nearly complete cranial human remains of an adult female.	1140	60479	PP	1550W0, se corner of wall & n of it, depth 4′, height–on bottom	Probably B&W	1	18–22	1	0	0	0	0	0	0	0	1	0	0	D c	F all features
24-4-10/60480.0	Nearly complete cranial and fragmentary postcranial human remains of an adult female.	1392	60480	PP	1200W150b, depth 2′6″, height 6″	Indet., early	1	18–24	1	0	0	0	0	0	0	0	1	0	0	D b2-c/c, T 1, AS 1, iliac crest stage 1 fusion	Female
24-4-10/60481.0	Nearly complete cranial and partial postcranial human remains of an adult female.	1401	60481	PP	1200W150b, depth 3′6″, height 6″	Indet., poss. Gl. 4 or 5	1	50–59	1	0	0	0	0	0	0	0	1	0	0	D h+/i+, AS 7	Female
24-4-10/60482.0	Nearly complete cranial and partial postcranial human remains of an adult male.	1568	60482	PP	1200W150, depth 16″, height 18–20″	Probably Gl. 2	1	45–49	1	0	0	0	0	0	1	0	0	0	0	D h/i, T 9, AS 6	Male
24-4-10/60483.0	Fragmentary postcranial human remains of an adult male.	1413	60483	PP	1100W150, depth 6″ in debris, height 2′ + on clay	Gl. 3	1	35–39	1	0	0	0	0	0	1	0	0	0	0	AS 4	Male
24-4-10/60483a	Nearly complete cranial human remains of an adult female.	1411	60483-a	PP	1200W150b, depth 3′6″, height–on bottom	Probably Gl. 3	1	35–45	1	0	0	0	0	0	0	0	1	0	0	D g/g-h	
24-4-10/60484.0	Nearly complete cranial and partial postcranial human remains of an adult female.	1398	60484	PP	1200W150b, depth 3′, height–on bottom	Indet., poss. Gl. 5	1	20–24	1	0	0	0	0	0	0	0	1	0	0	D b2-c/c, AS 2, prox. clavicle not fused, spheno-occipi-tal not fused	Female
24-4-10/60485.0	Fragmentary cranial and postcranial human remains of an adult male.	1556	60485	PP	1200W150, n exten-sion from b, depth 2′10″, height 2–3″	Gl. 1	1	40–49	1	0	0	0	0	0	1	0	0	0	0	D ant. H+	Male
24-4-10/60486.0	Fragmentary cranial human remains of a subadult of unknown sex.	1381	60486	PP	1100W150, ne of kiva 8, depth 4′6″, height–sunk 12″ in clay	Early B&W	1	16–20	0	1	0	0	0	0	0	0	0	0	1	D b1-b2/b2 (M3s just erupting), spheno-occipi-tal not fused	Indet.
24-4-10/60486.1	Fragmentary cranial human remains of an adult of unknown sex.	1381		PP	1100W150, ne of kiva 8, depth 4′6″, height–sunk 12″ in clay	Early B&W	1		1	0	0	0	0	0	0	0	0	0	1		Indet.
24-4-10/60487.0	Partial cranial and postcranial human remains of an adult that is probably female.	1563	60487	PP	1300W150, depth 4′, height 5′	Gl. 4	1	35–44	1	0	0	0	0	0	0	0	0	1	0	D e-f, AS 4-5	?Female
24-4-10/60488.0	Very fragmentary cranial and fragmentary postcranial human remains of an adult male.	1345	60488	PP	1050W200, depth 1′, height–on bottom	Probably B&W	1	45–49	1	0	0	0	0	0	1	0	0	0	0	T 9, AS 6	M all features
24-4-10/60489.0	Nearly complete cranial and fragmentary postcranial human remains of an adult female.	1571	60489	PP	1200W150, n extension of trench b, depth 2′, height 12″	Indet.	1	30–34	1	0	0	0	0	0	0	0	1	0	0	D g, AS 3	Female
24-4-10/60490.0	Fragmentary cranial and postcranial human remains of an adult that is probably male.	1570	60490	PP	Room 320, nw corner, under flr, depth 3′, height–on bottom	Probably Gl. 1	1	35–39	1	0	0	0	0	0	0	1	0	0	0	D i, AS 4	? Male (SN-M)
24-4-10/60491.0	Fragmentary postcranial human remains of an adult male.	1555	60491	PP	On wall between rooms 326 & 322, depth 3′, height 3′6″–4′	Gl. 3	1	35–39	1	0	0	0	0	0	1	0	0	0	0	AS 4	Male
24-4-10/60492.0	Nearly complete cranial human remains of a child of unknown sex.	1573	60492	PP	1200W150, n extension of trench b, depth 16″, height 2′	Indet.	1	3–4	0	0	1	0	0	0	0	0	0	0	1	D 4–5 ± 1 yrs	Indet.

Peabody number	Pathology type	Pathology description	Cranial mod.	Associated funerary objects	Field context	Notes
24-4-10/60477.1			Unobs.			Field card and provenience problematic
24-4-10/60479.0	Nutr.	Cribra orbitalia, porotic response on frontal and parietals	Occ. def.			Field card indicates 6 mos to 1 yr, but adult present, 1 M3 not erupted
24-4-10/60480.0	Nutr.	Enamel hypoplasia	Occ. def.			R humerus very robust
24-4-10/60481.0	Infection, arthritis, nutr., tumor	Abscessing on L maxilla, diploic expansion, osteoporosis, some DJD, button osteoma on frontal	Occ. def.			
24-4-10/60482.0	Infection, arthritis, nutr., tumor	Maxillary abscessing and resorption, ankylosed lumbar/sacral region, also at vertebrae/rib junctions and at sternum, spinal osteophytosis, enamel hypoplasia, osteoporosis, button osteoma frontal	Unobs.	2 broken bone flutes no. 2106 (73900), 10 modified obsidian stones (medicine outfit) no. 2107 (74633)		Perforated sternum
24-4-10/60483.0			No cran.	Gl. 3 bowl no. 1044 (72695), Gl. 3 bowl no. 1045 (72696)		
24-4-10/60483a			Occ. def.			
24-4-10/60484.0	Infection, nutr.	Dental abscesses, enamel hypoplasia	No cran. def.			
24-4-10/60485.0	Trauma, infection, nutr.	Healed fracture–midshaft L humerus, severe dental abscesses and tooth loss, osteoporosis	Unobs.			
24-4-10/60486.0	Infection, nutr.	Small frontal lesion, enamel hypoplasia	Unobs.	B&W and Gl. 1 red sherd, no numbers		
24-4-10/60486.1			Unobs.	B&W and Gl. 1 red sherd, no numbers		
24-4-10/60487.0			Occ. def.	Gl. 3 bowl no. 2101 (72755)		
24-4-10/60488.0	Infection, arthritis, nutr.	Cribra orbitalia and lytic lesion on frontal, thick diploe, DJD, spinal osteophytosis	Unobs.	Small B&W bowl no. 668 (72672)		Provenience is problematic
24-4-10/60489.0			?Occ. def.			
24-4-10/60490.0	Trauma, infection, arthritis, nutr., other	Healed fracture dist. R ulna (deformed and has cloaca) and L prox. humerus, vt/s4–5 look like healed fractures, sacrum is very asymmetrical, vt/c2–3 ankylosed, spinal osteophytosis and DJD, osteoporosis, extensive antemortem tooth loss	Unobs.			
24-4-10/60491.0			No cran.	Gl. 3 bowl no. 2032 (72749), blanket frags. and a turquoise bead no. 2050 (74814)		
24-4-10/60492.0	Nutr.	Cribra orbitalia	Slight occ. def.	Bisc. A bowl no. 2123 (72759)		

Peabody number	Object description	FN	Specimen label	Site	Intrasite provenience	Ceramic stage	MNI	Age (yrs.)	Ad	Su	Ch	In	Fe	Unk. age	M	M?	F	F?	Unk. sex	Age criteria	Sex criteria
24-4-10/60493.0	Partial cranial human remains of an adult female.	1562	60493	PP	1200W150, n extension from b, depth 2′10″, height 2–3″	Indet., early	1	35–39	1	0	0	0	0	0	0	0	1	0	0	D g	F all features
24-4-10/60494.0	Nearly complete cranial and fragmentary postcranial human remains of an adult male.	1581	60494	PP	Room e, under west wall, depth 4′, height 3′	Gl. 1R	1	35–49	1	0	0	0	0	0	1	0	0	0	0	D g-h	Male
24-4-10/60495.0	Partial cranial human remains of a child of unknown sex.	1592	60495	PP	Under porch of 326, depth 7′, height–on bottom	Indet.	1	7–9	0	0	1	0	0	0	0	0	0	0	1	D 9 ± 2 yrs (eruption)	Indet.
24-4-10/60496.0	Nearly complete cranial and postcranial human remains of an adult female.	1160	60496	PP	1250E100, e traverse of n trench	Gl. 3	1	40–49	1	0	0	0	0	0	0	0	1	0	0	D h/i, AS 5–6, T 10	
24-4-10/60497.0	Nearly complete cranial and fragmentary postcranial human remains of an adult female.	1410	60497	PP	1200W150b, depth 4′4″, height–on bottom	Gl. 3	1	25–29	1	0	0	0	0	0	0	0	1	0	0	D f, AS 2	F all features
24-4-10/60498.0	Very fragmentary cranial and fragmentary postcranial human remains of an adult of unknown sex.	1577	60498	PP	1250W150, south wall, depth 3′6″, height–on bottom	Gl. 1 or 2	1	40+	1	0	0	0	0	0	0	0	0	0	1	D i+	Indet.
24-4-10/60499.0	Fragmentary postcranial human remains of an adult of unknown sex.	1579	60499	PP	Under room e, depth 4′, height 3′+	Probably Gl. 1	1		1	0	0	0	0	0	0	0	0	0	1		Indet.
24-4-10/60500.0	Very fragmentary postcranial human remains of an adult female.	1572	60500	PP	Room 261, on n wall, depth 3′, height 3′ (2′ from orig. surface)	Indet.	1	40–49	1	0	0	0	0	0	0	0	1	0	0	AS 5, T 10	F all features
24-4-10/60501.0	Nearly complete cranial and postcranial human remains of an adult female.	1496	60501	PP	1250W200, room 262, depth 7′6″, height ?	Early, Probably Gl. 1	1	30–34	1	0	0	0	0	0	0	0	1	0	0	D f-g/g-h uneven, AS 3	Female
24-4-10/60502.0	Nearly complete cranial and partial postcranial human remains of an adult male.	1499	60502	PP	Room 214, depth 7, height 3′	Indet.	1	30–39	1	0	0	0	0	0	1	0	0	0	0	D e-f, AS 4	M all features
24-4-10/60503.0	Fragmentary cranial and postcranial human remains of an adult male.	1084a	60503	PP	1500W50	Probably Gl. 5	1	45–54	1	0	0	0	0	0	1	0	0	0	0	D i	M all features
24-4-10/60503a	Very fragmentary postcranial human remains of an adult of unknown sex.	1084a	60503a	PP	1500W50	Probably Gl. 5	1		1	0	0	0	0	0	0	0	0	0	1		Indet.
24-4-10/60504.0	Fragmentary cranial and postcranial human remains of an adult of unknown sex.	1070	60504	PP	1560W40	Probably Gl. 2	1	35–54	1	0	0	0	0	0	0	0	0	0	1	D g-h/i (uneven)	Indet.
24-4-10/60505.0	Nearly complete cranial and fragmentary postcranial human remains of an adult female.	1095	60505	PP	Room 83, depth 6–8″, height 2′6″	Probably Gl. 3	1	25–29	1	0	0	0	0	0	0	0	1	0	0	D e	Female
24-4-10/60507.0	Fragmentary cranial and postcranial human remains of an adult of unknown sex.	1302	60507	PP	Room 43, north wall, depth 3′, height 5′	Gl. 1 or 2	1	30–39	1	0	0	0	0	0	0	0	0	0	1	D e-h, very uneven	Indet.
24-4-10/60508.0	Fragmentary cranial and postcranial human remains of an adult female.	1288	60508	PP	1300W100, depth 4′6″, height 6″	Probably B&W or Gl. 1	1	45–49	1	0	0	0	0	0	0	0	1	0	0	D h, AS 6	Female
24-4-10/60509.0	Nearly complete cranial human remains of a child of unknown sex.	1001	60509/1001	PP	Room 50, depth 6′8″	Gl. 1	1	5–7	0	0	1	0	0	0	0	0	0	0	1	D 7 ± 2 yrs	Indet.
24-4-10/60510.0	Nearly complete cranial and fragmentary postcranial human remains of a subadult male.	1271	60510	PP	1350W100, west trench, depth 3′, height 2′6″	Probably Gl. 3	1	14–16	0	1	0	0	0	0	1	0	0	0	0	D b1, PC 14–16 yrs	Male
24-4-10/60511.0	Fragmentary postcranial human remains of an adult female.	1274	60511	PP	1350W100, west wall, depth 4′6″, height 2′	Probably Gl. 1	1		1	0	0	0	0	0	0	0	1	0	0		Female
24-4-10/60511.1	Fragmentary cranial human remains of a child of unknown sex.	1274	60511 (crossed out)	PP	1350W100, w wall, depth 4′6″, height 2′	Probably Gl. 1	1	6–8	0	0	1	0	0	0	0	0	0	0	1	D 8 ± 2 yrs	Indet.
24-4-10/60512.0	Partial cranial and fragmentary postcranial human remains of an adult female.	1275	60512	PP	1350W100, south wall, depth 3′6″, height 3′6″ above bottom	Indet.	1	40–49	1	0	0	0	0	0	0	0	1	0	0	D h/i, AS 5–6	Female

Peabody number	Pathology type	Pathology description	Cranial mod.	Associated funerary objects	Field context	Notes
24-4-10/60493.0			Not obs.			
24-4-10/60494.0	Infection	Lytic lesions on frontal	Not obs.	Gl. 1 red bowl no. 3002 (72762)		Unusual supraorbital torus, nasals are atypical–hook nose, occipital torus unusual, shoveled incisors
24-4-10/60495.0	Nutr.	Enamel hypoplasia	Not obs.	Gl. 1 yellow bowl no. 2186 (72770)	Wrapped in feather cloth	
24-4-10/60496.0	Arthritis	Minor spinal osteophytosis vt/l	Occ. def.	Frag. Gl. 3 bowl no. 1035 (71161), pipe piece, no number		
24-4-10/60497.0	Infection, nutr.	Extensive diploic expansion of parietals, enamel hypoplasia	No cran. def.	Frag. Gl. 3 bowl no. 1049 (72692)		
24-4-10/60498.0	Infection	Edentulous molars	Unobs.	Gl. 1 bowl no. 2132 (72760)		
24-4-10/60499.0			No cran.			
24-4-10/60500.0			No cran.	Gl. 3 bowl no. 2121 (72758)		
24-4-10/60501.0	Infection, nutr., other	Slight periosteal rxn L tibia, enamel hypoplasia, sacralization of vt/l5	Occ. def.			
24-4-10/60502.0	Infection, nutr.	Dental abscess and caries, enamel hypoplasia	No cran. def.	Gl. 4 bowl fragment no. 1547 (72727)		
24-4-10/60503.0	Infection, arthritis	Marked antemortem tooth loss, DJD	Unobs.	2 horns, no numbers		Time period assignment based on 1084 being Gl. 5
24-4-10/60503a			No cran.	2 horns, no numbers		Field card and provenience problematic; time period assignment based on 1084 being Gl. 5
24-4-10/60504.0	Infection, arthritis	Dental abscesses, marked tooth loss, spinal osteophytosis	Unobs.	Frag. of Gl. 3 bowl no. 685 (71125), bone awl (71766)		
24-4-10/60505.0	Arthritis, nutr.	Bone resorption/remodeling of femoral and tibial jts., cribra orbitalia, enamel hypoplasia	Not obs.	3 bone beads (72299)		
24-4-10/60507.0	Trauma, nutr.	Healed fracture R dist. radius, enamel hypoplasia	Unobs.	Gl. 1 yellow bowl no. 298 (72650)		
24-4-10/60508.0	Infection, arthritis, other	Dental abscessing mandible, marked tooth loss, spinal osteophytosis, dolichocephalic	Unobs.			
24-4-10/60509.0	Nutr.	Enamel hypoplasia, cribra orbitalia	Occ. def.	Potsherd of Gl. 1 bowl no. 228 (71074)		
24-4-10/60510.0	Nutr.	Cribra orbitalia	Occ. def.			
24-4-10/60511.0			No cran.	Frag. black pot no. 221 (72641)		
24-4-10/60511.1			Unobs.	Frag. black pot no. 221 (72641)		
24-4-10/60512.0	Infection, arthritis, nutr.	Dental abscesses, DJD, osteoporosis	?Occ. def.	Frag. Gl. 4 bowl no. 223 (72642), bone bead no. 222 (73801)		

Peabody number	Object description	FN	Specimen label	Site	Intrasite provenience	Ceramic stage	MNI	Age (yrs.)	Ad	Su	Ch	In	Fe	Unk. age	M	M?	F	F?	Unk. sex	Age criteria	Sex criteria
24-4-10/60513.0	Nearly complete cranial and postcranial human remains of an adult male.	1339	60513	PP	Room 110, depth 3′6″, height 2′6″	Gl. 4	1	40–49	1	0	0	0	0	0	1	0	0	0	0	D h/i, T-9, AS 5	Male
24-4-10/60514.0	Nearly complete cranial and postcranial human remains of an adult male.	1479	60514	PP	1150W100, room 3, depth 7′, height 5–6′	Probably Gl. 3	1	30–34	1	0	0	0	0	0	1	0	0	0	0	D e/f, AS 3–4	M all features
24-4-10/60515.0	Very fragmentary postcranial human remains of an adult of unknown sex.	1505	60515/f41	PP	1350W150, depth 3′6″, height 3′6″±	Gl. 3 or 4	1		1	0	0	0	0	0	0	0	0	0	1		Indet.
24-4-10/60516.0	Fragmentary cranial and postcranial human remains of an adult female.	1488	60516	PP	1350W150, depth and height not recorded	Probably Gl. 3	1	35–39	1	0	0	0	0	0	0	0	1	0	0	D g	Female
24-4-10/60517.0	Fragmentary cranial and postcranial human remains of an adult of unknown sex.	1492	60517	PP	Test hole 7, 3 w of wall, depth 3′10″, height–on bottom	Indet.	1	35–39	1	0	0	0	0	0	0	0	0	0	1	D g uneven	Indet.
24-4-10/60518.0	Nearly complete cranial and postcranial human remains of an adult male.	1167	60518	PP	950E150, baca-Ortega trench, depth 3′, height 2′	Gl. 3	1	45–49	1	0	0	0	0	0	1	0	0	0	0	D antemortem loss of molars, AS 6	Male
24-4-10/60519.0	Nearly complete cranial human remains of an adult male.	1490	60519	PP	1300W150, in kiva 10, depth 5′, height ?	Probably Gl. 1 or 2	1	40–49	1	0	0	0	0	0	1	0	0	0	0	D h	Male
24-4-10/60520.0	Nearly complete cranial and partial postcranial human remains of an adult female.	1172	60520	PP	1250W100, depth 5′, height 2′	Probably Gl. 4	1	40–49	1	0	0	0	0	0	0	0	1	0	0	D g, AS 6	Female
24-4-10/60521.0	Nearly complete cranial and fragmentary postcranial human remains of an adult female.	1497	60521/1497	PP	Room 221, under east wall, depth 5′, height–on bottom	Probably Gl. 1	1	40–49	1	0	0	0	0	0	0	0	1	0	0	D h? uneven	Female
24-4-10/60522.0	Nearly complete cranial and fragmentary postcranial human remains of an adult female.	1491	60522/1491	PP	1350W150, depth 18″, height 3′8″	Probably Gl. 3	1	25–29	1	0	0	0	0	0	0	0	1	0	0	D e	Female
24-4-10/60523.0	Nearly complete cranial human remains of a subadult of unknown sex.	1493	60523/1493	PP	1250W200, room 262, depth 4′4″, height ?	Probably Gl. 3	1	15–17	0	1	0	0	0	0	0	0	0	0	1	D b2, 15 ± 3 yrs	Indet.
24-4-10/60523.1	Very fragmentary cranial human remains of a subadult of unknown sex.	1493		PP	1250W200, room 262, depth 4 4″, height ?	Probably Gl. 3	1		0	1	0	0	0	0	0	0	0	0	1		Indet.
24-4-10/60524.0	Nearly complete cranial and fragmentary postcranial human remains of an adult female.	1500	60524/1500	PP	1350W150, depth 10″, height 3–4′	Probably Gl. 3	1	16–20	1	0	0	0	0	0	0	0	1	0	0	D b2, spheno-occipital fused	Female
24-4-10/60525.0	Nearly complete cranial and fragmentary postcranial human remains of an adult female.	1169	60525	PP	1250E100, depth 4′6″, height 3′6″	Probably Gl. 4 or 5	1	35–39	1	0	0	0	0	0	0	0	1	0	0	D h/i, AS 4	Female
24-4-10/60526.0	Fragmentary cranial and postcranial human remains of an adult female.	1446	60526/1446	PP	1200W200d, depth 4′, height–on bottom	Gl. 1 or 2	1	40–49	1	0	0	0	0	0	0	0	1	0	0	D i+, AS 5	F all features
24-4-10/60527.0	Nearly complete cranial and fragmentary postcranial human remains of an adult male.	1166	60527	PP	1250E100	Gl. 3	1	40–44	1	0	0	0	0	0	1	0	0	0	0	D g-h/h, T 8, AS 5	Male
24-4-10/60528.0	Nearly complete cranial and fragmentary postcranial human remains of an adult male.	1444	60528	PP	1250W200c, depth 2′, height 3′	Indet., poss. Gl. 4	1	45–49	1	0	0	0	0	0	1	0	0	0	0	D h/i, T 9	Male
24-4-10/60529.0	Nearly complete cranial and partial postcranial human remains of an adult female.	1472	60529	PP	Room 316, depth 5′, height 2′	Probably Gl. 3	1	20–29	1	0	0	0	0	0	0	0	1	0	0	D d-e, AS 2–3	Female
24-4-10/60530.0	Partial cranial and nearly complete postcranial human remains of an adult male.	1026	60530	PP	Room 50	Indet.	1	40–49	1	0	0	0	0	0	1	0	0	0	0	D h, AS 5–6, T 10	Male
24-4-10/60531.0	Very fragmentary cranial human remains of a child of unknown sex.	1537	60531	PP	1200W200, west extension, depth 3′6″, height–on bottom	Indet.	1	4–6	0	0	1	0	0	0	0	0	0	0	1	D 6 ± 2 yrs	Indet.
24-4-10/60532.0	Very fragmentary postcranial human remains of an adult that is probably female.	1443	60532	PP	1200W150b, depth 6′, height–on bottom	Gl. 1	1		1	0	0	0	0	0	0	0	0	1	0		?Female

Peabody number	Pathology type	Pathology description	Cranial mod.	Associated funerary objects	Field context	Notes
24-4-10/60513.0	Trauma, arthritis, nutr., other	Healed trauma to frontal, moderate spinal osteophytosis, moderate DJD, enamel hypoplasia, spina bifida vt/s4–5	No cran. def.	Gl. 1 bowl no. 669 (72671)		
24-4-10/60514.0	Infection, arthritis, nutr.	Periostitis R femur, tibia, fibula, sacrum, both innominates, slight DJD, enamel hypoplasia	?Occ. def.	Gl. 3 sherds no. 1408 (72720)		
24-4-10/60515.0			No cran.			
24-4-10/60516.0	Nutr., other	Enamel hypoplasia, bowing in fibulae	Unobs.			
24-4-10/60517.0	Trauma, nutr.	Healed fracture sacrum, enamel hypoplasia	Unobs.			
24-4-10/60518.0	Trauma, infection, arthritis ,other	Healed fracture prox. L tibia, antemortem loss of molars, moderate spinal osteophytosis, moderate DJD, spondylylosis	No cran. def.	Gl. 3 bowl no. 1032 (71165), bone awl no. 1129 (71386), broken effigy no. 1129 (71386), polished pot, no number		
24-4-10/60519.0	Infection	Lesion on frontal	Not obs.			
24-4-10/60520.0	Infection, arthritis, nutr.	Dental abscessing, slight DJD, osteoporosis	No cran. def.	Base of black jar no. 1073 (71166), 4 pottery counters (no number listed)		
24-4-10/60521.0	Infection, nutr., other	Dental abscessing, infection R mastoid, dental maleruption and impacted molar, 3 maxillary and 2 mandibular incisors only, enamel hypoplasia	Not obs.			
24-4-10/60522.0	Nutr.	Enamel hypoplasia	Not obs.			
24-4-10/60523.0	Nutr.	Enamel hypoplasia	Not obs.			
24-4-10/60523.1	Nutr.	Enamel hypoplasia	Unobs.			
24-4-10/60524.0	Infection, nutr., other	Expanded diploe frontal bone, enamel hypoplasia, partial metopic suture	Not obs.			
24-4-10/60525.0	Infection, arthritis	Severe dental abscesses, antemortem tooth loss, moderate DJD	Unobs.			
24-4-10/60526.0	Arthritis, nutr.	Slight DJD, slight osteoporosis,	Unobs.	Gl. 1 yellow bowl no. 1215 (72708)		
24-4-10/60527.0	Arthritis	Slight DJD	Occ. def.	Gl. 3 bowl no. 1031 (71164), stone knives nos. 98 (71893), 1126 (71894)		
24-4-10/60528.0	Arthritis, nutr., other	Slight DJD, enamel hypoplasia, spina bifida–sacrum	Unobs.			
24-4-10/60529.0	Nutr.	Enamel hypoplasia	Not obs.	Black pot (no number listed)		Very enlarged cranial foramina, odd postorbital constriction
24-4-10/60530.0	Infection, arthritis, nutr.	Impacted tooth in R maxilla near midline of palatal suture, moderate DJD, spinal osteophytosis, enamel hypoplasia	Unobs.	Gl. 1 red bowl no. 438 (71091)	Bark bindings at throat	
24-4-10/60531.0			Unobs.	Frag. Gl. 2 red bowl no. 1925 (72742)		Field card and ledger descriptions do not match
24-4-10/60532.0	Infection	Periostitis–R and L humerus, R tibia	No cran.	Gl. 1 red bowl no. 1216 (72706), half of painted mano, no number		

Peabody number	Object description	FN	Specimen label	Site	Intrasite provenience	Ceramic stage	MNI	Age (yrs.)	Ad	Su	Ch	In	Fe	Unk. age	M	M?	F	F?	Unk. sex	Age criteria	Sex criteria
24-4-10/60532a	Very fragmentary postcranial human remains of an adult that is probably female.	1443	60532a	PP	1200W150b, depth 6′, height–on bottom	Gl. 1	1		1	0	0	0	0	0	0	0	0	1	0		?Female
24-4-10/60533.0	Fragmentary postcranial human remains of an adult female.	1531	60533	PP	Room 219, depth 3′6″, height–on bottom	Indet.	1	40–49	1	0	0	0	0	0	0	0	1	0	0	T 10, AS 4–6	Female
24-4-10/60534.0	Partial cranial human remains of an adult male.	1419	60534/1419	PP	1200W150b, depth 4′6″, height–on bottom	Probably Gl. 2 or 3	1	35–39	1	0	0	0	0	0	1	0	0	0	0	D g	Male
24-4-10/60535.0	Partial cranial and fragmentary postcranial human remains of an adult of unknown sex.	974	60535/974	PP	1350W100, depth 2′	Gl. 3	1	18–22	1	0	0	0	0	0	0	0	0	0	1	D c, PC young adult	Indet.
24-4-10/60536.0	Nearly complete cranial and fragmentary postcranial human remains of an adult female.	1524	60536/1524	PP	1150W50, depth 7′, height ?	Probably B&W	1	25–29	1	0	0	0	0	0	0	0	1	0	0	D e	Female
24-4-10/60536.1	Fragmentary cranial human remains of an infant of unknown sex.	1524		PP	1150W50, depth 7′, height ?	Probably B&W	1	0–2	0	0	0	1	0	0	0	0	0	0	1	Infant, size	Indet.
24-4-10/60537.0	Fragmentary cranial and postcranial human remains of an adult of unknown sex.	975	60537/975	PP	1350W100, depth 2′	Gl. 3	1	30–39	1	0	0	0	0	0	0	0	0	0	1	D f-g	Indet.
24-4-10/60538.0	Fragmentary cranial and postcranial human remains of an adult female.	963	60538	PP	1400W100, depth 4′10″, height 2′11″	Gl. 3	1	16–20	1	0	0	0	0	0	0	0	1	0	0	D b2, PC young adult	Female
24-4-10/60539.0	Fragmentary postcranial human remains of an adult male.	990	60539	PP	1350W100, depth 42″, height 22″ above bottom	Probably Gl. 1	1		1	0	0	0	0	0	1	0	0	0	0		Male
24-4-10/60540.0	Fragmentary postcranial human remains of an adult of unknown sex.	1435	60540	PP	Room 213, through floor, depth 3′, height 2′	Probably Gl. 3	1		1	0	0	0	0	0	0	0	0	0	1		Indet.
24-4-10/60541.0	Partial cranial and fragmentary postcranial human remains of an adult male.	1408	60541	PP	1200W200A, depth 2′2″, height–on bottom	Indet.	1	35–45	1	0	0	0	0	0	1	0	0	0	0	D h+/i, T 8, AS 4	Male
24-4-10/60542.0	Partial cranial and fragmentary postcranial human remains of an adult female.	1199	60542	PP	1350W0, depth 44″, height 8–10″	Indet.	1	35–39	1	0	0	0	0	0	0	0	1	0	0	D g-h/i, AS 4	Female
24-4-10/60543.0	Nearly complete cranial and partial postcranial human remains of an adult female.	964	60543/964	PP	1400W50, depth 3′10″, height 1′11″	Probably Gl. 4	1	25–29	1	0	0	0	0	0	0	0	1	0	0	D e, T 6, AS 2	F all features
24-4-10/60543.1	Fragmentary postcranial human remains of an adult of unknown sex.	964		PP	1400W50, depth 3′10″, height 1′11″	Probably Gl. 4	1		1	0	0	0	0	0	0	0	0	0	1		Indet.
24-4-10/60544.0	Nearly complete cranial human remains of an adult female.	1540	60544/1540	PP	Test pit 3, depth 3′6″, height–on clay	Indet.	1	18–22	1	0	0	0	0	0	0	0	1	0	0	D c	Female
24-4-10/60545.0	Partial cranial and fragmentary postcranial human remains of an adult male.	1378	60545/1378	PP	1250W150, sw corner, depth 2′6″, height 1′	Probably Gl. 3	1	35–39	1	0	0	0	0	0	1	0	0	0	0	D e-i (uneven), AS 4	Male
24-4-10/60545a	Very fragmentary cranial human remains of an adult of unknown sex.	1378	60545-a/99 5	PP	1250W150, sw corner, depth 2′6″, height 1′	Probably Gl. 3	1	30–34	1	0	0	0	0	0	0	0	0	0	1	D f	Indet.
24-4-10/60546.0	Nearly complete cranial human remains of an adult female.	966	60546	PP	1400W100, depth 3′3″, height 4′1″	Probably Gl. 4	1	40–49	1	0	0	0	0	0	0	0	1	0	0	D h	Female
24-4-10/60547.0	Partial cranial and fragmentary postcranial human remains of an adult male.	969	60547	PP	1450W50, depth 4′5″, height1′2″	Probably Gl. 4	1	30–34	1	0	0	0	0	0	1	0	0	0	0	D f	Male
24-4-10/60548.0	Nearly complete cranial and fragmentary postcranial human remains of an adult male.	984	60548	PP	1400W100, depth 3′6″, height 3′6″	Gl. 3	1	35–54	1	0	0	0	0	0	1	0	0	0	0	D g-h/i antemortem loss	Male
24-4-10/60549.0	Partial cranial and postcranial human remains of an adult male.	1003	60549	PP	Room 50, depth 6′	Gl. 2	1	45–49	1	0	0	0	0	0	1	0	0	0	0	D h/i, AS 6	Male
24-4-10/60549.1	Very fragmentary postcranial human remains of an adult of unknown sex.	1003	60549	PP	Room 50, w wall, depth 6′	Gl. 2	1		1	0	0	0	0	0	0	0	0	0	1		Indet.
24-4-10/60550.0	Fragmentary postcranial human remains of an adult female.	1390	60550	PP	1250W150, depth 5′, height–on bottom	B&W	1	30–34	1	0	0	0	0	0	0	0	1	0	0	AS 3	Female

Peabody number	Pathology type	Pathology description	Cranial mod.	Associated funerary objects	Field context	Notes
24-4-10/60532a			No cran.	Gl. 1 red bowl no. 1216 (72706), half of painted mano, no number		
24-4-10/60533.0	Arthritis, nutr.	DJD, osteoporosis	No cran.	Gl. 1 bowl no. 1881 (72738)		
24-4-10/60534.0	Nutr.	Enamel hypoplasia	Occ. def.	2 bone beads no. 1017 (73803), 2 awls no. 1018 (73632), 1 tubular pipe frag. no. 1019 (73221)		
24-4-10/60535.0			Not obs.	Gl. 3 bowl no. 125 (71056)		
24-4-10/60536.0	Nutr., other	Enamel hypoplasia, peg 3rd molar	Not obs.	B&W bowl no. 1807 (72733)		
24-4-10/60536.1	Nutr.	Cribra orbitalia	Unobs.	B&W bowl no. 1807 (72733)		Does not match field card description
24-4-10/60537.0			Unobs.	Gl. 2 vessel frag. no. 124 (71057)		
24-4-10/60538.0	Nutr.	Enamel hypoplasia	Unobs.	Gl. 2 bowl no. 52 (71052), Gl. 3 olla no. 54 (71053), Gl. 2 olla no. 253 (71054), lump of hematite no. 316 (72270)		
24-4-10/60539.0	Trauma, infection	Fracture L ulna, periosteal inflammation R humerus	No cran.	Obsidian projectile point (no number listed)		
24-4-10/60540.0			No cran.			
24-4-10/60541.0	Arthritis, nutr., tumor, other	Slight DJD, enamel hypoplasia, button osteoma, spina bifida vt/s4–5	Unobs.	Gl. 5 olla no. 1048 (72690)		
24-4-10/60542.0	Trauma, infection, arthritis	Healed trauma to nasals?, dental abscess, L humerus–infection, lumbar vt–spondylytis deformans	No cran. def.	Gl. 1 yellow bowl no. 1265 (71171)		Very broad, flat nasals
24-4-10/60543.0	Infection, arthritis, nutr., tumor	Small lesion on L parietal, dental abscess, slight DJD, enamel hypoplasia, button osteoma	Occ. def.			
24-4-10/60543.1	Arthritis	Spondylytis deformans–lumbar vt	No cran.			Field card and provenience problematic
24-4-10/60544.0	Nutr.	Enamel hypoplasia incisors and canines	Occ. def.			
24-4-10/60545.0	Arthritis, nutr., other	Slight DJD, enamel hypoplasia, porotic hyperostosis on frontal and along sagittal suture of parietals, spina bifida sacrum	No cran. def.			
24-4-10/60545a			Unobs.			
24-4-10/60546.0	Nutr., tumor, other	Enamel hypoplasia, button osteoma, unusual dental development molars / premolars	Not obs.	Frag. black olla, worked round stone, no numbers		
24-4-10/60547.0	Nutr.	Enamel hypoplasia	Not obs.			
24-4-10/60548.0	Arthritis, nutr.	DJD appendicular, enamel hypoplasia	Not obs.			
24-4-10/60549.0	Trauma, infection, arthritis, nutr., other	Healed fracture sacrum, lesions on R parietal and small extranumerary bone at bregma, DJD, spondylytis deformans, osteoporosis	Slight occ. def.	Frag. of Gl. 2 bowl no. 346 (71075)		
24-4-10/60549.1	Infection	Periosteal rxn R tibia	No cran.	Frag. of Gl. 2 bowl no. 346 (71075)		
24-4-10/60550.0	Arthritis	Slight DJD	No cran.	B&W bowl no. 979 (72687)		

Peabody number	Object description	FN	Specimen label	Site	Intrasite provenience	Ceramic stage	MNI	Age (yrs.)	Ad	Su	Ch	In	Fe	Unk. age	M	M?	F	F?	Unk. sex	Age criteria	Sex criteria
24-4-10/60551.0	Fragmentary cranial and postcranial human remains of an adult of unknown sex.	1552	60551	PP	1250W150, e trench, depth 2′6″, height 2′+	Indet.	1	45–54	1	0	0	0	0	0	0	0	0	0	1	D i uneven	Indet.
24-4-10/60552.0	Nearly complete cranial and fragmentary postcranial human remains of an adult female.	1416	60552	PP	1100W150, depth 3′6″, height–cut in bottom of clay & rock	Probably B&W	1	25–29	1	0	0	0	0	0	0	0	1	0	0	D e	Female
24-4-10/60553.0	Nearly complete cranial human remains of a subadult female.	1164	60553	PP	1250E100, depth 4′6″, height 2′	Gl. 4	1	16–20	0	1	0	0	0	0	0	0	1	0	0	D b2	Female
24-4-10/60554.0	Fragmentary postcranial human remains of an adult male.	1162	60554	PP	1250E100, depth 6′, height 8″	Probably Gl. 3	1	25–34	1	0	0	0	0	0	1	0	0	0	0	T 6, AS 3, epiphyses youngish	Male
24-4-10/60555.0	Fragmentary cranial and postcranial human remains of an adult female.	1412	60555	PP	1250W150, n trench, depth 5′, height–on bottom	Indet.	1	35–39	1	0	0	0	0	0	0	0	1	0	0	D f-g, AS 4	Female
24-4-10/60556.0	Nearly complete cranial and partial postcranial human remains of an adult male.	1455	60556	PP	1350W150, depth 3′, height 4′	Probably Gl. 4 or 5	1	30–34	1	0	0	0	0	0	1	0	0	0	0	D g, T 6, AS 3	M, all features but small
24-4-10/60557.0	Fragmentary postcranial human remains of an adult of unknown sex.	1443 ?	60557/m16	PP	1200W150b, depth 6′, height–on bottom	Gl. 1	1		1	0	0	0	0	0	0	0	0	0	1		Indet.
24-4-10/60558.0	Partial cranial and postcranial human remains of an adult male.	1551	60558	PP	1350W150, e trench, depth 2′, height 2′+	Probably Gl. 4	1	15–17	0	1	0	0	0	0	1	0	0	0	0	D b2, PC 15–17 yrs (epiphyses)	Male (pelvis)
24-4-10/60559.0	Nearly complete cranial human remains of a subadult of unknown sex.	1495	60559/1495	PP	1300W200, room 320, depth 4′, height 1′	Gl. 3	1	16–18	0	1	0	0	0	0	0	0	0	0	1	D a	Indet.
24-4-10/60560.0	Nearly complete cranial and partial postcranial human remains of an adult male.	1405	60560	PP	1100W150, depth 2′, height 1′	Indet.	1	40–44	1	0	0	0	0	0	1	0	0	0	0	D h, T 8, AS 5	Male
24-4-10/60561.0	Nearly complete cranial and fragmentary postcranial human remains of an adult female.	1478	60561/1478	PP	Room 320, depth 5′, height–on bottom	Gl. 1	1	25–34	1	0	0	0	0	0	0	0	1	0	0	D e-f, AS 3	Female
24-4-10/60562.0	Nearly complete cranial and partial postcranial human remains of an adult male.	1345	60562/1345	PP	1050W200, depth 1′, height–on bottom	Probably B&W	1	40–49	1	0	0	0	0	0	1	0	0	0	0	D h/i, T 9, AS 6	Male
24-4-10/60563.0	Fragmentary cranial and postcranial human remains of an adult female.	1320	60563	PP	Room 43, depth 8′, height–on bottom	Probably B&W	1	35–45	1	0	0	0	0	0	0	0	1	0	0	D g, T 7–8, AS 4–5	Female
24-4-10/60564.0	Nearly complete cranial human remains of an adult female.	1331	60564	PP	Room 44, under s wall 3/5 from east, depth 4′, height 5′±	Probably Gl. 4	1	35–39	1	0	0	0	0	0	0	0	1	0	0	D g	Female
24-4-10/60565.0	Partial cranial and fragmentary postcranial human remains of an adult female.	1057	60565	PP	1450W100, depth 3′, height 1′	Gl. 3	1	50–59	1	0	0	0	0	0	0	0	1	0	0	D I, AS 7	Female
24-4-10/60566.0	Fragmentary cranial and postcranial human remains of a subadult female.	1056	60566	PP	1450W100, depth 6″, height 3′6″	Probably Gl. 5	1	18–20	0	1	0	0	0	0	0	0	1	0	0	AS < 1, PC epiphyses 18–20 yrs	Female
24-4-10/60567.0	Very fragmentary postcranial human remains of an adult of unknown sex.	1342	60567/m21	PP	Room 43, depth 5′6″, height 3′6″	Gl. 1 or 2	1		1	0	0	0	0	0	0	0	0	0	1		Indet.
24-4-10/60568.0	Nearly complete cranial and fragmentary postcranial human remains of an adult female.	1510	60568/1510	PP	Kiva 10, through s wall, depth 3′6″, height 2′6″	Probably Gl. 3	1	18–24	1	0	0	0	0	0	0	0	1	0	0	D b2, PC young adult	Female
24-4-10/60569.0	Nearly complete cranial and fragmentary postcranial human remains of an adult male.	1346	60569	PP	1050W150, near kiva 8, depth 2′–2′6″, height–on bottom	Gl. 6	1	45–49	1	0	0	0	0	0	1	0	0	0	0	D h, AS 6	Male
24-4-10/60570.0	Nearly complete cranial and fragmentary postcranial human remains of an adult female.	1072	60570/1072	PP	1550W50, depth 4′, height–on rock	Probably B&W	1	25–34	1	0	0	0	0	0	0	0	1	0	0	D d-e, AS 3	Female
24-4-10/60571.0	Nearly complete cranial and fragmentary postcranial human remains of an adult female.	1067	60571/ Pecos	PP	Room 19, against base of e wall, 5′ from se corner	Indet.	1	25–39	1	0	0	0	0	0	0	0	1	0	0	D e, AS 3–4	Female

Peabody number	Pathology type	Pathology description	Cranial mod.	Associated funerary objects	Field context	Notes
24-4-10/60551.0	Tumor	Bone spurs L tibia (counted as exostoses–tumors)	Unobs.	Gl. 1 yellow bowl no. 1982 (72747)		
24-4-10/60552.0	Trauma	Healed fracture sternum	Not obs.			Large sternal foramen
24-4-10/60553.0	Nutr., other	Enamel hypoplasia, small metopic suture, extra bone at bregma	Occ. def.	Gl. 5 olla bowl no. 1036 (71163)		
24-4-10/60554.0	Trauma	Healed fracture R fibula	No cran.			
24-4-10/60555.0	Infection, arthritis, nutr.	Dental abscess, slight DJD, enamel hypoplasia	Unobs.	Gl. 1 yellow bowl no. 1046 (72693), Gl. 1 yellow bowl frag. no. 1047 (72694)		
24-4-10/60556.0	Infection, arthritis, nutr., other	Lesion on frontal, dental abscessing, bone spurs on tibia, moderate DJD, severe enamel hypoplasia, bony tubercle L side inferior mandibular torus	Occ. def.			
24-4-10/60557.0	Arthritis	DJD	No cran.	Gl. 1 red bowl no. 1216 (72706), half of painted mano, no number		
24-4-10/60558.0	Nutr.	Enamel hypoplasia	Not obs.			
24-4-10/60559.0			Occ. def.	Gl. 3 bowls nos. 1464 (72725), 1465 (72726)		
24-4-10/60560.0	Trauma, infection, arthritis, nutr., other	Healed trauma to L parietal, lesion on occipital, DJD, enamel hypoplasia, spina bifida vt/s4–5	Occ. def.			Does not match field card description
24-4-10/60561.0	Arthritis, nutr.	Slight DJD, enamel hypoplasia	Occ. def.	2 Gl. 1 red bowls no. 1382 (72718), 1 B&W bowl no. 1393 (72719)		
24-4-10/60562.0	Trauma, infection, arthritis, nutr.	Healed fracture nasal and L fibula, heavy antemortem tooth loss, DJD, spondylytis deformans, osteoporosis	No cran. def.	Small B&W bowl no. 668 (72672)		
24-4-10/60563.0	Arthritis	DJD elbow	Unobs.			
24-4-10/60564.0	Other	Dental maleruption and rotation upper L M2–3	Unobs.			
24-4-10/60565.0	Infection, arthritis, nutr.	Slight periostitis on distal tibiae, moderate DJD, slight osteoporosis	No cran. def.	Gl. 3 bowl (no. 599 (71120)		
24-4-10/60566.0			Occ. def.	Pipe stem (late), no number listed		
24-4-10/60567.0	Arthritis, nutr.	DJD, osteoporosis	No cran.			
24-4-10/60568.0	Nutr.	Enamel hypoplasia	Not obs.	Fragment of cloth no. 1606 (74812)		
24-4-10/60569.0	Trauma, infection, arthritis, nutr.	Trauma to nasals, R clavicle–healed greenstick fracture?, periostitis distal R fibula, moderate DJD elbows, vt–spondylytis deformans, enamel hypoplasia	Occ. def.		Burial at length	
24-4-10/60570.0	Nutr.	Enamel hypoplasia	Lamb. def.			
24-4-10/60571.0	Other	Maleruption L upper C through max. sinus, developmentally only 2 lower incisors	Occ. def.			

Peabody number	Object description	FN	Specimen label	Site	Intrasite provenience	Ceramic stage	MNI	Age (yrs.)	Ad	Su	Ch	In	Fe	Unk. age	M	M?	F	F?	Unk. sex	Age criteria	Sex criteria
24-4-10/60572.0	Nearly complete cranial human remains of an adult female.	1101	60572/1101	PP	South east corner of room 89, depth 1′, height 2′	Probably Gl. 5	1	40–54	1	0	0	0	0	0	0	0	1	0	0	D h/i	Female
24-4-10/60573.0	Fragmentary postcranial human remains of an adult male.	1354	60573	PP	Room 110, depth 6′, height–on bottom	Probably Gl. 1	1	45–49	1	0	0	0	0	0	1	0	0	0	0	AS 6	Male
24-4-10/60574.0	Nearly complete cranial and fragmentary postcranial human remains of an adult female.	985	60574	PP	1400W100, depth 5′3″, height 6″ above bottom	Late B&W	1	50–59	1	0	0	0	0	0	0	0	1	0	0	D old, T 9, AS 7	Female
24-4-10/60575.0	Nearly complete cranial and fragmentary postcranial human remains of a subadult female.	986	60575/986	PP	1350W100, depth 4′10″, height 1′ above bottom	Gl. 2	1	16–20	0	1	0	0	0	0	0	0	1	0	0	D b2 M3s erupting, PC young adult, epiphyses just fused	Female
24-4-10/60576.0	Partial cranial and fragmentary postcranial human remains of an adult that is probably female.	971	60576/971	PP	1400W50, depth 3′6″, height 3′	Probably Gl. 4	1	< 35	1	0	0	0	0	0	0	0	0	1	0	Cranial sutures, relatively young adult	?Female
24-4-10/60577.0	Nearly complete cranial and fragmentary postcranial human remains of an adult male.	1516	60577	PP	1150W50, depth 4′6″, height indeterminate	Gl. 4 or 5	1	40–54	1	0	0	0	0	0	1	0	0	0	0	D h/i	Male
24-4-10/60578.0	Partial cranial and fragmentary postcranial human remains of an adult female.	1060	60578	PP	1450W100	Probably Gl. 1	1	27–34	1	0	0	0	0	0	0	0	1	0	0	D h, T 5, AS 3	Female
24-4-10/60579.0	Partial cranial and fragmentary postcranial human remains of an adult male.	1376	60579	PP	1350W150, depth 4′, height 3′	Indet.	1	40–54	1	0	0	0	0	0	1	0	0	0	0	D h/i	Male
24-4-10/60580.0	Nearly complete cranial and fragmentary postcranial human remains of an adult female.	1511	60580	PP	Kiva 10, depth 3′, height 4′	Indet.	1	50–59	1	0	0	0	0	0	0	0	1	0	0	D h/i, AS 7	F all features
24-4-10/60581.0	Nearly complete cranial and fragmentary postcranial human remains of an adult male.	1364	60581	PP	1250W150, depth 3′6″	Probably Gl. 5	1	35–49	1	0	0	0	0	0	1	0	0	0	0	D h/g	Male
24-4-10/60582.0	Nearly complete cranial and fragmentary postcranial human remains of an adult female.	1081	60582/1081	PP	1510W50, depth 3′	Indet.	1	18–24	1	0	0	0	0	0	0	0	1	0	0	D c-d/d, AS 2–3	Female
24-4-10/60583.0	Nearly complete cranial and fragmentary postcranial human remains of an adult male.	1377	60583/1377	PP	1100W150, ne of kiva 8, depth 4′, height–on clay	Probably B&W	1	45–55+	1	0	0	0	0	0	1	0	0	0	0	D h+/i+	Male
24-4-10/60584.0	Nearly complete cranial human remains of a child of unknown sex.	1396	60584/1396	PP	1200W150b, depth 2′, height–on rock	Indet., poss. Gl. 5	1	7–9	0	0	1	0	0	0	0	0	0	0	1	D 9 ± 2 yrs	Indet.
24-4-10/60585.0	Partial cranial human remains of a child of unknown sex.	1451	60585	PP	1150W200, north wall, depth 1′6″, height 1′6″	Gl. 1R	1	2–3	0	0	1	0	0	0	0	0	0	0	1	D 3 ± 1 yrs	Indet.
24-4-10/60586.0	Nearly complete cranial human remains of a child of unknown sex.	1537	60586/1537	PP	1200W200, w extension, depth 3′6″, height–on bottom	Indet.	1	3.5–5	0	0	1	0	0	0	0	0	0	0	1	D 5 ± 1.3 yrs	Indet.
24-4-10/60587.0	Nearly complete cranial human remains of a child of unknown sex.	1092	60587/1092	PP	1500W50, depth 2′, height–on bottom	Gl. 1	1	5–7	0	0	1	0	0	0	0	0	0	0	1	D 7 ± 2 yrs	Indet.
24-4-10/60588.0	Fragmentary cranial and postcranial human remains of an adult female.	1372	60588	PP	1300W150, depth 3′6″, height 3′6″	Indet., poss. Gl. 3	1	30–39	1	0	0	0	0	0	0	0	1	0	0	D d, T 7, AS 3	Female
24-4-10/60589.0	Nearly complete cranial and fragmentary postcranial human remains of an adult male.	1393	60589	PP	1200W200a, depth 3′, height–in crack	Gl. 1R	1	35–39	1	0	0	0	0	0	1	0	0	0	0	D g	Male
24-4-10/60590.0	Nearly complete cranial and partial postcranial human remains of an adult female.	1163	60590	PP	1250E100, n tren. e traverse, depth 6′, height–on bottom	Late Gl. 4	1	35–39	1	0	0	0	0	0	0	0	1	0	0	D f-g, T 7	Female

Peabody number	Pathology type	Pathology description	Cranial mod.	Associated funerary objects	Field context	Notes
24-4-10/60572.0	Infection, nutr., tumor	Lesion R parietal, enamel hypoplasia, exotoses on lingual mandible	Occ. def.			
24-4-10/60573.0	Arthritis,	DJD	No cran.			
24-4-10/60574.0	Arthritis, nutr.	Moderate DJD, slight osteoporosis	Occ. def.	Bisc. B bowl no. 83 (71061), smaller black vessel, no number		
24-4-10/60575.0	Nutr., other	Enamel hypoplasia, occipital swelling–hyperostosis	Not obs.	Partial Gl. 2 bowl no. 50 (71062)		
24-4-10/60576.0			Not obs.			Extremely robust app. elements, small mastoids, large brow ridge, L humerus markedly bowed–no evidence of trauma
24-4-10/60577.0	Infection, nutr.	Caries, enamel hypoplasia	No cran. def.	Gl. 4 sherds, no number		Highly retroflexed humerus
24-4-10/60578.0	Infection	Periostitis sacrum–deposition of woven bone	Not obs.	Frag. corrugated olla no. 706 (71121)		Preservation, dentition, and postcrania match although very different age estimates
24-4-10/60579.0	Nutr.	Enamel hypoplasia	Not obs.	Gl. 1 yellow bowl no. 983 (72680)		
24-4-10/60580.0	Trauma, arthritis, nutr.	Healed greenstick fracture on R ulna, DJD, osteoporosis	Occ. def.	Black pot with handles no. 1647 (72730)		
24-4-10/60581.0	Infection	Major infection R maxilla and antemortem molar tooth loss on left side	Not obs.			
24-4-10/60582.0	Infection, arthritis	Slight infection sacroiliac joint, slight DJD	Not obs.	Small black pot no. 698 (71127)		
24-4-10/60583.0	Infection, arthritis	Moderate DJD, severe dental abscesses	Not obs.	B&W bowl no. 1645 (72681)		
24-4-10/60584.0	Nutr.	Marked porosity on occipital bone near lambda	Not obs.	92 turquoise mosaic beads no. 984 (74723)		
24-4-10/60585.0	Nutr.	Cribra orbitalia	Not obs.	Snail shell no. 1246 (74714), Gl. 1 red frag. (no number)		
24-4-10/60586.0			Not obs.	Frag. Gl. 2 red bowl no. 1925 (72742)		Field card and ledger descriptions do not match
24-4-10/60587.0			Not obs.	Gl. 1 red bowl no. 717 (71132), old black olla scoop, no number		
24-4-10/60588.0	Infection, arthritis, nutr.	Infection in L prox. ulna, periosteal rxn dist. R fibula, slight DJD, enamel hypoplasia	Unobs.			Postcrania suggest older than dentition–problematic association?
24-4-10/60589.0	Arthritis, nutr.	DJD, enamel hypoplasia	Not obs.	Gl. 1 red bowl no. 982 (72688), projectile point no. 891 (74259)		
24-4-10/60590.0	Infection, arthritis, nutr., other	Dental abscess, DJD, enamel hypoplasia, spondylolysis vt/l5 neural arch not fused	Occ. def.	Gl. 3 bowl no. 1033 (71162)		

Peabody number	Object description	FN	Specimen label	Site	Intrasite provenience	Ceramic stage	MNI	Age (yrs.)	Ad	Su	Ch	In	Fe	Unk. age	M	M?	F	F?	Unk. sex	Age criteria	Sex criteria
24-4-10/60591.0	Fragmentary cranial and postcranial human remains of an adult female.	980	60591/980	PP	1350W100, depth 4′4″, height 8″	Probably Gl. 1	1	35–45	1	0	0	0	0	0	0	0	1	0	0	D h, T 7-8, AS 4	Female
24-4-10/60592.0	Nearly complete cranial and fragmentary postcranial human remains of an adult male.	968	60592	PP	1400W100, depth 7′4″, height–on bottom	Probably Gl. 1	1	45–55+	1	0	0	0	0	0	1	0	0	0	0	D h+/i+	Male
24-4-10/60592.1	Very fragmentary postcranial human remains of an adult of unknown sex.	1439	1439	PP	1200W200d, depth 3′6″, height–on bottom	Indet.	1	Adult	1	0	0	0	0	0	0	0	0	0	1		Indet.
24-4-10/60593.0	Nearly complete cranial and fragmentary postcranial human remains of an adult that is probably male.	1351	60593/1351	PP	1050W200, depth 2′, height 2′	Gl. 6	1	35–39	1	0	0	0	0	0	0	1	0	0	0	D h+/i+, AS 4	?Male
24-4-10/60594.0	Nearly complete cranial and fragmentary postcranial human remains of an adult female.	1098	60594	PP	1450W100, room 84, depth 3′, height 2′6″	Probably Gl. 4	1	35–45	1	0	0	0	0	0	0	0	1	0	0	D e-h, T 8, AS 4	Female
24-4-10/60595.0	Nearly complete cranial and fragmentary postcranial human remains of an adult that is probably male.	1088	60595/1088	PP	1450W50, depth 8″, height 12″	Gl. 4	1	40–49	1	0	0	0	0	0	0	1	0	0	0	D h	?Male
24-4-10/60596.0	Partial cranial and fragmentary postcranial human remains of a subadult female.	1032	60596	PP	50 (under base on n wall of room 50)	Indet., poss. Gl. 1R	1	15–19	0	1	0	0	0	0	0	0	1	0	0	D b2, T 1, AS < 1	Female
24-4-10/60597.0	Partial cranial and fragmentary postcranial human remains of an adult female.	1477	60597	PP	Room 320 by east wall, depth 5′, height 1′	Gl. 1 or 2	1	40–44	1	0	0	0	0	0	0	0	1	0	0	D f-h, AS 5	Female
24-4-10/60598.0	Nearly complete cranial and fragmentary postcranial human remains of an adult male.	1452	60598	PP	1200W100, b, room b-II, depth 9′, height–on bottom	Gl. 1R	1	45–49	1	0	0	0	0	0	1	0	0	0	0	D i+, AS 6	Male
24-4-10/60599.0	Partial cranial and fragmentary postcranial human remains of an adult male.	1017	60599	PP	Room 46, sw corner, depth 9′3″, height 1′	Gl. 1	1	30–34	1	0	0	0	0	0	1	0	0	0	0	D f, T 6, AS 3	Male
24-4-10/60600.0	Nearly complete cranial human remains of a subadult male.	1590	60600	PP	Room e, under w wall, depth 5′, height 2′	B&W or Gl. 1	1	18–20	0	1	0	0	0	0	1	0	0	0	0	D c	Male
24-4-10/60601.0	Nearly complete cranial and partial postcranial human remains of an adult that is probably male.	1593	60601/c	PP	Under porch 326, under rock, depth 7′, height–on bottom	Indet.	1	30–39	1	0	0	0	0	0	0	1	0	0	0	D g, T 6, AS 4	?Male (pelvis more M, cranium more F)
24-4-10/60602.0	Partial cranial and postcranial human remains of an adult male.	1300	60602	PP	Room 301, NE corner, depth 5′, height 1′ above bottom	Gl. 1R	1	50–59	1	0	0	0	0	0	1	0	0	0	0	D i, cranial sutures: very old, T 9, AS 7	Male
24-4-10/60603.0	Partial cranial and fragmentary postcranial human remains of an adult male.	1269	60603/1269	PP	1350W100, s wall, depth 1′, height 6′ above bottom	Probably Gl. 4	1	40–49	1	0	0	0	0	0	1	0	0	0	0	D h-i,uneven, AS 5	Male
24-4-10/60605.0	Nearly complete cranial and fragmentary postcranial human remains of an adult male.	1422	60605	PP	1200W150b, depth 4′, height 2′	Indet., poss. Gl. 3	1	45–55+	1	0	0	0	0	0	1	0	0	0	0	D h+/i+	Male
24-4-10/60606.0	Nearly complete cranial human remains of a child of unknown sex.	992	60606	PP	1450W50, depth 4′10″	Probably early Gl. 2	1	3–4	0	0	1	0	0	0	0	0	0	0	1	D 4 ± 1 yrs	Indet.
24-4-10/60607.0	Nearly complete cranial and fragmentary postcranial human remains of an adult male.	998	60607	PP	Room 46, depth 90″, height 15″ above bottom	Probably Gl. 3	1	35–39	1	0	0	0	0	0	1	0	0	0	0	D g	Male
24-4-10/60608.0	Nearly complete cranial and fragmentary postcranial human remains of an adult female.	1417	60608	PP	1200W150B, depth 3′6″, height 1′	Indet., poss. late	1	45–49	1	0	0	0	0	0	0	0	1	0	0	D f, cranial sutures suggest older, AS 6	Female

Peabody number	Pathology type	Pathology description	Cranial mod.	Associated funerary objects	Field context	Notes
24-4-10/60591.0	Infection, nutr.	Dental abscesses–severe with significant resorption of bone, osteoporosis	Unobs.	Gl. 1 bowl, no number		Highly polished teeth, very bowed L fibula,
24-4-10/60592.0	Infection	Severe antemortem tooth loss	Slight occ. def.	Flute, no number		
24-4-10/60592.1			No cran.	Gl. 4 bowl no. 1186 (72704)		1 femur, remainder of individual listed on field card is not present
24-4-10/60593.0	Infection, arthritis, nutr., tumor	Dental abscesses with antemortem tooth loss, moderate DJD, spinal osteophytosis, slight osteoporosis, enamel hypoplasia, tumor in R orbit encroaching into frontal sinus	Slight occ. def.		Burial at length	
24-4-10/60594.0	Infection, nutr.	Impacted L upper M3, enamel hypoplasia	No cran. def.	Black pot fragments no. 1061 (71133), no. 704 (71134)		
24-4-10/60595.0	Tumor	Osteoma on vault	Not obs.	Gl. 4 bowl no. 246 (72639), projectile point (in chest) no. 699 (71817), wrapped in buffalo hide, no number		
24-4-10/60596.0	Trauma, nutr.	Trauma to L parietal (2 indentations), slight porotic hyperostosis, enamel hypoplasia	Cran. def.	Gl. 1 red bowl no. 391 (71098), black duck pot no. 343 (71099), broken projectile point no. 344 (71815)		
24-4-10/60597.0	Infection, arthritis, nutr.	Dental abscesses, slight DJD, enamel hypoplasia	Occ. def.	Gl. 2 bowl no. 1367 (72717)		
24-4-10/60598.0	Infection, arthritis, nutr.	Substantial antemortem tooth loss, moderate DJD, osteoporosis	Occ. def.			
24-4-10/60599.0	Other	Spina bifida vt/s4–5, maleruption R upper canine caused dental crowding	Unobs.	Two thin bone objects, small bead (no number listed)		
24-4-10/60600.0	Trauma, nutr.	Scalping marks on parietals, enamel hypoplasia	Lamb. def.	Parts of B&W bowls no. 2190 (72769)		Facial shape different, shoveling not pronounced
24-4-10/60601.0	Infection, arthritis, nutr.	Dental abscesses, slight DJD, porotic hyperostosis	Occ. def.	B&W bowl no. 2188 (72771)		2 cutmarks on left and right side of frontal
24-4-10/60602.0	Infection, arthritis, nutr., tumor	Dental abscesses, marked DJD, marked spinal osteophytosis, vt/l–spondylytis deformans, cribra orbitalia–reworking massive brow ridge, bony tubercles–mandible	Occ. def.	Gl. 2 red bowl no. 293 (72649), crinoid (no number listed)		
24-4-10/60603.0	Trauma, infection, arthritis	Healed fractures to frontal bone, L maxilla, and slightly affecting R maxilla, affected L orbit, diploic expansion in response to trauma, DJD	Unobs.	Antler tip no. 47 (74106)		
24-4-10/60605.0	Infection, nutr.	Dental abscessing, extensive antemortem tooth loss, osteoporosis	Lamb. def.	Bone awl no. 1025 (73633), worked sherds no. 1024 (72937)		Facial shape different
24-4-10/60606.0	Trauma	Trauma to R parietal	Not obs.	Frag. Gl. 2 red bowl no. 82 (71066)		Previously assigned to 1423 but attached tag indicates field number was 992 (MEM 11/97)
24-4-10/60607.0	Nutr.	Enamel hypoplasia	Prob. occ. def.	Gl. 3 olla bowl no. 602 (71071), worked bone no. 254 (71691), turquoise and shell beads, no number listed		
24-4-10/60608.0	Infection, arthritis	Maxilla almost edentulous, vt–spondylytis deformans, moderate DJD, bony buildup prox. clavicle	Prob. occ. def.			

Peabody number	Object description	FN	Specimen label	Site	Intrasite provenience	Ceramic stage	MNI	Age (yrs.)	Ad	Su	Ch	In	Fe	Unk. age	M	M?	F	F?	Unk. sex	Age criteria	Sex criteria
24-4-10/60609.0	Nearly complete cranial and partial postcranial human remains of an adult female.	1021	60609	PP	Room 50, adjacent to w wall	Bisc. B?	1	40–49	1	0	0	0	0	0	0	0	1	0	0	D f-h , T 8 AS 5–6	Female
24-4-10/60610.0	Nearly complete cranial and partial postcranial human remains of an adult female.	1097	60610/1097	PP	1500W50, depth 18″, height–on bottom	Probably late Gl. 6	1	35–44	1	0	0	0	0	0	0	0	1	0	0	D g-h/g, T 7, AS 5	Female
24-4-10/60611.0	Nearly complete cranial and fragmentary postcranial human remains of an adult female.	1442	60611	PP	1250W150c, depth 3′6″, height–on bottom	B&W	1	35–39	1	0	0	0	0	0	0	0	1	0	0	D h/i, T 7, AS 4	Female
24-4-10/60612.0	Nearly complete cranial and fragmentary postcranial human remains of an adult female.	1597	60612	PP	Sw corner of 317, depth 8′, height 2′	Probably Gl. 3	1	35–45	1	0	0	0	0	0	0	0	1	0	0	D f, T 8, AS 4	Female
24-4-10/60613.0	Fragmentary postcranial human remains of an adult female.	962	60613	PP	1400W100, depth 4′1″, height 2′7″	Probably Gl. 1	1	25–39	1	0	0	0	0	0	0	0	1	0	0	PC general	Female (size pc)
24-4-10/60614.0	Fragmentary cranial and postcranial human remains of an adult female.	1514	606014	PP	1350W150, depth 3′8″, height 2′	Gl. 4	1	25–29	1	0	0	0	0	0	0	0	1	0	0	D d-e	Female
24-4-10/60615.0	Nearly complete cranial and fragmentary postcranial human remains of an adult female.	1125	606015/ 1125	PP	Room 53, s half	Probably Gl. 6	1	30–54	1	0	0	0	0	0	0	0	1	0	0	D f-g/i	Female (size pc)
24-4-10/60616.0	Partial cranial and postcranial human remains of an adult male.	1584	60616	PP	1250W150, s wall, depth 8′, height 3′	Probably Gl. 5	1	30–39	1	0	0	0	0	0	1	0	0	0	0	D g-h/h-i, AS 3–4	Male
24-4-10/60616a	Fragmentary cranial and postcranial human remains of an adult male.	1116	60616a, 60616	PP	Room 72, sw corner of subroom, depth 3′6″, height 5′	Probably Gl. 3 or 4	1	50–59	1	0	0	0	0	0	1	0	0	0	0	D i+	Male
24-4-10/60617.0	Fragmentary cranial human remains of an adult of unknown sex.	1116	60617/1116	PP	Room 72, sw corner of subroom, depth 3′6″, height 5′	Probably Gl. 3 or 4	1	40–49	1	0	0	0	0	0	0	0	0	0	1	D H	Indet.
24-4-10/60618.0	Fragmentary postcranial human remains of an adult that is probably female.	1076	60618	PP	1507W46, depth 4′6″, height–on bottom	Late B&W	1	Adult	1	0	0	0	0	0	0	0	0	1	0		?Female (size pc)
24-4-10/60620.0	Nearly complete cranial and fragmentary postcranial human remains of an adult female.	989	60620	PP	1400W100, depth 6′5″, height–on bottom	B&W	1	40–49	1	0	0	0	0	0	0	0	1	0	0	D g-h/i, T 10, AS 5–6	Female
24-4-10/60621.0	Nearly complete cranial human remains of a child of unknown sex.	1005	60621/ 60506	PP	Room 50, depth 5′	Probably Gl. 2	1	6–8	0	0	1	0	0	0	0	0	0	0	1	D 8 ± 2 yrs	Indet.
24-4-10/60622.0	Fragmentary cranial human remains of a subadult of unknown sex.	1053	60622/1053	PP	1450W100, depth 8–10″, height 3′6″	Probably Gl. 4	1	10–12	0	0	1	0	0	0	0	0	0	0	1	D 11–12 ± 30 mths (1 dm not erupted)	Indet.
24-4-10/60623.0	Nearly complete cranial and partial postcranial human remains of an adult female.	1159	606023	PP	Trench east of r 54 and r 55, depth 6′3″, height–in clay	B&W or Gl. 1	1	45–49	1	0	0	0	0	0	0	0	1	0	0	D h/i, AS 6	Female
24-4-10/N272.1	Fragmentary cranial and postcranial human remains of 1 adult of unknown sex.	1361	N272, 1361, 60977-10	PP	Room 43, under s wall, along ns low wall, depth 5′, height 4′	Probably Gl. 4	1	30–39	1	0	0	0	0	0	0	0	0	0	1	D f-g	Indet.
24-4-10/N500.0	Fragmentary cranial human remains of an adult of unknown sex.	995	N500/995	PP	1412W34, depth 20″, height 14″ above bottom	Gl. 3	1		1	0	0	0	0	0	0	0	0	0	1		Indet.
25-3-10/60626.0	Nearly complete cranial and fragmentary postcranial human remains of an a subadult female.	1795	60626	PP	Room e-vi, w side of e wall, depth 8′, height 4′ ?	Indet.	1	12–15	0	1	0	0	0	0	0	0	1	0	0	D B1, 15 ± 3 yrs; PC	Female
25-3-10/60627.0	Fragmentary postcranial human remains of a subadult of unknown sex.	1636	60627	PP	1300W0, trench b, depth 6′6″, height 1′6″ above large rock	Indet.	1	16–18	0	1	0	0	0	0	0	0	0	0	1	T 1, PC < 17–19 yrs	Indet.

Peabody number	Pathology type	Pathology description	Cranial mod.	Associated funerary objects	Field context	Notes
24-4-10/60609.0	Infection, arthritis	Dental abscessing, major infection left side mandible from lower L M3, slight DJD	Occ. def.	Bisc. B bowl no. 392 (71086), worked potsherd no. 360 (71288)		
24-4-10/60610.0	Trauma, infection, arthritis, nutr.	Scalping–frontal and parietal parallel cut marks on left and right side, dental abscessing lower M1s with mandibular bony abscesses, vt/l spondylytis deformans, severe enamel hypoplasia	Not obs.		Burial at length	Field number problematic
24-4-10/60611.0			Slight lamb. def.	Frag. B&W bowl no. 1217 (72705)		Different skull shape
24-4-10/60612.0	Infection, nutr.	Right ear infection, enamel hypoplasia	Not obs.	Rubbing stone no. 2208 (74492)		
24-4-10/60613.0	Infection, arthritis	Periostitis–prox. 1/3 R tibia, distal 1/4 L femur, prox. 1/3 L tibia, DJD	No cran.			
24-4-10/60614.0			Unobs.	Gl. 3 or 4 bowl no. 1692 (72731)		
24-4-10/60615.0	Infection, arthritis, nutr., tumor	Dental abscesses maxilla, vt/c 2–3 ankylosed–only vt present therefore cause difficult to assess, increased porosity orbits, frontal bone, enamel hypoplasia, button osteoma on frontal	Occ. def.	Piece of obsidian no. 887 (71928)	Burial at length	
24-4-10/60616.0	Infection, arthritis	Severe dental abscesses, DJD	Unobs.		Buried in top of oven	
24-4-10/60616a	Infection, arthritis	Dental abscesses, DJD	Unobs.	Large shell ornament no. 826 (72238), 8 beads (5 turquoise, 3 shell) no. 824 (72193), clay pipe no. 825 (71369)		
24-4-10/60617.0	Infection	Dental abscess, antemortem loss molars	Unobs.	Large shell ornament no. 826 (72238), 8 beads (5 turquoise, 3 shell) no. 824 (72193), clay pipe no. 825 (71369)	Burial at length	? Glaze and burial position
24-4-10/60618.0			No cran.	Frag. B&W bowl no. 684 (71126)		
24-4-10/60620.0	Infection, arthritis, nutr.	Dental abscesses, slight DJD, enamel hypoplasia	Not obs.	Black ware jar no. 81 (71064), B&W small handled jar no. 53 (71063)		Large female
24-4-10/60621.0	Nutr.	Cribra orbitalia, enamel hypoplasia	Not obs.			Does not match field card description
24-4-10/60622.0			Unobs.			
24-4-10/60623.0	Trauma, infection, arthritis, nutr., other	Trauma and infection lower R radius, L tibia, possible infection R clavicle, vt–moderate spinal osteophytosis, moderate DJD, enamel hypoplasia, cranial porosity, spheno-occipital has central hole?	Not obs.	Frag. early black pot no. 1037 (71160)		
24-4-10/N272.1			Unobs.	Black sherd no. 1804 (72964)		Does not match field card description
24-4-10/N500.0			Unobs.	Frag. of Gl. 3 bowl no. 51 (71069)		
25-3-10/60626.0	Trauma, nutr.	Trauma–scalping, no other trauma visible, cribra orbitalia and porotic hyperostosis	Not obs.		Burial at length	Assoc. pelvis with cranium dubious
25-3-10/60627.0			No cran.	Gl. 2 bowl no. 109 (74892)		

Peabody number	Object description	FN	Specimen label	Site	Intrasite provenience	Ceramic stage	MNI	Age (yrs.)	Ad	Su	Ch	In	Fe	Unk. age	M	M?	F	F?	Unk. sex	Age criteria	Sex criteria
25-3-10/60628a	Very fragmentary cranial human remains of a subadult female.	C	60628a	PP	Church	Indet.	1	16–20	0	1	0	0	0	0	0	0	1	0	0	D b1	Female
25-3-10/60628b	Very fragmentary cranial human remains of a child of unknown sex.	C	60628b	PP	Church	Indet.	1	5–7	0	0	1	0	0	0	0	0	0	0	1	7 ± 2 yrs	Indet.
25-3-10/60628c	Very fragmentary cranial human remains of an adult female.	C	60628c	PP	Church	Indet.	1	45–54	1	0	0	0	0	0	0	0	1	0	0	D I	Female
25-3-10/60628d	Very fragmentary cranial human remains of an adult male.	C	60628d	PP	Church	Indet.	1	35–39	1	0	0	0	0	0	1	0	0	0	0	D g	Male
25-3-10/60628e	Very fragmentary cranial human remains of an adult of unknown sex.	C	60628e	PP	Church	Indet.	1	30–39	1	0	0	0	0	0	0	0	0	0	1	D f-g	Indet.
25-3-10/60628f	Very fragmentary cranial human remains of an adult of unknown sex.	C	60628f	PP	Church	Indet.	1	20–24	1	0	0	0	0	0	0	0	0	0	1	D d	Indet.
25-3-10/60629.0	Fragmentary postcranial human remains of a subadult of unknown sex.	1795	60629	PP	Room e-vi, depth 8′, height 4′ ?	Indet.	1	12–16	0	1	0	0	0	0	0	0	0	0	1	D 13–17 yrs	Indet.
25-3-10/60629a	Very fragmentary postcranial human remains of an adult of unknown sex.	1795	60629a	PP	Room e-vi, west side of east wall, depth 8′, height 4′ ?	Indet.	1		1	0	0	0	0	0	0	0	0	0	1		Indet.
25-3-10/60630.0	Partial cranial and fragmentary postcranial human remains of an adult female.	1624	60630	PP	Location indeterminate, depth 3′6″, height 20″	Probably Gl. 3	1	40–44	1	0	0	0	0	0	0	0	1	0	0	D c, T 8, AS 5	Female
25-3-10/60631.0	Partial cranial and fragmentary postcranial human remains of an adult female.	1690	60631	PP	1300W0, trench a, depth 3′11″, height 4′	Probably Gl. 3	1	18–22	1	0	0	0	0	0	0	0	1	0	0	D c, PC young adult	Female
25-3-10/60631.1	Fragmentary postcranial human remains of an adult of unknown sex.	1690	60631	PP	1300W0, trench a, depth 3′11″, height 4′	Probably Gl. 3	1	Adult	1	0	0	0	0	0	0	0	0	0	1		Indet.
25-3-10/60632.0	Fragmentary postcranial human remains of an adult of unknown sex.	1615	60632	PP	1350W0, trench a, depth 4′, height–on clay	Indet.	1	Adult	1	0	0	0	0	0	0	0	0	0	1		Indet.
25-3-10/60633.0	Nearly complete cranial and partial postcranial human remains of an adult male.	C11	60633	PP	Church nave, depth 2′, height 3′10″	Indet.	1	25–34	1	0	0	0	0	0	1	0	0	0	0	D f-g, T 4–6, SB 3, AS 3–4	M all features
25-3-10/60634.0	Nearly complete cranial and postcranial human remains of an adult male.	C9	60634	PP	Church nave, depth 2′10″, height 1′2″	Indet.	1	35–49	1	0	0	0	0	0	1	0	0	0	0	D h, T 7–9, SB 3–4, AS 5–6	M all features
25-3-10/60651.0	Nearly complete cranial human remains of an adult male.	C46	60651	PP	Church nave, depth 3′6″, height 6″	Indet.	1	30–34	1	0	0	0	0	0	1	0	0	0	0	D f	M all features
25-3-10/60652.0	Nearly complete cranial human remains of an adult female.	C39	60652	PP	Church nave, depth 2′4″, height 1′8″	Indet.	1	16–20	1	0	0	0	0	0	0	0	1	0	0	D b2, spheno-occipit al fused	Female
25-3-10/60653.0	Nearly complete cranial human remains of an adult female.	C40	60653	PP	Church nave, depth 2′6″, height 1′6″	Indet.	1	<35	1	0	0	0	0	0	0	0	1	0	0	D not old adult	Female
25-3-10/60654.0	Nearly complete cranial human remains of an adult male.	C45	60654	PP	Church nave (under 44), depth 3′, height 1′	Indet.	1	50–59	1	0	0	0	0	0	1	0	0	0	0	D i+	Male
25-3-10/60655.0	Partial cranial human remains of an adult male.	1609	60655	PP	Room e, sub, depth 5′, height 2′	Gl. 1	1	40–54	1	0	0	0	0	0	1	0	0	0	0	D h/i	Male
25-3-10/60656.0	Nearly complete cranial human remains of an adult female.	1624	60656	PP	Location indeterminate, depth 3′6″, height 20″	Probably Gl. 3	1	30–34	1	0	0	0	0	0	0	0	1	0	0	D 30–34 yrs	Female
25-3-10/60657.0	Nearly complete cranial and very fragmentary postcranial human remains of an adult female.	1860	60657	PP	800W100, depth 3′, height 2′	Probably Gl. 3	1		1	0	0	0	0	0	0	0	1	0	0		Female
25-3-10/60661.0	Nearly complete cranial and postcranial human remains of an adult female.	1799	60661	PP	Room e-vii, depth 9′, height 4′	B&W or Gl. 1	1	60+	1	0	0	0	0	0	0	0	1	0	0	AS 8	Female
25-3-10/60662.0	Partial cranial and fragmentary postcranial human remains of an adult male.	1802	60662	PP	400E100, depth 10″, height ?	Indet.	1	25–34	1	0	0	0	0	0	1	0	0	0	0	D e, T 5, AS 3	M all features

Peabody number	Pathology type	Pathology description	Cranial mod.	Associated funerary objects	Field context	Notes
25-3-10/60628a	Nutr.	Enamel hypoplasia	Unobs.			
25-3-10/60628b			Unobs.			
25-3-10/60628c			Unobs.			
25-3-10/60628d	Other	Mandible–bony build up on lingual surface, left side of mandible pathological	Unobs.			
25-3-10/60628e	Nutr.	Enamel hypoplasia	Unobs.			
25-3-10/60628f			Unobs.			
25-3-10/60629.0			No cran.			Field card and provenience problematic
25-3-10/60629a			No cran.			Field card and provenience problematic
25-3-10/60630.0	Nutr.	Enamel hypoplasia	No cran. def.			Very different cranial shape, zygomatics tucked back, does not match field card description
25-3-10/60631.0	Trauma, nutr.	Healed trauma–occipital, enamel hypoplasia	Prob. cran. def.	Bone awl no. 328 (75480), bone bead no. 329 (75682)		
25-3-10/60631.1	Infection, arthritis	Severe DJD and infection prox. R ulna, vt/c spinal osteophytosis	No cran.	Awl and bead (no number listed)		Does not match field card description
25-3-10/60632.0			No cran.	Bisc. A bowl no. 244 (74878), antler tool, copper ore, cache stone tools, no numbers		Reassociated unnumbered R femur with 60632 (MEM 5/98)
25-3-10/60633.0	Infection, nutr.	Syphilis? (or tuberculosis?)–systemic infection, load bearing ligamentous and cranial regions, ribs, enamel hypoplasia	Cran. def.		Burial at length	Different population?
25-3-10/60634.0	Trauma, infection, arthritis, nutr.	Deformation of dist. ph/p 1st digit dental abscess, DJD, enamel hypoplasia	Not obs.	2 white opaque glass beads no. 1173 (76368)	Burial at length	Different cranial features / forehead, atypical coronal suture
25-3-10/60651.0	Trauma, infection	Severe trauma to R parietal, infection L mastoid	No cran. def.			
25-3-10/60652.0			Prob. cran. def.			
25-3-10/60653.0	Infection, arthritis	Marked antemortem dental loss, slight DJD at TMJ	Not obs.			Different orbital shape
25-3-10/60654.0	Infection	Maxillary antemortem tooth loss	No cran. def.			
25-3-10/60655.0	Infection	Dental abscesses, ectocranial porosis , right TMJ	Prob. occ. def.			
25-3-10/60656.0	Nutr.	Enamel hypoplasia	Prob. occ. def.			Does not match field card description
25-3-10/60657.0	Nutr.	Enamel hypoplasia	Occ. def.	Gl. 3 bowl no. 1487 (74975)		
25-3-10/60661.0	Infection, arthritis, tumor	Marked antemortem tooth loss, mild DJD, marked spondylytis deformans with severe degeneration vt/11, button osteomas on vault	Occ. def.	1 Olivella shell, no number		
25-3-10/60662.0	Trauma, nutr.	Healed torn ligament R prox. humerus, enamel hypoplasia	Occ. def.		Burial at length	

Peabody number	Object description	FN	Specimen label	Site	Intrasite provenience	Ceramic stage	MNI	Age (yrs.)	Ad	Su	Ch	In	Fe	Unk. age	M	M?	F	F?	Unk. sex	Age criteria	Sex criteria
25-3-10/60663.0	Nearly complete cranial and partial postcranial human remains of an adult male.	C56	60663	PP	Church nave, depth 3′6″, height 6″	Indet.	1	50–59+	1	0	0	0	0	0	1	0	0	0	0	SB 50–54, AS 8	M all features
25-3-10/60663.1	Fragmentary postcranial human remains of an adult of unknown sex.	C56		PP	Church nave, depth 3′6″, height 6″	Indet.	1		1	0	0	0	0	0	0	0	0	0	1		Indet.
25-3-10/60664.0	Fragmentary postcranial human remains of an adult female.	1679	60664	PP	1300W0, trench a, depth 5′2″, height 8″	Probably Gl. 1y or 2	1	30–34	1	0	0	0	0	0	0	0	1	0	0	AS 3	Female
25-3-10/60665.0	Fragmentary cranial and partial postcranial human remains of an adult male.	1705	60665	PP	Sw passage, sub, depth 5′, height 1′6″	Indet.	1	45–54	1	0	0	0	0	0	1	0	0	0	0	D h/h-I, T 10, AS 7	M all features
25-3-10/60666.0	Nearly complete cranial and postcranial human remains of an adult male.	1742	60666	PP	950E150, depth 6′, height 18″	Gl. 3 or poss. 4	1	35–44	1	0	0	0	0	0	1	0	0	0	0	D g/f-g,AS 5	Male
25-3-10/60667.0	Nearly complete cranial human remains of an adult that is probably male.	C26	60667	PP	Church nave, depth 3′	Indet.	1	18–24	1	0	0	0	0	0	0	1	0	0	0	D c-d/c	?Male
25-3-10/60668.0	Partial cranial and fragmentary postcranial human remains of an adult female.	1810	60668	PP	500W50, depth 3′, height–on flaky clay	Indet.	1	45–49	1	0	0	0	0	0	0	0	1	0	0	D h/i, AS 6	Female
25-3-10/60668.1	Very fragmentary postcranial human remains of an adult of unknown sex.	1810	60668	PP	500W50, depth 2′, height–on flaky clay	Indet.	1		1	0	0	0	0	0	0	0	0	0	1		Indet.
25-3-10/60668a	Nearly complete cranial human remains of an adult male.	1810	60668a	PP	500W50, depth 2′, height–on flaky clay	Indet.	1	40–54	1	0	0	0	0	0	1	0	0	0	0	D h/i	Male
25-3-10/60670.0	Nearly complete cranial and fragmentary postcranial human remains of an adult female.	1809	60670	PP	500W50, depth 3′, height–on bottom	Indet.	1	25–29	1	0	0	0	0	0	0	0	1	0	0	D d-e/e, AS 2	Female
25-3-10/60671.0	Partial postcranial human remains of an adult female.	1770	60671	PP	Outline trench, sw of 700 series, depth 4′, height 1′6″	Indet.	1	30–34	1	0	0	0	0	0	0	0	1	0	0	AS 3	F all features
25-3-10/60672.0	Fragmentary postcranial human remains of a subadult that is probably female.	C	60672	PP	Church	Indet.	1	14–16	0	1	0	0	0	0	0	0	0	1	0	PC 14–16 yrs	?Female
25-3-10/60673.0	Nearly complete cranial and fragmentary postcranial human remains of an adult female.	1641	60673	PP	Room 335, trench n, depth 3′, height 2′	Indet.	1	16–20	1	0	0	0	0	0	0	0	1	0	0	D b1	Female
25-3-10/60673.1	Fragmentary cranial and partial postcranial human remains of an adult male.	1641	60673	PP	Room 335, trench n, depth 3′, height 2′	Indet.	1	40–44	1	0	0	0	0	0	1	0	0	0	0	D g, T 8	Male
25-3-10/60673.2	Fragmentary postcranial human remains of an adult male.	1641	60673	PP	Room 335, trench n, depth 3′, height 2′	Indet.	1	Adult	1	0	0	0	0	0	1	0	0	0	0		Male (size pc)
25-3-10/60674.0	Nearly complete cranial human remains of an adult that is probably male.	1813	60674	PP	500W50, depth 6″, height 1′	Indet.	1	35–39	1	0	0	0	0	0	0	1	0	0	0	D g	?Male (pronounce d SR)
25-3-10/60674.1	Fragmentary postcranial human remains of a subadult that is probably female.	1813	60674	PP	500W50, depth 6″, height 1′	Indet.	1	12–18	0	1	0	0	0	0	0	0	0	1	0	PC 12–18 yrs	?Female
25-3-10/60674.2	Partial postcranial human remains of an adult female.	1813	60674	PP	500W50, depth 6″, height 1′	Indet.	1	20–29	1	0	0	0	0	0	0	0	1	0	0	Young adult	Female
25-3-10/60675.0	Nearly complete cranial and postcranial human remains of an adult male.	1819	60675	PP	500W50, depth 2′, height–slightly into clay	Indet.	1	25–29	1	0	0	0	0	0	1	0	0	0	0	D e, PC < 35 yrs	Male

Peabody number	Pathology type	Pathology description	Cranial mod.	Associated funerary objects	Field context	Notes
25-3-10/60663.0	Trauma, infection, arthritis	Healed fracture R ribs and nasals, fusion R innominate with sacrum, spinal osteophytosis, extensive DJD	Not obs.			
25-3-10/60663.1	Arthritis	Ankylosis vt/c6–7 with spinal osteophytosis	No cran.			Field card and provenience problematic
25-3-10/60664.0	Arthritis, other	Slight osteophytosis on vts, vt/l5 partially sacralized	No cran.			
25-3-10/60665.0	Trauma, infection, nutr., other	Trauma–broken nasals, systemic infection (possibly treponemal?)–osteomyelitis–long bone diaphyses L tibia, humerus, R humerus, and frontal with destruction of cortical bone and trabecular bone filling medullary, slight osteoporosis, spina bifida vt/s1, vt/s3–5	Unobs.	Gl. 3 bowl no. 578 (74929), Gl. 1 yellow bowl no. 564 (74928)		
25-3-10/60666.0	Trauma, infection, arthritis, nutr.	Trauma–healed fracture R ribs, Coll's fracture L radius, diploic expansion of skull, slight DJD, slight spinal osteophytosis of vt/l, slight osteoporosis, nonunion of olecranon process R ulna	Unobs.	2 Bisc. B bowls nos. 836 (74941), 857 (74942)		
25-3-10/60667.0	Nutr.	Enamel hypoplasia	Occ. def.			
25-3-10/60668.0	Infection, arthritis, nutr.	Extensive antemortem tooth loss, vt/l–s spinal osteophytosis, slight osteoporosis	Lamb. def.			
25-3-10/60668.1	Trauma, arthritis	Trauma–part of lithic projectile point embedded in spinous process of vt/l4, minor DJD	No cran.			
25-3-10/60668a			Occ. def.			Unusual facial shape orbits, does not match field card description,
25-3-10/60670.0	Arthritis	Slight DJD	Occ. def.			
25-3-10/60671.0	Arthritis	Slight DJD	No cran.	Gl. 1 red bowl no. 1007 (74952)		
25-3-10/60672.0			No cran.			
25-3-10/60673.0			Occ. def.	Bisc. B bowl no. 113 (74896)		
25-3-10/60673.1	Arthritis, nutr.	Moderate DJD, moderate osteoporosis	Unobs.	Bisc. B bowl no. 113 (74896)		
25-3-10/60673.2			No cran.	Bisc. B bowl no. 113 (74896)		
25-3-10/60674.0	Infection	Lesion on frontal	Not obs.	Gl. 5 bowl frag. no. 1223 (74964)		Postcrania recatalogued as 60674-2
25-3-10/60674.1			No cran.	Gl. 5 bowl frag. no. 1223 (74964)		Does not match field card description
25-3-10/60674.2	Other	Spina bifida vt/s3–5	No cran.	Gl. 5 bowl frag. no. 1223 (74964)		
25-3-10/60675.0	Infection, arthritis, nutr., tumor, other	Infection of auricular surface, DJD, mild spinal osteophytosis, enamel hypoplasia, vt/t12 bony growth off superior aspect of body ventral to rib facet, metopic suture fully present, swollen along sagittal suture	No cran. def.			Very small cranial capacity

Peabody number	Object description	FN	Specimen label	Site	Intrasite provenience	Ceramic stage	MNI	Age (yrs.)	Ad	Su	Ch	In	Fe	Unk. age	M	M?	F	F?	Unk. sex	Age criteria	Sex criteria
25-3-10/60676.0	Fragmentary cranial and postcranial human remains of an adult female.	1746	60676	PP	Extension trench, s 700 series, depth 5′, height 3′	Indet.	1	50–59	1	0	0	0	0	0	0	0	1	0	0	D d-e, T 9, SB 5, AS 6	Female
25-3-10/60677.0	Nearly complete cranial and postcranial human remains of an adult male.	1761	60677	PP	Trench n of test 3	Indet.	1	25–29	1	0	0	0	0	0	1	0	0	0	0	D e-f, T 5, AS 2	M all features
25-3-10/60677a	Partial cranial and fragmentary postcranial human remains of an adult that is probably female.	1761	60677a/1761/60677	PP	Trench n of test 3	Indet.	1	30–34	1	0	0	0	0	0	0	0	0	1	0	D f	?Female
25-3-10/60678.0	Partial postcranial human remains of an adult male.	1640	60678	PP	1300W0, trench a, depth 5′1″, height 1′+	Indet.	1	30–34	1	0	0	0	0	0	1	0	0	0	0	SB 3, AS e	M all features
25-3-10/60679.0	Partial cranial and postcranial human remains of an adult female.	1775	60679	PP	Room 37r next to west wall, depth 6′, height 4′+	Probably Gl. 3	1	25–29	1	0	0	0	0	0	0	0	1	0	0	D d-e, AS 1	Female
25-3-10/60679a	Partial cranial human remains of an adult male.	1775	60679-a	PP	Room 37r next to west wall, depth 6′, height 4′+	Probably Gl. 3	1	35–49	1	0	0	0	0	0	1	0	0	0	0	D g-h	Male
25-3-10/60680.0	Nearly complete cranial and partial postcranial human remains of a subadult that is probably male.	C34	60680	PP	Church nave, depth 2′, height 2′	Indet.	1	13.5–14.5	0	1	0	0	0	0	0	1	0	0	0	D b2 15 ± 3 yrs, PC 11.5–13.5 yrs	?Male
25-3-10/60681.0	Nearly complete cranial and partial postcranial human remains of an adult female.	C41	60681	PP	Church nave, depth 2′, height 2′	Indet.	1	35–39	1	0	0	0	0	0	0	0	1	0	0	D g, AS 4	Female (pelvis F, skull more M)
25-3-10/60683.0	Fragmentary postcranial human remains of an adult male.	C	60683	PP	Church	Indet.	1	45–49	1	0	0	0	0	0	1	0	0	0	0	AS 6	Male
25-3-10/60683a	Very fragmentary postcranial human remains of an adult that is probably female.	C	60683a	PP	Church	Indet.	1		1	0	0	0	0	0	0	0	0	1	0		?Female
25-3-10/60685.0	Partial cranial and postcranial human remains of an adult female.	C4	60685	PP	Church depth 2′, height 4′	Indet.	1	25–29	1	0	0	0	0	0	0	0	1	0	0	D b2-c, T 4, AS 2	Female
25-3-10/60685.1	Very fragmentary postcranial human remains of an adult of unknown sex.	C4		PP	Church	Indet.	1	Adult	1	0	0	0	0	0	0	0	0	0	1		Indet.
25-3-10/60685.2	Very fragmentary postcranial human remains of an adult of unknown sex.	C4		PP	Church	Indet.	1		1	0	0	0	0	0	0	0	0	0	1		Indet.
25-3-10/60686.0	Partial cranial and very fragmentary postcranial human remains of an adult male.	1832	60686	PP	Trench n of sp house, depth 5′	Gl. 3 or 4	1	30–39	1	0	0	0	0	0	1	0	0	0	0	D f-g/g	M all features
25-3-10/60687.0	Nearly complete cranial and postcranial human remains of an adult male.	1829	60687	PP	Trench n of sp house, depth 5′6″, height ?	Indet.	1	35–44	1	0	0	0	0	0	1	0	0	0	0	D h/i, T7, SB 4, AS 5	M all features
25-3-10/60688.0	Nearly complete cranial and postcranial human remains of a subadult male.	1820	60688	PP	500W50, depth 2′, height–slightly into clay	Indet.	1	18–19	0	1	0	0	0	0	1	0	0	0	0	D b2, T 1, SB 1, AS < 1	M all features
25-3-10/60689.0	Partial cranial and fragmentary postcranial human remains of an adult male.	1753	60689	PP	Trench e of test xv, depth 6′, height–on bottom	Indet.	1	40–44	1	0	0	0	0	0	0	0	1	0	0	D h, T 8, AS 5	Male
25-3-10/60691.0	Nearly complete cranial and postcranial human remains of an adult male.	C13	60691	PP	Church nave, depth 2′6″, height 1′6″	Indet.	1	45–59	1	0	0	0	0	0	1	0	0	0	0	D h+/i+, T 9, AS 7	Male, very robust
25-3-10/60692.0	Partial cranial and fragmentary postcranial human remains of a subadult of unknown sex.	1680	60692	PP	1250W50, s trench, depth 3′6″, height 5′6″	Indet.	1	16–20	0	1	0	0	0	0	0	0	0	0	1	D d 15 ± 3 yrs, PC epiphyseal lines visible	Indet.

Peabody number	Pathology type	Pathology description	Cranial mod.	Associated funerary objects	Field context	Notes
25-3-10/60676.0	Arthritis	Spinal osteophytosis	Unobs.	Gl. 1 red bowl no. 813 (74944)		Crania and postcrania may be separate individuals
25-3-10/60677.0	Arthritis, other	Vt/l5 spondylytis deformans, vt/s1 spina bifida, large occipital bun	Occ. def.		Burial at length	
25-3-10/60677a	Arthritis, nutr.	Arthritis–ribs 7–10, TMJ DJD on right side, severe enamel hypoplasia	Lamb. def.			
25-3-10/60678.0	Arthritis	DJD knee and elbow	No cran.	Frag. of black pot no. 139 (74895), Gl. 1 red bowl no. 138 (74894)		
25-3-10/60679.0	Infection, arthritis	Dental abscesses, slight appendicular DJD elbow, Schmorl's nodes–especially lumbar vt	Not obs.	Medicine cache of 12 pieces no. 640 (76185)		
25-3-10/60679a	Infection, nutr., tumor	Severe antemortem molar loss, increased porosity parietals, button osteoma–frontal	Not obs.	Medicine cache of 12 pieces no. 640 (76185)		
25-3-10/60680.0	Other	Early fusion L coronal suture with craniofacial distortion	Not obs.		Burial at length	
25-3-10/60681.0	Trauma, infection, arthritis, nutr.	Healed trauma to frontal–small depression fracture, antemortem tooth loss upper L molars, caries, severe appendicular DJD, osteoporosis	Not obs.		Burial at length	
25-3-10/60683.0	Arthritis	Spondylytis deformans on vt/l, DJD	No cran.			
25-3-10/60683a			No cran.			
25-3-10/60685.0	Infection, nutr.	Severe dental abscess lower L M3, enamel hypoplasia	No cran. def.			
25-3-10/60685.1			No cran.		Burial at length	
25-3-10/60685.2	Infection	Lytic lesions on 3 vt/t	No cran.			
25-3-10/60686.0	Infection	Healed periostitis on R tibia	Occ. def.	3 projectile points no. 1307 (75911)		
25-3-10/60687.0	Infection, arthritis, tumor, other	Dental caries right mandibular P1–M1, severe appendicular DJD especially elbow and knee, Schmorl's nodes, vt/l marked spinal osteophytosis, humeri significant enthesophytes near distal end, spina bifida vt/s4–5	Not obs.			
25-3-10/60688.0	Infection, nutr., other	Lytic lesion on 2 vt/t and vt/l3, enamel hypoplasia, vt/s4–5 neural arch unfused	Not obs.			
25-3-10/60689.0	Arthritis, other	Spondylytis deformans on vt/l, upper L P1 is double rooted	Not obs.	Gl. 1 red bowl no. 886 (74946)		
25-3-10/60691.0	Trauma, infection, arthritis, other	Trauma to spinous processes of vt/t1–2, healed fracture L clavicle, massive trauma to left side face, small lesion prox. L tibia, extensive antemortem tooth loss, moderate DJD, bony build up along dist. shaft R humerus, spinal osteophytosis, spondylytis deformans, spondylolysis of vt/l5	Not obs.	5 glass beads no. 1175 (76369)	Burial at length	
25-3-10/60692.0	Infection, nutr.	Impacted m3s, enamel hypoplasia	Not obs.			

Peabody number	Object description	FN	Specimen label	Site	Intrasite provenience	Ceramic stage	MNI	Age (yrs.)	Ad	Su	Ch	In	Fe	Unk. age	M	M?	F	F?	Unk. sex	Age criteria	Sex criteria
25-3-10/60693.0	Fragmentary postcranial human remains of an adult male.	1097	60693	PP	1500W50, depth 18″, height–on bottom	Probably Late Gl. 6	1	35–39	1	0	0	0	0	0	1	0	0	0	0	T 7, AS 4	Male
25-3-10/60694.0	Partial cranial and fragmentary postcranial human remains of an adult male.	1673	60694/1673	PP	1250W50, s trench, depth 6′, height 2′6″	Indet.	1	35–39	1	0	0	0	0	0	1	0	0	0	0	D g	Male
25-3-10/60695.0	Nearly complete cranial and postcranial human remains of an adult male.	1678	60695	PP	1300W0, tr. a, east extension, depth 4′6″, height–on bottom	Probably Gl. 1 or 2	1	40–49	1	0	0	0	0	0	1	0	0	0	0	D h+/i+. T 8, AS 5	Male
25-3-10/60696.0	Nearly complete cranial and postcranial human remains of an adult female.	C5	60696	PP	Church depth 2′, height 3′10″	Indet.	1	50–59+	1	0	0	0	0	0	0	0	1	0	0	D h very old, 60+ yrs, T 10, AS 7	Female
25-3-10/60697.0	Partial cranial human remains of an adult female.	1620	60697	PP	1250W100, depth 3′, height 2′	Indet.	1	35–39	1	0	0	0	0	0	0	0	1	0	0	D g	F all features
25-3-10/60698.0	Nearly complete cranial and postcranial human remains of an adult male.	1703	60698	PP	Sw passage, sub, depth 6′6″, height–on bottom	Indet.	1	40–49	1	0	0	0	0	0	1	0	0	0	0	D g/f, AS 6	Male
25-3-10/60699.0	Partial cranial and fragmentary postcranial human remains of an adult male.	1707	60699	PP	Sw passage, sub, depth 6′8″, height–on bottom	Indet.	1	55+	1	0	0	0	0	0	1	0	0	0	0	D 55+ yrs	Male
25-3-10/60700.0	Partial cranial and fragmentary postcranial human remains of an adult male.	1651	60700	PP	1300W50, trench b, depth 4′, height 5′	Indet.	1	18–22	1	0	0	0	0	0	1	0	0	0	0	PC 18–22 yrs	Male
25-3-10/60701.0	Nearly complete cranial human remains of an adult male.	1647	60701	PP	1300W50, trench b, depth 6′, height 3′	Probably Gl. 1 or 2	1	35–49	1	0	0	0	0	0	1	0	0	0	0	D g-h	Male
25-3-10/60702.0	Fragmentary cranial human remains of an adult of unknown sex.	1644	60702	PP	1300W50, trench b, depth 4′, height 5′	Probably Gl. 5	1	45–54	1	0	0	0	0	0	0	0	0	0	1	D h+/i	Indet.
25-3-10/60703.0	Nearly complete cranial and partial postcranial human remains of an adult female.	1717	60703	PP	Kiva 12, trench e, depth 7′6″, height ?	Indet.	1	30–39	1	0	0	0	0	0	0	0	1	0	0	D f-g/g	Female
25-3-10/60703a	Fragmentary cranial human remains of an adult that is probably female.	1692	60703	PP	1250W50, s trench, depth 7′, height about 2′	Indet.	1	20–29	1	0	0	0	0	0	0	0	0	1	0		?Female
25-3-10/60704.0	Nearly complete cranial and postcranial human remains of an adult female.	C7	60704	PP	Church, depth 2′4″, height 3′6″	Indet.	1	20–24	1	0	0	0	0	0	0	0	1	0	0	D e, AS 1, PC < 24 yrs based on epiphyses	Female
25-3-10/60704a	Partial cranial human remains of an adult female.	1769	60704a	PP	Kiva 12, trench E, depth 3′6″	Indet.	1	35–49	1	0	0	0	0	0	0	0	1	0	0	D g-h	Female
25-3-10/60706.0	Nearly complete cranial and partial postcranial human remains of an adult female.	C8	60706	PP	Church, depth 2′6″, height 3′4″	Indet.	1	50–59	1	0	0	0	0	0	0	0	1	0	0	D old, AS 7	Female
25-3-10/60708.0	Partial cranial and nearly complete postcranial human remains of a subadult female.	C23	60708	PP	Church nave, depth 3′, height 1′	Indet.	1	16–18	0	1	0	0	0	0	0	0	1	0	0	D b2, PC 16–19 yrs	Female
25-3-10/60709.0	Very fragmentary cranial human remains of an adult of unknown sex.	1177	60709/60679	PP	S end of plaza trench, e of rm 55, depth 2′6″, height 3′	Probably Gl. 2 or 3	1	35–54	1	0	0	0	0	0	0	0	0	0	1	D g-i+	Indet.
25-3-10/60710.0	Partial cranial and fragmentary postcranial human remains of an adult male.	1672	60710	PP	1300W50, trench b, depth 5′6″, height 4′6″	Indet.	1	35–39	1	0	0	0	0	0	1	0	0	0	0	D h/i+, T 7, AS 4	Male (pelvis M, cranium more F)

Peabody number	Pathology type	Pathology description	Cranial mod.	Associated funerary objects	Field context	Notes
25-3-10/60693.0	Arthritis, nutr.	Spondylytis deformans on vt/l, slight osteoporosis	No cran.		Burial at length	Field number problematic
25-3-10/60694.0	Infection, arthritis, nutr.	Dental abscess lower L M1, spondylytis deformans vt/l, ankylosis vt/c2–4, spinal osteophytosis, enamel hypoplasia	Lamb. def.	Gl. 1 yellow olla bottom, wrapped in mat or hide, no numbers		
25-3-10/60695.0	Trauma, infection, arthritis, tumor	Fracture prox. R clavicle with subsequent infection and fusion to manubrium, vt–spondylytis deformans, marked spinal osteophytosis, growth on R dist. femur	Not obs.			
25-3-10/60696.0	Arthritis, nutr.	Vt–spondylytis deformans, marked DJD, marked osteoporosis	Occ./ Lamb. def.		Burial at length	
25-3-10/60697.0			Occ. def.	Gl. 1 red bowl no. 107 (74880)		
25-3-10/60698.0	Nutr., other	Marked osteoporosis, spina bifida vt/s4–5	Lamb. def.	B&W bowl no. 546 (74927)		Very unusual muscle markings on L prox. humerus, bowed tibia
25-3-10/60699.0	Trauma, infection, arthritis	Trauma to frontal and parietals–several lesions with bony growth, dental abscesses and major antemortem tooth loss, marked DJD, spinal osteophytosis, vt/c2–3 fused, 2 v/t fused on right side with tumerous development, 4 vt/l exhibit spondylytis deformans	Lamb. def.	Bone awl no. 602 (75482), tubular steatite pipe no. 601 (76191)		
25-3-10/60700.0	Infection, arthritis, tumor	Frontal lesion, DJD at left TMJ, button osteoma on frontal	Lamb. def.	Fragments of hide wrapping, no number		Crania and postcrania may be separate individuals
25-3-10/60701.0			Not obs.	Bisc. B bowl no. 137 (74900)		
25-3-10/60702.0			Unobs.	Small whole pot no. 134 (74897)		
25-3-10/60703.0	Trauma, infection, nutr., other	L femur healed severe midshaft fracture with displacement of 7.5 cm, limb shortened 12.5+ cm, infection L upper C resulting in perforation of maxilla, palate also infected, caries upper R M2–3, porotic hyperostosis–parietals, sagittal suture obliterated, spina bifida–mild vt/s4	Not obs.			
25-3-10/60703a			Unobs.			
25-3-10/60704.0	Trauma	Healed fracture R clavicle	No cran. def.		Burial at length	
25-3-10/60704a	Infection	Severe dental abscesses molars	Lamb. def.			
25-3-10/60706.0	Infection, arthritis	Parietal ectocranial porosis?, vt–spondylytis deformans, moderate appendicular DJD	Not obs.			
25-3-10/60708.0	Infection, nutr.	Abscess lower L M2, major infection–R tibia, also ulnae, femora, fibulae, mild infection–R radius and L humerus, enamel hypoplasia	Not obs.	Bronze medal beads and copper cross no. 1195 (76400), frag. of silver and orange brocade no. 1193 (76360), frags. of cloth nos. 1194 (76359), 1192 (76358)	Burial at length	Field number problematic
25-3-10/60709.0	Infection,	Dental abscess	Unobs.			
25-3-10/60710.0	Arthritis, other	DJD, spina bifida vt/s4–5 unfused neural arch	Occ. def.	Gl. 3 olla bowl no. 270 (74912)		

Peabody number	Object description	FN	Specimen label	Site	Intrasite provenience	Ceramic stage	MNI	Age (yrs.)	Ad	Su	Ch	In	Fe	Unk. age	M	M?	F	F?	Unk. sex	Age criteria	Sex criteria
25-3-10/60711.0	Nearly complete cranial and postcranial human remains of an adult male.	C20	60711	PP	Church, depth 2′5″, height 1′6″	Indet.	1	25–29	1	0	0	0	0	0	1	0	0	0	0	D d, AS 2	M all features
25-3-10/60712.0	Nearly complete cranial human remains of a child of unknown sex. Individual is of Caucasian ancestry.	C27	60712	PP	Church nave, depth 2′, height 1′	Indet.	1	6–8	0	0	1	0	0	0	0	0	0	0	1	D 8 ± 2 yrs	Indet.
25-3-10/60713.0	Fragmentary cranial and postcranial human remains of an adult female.	1649	70713	PP	1300W0, trench b, depth 3′4″, height 3′	Probably Gl. 3	1	40–44	1	0	0	0	0	0	0	0	1	0	0	D g-h, AS 5	Female
25-3-10/60713.1	Fragmentary cranial human remains of a subadult of unknown sex.	1649		PP	1300W0, trench b, depth 3′4″, height 3′	Probably Gl. 3	1	12–15	0	1	0	0	0	0	0	0	0	0	1	D 15 ± 3 yrs	Indet.
25-3-10/60714.0	Nearly complete cranial and postcranial human remains of an adult male.	C17	60714	PP	Church nave, depth 2′6″, height 1′6″	Indet.	1	60+	1	0	0	0	0	0	1	0	0	0	0	D 60+ yrs	M all features
25-3-10/60715.0	Nearly complete cranial and partial postcranial human remains of an adult female.	1833	60715	PP	700E150, trench e of s house, depth 3′, height–on clay	Indet.	1	25–34	1	0	0	0	0	0	0	0	1	0	0	D d-e/d, AS 3	Female (pelvis F, cranium more M)
25-3-10/60716.0	Nearly complete cranial and postcranial human remains of a subadult male.	1713	60716	PP	Trench e of kiva 12, depth 7′, height–on bottom	Indet.	1	17–19	0	1	0	0	0	0	1	0	0	0	0	D b2, AS <1	Male (pelvis M, cranium more F)
25-3-10/60717.0	Nearly complete cranial and postcranial human remains of an adult male.	C19	60717	PP	Church nave, depth 2′6″, height 1′6″	Indet.	1	27–34	1	0	0	0	0	0	1	0	0	0	0	D e, T 5, AS 3	M all features
25-3-10/60718.0	Nearly complete cranial human remains of an adult male.	C14	60718	PP	Church nave, depth 2′, height 2′	Indet.	1	30–39	1	0	0	0	0	0	1	0	0	0	0	D f-g/g	Male
25-3-10/60719.0	Nearly complete cranial human remains of a subadult that is probably female.	C18	60719	PP	Church nave, depth 2′, height 2′	Indet.	1	10–12	0	1	0	0	0	0	0	0	0	1	0	D b2, spheno-occipital not fused	?Female
25-3-10/60720.0	Nearly complete cranial human remains of an adult that is probably male.	C41	60720, 60681	PP	Church nave, depth 2′, height 2′	Indet.	1	50–59+	1	0	0	0	0	0	0	1	0	0	0	D h/i	?Male
25-3-10/60721.0	Partial cranial human remains of an adult male.	C38	60721	PP	Church nave, depth 2′6″, height 1′6″	Indet.	1	25–29	1	0	0	0	0	0	1	0	0	0	0	D e	Male
25-3-10/60722.0	Nearly complete cranial and postcranial human remains of an adult female.	1834	60722	PP	Trench n of sp house, depth 6′	Indet.	1	25–44	1	0	0	0	0	0	0	0	1	0	0	D d, T 8, SB 5, AS 5	F all features
25-3-10/60723.0	Nearly complete cranial human remains of an adult female.	C16	60723	PP	Church nave, depth 2′4″, height 1′6″	Indet.	1	45–54+	1	0	0	0	0	0	0	0	1	0	0	D elderly	Female
25-3-10/60724.0	Nearly complete cranial human remains of an adult male.	C30	60724	PP	Church nave, depth 3′, height 1′	Indet.	1	25–29	1	0	0	0	0	0	1	0	0	0	0	D e	Male
25-3-10/60725.0	Fragmentary postcranial human remains of an adult male.	C6	60725	PP	Church depth 2′6″, height 3′4″	Indet.	1	35–39	1	0	0	0	0	0	1	0	0	0	0	AS 4	Male
25-3-10/60726.0	Nearly complete cranial human remains of a child of unknown sex.	1633	60726	PP	1300W0, trench b, depth 4′, height ?	Indet.	1	7–9	0	0	1	0	0	0	0	0	0	0	1	D 9–10 ± 2.5 yrs	Indet.
25-3-10/60727.0	Nearly complete cranial and partial postcranial human remains of an adult female.	1517	60727	PP	Room c, depth 5′6″, height ?	Gl. 3, 4, or 5	1	50–59	1	0	0	0	0	0	0	0	1	0	0	D h+/i+, AS 7	Female
25-3-10/60728.0	Partial postcranial human remains of an adult that is probably female.	C25	60728	PP	Church nave, depth 2′8″, height 1′2″	Indet.	1	50–59	1	0	0	0	0	0	0	0	0	1	0	AS 7	?Female
25-3-10/60729.0	Fragmentary postcranial human remains of an adult male.	1800	60729	PP	400E100, depth 1′, height 1′	Indet.	1	50–59	1	0	0	0	0	0	1	0	0	0	0	T 10, AS 7	Male

Peabody number	Pathology type	Pathology description	Cranial mod.	Associated funerary objects	Field context	Notes
25-3-10/60711.0	Trauma, arthritis, other	Trauma–scalped–marks on frontal, severe healed trauma to midshaft R humerus and radius, fusion of sacrum to R innom. with arthritis, vt/s4–6 neural arches not fused, spondylolysis vt/l5, R femur shows evidence of stab wound	Not obs.		Burial at length	
25-3-10/60712.0			Not obs.			
25-3-10/60713.0	Infection	Dental abscess–lower L M2	Unobs.			
25-3-10/60713.1			Unobs.			Does not match field card description
25-3-10/60714.0	Infection, arthritis, nutr.	Edentulous, debilitating DJD all limb joints especially knee, ankle, elbow, vt/l spondylytis deformans, severe osteoporosis	Occ. def.		Burial at length	
25-3-10/60715.0	Arthritis, nutr.	Slight DJD, enamel hypoplasia	Occ. def.	Cloth fragments no. 1284 (76357), hide garment, no numbers	Grave lined with bark, painted green, cloth garment, hide garment	Copper staining–8 limb bones, marked bowing fibula,
25-3-10/60716.0			Occ. def.	Gl. 1 yellow bowl no. 694 (74933)		
25-3-10/60717.0	Trauma, infection, arthritis, nutr., other	Trauma death and scalping, dental abscesses slight DJD, increased porosity parietals along sagittal suture, enamel hypoplasia, spina bifida–vt/s4 incomplete fusion, vt/s5 not fused	Occ. def.		Burial at length	
25-3-10/60718.0	Trauma, infection, nutr., other	Trauma–along left temporal line, porotic hyperostosis, molarized upper R P1	Not obs.			
25-3-10/60719.0	Trauma, nutr.	Trauma–partially healed fracture nasal cavity, cribra orbitalia	Cran. def.			
25-3-10/60720.0	Infection, other	Severe dental abscess–lower L M2, malocclusion in left side mandible due to overcrowding	Not obs.			Reassociated human remains from 60681.1 (MEM)
25-3-10/60721.0			Not obs.			
25-3-10/60722.0	Arthritis, nutr.	Vt/l Schmorl's nodes, enamel hypoplasia, developing osteoporosis	Lamb. def.		Mat wrapped	Unusually deep trochlea dist. humerus
25-3-10/60723.0			Not obs.			Edentulous antemortem
25-3-10/60724.0			Not obs.			
25-3-10/60725.0	Arthritis, other	Slight DJD, vt/s4–5 not fused	No cran.			
25-3-10/60726.0	Nutr.	Cribra orbitalia	Not obs.	Gl. 3 olla bowl no. 111 (74889)		
25-3-10/60727.0	Infection, arthritis, nutr., tumor, other	Extensive antemortem tooth loss, extensive DJD appendicular and axial, Schmorl's nodes, developing osteoporosis, button osteomas on vault, vt/s4–5 not fused	Not obs.			Field card number and description problematic
25-3-10/60728.0	Trauma, arthritis	Trauma induced fusion R Mc3 with capitate and hamate, DJD elbow and prox. ribs	No cran.			
25-3-10/60729.0	Trauma, infection, arthritis	L femur healed greenstick fracture, L humerus periosteal infection, spinal osteophytosis, DJD appendicular	No cran.			

Peabody number	Object description	FN	Specimen label	Site	Intrasite provenience	Ceramic stage	MNI	Age (yrs.)	Ad	Su	Ch	In	Fe	Unk. age	M	M?	F	F?	Unk. sex	Age criteria	Sex criteria
25-3-10/60730.0	Nearly complete cranial human remains of a child of unknown sex.	C	60730	PP	Church?	Indet.	1	6–8	0	0	1	0	0	0	0	0	0	0	1	D 8 ± 2 yrs	Indet.
25-3-10/60731.0	Nearly complete cranial and fragmentary postcranial human remains of an adult female.	1688	60731	PP	1250W50, s trench, depth 3′, height 8′	Indet.	1	50–59	1	0	0	0	0	0	0	0	1	0	0	D g/i+, AS 7	Female
25-3-10/60732.0	Fragmentary postcranial human remains of a subadult female.	1676	60732	PP	1350W50, trench b, depth 6′6″, height 3′6″	Indet.	1	18–19	0	1	0	0	0	0	0	0	1	0	0	T 1, SB 1, iliac crest not fused	Female
25-3-10/60733.0	Nearly complete cranial and partial postcranial human remains of an adult male.	1671	60733	PP	1300W0, trench a, depth 6′, height–on bottom	Indet.	1	45–49	1	0	0	0	0	0	1	0	0	0	0	T 9, AS 6	M all features
25-3-10/60735.0	Nearly complete cranial and postcranial human remains of a subadult male. Individual is of Caucasian ancestry.	C42	60735	PP	Church nave, depth 3′, height 1′	Indet.	1	14–16	0	1	0	0	0	0	1	0	0	0	0	D b1, 15 ± 3 yrs, T 1, AS < 1	Male
25-3-10/60735a	Very fragmentary postcranial human remains of a subadult of unknown sex.	C42	60735a	PP	Church nave, depth 3′, height 1′	Indet.	1		0	1	0	0	0	0	0	0	0	0	1		
25-3-10/60735b	Very fragmentary postcranial human remains of an adult male.	C42	60735b	PP	Church nave, depth 3′, height 1′	Indet.	1	50–59	1	0	0	0	0	0	1	0	0	0	0	T 10, AS 7	Male
25-3-10/60736.0	Nearly complete cranial and postcranial human remains of an adult female.	1827	60736	PP	500W50, depth 2′6″, height–in clay 1′	Gl. 5	1	50–59	1	0	0	0	0	0	0	0	1	0	0	D h, AS 7	Female
25-3-10/60737.0	Nearly complete cranial and partial postcranial human remains of an adult male.	C28	60737	PP	Church nave, depth 2′2″, height 1′10″	Indet.	1	25–29	1	0	0	0	0	0	1	0	0	0	0	D e, T 5, AS 2	Male
25-3-10/60738.0	Nearly complete cranial human remains of an adult female.	1631	60738	PP	130W0, room 40, depth 6′5″, height–in clay	Probably Gl. 3	1	40–54	1	0	0	0	0	0	0	0	1	0	0	D h	Female
25-3-10/60739.0	Partial postcranial human remains of an adult female.	1824	60739	PP	Sw passage, 2′ from street surface, height 4′	Indet.	1	50–59	1	0	0	0	0	0	0	0	1	0	0	AS 7	F all features
25-3-10/60740.0	Partial postcranial human remains of an adult female.	1826	60740	PP	37s, depth 8′, height 1′ from top of penasco	Probably B&W	1	>35	1	0	0	0	0	0	0	0	1	0	0		Female
25-3-10/60741.0	Nearly complete cranial and postcranial human remains of an adult male.	1813	60741	PP	500W50, depth 6″, height 1′	Indet.	1	30–39	1	0	0	0	0	0	1	0	0	0	0	D d, AS 4	M all features
25-3-10/60742.0	Fragmentary postcranial human remains of a subadult male.	1801	60742	PP	400E100, depth 1′, height–on venarco ?	Indet.	1	14–17	0	1	0	0	0	0	1	0	0	0	0	PC 14–17 yrs	Male
25-3-10/60743.0	Nearly complete cranial and postcranial human remains of an adult male.	1825	60743	PP	500W50, depth 2′6″, height–in clay	Probably Gl. 5	1	40–50+	1	0	0	0	0	0	1	0	0	0	0	D g, T 10, SB 6, AS 5–6	M all features
25-3-10/60744.0	Nearly complete cranial and partial postcranial human remains of an adult female.	1805	60744	PP	500W50, depth 18″, height–on bottom	Indet.	1	45–50+	1	0	0	0	0	0	0	0	1	0	0	D h+/i+, T 10, SB 6, AS 6	F all features
25-3-10/60745.0	Fragmentary postcranial human remains of an adult male.	1621	60745	PP	1350W0, trench a, depth 5′, height–sunk in clay	Probably B&W	1		1	0	0	0	0	0	1	0	0	0	0		Male
25-3-10/60746.0	Nearly complete cranial and partial postcranial human remains of an adult female.	1612	60746, 60755	PP	1300W0, depth 4′, height–on clay	Indet.	1	45–50+	1	0	0	0	0	0	0	0	1	0	0	D g/i, T 10, SB 5, AS 6	F all features
25-3-10/60747.0	Partial cranial human remains of an adult male.	C33	60747	PP	Church nave, depth 2′6″, height 1′6″	Indet.	1	35–39	1	0	0	0	0	0	1	0	0	0	0	D g	Male
25-3-10/60748.0	Nearly complete cranial human remains of an adult male.	C	60748	PP	Church	Indet.	1	40–49	1	0	0	0	0	0	1	0	0	0	0	D h	Male
25-3-10/60749.0	Nearly complete cranial human remains of an adult male.	C15	60749	PP	Church nave, depth 2′, height 2′	Indet.	1	20–24	1	0	0	0	0	0	1	0	0	0	0	D d	Male

Peabody number	Pathology type	Pathology description	Cranial mod.	Associated funerary objects	Field context	Notes
25-3-10/60730.0	Nutr.	Enamel hypoplasia	Not obs.			
25-3-10/60731.0	Trauma, infection, arthritis, nutr.	Perimortem trauma to right side frontal, severe dental abscesses, impacted upper M3, severe DJD and osteoporosis	Occ. def.	Gl. 3 bowl no. 327 (74919)		Robust female, density of crania and postcrania very different but age and preservation similar
25-3-10/60732.0			No cran.	Gl. 1 red bowl no. 269 (74914)		Copper staining L tibia
25-3-10/60733.0	Arthritis, nutr.	Vt/l spinal osteophytosis, enamel hypoplasia	Not obs.	Gl. 4 bowls no. 278 (74909), no. 264 (74911), ceremonial jar no. 223 (74910), bead, no number		
25-3-10/60735.0	Arthritis, other	Vt/t and vt/l Schmorl's nodes, spondylolysis of vt/l5	Not obs.			Reassociated human remains from 60734 (MEM)
25-3-10/60735a			No cran.	Buttons(?), projectile point in chest, no numbers		Field card number and provenience problematic
25-3-10/60735b			No cran.	Buttons(?), projectile point in chest, no numbers		Field card number and provenience problematic
25-3-10/60736.0	Infection, arthritis	Lower molars lost antemortem, vt/t Schmorl's nodes	Occ. def.			
25-3-10/60737.0	Infection	Healed ectocranial porosis–frontal and parietal	No cran. def.			Small teeth
25-3-10/60738.0	Infection, nutr.	Dental abscess L maxilla, enamel hypoplasia	Occ. def.			
25-3-10/60739.0	Arthritis, nutr.	DJD, vt/l spinal osteophytosis, osteoporosis	No cran.	Early Gl. 3 sherds, no number		
25-3-10/60740.0	Nutr.	Mild osteoporosis	Not obs.			Field card description does not match human remains
25-3-10/60741.0	Arthritis, nutr.	Vt/c1 fused to cranium with arthritis, enamel hypoplasia	Not obs.	Gl. 5 bowl frag. no. 1223 (74964)		Does not match field card description
25-3-10/60742.0			No cran.		Burial at length	
25-3-10/60743.0	Trauma, infection, arthritis, other	Ankylosis 2 vt/t ant. and post., dental abscesses, 5 healed lesions on parietal, 1 on frontal, DJD appendicular with severe DJD dist. R radius and ulna–very restricted movement ,spina bifida vt/s4–5 neural arch not fused	Not obs.			Bit of flint between two fused dorsal vertebra
25-3-10/60744.0	Arthritis, nutr.	Appendicular DJD, osteoporosis, enamel hypoplasia	Slight occ. def.			
25-3-10/60745.0			No cran.	B&W sherds, no number	Burial at length	Drawn at length but field card says flexed, field number problematic
25-3-10/60746.0	Infection, arthritis	Several molar abscesses affecting lower L M1, upper R P2–M3, all upper L molars lost antemortem, vt–Schmorl's nodes	Occ. def.	Pot, no number		Reassociated human remains from 60755 (MEM 5/98)
25-3-10/60747.0	Nutr.	Enamel hypoplasia	Not obs.			
25-3-10/60748.0	Trauma, nutr.	Trauma–scalped, 3 healed depression fractures on frontal, enamel hypoplasia	Not obs.			
25-3-10/60749.0	Trauma, infection, nutr.	Healed depression fracture L parietal, healed fracture nasals, healed ectocranial porosis frontal and R parietal, enamel hypoplasia	Not obs.			

Peabody number	Object description	FN	Specimen label	Site	Intrasite provenience	Ceramic stage	MNI	Age (yrs.)	Ad	Su	Ch	In	Fe	Unk. age	M	M?	F	F?	Unk. sex	Age criteria	Sex criteria
25-3-10/60750.0	Nearly complete cranial and fragmentary postcranial human remains of an adult female.	C	60750, 60959	PP	Church	Indet.	1	20–24	1	0	0	0	0	0	0	0	1	0	0	D d	Female
25-3-10/60750b	Very fragmentary cranial human remains of an adult that is probably male.	C	60750b	PP	Church	Indet.	1	35–39	1	0	0	0	0	0	0	1	0	0	0	D g	?Male
25-3-10/60751.0	Nearly complete cranial human remains of an adult male.	C	60751, 60771a	PP	Church	Indet.	1	25–29	1	0	0	0	0	0	1	0	0	0	0	D e	Male
25-3-10/60752.0	Nearly complete cranial and fragmentary postcranial human remains of an adult female.	1739	60752	PP	950E150, e side, depth 5 6″, height 6″	Probably early Gl. 3	1	40–55+	1	0	0	0	0	0	0	0	1	0	0	D h/i+	F all features
25-3-10/60752a	Fragmentary postcranial human remains of a subadult that is probably female.	1739	60752a	PP	950E150, e side, depth 5′6″, height 6″	Probably early Gl. 3	1	16–18	0	1	0	0	0	0	0	0	0	1	0	PC 16–19 yrs	?Female
25-3-10/60753.0	Nearly complete cranial human remains of an adult male.	C30	60753	PP	Church nave	Indet.	1	30–34	1	0	0	0	0	0	1	0	0	0	0	D f-g/g	Male
25-3-10/60754.0	Partial postcranial human remains of an adult male.	C36	60754	PP	Church nave, depth 2′7″, height 1′5″	Indet.	1	40–50+	1	0	0	0	0	0	1	0	0	0	0	T 10, SB 6, AS 5	M all features
25-3-10/60754a	Very fragmentary postcranial human remains of an adult of unknown sex.	C36	60754a	PP	Church nave, depth 2′7″, height 1′5″	Indet.	1		1	0	0	0	0	0	0	0	0	0	1		Indet.
25-3-10/60756.0	Fragmentary postcranial human remains of an adult that is probably female.	1755	60756, 60755	PP	950E150, depth 6′, height ?	Indet.	1	>35	1	0	0	0	0	0	0	0	0	1	0	Postcrania	?Female
25-3-10/60757.0	Nearly complete cranial human remains of an adult male.	C49	60757	PP	Church nave, depth 3′, height 1′	Indet.	1	18–22	1	0	0	0	0	0	1	0	0	0	0	D c	Male
25-3-10/60758.0	Nearly complete cranial human remains of a child of unknown sex.	C52	60758/60750a	PP	Church nave, depth 2′6″, height 1′6″	Indet.	1	7–9	0	0	1	0	0	0	0	0	0	0	1	D 9–10 ± 2 yrs	Indet.
25-3-10/60759.0	Nearly complete cranial human remains of an adult female.	C51	60759	PP	Church nave, depth 4′, height–penasco	Indet.	1	35–49	1	0	0	0	0	0	0	0	1	0	0	D g-h	Female
25-3-10/60760.0	Partial cranial human remains of a subadult female.	C	60760	PP	Church	Indet.	1	10–12	0	1	0	0	0	0	0	0	1	0	0	D 12 ± 2.5 yrs	Female
25-3-10/60761.0	Partial cranial and fragmentary postcranial human remains of an adult male.	C47	60761, 60963, 60959	PP	Church nave, depth 2′6″	Indet.	1	50–59	1	0	0	0	0	0	1	0	0	0	0	D h, old	Male
25-3-10/60762.0	Fragmentary cranial human remains of an adult that is probably female.	C44	60762	PP	Church nave, depth 3′, height 1′	Indet.	1	30–34	1	0	0	0	0	0	0	0	0	1	0		?Female
25-3-10/60763.0	Nearly complete cranial human remains of an adult male.	1684	60763, 1684	PP	1300W50, trench b, depth 5′6″, height 6?	Indet.	1	35–49	1	0	0	0	0	0	1	0	0	0	0	D g-h	Male
25-3-10/60764.0	Nearly complete cranial human remains of an adult male.	C	60764	PP	Church	Indet.	1	18–24	1	0	0	0	0	0	1	0	0	0	0	D c-d	Male
25-3-10/60765.0	Nearly complete cranial human remains of an adult female.	C	60765	PP	Church	Indet.	1	45–55+	1	0	0	0	0	0	0	0	1	0	0	D h+/i+	Female
25-3-10/60766.0	Partial cranial human remains of an adult male.	C53	60766	PP	Church nave, depth 2′6″, height 1′6″	Indet.	1	25–34	1	0	0	0	0	0	1	0	0	0	0	D e uneven	Male
25-3-10/60767.0	Nearly complete cranial human remains of an adult female.	C	60767, 60771a	PP	Church	Indet.	1	35–49	1	0	0	0	0	0	0	0	1	0	0	D h/g	Female
25-3-10/60768.0	Nearly complete cranial human remains of an adult male.	C32	60768	PP	Church nave, depth 3′, height 1′	Indet.	1	30–39	1	0	0	0	0	0	1	0	0	0	0	D f-g/g	Male
25-3-10/60769.0	Nearly complete cranial human remains of an adult male. Individual is of Caucasian ancestry.	C	60769	PP	Church	Indet.	1	20–24	1	0	0	0	0	0	1	0	0	0	0	D d	Male
25-3-10/60770.0	Nearly complete cranial human remains of an adult male.	C31	60770	PP	Church nave, depth 3′, height 1′	Indet.	1	35–39	1	0	0	0	0	0	1	0	0	0	0	D g	Male
25-3-10/60771.0	Fragmentary cranial human remains of an adult female.	C	60771c	PP	Church	Indet.	1	40–45	1	0	0	0	0	0	0	0	1	0	0	D h	Female
25-3-10/60771.1	Fragmentary cranial human remains of a child of unknown sex.	C	60771e	PP	Church	Indet.	1	3–4	0	0	1	0	0	0	0	0	0	0	1	4 ± 1 yrs	Indet.

Peabody number	Pathology type	Pathology description	Cranial mod.	Associated funerary objects	Field context	Notes
25-3-10/60750.0	Trauma	Trauma death from blow to frontal	Not obs.			Examined by Owsley team 1/98, reassociated
25-3-10/60750b	Nutr.	Enamel hypoplasia	Unobs.			
25-3-10/60751.0	Trauma, other	Trauma death–scalped, hit with blunt instrument on back of skull, dental pathology–supernumerary upper incisor	No cran. def.			Examined by Owsley team 1/98, reassociated mandible.
25-3-10/60752.0	Trauma, infection, arthritis	Severe healed fracture L dist. radius–Coll's, periodontal disease, DJD	Not obs.		Mat wrapped, with burial no. 1758 (orig no.) infant burial	
25-3-10/60752a	Tumor	Exotoses R ulna	No cran.		With adult burial no.	Field card description does not match human remains
25-3-10/60753.0			Occ. def.			
25-3-10/60754.0	Arthritis	DJD, spinal osteophytosis	No cran.	Wooden cups, no number		
25-3-10/60754a			No cran.	Wooden cups, no number		Field number and provenience doubtful
25-3-10/60756.0	Trauma, arthritis, nutr.	Healed fracture L ulna, DJD, osteoporosis	No cran.	Gl. 2 bowl no. 922 (74948)		
25-3-10/60757.0	Trauma	Trauma death–perimortem trauma L and right temporal region, also to left side face	Occ. def.			
25-3-10/60758.0	Trauma	Trauma death–blows to head	Not obs.			8 dm, 1st molars erupted, 2nd not erupted
25-3-10/60759.0	Infection	Severe abscess R canine	No cran. def.			
25-3-10/60760.0	Trauma	Trauma death–scalped	Not obs.			
25-3-10/60761.0	Trauma	Trauma to frontal	Not obs.			Examined by Owsley team 1/98, mandible and postcrania reassociated
25-3-10/60762.0	Trauma, arthritis, tumor	Perimortem trauma to L parietal, DJD at TMJ, button osteomas on frontal	Unobs.			
25-3-10/60763.0	Trauma	Trauma–healed fracture L parietal	Occ. def.	Late Gl. 1 yellow bowl no. 326 (74917)		
25-3-10/60764.0			Not obs.			Cranium looks different, Plains?
25-3-10/60765.0	Infection	All molars lost antemortem, 2 premolars abscessed	No cran. def.			
25-3-10/60766.0			Not obs.			
25-3-10/60767.0	Infection, nutr.	Periodontal disease, mild ectocranial porosis	Not obs.			Examined by Owsley team 1/98, mandible 60771a reassociated by sort.
25-3-10/60768.0	Infection	Infection–frontal and parietal region	Not obs.			
25-3-10/60769.0			Not obs.			
25-3-10/60770.0	Infection	Frontal lesion	?Cran. def.			
25-3-10/60771.0			Unobs.			
25-3-10/60771.1			Unobs.			

Peabody number	Object description	FN	Specimen label	Site	Intrasite provenience	Ceramic stage	MNI	Age (yrs.)	Ad	Su	Ch	In	Fe	Unk. age	M	M?	F	F?	Unk. sex	Age criteria	Sex criteria
25-3-10/60772.0	Nearly complete cranial and fragmentary postcranial human remains of an adult female.	1734	60772	PP	950E150, e side, depth 3′, height 2′	Indet.	1	25–29	1	0	0	0	0	0	0	0	1	0	0	D d, AS 2, iliac crest stage 1 fusion	F all features
25-3-10/60773.0	Nearly complete cranial and partial postcranial human remains of an adult that is probably male.	1730	60773	PP	950E150, e side, depth 3′, height 3′	Indet.	1	40–54	1	0	0	0	0	0	0	1	0	0	0	D g-h, T 8, SB 4–5, AS 7	?Male (Pelvis M, Cranium-all but NC F)
25-3-10/60774.0	Nearly complete cranial and fragmentary postcranial human remains of an adult female.	1621	60774	PP	1350W0, trench a, depth 5′, height–sunk in clay	Probably B&W	1	40–49	1	0	0	0	0	0	0	0	1	0	0	D h	Female
25-3-10/60775.0	Partial cranial and fragmentary postcranial human remains of an adult male.	1771	60775	PP	Outline trench, s of w 700 series, depth 3′, height 3′	Indet.	1	25–29	1	0	0	0	0	0	1	0	0	0	0	D e	M all features
25-3-10/60777.0	Fragmentary postcranial human remains of an adult of unknown sex.	1772	60777	PP	Kiva xii, trench e of kiva, depth 6′, height 1′	Indet.	1		1	0	0	0	0	0	0	0	0	0	1		Indet.
25-3-10/60778.0	Very fragmentary cranial and partial postcranial human remains of a subadult male.	1738	60778	PP	Room 703, under glaze 2 floor, depth 7′6″, height 1′6″	Indet.	1	16–20	0	1	0	0	0	0	1	0	0	0	0	PC 16–20 yrs	Male
25-3-10/60779.0	Nearly complete cranial and postcranial human remains of an adult female.	C2	60779	PP	Church, depth 1′6″, height 4′4″	Indet.	1	25–34	1	0	0	0	0	0	0	0	1	0	0	D f-h, SB 3, AS 2–3	Female
25-3-10/60781.0	Nearly complete cranial human remains of an adult male.	1854	60781 (60644 crossed out)	PP	Room s-i, depth 4′	Indet.	1	20–29	1	0	0	0	0	0	1	0	0	0	0	D e/d	Male
25-3-10/60781.1	Fragmentary postcranial human remains of an adult female.	1854	60781 (60644 crossed out)	PP	Room s-i, depth 4′	Indet.	1	50–59	1	0	0	0	0	0	0	0	1	0	0	PC 50+ yrs	Female
25-3-10/60782.0	Fragmentary postcranial human remains of an adult female.	C	60782	PP	Church	Indet.	1	22–27	1	0	0	0	0	0	0	0	1	0	0	T 3, SB 2, AS 2	F all features
25-3-10/61122.0	Nearly complete cranial human remains of an adult female.	C	61122/ church	PP	Church	Indet.	1	>35	1	0	0	0	0	0	0	0	1	0	0	Cranial, 30–40 yrs	Female
25-3-10/61152.0	Nearly complete cranial human remains of an adult female.	Bb65	61152/bb65	FL	Just w of rm 19, dpth 2′, height about 1′ above adobe floor	Probably B&W	1	18–24	1	0	0	0	0	0	0	0	1	0	0	D c-d/d	Female
25-3-10/61153.0	Nearly complete cranial human remains of an adult female.	Bb87	61153/ BB87	FL	Room 75, depth 36″, under floor	Probably B&W	1	35–39	1	0	0	0	0	0	0	0	1	0	0	D g/g	Female
25-3-10/61154.0	Nearly complete cranial human remains of an adult male.	Bb98	61154/ BB98	FL	Trench l-c, depth 47″, height ?	Probably B&W	1	40–49	1	0	0	0	0	0	1	0	0	0	0	D h	Male
25-3-10/61155.0	Nearly complete cranial human remains of an adult female.	C	61155/ church	PP	Church	Indet.	1	50–59	1	0	0	0	0	0	0	0	1	0	0	Cranial 50+ yrs	Female
25-3-10/61172.0	Nearly complete cranial human remains of a child of unknown sex.	Bb58	61172, bb58	FL	Room 50, depth head 14″, height 10″ above room floor	Probably B&W	1	8–10	0	0	1	0	0	0	0	0	0	0	1	D 10 ± 2.5 yrs	Indet.
26-3-10/61179.0	Fragmentary cranial and nearly complete postcranial human remains of an adult female.	Bb115	61179/ bb115	FL	Trench n, depth 34″, height ?	Probably B&W	1	30–39	1	0	0	0	0	0	0	0	1	0	0	D e-f, T 5, SB 3, AS 4	F all features
26-3-10/61180.0	Nearly complete postcranial human remains of an adult female.	Bb23		FL	Trench c, depth 3′, height 1′	Probably B&W	1	25–29	1	0	0	0	0	0	0	0	1	0	0	AS 2	Female
26-3-10/61181.0	Nearly complete cranial and partial postcranial human remains of an adult male.	Bb52	61181/ BB52	FL	N15W30, depth 4′, height 8–10″	Probably B&W	1	40–54	1	0	0	0	0	0	1	0	0	0	0	D h/i+	Male
26-3-10/61182.0	Partial cranial human remains of an adult female.	Bb34	61182, bb34	FL	Depth 20″, height 16″	Probably B&W	1		1	0	0	0	0	0	0	0	1	0	0		Female
26-3-10/61183.0	Fragmentary cranial and postcranial human remains of an adult male.	Bb15	Bb15	FL	Room bb3, depth 8″, height 2′6″ above room floor	B&W	1	>45	1	0	0	0	0	0	1	0	0	0	0		Male

Peabody number	Pathology type	Pathology description	Cranial mod.	Associated funerary objects	Field context	Notes
25-3-10/60772.0	Arthritis, nutr., other	Vt/c2–3 ankylosed with arthritis, cribra orbitalia, peg upper L M3, upper R P1 misoriented longitudinally	Occ. def.	Gl. 3 olla no. 1009 (74937)		
25-3-10/60773.0	Arthritis, nutr., other	DJD, osteoporosis onset, enamel hypoplasia, vt/s4–5 neural arch not fused	Not obs.	Bisc. B bowl no. 811 (74936), bone awl no. 877 (75483), 2 stone knives no. 876 (75991), stone effigy no. 878 (76197)		
25-3-10/60774.0	Infection, arthritis	Antemortem molar loss, DJD	Occ. def.	B&W sherds, no number		Field number problematic
25-3-10/60775.0	Trauma, infection, nutr., other	Perimortem trauma to L frontal with radiating fracture, unusual increased porosity frontal, parietal, occipital, enamel hypoplasia, maleruption R premolar	Occ. def.	Gl. 2 bowl frags. no. 968 (74953)		
25-3-10/60777.0	Arthritis	DJD–especially shoulder, elbow, hip, knee, and ankle	No cran.			
25-3-10/60778.0	Trauma	Healed fracture neck of L femur with marked displacement	Occ. def.		Floor slabs lay above burial	
25-3-10/60779.0	Infection, arthritis, nutr.	Possible tuberculosis, vt/t8–10 fused, vt/t12 healed fracture, abscess upper R M3 and L M2, osteoporosis-very thin scapulae	Occ. def.		Burial at length	
25-3-10/60781.0			Occ. def.	Gl. 3 bowl frag. no. 3827 (74971)		
25-3-10/60781.1	Nutr.	Osteoporosis	No cran.			Field number problematic
25-3-10/60782.0	Other	Vt/s4–5 neural arch not fused	No cran.			
25-3-10/61122.0	Trauma, infection	Perimortem trauma to frontal, L maxilla, some antemortem loss molars	Occ. def.			
25-3-10/61152.0	Infection, nutr.	Dental abscess upper R M1–2, enamel hypoplasia	Cr. def.			Different cranial shape
25-3-10/61153.0	Infection, nutr.	Lesion on left TMJ, lower L M2 abscessed, enamel hypoplasia	Not obs.			
25-3-10/61154.0			Not obs.			
25-3-10/61155.0	Infection	Extensive antemortem tooth loss	Not obs.			
25-3-10/61172.0	Nutr.	Cribra orbitalia, enamel hypoplasia	Not obs.			
26-3-10/61179.0	Arthritis, nutr.	Slight spinal osteophytosis, enamel hypoplasia	Unobs.			
26-3-10/61180.0			Not obs.			
26-3-10/61181.0	Arthritis, other	DJD, spinal osteophytosis, ankylosis of 4 vt/t and 2 vt/l, ankylosis of sacrum to R innominate, sacralization of vt/l5	Not obs.			Reassociated human remains from 61274 (MEM 5/98)
26-3-10/61182.0			Lamb. def.			
26-3-10/61183.0	Arthritis	DJD at TMJ	Unobs.	1 Olivella shell, no number		Reassociated human remains, formerly cat. no. 61278 (MEM 4/98)

Peabody number	Object description	FN	Specimen label	Site	Intrasite provenience	Ceramic stage	MNI	Age (yrs.)	Ad	Su	Ch	In	Fe	Unk. age	M	M?	F	F?	Unk. sex	Age criteria	Sex criteria
26-3-10/61184.0	Partial cranial and fragmentary postcranial human remains of an adult female.	Bb95	Bb95	FL	Trench d, depth 69″, height ?	Probably B&W	1	20–24	1	0	0	0	0	0	0	0	1	0	0	D d	Female
26-3-10/61185.0	Very fragmentary cranial and postcranial human remains of a child of unknown sex.	Bb56		FL	Room 50, se corner, depth 3′, height 1′	Probably B&W	1		0	0	1	0	0	0	0	0	0	0	1	Child	Indet.
26-3-10/61186.0	Fragmentary cranial human remains of a child of unknown sex.	Bb70		FL	Room 64, depth 8″, height 3′± from floor	Probably B&W	1		0	0	1	0	0	0	0	0	0	0	1	Child	Indet.
26-3-10/61187.0	Fragmentary cranial and postcranial human remains of an adult male.	Bb61		FL	Room 52, se corner, depth 3′5″, on floor	Probably B&W	1	30–49	1	0	0	0	0	0	1	0	0	0	0	D f-h	Male
26-3-10/61187.2	Fragmentary cranial human remains of a subadult of unknown sex.	Bb61		FL	Room 52, se corner, depth 3′5″, on floor	Probably B&W	1	16–20	0	1	0	0	0	0	0	0	0	0	1	D b2, subadult	Indet.
26-3-10/61188.0	Fragmentary cranial and postcranial human remains of a child of unknown sex.	Bb63		FL	Rm 48, agnst w wall 2' n frm sw crnr, dpth 3′4″, blw flr in sbsoil	Probably B&W	1		0	0	1	0	0	0	0	0	0	0	1	Child	Indet.
26-3-10/61191.0	Fragmentary cranial and postcranial human remains of an adult of unknown sex.	Bb68		FL	Room 50, depth 3′6″, height 1′	Probably B&W	1		1	0	0	0	0	0	0	0	0	0	1		Indet.
26-3-10/61197.0	Nearly complete cranial and postcranial human remains of a subadult male.	C	61197/ church	PP	Church	Indet.	1	18–20	0	1	0	0	0	0	1	0	0	0	0	D d but M3s not erupted, T 1, SB 1	M all features
26-3-10/61198.0	Nearly complete cranial and postcranial human remains of an adult female.	Bb102	61198, bb102	FL	Room 79, nw corner, depth 72″, height–in grave cut into	Probably B&W	1	40–49	1	0	0	0	0	0	0	0	1	0	0	D f-i, very fragmentary, AS 6	Female
26-3-10/61199.0	Fragmentary cranial and postcranial human remains of a child of unknown sex.	Bb75	Bb75	FL	3' n of 74, depth 1′4″, height 3′6″	Probably B&W	1	1–2	0	0	1	0	0	0	0	0	0	0	1	D 2 yrs, PC 1.5 yrs	Indet.
26-3-10/61200.0	Fragmentary cranial and postcranial human remains of a child of unknown sex.	Bb86	61200/ ?BB86	FL	Room 74, depth 40″, on room floor	Probably B&W	1	7–8	0	0	1	0	0	0	0	0	0	0	1	PC 7.5–8.5 yrs	Indet.
26-3-10/61201.0	Fragmentary postcranial human remains of an adult of unknown sex.	Bb67		FL	W of room 51, depth 2′8″, height ?	Probably B&W	1		1	0	0	0	0	0	0	0	0	0	1		Indet.
26-3-10/61202.0	Fragmentary cranial and postcranial human remains of an infant of unknown sex.	Bb76	Bb76	FL	S5W18, depth 4′6″, height 1′ + (about 2′)	Probably B&W	1	1–1.5	0	0	0	1	0	0	0	0	0	0	1	D 1.5 ± 0.5 yrs, PC 0.5–1.5 yrs	Indet.
26-3-10/61203.0	Fragmentary cranial and very fragmentary postcranial human remains of an adult male.	Bb83		FL	Trench w of 18, depth 4′10″, on bottom	Probably B&W	1		1	0	0	0	0	0	1	0	0	0	0		Male
26-3-10/61204.0	Fragmentary cranial and postcranial human remains of an adult that is probably male.	Bb74	Bb74	FL	Trench west of room 18, depth 5′, scoop in bottom	Probably B&W	1	35–45	1	0	0	0	0	0	0	1	0	0	0	D g/g	?Male
26-3-10/61216.0	Partial postcranial human remains of an adult female.	Bb110		FL	Trench n, depth 39″, height ?	Probably B&W	1	50–59	1	0	0	0	0	0	0	0	1	0	0	T 10, SB 6, AS 7	F all features
27-1-10/60684.0	Nearly complete cranial human remains of a subadult female.	Bb94	60684/bb94	FL	W173N78, depth 4″, height ?	Probably B&W	1	16–20	0	1	0	0	0	0	0	0	1	0	0	M3s just erupting	Female
27-1-10/61261.1	Very fragmentary cranial human remains of a subadult male.	C		PP	Church	Indet.	1	16–20	0	1	0	0	0	0	1	0	0	0	0	D b2, subadult	Male
27-1-10/61261.2	Very fragmentary cranial human remains of an adult that is probably female.	C	Floater	PP	Church	Indet.	1	>50	1	0	0	0	0	0	0	0	0	1	0	D old, >50 yrs	?Female
27-1-10/61261.3	Very fragmentary cranial human remains of an adult that is probably male.	C	Floater	PP	Church	Indet.	1	30–34	1	0	0	0	0	0	0	1	0	0	0	D f	?Male
27-1-10/61262.0	Fragmentary postcranial human remains of an adult of unknown sex.	Bb88		FL	Room w1, depth 3′6″, height 9″	Probably B&W	1		1	0	0	0	0	0	0	0	0	0	1		Indet.

Peabody number	Pathology type	Pathology description	Cranial mod.	Associated funerary objects	Field context	Notes
26-3-10/61184.0			Occ. def.	Medicine outfit, 21 pcs (bone frags. and stone), no. 270 (77094)		
26-3-10/61185.0			Unobs.			
26-3-10/61186.0			Unobs.			
26-3-10/61187.0	Nutr.	Enamel hypoplasia	Unobs.	Frag. of B&W bowl, no number		Reassociated human remains from 61187-1 and 61187.3 (MEM 5/98)
26-3-10/61187.2	Nutr.	Enamel hypoplasia	Unobs.	Frag. of B&W bowl, no number		Reassociated human remains from 61887-3 (MEM 5/98). Field card number and provenience problematic.
26-3-10/61188.0			Unobs.			
26-3-10/61191.0			Unobs.			? Association of crania and postcrania– preservation very different
26-3-10/61197.0	Trauma, infection, arthritis	Trauma–L temporal region–radiating fracture L parietal, partial crack along coronal suture but skull badly damaged, vt/t 8–9 and vt/c5–6 ankylosed, fracture of coccyx, infection L mastoid	Not obs.			Field number problematic
26-3-10/61198.0	Infection, arthritis	Maxilla edentulous and no mandibular molars, spinal osteophytosis, some DJD	Lamb. def.			
26-3-10/61199.0			Unobs.			Crania and postcrania may be from separate individuals
26-3-10/61200.0			Unobs.			
26-3-10/61201.0	Arthritis	Spinal osteophytosis	No cran.			
26-3-10/61202.0			Unobs.			
26-3-10/61203.0			Unobs.	Frags. black jar, no number		
26-3-10/61204.0	Nutr., other	Enamel hypoplasia, upper L P1 rotated 90° clockwise	Unobs.			
26-3-10/61216.0	Arthritis	DJD, spinal osteophytosis	Not obs.			
27-1-10/60684.0			Not obs.			
27-1-10/61261.1	Infection, nutr.	Abscess in lower R M2, enamel hypoplasia	Unobs.			
27-1-10/61261.2			Unobs.			
27-1-10/61261.3			Unobs.			
27-1-10/61262.0			No cran.			Field number and provenience problematic

Peabody number	Object description	FN	Specimen label	Site	Intrasite provenience	Ceramic stage	MNI	Age (yrs.)	Ad	Su	Ch	In	Fe	Unk. age	M	M?	F	F?	Unk. sex	Age criteria	Sex criteria
27-1-10/61265.0	Fragmentary cranial and partial postcranial human remains of an infant of unknown sex.	Bb78		FL	S40W75, depth 2′10″, scoop in bottom	Probably B&W	1	0.5–1.0	0	0	0	1	0	0	0	0	0	0	1	D 0.75 ± 0.25 yrs, PC 0.5–1.5 yrs	Indet.
27-1-10/61266.0	Very fragmentary postcranial human remains of an adult of unknown sex.	Bb92		FL	Trench k, depth 3′, height ?	Probably B&W	1		1	0	0	0	0	0	0	0	0	0	1		Indet.
27-1-10/61268.0	Partial cranial and fragmentary postcranial human remains of an infant of unknown sex.	Bb85	Bb85	FL	Room 76, 1′ s of n wall, middle, depth 3′, height ?	Probably B&W	1	0.25–0.75	0	0	0	1	0	0	0	0	0	0	1	D 0.75 ± 0.25 yrs, PC 0–0.5 yrs	Indet.
27-1-10/61269.0	Fragmentary cranial and partial postcranial human remains of an infant of unknown sex.	Bb79		FL	Room 68, ne corner, depth 5′, on bottom	Probably B&W	1	0.5–1.0	0	0	0	1	0	0	0	0	0	0	1	PC 0.5–1.5 yrs	Indet.
27-1-10/61270.0	Nearly complete cranial and fragmentary postcranial human remains of an infant of unknown sex.	Bb73	61268	FL	Ash bed w of room 55, depth 4′, on bottom	Probably B&W	1	0–0.5	0	0	0	1	0	0	0	0	0	0	1	D 0 ± 0.2 yrs, PC 0–0.5 yrs	Indet.
27-1-10/61271.0	Fragmentary cranial human remains of an infant of unknown sex.	Bb83		FL	Trench w of 18, depth 4′10″, on bottom	Probably B&W	1	0–2.5	0	0	0	1	0	0	0	0	0	0	1		Indet.
27-1-10/61272.0	Fragmentary cranial and postcranial human remains of an infant of unknown sex.	Bb80		FL	Room 72, depth 4′, on or near bottom	Probably B&W	1	0.25–0.75	0	0	0	1	0	0	0	0	0	0	1	D 0.75 ± 0.25 yrs, PC 0–0.5 yrs	Indet.
27-1-10/61276.0	Partial cranial human remains of an adult that is probably male.	Bb84	61276	FL	Trench j, depth 5′, on bottom	Probably B&W	1	25–29	1	0	0	0	0	0	0	1	0	0	0	D e	?Male
27-1-10/61277.0	Fragmentary postcranial human remains of an adult of unknown sex.	Bb106		FL	Trench n, depth 60″, height–on or scooped into	Probably B&W	1		1	0	0	0	0	0	0	0	0	0	1		Indet.
27-1-10/61280.0	Partial cranial and postcranial human remains of an adult of unknown sex.	Bb104	Bb104	FL	N83W59, depth 32″, height ?	Probably B&W	1	35–45	1	0	0	0	0	0	0	0	0	0	1	D g/h	Indet.
27-1-10/61280a	Fragmentary cranial human remains of an adult of unknown sex.	Bb104		FL	N83W59, depth 32″, height ?	Probably B&W	1	25–29	1	0	0	0	0	0	0	0	0	0	1	D e	Indet.
27-1-10/61281.0	Partial postcranial human remains of an adult of unknown sex.	Bb64		FL	Room 57, se corner, depth 2′3″, on floor	Probably B&W	1		1	0	0	0	0	0	0	0	0	0	1		Indet.
27-1-10/61282.0	Fragmentary postcranial human remains of an adult male.	Bb9	61282 (labeled, MEM 11/96)	FL	Kidder creek trench, depth 5–6 ″, lies on ash layer	B&W	1	45–49	1	0	0	0	0	0	1	0	0	0	0	PC 45–49 yrs	Male
27-1-10/61282.1	Very fragmentary postcranial human remains of an adult of unknown sex.	Bb9	61282 (labeled, MEM 11/96)	FL	Kidder crk tr/rm16, s wall, bb9: d6-5″, on ash, bb48: d34″, h ?	B&W	1		1	0	0	0	0	0	0	0	0	0	1		Indet.
27-1-10/61284.0	Fragmentary cranial and postcranial human remains of an infant of unknown sex.	Bb107	61284	FL	N50E20, depth 3′, height 18″	Probably B&W	1	0–0.5	0	0	0	1	0	0	0	0	0	0	1	0–0.5 yrs	Indet.
27-1-10/61285.0	Nearly complete postcranial human remains of a child of unknown sex.	Bb54	61285 (labeled, MEM 11/96)	FL	Rm 43, sw corner, dpth pelvis 12″ skull 14″, face 20″ abv flr	B&W	1	6.5–7.5	0	0	1	0	0	0	0	0	0	0	1	PC 6.5–7.5 yrs, no teeth	Indet.
27-1-10/61286.0	Fragmentary postcranial human remains of an adult of unknown sex.	Bb131	61286 (labeled, MEM 11/96)	FL	Trench j, west extension	Probably B&W	1		1	0	0	0	0	0	0	0	0	0	1		Indet.
27-1-10/61287.0	Fragmentary cranial and postcranial human remains of an adult male.	Bb34	61287 (labeled, MEM 11/96)	FL	Depth 20″, height 16″	Probably B&W	1	40–49	1	0	0	0	0	0	1	0	0	0	0	D h/i+, AS 5–6	Male
27-1-10/61288.0	Fragmentary postcranial human remains of an adult female.	Bb81	61288 (labeled, MEM 11/96)	FL	Below 80, depth 5′, on bottom	Probably B&W	1		1	0	0	0	0	0	0	0	1	0	0		Female

Peabody number	Pathology type	Pathology description	Cranial mod.	Associated funerary objects	Field context	Notes
27-1-10/61265.0			Unobs.	6 bone beads no. 144 (76883)		
27-1-10/61266.0			No cran.	Seed bead no. 219 (77055)		Field number and provenience problematic
27-1-10/61268.0			Not obs.			
27-1-10/61269.0			Unobs.			
27-1-10/61270.0			Not obs.			
27-1-10/61271.0			Unobs.	Frags. black jar, no number		Field number and provenience problematic
27-1-10/61272.0			Unobs.			
27-1-10/61276.0			Not obs.			Field number and provenience problematic
27-1-10/61277.0			No cran.			
27-1-10/61280.0			Not obs.			Field number and provenience problematic
27-1-10/61280a	Nutr.	Enamel hypoplasia	Unobs.			
27-1-10/61281.0	Arthritis	Slight DJD	Not obs.			Field number and provenience problematic
27-1-10/61282.0	Arthritis, nutr.	Slight DJD, slight osteoporosis	No cran.	B&W frag, black corrugated jar, no number		
27-1-10/61282.1			No cran.	B&W frag, black corrugated jar, no number		Does not match field card description
27-1-10/61284.0			Unobs.			
27-1-10/61285.0			Not obs.	2 pcs broken manos, no number		Field number and provenience problematic
27-1-10/61286.0	Nutr.	Osteoporosis	No cran.			
27-1-10/61287.0	Arthritis	Spinal osteophytosis	Unobs.			Field number and provenience problematic
27-1-10/61288.0	Trauma	Healed fracture 1 phalange	No cran.			

Peabody number	Object description	FN	Specimen label	Site	Intrasite provenience	Ceramic stage	MNI	Age (yrs.)	Ad	Su	Ch	In	Fe	Unk. age	M	M?	F	F?	Unk. sex	Age criteria	Sex criteria
27-1-10/61289.0	Partial cranial human remains of an adult male.	C23	61289/ church	PP	Church nave, depth 3′, height 1′	Indet.	1	35–44	1	0	0	0	0	0	1	0	0	0	0	Cranium, 35–44 yrs	Male
27-1-10/61290.0	Nearly complete cranial human remains of an adult female.	Bb110	61290/ BB110	FL	Trench n, depth 39″, height ?	Probably B&W	1	30–39	1	0	0	0	0	0	0	0	1	0	0	D f-g	Female
27-1-10/61291.0	Nearly complete cranial human remains of an adult female.	Bb111	61291/ bb111	FL	N76W53, depth 2′, height± 3′	Probably B&W	1	45–54+	1	0	0	0	0	0	0	0	1	0	0	D h, sutures fused	Female
27-1-10/61293.0	Partial cranial human remains of a child of unknown sex.	Bb130	61293	FL	Trench i, depth 30″, height 12″	Probably B&W	1		0	0	1	0	0	0	0	0	0	0	1		Indet.
27-1-10/61294.0	Partial cranial and postcranial human remains of a child of unknown sex.	Bb101		FL	Trench w of 55, depth 52″, height 4″	Probably B&W	1	4.5–5.5	0	0	1	0	0	0	0	0	0	0	1	D 5 ± 1.3 yrs, PC 4.5–5.5 yrs	Indet.
27-1-10/61295.0	Partial cranial human remains of an adult female.	Bb124	61295	FL	Trench n, depth 30″, height ?	Probably B&W	1		1	0	0	0	0	0	0	0	1	0	0		Female
27-1-10/61295.1	Very fragmentary cranial human remains of a subadult of unknown sex.	Bb124	Bb124	FL	Trench n, depth 30″, height ?	Probably B&W	1	12–15	0	1	0	0	0	0	0	0	0	0	1	D 15 ± 3 yrs	Indet.
27-1-10/61296.0	Fragmentary cranial human remains of an adult of unknown sex.	Bb59	61296 (la-beled, MEM 11/96)	FL	Room 53, se corner, at surface partly denuded	Probably B&W	1		1	0	0	0	0	0	0	0	0	0	1		Indet.
27-1-10/61297.0	Fragmentary cranial human remains of a child of unknown sex.	Bb106	61297 (la-beled, MEM 11/96)	FL	Trench n, depth 60″, height–on or scooped into	Probably B&W	1		0	0	1	0	0	0	0	0	0	0	1		Indet.
27-1-10/61298.0	Fragmentary cranial and postcranial human remains of an adult of unknown sex.	Bb109	Bb109, 61298 (MEM 11/96)	FL	Trench n	Probably B&W	1		1	0	0	0	0	0	0	0	0	0	1		Indet.
27-1-10/61299.0	Fragmentary cranial human remains of an adult that is probably female.	Bb129	91299 (la-beled, MEM 11/96)	FL	Trench I, depth 30″, height 8″	Probably B&W	1	40–50+	1	0	0	0	0	0	0	0	1	0	0	D h+	?Female
27-1-10/61300.0	Fragmentary postcranial human remains of an adult male.	Bb34	61300 (la-beled, MEM 11/96)	FL	Depth 20″, height 16″	Probably B&W	1		1	0	0	0	0	0	1	0	0	0	0		Male
27-1-10/61304.0	Nearly complete cranial human remains of an adult male.	Bb131	61304 (la-beled, MEM 11/96)	FL	Trench j, w extension, depth 51″, height–on bottom	Probably B&W	1	40–54	1	0	0	0	0	0	1	0	0	0	0	D h/h-I	Male
27-1-10/61309.0	Fragmentary postcranial human remains of a subadult of unknown sex.	Bb112	61309 (la-beled, MEM 11/96)	FL	Trench n, depth 35″, height ?	Probably B&W	1	18–22	0	1	0	0	0	0	0	0	0	0	1	PC iliac crest unfused	Indet.
27-1-10/61310.0	Nearly complete cranial and very fragmentary postcranial human remains of an adult female.	Bb118	61310/ BB118	FL	Trench p, depth 39″, height–on bottom	Probably B&W	1	35–39	1	0	0	0	0	0	0	0	1	0	0	D g	Female
27-1-10/61312.0	Nearly complete cranial human remains of a subadult female.	Bb112	61312/ BB112	FL	Trench n, depth 35″, height ?	Probably B&W	1	13–16	0	1	0	0	0	0	0	0	1	0	0	D c/c-d, M3s partially erupted	Female
27-1-10/61314.0	Nearly complete cranial and very fragmentary postcranial human remains of a subadult of unknown sex.	Bb126	61314/ BB126	FL	N60E50, depth 49″, height ?	Probably B&W	1	16–22	0	1	0	0	0	0	0	0	0	0	1	D b2-c/b2, M3s erupting	Indet.
27-1-10/61315.0	Partial cranial and fragmentary postcranial human remains of an adult male.	Bb100	61315/ BB100	FL	Trench l-k, depth 12″, height 41″	Probably B&W	1	20–29	1	0	0	0	0	0	1	0	0	0	0	D d-e	Male
27-1-10/61322.0	Fragmentary cranial and postcranial human remains of an adult that is probably female.	Bb114	Bb114	FL	Room 81, trench p, depth 3′6″, height–scoop into bottom	Probably B&W	1		1	0	0	0	0	0	0	0	0	1	0		?Female

Peabody number	Pathology type	Pathology description	Cranial mod.	Associated funerary objects	Field context	Notes
27-1-10/61289.0	Trauma, tumor	Trauma death–scalped, small tumor in frontal sinus	Not obs.	Bronze medal beads and copper cross no. 1195 (76400), frag. of silver and orange brocade no. 1193 (76360), frags. of cloth no. 1194 (76359), no. 1192 (76358)		Field number and provenience problematic
27-1-10/61290.0	Infection	Antemortem tooth loss	Not obs.			
27-1-10/61291.0	Infection	Upper dentition lost antemortem	Not obs.			
27-1-10/61293.0			Not obs.			
27-1-10/61294.0	Nutr.	Cribra orbitalia	Not obs.			Reassociated human remains from 61267 (MEM 4/98)
27-1-10/61295.0			Not obs.	2 fossil brachiopods and 1 fossilized plant frag. (?medicine outfit), no. 351 (77095)		
27-1-10/61295.1	Nutr.	Enamel hypoplasia	Unobs.	2 fossil brachiopods and 1 fossilized plant frag. (?medicine outfit), no. 351 (77095)		
27-1-10/61296.0			Unobs.	Frags. black bowl no. 184 (76656)	trauma-fracture tibia, arrowhead just above ribs	No postcrania present
27-1-10/61297.0			Unobs.			
27-1-10/61298.0			Unobs.	Medicine outfit (8 stones, 1 bone, 1 bead) no. 380 (77093)		
27-1-10/61299.0			Unobs.	18 beads, no number		
27-1-10/61300.0			No cran.			Large sternal foramina, field number and provenience problematic
27-1-10/61304.0	Infection, arthritis, nutr., tumor	Extensive antemortem tooth loss (pronounced calculus on molars), ankylosis vt/c2–3, enamel hypoplasia, multiple osteomas on frontal	Occ. def.			
27-1-10/61309.0			No cran.	2 strands of shell beads no. 334 (77068)		Description does not match field card
27-1-10/61310.0	Infection, nutr.	Extensive antemortem tooth loss, enamel hypoplasia, cribra orbitalia	Not obs.			Reassociated humerus from 61214 MEM 4/98
27-1-10/61312.0	Nutr.	Enamel hypoplasia	Occ. def.	2 strands of shell beads no. 334 (77068)		
27-1-10/61314.0	Nutr., other	Enamel hypoplasia, scaphocephaly				
27-1-10/61315.0	Arthritis, nutr.	DJD, enamel hypoplasia	Occ. def.	Small black olla no. 282 (76640)		
27-1-10/61322.0	Nutr.	Enamel hypoplasia	Unobs.			

Peabody number	Object description	FN	Specimen label	Site	Intrasite provenience	Ceramic stage	MNI	Age (yrs.)	Ad	Su	Ch	In	Fe	Unk. age	M	M?	F	F?	Unk. sex	Age criteria	Sex criteria
27-1-10/61323.0	Fragmentary cranial human remains of an adult of unknown sex.	Bb142	61323 (labeled, MEM 11/96)	FL	Room 104, just outside ne corner rm 101, depth 3′4″	Probably B&W	1		1	0	0	0	0	0	0	0	0	0	1		Indet.
27-1-10/61327.0	Fragmentary cranial and postcranial human remains of an infant of unknown sex.	Bb108	61327 (labeled, MEM 11/96)	FL	Room 74, depth 3′6″, height 1′	Probably B&W	1	0.5–2.5	0	0	0	1	0	0	0	0	0	0	1	Overall size	Indet.
27-1-10/61328.0	Very fragmentary cranial and postcranial human remains of an adult of unknown sex.	Bb59	61329 (labeled, MEM 11/96)	FL	Room 53, se corner, at surface partly denuded	Probably B&W	1		1	0	0	0	0	0	0	0	0	0	1		Indet.
27-1-10/61329.0	Very fragmentary cranial and nearly complete postcranial human remains of an infant of unknown sex.	Bb103	15	FL	Room 80, depth 53″, height–under floor	Probably B&W	1	0.5–1.0	0	0	0	1	0	0	0	0	0	0	1	PC 0.5–1.5 yrs	Indet.
27-6-10/60889.0	Nearly complete cranial human remains of an adult male.	Bb76	60889/bb76	FL	S5W18, depth 4′6″, height 1′+ (about 2′)	Probably B&W	1	45–55+	1	0	0	0	0	0	1	0	0	0	0	Dh+/i+	Male
27-6-10/60900.0	Nearly complete cranial human remains of a subadult of unknown sex.	C	60900, church	PP	Church	Indet.	1	15–17	0	1	0	0	0	0	0	0	0	0	1	D 15 ± 3 yrs, spheno-occipital not fused	Indet.
27-6-10/60905.0	Nearly complete cranial human remains of a subadult female.	C	60905/church	PP	Church	Indet.	1	12–15	0	1	0	0	0	0	0	0	1	0	0	D 15 ± 3 yrs	Female
27-6-10/60906.0	Partial cranial and fragmentary postcranial human remains of an adult female.	Bb88	60906, bb88	FL	Room w1, depth 3′6″, height 9″	Probably B&W	1	40–54	1	0	0	0	0	0	0	0	1	0	0	Dh/i	Female
27-6-10/60910.0	Fragmentary cranial human remains of an adult male.	Bb141	60910/141	FL	Room 101, e wall, depth 2′10″, height–on bottom	Probably B&W	1	40–54	1	0	0	0	0	0	1	0	0	0	0	Dh/i	Male
29-2-10/60888.0	Nearly complete cranial and postcranial human remains of an adult female.	2011	60888/2011	PP	1250W0, plaza trench, depth 4′6″	Indet.	1	60+	1	0	0	0	0	0	0	0	1	0	0	Dh+/i+, SB 6, AS 8	F all features
29-2-10/60890.0	Fragmentary cranial human remains of a child of unknown sex.	Bb146		FL	Ne bank of an arroyo, depth 5′6″, height–on bottom	Probably B&W	1	4–6	0	0	1	0	0	0	0	0	0	0	1	D 6–7 ± 2 yrs	Indet.
29-2-10/60891.0	Partial cranial and postcranial human remains of an infant of unknown sex.	Bb149		FL	Room 123, depth 4′, height–on bottom	Probably B&W	1	0.75–1.5	0	0	0	1	0	0	0	0	0	0	1	D 1.5 ± 0.5 yrs, PC 0.5–1.5 yrs	Indet.
29-2-10/60892.0	Fragmentary cranial human remains of a child of unknown sex.	Bb147		FL	Trench n of room 124, depth 18″, height 2′	Probably B&W	1	3–4	0	0	1	0	0	0	0	0	0	0	1	D 4 ± 1 yrs	Indet.
29-2-10/60893.0	Fragmentary cranial and postcranial human remains of a child of unknown sex.	Bb150		FL	Room 122, se corner, depth 3′, height–on lowest floor level	Probably B&W	1	1–2	0	0	1	0	0	0	0	0	0	0	1	D 3 ± 1yrs, PC 0.5–1.5 yrs	Indet.
29-2-10/60894.0	Nearly complete cranial and fragmentary postcranial human remains of an adult male.	2059	60894/2059	PP	1250E0, plaza, test 20, depth 7′, height 3′	Indet.	1	45–55+	1	0	0	0	0	0	1	0	0	0	0	Dh+/h	Male
29-2-10/60895.0	Partial cranial and postcranial human remains of a subadult female.	2013		PP	950E150, depth 3′, height–on botttom	Indet.	1	14–16	0	1	0	0	0	0	0	0	1	0	0	Db2, M3s just erupting, T 1, SB 1, AS <1, PC 13–15 yrs	Female
29-2-10/60896.0	Fragmentary cranial and postcranial human remains of a child of unknown sex.	Bb151		FL	Room 123, nw corner, depth 2′, height–on room floor	Probably B&W	1	7–8	0	0	1	0	0	0	0	0	0	0	1	D 7 ± 2 yrs, PC 7.5–8.5 yrs	Indet.
29-2-10/60897.0	Fragmentary cranial human remains of a subadult of unknown sex.	Bb152		FL	Room, through subfloor refuse, depth 6′, height–on bottom	Probably B&W	1	16–20	0	1	0	0	0	0	0	0	0	0	1	Db1	Indet.
29-2-10/60898.0	Nearly complete cranial and postcranial human remains of an infant of unknown sex.	2057		PP	1250E0, plaza, below glaze 4 level, depth 4′±, height 5′±	Indet.	1	1–2	0	0	0	1	0	0	0	0	0	0	1	Overall 2 ± 0.66 yrs	Indet.

Peabody number	Pathology type	Pathology description	Cranial mod.	Associated funerary objects	Field context	Notes
27-1-10/61323.0	Nutr.	Enamel hypoplasia	Unobs.			
27-1-10/61327.0			Unobs.			
27-1-10/61328.0			Unobs.	Frags. black bowl no. 184 (76656)		
27-1-10/61329.0			Unobs.			
27-6-10/60889.0	Infection, tumor	Extensive antemortem tooth loss, button osteoma on frontal	Occ. def.			Description does not match field card
27-6-10/60900.0			No cran. def.			
27-6-10/60905.0	Trauma	Trauma to cranium associated with death	Not obs.			
27-6-10/60906.0	Infection, nutr.	Extensive antemortem tooth loss, R mandibular abscesses M1–M3, caries in lower P1s, upper L I2, enamel hypoplasia	Not obs.			
27-6-10/60910.0	Trauma	Healed lesion to lambda	Unobs.			
29-2-10/60888.0	Trauma, arthritis, nutr.	Healed fracture dist. L radius, extensive DJD, spinal osteophytosis, osteoporosis	Occ. def.	2 Gl. 4 bowls nos. 88 (77178), 375 (77183), 4 bone awls no. 376 (77364), bone bead no. 377 (77398)		
29-2-10/60890.0	Nutr.	Enamel hypoplasia	Unobs.			
29-2-10/60891.0			Not obs.			
29-2-10/60892.0			Unobs.	Corrugated pot no. F37 (77719)		
29-2-10/60893.0	Other	Space for only 4 anterior lower teeth	Unobs.	Turtle carapace no. F38 (77762)		
29-2-10/60894.0	Infection, nutr.	Multiple dental abscesses, enamel hypoplasia	No cran. def.			
29-2-10/60895.0	Nutr., other	Enamel hypoplasia, vt/s1 and vt/s3–5 neural arches not fused	Not obs.	2 bone awls no. 174 (77366), frag. horn flaker no. 175 (77676), small polishing stone no. 173 (77539), fragments of clay pipe, no number	Burial at length	Reassociated 60920 with 60895 (MEM)
29-2-10/60896.0			Unobs.	Antler prong (no number)		
29-2-10/60897.0	Nutr.	Enamel hypoplasia	Unobs.	2 obsidian projectile points no. F40a (77749)		
29-2-10/60898.0			Not obs.			

Peabody number	Object description	FN	Specimen label	Site	Intrasite provenience	Ceramic stage	MNI	Age (yrs.)	Ad	Su	Ch	In	Fe	Unk. age	M	M?	F	F?	Unk. sex	Age criteria	Sex criteria
29-2-10/60899.0	Fragmentary cranial and postcranial human remains of an infant of unknown sex.	2033		PP	1250E0, depth 4′, height 5′	Indet.	1	0.25–0.75	0	0	0	1	0	0	0	0	0	0	1	Overall 0.5 yrs	Indet.
29-2-10/60901.0	Partial cranial and postcranial human remains of a subadult male.	2039		PP	950E150, depth 4′, height–on clay	Indet.	1	12.5–15.5	0	1	0	0	0	0	1	0	0	0	0	D 15 ± 3 yrs, PC 11–18 yrs, T 1, SB 1, AS < 1	Male
29-2-10/60902.0	Nearly complete cranial and fragmentary postcranial human remains of an adult male.	2039	60902	PP	950E150, depth 4′, height 1′	Indet.	1	35–39	1	0	0	0	0	0	1	0	0	0	0	D F, T 7, SB 4, AS 5	M all features
29-2-10/60903.0	Nearly complete cranial and partial postcranial human remains of an adult female.	Bb148	60903/BB148	FL	Trench s of room 124, depth 2′, height–on bottom	Probably B&W	1	45–49	1	0	0	0	0	0	0	0	1	0	0	T 9, SB 5, AS 6	Female (Cranium F, Pelvis indet.)
29-2-10/60907.0	Nearly complete cranial human remains of an adult male.	2049	60907	PP	950E150, depth 5′, height–in clay	Indet.	1	30–34	1	0	0	0	0	0	1	0	0	0	0	D f	Male
29-2-10/60908.0	Fragmentary cranial human remains of an adult female.	2032	60908/2049	PP	Room e12, west wall, depth 6′6″, height 3′	Indet.	1	25–34	1	0	0	0	0	0	0	0	1	0	0	D e? uneven	Female
29-2-10/60909.0	Fragmentary postcranial human remains of an adult male.	2046	60909	PP	985E155, depth 5′6″, height–on clay bottom	Indet.	1	45–59	1	0	0	0	0	0	1	0	0	0	0	T 8–9, SB 5, AS 6–7	M all features
29-2-10/60911.0	Nearly complete cranial and partial postcranial human remains of an adult female.	2043	60911	PP	950E150, depth 7′, height–in clay	Indet.	1	50–59+	1	0	0	0	0	0	0	0	1	0	0	D h+/i+, T 10+, SB 6+, AS 8	F all features
29-2-10/60912.0	Partial cranial and postcranial human remains of an adult male.	2034	60912	PP	950E150, depth 4′, height–on bottom	Indet.	1	40–44	1	0	0	0	0	0	1	0	0	0	0	D f-i, T 8, SB 5, AS 4–5	M all features
29-2-10/60913.0	Partial cranial and fragmentary postcranial human remains of an infant of unknown sex.	2012		PP	950E150, depth 3′6″, height–on clay	Indet.	1	0–0.5	0	0	0	1	0	0	0	0	0	0	1	D 0.75 ± 0.25 yrs, PC 0–0.5 yrs	Indet.
29-2-10/60914.0	Very fragmentary cranial human remains of an infant of unknown sex.	2054		PP	950E150, depth 2′, height–on bottom	Indet.	1	2–3	0	0	0	1	0	0	0	0	0	0	1	D 2.5 ± 0.66 yrs	Indet.
29-2-10/60915.0	Fragmentary postcranial human remains of an adult male.	2030		PP	E10, against black &white wall, in fill, depth 15′, height 1′	Indet.	1	45–49	1	0	0	0	0	0	1	0	0	0	0	T 9, SB 4–5	Male
29-2-10/60916.0	Fragmentary cranial and postcranial human remains of an adult that is probably male.	2042		PP	Room e12, depth 6′6″, height ?	Gl. 4	1	45–54	1	0	0	0	0	0	0	1	0	0	0	D i	?Male
29-2-10/60917.0	Nearly complete cranial and fragmentary postcranial human remains of a child of unknown sex.	2055	60917	PP	950E150, trench, depth 5′, height–on bottom	Indet.	1	8–9	0	0	1	0	0	0	0	0	0	0	1	D 11 ± 2.5 yrs, PC 7.5–8.5 yrs	Indet.
29-2-10/60918.0	Partial cranial and fragmentary postcranial human remains of an adult male.	1997	60918	PP	1300E100, trench, depth 2′6″, height–bottom not known	Indet.	1	18–22	1	0	0	0	0	0	1	0	0	0	0	D c	M all features
29-2-10/60919.0	Fragmentary cranial and postcranial human remains of an adult female.	2013		PP	950E150, depth–grass roots, depth 2′6″	Indet.	1	35–54	1	0	0	0	0	0	0	0	1	0	0	D g-i	F all features
29-2-10/60921.0	Fragmentary cranial and postcranial human remains of a fetus of unknown sex.	2041		PP	Room e12, center of s wall, in cellar refuse	Indet.	1	Fetal 5–9 mths	0	0	0	0	1	0	0	0	0	0	1	D 7 ± 2 mths in utero, mandible not fused, PC < NB	Indet.
29-2-10/60922.0	Very fragmentary cranial and postcranial human remains of an infant of unknown sex.	2004		PP	E12, 2′ from floor 1, near north wall	Indet.	1	0–2.5	0	0	0	1	0	0	0	0	0	0	1	Size	Indet.
29-2-10/60923.0	Fragmentary cranial human remains of an infant of unknown sex.	2029		PP	950E150, lower tren., refuse, depth 6′, height–on red clay	Indet.	1	1–1.5	0	0	0	1	0	0	0	0	0	0	1	D 1.5 ± 0.5 yrs	Indet.
29-2-10/60924.0	Partial cranial and nearly complete postcranial human remains of a fetus of unknown sex.	2047		PP	Room e12, center of w wall, almost under post 4	Indet.	1	Fetal 7–9 mths	0	0	0	0	1	0	0	0	0	0	1	D 7 ± 2 mths in utero, PC < NB	Indet.

Peabody number	Pathology type	Pathology description	Cranial mod.	Associated funerary objects	Field context	Notes
29-2-10/60899.0	Nutr.	Mild cribra orbitalia	Unobs.			
29-2-10/60901.0	Nutr.	Enamel hypoplasia	Not obs.	Bowl sherds no. 335 (77267)		
29-2-10/60902.0	Arthritis, nutr.	Appendicular DJD, enamel hypoplasia	Occ. def.			
29-2-10/60903.0	Arthritis, nutr.	Marked DJD, vt/l osteophytes, enamel hypoplasia	Not obs.			
29-2-10/60907.0	Nutr.	Enamel hypoplasia	Not obs.	Gl. 2(?) bowl no. 374 (77199)		
29-2-10/60908.0			Unobs.	Bowl frags. no. 514 (77200)		
29-2-10/60909.0	Arthritis, nutr., other	Marked spinal osteophytosis, osteoporosis, vt/s5 not fused	No cran.	Gl. 3 bowl no. 288 (77176), blind corrugated sherds no. 334 (77260), Bisc. A bowl no. 405 (77161)		
29-2-10/60911.0	Infection, arthritis	Dental abscess upper R M1, loss of M2s, extensive DJD, vt–marked osteophyte formation, spondylytis deformans	Occ. def.	Frags. of 2 Gl. 2 bowls nos. 441 (77168), 442 (77189)		
29-2-10/60912.0	Infection, arthritis, nutr.	Upper R M3–deep caries, huge incisive canal–infection?, marked DJD, enamel hypoplasia	Not obs.			
29-2-10/60913.0			Not obs.			
29-2-10/60914.0			Unobs.	Bone awl no. 145 (77365), obsidian projectile point no. 146 (77486)		
29-2-10/60915.0			No cran.	Medicine outfit 21 pieces of crystal, glass, pebble, gypsum (77763)		
29-2-10/60916.0	Infection	Lower molars lost antemortem	Unobs.	Gl. 4 bowl no. 308 (77179), horn flaker no. 304 (77475)		
29-2-10/60917.0	Nutr.	Enamel hypoplasia	Slight lamb. def.	Gl. 2 bowl no. 399 (77170)		
29-2-10/60918.0	Nutr.	Enamel hypoplasia	Not obs.			Fm flattened, teeth not very shovel-shaped
29-2-10/60919.0	Infection, nutr.	Thickening skull vault, enamel hypoplasia	Unobs.			
29-2-10/60921.0			Unobs.			
29-2-10/60922.0			Unobs.			
29-2-10/60923.0			Unobs.			Description does not match field card
29-2-10/60924.0			Not obs.	Blind corrugated olla (nonumber)		

Peabody number	Object description	FN	Specimen label	Site	Intrasite provenience	Ceramic stage	MNI	Age (yrs.)	Ad	Su	Ch	In	Fe	Unk. age	M	M?	F	F?	Unk. sex	Age criteria	Sex criteria
29-2-10/60925.0	Nearly complete cranial and fragmentary postcranial human remains of an infant of unknown sex.	2028		PP	950E150, depth 3′, height 2′6″	Gl. 3	1	0–0.5	0	0	0	1	0	0	0	0	0	0	1	D 0.5 ± 0.25 yrs, PC NB–0.5 yrs	Indet.
29-2-10/60926.0	Fragmentary cranial and partial postcranial human remains of an infant of unknown sex.	2020		PP	950E150, depth 2′6″, height–few inches above ledge	Indet.	1	0.5–1.0	0	0	0	1	0	0	0	0	0	0	1	D 0.75 ± 0.25 yrs, PC 0.5–1.5 yrs	Indet.
29-2-10/60927.0	Nearly complete cranial and postcranial human remains of a child of unknown sex.	2021		PP	950E150, depth 1′6″, height 1′	Indet.	1	0.75–2.0	0	0	1	0	0	0	0	0	0	0	1	D 2 ± 0.66 yrs, PC 0.5–1.5 yrs	Indet.
29-2-10/60928.0	Partial cranial and postcranial human remains of a child of unknown sex.	2026		PP	950E150, depth 2′6″, height 6″	Indet.	1	1.5–2.5	0	0	1	0	0	0	0	0	0	0	1	D 3 ± 1 yrs, PC 2.5–3.5 yrs	Indet.
29-2-10/60929.0	Nearly complete cranial and postcranial human remains of an infant of unknown sex.	2001		PP	950E150, depth 4′, height 1′ in red clay	Transi-tional	1	0.75–1.25	0	0	0	1	0	0	0	0	0	0	1	D 1 ± 0.33 yrs, PC 0.5–1.5 yrs	Indet.
29-2-10/60930.0	Nearly complete cranial and postcranial human remains of an adult female.	2005	60930	PP	950E150, depth 4′ below surface, height–on bottom	Indet.	1	35–44	1	0	0	0	0	0	0	0	1	0	0	D f-g, T 7, SB 3, AS 5	F all features
29-2-10/60931.0	Partial cranial and fragmentary postcranial human remains of an adult female.	2004	60931/2005	PP	950E150, depth 4′6″, height–on clay	Indet.	1	50–59	1	0	0	0	0	0	0	0	1	0	0	D h+/i+, T 10, SB 6, AS 7	F all features
29-2-10/60932.0	Nearly complete cranial and partial postcranial human remains of an adult male.	1998	60932/2004	PP	950E150, lower tren., refuse depth 6′, height–on red clay	Indet.	1	50–59+	1	0	0	0	0	0	1	0	0	0	0	D h/i, AS 8	M all features
29-2-10/60933.0	Partial cranial and fragmentary postcranial human remains of an adult female.	2006	60933/1998	PP	985E178, depth–grass roots, height 2′6″	Indet.	1	18–22	1	0	0	0	0	0	0	0	1	0	0	D b2, PC young adult	F all features
29-2-10/60934.0	Fragmentary cranial and postcranial human remains of a child of unknown sex.	2052		PP	950E150, depth 3′, height–on bottom	Indet.	1	4.5–5.5	0	0	1	0	0	0	0	0	0	0	1	D 6 ± 2 yrs, PC 4.5–5.5 yrs	Indet.
29-2-10/60935.0	Partial cranial and postcranial human remains of an adult male.	2022	60935	PP	950E150, depth 4′2″, height–on floor	Indet.	1	20–24	1	0	0	0	0	0	1	0	0	0	0	D d, T 4, SB 2, AS 1	M all features
29-2-10/60936.0	Nearly complete cranial and postcranial human remains of an adult female.	1999	2022	PP	950E150, depth–about 3′, height–sunk in red clay	Indet.	1	60+	1	0	0	0	0	0	0	0	1	0	0	D i+, AS 8	Female (Pelvis F, Cranium F except SM, SR)
29-2-10/60937.0	Fragmentary cranial and postcranial human remains of a child of unknown sex.	2045	60937	PP	980E177, depth 1′6″, height 2′6″	Indet.	1	0.75–1.5	0	0	1	0	0	0	0	0	0	0	1	D 1–1.5 yrs ± 0.5 yrs, PC 0.5–1.5 yrs	Indet.
29-2-10/60938.0	Fragmentary cranial and partial postcranial human remains of a child of unknown sex.	2044		PP	Room e12, depth 9′±, height 1′1″	Indet.	1	1–2	0	0	1	0	0	0	0	0	0	0	1	D 3 ± 1 yrs, PC 0.5–1.5 yrs	Indet.
29-2-10/60939.0	Partial cranial and postcranial human remains of an infant of unknown sex.	1996		PP	Room e 12, depth 7′2″, height 2′3″	Indet.	1	1–1.5	0	0	0	1	0	0	0	0	0	0	1	D 1.5 ± 0.5 yrs, PC 0.5–1.5 yrs	Indet.
29-2-10/60940.0	Very fragmentary cranial and fragmentary postcranial human remains of an infant of unknown sex.	2003	60940	PP	950E150, depth 10″	Indet.	1	0–0.5	0	0	0	1	0	0	0	0	0	0	1	D 0.5 ± 0.25 yrs, PC NB–0.5 yrs	Indet.
29-2-10/60941.0	Fragmentary cranial and postcranial human remains of a child of unknown sex.	2016		PP	950E150, depth 2′, height 2′	Indet.	1	1.5–2.5	0	0	1	0	0	0	0	0	0	0	1	D 3 ± 1 yrs, PC 1.5–2.5 yrs	Indet.
29-2-10/60978.0	Fragmentary cranial and postcranial human remains of an adult male.	2015	60978/2016	PP	Plaza trench, depth 5′	Indet.	1	45–59	1	0	0	0	0	0	1	0	0	0	0	D h+/i+. T 9, SB 4–5, AS 7	M all features
29-2-10/60979.0	Very fragmentary cranial human remains of a child of unknown sex.	2032		PP	950E150, depth 2′, height 1′	Indet.	1	3–4	0	0	1	0	0	0	0	0	0	0	1	D 4 ± 1 yrs	Indet.

Peabody number	Pathology type	Pathology description	Cranial mod.	Associated funerary objects	Field context	Notes
29-2-10/60925.0			Not obs.			
29-2-10/60926.0			Unobs.	Projectile point no. 260 (10447)		
29-2-10/60927.0			Not obs.	Mixture of pot sherds, Gl. 3–4, Bisc. A, no number		
29-2-10/60928.0	Infection, arthritis	Periostitis fibula and L orbit, pseudarthrosis 3 vt/t and R ribs	Not obs.	49 turquoise beads and 2 pendants no. 234 (77647), Gl. 3 and 4 potsherds no number	Burial at length	
29-2-10/60929.0			Not obs.	Gl. 3 (?)transitional (?) bowl no. 403 (77175), 2 turquoise beads no. 233 (77648)		
29-2-10/60930.0	Other	Sacralization vt/l5	Occ. def.	Bisc. B bowl no. 91 (77162), worked pottery frag. no. 110 (77238), frag. of bone pendant no. 109 (77452)	Partially flexed burial	
29-2-10/60931.0	Trauma, infection, arthritis, nutr.	Healed fracture R femur, multiple dental abscesses incisors, canines, premolar, molars, DJD, slight osteoporosis	Occ. def.	Bisc. B bowl no. 87 (77163), clay frag. with textile impression no. 38 (77335)		
29-2-10/60932.0	Infection, arthritis, nutr., other	Dental abscesses upper L P2, lower R premolar and canine, DJD at right TMJ, spinal osteophytosis, enamel hypoplasia, spina bifida vt/s4–5	Occ. def.			Reassociated 29-2-10/60982.0 with 29-2-10/60932.0. These match burial card no. 2004 (MEM 4/98).
29-2-10/60933.0	Infection	Impacted M3	Not obs.			Reassociated human remains from 60932-1 (MEM 5/98)
29-2-10/60934.0	Other	Upper permanent Is erupting through maxilla, upper I2s also malpositioned	Unobs.			
29-2-10/60935.0	Infection, other	Abscess upper L C and P1–severe infection, very heavy M1 wear, vt/s1 lumbarization, vt/s4 incomplete fusion	Not obs.	Frags. Gl. 3 bowl no. 513 (77202), Gl. 4 sherds, no number		
29-2-10/60936.0	Infection, arthritis, tumor	All molars lost antemortem, DJD, spinal osteophytosis, exotosis L tibia	Occ. def.	Gl. 3, 4, and Bisc. B sherds, no number		
29-2-10/60937.0			Unobs.	Uncharred wood frags., no number		
29-2-10/60938.0			Unobs.	Frags. of corrugated pot with B&W interior decoration no. 432 (77193)		
29-2-10/60939.0			Not obs.	1 projectile point, Bisc. A sherd, no number		
29-2-10/60940.0			Unobs.	3 small worked sherds no. 23 (77237)		
29-2-10/60941.0			Unobs.			May be mixing with 0-60939 and 0-60938
29-2-10/60978.0	Infection, arthritis	Multiple abscesses and caries, antemortem loss upper R P2–M2, DJD, spinal osteophytosis	Occ. def.			
29-2-10/60979.0			Unobs.			

Peabody number	Object description	FN	Specimen label	Site	Intrasite provenience	Ceramic stage	MNI	Age (yrs.)	Ad	Su	Ch	In	Fe	Unk. age	M	M?	F	F?	Unk. sex	Age criteria	Sex criteria
29-2-10/60980.0	Nearly complete cranial and fragmentary postcranial human remains of an adult male.	2000		PP	985E155, depth 5′6″, height–on clay bottom	Indet.	1	45–55+	1	0	0	0	0	0	1	0	0	0	0	D h+/i+	M all features
29-2-10/60981.0	Very fragmentary cranial and postcranial human remains of a child of unknown sex.	2007		PP	950E150, depth 16″, height 2′6″	Indet.	1	1.5–2.5	0	0	1	0	0	0	0	0	0	0	1	D 3 ± 1 yrs, PC 1.5–2.5 yrs	Indet.
29-2-10/60983.0	Nearly complete cranial human remains of an adult that is probably female.	2025	60983, 2025 (on skull)	PP	950E150, depth 4′, height–entirely in red clay bottom	Transi-tional, Bisc. A&B	1	18–22	1	0	0	0	0	0	0	0	0	1	0	D c	?Female (quite robust, prog-nathic)
29-2-10/60984.0	Fragmentary cranial and postcranial human remains of an adult female.	2010	60984	PP	950E150, trench, depth 18″, height unknown	Indet.	1	40–44	1	0	0	0	0	0	0	0	1	0	0	D h, T 8, SB 4, AS 5	Female
29-2-10/60985.0	Nearly complete cranial and postcranial human remains of a child of unknown sex.	2009	60985/2009	PP	950E150 trench, depth 3′, height–on bottom	Gl. 3	1	5–6	0	0	1	0	0	0	0	0	0	0	1	D 6-7 ± 2 yrs, PC 5–6 yrs	Indet.
29-2-10/60986.0	Nearly complete cranial and postcranial human remains of an adult female.	2051	60986/2051	PP	950E150, depth 4′, height–on floor	Gl. 1	1	16–20	1	0	0	0	0	0	0	0	1	0	0	D b1, spheno-occipit al incomplete fusion, T 1, SB 1, AS 1	F all features
29-2-10/60987.0	Nearly complete cranial and postcranial human remains of an adult that is probably male.	2048	2048	PP	950E150, depth 4′, height 1′6″	Indet.	1	50–59	1	0	0	0	0	0	0	1	0	0	0	D i+, AS 7	?Male (Cranium-p ath., Pelvis more M)
29-2-10/60988.0	Fragmentary cranial and postcranial human remains of an adult male.	2040		PP	Room 12, depth 6′, height 4′	Indet.	1	35–45	1	0	0	0	0	0	1	0	0	0	0	D g-h/h	Male
29-2-10/60989.0	Nearly complete cranial and partial postcranial human remains of an adult male.	2038	60989	PP	950E150, depth 4′4″, height–on clay	Indet.	1	50–59+	1	0	0	0	0	0	1	0	0	0	0	D i+, AS 7–8	M, all features
29-2-10/60990.0	Partial cranial and postcranial human remains of an adult that is probably female.	2037	60990	PP	Room e12, depth 6′1″	Indet.	1	16–20	1	0	0	0	0	0	0	0	0	1	0	D b2, T 1, SB 1, AS 1	?Female (cranium-NC M, MP F, ME 3, pelvis- PS 3, SN 3)
29-3-10/60993.0	Nearly complete cranial and postcranial human remains of a child of unknown sex.	2002	60993/2002	PP	950E150, depth 3′6″, height–sunk in bottom	Indet.	1	11.5–13.5	0	0	1	0	0	0	0	0	0	0	1	D 10–11 ± 2.5 yrs, PC 11.5–13.5 yrs	Indet.
29-3-10/60994.0	Nearly complete cranial and partial postcranial human remains of an adult male.	2018	60994/2018	PP	950E150, depth 2′6″, height 1′6″	Indet.	1	27–34	1	0	0	0	0	0	1	0	0	0	0	D d-e/e, T 5, SB 3, AS 3	M all features
36-116-10/N1472.0	Fragmentary cranial human remains of an adult that is probably female.	2056	N1472/2056	PP	1250E0, test 20, layer 6, depth 4′, height 6′	Indet.	1	35–45	1	0	0	0	0	0	0	0	0	1	0	D g/h	?Female (SR F, SM3)
36-95-10/N1466.0	Fragmentary cranial human remains of a child of unknown sex.	Bb145	N1466	FL	Pit in trench d, n edge, depth 8′, height 2′	Probably B&W	1	4–6	0	0	1	0	0	0	0	0	0	0	1	D 6 ± 2 yrs	Indet.
45-1-10/N3986.2	Fragmentary cranial human remains of an infant of unknown sex.	1251	N3986/1251	PP		Indet.	1	0.5–1.0	0	0	0	1	0	0	0	0	0	0	1	D 1 ± 0.3 yrs	Indet.
985-27-10/66196.0	Very fragmentary postcranial human remains of an adult of unknown sex.	33526	33526	PP	NE midden edge just outside fence, 30 ft. N up slope, tr. 3 ft. deep 15 ft. wide	1400–1838	1		1	0	0	0	0	0	0	0	0	0	1		Indet.

Peabody number	Pathology type	Pathology description	Cranial mod.	Associated funerary objects	Field context	Notes
29-2-10/60980.0	Trauma, infection, arthritis	Healed nasal fracture, severe dental abscesses–loss of upper L P2–M2, lower R M1 and M3, and lower LM3 due to abscesses, upper R M1 abscess, DJD	Occ. def.	Gl. 3 bowl no. 288 (77176), blind corrugated sherds no. 334 (77260), Bisc. A bowl no. 405 (77161)		
29-2-10/60981.0			Unobs.			
29-2-10/60983.0	Nutr.	Cribra orbitalia	Occ. def.	Bone awl no. 236 (77367), worked sherd no. 235 (77239), sherd frags., no number		Very prognathic
29-2-10/60984.0	Infection, arthritis, other	Abscess upper L M1 and lower R M2, DJD, vt/s4–5 not fused	Unobs.	Gl. 4 bowl no. 90 (78051), prismatic polishing stone (?) no. 112 (77538), worked sherds and 1 Hopi sherd no. 111 (77272), frags. of 2 Gl. 3 bowls nos. 516 (77177), 89 (77173), Gl. 3 (?) bowl no. 127 (77174)		
29-2-10/60985.0	Other	Spina bifida vt/s4–5 wide open, only vt/s1 fused (dolichocephalic)	Not obs.	Bisc. B (?) frag. no. 128 (77164)		
29-2-10/60986.0	Nutr.	Enamel hypoplasia	Occ. def.	Bowl frags. no. 515 (77201), Gl. 1 yellow and Bisc. A sherds, no number		
29-2-10/60987.0	Infection, arthritis, other	Complete antemortem loss maxillary dentition, extensive loss mandibular teeth, osteomyelitis cervical vt, DJD, spinal osteophytosis, ankylosis vt/c5–6, spondylytis deformans, scaphocephaly	Not obs.	Flint scraper no. 484 (77588)		
29-2-10/60988.0	Infection, arthritis, nutr.	Abscesses and caries, DJD, enamel hypoplasia	Unobs.	Transitional B&W–Bisc. A bowl no. 394 (77165), cooking pot no. 395 (77191), B&W bowl frags. no. 396 (77166)		Sciatic notch, overall size, mental eminence and mastoid–male
29-2-10/60989.0	Infection, arthritis	Dental abscesses, severe antemortem tooth loss, spinal osteophytosis, DJD	Not obs.			Unusual perforation in scapula near coracoid
29-2-10/60990.0	Trauma, infection, nutr.	Healed trauma R humeral shaft, infection R ulna and radius, enamel hypoplasia	Occ. def.	Gl. 4 bowl no. 400 (77180)		
29-3-10/60993.0	Other	No upper lateral incisors	Occ. def.	Projectile point no. 29 (77485)		
29-3-10/60994.0	Arthritis, other	DJD, vt/s4–5 not fused	Not obs.	Bowl sherds no. 402 (77285)		Dome-shaped individual–looks very different
36-116-10/N1472.0	Tumor	Button osteoma on R parietal	Unobs.	Carbonized, twilled tube mat, no number		Field card notes occipital deformation
36-95-10/N1466.0			Cran. def.			
45-1-10/N3986.2			Unobs.			
985-27-10/66196.0	Arthritis	Spinal osteophytosis	No cran.			

Peabody number	Object description	FN	Specimen label	Site	Intrasite provenience	Ceramic stage	MNI	Age (yrs.)	Ad	Su	Ch	In	Fe	Unk. age	M	M?	F	F?	Unk. sex	Age criteria	Sex criteria
985-27-10/N10649.0	Fragmentary postcranial human remains of an adult of unknown sex.	33524	33524	PP	NE midden edge just outside fence, 30 ft. N up slope, tr. 3 ft. deep 15 ft. wide	1400–1838	1		1	0	0	0	0	0	0	0	0	0	1		Indet.
985-27-10/N10650.0	Fragmentary postcranial human remains of an adult of unknown sex.	33525	33525	PP	NE midden edge just outside fence, 30 ft. N up slope, tr. 3 ft. deep 15 ft. wide	1400–1838	1		0	0	1	0	0	0	0	0	0	0	1		Indet.
997-27-10/N10668.0	Fragmentary postcranial human remains of an infant of unknown sex.	Bb155	N10668/ 997-27-10	FL		Probably B&W	1	0.75–1.25	0	0	0	1	0	0	0	0	0	0	1	PC 0.5–1.5 yrs	Indet.
997-27-10/N10669.0	Fragmentary cranial and postcranial human remains of an adult of unknown sex.	Bb154	N10669/ 997-27-10	FL		Probably B&W	1	16–20	1	0	0	0	0	0	0	0	0	0	1	D b2, PC < 35 yrs	Indet.
997-27-10/N10670.0	Fragmentary cranial human remains of an adult of unknown sex.	Bb153	N10670/ 997-27-10	FL	Burnt house layer, trench d	Probably B&W	1	16–22	1	0	0	0	0	0	0	0	0	0	1	D b2-c	Indet.

Peabody number	Pathology type	Pathology description	Cranial mod.	Associated funerary objects	Field context	Notes
985-27-10/N10649.0			No cran.			
985-27-10/N10650.0			No cran.			
997-27-10/N10668.0			No cran.	Left femur of an immature lagomorph		Remarks:Transferred from RSPM.
997-27-10/N10669.0			Unobs.			Remarks:Transferred from RSPM.
997-27-10/N10670.0			Unobs.			Bone has been burnt

Appendix B

Information on Human Remains from the Upper Pecos Valley That Were Not Used in the Chapter 4 Analysis

Michèle E. Morgan

The data for each Peabody Number continue from the left page to the right page.

Key to Column Heads	
Peabody number	Unique number assigned by the Peabody Museum of Archaeology and Ethnology
Object description	Brief, standardized description of human remains
FN	Field number
Specimen label	All numbers, words, etc., recorded on human remains. Notations in the form of "(MEM 11/96)" indicate that the author labeled the specimens on the given date.
Site	DR (Dick's Ruin), FL (Forked Lightning), LL (Loma Lothrup), PP (Pecos Pueblo), R (Rowe)
Intrasite provenience	Information quoted directly from field cards and other primary sources
Ceramic stage	Abbreviated, e.g., "B&W" (Black-on-white), "Pres. Gl. 2 or 3" (presumably Glaze 2 or 3).
MNI	Minimum number of individuals
Age (yrs.)	Age estimate in years
Ad	Adult. Describes developmental age on the basis of physical maturation of the remains determined by full skeletal maturation, which could include eruption of third molars and fusion of epiphyses and the basilar suture.
Su	Subadult. Describes developmental age on the basis of physical maturation of the remains determined by the eruption of the second permanent premolar or a postcranial age estimate of ≥ 12 years.
Ch	Child. Describes developmental age on the basis of physical maturation of the remains determined by the presence of complete deciduous dentition or a postcranial age estimate of ≥ 2.5 years.

In	Infant. Describes developmental age on the basis of physical maturation of the remains determined by long bone measurements and tooth maturation.
Fe	Fetus. Describes developmental age on the basis of physical maturation of the remains determined by long bone measurements and tooth maturation.
Unk. age	Unknown age. An insufficient number of features were present to allow determination of the physical maturation of the remains.
M	Identified as male.
M?	Identified as probably male.
F	Identified as female.
F?	Identified as probably female.
Unk. sex	Undetermined sex. An insufficient number of features were present to allow a confident determination of the sex of the individual.
Age criteria	Criteria used to assess age: D (dental stage as reproduced in White 1991), T (Todd), S (Suchey-Brooks), AS (auricular surface), PC (postcranial age based on epiphyseal fusion and/or metric data)
Sex criteria	Female (indicates all or nearly all criteria support female); F all features (all criteria indicate female); ?Female (majority of criteria support female); Male (indicates all or nearly all criteria support male); M all features (all criteria indicate male); ?Male (majority of criteria support male); Indet. (indicates that sex could not be confidently determined). Pelvic criteria abbreviated, "VA" (ventral arc), "SPC" (subpubic cavity), "IPR" (ischiopubic ramus), "SN" (sciatic notch), "PS" (preauricular sulcus); cranial criteria abbreviated, "NC" (nuchal crest), "MP" (mastoid process), "SM" (supraorbital margin), "SR" (supraorbital ridge), "ME" (mental eminence)
Pathology type	Presence of any of six categories of pathology: trauma, infection, arthritis, tumor, nutritional/physiological stress ("Nutr."), other (includes neurological and all other pathologies)
Pathology description	Brief notes describing pathology, abbreviated, e.g., "DJD" (degenerative joint disease), "TMJ" (temporal mandibular joint)
Cranial mod.	Cranial modification, abbreviated, e.g., "Sl. occ. def." (slight occipital deformation), "Lamb." (lambdoidal), "Unobs." (unobservable), "Not obs." (not observed).
Associated funerary objects	Objects listed on NAGPRA inventory, quoting directly from field cards (FCs). ID numbers preceded by "no." are field numbers; those in parentheses are Peabody Andover numbers.
Field context	Notes recorded on field card that pertain to context. "Agua Fria," "Pajaritan," and "Frijolito" refer to categories of pottery.
Notes	Additional descriptive notes

Peabody number	Object description	FN	Specimen label	Site	Intrasite provenience	Ceramic stage	MNI	Age (yrs.)	Ad	Su	Ch	In	Fe	Unk. age	M	M?	F	F?	Unk. sex	Age criteria	Sex criteria
00-0-10/N9468.35	Very fragmentary cranial human remains of 2 infants, 3 children, 1 subadult, and 6 adults, all of unknown sex.	1716/2b and others	9468-35 (MEM 11/96)	PP	Trench n of sp house, depth 9′, height 2′, and other locations	Indet.	12		6	1	3	2	0	0	0	0	0	0	12		Indet.
00-0-10/N9468.51	Very fragmentary cranial human remains of 1 child and 1 adult, both of unknown sex.	Bb	9468-51 (MEM 11/96)	FL		Probably B&W	2		1	0	1	0	0	0	0	0	0	0	2		Indet.
00-0-10/N9468.64	Fragmentary cranial human remains of an infant of unknown sex.	Dr7	Dr7	DR	Room 31	Indet.	1	0–0.5	0	0	0	1	0	0	0	0	0	0	1	D 0.5 ± .25 yrs	Indet.
19-6-10/59839a	Very fragmentary cranial and postcranial human remains of an individual of unknown age and sex.	478	59839a	PP	1000E150, depth 30″, height–above red clay 44″	Probably Gl. 4 or 5	1		0	0	0	0	0	1	0	0	0	0	1		Indet.
19-6-10/59845.1	Reassociated with 59849.	552		PP	1550W50	B&W or Gl. 1	0		0	0	0	0	0	0	0	0	0	0	0		
19-6-10/59878.0	Very fragmentary postcranial human remains of 3 adults of unknown sex.	C	59878	PP	Church	Indet.	3		3	0	0	0	0	0	0	0	0	0	3		Indet.
19-6-10/59878.2	Very fragmentary postcranial human remains of an adult of unknown sex.	C	59878-3	PP	Church	Indet.	1		1	0	0	0	0	0	0	0	0	0	1		Indet.
19-6-10/59878.5	Very fragmentary postcranial human remains of 4 adults of unknown sex.	C	59878	PP	Church	Indet.	4		4	0	0	0	0	0	0	0	0	0	4		Indet.
19-6-10/59879.0	Very fragmentary postcranial human remains of 4 adults of unknown sex.	C	59879-8, -12, -11, -4, -9, -10. -6	PP	Church	Indet.	4		4	0	0	0	0	0	0	0	0	0	4		Indet.
19-6-10/59880.0	Very fragmentary postcranial human remains of 2 adults of unknown sex.	C	59880-9, -10	PP	Church	Indet.	2		2	0	0	0	0	0	0	0	0	0	2		Indet.
19-6-10/59882.0	Very fragmentary postcranial human remains of 3 adult males.	C	59882	PP	Church	Indet.	3		3	0	0	0	0	0	3	0	0	0	0		Male
19-6-10/59884.0	Very fragmentary postcranial human remains of 1 subadult and 10 adults, all of unknown sex.	C	59884	PP	Church	Indet.	11		10	1	0	0	0	0	0	0	0	0	11		Indet.
19-6-10/59885.0	Reassociated.	C	59885	PP	Church	Indet.	0		0	0	0	0	0	0	0	0	0	0	0		
19-6-10/59886.0	Reassociated.	C	59886	PP	Church	Indet.	0		0	0	0	0	0	0	0	0	0	0	0		
19-6-10/59887.0	Very fragmentary postcranial human remains of 3 adults of unknown sex.	C	59987	PP	Church	Indet.	3		3	0	0	0	0	0	0	0	0	0	3		Indet.
19-6-10/60024.0	Partial cranial and postcranial human remains of an adult that is probably male.	Rowe 11	60024	R	N130E40	Indet.	1	45–54	1	0	0	0	0	0	0	1	0	0	0	D i	?Male
19-6-10/60037.0	Nearly complete cranial and fragmentary postcranial human remains of a subadult male.	Rowe 10	60037	R	Room viii, north side, depth 7′9″	Indet.	1	16–20	0	1	0	0	0	0	1	0	0	0	0	D b2, PC 16–20	M all features
19-6-10/60069.0	Partial cranial and postcranial human remains of a child of unknown sex.	Rowe 15	60069	R	N130E30, 43" w of w wall middle ix, depth 87″	Indet.	1	5–6	0	0	1	0	0	0	0	0	0	0	1	D 7 ± 2 yrs, PC 4.5–6.5 yrs	Indet.
19-6-10/60105.0	Nearly complete cranial human remains of an adult male.	LL2	60105, LL2	LL	Under north wall, depth 3′ underground, 1′ under flr level	Indet.	1	40–50+	1	0	0	0	0	0	1	0	0	0	0	D h+	Male, cranium–all but MP indicate male
19-6-10/60110.0	Fragmentary postcranial human remains of 3 adults of unknown sex.	C	60110	PP	Church	Indet.	3		3	0	0	0	0	0	0	0	0	0	3		Indet.
19-6-10/60111.2	Very fragmentary postcranial human remains of 3 adults of unknown sex.	C	60111	PP	Church	Indet.	3		3	0	0	0	0	0	0	0	0	0	3		Indet.
19-6-10/60114.0	Fragmentary postcranial human remains of 3 adults of unknown sex.	C	60114-1, -2, -3, -4, -6	PP	Church	Indet.	3		3	0	0	0	0	0	0	0	0	0	3		Indet.

Peabody number	Pathology type	Pathology description	Cranial mod.	Associated funerary objects	Field context	Notes
00-0-10/N9468.35	Nutr., other	Enamel hypoplasia, malerupted M3	Unobs.	Undecorated red bowl no. 1360 (74934)		Primarily teeth, field number problematic
00-0-10/N9468.51	Nutr.	Enamel hypoplasia	Unobs.			May be mixing with N9468-49 and N9468-50
00-0-10/N9468.64			Unobs.			
19-6-10/59839a			Unobs.	Bone awl no. 1563 (67842)	Pottery distinctly Pajaritan	
19-6-10/59845.1						Reassociated with 59849 (MEM 10/96)
19-6-10/59878.0			No cran.			3 pairs femora
19-6-10/59878.2			No cran.			Reassociated by Owsley team as BB (1/98)
19-6-10/59878.5			No cran.			Femora
19-6-10/59879.0			No cranium			
19-6-10/59880.0			No cranium			Reassociated 3 fibulae with tibiae and recatalogued as 59879-1, 59879-2, 59879-3
19-6-10/59882.0			No cranium			
19-6-10/59884.0	Trauma, arthritis	Healed fracture, DJD	No cranium			
19-6-10/59885.0						Reassociated the 4 radii with 4 ulnae from 59886, renumbered as 59885-1, 59885-3, 59885-4, 59885-5 (MEM 12/96)
19-6-10/59886.0						Reassociated 4 left ulnae with radii from 59885, renumbered as 59885-1, 59885-3, 59885-4, 59885-5. (MEM 12/96)
19-6-10/59887.0			No cranium			
19-6-10/60024.0	Arthritis, nutr.	Severe DJD–especially vertebrae, osteoporosis	Lamb. def.	2 B&W bowls nos. 108 (66636), 109 (66637), clay pipe frags. no. 107 (66708)	Imprint of blanket on ground	
19-6-10/60037.0	Nutr.	Enamel hypoplasia	Occ. def.	3 B&W bowls nos. 88 (66633), 89 (66634), 103 (66635)	Burial at length	Field number problematic
19-6-10/60069.0			Not obs.	Sherd from bottom of black corrugated pot, no number		
19-6-10/60105.0	Infection	Periodontal disease	Not obs.			
19-6-10/60110.0			No cranium			
19-6-10/60111.2	Trauma	Healed fracture on 60111-2	No cranium			Tibia
19-6-10/60114.0			No cranium			

Peabody number	Object description	FN	Specimen label	Site	Intrasite provenience	Ceramic stage	MNI	Age (yrs.)	Ad	Su	Ch	In	Fe	Unk. age	M	M?	F	F?	Unk. sex	Age criteria	Sex criteria
19-6-10/60115.0	Fragmentary postcranial human remains of 2 adults of unknown sex.	C	60115-1, 60115-2	PP	Church	Indet.	2		2	0	0	0	0	0	0	0	0	0	2		Indet.
19-6-10/60116.0	Fragmentary postcranial human remains of 2 adults of unknown sex.	C	60116	PP	Church	Indet.	2		2	0	0	0	0	0	0	0	0	0	2		Indet.
19-6-10/60120.0	Fragmentary postcranial human remains of 1 child and 1 adult, both of unknown sex.	C	60120G, 60120 3, 60120 2	PP	Church	Indet.	2		1	0	1	0	0	0	0	0	0	0	2	mixed	Indet.
19-6-10/60139.0	Partial cranial and fragmentary postcranial human remains of an adult female.	LL4	60139	LL	North wall, under floor level	Indet.	1	27–34	1	0	0	0	0	0	0	0	1	0	0	D g?, T 5, SB 3, AS 3–4	F all features
19-6-10/60141.0	Partial cranial human remains of an adult female.	LL1	60141, LL1	LL	2′ from north wall	Indet.	1	18–22	1	0	0	0	0	0	0	0	1	0	0	D c	Female
19-6-10/60141.1	Fragmentary cranial and postcranial human remains of an adult male.	LL1	60141	LL	2′ from north wall	Indet.	1	16–20	1	0	0	0	0	0	1	0	0	0	0	D b2	Male
19-6-10/60160.0	Fragmentary cranial and partial postcranial human remains of an adult female.	Rowe 14	60160	R	N130E30 (base 3rd wall, just w of w wall of mid ix), depth 69″	Indet.	1	35–39	1	0	0	0	0	0	0	0	1	0	0	D g	F all features
19-6-10/60186.1	Fragmentary postcranial human remains of 1 fetus, 2 infants, 5 children and 1 adult, all of unknown sex			PP	North terrace	Indet.	9		1	0	5	2	1	0	0	0	0	0	9		Indet.
19-6-10/60187.0	Fragmentary cranial human remains of 3 children of unknown sex.		60187-11	PP		Indet.	3	2–3,2–3, 3–4	0	0	3	0	0	0	0	0	0	0	3	D 2 are 3 ± 1 yrs, 1 is 4 ± 1 yrs	Indet.
19-6-10/60625.0	Very fragmentary postcranial human remains of an adult of unknown sex.		60625	PP		Indet.	1		1	0	0	0	0	0	0	0	0	0	1		Indet.
19-6-10/60645.0	Reassociated.	C	60645	PP	Church	Indet.	0		0	0	0	0	0	0	0	0	0	0	0		
19-6-10/60645.1	Problematic. Partial cranial human remains of an adult of unknown sex.			PP		Indet.	1	40–49	1	0	0	0	0	0	0	0	0	0	1	D h	
19-6-10/60645.2	Reassociated.			PP?			0		0	0	0	0	0	0	0	0	0	0	0	D b2	
19-6-10/60957.0	Fragmentary cranial and postcranial human remains of 1 adult female and 2 adult males.	1697	1697/60957	PP	1250E50, east trench, depth 2′6″, height 1″	Indet.	3	25–39 for all	3	0	0	0	0	0	2	0	1	0	0	Mixed	2 Male, 1 Female
19-6-10/60958.1	Very fragmentary cranial and postcranial human remains of 1 child and 1 subadult, both of unknown sex.	C	60958	PP	Church	Indet.	2		0	1	1	0	0	0	0	0	0	0	2	Child, subadult	Indet.
19-6-10/60959.0	Very fragmentary postcranial human remains of 1 subadult and 4 adults, all of unknown sex.	C		PP	Church	Indet.	5		4	1	0	0	0	0	0	0	0	0	5		Indet.
19-6-10/60960.0	Very fragmentary postcranial human remains of 8 adults of unknown sex.	C		PP	Church	Indet.	8		8	0	0	0	0	0	0	0	0	0	8		Indet.
19-6-10/60961.0	Very fragmentary postcranial human remains of 4 adults of unknown sex.	C		PP	Church	Indet.	4		4	0	0	0	0	0	0	0	0	0	4		Indet.
19-6-10/60962.0	Very fragmentary postcranial human remains of 7 adults of unknown sex.	C		PP	Church	Indet.	7		7	0	0	0	0	0	0	0	0	0	7		Indet.
19-6-10/60963.0	Very fragmentary postcranial human remains of 1 subadult and 2 adults of unknown sex.	C		PP	Church	Indet.	3		2	1	0	0	0	0	0	0	0	0	3		Indet.
19-6-10/60964.0	Very fragmentary postcranial human remains of 1 subadult male and 1 adult female.	C	60964	PP	Church	Indet.	2	18–19, 50+	1	1	0	0	0	0	1	0	1	0	0	Subadult T 1, SB 1, AS 1, Adult–AS 7	Mixed
19-6-10/60968.0	Very fragmentary postcranial human remains of 1 adult female and 1 adult male.	C	60968	PP	Church	Indet.	2	F 35–39, M 60+	2	0	0	0	0	0	1	0	1	0	0	M–T 10, SB 6, AS 8, F–AS 4	Male, Female

Peabody number	Pathology type	Pathology description	Cranial mod.	Associated funerary objects	Field context	Notes
19-6-10/60115.0			No cranium			
19-6-10/60116.0			No cranium			Reassociated 2 pairs humeri 60116-2, -5, and 60116-7, -6; removed and recat. 3 humeri as 60116-1 and 60116-2, MEM 2/97
19-6-10/60120.0	Trauma, arthritis	Healed fracture, spinal osteophytosis	No cranium			2 sacra
19-6-10/60139.0	Arthritis, other	Arthritis, very large maxillary incisive canal	No cran. def.			
19-6-10/60141.0			Not obs.	Bone awl LL5 (66465), frags. Gl. D bowl no. LL28 (66488)		
19-6-10/60141.1			Unobs.	Bone awl LL5 (66465), frags. Gl. D bowl no. LL28 (66488)		Field number problematic
19-6-10/60160.0	Nutr.	Osteoporosis	Unobs.			
19-6-10/60186.1			No cranium			Provenience information copied from 60186 and is uncertain for 60186-1
19-6-10/60187.0			Unobs.			
19-6-10/60625.0	Infection	Osteomyelitis on R ulna	No cranium			R ulna only present
19-6-10/60645.0						Reassociated and recatalogued as 60977-11 (MEM 3/98)
19-6-10/60645.1						Pronounced occipital bun, very long low cranium with sagital crest. Very rectangular orbits, small zygomatic arches tucked back, very small teeth. Probably not Pecos.
19-6-10/60645.2						Reassociated
19-6-10/60957.0	Infection, arthritis, other	Frontal lesion, spinal osteophytosis, 1 vt/c spondylitis deformans, 1 sacrum–vt/s3–5 unfused and lumbarization of vt/s1	Unobs.	Sherds, no number		1 older female, 1 young male, 1 male, field card description problematic
19-6-10/60958.1			Unobs.			
19-6-10/60959.0			No cranium			Humeri
19-6-10/60960.0	Arthritis	DJD	No cranium			Fibulae only
19-6-10/60961.0			No cranium			Clavicles only
19-6-10/60962.0			No cranium			Radii only
19-6-10/60963.0	Arthritis	Slight DJD	No cranium			Ulnae only
19-6-10/60964.0			No cranium			2 innominates
19-6-10/60968.0			No cranium			2 innominates

Peabody number	Object description	FN	Specimen label	Site	Intrasite provenience	Ceramic stage	MNI	Age (yrs.)	Ad	Su	Ch	In	Fe	Unk. age	M	M?	F	F?	Unk. sex	Age criteria	Sex criteria
19-6-10/60969.0	Very fragmentary postcranial human remains of 1 subadult and 4 adults, all of unknown sex.	C		PP	Church	Indet.	5		4	1	0	0	0	0	0	0	0	0	5		Indet.
19-6-10/60970.0	Very fragmentary postcranial human remains of 2 adult males.	C		PP	Church	Indet.	2		2	0	0	0	0	0	2	0	0	0	0		Males
19-6-10/60970.1	Very fragmentary postcranial human remains of 2 adults of unknown sex.	C		PP	Church	Indet.	2		2	0	0	0	0	0	0	0	0	0	2		Indet.
19-6-10/60970.2	Very fragmentary postcranial human remains of 2 adults of unknown sex.	C		PP	Church	Indet.	2		2	0	0	0	0	0	0	0	0	0	2		Indet.
19-6-10/60970.3	Very fragmentary postcranial human remains of 1 subadult and 5 adults, all of unknown sex.	C		PP	Church	Indet.	6		5	1	0	0	0	0	0	0	0	0	6		Indet.
19-6-10/60971.0	Reassociated.	C		PP	Church	Indet.	0		0	0	0	0	0	0	0	0	0	0	0		
19-6-10/60971.13	Very fragmentary postcranial human remains of 1 subadult and 7 adults, all of unknown sex.	C		PP	Church	Indet.	8		7	1	0	0	0	0	0	0	0	0	8		Indet.
19-6-10/60972.0	Reassociated.	C		PP	Church	Indet.	0		0	0	0	0	0	0	0	0	0	0	0		
19-6-10/60972.4	Very fragmentary postcranial human remains of 7 adults of unknown sex.	C		PP	Church	Indet.	7		7	0	0	0	0	0	0	0	0	0	7		Indet.
19-6-10/60973.0	Reassociated.	C		PP	Church	Indet.	0		0	0	0	0	0	0	0	0	0	0	0		
19-6-10/60973.1	Reassociated.	C		PP	Church	Indet.	0		0	0	0	0	0	0	0	0	0	0	0		
19-6-10/60974.0	Reassociated.	C		PP	Church	Indet.	0		0	0	0	0	0	0	0	0	0	0	0		
19-6-10/60974.1	Very fragmentary postcranial human remains of 8 adults of unknown sex.	C		PP	Church	Indet.	8		8	0	0	0	0	0	0	0	0	0	8		Indet.
19-6-10/60975.0	Reassociated.	C		PP	Church	Indet.	0		0	0	0	0	0	0	0	0	0	0	0		
19-6-10/60975.1	Very fragmentary postcranial human remains of 6 adults of unknown sex.	C		PP	Church	Indet.	6		6	0	0	0	0	0	0	0	0	0	6		Indet.
19-6-10/60976.0	Reassociated.	C		PP	Church	Indet.	0		0	0	0	0	0	0	0	0	0	0	0		
19-6-10/60976.1	Very fragmentary postcranial human remains of 4 subadults and 13 adults, all of unknown sex.	C		PP	Church	Indet.	17		13	4	0	0	0	0	0	0	0	0	17		Indet.
19-6-10/60977.0	Reassociated.	C	60977	PP	Church	Indet.	0		0	0	0	0	0	0	0	0	0	0	0		
19-6-10/60977.1	Fragmentary cranial and postcranial human remains of 1 adult male and 1 adult of unknown sex.	C	60977	PP?	Church	Indet.	2	35–39	2	0	0	0	0	0	1	0	0	0	1	D g	Male, Indet.
19-6-10/60977.2	Very fragmentary cranial and fragmentary postcranial human remains of a subadult of unknown sex.	C		PP?	Church	Indet.	1	16–20	0	1	0	0	0	0	0	0	0	0	1	D b2	Indet.
19-6-10/60995.0	Nearly complete cranial human remains of an adult male.			PP		Indet.	1	18–22	1	0	0	0	0	0	1	0	0	0	0	D c	M all features
19-6-10/60996.0	Fragmentary cranial and very fragmentary postcranial human remains of an adult that is probably female.			PP		Indet.	1	50–59	1	0	0	0	0	0	0	0	0	1	0	D very old	?Female
19-6-10/60997.0	Fragmentary cranial and very fragmentary postcranial human remains of an adult male.			PP		Indet.	1	>50+	1	0	0	0	0	0	1	0	0	0	0	D old	Male
19-6-10/60998.0	Partial cranial and postcranial human remains of a child of unknown sex.			PP		Indet.	1	5.5–6.5	0	0	1	0	0	0	0	0	0	0	1	D 7 ± 2 yrs; PC 5.5–6.5 yrs	Indet.

Peabody number	Pathology type	Pathology description	Cranial mod.	Associated funerary objects	Field context	Notes
19-6-10/60969.0	Arthritis	DJD, spinal osteophytosis	No cranium			
19-6-10/60970.0			No cranium			
19-6-10/60970.1			No cranium			2 humeri
19-6-10/60970.2			No cranium			2 radii
19-6-10/60970.3			No cranium			Ulnae
19-6-10/60971.0						Reassociated with other elements from the church, MEM 4/98.
19-6-10/60971.13			No cranium			Femora
19-6-10/60972.0						Reassociated with other elements from the church, MEM 4/98.
19-6-10/60972.4	Infection, nutr., other	1 tibia has infection at midshaft, 1 tibia has osteoporosis, 1 tibia has rodent gnawing	No cranium			Tibia
19-6-10/60973.0						Reassociated with other elements from the church, MEM 4/98.
19-6-10/60973.1						Reassociated with other elements from the church, MEM 4/98.
19-6-10/60974.0						Reassociated with other elements from the church, MEM 4/98.
19-6-10/60974.1			No cranium			Ulnae
19-6-10/60975.0						Reassociated with other elements from the church, MEM 4/98.
19-6-10/60975.1			No cranium			Clavicles
19-6-10/60976.0						Reassociated with other elements from the church, MEM 4/98.
19-6-10/60976.1			No cranium			Fibulae
19-6-10/60977.0						Reassociated with other elements from the church, MEM 4/98.
19-6-10/60977.1	Arthritis	Spinal osteophytosis	Unobs.			
19-6-10/60977.2	Infection	Periosteal reaction on mandible	Unobs.			
19-6-10/60995.0	Nutr., other	Enamel hypoplasia, extra bud on upper L M3	Not obs.			Very different look but is definitely Native American
19-6-10/60996.0	Infection	Nearly edentulous	Unobs.			
19-6-10/60997.0	Infection	Nearly edentulous?	Unobs.			
19-6-10/60998.0	Other	Sacral spina bifida	Not obs.			

Peabody number	Object description	FN	Specimen label	Site	Intrasite provenience	Ceramic stage	MNI	Age (yrs.)	Ad	Su	Ch	In	Fe	Unk. age	M	M?	F	F?	Unk. sex	Age criteria	Sex criteria
19-6-10/60999.0	Fragmentary postcranial human remains of an adult male.			PP		Indet.	1	40–44	1	0	0	0	0	0	1	0	0	0	0	AS 5	Male
19-6-10/61000.0	Fragmentary postcranial human remains of an adult female.			PP		Indet.	1	30–34	1	0	0	0	0	0	0	0	1	0	0	T 6, SB 3–4	Female (pelvis)
19-6-10/N96.0	Fragmentary cranial human remains of an adult of unknown sex.		N-96	PP		Early 3rd cent. A.D.	1		1	0	0	0	0	0	0	0	0	0	1		Indet.
20-6-10/60244a	Fragmentary postcranial human remains of 2 adults of unknown sex.	769	60244a	PP	1550E100, depth 4′2″, height 3′	Gl. 3	2		2	0	0	0	0	0	0	0	0	0	2		Indet.
20-6-10/60322.1	Fragmentary cranial human remains of 3 adults of unknown sex.	854	60322	PP	1280E130, depth 3′, height 3′	Indet.	3		3	0	0	0	0	0	0	0	0	0	3		Indet.
20-6-10/60323.0	Reassociated.	846		PP	1310E130, depth 6′, height 2′	Gl. 3 or 4	0		0	0	0	0	0	0	0	0	0	0	0		
20-6-10/60324.0	Fragmentary postcranial human remains of 20 adults of unknown sex.		60324 (2–22 except 7), -21 also labeled 60236	PP	General digging	Indet.	20		20	0	0	0	0	0	0	0	0	0	20		Indet.
20-6-10/60325.0	Fragmentary postcranial human remains of 1 subadult and 18 adults, all of unknown sex.		60325/1–19, -18 also 60236, -3 also 59986, -8 also 60083	PP	General digging	Indet.	19		18	1	0	0	0	0	0	0	0	0	19		Indet.
20-6-10/60326.0	Fragmentary postcranial human remains of 10 adults of unknown sex.		60326 (2–12, except 10)	PP	General digging	Indet.	10		10	0	0	0	0	0	0	0	0	0	10		Indet.
20-6-10/60327.0	Fragmentary postcranial human remains of 7 adults of unknown sex.		60327 (1–8, except 6)	PP	General digging	Indet.	7		7	0	0	0	0	0	0	0	0	0	7		Indet.
20-6-10/60328.0	Fragmentary postcranial human remains of 6 adults of unknown sex.		60328 (1-6), -4 also f38, -5 also f42	PP	General digging	Indet.	6		6	0	0	0	0	0	0	0	0	0	6		Indet.
20-6-10/60329.0	Fragmentary postcranial human remains of 7 adults of unknown sex.		60329 (1–7), -3 also f43, -7 also m19	PP	General digging	Indet.	7		7	0	0	0	0	0	0	0	0	0	7		Indet.
20-6-10/60330.0	Fragmentary postcranial human remains of 11 adults of unknown sex.		60330 (1–11)	PP	General digging	Indet.	11		11	0	0	0	0	0	0	0	0	0	11		Indet.
20-6-10/60331.0	Fragmentary postcranial human remains of 10 adults of unknown sex.		60321 (1–8, 11, 12)	PP	General digging	Indet.	10		10	0	0	0	0	0	0	0	0	0	10		Indet.
20-6-10/60332.0	Fragmentary postcranial human remains of 6 adults of unknown sex.		60332 (1, 3, 4, 6–8)	PP	General digging	Indet.	6		6	0	0	0	0	0	0	0	0	0	6		Indet.
20-6-10/60333.0	Fragmentary postcranial human remains of 8 adults of unknown sex.		60333 (-2, 4–10)	PP	General digging (?) (MEM 7/97)	Indet.	8		8	0	0	0	0	0	0	0	0	0	8		Indet.
20-6-10/60336.0	Fragmentary postcranial human remains of 13 adults of unknown sex.		60336 (1–3, 5, 7-15)	PP	General digging	Indet.	13		13	0	0	0	0	0	0	0	0	0	13		Indet.
20-6-10/60336.2	Fragmentary postcranial human remains of an adult of unknown sex.		60336-6	PP	Near fibrolithic ax, see notes 6/21/16	Indet.	1		1	0	0	0	0	0	0	0	0	0	1		Indet.
20-6-10/60337.0	Fragmentary postcranial human remains of 11 adults of unknown sex.		60337 (1–3, 5–10, 12, 13)	PP	General digging	Indet.	11		11	0	0	0	0	0	0	0	0	0	11		Indet.
20-6-10/60338.0	Fragmentary postcranial human remains of 7 adults of unknown sex.		60338 (2–8)	PP	General digging	Indet.	7		7	0	0	0	0	0	0	0	0	0	7		Indet.
20-6-10/60339.0	Fragmentary postcranial human remains of 6 adults of unknown sex.		60339 (1, 2, 4–7)	PP	General digging	Indet.	6		6	0	0	0	0	0	0	0	0	0	6		Indet.
20-6-10/60340.0	Fragmentary postcranial human remains of 7 adults of unknown sex.		60340 (1–7)	PP	General digging	Indet.	7		7	0	0	0	0	0	0	0	0	0	7		Indet.
20-6-10/60341.0	Fragmentary postcranial human remains of 6 adults of unknown sex.		60341 (1–6)	PP	General digging	Indet.	6		6	0	0	0	0	0	0	0	0	0	6		Indet.
20-6-10/60342.0	Nearly complete cranial and fragmentary postcranial human remains of an adult female.	Rowe	60342, 1350-150e (on extra element)	R		B&W	1	40–49	1	0	0	0	0	0	0	0	1	0	0	D h	Female

Peabody number	Pathology type	Pathology description	Cranial mod.	Associated funerary objects	Field context	Notes
19-6-10/60999.0			No cranium			
19-6-10/61000.0			No cranium			
19-6-10/N96.0			Unobs.			
20-6-10/60244a	Other	Vt/l spondylolysis	No cranium	Gl. 3 bowl no. 869, scraper no number		
20-6-10/60322.1			Unobs.	Pipe, deer horn, 2 cut bones, no numbers listed		Description does not match field card
20-6-10/60323.0						Reassociated human remains with 60216 (MEM)
20-6-10/60324.0			No cranium			
20-6-10/60325.0	Arthritis, nutr.	Mild DJD, osteoporosis	No cranium			
20-6-10/60326.0	Arthritis	DJD	No cranium			
20-6-10/60327.0			No cranium			
20-6-10/60328.0			No cranium			
20-6-10/60329.0	Arthritis	DJD	No cranium			
20-6-10/60330.0			No cranium			
20-6-10/60331.0			No cranium			
20-6-10/60332.0	Nutr.	Osteoporosis	No cranium			
20-6-10/60333.0	Nutr.	Osteoporosis	No cranium			
20-6-10/60336.0	Arthritis, nutr.	DJD, osteoporosis	No cranium			
20-6-10/60336.2			No cranium			
20-6-10/60337.0	Arthritis	DJD	No cranium			
20-6-10/60338.0	Arthritis	DJD	No cranium			
20-6-10/60339.0	Nutr.	1 individual has osteoporosis	No cranium			
20-6-10/60340.0			No cranium			
20-6-10/60341.0			No cranium			
20-6-10/60342.0	Trauma, infection, other	Clavicle–lytic infection resulting from trauma, dental abcesses, extensive exotoses on lingual side mandible, extensive antemortem tooth loss, ?cut marks on nasals	Occ. def.			Intrasite location on basis of specimen label

Peabody number	Object description	FN	Specimen label	Site	Intrasite provenience	Ceramic stage	MNI	Age (yrs.)	Ad	Su	Ch	In	Fe	Unk. age	M	M?	F	F?	Unk. sex	Age criteria	Sex criteria
20-6-10/60342.1	Fragmentary postcranial human remains of an adult of unknown sex.	Rowe		R		B&W	1		1	0	0	0	0	0	0	0	0	0	1		Indet.
20-6-10/60342.2	Fragmentary postcranial human remains of an adult of unknown sex.	Rowe		R	1350E100	B&W	1		1	0	0	0	0	0	0	0	0	0	1		Indet.
20-6-10/60342.3	Fragmentary postcranial human remains of an adult of unknown sex.	Rowe		R		B&W	1		1	0	0	0	0	0	0	0	0	0	1		Indet.
20-6-10/60342.4	Fragmentary postcranial human remains of a subadult of unknown sex.	Rowe		R		B&W	1		0	1	0	0	0	0	0	0	0	0	1		Indet.
20-6-10/60343.0	Fragmentary postcranial human remains of 1 subadult and 19 adults, all of unknown sex.		60343 (1–22, except 2, 15)	PP	General digging	Indet.	20		19	1	0	0	0	0	0	0	0	0	20		Indet.
20-6-10/60344.0	Fragmentary postcranial human remains of 2 subadults of unknown sex, 7 adult females, and 7 adult males.		60344 (1–30, except 7) (-29 also 59912)	PP	General digging	Indet.	16		14	2	0	0	0	0	7	0	7	0	2		Indet.
20-6-10/60345.0	Fragmentary postcranial human remains of 17 adults of unknown sex.		60345- (1–34)	PP	General digging	Indet.	17		17	0	0	0	0	0	0	0	0	0	17		Indet.
20-6-10/60346.0	Fragmentary postcranial human remains of 22 adults of unknown sex.		60346- (1–44, except 20)	PP	General digging	Indet.	22		19	3	0	0	0	0	0	0	0	0	22		Indet.
20-6-10/60347.0	Fragmentary postcranial human remains of 7 adults of unknown sex.		60347- (1–16 except 3, 6, 10)	PP	General digging	Indet.	7		7	0	0	0	0	0	0	0	0	0	7		Indet.
20-6-10/60348.0	Fragmentary postcranial human remains of 15 adults of unknown sex.		60348 (1–5, 7–16, some unnumbered)	PP	General digging	Indet.	15		15	0	0	0	0	0	0	0	0	0	15		Indet.
20-6-10/60349.0	Fragmentary postcranial human remains of 2 infants, 4 children, and 4 adults, all of unknown sex.		60349- (2, 3, 11, 13, 14, 16, 17, 19, 22, 23, 24, 26, 27, 28, 29)	PP	General digging	Indet.	10		4	0	4	2	0	0	0	0	0	0	10		Indet.
20-6-10/60350.0	Fragmentary cranial human remains of 9 children, 1 subadult, and 14 adults, all of unknown sex.		60350- (1, 3, 5, 6, 12, 13, 14, 17, 18, 20, 21, 24, 27, 29, 30, 31, 32, 33, 34)	PP	General digging, lay 20″ below surface, 3′ above red	Probably Gl. 6	24		14	1	9	0	0	0	0	0	0	0	24		Indet.
20-6-10/60350.2	Fragmentary cranial human remains of 3 adults of unknown sex.		60350 (9, 23, 25), gen. dig 1400/E150	PP	1400E150	Indet.	3		3	0	0	0	0	0	0	0	0	0	3		Indet.
20-6-10/60350.3	Fragmentary cranial human remains of an adult of unknown sex.		60350-7, gen. dig 1450/E150	PP	1400E150	Indet.	1		1	0	0	0	0	0	0	0	0	0	1		Indet.
20-6-10/60358.0	Fragmentary cranial and postcranial human remains of 4 adults of unknown sex.		60358- (1, 2, 3, 5, 6, 8, 9, 10, 12, 14, 16, 17)	PP	General digging	Indet.	4		4	0	0	0	0	0	0	0	0	0	4		Indet.
20-6-10/60991.0	Very fragmentary postcranial human remains of 6 adults of unknown sex.		60991- (1–7), Pecos	PP		Indet.	6		6	0	0	0	0	0	0	0	0	0	6		Indet.
20-6-10/60992.0	Nearly complete cranial human remains of an adult female.		60992, ?, x	PP		Indet.	1	16–20	1	0	0	0	0	0	0	0	1	0	0	D b2	Female
2000.1.856.0	Very fragmentary cranial human remains of 6 adults of unknown sex.		Pecos teeth, 2000.1.856.0	PP		Pueblo III or later	6		6	0	0	0	0	0	0	0	0	0	6		Indet.
2000.1.857.0	Very fragmentary postcranial human remains of 2 adults of unknown sex.		Pecos, 2549, 2000.1.857.0	PP		Pueblo III or later	2		2	0	0	0	0	0	0	0	0	0	2		Indet.
24-4-10/60478.0	Nearly complete cranial human remains of an adult female.		60478	PP		Indet.	1	20–29	1	0	0	0	0	0	0	0	1	0	0	D d-e/e	
24-4-10/60604.0	Fragmentary cranial and postcranial human remains of an adult male.		60604	PP		Indet.	1	50–59	1	0	0	0	0	0	1	0	0	0	0	D i+	Male

Peabody number	Pathology type	Pathology description	Cranial mod.	Associated funerary objects	Field context	Notes
20-6-10/60342.1			No cranium			
20-6-10/60342.2			No cranium			
20-6-10/60342.3			No cranium			
20-6-10/60342.4			No cranium			
20-6-10/60343.0	Arthritis, other	Some spinal osteophytosis, 1 of 20 sacra has spina bifida	No cranium			
20-6-10/60344.0			No cranium			
20-6-10/60345.0	Trauma	Trauma on 1 L radius (60345-26)	No cranium			
20-6-10/60346.0	Arthritis	Slight DJD on some elements	No cranium			60346-20 renumbered as 60333-1
20-6-10/60347.0			No cranium			
20-6-10/60348.0	Arthritis	DJD on some elements	No cranium			
20-6-10/60349.0			No cranium			
20-6-10/60350.0	Infection, arthritis, nutr.	Some periodontal disease, some DJD at TMJ, some enamel hypoplasia	Unobs.	Pecos white pottery, 3 projectile points no. 1324 (68428), bone bead no. 1325 (68017), stone drill no. 1327 (68660)		
20-6-10/60350.2	Arthritis	DJD at TMJ	Unobs.			
20-6-10/60350.3	Nutr.	Enamel hypoplasia	Unobs.			
20-6-10/60358.0	Trauma, infection	Healed fracture on 1 femur, infection on 1 parietal, other long bones also indicate infection	Unobs.			
20-6-10/60991.0			No cranium			Humeri
20-6-10/60992.0			Not obs.			
2000.1.856.0	Nutr.	Enamel hypoplasia	Unobs.			Incisors
2000.1.857.0			No cranium			2 postcranial elements
24-4-10/60478.0	Infection, nutr.	Interesting healed lesion on frontal, mild cribra orbitalia	Not obs.			
24-4-10/60604.0	Nutr.	Enamel hypoplasia	Unobs.			

Peabody number	Object description	FN	Specimen label	Site	Intrasite provenience	Ceramic stage	MNI	Age (yrs.)	Ad	Su	Ch	In	Fe	Unk. age	M	M?	F	F?	Unk. sex	Age criteria	Sex criteria
24-4-10/60619.0	Fragmentary postcranial human remains of an adult female.		60619	PP		Indet.	1	40–44	1	0	0	0	0	0	0	0	1	0	0	T 8, AS 5	Female
24-4-10/61702.0	Partial cranial and very fragmentary postcranial human remains of an adult male.		61702	PP		Indet.	1	45–55+	1	0	0	0	0	0	1	0	0	0	0	D h+/i++	Male
24-4-10/61704.0	Fragmentary postcranial human remains of an individual of unknown age and sex.			PP		Indet.	1		0	0	0	0	0	1	0	0	0	0	1		Indet.
24-4-10/N263.0	Fragmentary cranial and postcranial human remains of 1 child, 1 subadult and 10 adults, all of unknown sex.		N263	PP		Indet.	12		10	1	1	0	0	0	0	0	0	0	12	D 3 dm present	Indet.
24-4-10/N272.0	Fragmentary cranial and postcranial human remains of 1 child, 4 subadults and 7 adults, all of unknown sex.		N272	PP		Indet.	12		7	4	1	0	0	0	0	0	0	0	12	D 1 dm present	Indet.
24-4-10/N272.2	Fragmentary cranial and postcranial human remains of 1 adult of unknown sex.		N272, 60441	PP		Indet.	1		1	0	0	0	0	0	0	0	0	0	1		Indet.
24-4-10/N331.0	Fragmentary cranial and postcranial human remains of an adult male.		N331	PP		Indet.	1	45–50+	1	0	0	0	0	0	1	0	0	0	0	D h+	Male
25-3-10/60669.0	Nearly complete cranial and postcranial human remains of an adult male.	Mesa	60669	PP	Mesa	Indet.	1	20–24	1	0	0	0	0	0	1	0	0	0	0	D c, T 2, AS 1	Male
25-3-10/60672.1	Reassociated.	C	60672	PP	Church	Indet.	0		0	0	0	0	0	0	0	0	0	0	0		
25-3-10/60681.1	Reassociated.		60681	PP	Church nave	Indet.	0	50–59	0	0	0	0	0	0	0	0	0	0	0	D i+	
25-3-10/60682.0	Very fragmentary postcranial human remains of 4 adults, including 2 male and 2 female.	1746	60682	PP	Extension trench S700 series, depth 5′, height 3′	Indet.	4		4	0	0	0	0	0	2	0	2	0	0		Mixed
25-3-10/60682.1	Very fragmentary postcranial human remains of 1 child and 3 adults, all of unknown sex.	1746	60682 (a–d)	PP	Extension trench S of 700 series, depth 5′, height 3′	Indet.	4		3	0	1	0	0	0	0	0	0	0	4		
25-3-10/60734.0	Reassociated.	Exc. on viii.25	60734	PP	Room sp-II	Indet.	0		0	0	0	0	0	0	0	0	0	0	0		
25-3-10/60750a	Reassociated.	C	60750a	PP	Church	Indet.	0		0	0	0	0	0	0	0	0	0	0	0		
25-3-10/60755.0	Reassociated.	1612	60755	PP	1300W0	Indet.	0		0	0	0	0	0	0	0	0	0	0	0		
25-3-10/60776.0	Very fragmentary postcranial human remains of 13 adults of unknown sex.		60776	PP		Indet.	13		13	0	0	0	0	0	0	0	0	0	13		Indet.
25-3-10/60780.0	Very fragmentary postcranial human remains of 8 adults of unknown sex.		60780	PP		Indet.	8		8	0	0	0	0	0	0	0	0	0	8		Indet.
25-3-10/61116.0	Very fragmentary postcranial human remains of 1 child of unknown sex, 6 adult females, and 3 adult males.			PP		Indet.	10		9	0	1	0	0	0	3	0	6	0	1		Mixed
25-3-10/61117.0	Very fragmentary postcranial human remains of 6 adults of unknown sex.			PP		Indet.	6		6	0	0	0	0	0	0	0	0	0	6		
25-3-10/61118.0	Very fragmentary postcranial human remains of 1 child, 1 subadult, and 13 adults, including 4 females, 10 males and 1 individual of unknown sex.		61118	PP		Indet.	15		13	1	1	0	0	0	10	0	4	0	1		Mixed
25-3-10/61119.0	Very fragmentary postcranial human remains of 1 subadult and 5 adults, all of unknown sex.		61119 a–f, 2 unlabeled	PP		Indet.	6		5	1	0	0	0	0	0	0	0	0	6		
26-3-10/60640a	Very fragmentary cranial human remains of an adult of unknown sex.		60640-a	PP?		Indet.	1	25–34	1	0	0	0	0	0	0	0	0	0	1	D e-f	Indet.
26-3-10/61187.1	Reassociated.	Bb61		FL	Room 52, se corner, depth 3′5″, on floor	Probably B&W	0		0	0	0	0	0	0	0	0	0	0	0		
26-3-10/61187.3	Reassociated.	Bb61		FL	Room 52, se corner, depth 3′5″, on floor	Probably B&W	0		0	0	0	0	0	0	0	0	0	0	0		

Peabody number	Pathology type	Pathology description	Cranial mod.	Associated funerary objects	Field context	Notes
24-4-10/60619.0	Infection, arthritis, nutr.	Periosteal infection R proximal femur, moderate DJD, osteoporosis	No cranium			
24-4-10/61702.0	Infection	Extensive antemortem tooth loss	Not obs.			
24-4-10/61704.0			No cranium			Not observable due to matrix.
24-4-10/N263.0	Nutr.	Enamel hypoplasia	Unobs.			A few pairs of postcranial elements could be associated as individuals
24-4-10/N272.0	Nutr.	Enamel hypoplasia	Unobs.			
24-4-10/N272.2			Unobs.			
24-4-10/N331.0			Occ. def.			
25-3-10/60669.0	Nutr.	Enamel hypoplasia	Occ. def.			
25-3-10/60672.1						Reassociated with human remains from the church and recatalogued, MEM 2/98
25-3-10/60681.1						Reassociated with catalog number 60720.
25-3-10/60682.0			No cranium	Gl. 1 red bowl no. 813 (74944)		Description does not match field card
25-3-10/60682.1			No cranium	Gl. 1 red bowl no. 813 (74944)		Description does not match field card
25-3-10/60734.0						Reassociated with 60735. may not be Native American, provenience problematic
25-3-10/60750a						Reassociated with other church human remains, MEM 4/98
25-3-10/60755.0						Reassociated with 60746 (MEM 3/98).
25-3-10/60776.0	Trauma, infection	Trauma on 60776-7, embedded stone projectile with accompanying infection	No cranium			Paired tibiae only
25-3-10/60780.0			No cranium			Paired femora only, 60780-8 unusual "cut" marks on femoral heads
25-3-10/61116.0			No cranium			Left innominates only
25-3-10/61117.0	Arthritis, other	2 of 6 sacra have unfused vt/s4–5	No cranium			6 sacra
25-3-10/61118.0			No cranium			14 right innominates
25-3-10/61119.0			No cranium			Scapulae only
26-3-10/60640a			Unobs.			
26-3-10/61187.1						Reassociated with 61187-0, MEM 4/98.
26-3-10/61187.3						Reassociated with 61187-0 and 61187-2, MEM 4/98.

Peabody number	Object description	FN	Specimen label	Site	Intrasite provenience	Ceramic stage	MNI	Age (yrs.)	Ad	Su	Ch	In	Fe	Unk. age	M	M?	F	F?	Unk. sex	Age criteria	Sex criteria
26-3-10/61189.0	Fragmentary postcranial human remains of an adult of unknown sex.	Dr1		DR	On floor of 10	Indet.	1	< 35	1	0	0	0	0	0	0	0	0	0	1	PC < 35 yrs	Indet.
26-3-10/61190.0	Fragmentary cranial human remains of an adult that is probably female.	Dr4		DR	Room 24, 7" below floor	B&W	1		1	0	0	0	0	0	0	0	0	1	0		?Female
26-3-10/61192.0	Very fragmentary postcranial human remains of 2 children and 6 adults, all of unknown sex.			PP		Indet.	8		6	0	2	0	0	0	0	0	0	0	8	Children, 4–5 yrs, 7.5–8.5 yrs	
26-3-10/61193.0	Very fragmentary postcranial human remains of 1 infant, 2 subadults, and 12 adults, all of unknown sex.	C		PP	Church?	Indet.	15		12	2	0	1	0	0	0	0	0	0	15		Indet.
26-3-10/61194.0	Very fragmentary postcranial human remains of 2 subadults and 9 adults, all of unknown sex.		61194	PP		Indet.	11		9	2	0	0	0	0	0	0	0	0	11		Indet.
26-3-10/61195.0	Very fragmentary postcranial human remains of 2 adults of unknown sex.			PP		Indet.	2		2	0	0	0	0	0	0	0	0	0	2		Indet.
26-3-10/61196.0	Fragmentary cranial and postcranial human remains of an adult male.		61196	PP		Indet.	1		1	0	0	0	0	0	1	0	0	0	0		Male
26-3-10/61196.1	Very fragmentary cranial and postcranial human remains of a child of unknown sex.			PP?		Indet.	1		0	0	1	0	0	0	0	0	0	0	1		Indet.
26-3-10/61205.0	Very fragmentary postcranial human remains of 3 children of unknown sex, 6 adult females, and 4 adult males.			PP		Indet.	13	16–20	10	0	3	0	0	0	4	0	6	0	3		Mixed
26-3-10/61206.0	Very fragmentary postcranial human remains of 3 children, 1 subadult, and 9 adults, all of unknown sex.		61206	PP		Indet.	13		9	1	3	0	0	0	0	0	0	0	13		Indet.
26-3-10/61207.0	Very fragmentary postcranial human remains of 1 child, 1 subadult and 6 adults, all of unknown sex.			PP		Indet.	8		6	1	1	0	0	0	0	0	0	0	8		Indet.
26-3-10/61208.0	Very fragmentary postcranial human remains of 2 children, 1 subadult, and 9 adults, all of unknown sex.			PP		Indet.	12		9	1	2	0	0	0	0	0	0	0	12		Indet.
26-3-10/61209.0	Very fragmentary postcranial human remains of 6 adults of unknown sex.			PP		Indet.	6		4	1	1	0	0	0	0	0	0	0	6		Indet.
26-3-10/61210.0	Very fragmentary postcranial human remains of 8 adults of unknown sex.			PP		Indet.	8		8	0	0	0	0	0	0	0	0	0	8		Indet.
26-3-10/61211.0	Very fragmentary postcranial human remains of 3 children and 10 adults, all of unknown sex.			PP		Indet.	13		10	0	3	0	0	0	0	0	0	0	13		Indet.
26-3-10/61212.0	Very fragmentary postcranial human remains of 4 subadults and 5 adults, all of unknown sex.			PP		Indet.	9		5	4	0	0	0	0	0	0	0	0	9		Indet.
26-3-10/61213.0	Very fragmentary cranial human remains of an infant of unknown sex.			PP		Indet.	1		0	0	0	1	0	0	0	0	0	0	1	Size	Indet.
26-3-10/61214.0	Reassociated with 27-1-10/61310.	Bb118		FL	Trench p	Probably B&W	0	>35	0	0	0	0	0	0	0	0	0	0	0		Indet.
26-3-10/61215.0	Very fragmentary postcranial human remains of 3 adults of unknown sex.			PP		Indet.	3		3	0	0	0	0	0	0	0	0	0	3		Indet.
27-1-10/61259.0	Fragmentary cranial and postcranial human remains of an adult female.			PP		Indet.	1	20–29	1	0	0	0	0	0	0	0	1	0	0	D d/e	Fall features
27-1-10/61260.0	Fragmentary postcranial human remains of an adult male.			PP		Indet.	1		1	0	0	0	0	0	1	0	0	0	0		Male
27-1-10/61261.0	Very fragmentary postcranial human remains of 2 adults of unknown sex.	C	Floater	PP	Church	Indet.	2		2	0	0	0	0	0	0	0	0	0	2		Indet.
27-1-10/61263.0	Fragmentary postcranial human remains of an adult that is probably female.			PP		Indet.	1		1	0	0	0	0	0	0	0	0	1	0		?Female

Peabody number	Pathology type	Pathology description	Cranial mod.	Associated funerary objects	Field context	Notes
26-3-10/61189.0			No cranium	Frags. Gl. 1 red bowl, no number		
26-3-10/61190.0			Unobs.	B&W bowl, B&W frag., bone tube no. 250 (76647)		
26-3-10/61192.0			No cranium			Femora and tibia
26-3-10/61193.0			No cranium			Humeri
26-3-10/61194.0	Trauma	1 femur has fracture	No cranium			Femora
26-3-10/61195.0			No cranium			Clavicles
26-3-10/61196.0	Tumor	Button osteoma on parietal	Unobs.			
26-3-10/61196.1			Unobs.			
26-3-10/61205.0			No cranium			Innominates only
26-3-10/61206.0			No cranium			Scapulae
26-3-10/61207.0	Other	Ankylosis of 2 vt/1	No cranium			Vertebrae
26-3-10/61208.0			No cranium			3 patellae, Mc, Mt, tarsals and phalanges
26-3-10/61209.0			No cranium			Sacra
26-3-10/61210.0			No cranium			Radii
26-3-10/61211.0	Trauma	1 healed midshaft fracture on R ulna	No cranium			Ulnae
26-3-10/61212.0	Arthritis	DJD	No cranium			Clavicles
26-3-10/61213.0			Unobs.			
26-3-10/61214.0			No cranium			Reassociated with 61310, MEM 4/98
26-3-10/61215.0			No cranium			Fibulae, 1 patella, 1 sternum
27-1-10/61259.0	Arthritis, nutr.	Spinal osteophytosis, enamel hypoplasia	Unobs.			
27-1-10/61260.0			No cranium			
27-1-10/61261.0			No cranium			Clavicles
27-1-10/61263.0			No cranium			

Peabody number	Object description	FN	Specimen label	Site	Intrasite provenience	Ceramic stage	MNI	Age (yrs.)	Ad	Su	Ch	In	Fe	Unk. age	M	M?	F	F?	Unk. sex	Age criteria	Sex criteria
27-1-10/61264.0	Fragmentary postcranial human remains of an adult that is probably female.			PP		Indet.	1	40–44	1	0	0	0	0	0	0	0	0	1	0	AS 5	?Female
27-1-10/61267.0	Reassociated.	Bb101		FL	Trench w of 55	Probably B&W	0		0	0	0	0	0	0	0	0	0	0	0		
27-1-10/61273.0	Fragmentary postcranial human remains of an adult that is probably male.			PP		Indet.	1		1	0	0	0	0	0	0	1	0	0	0		?Male
27-1-10/61274.0	Reassociated.	Bb52	61274	FL	N15W30, depth 4′, height 8–10″	Probably B&W	0		0	0	0	0	0	0	0	0	0	0	0		
27-1-10/61275.0	Very fragmentary cranial and postcranial human remains of 1 subadult and 2 adults, all of unknown sex.	Bb82	Pecos	FL	Room 74, depth 2′, height about 1′6″ above room floor	Probably B&W	3		2	1	0	0	0	0	0	0	0	0	3		Indet.
27-1-10/61278.0	Reassociated.	Bb15		FL	Room bb 3, depth 8″, height 2′6″ above room floor	Probably B&W	0		0	0	0	0	0	0	0	0	0	0	0		
27-1-10/61279.0	Nearly complete cranial human remains of a child of unknown sex.	Dr6		DR	East of room 14, 3′ from surface	Indet.	1	2–3	0	0	1	0	0	0	0	0	0	0	1	D 3±1 yr	Indet.
27-1-10/61283.0	Very fragmentary postcranial human remains of an adult of unknown sex.		61283 (labeled, MEM 11/96)	PP		Indet.	1		1	0	0	0	0	0	0	0	0	0	1		Indet.
27-1-10/61301.0	Fragmentary cranial and postcranial human remains of an infant of unknown sex.		61301 (labeled femur, MEM 12/96)	PP		Indet.	1		0	0	0	1	0	0	0	0	0	0	1	Size	Indet.
27-1-10/61301.1	Fragmentary cranial and postcranial human remains of an infant of unknown sex.		61301-1 (labeled ilium, MEM 12/96)	PP		Indet.	1		0	0	0	1	0	0	0	0	0	0	1		Indet.
27-1-10/61302.0	Fragmentary cranial and postcranial human remains of an infant of unknown sex.		61302 (labeled, MEM 11/96)	PP		Indet.	1		0	0	0	1	0	0	0	0	0	0	1		Indet.
27-1-10/61303.0	Fragmentary cranial human remains of an adult of unknown sex.		61303 (labeled, MEM 11/96)	PP		Indet.	1		1	0	0	0	0	0	0	0	0	0	1		Indet.
27-1-10/61305.0	Partial cranial human remains of an adult of unknown sex.	Pr389	61305 (labeled, MEM 11/96)	PP		Indet.	1		1	0	0	0	0	0	0	0	0	0	1		Indet.
27-1-10/61306.0	Fragmentary cranial and postcranial human remains of a child of unknown sex.	Pr238	61306 (labeled, MEM 11/96)	PP		Indet.	1		0	0	1	0	0	0	0	0	0	0	1		Indet.
27-1-10/61307.0	Nearly complete cranial and fragmentary postcranial human remains of an adult female.	Pr132	61307	PP		Indet.	1	30–39	1	0	0	0	0	0	0	0	1	0	0	D h, T 7, AS 3	Female
27-1-10/61308.0	Very fragmentary postcranial human remains of 1 adult female and 1 adult male.	Pr123 or pr389		PP		Indet.	2		2	0	0	0	0	0	1	0	1	0	0	T 1	Mixed
27-1-10/61311.0	Partial cranial human remains of an adult male.		61311	PP		Indet.	1	25–39	1	0	0	0	0	0	1	0	0	0	0	D e-g	Male
27-1-10/61313.0	Partial cranial human remains of an adult of unknown sex.		61313 (labeled, MEM 11/96)	PP		Indet.	1	40–49	1	0	0	0	0	0	0	0	0	0	1	D h	Indet.

Peabody number	Pathology type	Pathology description	Cranial mod.	Associated funerary objects	Field context	Notes
27-1-10/61264.0	Arthritis	Spinal osteophytosis	No cranium			
27-1-10/61267.0						Reassociated with catalog number 61294 (MEM 4/98)
27-1-10/61273.0			No cranium			
27-1-10/61274.0						Reassociated with catalog number 61181 (MEM 4/98)
27-1-10/61275.0	Other	1 individual has spina bifida vt/s3–5	Unobs.			Description does not match field card
27-1-10/61278.0						Reassociated with catalog number 61183 (MEM 4/98)
27-1-10/61279.0			Not obs.	B&W bowl, no number		
27-1-10/61283.0			No cranium			
27-1-10/61301.0			Unobs.			
27-1-10/61301.1			Unobs.			
27-1-10/61302.0			Unobs.			
27-1-10/61303.0			Unobs.			
27-1-10/61305.0			Not obs.			
27-1-10/61306.0			Unobs.			
27-1-10/61307.0	Infection	Extensive antemortem molar tooth loss	Not obs.			
27-1-10/61308.0			No cranium			
27-1-10/61311.0	Infection	Antemortem tooth loss	Occ. def.			
27-1-10/61313.0			Not obs.			

Peabody number	Object description	FN	Specimen label	Site	Intrasite provenience	Ceramic stage	MNI	Age (yrs.)	Ad	Su	Ch	In	Fe	Unk. age	M	M?	F	F?	Unk. sex	Age criteria	Sex criteria
27-1-10/61316.0	Very fragmentary cranial and postcranial human remains of 1 child and 1 subadult, both of unknown sex.		61316 (labeled, MEM 11/96)	PP		Indet.	2		0	1	1	0	0	0	0	0	0	0	2		Indet.
27-1-10/61317.0	Fragmentary postcranial human remains of a child of unknown sex.		61317 (labeled, MEM 11/96)	PP		Indet.	1		0	0	1	0	0	0	0	0	0	0	1		Indet.
27-1-10/61318.0	Fragmentary postcranial human remains of an adult of unknown sex.			PP		Indet.	1		1	0	0	0	0	0	0	0	0	0	1		Indet.
27-1-10/61319.0	Very fragmentary postcranial human remains of an adult of unknown sex.			PP		Indet.	1		1	0	0	0	0	0	0	0	0	0	1		Indet.
27-1-10/61320.0	Fragmentary cranial and postcranial human remains of an infant of unknown sex.		61320 (labeled, MEM 11/96)	PP		Indet.	1	0–2.5	0	0	0	1	0	0	0	0	0	0	1	Size	Indet.
27-1-10/61321.0	Fragmentary cranial human remains of an infant of unknown sex.		61321 (labeled, MEM 11/96)	PP		Indet.	1	0.5–2.5	0	0	0	1	0	0	0	0	0	0	1	Size	Indet.
27-1-10/61324.0	Fragmentary postcranial human remains of an adult of unknown sex.		61324 (labeled, MEM 11/96)	PP		Indet.	1		1	0	0	0	0	0	0	0	0	0	1		Indet.
27-1-10/61325.0	Fragmentary postcranial human remains of an adult of unknown sex.		61325 (labeled, MEM 11/96)	PP		Indet.	1		1	0	0	0	0	0	0	0	0	0	1		Indet.
27-1-10/61326.0	Very fragmentary cranial human remains of 1 subadult and 3 adults, all of unknown sex.		61326 (labeled, MEM 11/96)	PP		Indet.	4		4	0	0	0	0	0	0	0	0	0	4		Indet.
27-1-10/61326.1	Very fragmentary cranial human remains of a child of unknown sex.		61326 (labeled, MEM 11/96)	PP?		Indet.	1	3–11	0	0	1	0	0	0	0	0	0	0	1	Size	Indet.
27-1-10/61326.2	Fragmentary cranial human remains of a child of unknown sex.		61326 (labeled, MEM 11/96)	PP?		Indet.	1	3–11	0	0	1	0	0	0	0	0	0	0	1	Size	Indet.
27-1-10/61330.0	Very fragmentary postcranial human remains of 2 adults of unknown sex.		61330 (labeled, MEM 11/96)	PP		Indet.	2		2	0	0	0	0	0	0	0	0	0	2		Indet.
27-1-10/61331.0	Fragmentary cranial and postcranial human remains of an adult male.		61331 (labeled 1 femur, MEM 11/96)	PP		Indet.	1	> 35	1	0	0	0	0	0	1	0	0	0	0	D >35 yrs	Male
27-1-10/61332.0	Fragmentary cranial and partial postcranial human remains of an adult of unknown sex.		61332 (labeled mandible, MEM 11/96)	PP		Indet.	1	20–29	1	0	0	0	0	0	0	0	0	0	1	D and PC 20–30 yrs	Indet.
27-1-10/61333.0	Fragmentary cranial and postcranial human remains of an adult that is probably female.		61333 (labeled mandible, MEM 11/96)	PP		Indet.	1	20–24	1	0	0	0	0	0	0	0	0	1	0	D d, epiphyseal lines still visible	?Female

Peabody number	Pathology type	Pathology description	Cranial mod.	Associated funerary objects	Field context	Notes
27-1-10/61316.0			Unobs.			
27-1-10/61317.0			No cranium			
27-1-10/61318.0	Arthritis	DJD, spinal osteophytosis	No cranium			
27-1-10/61319.0			No cranium			
27-1-10/61320.0			Unobs.			
27-1-10/61321.0			Unobs.			
27-1-10/61324.0	Arthritis	DJD, spinal osteophytosis	No cranium			
27-1-10/61325.0			No cranium			
27-1-10/61326.0			Unobs.			
27-1-10/61326.1			Unobs.			
27-1-10/61326.2			Unobs.			Doubtful association of cranium and mandible
27-1-10/61330.0			No cranium			
27-1-10/61331.0	Nutr.	Osteoporosis	Unobs.			Looks burned
27-1-10/61332.0			Unobs.			
27-1-10/61333.0	Other	Partial sacralization of vt/15, socket for supernumerary L upper incisor	Unobs.			

Peabody number	Object description	FN	Specimen label	Site	Intrasite provenience	Ceramic stage	MNI	Age (yrs.)	Ad	Su	Ch	In	Fe	Unk. age	M	M?	F	F?	Unk. sex	Age criteria	Sex criteria
27-1-10/61334.0	Very fragmentary cranial and postcranial human remains of an adult of unknown sex.		61334 (labeled femur, MEM 11/96)	PP		Indet.	1	>35	1	0	0	0	0	0	0	0	0	0	1	>35 yrs	Indet.
27-1-10/61335.0	Fragmentary postcranial human remains of an adult of unknown sex.		61335 (labeled l humerus, MEM 11/96)	PP		Indet.	1		1	0	0	0	0	0	0	0	0	0	1		Indet.
27-1-10/61336.0	Very fragmentary cranial and fragmentary postcranial human remains of an adult of unknown sex.		61336 (MEM 11/96)	PP		Indet.	1		1	0	0	0	0	0	0	0	0	0	1		Indet.
27-1-10/N502.0	Fragmentary postcranial human remains of 2 children and 3 adults, all of unknown sex.		N502	PP		Indet.	5		3	0	0	2	0	0	0	0	0	0	5		Indet.
29-2-10/60920.0	Reassociated.	2027	2013	PP	950E150	Biscuit B	0		0	0	0	0	0	0	0	0	0	0	0		
29-2-10/60932.1	Reassociated.	1998	60932	PP	985E180	Probably Gl. 3	0		0	0	0	0	0	0	0	0	0	0	0		
29-2-10/60982.0	Reassociated.	2011		PP	950E150		0		0	0	0	0	0	0	0	0	0	0	0		
29-2-10/60983.1	Fragmentary postcranial human remains of an adult female.			PP		Indet.	1	50–59	1	0	0	0	0	0	0	0	1	0	0	AS 7	Female
36-30-10/N1396.0	Fragmentary postcranial human remains of an individual of unknown age and sex.		N,1396	PP		Indet.	1		0	0	0	0	0	1	0	0	0	0	1		Indet.
42-19-10/N3948.3	Fragmentary postcranial human remains of an adult of unknown sex.		N3948-b-4, Pecos			Undet.	1		1	0	0	0	0	0	0	0	0	0	1		Indet.
42-19-10/N3950.1	Fragmentary postcranial human remains of an adult of unknown sex.		N3950-6, Pecos			Undet.	1		1	0	0	0	0	0	0	0	0	0	1		Indet.
80-60-10/23920.0	Fragmentary postcranial human remains of an infant of unknown sex.		23920		62 meters SSW of projecting wall	Undet.	1	0.0–0.5	0	0	0	1	0	0	0	0	0	0	1		Indet.
985-27-10/N10651.0	Fragmentary postcranial human remains of an adult of unknown sex.		10651 (MEM 11/96)	PP	NE midden edge just outside fence, 30 ft. N up slope, tr. 3 ft. deep 15 ft. wide	A.D. 1400-1838	1		1	0	0	0	0	0	0	0	0	0	1		Indet.
997-27-10/N10666.0	Fragmentary postcranial human remains of an adult of unknown sex.	Rowe	10, N10666	R	Provenience unknown	Indet.	1		1	0	0	0	0	0	0	0	0	0	1		Indet.
997-27-10/N10667.0	Human hair of an individual of unknown age and unknown sex.	Rowe	N10667, 997-27-10	R	Grave 11	Indet.	1		0	0	0	0	0	1	0	0	0	0	1		Indet.

Peabody number	Pathology type	Pathology description	Cranial mod.	Associated funerary objects	Field context	Notes
27-1-10/61334.0	Nutr.	Osteoporosis	Unobs.			
27-1-10/61335.0			No cranium			
27-1-10/61336.0	Arthritis	DJD	Unobs.			
27-1-10/N502.0			No cranium			
29-2-10/60920.0						Reassociated with catalog number 60895.
29-2-10/60932.1						Reassociated with catalog number 60933, MEM 4/98
29-2-10/60982.0						Reassociated with 29-2-10/60932.0
29-2-10/60983.1	Arthritis, nutr.	DJD, osteoporosis	No cranium	Frags. cooking pot, bone awl no. 236, worked sherd no. 235		
36-30-10/N1396.0	Arthritis	Extreme spondylitis (vt/t and vt/l)	No cranium			Fabric adhering to spine
42-19-10/N3948.3	Infection, nutr.	Some periosteal infection, some osteoporosis	No cranium			
42-19-10/N3950.1	Infection	Periosteal reaction in fibula	No cranium			
80-60-10/23920.0			No cranium			
985-27-10/N10651.0			No cranium			
997-27-10/N10666.0			No cranium			Remarks: Transferred from RSPM
997-27-10/N10667.0			No cran.			Remarks: Transferred from RSPM

Appendix C

Quantification of Pathology Data Presented in Appendix A

Michèle E. Morgan

Key to Column Heads	
Peabody number	Unique number assigned by the Peabody Museum of Archaeology and Ethnology
CS	Ceramic Stage: IN, indeterminate; P, problematic; B, Black-on-white; 1, Glaze 1; 1A, Glaze 1 or 2; 2, Glaze 2; 2A, Glaze 2 or 3; 3, Glaze 3; 3A, Glaze 3 or 4; 4, Glaze 4; 4A, Glaze 4 or 5; 5, Glaze 5; 5A, Glaze 5 or 6; 6, Glaze 6; 6A, Glaze 6 or Late; 7, Late
MNI	Minimum number of individuals
Age	A, adult; J, juvenile, which includes all other age categories
Sex	M, male or probably male; F, female or probably female; U, undetermined. Entries record presence (1) or absence (0).
AFO	Associated funerary objects: Y, yes; N, no
Cr. mod.	Cranial modification: N, no cranial modification; Q, questionable (not observed because of lack of recording and/or relevant anatomy was insufficiently present to record); U, unobservable (relevant anatomy not present); Y, cranial modification positively recorded
Trauma	All entries given as presence (1) or absence (0)
T cr	Cranial trauma
T pc	Postcranial trauma
T h	Healed trauma
T d	Death: perimortem trauma
Infection	All entries given as presence (1) or absence (0)
I de	Dental: dental-related infection including caries and abscesses

I di	Disease: disease-related infection
I tr	Trauma: trauma-related infection
I g	General: infection not clearly attributable to one of the previous three categories
Arthritis	All entries given as presence (1) or absence (0)
A sl	Slight degenerative joint disease
A m	Medium degenerative joint disease
A s	Severe degenerative joint disease
A sp	Spinal osteophytosis
A o	Other arthritis
Physiological stress	All entries given as presence (1) or absence (0)
EH	Enamel hypoplasia
CO	Cribra orbitalia
PH	Porotic hyperostosis
O	Osteoporosis
Neurological	All entries given as presence (1) or absence (0)
Sb sl	Spina bifida, slight
Sb m/s	Spina bifida, medium to severe
O na	Other neurological anomalies
Tumor	All entries given as presence (1) or absence (0)
T bo	Button osteoma
T o	Other tumor
Other path.	All entries given as presence (1) or absence (0)
Op d	Other dental pathology
Op c/pc	Other cranial and/or postcranial pathology

Peabody number	CS	MNI	Age	Sex			AFO	Cr mod	Trauma				Infection				Arthritis					Physiological stress				Neurological			Tumor		Other path.	
				M	F	U			T c	T pc	T h	T d	I de	I di	I tr	I g	A sl	A m	A s	A sp	A o	EH	CO	PH	O	Sb sl	Sb m/s	O na	T bo	T o	OP d	OP c/pc
00-0-10/N9468.101	B	1	J	0	0	1	N	U	0	0	0	0	0	0	0	0	0	0	0	0	0	0	0	0	0	0	0	0	0	0	0	0
00-0-10/N9468.38	2A	1	A	0	0	1	Y	U	0	0	0	0	0	0	0	0	0	0	0	0	0	1	0	0	0	0	0	0	0	0	0	0
00-0-10/N9468.40	B	1	A	0	0	1	N	U	0	0	0	0	0	0	0	0	0	0	0	0	0	0	0	0	0	0	0	0	0	0	0	0
00-0-10/N9468.42	B	1	J	0	0	1	N	U	0	0	0	0	0	0	0	0	0	0	0	0	0	0	0	0	0	0	0	0	0	0	0	0
00-0-10/N9468.43	B	1	A	0	0	1	N	U	0	0	0	0	0	0	0	0	0	0	0	0	0	0	0	0	0	0	0	0	0	0	0	0
00-0-10/N9468.44	B	1	J	0	0	1	N	U	0	0	0	0	0	0	0	0	0	0	0	0	0	0	0	0	0	0	0	0	0	0	0	0
00-0-10/N9468.45	B	1	J	0	0	1	N	U	0	0	0	0	0	0	0	0	0	0	0	0	0	0	0	0	0	0	0	0	0	0	0	0
00-0-10/N9468.46	B	1	J	0	0	1	N	U	0	0	0	0	0	0	0	0	0	0	0	0	0	0	0	0	0	0	0	0	0	0	0	0
00-0-10/N9468.47	B	1	A	0	0	1	Y	U	0	0	0	0	0	0	0	0	0	0	0	0	0	0	0	0	0	0	0	0	0	0	0	0
00-0-10/N9468.48	B	1	J	0	0	1	N	U	0	0	0	0	0	0	0	0	0	0	0	0	0	0	0	0	0	0	0	0	0	0	0	0
00-0-10/N9468.49	B	1	A	0	0	1	N	U	0	0	0	0	0	0	0	0	0	0	0	0	0	0	0	0	0	0	0	0	0	0	0	0
00-0-10/N9468.50	B	1	A	0	0	1	Y	U	0	0	0	0	0	0	0	0	0	0	0	0	0	0	0	0	0	0	0	0	0	0	0	0
00-0-10/N9468.52	B	1	A	0	0	1	N	U	0	0	0	0	0	0	0	0	0	0	0	0	0	1	0	0	0	0	0	0	0	0	0	0
00-0-10/N9468.53	B	1	A	0	0	1	N	U	0	0	0	0	0	0	0	0	0	0	0	0	0	1	0	0	0	0	0	0	0	0	0	0
00-0-10/N9468.54	B	1	A	0	0	1	N	U	0	0	0	0	0	0	0	0	0	0	0	0	0	1	0	0	0	0	0	0	0	0	0	0
00-0-10/N9468.55	B	1	J	0	0	1	N	U	0	0	0	0	0	0	0	0	0	0	0	0	0	0	0	0	0	0	0	0	0	0	0	0
00-0-10/N9468.56	B	1	J	0	0	1	N	U	0	0	0	0	0	0	0	0	0	0	0	0	0	1	0	0	0	0	0	0	0	0	0	0
00-0-10/N9468.57	B	1	J	0	0	1	N	U	0	0	0	0	0	0	0	0	0	0	0	0	0	0	0	0	0	0	0	0	0	0	0	0
00-0-10/N9468.58	B	1	J	0	0	1	N	U	0	0	0	0	0	0	0	0	0	0	0	0	0	1	0	0	0	0	0	0	0	0	0	0
00-0-10/N9468.59	B	1	A	0	0	1	Y	U	0	0	0	0	0	0	0	0	0	0	0	0	0	0	0	0	0	0	0	0	0	0	0	0
00-0-10/N9468.60	B	1	A	0	0	1	Y	U	0	0	0	0	0	0	0	0	0	0	0	0	0	0	0	0	0	0	0	0	0	0	0	0
00-0-10/N9468.61	B	1	A	0	0	1	Y	U	0	0	0	0	0	0	0	0	0	0	0	0	0	1	0	0	0	0	0	0	0	0	0	0
00-0-10/N9468.63	IN	1	J	0	0	1	Y	U	0	0	0	0	0	0	0	0	0	0	0	0	0	0	0	0	0	0	0	0	0	0	0	0
00-0-10/N9468a.12	IN	1	A	0	0	1	Y	U	0	0	0	0	0	0	0	0	0	0	0	0	0	0	0	0	0	0	0	0	0	0	0	0
19-6-10/59801.0	IN	1	A	1	0	0	Y	Q	0	0	0	0	0	0	0	0	0	0	0	0	0	0	0	0	0	0	0	0	0	0	0	0
19-6-10/59802.0	B	1	A	0	1	0	Y	Q	0	0	0	0	0	1	0	0	0	0	0	0	0	0	0	0	0	0	1	0	0	0	0	0
19-6-10/59803.0	IN	1	A	0	1	0	Y	Y	1	0	1	0	1	0	0	0	0	1	0	0	0	0	0	0	1	0	0	0	0	0	0	0
19-6-10/59804.0	1A	1	A	0	1	0	Y	Q	0	0	0	0	0	0	0	0	0	0	0	0	0	0	0	0	0	0	0	0	0	0	0	0
19-6-10/59805.0	5	1	A	1	0	0	Y	Q	0	0	0	0	0	0	0	1	0	0	0	1	0	0	0	0	0	0	0	0	0	0	1	0
19-6-10/59806.0	2A	1	A	1	0	0	N	Q	0	0	0	0	0	0	0	0	1	0	0	1	0	0	0	0	1	0	0	0	0	0	0	0
19-6-10/59807.0	3	1	J	1	0	0	Y	Q	0	0	0	0	0	0	0	1	0	0	0	0	0	0	0	0	0	0	0	0	0	0	0	0
19-6-10/59808.0	IN	1	J	1	0	0	Y	Q	0	0	0	0	0	0	0	0	0	0	0	0	0	0	0	0	0	0	0	0	0	0	0	0
19-6-10/59809.0	IN	1	A	0	1	0	Y	Q	0	0	0	0	0	0	0	0	0	0	0	0	0	0	0	0	0	0	0	0	0	0	0	0
19-6-10/59810.0	6	1	A	0	1	0	Y	Q	0	0	0	0	0	0	0	0	0	0	0	1	0	0	0	0	0	0	0	0	0	0	0	0
19-6-10/59811.0	4	1	A	0	1	0	Y	Q	0	0	0	0	0	0	0	0	0	0	0	1	0	0	0	0	1	0	0	0	0	0	0	0
19-6-10/59812.0	1A	1	A	1	0	0	Y	Q	0	0	0	0	0	0	0	0	0	0	0	0	0	0	0	0	0	0	0	0	0	0	0	0
19-6-10/59813.0	6	1	A	0	1	0	Y	Q	0	0	0	0	0	0	0	0	0	0	0	1	0	0	0	0	0	0	0	0	0	0	0	0
19-6-10/59814.0	3	1	A	0	1	0	Y	Q	0	0	0	0	0	1	0	0	1	0	0	0	0	0	0	0	0	0	0	0	0	1	0	0
19-6-10/59814.1	3	1	A	0	0	1	Y	U	0	0	0	0	0	0	0	0	0	0	0	0	0	0	0	0	0	0	0	0	0	0	0	0
19-6-10/59815.0	4A	1	A	0	1	0	Y	Q	0	0	0	0	0	0	0	0	1	0	0	0	0	0	0	0	0	0	0	0	0	0	0	0
19-6-10/59816.0	5	1	A	1	0	0	Y	Q	1	0	1	0	0	0	0	0	0	0	0	0	0	0	0	0	1	0	0	0	0	0	1	0
19-6-10/59817.0	IN	1	A	0	1	0	Y	Q	0	0	0	0	0	0	0	0	0	0	0	1	1	0	0	0	0	0	0	0	0	0	0	0
19-6-10/59818.0	IN	1	A	0	1	0	Y	U	0	0	0	0	0	0	0	0	0	0	0	0	0	0	0	0	0	0	0	0	0	0	0	0
19-6-10/59818.1	IN	1	J	0	0	1	Y	U	0	0	0	0	0	0	0	0	0	0	0	0	0	0	0	0	0	0	0	0	0	0	0	0

Peabody number	CS	MNI	Age	Sex			AFO	Cr mod	Trauma				Infection				Arthritis					Physiological stress				Neurological			Tumor		Other path.	
				M	F	U			T c	T pc	T h	T d	I de	I di	I tr	I g	A sl	A m	A s	A sp	A o	EH	CO	PH	O	Sb sl	Sb m/s	O na	T bo	T o	OP d	OP c/pc
19-6-10/59819.0	P	1	A	0	1	0	N	U	0	0	0	0	0	0	0	1	0	0	0	0	0	0	0	0	1	0	0	0	0	0	0	0
19-6-10/59820.0	IN	1	A	0	1	0	Y	Q	1	0	1	0	1	0	0	0	1	0	0	1	0	0	0	0	0	0	1	0	0	0	0	0
19-6-10/59821.0	IN	1	J	0	0	1	Y	Q	0	0	0	0	1	0	0	0	0	0	0	0	0	0	0	0	0	0	0	0	0	0	0	0
19-6-10/59822.0	6	1	J	1	0	0	Y	Q	0	0	0	0	0	0	0	1	0	0	0	0	0	0	0	0	0	0	0	0	0	0	0	0
19-6-10/59823.0	4	1	A	1	0	0	Y	Q	0	0	0	0	0	0	0	1	0	0	0	0	1	0	0	0	0	0	0	0	0	1	0	0
19-6-10/59824.0	5	1	A	0	1	0	N	Q	0	1	1	0	0	0	0	0	0	0	0	1	1	0	0	0	0	0	0	0	0	1	0	0
19-6-10/59825.0	4	1	A	1	0	0	Y	Q	0	0	0	0	1	0	0	0	0	0	0	1	0	1	0	0	0	0	0	1	0	0	0	0
19-6-10/59825.1	4	1	A	0	0	1	N	U	0	0	0	0	0	0	0	0	0	0	0	0	0	0	0	0	0	0	0	0	0	0	0	0
19-6-10/59826.0	4	1	A	0	1	0	Y	Q	0	0	0	0	0	0	0	0	0	0	0	0	0	0	0	0	0	0	0	0	0	0	0	0
19-6-10/59827.0	IN	1	A	1	0	0	Y	Q	0	0	0	0	0	0	0	0	0	0	0	1	0	0	0	0	0	0	0	0	0	0	0	0
19-6-10/59828.0	5	1	A	0	1	0	Y	Q	0	0	0	0	1	0	0	0	0	0	0	0	0	0	0	0	0	0	0	0	0	0	0	0
19-6-10/59829.0	1	1	A	0	1	0	Y	Q	0	0	0	0	0	0	0	0	0	0	0	0	0	0	0	0	0	0	0	0	0	0	0	0
19-6-10/59830.0	6	1	A	1	0	0	Y	Q	0	0	0	0	1	0	0	0	0	0	0	0	0	0	0	0	0	0	0	0	0	0	0	0
19-6-10/59830.1	IN	1	A	0	0	1	N	U	0	0	0	0	0	0	0	0	0	0	0	0	1	1	0	0	0	0	0	0	0	0	0	0
19-6-10/59832.0	IN	1	A	1	0	0	Y	Q	0	0	0	0	1	0	0	0	0	0	0	0	0	0	0	0	0	0	0	0	0	0	0	0
19-6-10/59833.0	IN	1	A	1	0	0	Y	U	0	0	0	0	0	0	0	0	0	0	0	0	0	0	0	0	0	0	0	0	0	0	0	0
19-6-10/59834.0	3	1	A	0	1	0	Y	Q	0	1	1	0	0	1	0	0	0	0	0	0	0	0	0	0	0	0	0	0	0	1	0	0
19-6-10/59835.0	3	1	A	0	1	0	Y	Q	0	0	0	0	0	0	0	0	0	0	0	0	0	0	0	0	0	0	1	0	0	0	0	0
19-6-10/59836.0	4	1	A	0	1	0	N	Q	0	0	0	0	0	0	0	1	0	0	1	1	0	0	0	0	1	0	0	0	0	0	0	0
19-6-10/59837.0	5	1	A	1	0	0	Y	Q	0	0	0	0	1	0	0	0	0	0	0	0	0	0	0	0	0	0	0	0	0	1	0	0
19-6-10/59838.0	2A	1	A	1	0	0	N	Q	0	0	0	0	1	0	0	0	0	1	0	0	1	0	0	0	0	0	0	0	0	1	0	0
19-6-10/59839.0	4A	1	A	0	1	0	Y	Q	0	0	0	0	1	0	0	0	0	0	0	1	0	0	0	0	1	0	0	0	0	1	0	0
19-6-10/59841.0	IN	1	A	0	1	0	Y	Q	0	1	1	0	0	0	0	0	0	0	0	1	0	0	0	0	0	0	0	0	0	0	0	0
19-6-10/59842.0	1A	1	A	1	0	0	Y	Q	0	1	1	0	1	0	0	0	1	0	0	0	0	0	0	0	1	0	0	0	0	0	0	0
19-6-10/59843.0	IN	1	A	0	1	0	Y	Q	0	0	0	0	1	0	0	0	1	0	0	0	0	0	0	0	0	0	0	0	0	0	0	0
19-6-10/59844.0	BA	1	A	0	0	1	Y	U	0	1	1	0	0	0	0	0	0	0	0	0	0	0	0	0	0	0	0	0	0	0	0	0
19-6-10/59845.0	BA	1	A	0	1	0	Y	Q	1	0	0	1	0	0	0	1	0	0	0	1	0	1	0	0	0	0	0	0	1	0	0	0
19-6-10/59846.0	IN	1	A	0	1	0	Y	Q	1	0	1	0	1	0	0	0	0	1	0	1	0	0	0	0	0	1	0	0	0	0	0	0
19-6-10/59847.0	3	1	J	0	1	0	N	Q	0	0	0	0	0	0	0	0	0	0	0	0	0	0	0	0	0	0	0	0	0	0	0	0
19-6-10/59848.0	IN	1	J	0	0	1	Y	Q	0	0	0	0	0	0	0	0	0	0	0	0	0	0	0	0	0	0	0	0	0	0	0	0
19-6-10/59849.0	IN	1	A	0	1	0	Y	Q	0	0	0	0	0	0	0	0	0	0	0	0	0	0	0	0	0	0	0	0	0	0	0	0
19-6-10/59850.0	IN	1	A	0	1	0	Y	Q	0	0	0	0	0	0	0	0	0	0	0	0	0	0	0	0	0	0	0	0	0	0	0	0
19-6-10/59851.0	1	1	A	1	0	0	Y	Q	0	0	0	0	0	0	0	0	0	0	1	1	0	0	0	0	1	0	0	0	0	0	0	0
19-6-10/59852.0	IN	1	A	1	0	0	Y	N	0	0	0	0	0	0	0	0	0	0	0	0	0	1	0	0	0	0	0	0	0	0	0	0
19-6-10/59853.0	4	1	A	1	0	0	N	Q	0	0	0	0	1	0	0	0	1	0	0	0	0	0	0	0	1	0	0	0	0	0	0	0
19-6-10/59854.0	2	1	A	0	1	0	N	Y	1	0	1	0	0	0	1	0	0	1	0	1	0	0	0	0	1	0	0	0	0	0	0	0
19-6-10/59855.0	6	1	A	0	0	1	N	U	0	0	0	0	0	0	0	0	0	0	0	0	0	0	0	0	0	0	0	0	0	0	0	0
19-6-10/59856.0	IN	1	A	0	1	0	Y	Q	0	0	0	0	0	0	0	0	0	0	0	1	1	1	0	0	0	0	0	0	0	0	0	0
19-6-10/59857.0	5	1	A	0	1	0	N	Q	0	1	1	0	1	0	1	0	0	0	0	1	0	0	0	0	0	0	0	0	0	0	0	0
19-6-10/59859.0	1	1	A	0	1	0	Y	Y	0	0	0	0	0	0	0	0	1	0	0	0	0	0	0	0	1	1	0	0	0	0	0	0
19-6-10/59861.0	IN	1	A	0	1	0	Y	Q	0	0	0	0	0	0	0	0	0	0	1	1	0	0	0	0	1	0	0	0	0	0	0	0
19-6-10/59862.0	IN	1	A	1	0	0	Y	Y	0	0	0	0	0	0	0	1	0	0	0	0	0	0	0	0	0	0	0	0	0	0	1	0
19-6-10/59863.0	3	1	A	1	0	0	Y	Y	0	0	0	0	0	0	0	0	1	0	0	1	0	0	0	0	0	0	0	0	0	0	1	0
19-6-10/59864.0	IN	1	A	0	1	0	Y	Q	0	0	0	0	0	1	0	0	0	0	0	0	0	0	0	0	0	0	0	0	0	0	0	0
19-6-10/59865.0	3	1	A	1	0	0	Y	Q	0	0	0	0	0	0	0	0	1	0	0	1	0	1	0	0	0	0	0	0	0	0	0	0

Peabody number	CS	MNI	Age	Sex			AFO	Cr mod	Trauma				Infection				Arthritis					Physiological stress				Neurological			Tumor		Other path.	
				M	F	U			T c	T pc	T h	T d	I de	I di	I tr	I g	A sl	A m	A s	A sp	A o	EH	CO	PH	O	Sb sl	Sb m/s	O na	T bo	T o	OP d	OP c/pc
19-6-10/59866.0	6	1	A	1	0	0	Y	N	0	0	0	0	0	0	0	0	1	0	0	0	0	0	0	0	0	0	0	0	0	0	0	0
19-6-10/59867.0	4	1	J	0	1	0	Y	U	0	0	0	0	0	0	0	0	0	0	0	0	0	0	0	0	0	0	0	0	0	0	0	0
19-6-10/59868.0	3A	1	A	1	0	0	Y	Y	0	0	0	0	0	0	0	0	0	0	1	1	0	0	0	0	0	0	0	0	0	0	0	0
19-6-10/59869.0	4	1	A	0	0	1	Y	U	0	0	0	0	0	0	0	0	0	0	0	0	0	1	0	0	0	0	0	0	0	0	0	0
19-6-10/59870.0	4	1	A	1	0	0	N	Y	0	0	0	0	0	0	0	0	1	0	0	0	0	0	0	0	1	0	1	0	0	0	0	0
19-6-10/59871.0	6	1	A	0	1	0	Y	Y	0	0	0	0	0	0	0	0	1	0	0	0	0	0	0	0	0	0	0	0	0	0	0	0
19-6-10/59871.1	2	1	A	1	0	0	Y	Q	0	0	0	0	0	0	0	0	0	0	0	0	0	1	0	0	0	0	0	0	0	0	0	0
19-6-10/59873.0	5	1	J	1	0	0	Y	U	0	0	0	0	0	0	0	0	0	0	0	0	0	0	0	0	0	0	0	0	0	0	0	0
19-6-10/59874.0	IN	1	A	1	0	0	Y	U	0	0	0	0	0	0	0	0	0	0	0	1	0	0	0	0	1	0	0	0	0	0	0	0
19-6-10/59874.1	IN	1	A	0	0	1	Y	U	0	0	0	0	0	0	0	0	0	0	0	0	0	0	0	0	0	0	0	0	0	0	0	0
19-6-10/59875.0	3	1	A	1	0	0	N	Q	0	0	0	0	0	0	0	0	1	0	0	1	0	0	0	0	1	0	0	0	0	0	0	0
19-6-10/59876.0	5A	1	A	1	0	0	Y	N	0	0	0	0	0	0	0	1	1	0	0	1	0	0	0	0	0	1	0	0	0	0	0	0
19-6-10/59877.0	6	1	A	0	1	0	Y	U	0	0	0	0	0	0	0	0	1	0	0	0	0	0	0	0	0	0	0	0	0	0	0	0
19-6-10/59877.1	6	1	A	1	0	0	Y	U	0	0	0	0	0	0	0	0	0	0	0	0	1	0	0	0	0	0	0	0	0	0	0	0
19-6-10/59878.1	CH	1	A	1	0	0	N	U	0	0	0	0	0	0	0	0	0	0	0	0	0	0	0	0	0	0	0	0	0	0	0	1
19-6-10/59878.3	CH	1	A	0	0	1	N	U	0	0	0	0	0	0	0	0	0	0	0	0	0	0	0	0	0	0	0	0	0	0	0	0
19-6-10/59878.4	CH	1	A	0	0	1	N	U	0	0	0	0	0	0	0	0	0	0	0	0	0	0	0	0	0	0	0	0	0	0	0	0
19-6-10/59879.1	CH	1	A	0	0	1	N	U	0	0	0	0	0	0	0	0	0	0	0	0	0	0	0	0	0	0	0	0	0	0	0	0
19-6-10/59879.2	CH	1	A	1	0	0	N	U	0	1	1	0	0	0	0	0	0	0	0	0	0	0	0	0	0	0	0	0	0	0	0	0
19-6-10/59879.3	CH	1	J	1	0	0	N	U	0	0	0	0	0	0	0	0	0	0	0	0	0	0	0	0	0	0	0	0	0	0	0	0
19-6-10/59879.4	CH	1	A	0	0	1	N	U	0	0	0	0	0	0	0	0	0	0	0	0	0	0	0	0	0	0	0	0	0	0	0	0
19-6-10/59879.5	CH	1	A	0	0	1	N	U	0	0	0	0	0	0	0	0	0	0	0	0	0	0	0	0	0	0	0	0	0	0	0	0
19-6-10/59879.6	CH	1	A	0	0	1	N	U	0	0	0	0	0	0	0	0	0	0	0	0	0	0	0	0	0	0	0	0	0	0	0	0
19-6-10/59879.7	CH	1	A	0	1	0	N	U	0	0	0	0	0	0	0	0	0	0	0	0	0	0	0	0	0	0	0	0	0	0	0	0
19-6-10/59881.0	CH	1	A	1	0	0	N	U	0	0	0	0	0	0	0	0	0	0	0	0	0	0	0	0	0	0	0	0	0	0	0	0
19-6-10/59881.1	CH	1	A	1	0	0	N	U	0	0	0	0	0	0	0	0	0	0	0	0	0	0	0	0	0	0	0	0	0	0	0	0
19-6-10/59881.2	CH	1	A	0	1	0	N	U	0	0	0	0	0	0	0	0	0	0	0	0	0	0	0	0	0	0	0	0	0	0	0	0
19-6-10/59882.1	CH	1	A	1	0	0	N	U	0	0	0	0	0	0	0	0	0	0	0	0	0	0	0	0	0	0	0	0	0	0	0	0
19-6-10/59883.0	CH	1	A	0	0	1	N	U	0	0	0	0	0	0	0	0	0	0	0	0	0	0	0	0	0	0	0	0	0	0	0	0
19-6-10/59885.1	CH	1	A	0	0	1	N	U	0	0	0	0	0	0	0	0	0	0	0	0	0	0	0	0	0	0	0	0	0	0	0	0
19-6-10/59885.3	CH	1	A	0	0	1	N	U	0	0	0	0	0	0	0	0	0	0	0	0	0	0	0	0	0	0	0	0	0	0	0	0
19-6-10/59885.4	CH	1	A	0	0	1	N	U	0	0	0	0	0	0	0	0	0	0	0	0	0	0	0	0	0	0	0	0	0	0	0	0
19-6-10/59885.5	CH	1	A	0	0	1	N	U	0	0	0	0	0	0	0	0	0	0	0	0	0	0	0	0	0	0	0	0	0	0	0	0
19-6-10/59888.0	CH	1	J	0	1	0	N	U	0	0	0	0	0	0	0	0	0	0	0	0	0	0	0	0	0	0	0	0	0	0	0	0
19-6-10/59889.0	CH	1	A	0	1	0	N	Q	0	0	0	0	1	0	0	0	1	0	0	0	0	0	0	0	1	0	0	0	0	0	0	0
19-6-10/59890.0	CH	1	A	0	1	0	N	Q	0	0	0	0	0	0	0	0	0	0	0	0	0	0	0	0	0	0	0	0	0	0	0	1
19-6-10/59891.0	CH	1	A	1	0	0	N	Q	0	0	0	0	0	0	0	0	1	0	0	1	0	0	0	0	0	0	0	0	0	0	0	0
19-6-10/59892.0	CH	1	A	1	0	0	N	Q	1	0	1	0	1	0	0	1	0	0	0	0	0	0	0	0	0	0	0	0	0	0	0	1
19-6-10/59893.0	CH	1	A	1	0	0	N	Q	0	0	0	0	0	0	0	0	0	0	0	0	0	0	0	0	0	0	0	0	0	0	0	1
19-6-10/59894.0	CH	1	J	0	1	0	N	Q	0	0	0	0	0	0	0	0	0	0	0	0	0	0	0	0	0	0	0	0	0	0	0	0
19-6-10/59895.0	CH	1	A	0	1	0	N	Q	0	0	0	0	0	0	0	0	0	0	0	0	0	0	0	0	0	0	0	0	0	0	0	1
19-6-10/59896.0	CH	1	J	0	0	1	N	Q	1	0	1	1	0	0	0	0	0	0	0	0	0	0	1	0	0	0	0	0	0	0	0	1
19-6-10/59897.0	CH	1	A	0	1	0	N	Q	0	0	0	0	0	0	0	0	0	0	0	0	1	0	0	0	0	0	0	0	0	0	0	0
19-6-10/59898.0	CH	1	A	0	1	0	N	Q	1	0	0	0	0	0	0	1	0	0	0	0	0	0	0	0	0	0	0	0	0	0	0	0
19-6-10/59899.0	CH	1	A	1	0	0	N	Q	0	0	0	0	0	0	0	0	0	0	0	0	0	0	0	0	0	0	0	0	0	0	0	0

Peabody number	CS	MNI	Age	Sex			AFO	Cr mod	Trauma				Infection				Arthritis					Physiological stress				Neurological			Tumor		Other path.	
				M	F	U			T c	T pc	T h	T d	I de	I di	I tr	I g	A sl	A m	A s	A sp	A o	EH	CO	PH	O	Sb sl	Sb m/s	O na	T bo	T o	OP d	OP c/pc
19-6-10/59900.0	CH	1	J	0	0	1	N	Y	0	0	0	0	0	0	0	0	0	0	0	0	0	0	1	0	0	0	0	0	0	0	0	0
19-6-10/59901.0	CH	1	A	1	0	0	N	Q	0	0	0	0	0	0	0	1	0	0	0	0	0	0	0	0	0	0	0	0	0	0	0	0
19-6-10/59902.0	CH	1	A	1	0	0	N	Q	1	0	0	1	0	0	0	0	0	0	0	0	0	0	0	0	0	0	0	0	0	0	0	0
19-6-10/59903.0	CH	1	A	0	1	0	N	U	0	0	0	0	0	0	0	0	0	0	0	0	0	0	0	0	0	0	0	0	0	0	0	0
19-6-10/59904.0	CH	1	A	1	0	0	N	Q	1	0	0	1	0	1	0	0	0	0	0	0	0	0	0	0	0	0	0	0	0	0	0	0
19-6-10/59905.0	CH	1	A	1	0	0	N	Q	0	0	0	0	0	0	0	0	0	0	0	0	0	0	0	0	0	0	0	0	0	0	0	0
19-6-10/59906.0	CH	1	A	1	0	0	N	Q	1	0	1	1	1	0	0	0	0	0	0	0	0	0	0	0	0	0	0	0	0	0	1	0
19-6-10/59907.0	5	1	A	1	0	0	Y	Q	0	0	0	0	0	0	0	0	1	0	0	0	0	0	0	0	0	0	0	1	0	0	1	1
19-6-10/59908.0	IN	1	A	0	1	0	Y	Q	1	0	0	0	1	0	0	0	0	0	0	0	0	0	0	0	0	0	0	0	0	0	0	0
19-6-10/59909.0	P	1	J	0	0	1	N	Q	0	0	0	0	0	0	0	0	0	0	0	0	0	1	0	0	0	0	0	0	0	0	0	0
19-6-10/59910.0	IN	1	A	0	1	0	Y	Q	0	0	0	0	1	0	0	0	0	0	0	0	0	0	0	0	0	0	0	0	0	1	0	0
19-6-10/59911.0	IN	1	A	0	1	0	Y	Y	0	0	0	0	0	0	0	0	0	0	0	0	0	0	0	0	1	0	0	0	0	0	0	0
19-6-10/59912.0	IN	1	A	0	1	0	Y	Y	0	0	0	0	1	0	0	0	1	0	0	0	0	0	0	0	1	0	0	0	0	0	0	0
19-6-10/59913.0	IN	1	A	0	1	0	Y	Y	0	0	0	0	0	0	0	0	1	0	0	0	0	0	0	0	1	0	0	0	0	0	0	0
19-6-10/59914.0	3	1	A	0	1	0	N	Q	0	0	0	0	0	0	0	1	1	0	0	0	0	0	0	0	0	0	0	0	0	0	0	0
19-6-10/59915.0	5	1	A	1	0	0	Y	Y	0	1	1	0	0	0	0	0	0	0	0	1	0	1	0	0	0	0	1	0	1	0	0	0
19-6-10/59916.0	5	1	A	0	1	0	Y	Q	0	0	0	0	1	0	0	0	0	0	0	1	0	0	0	0	0	0	0	0	0	0	0	1
19-6-10/59917.0	2	1	A	0	1	0	Y	Y	0	0	0	0	0	0	0	0	1	0	0	0	0	0	0	0	0	0	0	0	0	0	0	0
19-6-10/59918.0	5	1	A	1	0	0	Y	Y	0	0	0	0	0	0	0	0	1	0	0	0	0	0	0	0	0	0	0	0	0	0	0	1
19-6-10/59919.0	P	1	A	1	0	0	Y	Y	0	0	0	0	0	0	0	0	1	0	0	0	0	0	0	0	0	0	0	0	0	0	0	0
19-6-10/59920.0	3A	1	A	0	1	0	N	Y	0	0	0	0	0	0	0	0	1	0	0	1	0	1	0	0	1	0	0	0	0	0	0	0
19-6-10/59921.0	4A	1	A	1	0	0	N	Y	0	0	0	0	0	0	0	0	0	0	0	1	0	1	0	0	0	0	1	0	0	0	0	0
19-6-10/59922.0	4	1	A	0	1	0	Y	Q	0	1	0	0	0	0	0	0	1	0	0	0	0	0	0	0	0	0	0	0	0	0	0	0
19-6-10/59923.0	P	1	A	0	1	0	N	Y	0	0	0	0	0	0	0	0	1	0	0	0	0	0	0	0	0	0	0	0	0	0	0	0
19-6-10/59924.0	6	1	A	1	0	0	Y	Q	0	0	0	0	0	0	0	0	0	0	0	1	0	0	0	0	0	0	0	0	0	0	0	0
19-6-10/59925.0	IN	1	A	0	1	0	Y	Y	0	0	0	0	0	0	0	0	0	0	0	0	0	0	0	0	0	0	0	0	0	0	0	0
19-6-10/59951.0	3	1	A	1	0	0	Y	Q	0	0	0	0	0	0	0	0	0	0	0	1	0	0	0	0	0	0	0	0	0	0	1	0
19-6-10/59952.0	IN	1	A	0	1	0	Y	Y	0	0	0	0	0	0	0	0	1	0	0	0	0	0	0	0	1	0	0	0	1	0	0	0
19-6-10/59953.0	4	1	A	1	0	0	Y	Y	0	0	0	0	0	0	0	1	1	0	0	0	0	1	0	0	0	0	0	0	0	0	0	0
19-6-10/59954.0	2	1	A	1	0	0	N	Q	0	0	0	0	1	0	0	0	1	0	0	0	0	0	0	1	0	0	0	0	0	0	1	0
19-6-10/59955.0	3A	1	A	1	0	0	N	Q	0	0	0	0	0	0	0	0	0	0	0	0	0	0	0	0	0	0	0	0	0	0	1	0
19-6-10/59956.0	3	1	A	1	0	0	Y	Y	0	0	0	0	0	0	0	0	1	0	0	0	0	0	0	0	0	0	0	0	0	0	0	0
19-6-10/59957.0	P	1	A	1	0	0	N	N	0	0	0	0	0	0	0	0	0	1	0	0	0	0	0	0	0	0	0	0	0	0	1	1
19-6-10/59958.0	4A	1	A	1	0	0	N	Y	0	0	0	0	0	0	0	0	0	0	1	1	0	0	0	0	0	0	0	0	0	0	0	0
19-6-10/59959.0	IN	1	A	0	1	0	Y	Y	0	0	0	0	0	0	0	0	1	0	0	0	0	0	0	0	0	0	0	0	0	0	0	0
19-6-10/59960.0	IN	1	A	0	1	0	N	Q	0	0	0	0	0	0	0	0	0	0	0	0	0	0	0	0	0	0	0	0	0	0	0	0
19-6-10/59961.0	2A	1	J	0	0	1	N	Q	0	0	0	0	0	0	0	0	0	0	0	0	0	0	0	0	0	0	0	0	0	0	0	0
19-6-10/59962.0	IN	1	A	1	0	0	Y	Q	0	0	0	0	0	0	0	0	1	0	0	1	0	0	0	0	0	0	0	0	0	0	0	0
19-6-10/59963.0	4	1	A	1	0	0	Y	Y	0	0	0	0	0	0	0	0	0	1	0	0	0	0	0	0	1	0	0	0	0	0	0	0
19-6-10/59963.1	4	1	A	0	0	1	Y	U	0	0	0	0	0	0	0	0	0	0	0	0	0	0	0	0	0	0	0	0	0	0	0	0
19-6-10/59964.0	3A	1	A	0	1	0	N	Q	0	0	0	0	0	0	0	0	0	0	0	0	0	0	0	0	0	0	0	0	0	0	0	0
19-6-10/59965.0	B	1	A	0	1	0	N	Q	0	0	0	0	0	0	0	1	1	0	0	0	0	0	0	0	0	1	0	0	0	0	0	0
19-6-10/59966.0	IN	1	A	1	0	0	Y	Y	0	1	1	0	1	0	0	0	1	0	0	1	0	1	0	0	0	0	0	0	0	0	0	0
19-6-10/59967.0	IN	1	A	1	0	0	Y	Q	0	1	1	0	0	0	0	0	0	0	1	1	0	0	0	0	1	0	1	0	0	0	0	0
19-6-10/59968.0	IN	1	J	0	0	1	N	Q	0	0	0	0	0	1	0	0	0	0	0	0	1	0	0	1	0	0	0	0	0	0	0	0

				Sex					Trauma				Infection				Arthritis					Physiological stress				Neurological			Tumor		Other path.	
Peabody number	CS	MNI	Age	M	F	U	AFO	Cr mod	Tc	Tpc	Th	Td	Ide	Idi	Itr	Ig	Asl	Am	As	Asp	Ao	EH	CO	PH	O	Sb sl	Sb m/s	Ona	Tbo	To	OP d	OP c/pc
19-6-10/59969.0	4	1	A	0	1	0	Y	Y	0	0	0	0	0	0	0	0	0	0	0	0	1	1	0	0	0	0	0	0	0	0	0	0
19-6-10/59970.0	BA	1	A	1	0	0	N	Q	0	0	0	0	1	0	0	0	0	0	0	0	0	0	0	0	0	0	0	0	0	0	0	0
19-6-10/59971.0	P	1	A	0	1	0	N	Q	0	0	0	0	0	0	0	0	0	1	0	0	0	1	0	0	0	0	0	0	0	0	0	0
19-6-10/59972.0	4A	1	A	1	0	0	Y	Q	0	1	1	0	0	0	0	0	0	0	0	1	0	1	0	0	1	0	0	0	0	0	0	0
19-6-10/59973.0	IN	1	J	0	0	1	Y	Q	0	0	0	0	0	0	0	0	0	0	0	0	0	0	0	0	0	0	0	0	0	0	1	0
19-6-10/59974.0	4	1	J	0	1	0	Y	Y	0	0	0	0	0	0	0	0	0	0	0	0	0	1	0	0	0	0	0	0	0	0	0	0
19-6-10/59975.0	IN	1	A	1	0	0	Y	Q	0	0	0	0	0	0	0	0	0	0	0	1	0	0	0	0	1	0	0	0	0	0	0	0
19-6-10/59976.0	IN	1	A	0	1	0	Y	Y	1	1	1	0	0	0	0	0	1	0	0	1	1	0	0	0	1	0	0	0	0	0	0	0
19-6-10/59977.0	5	1	A	0	1	0	Y	Q	0	0	0	0	1	0	0	0	1	0	0	0	0	1	0	0	0	0	0	0	0	0	0	0
19-6-10/59978.0	4A	1	A	0	1	0	Y	Y	0	0	0	0	1	0	0	0	1	0	0	0	0	0	0	0	0	0	0	0	0	0	0	0
19-6-10/59979.0	4	1	A	0	1	0	N	Q	0	0	0	0	0	0	0	0	0	1	0	1	0	1	0	0	0	0	0	0	0	0	0	0
19-6-10/59980.0	6	1	A	0	1	0	Y	Y	0	0	0	0	1	0	0	0	1	0	0	1	0	0	0	0	0	0	0	0	0	0	0	0
19-6-10/59981.0	IN	1	A	1	0	0	Y	Y	0	0	0	0	0	0	0	0	0	0	0	0	0	1	0	0	0	0	0	0	0	0	0	0
19-6-10/59982.0	6	1	A	1	0	0	Y	Y	0	0	0	0	0	0	0	0	1	0	0	0	0	1	0	0	0	0	0	0	0	0	0	0
19-6-10/59983.0	5	1	J	1	0	0	N	Q	0	0	0	0	0	0	0	1	0	0	0	0	0	0	0	0	0	0	0	0	0	0	0	0
19-6-10/59984.0	3A	1	A	1	0	0	Y	Y	0	0	0	0	1	0	0	0	0	0	1	1	1	1	0	0	0	0	0	0	0	0	0	0
19-6-10/59985.0	IN	1	A	1	0	0	Y	Q	0	0	0	0	0	0	0	0	1	0	0	0	0	0	0	0	0	0	0	0	0	0	0	1
19-6-10/59986.0	3	1	A	0	1	0	Y	Y	0	1	0	0	1	0	0	0	0	1	0	1	0	1	0	0	1	0	0	0	0	0	0	0
19-6-10/59986.1	3	1	A	1	0	0	Y	U	0	1	0	0	0	0	0	0	0	0	0	0	0	0	0	0	0	0	0	0	0	0	0	0
19-6-10/59987.0	2A	1	A	1	0	0	Y	Y	0	0	0	0	0	0	1	0	0	0	0	0	0	1	0	0	0	0	0	1	0	0	0	0
19-6-10/59988.0	IN	1	A	0	1	0	Y	U	0	0	0	0	0	0	0	0	1	0	0	0	0	1	0	0	0	0	0	0	0	0	0	0
19-6-10/59989.0	IN	1	A	0	1	0	Y	Y	0	0	0	0	1	0	0	0	0	1	0	0	0	0	0	0	1	0	0	0	0	0	0	0
19-6-10/59990.0	3	1	A	1	0	0	Y	Q	0	1	0	0	0	0	0	0	1	0	0	1	0	0	0	0	0	0	0	1	0	0	0	0
19-6-10/59991.0	4	1	A	0	1	0	Y	Y	0	1	1	0	1	0	0	0	1	0	0	1	0	0	0	0	1	0	0	0	0	0	0	0
19-6-10/59992.0	5	1	J	0	0	1	Y	Q	0	0	0	0	0	0	0	0	0	0	0	0	0	0	0	0	0	0	0	0	0	0	0	0
19-6-10/59993.0	IN	1	A	0	1	0	Y	Y	0	0	0	0	0	0	0	0	0	1	0	1	0	1	0	0	1	0	0	0	0	0	0	0
19-6-10/59994.0	IN	1	A	1	0	0	N	Y	0	0	0	0	0	0	0	0	0	0	0	1	0	0	0	0	0	0	1	0	0	0	0	0
19-6-10/59995.0	4	1	A	0	1	0	Y	Y	1	0	1	0	0	0	0	0	1	0	0	1	0	0	0	0	0	0	0	0	1	0	0	0
19-6-10/59996.0	3	1	A	0	1	0	Y	Y	0	0	0	0	0	0	0	0	0	0	0	0	0	0	0	0	0	1	0	0	0	0	0	0
19-6-10/59997.0	IN	1	A	1	0	0	Y	Q	0	0	0	0	0	0	0	0	0	0	0	1	0	0	0	0	0	0	0	0	0	0	0	0
19-6-10/59998.0	IN	1	A	1	0	0	Y	Y	0	0	0	0	0	0	0	0	0	0	0	0	0	1	0	0	0	0	0	0	0	0	0	0
19-6-10/59999.0	IN	1	A	0	1	0	Y	Y	0	0	0	0	1	0	0	0	1	0	0	0	0	0	0	0	1	0	0	0	0	0	0	0
19-6-10/60000.0	4	1	A	1	0	0	Y	Y	0	0	0	0	0	0	0	1	0	0	0	0	0	0	0	0	0	0	0	0	0	0	1	0
19-6-10/60001.0	IN	1	A	0	1	0	Y	Y	0	0	0	0	0	0	0	0	1	0	0	1	0	0	0	0	1	0	0	0	0	0	0	0
19-6-10/60002.0	IN	1	A	1	0	0	Y	Y	0	0	0	0	0	0	0	0	1	0	0	0	0	1	0	0	1	0	1	0	0	0	0	0
19-6-10/60003.0	1A	1	A	0	1	0	N	N	0	0	0	0	0	0	0	0	0	0	0	0	0	1	0	0	0	1	0	0	0	0	0	1
19-6-10/60004.0	IN	1	A	0	1	0	Y	Y	0	0	0	0	0	0	0	0	1	0	0	1	0	1	0	0	0	1	0	0	0	0	0	0
19-6-10/60005.0	4A	1	A	0	1	0	Y	Y	0	0	0	0	0	0	0	0	0	0	1	0	0	0	0	0	1	0	0	0	0	0	0	0
19-6-10/60006.0	4A	1	A	0	1	0	Y	Q	0	0	0	0	1	0	0	0	0	0	1	1	0	1	0	0	1	0	0	0	0	0	0	0
19-6-10/60007.0	IN	1	A	0	1	0	Y	Y	0	0	0	0	0	0	0	0	1	0	0	1	0	0	0	0	0	0	0	0	0	0	0	0
19-6-10/60008.0	4	1	A	0	1	0	N	Y	0	0	0	0	1	0	0	0	1	0	0	1	0	1	0	0	0	0	0	0	0	0	0	1
19-6-10/60009.0	6	1	A	0	1	0	Y	Y	0	0	0	0	1	0	0	0	0	0	1	1	0	0	0	0	1	0	0	1	0	0	0	0
19-6-10/60010.0	IN	1	A	0	1	0	Y	Q	0	0	0	0	1	0	0	1	0	0	0	1	1	0	0	0	0	0	0	0	0	0	0	0
19-6-10/60011.0	3A	1	A	0	1	0	N	Y	0	0	0	0	0	0	0	0	1	0	0	0	0	1	0	0	0	0	0	0	0	0	0	0
19-6-10/60012.0	IN	1	A	0	1	0	Y	Y	0	1	0	0	0	0	0	0	1	0	0	0	0	0	0	0	1	0	0	0	0	0	0	0

				Sex					Trauma				Infection				Arthritis					Physiological stress				Neurological			Tumor		Other path.	
Peabody number	CS	MNI	Age	M	F	U	AFO	Cr mod	T c	T pc	T h	T d	I de	I di	I tr	I g	A sl	A m	A s	A sp	A o	EH	CO	PH	O	Sb sl	Sb m/s	O na	T bo	T o	OP d	OP c/pc
19-6-10/60013.0	B	1	A	0	1	0	Y	Q	0	0	0	0	0	0	0	0	0	0	0	0	0	1	1	0	1	0	0	0	0	0	0	0
19-6-10/60014.0	4A	1	J	0	0	1	N	Q	0	0	0	0	0	0	0	0	0	0	0	0	0	0	0	0	0	0	0	0	0	0	0	0
19-6-10/60016.0	2	1	A	1	0	0	Y	Y	0	0	0	0	0	0	0	0	1	0	0	0	0	0	0	0	0	0	0	0	0	0	0	0
19-6-10/60017.0	B	1	A	0	1	0	N	Y	0	0	0	0	0	0	0	0	1	0	0	0	0	0	0	0	1	0	0	0	0	0	0	0
19-6-10/60018.0	2	1	A	1	0	0	Y	N	0	1	1	0	0	0	0	0	1	0	0	1	0	0	0	0	0	0	0	0	0	0	0	0
19-6-10/60018.1	2	1	A	0	1	0	Y	Y	0	0	0	0	0	0	0	0	0	0	0	0	0	0	0	0	1	0	0	0	0	0	0	0
19-6-10/60019.0	IN	1	A	1	0	0	Y	Y	0	0	0	0	0	0	0	0	0	0	0	1	0	1	1	0	0	0	0	0	0	0	0	0
19-6-10/60020.0	IN	1	A	0	1	0	Y	U	0	0	0	0	0	0	0	0	0	0	0	0	0	0	0	0	0	1	0	0	0	0	0	0
19-6-10/60020.1	6	1	A	0	1	0	Y	Y	0	0	0	0	1	0	0	0	0	0	0	0	0	0	0	0	0	0	0	0	0	0	0	0
19-6-10/60021.0	4	1	A	0	1	0	N	Y	0	0	0	0	1	0	0	0	0	0	0	0	0	0	0	0	0	0	0	0	0	0	0	0
19-6-10/60022.0	IN	1	A	1	0	0	Y	Q	0	1	0	0	0	0	0	0	0	0	0	1	0	0	0	0	1	0	0	0	0	0	0	0
19-6-10/60022.1	IN	1	A	0	0	1	Y	U	0	0	0	0	1	0	0	0	0	0	0	0	0	0	0	0	0	0	0	0	0	0	0	0
19-6-10/60023.0	5	1	J	0	1	0	Y	Y	0	0	0	0	0	0	0	1	0	0	0	0	0	0	0	0	0	0	0	0	0	0	1	0
19-6-10/60025.0	2	1	A	0	1	0	Y	Y	0	1	1	0	0	0	0	1	0	0	0	1	0	1	0	0	1	0	0	0	0	0	0	0
19-6-10/60026.0	6	1	A	1	0	0	Y	N	0	0	0	0	0	0	0	0	0	0	0	0	0	0	0	0	0	0	0	0	0	0	0	0
19-6-10/60027.0	6	1	A	1	0	0	N	Y	0	0	0	0	0	0	0	0	1	0	0	0	0	1	0	0	0	0	0	0	0	0	0	0
19-6-10/60028.0	6	1	A	1	0	0	N	Y	0	0	0	0	0	0	1	0	0	0	1	1	0	0	0	0	1	0	0	0	0	0	0	0
19-6-10/60029.0	3A	1	A	0	1	0	Y	Y	0	0	0	0	0	0	0	0	0	0	0	0	0	0	0	0	0	0	0	0	0	0	1	0
19-6-10/60030.0	4	1	A	1	0	0	Y	Y	0	0	0	0	0	0	0	0	1	0	0	1	0	0	0	0	1	0	0	0	0	0	0	0
19-6-10/60031.0	1A	1	A	0	1	0	N	N	0	0	0	0	0	0	0	0	0	0	0	0	0	0	0	0	0	0	0	0	0	0	0	0
19-6-10/60032.0	BA	1	A	0	1	0	N	Y	1	0	1	0	0	0	1	0	0	0	0	0	0	0	0	0	0	0	0	0	0	0	0	0
19-6-10/60033.0	2A	1	A	0	1	0	N	Q	0	0	0	0	0	0	0	0	0	0	0	0	0	0	0	0	0	0	0	0	0	0	0	0
19-6-10/60034.0	4	1	A	1	0	0	Y	Y	0	0	0	0	1	0	0	0	0	0	0	0	0	0	0	0	0	0	0	0	0	0	0	0
19-6-10/60035.0	3	1	A	0	1	0	N	Y	0	0	0	0	0	0	0	1	0	1	0	1	0	0	0	0	1	0	0	0	0	0	0	0
19-6-10/60036.0	P	1	A	0	1	0	N	Y	0	0	0	0	0	0	0	0	0	0	0	0	0	0	0	0	0	0	0	0	0	0	0	0
19-6-10/60038.0	3	1	A	1	0	0	Y	Y	1	0	1	0	0	1	0	0	0	0	0	1	0	0	0	0	1	0	0	0	0	0	0	1
19-6-10/60039.0	4A	1	A	1	0	0	N	Q	0	0	0	0	1	0	0	0	0	0	0	1	0	0	0	0	1	0	0	0	1	1	0	0
19-6-10/60040.0	5	1	A	0	1	0	Y	Y	1	0	0	0	0	0	0	0	0	0	0	1	0	0	0	0	0	0	0	0	0	0	0	0
19-6-10/60041.0	4	1	A	0	1	0	Y	N	0	0	0	0	0	0	0	0	0	0	0	0	0	1	0	0	0	0	0	0	0	0	0	0
19-6-10/60042.0	3	1	A	1	0	0	Y	Y	0	0	0	0	1	0	0	0	0	0	0	1	1	0	0	0	1	0	0	0	0	0	0	0
19-6-10/60043.0	IN	1	J	0	1	0	Y	Q	0	0	0	0	0	0	0	0	0	0	0	0	0	0	0	0	0	0	0	0	0	0	0	0
19-6-10/60044.0	2	1	A	1	0	0	N	N	0	0	0	0	0	0	0	1	0	0	0	0	0	0	0	1	0	0	0	0	0	0	1	0
19-6-10/60045.0	IN	1	J	0	1	0	Y	Y	0	0	0	0	0	0	0	0	0	0	0	0	0	0	0	0	0	0	0	0	0	1	0	0
19-6-10/60046.0	5A	1	A	1	0	0	N	Q	0	0	0	0	1	0	0	0	0	0	0	1	0	0	0	0	0	0	0	0	0	0	0	0
19-6-10/60047.0	1A	1	A	0	1	0	Y	N	0	0	0	0	0	0	0	0	0	0	0	0	0	0	0	0	0	0	0	0	0	0	0	0
19-6-10/60048.0	5A	1	A	1	0	0	Y	Q	0	0	0	0	1	0	0	0	0	0	0	0	0	0	0	0	1	0	0	0	0	0	0	0
19-6-10/60049.0	4	1	A	0	1	0	Y	N	0	0	0	0	1	0	0	1	0	1	0	1	0	0	0	0	1	0	0	0	0	0	0	0
19-6-10/60050.0	BA	1	A	1	0	0	Y	Y	0	0	0	0	0	0	0	0	0	0	1	0	0	0	0	0	1	0	1	0	0	0	0	0
19-6-10/60051.0	IN	1	A	1	0	0	Y	Y	0	0	0	0	0	0	0	1	0	0	0	1	0	1	0	0	0	0	0	0	0	1	0	0
19-6-10/60052.0	2	1	A	0	1	0	Y	N	0	0	0	0	0	0	0	0	0	0	0	0	0	1	0	0	0	0	0	0	0	0	0	0
19-6-10/60053.0	P	1	A	1	0	0	N	N	0	0	0	0	1	0	0	0	0	0	0	0	0	0	0	0	0	0	0	0	0	0	0	0
19-6-10/60054.0	4A	1	A	0	1	0	Y	Y	0	0	0	0	0	0	0	0	1	0	0	1	1	1	0	0	1	0	0	0	0	0	0	0
19-6-10/60055.0	3A	1	A	0	1	0	Y	Y	0	0	0	0	0	0	0	0	1	0	0	0	0	1	0	0	0	0	0	0	0	1	0	0
19-6-10/60056.0	1A	1	J	0	0	1	Y	Q	0	0	0	0	0	0	0	0	0	0	0	0	0	0	0	0	0	0	0	0	0	0	0	0
19-6-10/60057.0	P	1	A	0	1	0	Y	Q	0	1	1	0	1	0	0	0	0	1	0	1	0	0	0	0	0	0	0	0	1	0	0	0

Peabody number	CS	MNI	Age	Sex			AFO	Cr mod	Trauma				Infection				Arthritis					Physiological stress				Neurological			Tumor		Other path.	
				M	F	U			T c	T pc	T h	T d	I de	I di	I tr	I g	A sl	A m	A s	A sp	A o	EH	CO	PH	O	Sb sl	Sb m/s	O na	T bo	T o	OP d	OP c/pc
19-6-10/60058.0	2	1	A	1	0	0	N	Y	0	0	0	0	0	0	0	0	0	0	1	1	0	0	0	0	1	0	0	0	0	0	0	0
19-6-10/60059.0	3	1	A	0	1	0	N	N	0	0	0	0	1	0	0	0	1	0	0	0	0	1	0	0	0	0	0	1	0	0	0	0
19-6-10/60060.0	5	1	A	0	1	0	Y	Y	0	0	0	0	1	0	0	0	1	0	0	0	0	0	0	0	0	0	0	0	0	0	0	0
19-6-10/60061.0	6	1	A	0	1	0	Y	Y	0	1	0	0	1	1	0	0	0	0	1	1	1	0	0	0	0	0	0	0	0	0	0	0
19-6-10/60062.0	P	1	A	0	1	0	N	Q	0	0	0	0	1	0	0	0	1	0	0	1	0	1	0	0	0	0	0	0	0	0	0	0
19-6-10/60063.0	IN	1	A	0	1	0	Y	Y	0	1	1	0	1	0	0	0	0	1	0	1	0	0	0	0	0	0	0	0	1	0	0	0
19-6-10/60064.0	4A	1	A	1	0	0	Y	Y	0	0	0	0	0	0	0	0	1	0	0	1	0	1	0	0	0	0	0	0	1	0	0	0
19-6-10/60065.0	5	1	A	1	0	0	N	N	0	0	0	0	1	0	0	0	0	1	0	0	0	1	0	0	1	0	0	0	1	0	0	0
19-6-10/60066.0	IN	1	A	0	1	0	Y	N	0	0	0	0	0	0	0	0	0	1	0	1	0	0	0	0	1	0	0	0	0	0	0	0
19-6-10/60067.0	IN	1	J	0	1	0	Y	Y	1	1	1	0	0	1	1	0	0	0	0	0	1	0	0	0	0	0	0	0	0	1	0	0
19-6-10/60068.0	4A	1	A	0	1	0	N	Y	0	0	0	0	0	0	0	0	0	0	0	0	0	0	1	0	0	0	0	0	0	0	0	0
19-6-10/60070.0	3	1	A	0	1	0	Y	Y	0	1	0	0	0	0	0	0	0	0	1	1	1	0	0	0	1	0	0	0	0	1	0	0
19-6-10/60071.0	IN	1	A	1	0	0	Y	N	0	0	0	0	1	0	0	0	0	1	0	1	0	0	0	0	1	0	0	0	0	0	0	0
19-6-10/60072.0	4A	1	A	0	1	0	Y	N	0	0	0	0	1	0	0	0	0	1	0	1	0	0	0	0	1	0	0	0	0	0	0	0
19-6-10/60073.0	6	1	A	0	1	0	N	U	0	0	0	0	0	0	0	0	1	0	0	1	0	0	0	0	0	0	0	0	0	0	0	0
19-6-10/60074.0	3	1	A	1	0	0	N	Y	0	0	0	0	0	0	0	0	0	0	0	0	0	1	0	0	0	0	0	0	0	0	0	0
19-6-10/60075.0	1	1	A	1	0	0	Y	N	0	1	1	0	0	0	0	0	0	1	0	1	1	0	0	0	0	0	0	0	0	0	0	0
19-6-10/60076.0	IN	1	A	0	1	0	Y	Y	0	0	0	0	0	0	0	0	1	0	0	0	1	1	0	0	1	0	0	0	0	0	0	0
19-6-10/60077.0	B	1	J	0	1	0	N	Y	0	0	0	0	0	0	0	1	0	0	0	0	0	0	0	0	0	0	0	0	0	0	0	0
19-6-10/60078.0	4A	1	J	0	0	1	N	Q	0	0	0	0	0	0	0	0	0	0	0	0	0	0	0	0	0	0	0	0	0	0	0	0
19-6-10/60079.0	IN	1	J	0	0	1	N	Q	0	0	0	0	0	0	0	0	0	0	0	0	0	0	0	0	0	0	0	0	0	0	0	0
19-6-10/60080.0	1	1	A	1	0	0	Y	Y	0	0	0	0	0	0	0	0	1	0	0	1	0	0	0	0	0	0	0	0	0	0	0	0
19-6-10/60081.0	IN	1	A	0	1	0	Y	Y	0	0	0	0	1	0	0	0	1	0	0	0	0	0	0	0	0	0	0	0	0	0	0	0
19-6-10/60082.0	P	1	A	0	1	0	Y	Y	0	1	1	0	1	1	0	0	1	0	0	0	0	0	0	0	0	0	0	0	0	0	0	0
19-6-10/60083.0	4A	1	A	0	1	0	N	N	0	0	0	0	1	0	0	1	1	0	0	1	0	0	0	0	0	0	0	0	0	0	0	0
19-6-10/60084.0	6	1	A	0	1	0	N	Y	0	0	0	0	1	0	0	0	0	0	1	1	0	0	0	0	1	0	0	0	0	0	0	0
19-6-10/60085.0	5	1	A	1	0	0	Y	Q	0	0	0	0	0	0	0	0	0	0	0	0	0	1	0	0	0	0	0	0	0	0	0	0
19-6-10/60086.0	5	1	A	1	0	0	Y	Y	0	0	0	0	1	0	0	0	1	0	0	1	0	1	0	0	0	0	0	0	0	0	0	0
19-6-10/60088.0	6	1	A	1	0	0	Y	Y	0	0	0	0	1	0	0	0	1	0	0	1	0	0	0	0	0	0	0	0	0	0	0	0
19-6-10/60089.0	4A	1	A	0	1	0	Y	Y	0	0	0	0	0	0	0	0	0	1	0	1	0	1	0	0	1	0	0	0	0	0	0	0
19-6-10/60091.0	3A	1	A	0	1	0	N	Y	0	0	0	0	1	0	0	0	0	1	0	1	0	0	0	0	1	0	0	0	0	0	0	0
19-6-10/60092.0	1	1	A	1	0	0	Y	Y	0	0	0	0	0	0	0	0	1	0	0	0	0	0	0	0	1	0	0	0	0	0	1	0
19-6-10/60093.0	1	1	A	0	1	0	Y	Y	0	0	0	0	0	0	0	0	1	0	0	0	0	0	0	0	0	0	0	0	0	0	0	0
19-6-10/60094.0	3A	1	A	0	1	0	Y	Y	0	0	0	0	1	0	0	0	0	0	0	0	0	1	0	0	0	0	0	0	0	0	0	0
19-6-10/60095.0	IN	1	A	0	1	0	Y	Y	0	0	0	0	0	0	0	0	1	0	0	0	0	0	0	0	1	0	0	0	0	0	0	0
19-6-10/60096.0	IN	1	A	1	0	0	Y	Q	0	1	1	0	0	0	1	0	1	0	0	1	0	1	0	0	0	0	0	0	0	0	0	0
19-6-10/60097.0	IN	1	A	0	1	0	Y	Y	0	0	0	0	1	0	0	1	0	0	0	0	0	0	0	0	0	0	0	0	0	0	0	0
19-6-10/60098.0	IN	1	A	0	1	0	Y	Y	0	0	0	0	0	0	0	0	0	0	0	0	0	0	0	0	0	0	0	0	0	0	0	0
19-6-10/60099.0	5	1	A	0	1	0	Y	Y	0	0	0	0	0	0	0	0	0	0	0	0	0	0	0	0	0	0	0	0	0	0	0	0
19-6-10/60100.0	6	1	A	0	1	0	N	Q	0	0	0	0	0	1	0	0	0	0	0	0	0	0	0	0	0	0	0	0	0	0	0	0
19-6-10/60101.0	5	1	A	0	1	0	Y	Q	0	1	1	0	1	0	0	0	0	0	0	0	1	0	0	0	0	0	0	0	0	0	0	0
19-6-10/60102.0	IN	1	A	1	0	0	Y	Q	0	0	0	0	0	0	0	0	0	0	0	0	0	1	0	0	0	0	0	0	0	0	1	0
19-6-10/60103.0	IN	1	A	1	0	0	Y	Q	0	1	0	0	0	0	0	0	0	0	0	0	1	0	0	0	0	0	0	0	0	0	0	0
19-6-10/60104.0	B	1	A	1	0	0	N	Q	0	0	0	0	1	0	0	0	0	0	0	0	0	0	0	0	0	0	0	0	0	0	0	0
19-6-10/60106.0	3A	1	A	1	0	0	Y	Q	0	0	0	0	0	0	0	0	0	0	0	0	0	0	0	0	0	0	0	0	0	0	0	0

Peabody number	CS	MNI	Age	Sex			AFO	Cr mod	Trauma				Infection				Arthritis					Physiological stress				Neurological			Tumor		Other path.	
				M	F	U			T c	T pc	T h	T d	I de	I di	I tr	I g	A sl	A m	A s	A sp	A o	EH	CO	PH	O	Sb sl	Sb m/s	O na	T bo	T o	OP d	OP c/pc
19-6-10/60107.0	6	1	J	1	0	0	Y	Q	0	0	0	0	0	0	0	0	0	0	0	0	0	0	0	0	0	0	0	1	0	0	0	0
19-6-10/60108.0	4	1	A	0	1	0	N	Y	0	0	0	0	0	0	0	0	0	0	0	0	0	1	0	0	0	0	0	0	0	0	1	0
19-6-10/60109.0	3	1	A	1	0	0	Y	Y	0	0	0	0	0	0	0	0	1	0	0	0	0	1	0	0	0	0	0	0	0	0	0	0
19-6-10/60110.1	CH	1	A	0	0	1	N	U	0	0	0	0	0	0	0	0	0	0	0	0	0	0	0	0	0	0	0	0	0	0	0	0
19-6-10/60110.2	CH	1	A	0	0	1	N	U	0	0	0	0	0	0	0	0	0	0	0	0	0	0	0	0	0	0	0	0	0	0	0	0
19-6-10/60111.1	CH	1	A	0	1	0	N	U	0	0	0	0	0	0	0	0	0	0	0	0	0	0	0	0	0	0	0	0	0	0	0	0
19-6-10/60112.0	CH	1	A	0	0	1	N	U	0	1	1	0	0	0	0	0	0	0	0	0	0	0	0	0	0	0	0	0	0	0	0	0
19-6-10/60113.0	CH	1	A	1	0	0	N	U	0	0	0	0	0	0	0	0	0	0	0	0	0	0	0	0	0	0	0	0	0	0	0	0
19-6-10/60113.1	CH	1	A	0	1	0	N	U	0	0	0	0	0	0	0	0	0	0	0	0	0	0	0	0	0	0	0	0	0	0	0	0
19-6-10/60113.2	CH	1	A	1	0	0	N	U	0	0	0	0	0	0	0	0	0	0	0	0	0	0	0	0	0	0	0	0	0	0	0	0
19-6-10/60116.1	CH	1	A	0	0	1	N	U	0	0	0	0	0	0	0	0	0	0	0	0	0	0	0	0	0	0	0	0	0	0	0	0
19-6-10/60116.2	CH	1	A	0	0	1	N	U	0	0	0	0	0	0	0	0	0	0	0	0	0	0	0	0	0	0	0	0	0	0	0	0
19-6-10/60117.0	CH	1	A	0	0	1	N	U	0	0	0	0	0	0	0	0	0	0	0	0	0	0	0	0	0	0	0	0	0	0	0	0
19-6-10/60118.0	CH	1	A	1	0	0	N	U	0	1	1	0	0	0	0	0	0	0	0	0	0	0	0	0	0	0	0	0	0	0	0	0
19-6-10/60119.0	CH	1	A	0	0	1	N	U	0	0	0	0	0	0	0	0	0	0	0	0	0	0	0	0	0	0	0	0	0	0	0	0
19-6-10/60133.0	3A	1	A	1	0	0	Y	U	0	0	0	0	0	0	0	0	0	0	0	0	0	1	0	0	0	0	0	0	0	0	0	0
19-6-10/60134.0	P	1	A	0	1	0	N	U	0	0	0	0	0	0	0	0	0	0	0	0	0	0	0	0	0	0	0	0	0	0	0	0
19-6-10/60135.0	IN	1	A	1	0	0	Y	Q	0	0	0	0	0	0	0	0	1	0	0	0	0	0	0	0	0	0	1	0	0	0	0	0
19-6-10/60136.0	B	1	A	1	0	0	N	U	0	0	0	0	0	0	0	0	0	0	0	0	0	0	0	0	0	0	0	0	0	0	0	1
19-6-10/60137.0	4	1	A	0	1	0	N	Q	0	0	0	0	0	0	0	0	1	0	0	1	0	0	0	0	0	0	0	0	0	0	0	1
19-6-10/60138.0	IN	1	A	0	1	0	Y	U	0	0	0	0	0	0	0	0	1	0	0	0	0	0	0	0	0	0	0	0	0	0	0	0
19-6-10/60140.0	3	1	J	0	0	1	N	Q	0	0	0	0	0	0	0	0	0	0	0	0	0	0	0	0	0	0	0	0	0	0	0	0
19-6-10/60142.0	IN	1	J	0	0	1	N	Y	0	0	0	0	0	0	0	0	0	0	0	0	0	1	0	0	0	0	0	0	0	0	1	0
19-6-10/60143.0	IN	1	A	0	1	0	Y	Q	0	0	0	0	1	0	0	0	0	0	0	0	0	0	0	0	0	1	0	0	0	0	0	0
19-6-10/60144.0	IN	1	A	0	1	0	Y	Q	0	1	1	0	0	0	0	0	0	0	0	0	0	0	0	0	0	0	0	0	0	0	0	0
19-6-10/60145.0	1	1	A	0	1	0	Y	Y	0	0	0	0	0	0	0	1	0	1	0	0	0	0	0	0	0	0	1	0	0	0	0	0
19-6-10/60146.0	IN	1	A	0	1	0	Y	U	0	0	0	0	0	0	0	0	0	0	0	0	0	0	0	0	0	0	0	0	0	0	0	0
19-6-10/60146a	IN	1	J	0	0	1	Y	U	0	0	0	0	0	0	0	0	0	0	0	0	0	0	0	0	0	0	0	0	0	0	0	0
19-6-10/60147.0	2A	1	A	0	1	0	Y	U	0	0	0	0	0	0	0	0	1	0	0	0	1	0	0	0	1	0	1	0	0	0	0	0
19-6-10/60148.0	P	1	A	1	0	0	Y	Q	0	1	1	0	0	0	0	0	0	0	0	1	0	1	0	0	0	0	0	0	0	0	0	0
19-6-10/60149.0	6	1	J	0	0	1	N	Q	0	0	0	0	0	0	0	0	0	0	0	0	0	1	0	0	0	0	0	0	0	0	0	0
19-6-10/60150.0	3	1	J	0	0	1	N	Q	0	0	0	0	1	0	0	0	0	0	0	0	0	1	0	0	0	0	0	0	0	0	0	0
19-6-10/60151.0	IN	1	J	0	0	1	Y	Q	0	0	0	0	0	0	0	0	0	0	0	0	0	0	0	0	0	0	0	0	0	0	0	0
19-6-10/60152.0	IN	1	A	0	1	0	Y	U	0	0	0	0	0	0	0	0	0	0	0	0	0	1	0	0	0	0	0	0	0	0	0	1
19-6-10/60154.0	IN	1	A	1	0	0	Y	U	0	0	0	0	0	0	0	0	0	0	0	0	0	0	0	0	0	0	0	0	0	0	0	0
19-6-10/60155.0	3	1	J	0	0	1	N	U	0	0	0	0	0	0	0	0	0	0	0	0	0	1	0	0	0	0	0	0	0	0	0	1
19-6-10/60156.0	IN	1	A	1	0	0	Y	U	0	0	0	0	0	0	0	1	1	0	0	1	0	0	0	0	1	0	0	0	0	0	0	0
19-6-10/60157.0	4	1	J	1	0	0	Y	U	0	0	0	0	0	0	0	0	0	0	0	0	1	1	0	0	0	0	0	0	0	0	0	0
19-6-10/60158.0	5	1	A	1	0	0	Y	U	0	0	0	0	1	0	0	0	0	0	0	1	0	0	0	0	1	0	0	0	0	0	0	0
19-6-10/60159.0	B	1	A	0	1	0	N	U	0	0	0	0	0	0	0	1	0	0	0	1	0	0	0	0	1	0	0	0	0	0	0	0
19-6-10/60161.0	IN	1	A	1	0	0	Y	U	0	0	0	0	0	0	0	0	0	0	0	0	0	0	0	0	0	0	0	0	0	0	0	0
19-6-10/60162.0	3	1	J	0	0	1	Y	Q	0	0	0	0	0	0	0	0	0	0	0	0	0	0	0	0	0	0	0	0	0	0	0	0
19-6-10/60163.0	IN	1	A	1	0	0	Y	U	0	0	0	0	0	0	0	1	0	0	1	1	0	0	0	0	1	0	0	0	0	0	0	0
19-6-10/60164.0	4A	1	A	0	1	0	Y	Q	0	0	0	0	0	0	0	0	0	0	0	0	0	1	0	0	0	0	0	0	0	0	0	0
19-6-10/60165.0	IN	1	A	0	1	0	Y	Q	0	0	0	0	0	0	0	0	1	0	0	0	0	1	0	0	1	0	0	0	0	0	0	0

				Sex					Trauma				Infection				Arthritis					Physiological stress				Neurological			Tumor		Other path.	
Peabody number	CS	MNI	Age	M	F	U	AFO	Cr mod	T c	T pc	T h	T d	I de	I di	I tr	I g	A sl	A m	A s	A sp	A o	EH	CO	PH	O	Sb sl	Sb m/s	O na	T bo	T o	OP d	OP c/pc
19-6-10/60166.0	3A	1	A	1	0	0	N	U	0	0	0	0	0	0	0	0	0	0	0	0	0	0	0	0	0	0	0	0	0	0	0	0
19-6-10/60167.0	3A	1	A	0	0	1	N	U	0	1	0	0	0	0	0	0	0	0	0	1	0	0	0	0	1	0	0	0	0	0	0	0
19-6-10/60168.0	2	1	A	0	1	0	Y	U	0	0	0	0	0	0	0	0	1	0	0	1	0	0	0	0	1	0	0	0	0	0	0	0
19-6-10/60169.0	IN	1	A	0	1	0	Y	Q	0	0	0	0	0	0	0	0	0	0	0	1	0	0	0	0	1	0	0	0	0	0	0	0
19-6-10/60171.0	2	1	A	0	1	0	N	U	0	1	0	0	0	0	0	0	0	0	0	0	0	0	0	1	0	0	0	0	0	0	0	1
19-6-10/60172.0	6	1	A	1	0	0	N	U	0	0	0	0	0	0	0	0	0	0	0	1	0	0	0	0	1	0	0	0	0	0	0	0
19-6-10/60173.0	IN	1	A	0	1	0	Y	U	0	0	0	0	0	0	0	0	0	0	0	1	0	0	0	0	0	0	0	0	0	0	0	0
19-6-10/60174.0	4A	1	A	0	1	0	Y	Q	0	0	0	0	0	0	0	1	1	0	0	0	0	0	0	0	1	0	0	0	0	0	0	0
19-6-10/60175.0	2	1	A	0	1	0	N	Q	0	0	0	0	0	0	0	0	0	0	0	0	0	0	0	0	0	0	0	0	0	0	0	0
19-6-10/60176.0	4A	1	A	1	0	0	N	Q	0	0	0	0	0	0	0	0	1	0	0	0	0	0	0	0	1	0	0	0	0	0	0	0
19-6-10/60177.0	IN	1	A	1	0	0	Y	Q	0	0	0	0	0	0	0	0	0	0	0	0	0	0	0	0	0	0	0	0	0	0	0	0
19-6-10/60178.0	6	1	A	1	0	0	N	U	0	0	0	0	0	0	0	0	0	0	0	0	0	0	0	0	0	0	0	0	0	0	0	0
19-6-10/60179.0	IN	1	A	0	1	0	Y	U	0	0	0	0	0	0	0	0	1	0	0	0	0	1	0	0	0	0	0	0	0	0	0	0
19-6-10/60180.0	4A	1	A	1	0	0	Y	U	0	0	0	0	0	0	0	0	1	0	0	0	0	1	0	0	0	0	0	0	0	0	1	0
19-6-10/60181.0	IN	1	A	1	0	0	Y	Q	0	0	0	0	0	0	0	0	1	0	0	0	0	1	0	0	0	0	0	0	0	0	0	0
19-6-10/60182.0	P	1	A	1	0	0	Y	U	0	0	0	0	0	0	0	0	1	0	0	0	0	0	0	0	1	0	0	0	0	0	0	0
19-6-10/60183.0	5	1	J	0	0	1	Y	Y	0	0	0	0	0	0	0	0	0	0	0	0	0	0	1	0	0	0	0	0	0	0	0	0
19-6-10/60183.1	5	1	A	0	1	0	Y	Y	0	0	0	0	0	0	0	0	0	0	0	0	0	0	0	0	0	0	0	0	0	0	0	0
19-6-10/60184.0	5	1	J	0	0	1	N	Q	0	0	0	0	0	0	0	0	0	0	0	0	0	0	1	1	0	0	0	0	0	0	0	0
19-6-10/60185.0	IN	1	J	0	0	1	Y	U	0	0	0	0	0	0	0	0	0	0	0	0	0	0	0	0	0	0	0	0	0	0	0	0
19-6-10/60185a	IN	1	J	0	0	1	Y	U	0	0	0	0	0	0	0	0	0	0	0	0	0	0	0	1	0	0	0	0	0	0	0	0
19-6-10/60186.0	P	1	J	0	0	1	Y	U	0	0	0	0	0	0	0	1	0	0	0	0	0	0	0	0	0	0	0	0	0	0	0	0
19-6-10/60187.1	5	1	J	0	0	1	N	U	0	0	0	0	0	0	0	0	0	0	0	0	0	0	0	0	0	0	0	0	0	0	0	0
19-6-10/60187.10	5	1	J	0	0	1	N	U	0	0	0	0	0	0	0	0	0	0	0	0	0	0	0	0	0	0	0	0	0	0	0	0
19-6-10/60187.2	4	1	J	0	0	1	Y	U	0	0	0	0	0	0	0	0	0	0	0	0	0	0	0	0	0	0	0	0	0	0	0	0
19-6-10/60187.3	4	1	J	0	0	1	Y	U	0	0	0	0	0	0	0	0	0	0	0	0	0	0	0	0	0	0	0	0	0	0	0	0
19-6-10/60187.4	2	1	J	0	0	1	N	U	0	0	0	0	0	0	0	0	0	0	0	0	0	0	0	0	0	0	0	0	0	0	0	0
19-6-10/60187.5	P	1	J	0	0	1	N	U	0	0	0	0	0	0	0	0	0	0	0	0	0	0	0	0	0	0	0	0	0	0	0	0
19-6-10/60187.6	P	1	J	0	0	1	N	U	0	0	0	0	0	0	0	0	0	0	0	0	0	0	0	0	0	0	0	0	0	0	0	0
19-6-10/60187.7	1A	1	J	0	0	1	N	U	0	0	0	0	0	0	0	0	0	0	0	0	0	0	0	0	0	0	0	0	0	0	0	0
19-6-10/60187.8	IN	1	J	0	0	1	N	U	0	0	0	0	0	0	0	0	0	0	0	0	0	0	0	0	0	0	0	0	0	0	0	0
19-6-10/60187.9	IN	1	J	0	0	1	Y	U	0	0	0	0	0	0	0	0	0	0	0	0	0	0	0	0	0	0	0	0	0	0	0	0
19-6-10/60189.0	2	1	A	1	0	0	Y	U	0	1	1	0	0	0	0	0	0	1	0	0	1	0	0	0	1	0	0	1	0	1	0	0
19-6-10/60190.0	1A	1	J	0	0	1	Y	Y	0	0	0	0	0	0	0	0	0	0	0	0	0	1	0	0	0	0	0	0	0	0	0	0
19-6-10/60191.0	4A	1	A	0	1	0	Y	Q	0	0	0	0	1	0	0	0	1	0	0	0	0	0	0	0	0	0	0	0	0	0	0	0
19-6-10/60192.0	B	1	A	0	1	0	Y	U	0	0	0	0	0	0	0	1	0	0	0	0	1	0	0	0	0	1	0	0	0	0	0	0
19-6-10/60192.1	B	1	J	0	0	1	Y	U	0	0	0	0	0	0	0	0	0	0	0	0	0	0	0	0	0	0	0	0	0	0	0	0
19-6-10/60193.0	3	1	A	0	1	0	N	Y	0	0	0	0	1	0	0	0	1	0	0	0	0	0	0	0	0	0	0	0	0	0	0	0
19-6-10/60624.0	1	1	A	0	0	1	N	U	0	1	1	0	0	1	0	0	0	0	0	0	0	0	0	0	0	0	0	0	0	0	0	0
19-6-10/60646.0	IN	1	A	0	1	0	Y	Y	0	0	0	0	0	0	0	0	1	0	0	0	0	1	0	0	0	0	0	0	0	0	0	0
19-6-10/60647.0	IN	1	J	0	1	0	N	Q	0	0	0	0	1	0	0	0	0	0	0	0	0	0	0	0	0	0	1	0	0	0	0	0
19-6-10/60648.0	CH	1	A	1	0	0	Y	Q	0	0	0	0	1	0	0	1	0	0	0	0	0	0	0	0	0	0	0	0	0	0	0	0
19-6-10/60648.1	CH	1	A	0	0	1	Y	U	0	0	0	0	0	0	0	0	0	0	0	0	0	0	0	0	0	0	0	0	0	0	0	0
19-6-10/60649.0	CH	1	A	1	0	0	N	N	0	0	0	0	0	0	0	1	0	0	0	0	0	0	0	0	0	0	0	0	0	0	0	0
19-6-10/60650.0	IN	1	A	1	0	0	Y	Y	0	0	0	0	1	0	0	0	0	0	0	0	0	0	0	0	0	0	0	0	0	0	0	0

				Sex					Trauma				Infection				Arthritis					Physiological stress				Neurological			Tumor		Other path.	
Peabody number	CS	MNI	Age	M	F	U	AFO	Cr mod	T c	T pc	T h	T d	I de	I di	I tr	I g	A sl	A m	A s	A sp	A o	EH	CO	PH	O	Sb sl	Sb m/s	O na	T bo	T o	OP d	OP c/pc
19-6-10/60958.0	CH	1	A	0	1	0	N	U	0	0	0	0	0	0	0	0	0	0	0	0	0	0	0	0	0	0	0	0	0	0	0	0
19-6-10/60959.1	CH	1	A	1	0	0	N	U	0	1	0	1	0	0	0	0	0	0	0	0	0	0	0	0	0	0	0	0	0	0	0	0
19-6-10/60959.2	CH	1	A	0	0	1	N	U	0	0	0	0	0	0	0	0	0	0	0	0	0	0	0	0	0	0	0	0	0	0	0	0
19-6-10/60959.3	CH	1	A	0	0	1	N	U	0	0	0	0	0	0	0	0	0	0	0	0	0	0	0	0	0	0	0	0	0	0	0	0
19-6-10/60959.4	CH	1	A	0	0	1	N	U	0	0	0	0	0	0	0	0	0	0	0	0	0	0	0	0	0	0	0	0	0	0	0	0
19-6-10/60959.5	CH	1	A	0	0	1	N	U	0	0	0	0	0	0	0	0	0	0	0	0	0	0	0	0	0	0	0	0	0	0	0	0
19-6-10/60962.1	CH	1	A	0	0	1	N	U	0	0	0	0	0	0	0	0	0	0	0	0	0	0	0	0	0	0	0	0	0	0	0	0
19-6-10/60962.2	CH	1	A	0	0	1	N	U	0	0	0	0	0	0	0	0	0	0	0	0	0	0	0	0	0	0	0	0	0	0	0	0
19-6-10/60962.3	CH	1	A	0	0	1	N	U	0	0	0	0	0	0	0	0	0	0	0	0	0	0	0	0	0	0	0	0	0	0	0	0
19-6-10/60962.4	CH	1	J	0	0	1	N	U	0	0	0	0	0	0	0	0	0	0	0	0	0	0	0	0	0	0	0	0	0	0	0	0
19-6-10/60963.1	CH	1	A	0	0	1	N	U	0	0	0	0	0	0	0	0	0	0	0	0	0	0	0	0	0	0	0	0	0	0	0	0
19-6-10/60963.2	CH	1	A	0	0	1	N	U	0	0	0	0	0	0	0	0	0	0	0	0	0	0	0	0	0	0	0	0	0	0	0	0
19-6-10/60965.0	CH	1	J	0	1	0	N	U	0	0	0	0	0	0	0	0	0	0	0	0	0	0	0	0	0	0	0	0	0	0	0	0
19-6-10/60966.0	CH	1	A	0	1	0	N	U	0	0	0	0	0	0	0	0	0	0	0	0	0	1	0	0	0	0	0	0	0	0	0	0
19-6-10/60967.0	CH	1	A	1	0	0	N	U	0	0	0	0	0	0	0	0	0	0	0	0	0	0	0	0	0	0	0	0	0	0	0	0
19-6-10/60970.4	CH	1	A	0	0	1	N	U	0	0	0	0	0	0	0	0	0	0	0	0	0	0	0	0	0	0	0	0	0	0	0	0
19-6-10/60970.5	CH	1	A	0	0	1	N	U	0	0	0	0	0	0	0	0	0	0	0	0	0	0	0	0	0	0	0	0	0	0	0	0
19-6-10/60971.1	CH	1	A	0	1	0	N	U	0	0	0	0	0	0	0	0	0	0	0	0	0	0	0	0	0	0	0	0	0	0	0	0
19-6-10/60971.10	CH	1	A	0	0	1	N	U	0	0	0	0	0	0	0	0	0	0	0	0	0	0	0	0	0	0	0	0	0	0	0	0
19-6-10/60971.11	CH	1	A	0	0	1	N	U	0	0	0	0	0	0	0	0	0	0	0	0	0	0	0	0	0	0	0	0	0	0	0	0
19-6-10/60971.12	CH	1	A	0	0	1	N	U	0	0	0	0	0	0	0	0	0	0	0	0	0	0	0	0	0	0	0	0	0	0	0	0
19-6-10/60971.2	CH	1	A	0	1	0	N	U	0	0	0	0	0	0	0	0	0	0	0	0	0	0	0	0	1	0	0	0	0	0	0	0
19-6-10/60971.3	CH	1	A	0	1	0	N	U	0	0	0	0	0	0	0	0	0	0	0	0	0	0	0	0	0	0	0	0	0	0	0	0
19-6-10/60971.4	CH	1	A	0	1	0	N	U	0	0	0	0	0	0	0	0	0	0	0	0	0	0	0	0	0	0	0	0	0	0	0	0
19-6-10/60971.5	CH	1	A	1	0	0	N	U	0	0	0	0	0	0	0	0	1	0	0	0	0	0	0	0	0	0	0	0	0	0	0	0
19-6-10/60971.6	CH	1	A	1	0	0	N	U	0	0	0	0	0	0	0	0	0	0	0	0	0	0	0	0	0	0	0	0	0	0	0	0
19-6-10/60971.7	CH	1	J	0	0	1	N	U	0	0	0	0	0	0	0	0	0	0	0	0	0	0	0	0	0	0	0	0	0	0	0	0
19-6-10/60971.8	CH	1	J	1	0	0	N	U	0	0	0	0	0	0	0	0	0	0	0	0	0	0	0	0	0	0	0	0	0	0	0	0
19-6-10/60971.9	CH	1	A	0	0	1	N	U	0	0	0	0	0	0	0	0	0	0	0	0	0	0	0	0	0	0	0	0	0	0	0	0
19-6-10/60972.1	CH	1	A	0	0	1	N	U	0	0	0	0	0	0	0	0	0	0	0	0	0	0	0	0	0	0	0	0	0	0	0	0
19-6-10/60972.2	CH	1	A	0	0	1	N	U	0	0	0	0	0	0	0	0	0	0	0	0	0	0	0	0	0	0	0	0	0	0	0	0
19-6-10/60972.3	CH	1	A	0	1	0	N	U	0	0	0	0	0	0	0	0	0	0	0	0	0	0	0	0	0	0	0	0	0	0	0	0
19-6-10/60972.5	CH	1	A	1	0	0	N	U	0	0	0	0	0	0	0	0	0	0	0	0	0	0	0	0	0	0	0	0	0	0	0	0
19-6-10/60972.6	CH	1	J	0	0	1	N	U	0	0	0	0	0	0	0	0	0	0	0	0	0	0	0	0	0	0	0	0	0	0	0	0
19-6-10/60975.2	CH	1	A	0	0	1	N	U	0	0	0	0	0	0	0	0	0	0	0	0	0	0	0	0	0	0	0	0	0	0	0	0
19-6-10/60977.10	CH	1	A	0	0	1	N	U	0	0	0	0	0	0	0	0	0	0	0	0	0	0	0	0	0	0	0	0	0	0	0	0
19-6-10/60977.11	CH	1	A	0	1	0	N	U	0	0	0	0	0	0	0	1	0	0	0	0	0	0	0	0	0	0	0	0	0	0	0	0
19-6-10/60977.12	CH	1	A	0	1	0	N	U	0	0	0	0	0	0	0	0	0	0	0	0	0	0	0	0	0	0	0	0	0	0	0	0
19-6-10/60977.3	1A	1	A	0	0	1	Y	U	0	0	0	0	0	0	0	0	1	0	0	0	0	0	0	0	0	0	0	0	0	0	0	0
19-6-10/60977.4	3A	1	A	0	1	0	N	U	0	0	0	0	0	0	0	0	0	0	0	0	0	0	0	0	1	0	0	0	0	0	0	0
19-6-10/60977.5	CH	1	A	0	0	1	N	U	0	0	0	0	0	0	0	0	0	0	0	0	0	0	0	0	0	0	0	0	0	0	0	0
19-6-10/60977.6	CH	1	A	0	0	1	N	U	0	0	0	0	0	0	0	0	0	0	0	0	0	0	0	0	0	0	0	0	0	0	0	0
19-6-10/60977.7	CH	1	A	0	0	1	N	U	0	0	0	0	0	0	0	0	0	0	0	0	0	0	0	0	0	0	0	0	0	0	0	0
19-6-10/60977.8	CH	1	A	0	0	1	N	U	0	0	0	0	0	0	0	0	0	0	0	0	0	0	0	0	0	0	0	0	0	0	0	0
19-6-10/60977.9	CH	1	J	0	0	1	N	U	0	0	0	0	0	0	0	0	0	0	0	0	0	0	0	0	0	0	0	0	0	0	0	0

Peabody number	CS	MNI	Age	Sex			AFO	Cr mod	Trauma				Infection				Arthritis					Physiological stress				Neurological			Tumor		Other path.	
				M	F	U			T c	T pc	T h	T d	I de	I di	I tr	I g	A sl	A m	A s	A sp	A o	EH	CO	PH	O	Sb sl	Sb m/s	O na	T bo	T o	OP d	OP c/pc
19-6-10/61111.0	B	1	J	0	0	1	Y	Y	0	0	0	0	0	0	0	1	0	0	0	0	0	0	0	0	0	0	0	0	0	0	1	0
19-6-10/61112.0	B	1	A	1	0	0	N	Y	0	0	0	0	1	0	0	1	1	0	0	1	0	1	0	0	0	0	0	0	0	0	0	0
19-6-10/61113.0	1	1	J	0	0	1	N	Q	0	0	0	0	0	0	0	0	0	0	0	0	0	0	0	0	0	0	0	0	0	0	0	0
19-6-10/61114.0	B	1	A	0	1	0	Y	Y	0	0	0	0	1	0	0	0	0	0	0	0	0	0	0	0	0	0	0	0	0	0	0	0
19-6-10/61120.0	B	1	A	0	1	0	N	Y	0	0	0	0	0	0	0	0	0	0	0	0	0	0	0	0	0	0	0	0	0	0	0	0
19-6-10/N687.0	B	1	A	0	0	1	N	U	0	0	0	0	0	0	0	0	0	0	0	0	0	0	0	0	0	0	0	0	0	0	0	0
19-6-10/N690.0	B	1	A	1	0	0	N	U	0	0	0	0	0	0	0	0	0	0	0	0	0	1	0	0	0	0	0	0	0	0	0	0
20-6-10/60194.0	5	1	A	1	0	0	Y	Q	0	0	0	0	1	0	0	0	1	0	0	0	0	0	0	0	0	0	0	0	1	0	0	0
20-6-10/60195.0	3	1	J	1	0	0	N	Q	0	1	1	0	0	0	1	0	0	0	0	0	0	0	0	0	0	0	0	0	0	0	0	0
20-6-10/60196.0	P	1	A	0	1	0	N	Q	0	0	0	0	1	1	0	0	1	0	0	0	0	0	0	0	1	0	0	0	0	0	0	0
20-6-10/60197.0	5	1	A	1	0	0	N	Q	0	0	0	0	0	0	0	0	0	0	0	0	0	0	0	0	0	0	0	0	0	0	0	0
20-6-10/60198.0	4A	1	A	1	0	0	Y	Y	0	0	0	0	1	0	0	0	0	1	0	0	0	0	0	0	0	0	0	0	0	0	0	0
20-6-10/60199.0	B	1	A	0	1	0	Y	Y	0	0	0	0	0	0	0	1	0	0	0	0	0	1	0	0	0	0	0	0	0	0	1	0
20-6-10/60200.0	5	1	A	1	0	0	N	Q	0	0	0	0	1	0	0	0	0	0	0	0	0	1	0	0	0	1	0	0	0	0	0	0
20-6-10/60201.0	3	1	A	1	0	0	Y	Q	0	0	0	0	0	0	0	0	1	0	0	0	0	1	0	0	0	0	0	0	0	0	0	0
20-6-10/60202.0	2	1	A	1	0	0	Y	U	0	0	0	0	0	0	0	0	1	0	0	1	0	0	0	0	0	0	0	0	0	0	0	0
20-6-10/60203.0	4	1	A	1	0	0	N	U	0	0	0	0	0	0	0	0	0	0	0	0	0	0	0	0	0	0	0	0	0	0	0	0
20-6-10/60204.0	3	1	J	0	0	1	N	U	0	0	0	0	0	0	0	0	0	0	0	0	0	0	0	0	0	0	0	0	0	0	1	0
20-6-10/60205.0	4	1	J	0	1	0	Y	Y	0	0	0	0	0	0	0	0	0	0	0	0	0	1	0	0	0	0	0	0	0	0	0	0
20-6-10/60206.0	3	1	J	0	0	1	Y	Q	0	0	0	0	0	0	0	0	0	0	0	0	0	0	0	0	0	0	0	0	0	0	0	0
20-6-10/60207.0	7	1	A	0	1	0	N	Q	0	0	0	0	1	0	0	0	0	1	0	0	0	0	0	0	0	0	0	0	1	0	0	0
20-6-10/60208.0	5	1	J	0	0	1	Y	Q	0	0	0	0	0	0	0	0	0	0	0	0	0	0	0	0	0	0	0	0	0	0	0	0
20-6-10/60209.0	IN	1	J	0	0	1	N	U	0	0	0	0	0	0	0	0	0	0	0	0	0	0	0	0	0	0	0	0	0	0	0	0
20-6-10/60210.0	1	1	A	1	0	0	Y	U	0	0	0	0	0	0	0	0	0	0	0	0	0	0	0	0	0	0	0	0	0	0	0	0
20-6-10/60211.0	3	1	A	0	1	0	Y	Q	0	0	0	0	0	0	0	0	1	0	0	1	0	0	0	0	0	0	0	0	0	0	0	0
20-6-10/60212.0	4	1	A	1	0	0	Y	U	0	1	1	0	1	0	0	1	1	0	0	1	0	0	0	0	0	0	0	0	0	0	0	1
20-6-10/60213.0	2	1	A	0	1	0	Y	Q	0	0	0	0	0	0	0	0	0	0	0	0	0	0	0	0	0	0	0	0	0	0	0	0
20-6-10/60214.0	3	1	J	0	1	0	Y	U	0	0	0	0	0	0	0	0	0	0	0	0	0	0	0	0	0	0	0	0	0	0	0	0
20-6-10/60215.0	3	1	A	1	0	0	Y	Q	0	0	0	0	0	0	0	0	0	0	0	0	0	1	0	0	0	0	0	0	0	0	0	0
20-6-10/60216.0	3A	1	A	0	1	0	Y	Q	0	0	0	0	1	0	0	0	0	0	0	1	0	0	0	0	1	0	0	0	0	0	0	0
20-6-10/60217.0	4A	1	J	0	0	1	Y	Y	0	0	0	0	0	0	0	0	0	0	0	0	0	0	0	0	0	0	0	0	0	0	1	0
20-6-10/60218.0	5	1	A	1	0	0	Y	Y	0	0	0	0	1	0	0	1	0	0	0	0	0	1	0	0	0	1	0	0	1	0	0	0
20-6-10/60219.0	3	1	A	1	0	0	Y	Y	0	0	0	0	0	0	0	0	0	0	0	0	0	0	0	0	0	0	0	0	0	0	0	0
20-6-10/60220.0	3	1	A	1	0	0	Y	Y	0	0	0	0	0	0	0	0	1	0	0	0	0	1	0	0	0	0	0	0	0	0	0	0
20-6-10/60221.0	IN	1	A	0	0	1	N	U	0	0	0	0	0	0	0	0	0	0	0	0	0	0	0	0	0	0	0	0	0	0	1	0
20-6-10/60222.0	1	1	A	0	1	0	Y	Y	0	0	0	0	0	0	0	1	0	0	0	0	0	1	0	0	0	0	0	0	0	0	0	0
20-6-10/60223.0	3A	1	A	0	0	1	N	U	0	0	0	0	0	0	0	0	0	0	0	0	0	0	0	0	0	0	0	0	0	0	0	0
20-6-10/60224.0	4	1	A	1	0	0	Y	Q	0	0	0	0	1	0	0	0	0	0	0	1	0	1	0	0	0	0	0	0	0	0	0	0
20-6-10/60225.0	4	1	A	0	1	0	N	Q	0	0	0	0	0	0	0	0	1	0	0	0	0	0	0	0	0	0	0	0	1	0	1	1
20-6-10/60226.0	P	1	A	1	0	0	N	Q	0	0	0	0	0	0	0	1	1	0	0	1	0	0	0	0	0	0	1	0	0	0	0	0
20-6-10/60227.0	1	1	A	0	1	0	N	Y	0	0	0	0	0	0	0	0	0	0	0	0	0	0	0	0	0	0	0	0	0	0	0	0
20-6-10/60228.0	4	1	A	0	1	0	Y	Q	0	1	1	0	0	0	0	1	0	0	0	1	0	1	0	0	0	0	0	0	0	0	0	0
20-6-10/60229.0	5	1	A	0	1	0	Y	Q	0	0	0	0	0	1	0	0	1	0	0	1	1	0	0	0	1	0	0	0	0	0	0	0
20-6-10/60230.0	4	1	A	1	0	0	Y	Q	0	0	0	0	0	0	0	0	1	0	0	1	0	0	0	0	1	0	0	0	0	0	0	0
20-6-10/60231.0	3	1	A	0	1	0	N	Q	0	0	0	0	1	0	0	0	0	0	0	1	0	0	0	0	1	0	0	0	0	0	0	1

Peabody number	CS	MNI	Age	Sex			AFO	Cr mod	Trauma				Infection				Arthritis					Physiological stress				Neurological			Tumor		Other path.	
				M	F	U			T c	T pc	T h	T d	I de	I di	I tr	I g	A sl	A m	A s	A sp	A o	EH	CO	PH	O	Sb sl	Sb m/s	O na	T bo	T o	OP d	OP c/pc
20-6-10/60232.0	1	1	A	1	0	0	N	Q	0	0	0	0	0	0	0	0	1	0	0	0	0	0	0	0	1	0	0	0	0	0	0	0
20-6-10/60233.0	1	1	A	0	1	0	Y	U	0	0	0	0	1	0	0	0	0	0	1	0	0	0	0	0	1	0	0	0	0	0	0	0
20-6-10/60234.0	2	1	A	0	1	0	Y	Y	0	0	0	0	1	0	0	0	0	0	0	0	0	1	0	0	1	0	0	0	0	0	0	0
20-6-10/60235.0	1	1	A	1	0	0	Y	N	0	0	0	0	0	0	0	1	0	0	0	0	0	0	0	1	0	1	0	0	0	0	1	0
20-6-10/60236.0	2	1	A	1	0	0	N	Y	0	0	0	0	1	0	0	0	1	0	0	0	0	0	0	0	0	0	0	0	0	0	0	0
20-6-10/60237.0	5	1	A	1	0	0	Y	Y	0	0	0	0	1	0	0	0	0	0	0	1	0	0	0	0	0	0	0	0	0	0	0	0
20-6-10/60238.0	3	1	A	1	0	0	Y	Q	0	0	0	0	0	0	0	0	0	0	0	0	0	1	0	0	0	0	0	0	0	0	0	1
20-6-10/60239.0	2	1	A	0	1	0	N	Q	0	0	0	0	1	0	0	0	0	0	0	0	0	1	0	1	0	0	0	0	0	0	0	1
20-6-10/60240.0	1	1	A	0	1	0	Y	U	0	1	1	0	1	0	0	0	1	0	0	1	0	0	0	0	1	0	0	0	0	0	0	0
20-6-10/60241.0	3	1	A	0	1	0	N	Y	0	1	1	0	1	0	0	0	1	0	0	0	0	0	0	0	0	1	0	0	0	0	0	0
20-6-10/60242.0	4	1	A	1	0	0	Y	Y	0	1	1	0	0	0	1	0	0	0	1	1	0	0	0	0	0	0	0	0	0	0	0	0
20-6-10/60243.0	4	1	A	1	0	0	N	Y	0	0	0	0	1	0	0	0	0	0	0	1	0	0	0	0	0	1	0	0	0	0	0	0
20-6-10/60244.0	3	1	A	1	0	0	Y	Q	0	1	1	0	0	0	0	0	1	0	0	0	0	1	0	0	0	1	0	1	0	0	0	0
20-6-10/60245.0	5	1	J	0	0	1	N	Q	0	0	0	0	0	0	0	0	0	0	0	0	0	1	0	0	0	0	0	0	0	0	0	0
20-6-10/60246.0	2A	1	A	0	1	0	Y	Y	0	0	0	0	1	0	0	0	1	0	0	0	1	0	0	0	1	0	0	0	0	0	0	0
20-6-10/60247.0	2	1	A	0	1	0	Y	Q	0	0	0	0	1	0	0	0	0	0	0	0	0	0	0	0	0	0	0	0	0	0	1	0
20-6-10/60248.0	2	1	A	0	1	0	Y	Q	0	0	0	0	1	0	0	0	1	0	0	0	0	0	0	0	1	0	0	0	0	0	0	0
20-6-10/60249.0	1	1	A	0	1	0	N	Y	0	1	0	0	1	0	0	0	0	0	0	0	0	1	0	0	1	0	0	0	0	0	0	0
20-6-10/60250.0	5	1	A	0	0	1	N	U	0	0	0	0	0	0	0	0	0	0	0	0	0	0	0	0	0	0	0	0	0	0	0	0
20-6-10/60251.0	IN	1	A	0	1	0	N	U	0	1	1	0	1	0	0	1	0	0	0	0	0	1	0	0	0	0	0	0	0	0	1	0
20-6-10/60252.0	3	1	A	0	1	0	N	Q	0	0	0	0	0	0	0	0	0	0	0	0	0	1	0	0	0	1	0	0	0	0	0	0
20-6-10/60253.0	3	1	A	1	0	0	Y	U	0	0	0	0	1	0	0	0	0	0	0	0	0	0	0	0	0	0	0	0	0	0	0	0
20-6-10/60254.0	1	1	A	1	0	0	Y	U	0	0	0	0	0	0	0	0	0	0	1	1	1	0	0	0	0	0	0	0	0	0	0	0
20-6-10/60255.0	3	1	A	0	1	0	Y	Y	0	1	0	0	0	0	1	0	1	0	0	0	0	0	0	0	1	0	0	0	0	0	0	0
20-6-10/60256.0	1	1	A	1	0	0	Y	Q	0	0	0	0	0	0	0	0	0	0	0	0	0	0	0	1	0	0	0	0	0	0	0	0
20-6-10/60257.0	3	1	A	0	1	0	N	Y	0	1	1	0	1	0	0	0	0	1	0	0	0	0	0	0	0	0	0	0	0	0	0	0
20-6-10/60258.0	3	1	A	0	1	0	Y	Q	0	0	0	0	1	0	0	1	0	0	0	0	0	1	0	0	0	0	0	0	0	0	0	0
20-6-10/60259.0	1	1	A	0	1	0	Y	Y	0	0	0	0	0	0	0	1	0	0	0	0	1	0	0	0	0	1	0	0	0	0	0	0
20-6-10/60260.0	4	1	A	1	0	0	Y	Q	0	0	0	0	0	1	0	0	0	1	0	1	1	1	0	0	1	0	0	0	0	0	0	0
20-6-10/60261.0	2	1	A	1	0	0	N	Q	0	0	0	0	1	0	0	0	1	0	0	0	0	1	0	0	0	0	0	0	0	0	0	1
20-6-10/60262.0	2	1	A	0	1	0	N	Q	0	0	0	0	0	0	0	0	0	0	0	0	0	1	0	0	0	0	0	0	0	0	0	0
20-6-10/60263.0	3A	1	J	0	0	1	Y	Q	0	0	0	0	0	0	0	0	0	0	0	0	0	0	0	0	0	0	0	0	0	0	0	0
20-6-10/60264.0	5	1	A	0	1	0	Y	U	0	0	0	0	0	0	1	0	1	0	0	0	0	1	0	0	0	0	0	0	0	0	0	0
20-6-10/60264.1	5	1	J	0	0	1	Y	U	0	0	0	0	0	0	0	0	0	0	0	0	0	0	0	0	0	0	0	0	0	0	0	0
20-6-10/60265.0	3	1	A	0	1	0	N	U	0	0	0	0	0	0	0	0	0	0	0	0	0	0	0	0	0	0	0	0	0	0	0	0
20-6-10/60266.0	B	1	A	1	0	0	Y	Y	0	1	1	0	0	0	0	0	1	0	0	1	0	1	0	0	0	0	0	0	0	0	0	0
20-6-10/60267.0	3	1	A	1	0	0	Y	Q	0	0	0	0	1	0	0	0	1	1	0	1	0	0	0	0	0	0	1	0	0	0	0	0
20-6-10/60268.0	3	1	A	1	0	0	Y	U	0	1	1	0	0	0	0	1	0	0	0	0	0	1	0	0	0	0	0	0	0	0	0	0
20-6-10/60269.0	7	1	J	0	0	1	Y	U	0	0	0	0	0	0	0	0	0	0	0	0	0	0	0	0	0	0	0	0	0	0	0	0
20-6-10/60270.0	7	1	J	0	0	1	N	Q	0	0	0	0	0	0	0	0	0	0	0	0	1	0	1	0	0	0	0	0	0	0	0	1
20-6-10/60271.0	4A	1	A	0	1	0	N	Y	0	0	0	0	1	0	0	0	0	0	0	0	0	0	0	0	0	0	0	0	0	0	0	0
20-6-10/60272.0	3	1	A	0	1	0	Y	Y	0	1	1	0	0	0	0	0	0	0	0	0	0	0	0	0	0	0	0	0	0	0	0	0
20-6-10/60273.0	2	1	A	1	0	0	N	U	0	1	1	0	0	0	0	0	1	0	0	1	0	0	0	0	0	0	0	0	0	0	0	0
20-6-10/60274.0	1	1	A	1	0	0	Y	Q	0	0	0	0	1	0	0	0	1	0	0	1	0	0	0	0	1	0	0	0	0	0	0	0
20-6-10/60275.0	1	1	A	1	0	0	Y	U	0	0	0	0	0	0	0	0	0	0	0	0	0	0	0	0	0	0	0	0	0	0	0	1

Peabody number	CS	MNI	Age	Sex			AFO	Cr mod	Trauma				Infection				Arthritis					Physiological stress				Neurological			Tumor		Other path.	
				M	F	U			T c	T pc	T h	T d	I de	I di	I tr	I g	A sl	A m	A s	A sp	A o	EH	CO	PH	O	Sb sl	Sb m/s	O na	T bo	T o	OP d	OP c/pc
20-6-10/60276.0	2A	1	A	1	0	0	N	Y	0	0	0	0	0	0	0	1	0	0	0	1	0	0	0	0	1	1	0	0	0	0	0	0
20-6-10/60277.0	3	1	A	1	0	0	N	Q	0	0	0	0	0	0	0	0	0	1	0	0	1	0	0	0	0	0	0	0	0	0	1	0
20-6-10/60278.0	3	1	A	1	0	0	Y	Q	0	1	1	0	0	0	0	0	1	0	0	1	0	0	0	0	1	1	0	0	0	0	0	0
20-6-10/60279.0	3	1	A	1	0	0	N	Y	0	0	0	0	1	0	0	0	1	0	0	0	0	1	0	0	0	0	0	0	0	0	0	0
20-6-10/60280.0	4	1	A	0	1	0	N	Y	0	0	0	0	1	0	0	0	0	1	0	0	0	1	0	0	0	0	0	0	0	0	0	0
20-6-10/60281.0	2	1	A	0	1	0	Y	Y	0	0	0	0	0	0	0	0	1	0	0	1	1	0	0	0	0	0	0	0	0	0	0	0
20-6-10/60282.0	5	1	A	0	1	0	Y	Y	0	0	0	0	0	0	0	0	1	0	0	0	0	1	0	0	0	0	0	0	0	0	0	0
20-6-10/60283.0	4A	1	A	1	0	0	Y	Q	0	0	0	0	1	0	0	1	1	0	0	0	0	1	0	0	0	0	1	0	0	0	0	0
20-6-10/60284.0	3	1	A	0	1	0	N	U	0	0	0	0	0	0	0	0	0	0	0	0	0	0	0	0	0	0	0	0	0	0	0	0
20-6-10/60285.0	3	1	A	0	1	0	Y	U	0	0	0	0	0	0	0	0	1	0	0	0	0	1	0	0	0	0	0	0	0	0	0	0
20-6-10/60286.0	4	1	A	1	0	0	Y	Y	1	0	1	0	0	0	1	0	0	1	0	1	1	0	0	0	0	0	0	0	0	0	1	0
20-6-10/60287.0	3A	1	A	0	1	0	N	Q	0	0	0	0	1	0	0	0	0	0	1	0	0	0	0	0	1	0	0	0	0	0	0	0
20-6-10/60288.0	3	1	A	0	1	0	N	Q	0	0	0	0	1	0	0	1	0	0	0	0	0	0	0	0	0	0	0	0	0	0	0	0
20-6-10/60289.0	3	1	A	0	1	0	N	U	0	0	0	0	0	0	0	1	0	0	0	0	0	0	0	0	0	0	0	0	0	0	0	0
20-6-10/60290.0	5	1	A	0	1	0	Y	U	0	0	0	0	0	0	0	0	1	0	0	0	0	0	0	0	1	0	0	0	0	0	0	0
20-6-10/60291.0	3	1	A	0	1	0	N	U	0	0	0	0	1	0	0	0	0	0	0	0	0	1	0	0	0	0	0	0	0	0	0	0
20-6-10/60292.0	P	1	A	0	1	0	Y	Y	0	1	0	0	0	0	1	1	0	0	0	0	1	0	0	1	0	0	0	0	0	0	0	1
20-6-10/60293.0	3	1	A	1	0	0	Y	Q	1	0	0	0	0	0	0	0	0	0	0	0	0	1	0	0	0	0	0	0	0	0	0	0
20-6-10/60294.0	3	1	A	1	0	0	N	Q	0	1	1	0	1	0	0	0	1	0	0	1	0	0	0	0	0	0	0	0	0	0	0	0
20-6-10/60295.0	3	1	A	0	1	0	N	U	0	0	0	0	1	1	0	0	1	0	0	0	1	0	0	0	0	0	0	0	0	0	0	0
20-6-10/60296.0	B	1	A	1	0	0	Y	Q	0	0	0	0	0	0	0	0	1	0	0	0	0	1	0	0	0	0	0	0	0	0	0	0
20-6-10/60297.0	5	1	A	0	1	0	Y	Y	0	0	0	0	0	0	0	0	0	0	0	0	0	1	0	0	0	0	0	0	0	0	0	0
20-6-10/60298.0	4	1	A	0	1	0	Y	Q	0	0	0	0	0	0	0	1	1	0	0	0	0	1	0	0	0	1	0	0	0	0	0	0
20-6-10/60299.0	4	1	A	1	0	0	N	Q	0	0	0	0	0	0	0	0	1	0	0	0	0	0	0	0	0	0	0	0	0	0	1	0
20-6-10/60300.0	4	1	A	0	1	0	N	Q	0	0	0	0	0	1	0	0	0	0	0	0	1	0	0	0	0	0	0	0	0	0	0	0
20-6-10/60301.0	5	1	A	0	1	0	N	Q	0	1	1	0	1	1	0	0	1	0	0	0	0	0	0	0	1	0	0	0	0	0	0	0
20-6-10/60302.0	3	1	A	0	1	0	Y	Q	0	0	0	0	1	0	0	0	1	0	0	1	0	1	0	0	0	0	0	0	0	0	0	0
20-6-10/60303.0	7	1	A	0	1	0	N	Y	0	0	0	0	0	0	0	0	1	0	0	0	0	1	0	0	0	0	0	0	0	0	0	0
20-6-10/60304.0	5	1	J	0	1	0	Y	Y	0	0	0	0	0	1	0	0	0	0	0	0	0	1	0	0	0	0	0	0	0	0	0	0
20-6-10/60305.0	3	1	J	1	0	0	Y	Q	0	0	0	0	0	0	0	0	0	0	0	0	0	1	0	0	0	0	0	0	0	0	0	0
20-6-10/60306.0	4	1	A	1	0	0	Y	Y	0	0	0	0	0	0	0	0	1	0	0	0	1	1	0	0	0	0	0	0	0	0	0	0
20-6-10/60307.0	1	1	A	0	1	0	Y	Y	0	0	0	0	1	0	0	0	1	0	0	0	0	1	0	0	0	0	0	0	0	0	0	0
20-6-10/60308.0	3	1	A	0	1	0	Y	Q	0	0	0	0	0	0	0	1	1	0	0	0	0	1	0	0	0	0	0	0	0	0	0	0
20-6-10/60309.0	5	1	A	0	1	0	Y	Y	0	1	1	0	0	0	0	0	0	0	1	1	0	0	0	0	1	0	0	0	0	0	0	0
20-6-10/60310.0	5	1	A	0	1	0	Y	U	0	0	0	0	0	0	0	0	0	0	0	0	0	0	0	0	0	0	0	0	0	0	0	0
20-6-10/60311.0	3	1	A	0	1	0	Y	Q	1	1	1	1	0	0	1	0	0	0	0	0	0	0	0	0	0	0	0	0	0	0	0	1
20-6-10/60312.0	2A	1	A	0	1	0	Y	Y	0	0	0	0	0	0	0	0	0	0	0	0	0	0	0	0	0	0	0	0	0	0	0	0
20-6-10/60313.0	3	1	A	1	0	0	Y	Y	0	1	1	0	1	0	0	0	1	0	0	1	0	0	0	0	1	0	1	0	0	0	0	0
20-6-10/60314.0	3	1	A	1	0	0	N	Y	0	1	1	0	1	0	0	0	1	0	0	1	0	1	0	1	0	0	0	0	0	0	0	0
20-6-10/60315.0	3	1	A	1	0	0	Y	N	0	0	0	0	1	0	0	0	0	0	0	0	0	0	0	0	1	0	0	0	0	0	0	0
20-6-10/60316.0	5	1	A	1	0	0	Y	Q	0	0	0	0	0	0	0	0	0	0	0	1	0	0	0	0	1	0	0	0	0	0	0	0
20-6-10/60317.0	3	1	A	1	0	0	Y	Q	0	1	1	0	1	1	0	0	0	0	0	0	0	0	0	1	1	0	0	0	0	1	0	0
20-6-10/60318.0	P	1	A	1	0	0	Y	Q	0	0	0	0	1	1	0	0	0	0	0	0	0	1	0	0	0	0	0	0	0	0	0	0
20-6-10/60319.0	3	1	A	1	0	0	Y	Q	0	0	0	0	0	0	0	0	1	0	0	0	0	1	0	0	0	0	0	0	0	0	0	0
20-6-10/60320.0	2	1	A	0	0	1	N	Y	0	0	0	0	0	0	0	0	0	0	0	1	0	1	0	0	0	0	0	0	0	0	0	0

Peabody number	CS	MNI	Age	Sex			AFO	Cr mod	Trauma				Infection				Arthritis					Physiological stress				Neurological			Tumor		Other path.	
				M	F	U			T c	T pc	T h	T d	I de	I di	I tr	I g	A sl	A m	A s	A sp	A o	EH	CO	PH	O	Sb sl	Sb m/s	O na	T bo	T o	OP d	OP c/pc
20-6-10/60321.0	2	1	A	0	1	0	N	U	0	0	0	0	0	0	0	0	0	0	0	0	0	0	0	0	0	0	0	0	0	0	0	0
20-6-10/60322.0	IN	1	A	0	0	1	Y	U	0	0	0	0	0	0	0	0	0	0	0	0	0	0	0	0	0	0	0	0	0	0	0	0
20-6-10/60324.1	IN	1	A	0	0	1	N	U	0	0	0	0	0	0	0	0	0	0	0	0	0	0	0	0	0	0	0	0	0	0	0	0
20-6-10/60327.1	IN	1	A	0	0	1	Y	U	0	0	0	0	0	0	0	0	0	0	0	0	0	0	0	0	0	0	0	0	0	0	0	0
20-6-10/60332.1	IN	1	A	0	0	1	Y	U	0	0	0	0	0	0	0	0	1	0	0	0	0	0	0	0	0	0	0	0	0	0	0	0
20-6-10/60333.1	IN	1	A	0	0	1	Y	U	0	0	0	0	0	0	0	0	1	0	0	0	0	0	0	0	0	0	0	0	0	0	0	0
20-6-10/60336.1	IN	1	A	0	0	1	Y	U	0	0	0	0	0	0	0	0	0	0	0	0	0	0	0	0	0	0	0	0	0	0	0	0
20-6-10/60337.1	IN	1	A	0	0	1	Y	U	0	0	0	0	0	0	0	0	1	0	0	0	0	0	0	0	0	0	0	0	0	0	0	0
20-6-10/60337.2	2	1	A	0	0	1	N	U	0	0	0	0	0	0	0	0	0	0	0	0	0	0	0	0	0	0	0	0	0	0	0	0
20-6-10/60337.3	1	1	A	0	0	1	N	U	0	0	0	0	0	0	0	0	1	0	0	0	0	0	0	0	0	0	0	0	0	0	0	0
20-6-10/60339.1	P	1	A	0	0	1	Y	U	0	0	0	0	0	0	0	0	0	0	0	0	0	0	0	0	0	0	0	0	0	0	0	0
20-6-10/60343.1	4A	1	A	0	0	1	Y	U	0	0	0	0	0	0	0	0	0	0	0	0	0	0	0	0	0	0	0	0	0	0	0	0
20-6-10/60344.1	IN	1	A	1	0	0	Y	U	0	0	0	0	0	0	0	0	1	0	0	0	0	0	0	0	0	0	0	0	0	0	0	0
20-6-10/60350.1	IN	1	J	0	0	1	Y	U	0	0	0	0	0	0	0	0	0	0	0	0	0	1	0	0	0	0	0	0	0	0	0	0
20-6-10/60350.10	3A	1	J	0	0	1	Y	U	0	0	0	0	0	0	0	0	0	0	0	0	0	1	0	0	0	0	0	0	0	0	0	0
20-6-10/60350.11	6	1	J	0	0	1	Y	U	0	0	0	0	0	0	0	0	0	0	0	0	0	0	0	0	0	0	0	0	0	0	0	0
20-6-10/60350.4	3	1	A	0	0	1	Y	U	0	0	0	0	0	0	0	0	0	0	0	0	0	0	0	0	0	0	0	0	0	0	0	0
20-6-10/60350.5	3	1	A	0	0	1	N	U	0	0	0	0	0	0	0	0	0	0	0	0	0	0	0	0	0	0	0	0	0	0	0	0
20-6-10/60350.6	4A	1	A	0	0	1	N	U	0	0	0	0	0	0	0	0	0	0	0	0	0	0	0	0	0	0	0	0	0	0	0	0
20-6-10/60350.7	2	1	A	0	0	1	Y	U	0	0	0	0	0	0	0	0	0	0	0	0	0	0	0	0	0	0	0	0	0	0	1	0
20-6-10/60350.8	4	1	J	0	0	1	Y	U	0	0	0	0	0	0	0	0	0	0	0	0	0	1	0	0	0	0	0	0	0	0	0	0
20-6-10/60350.9	IN	1	A	0	0	1	N	U	0	0	0	0	0	0	0	0	0	0	0	0	0	0	0	0	0	0	0	0	0	0	0	0
20-6-10/60358.1	IN	1	A	0	0	1	Y	U	0	1	1	0	0	0	0	0	0	0	0	0	0	0	0	0	0	0	0	0	0	0	0	0
20-6-10/60358.2	P	1	J	0	0	1	N	U	0	1	1	0	0	0	1	0	0	0	0	0	0	0	0	0	0	0	0	0	0	0	0	0
20-6-10/60358.3	BA	1	A	0	0	1	Y	U	0	0	0	0	0	0	0	1	0	0	0	0	0	0	0	0	0	0	0	0	0	0	0	0
20-6-10/60358.4	IN	1	A	0	0	1	Y	U	0	0	0	0	0	0	0	0	0	0	0	0	0	0	0	0	0	0	0	0	0	0	0	0
24-4-10/60426.0	4	1	A	1	0	0	Y	Q	0	1	1	0	0	0	0	0	0	0	0	0	0	0	0	0	0	0	0	0	0	0	0	0
24-4-10/60427.0	3	1	A	0	0	1	N	U	0	1	1	0	0	0	0	0	0	0	0	0	0	0	0	0	0	0	0	0	0	0	0	0
24-4-10/60427a	3	1	J	0	0	1	N	Q	0	0	0	0	0	0	0	0	1	0	0	0	0	0	0	0	0	0	0	0	0	0	0	0
24-4-10/60428.0	IN	1	A	0	1	0	Y	Y	0	0	0	0	0	0	0	0	1	0	0	0	0	1	0	0	0	0	0	0	0	0	0	0
24-4-10/60429.0	4	1	A	1	0	0	Y	N	0	0	0	0	1	0	0	0	0	0	0	0	0	0	0	0	0	0	0	0	0	0	0	0
24-4-10/60430.0	1A	1	A	0	1	0	Y	Q	0	0	0	0	1	0	0	0	0	0	0	0	0	1	0	0	0	0	0	0	0	0	0	0
24-4-10/60431.0	1	1	J	0	0	1	N	Y	0	0	0	0	0	0	0	0	0	0	0	0	0	0	1	0	0	0	0	0	0	0	0	0
24-4-10/60432.0	3	1	A	1	0	0	Y	U	0	0	0	0	0	0	0	0	0	0	0	0	0	1	0	0	0	0	0	0	0	0	0	0
24-4-10/60433.0	5	1	A	0	1	0	N	Q	0	0	0	0	0	0	0	1	1	0	0	1	0	0	0	0	0	0	0	0	0	0	0	0
24-4-10/60434.0	2A	1	A	0	1	0	Y	Q	0	0	0	0	1	0	0	0	1	0	0	0	1	0	0	0	1	0	0	0	0	0	0	0
24-4-10/60435.0	3A	1	A	0	1	0	N	U	0	0	0	0	0	0	0	0	0	0	0	0	0	0	0	0	0	0	0	0	0	0	0	0
24-4-10/60436.0	P	1	A	0	1	0	N	U	0	1	1	0	1	0	0	0	0	0	1	1	0	0	0	0	1	0	0	0	0	0	0	0
24-4-10/60437.0	4	1	A	0	1	0	N	Q	0	0	0	0	0	0	0	0	1	0	0	1	0	0	0	0	1	0	0	0	0	0	0	0
24-4-10/60438.0	5	1	A	0	1	0	Y	U	0	0	0	0	0	0	0	0	0	0	0	0	0	0	0	0	0	0	0	0	0	0	0	0
24-4-10/60439.0	1A	1	A	0	1	0	Y	Q	0	0	0	0	0	0	0	0	0	0	0	0	0	0	0	0	0	0	0	0	0	0	0	0
24-4-10/60440.0	4	1	A	0	1	0	Y	Q	1	0	1	0	1	0	0	0	0	0	0	0	0	1	0	0	0	0	0	0	0	0	1	0
24-4-10/60441.0	B	1	A	0	0	1	N	U	0	0	0	0	1	0	0	0	0	0	0	0	0	0	0	0	1	0	0	0	0	0	0	0
24-4-10/60442.0	1	1	A	0	1	0	Y	Q	0	0	0	0	0	0	0	0	0	0	0	0	0	1	0	0	0	0	0	0	0	0	0	0
24-4-10/60443.0	3	1	A	0	1	0	Y	Q	0	0	0	0	0	0	0	0	0	0	0	0	0	0	0	0	1	0	0	0	0	0	1	0

Peabody number	CS	MNI	Age	Sex			AFO	Cr mod	Trauma				Infection				Arthritis					Physiological stress				Neurological			Tumor		Other path.	
				M	F	U			Tc	Tpc	Th	Td	Ide	Idi	Itr	Ig	Asl	Am	As	Asp	Ao	EH	CO	PH	O	Sb sl	Sb m/s	Ona	Tbo	To	OP d	OP c/pc
24-4-10/60444.0	1A	1	A	1	0	0	Y	U	0	1	1	0	1	0	0	0	1	0	0	0	0	0	0	0	0	0	0	0	0	0	0	0
24-4-10/60445.0	1	1	A	1	0	0	Y	Q	0	1	1	0	0	0	0	0	1	0	0	0	0	0	0	0	0	0	0	0	0	0	0	0
24-4-10/60446.0	1	1	A	0	1	0	N	Y	0	0	0	0	0	0	0	0	0	0	0	1	0	0	0	0	0	0	0	0	0	0	0	0
24-4-10/60447.0	1	1	J	0	0	1	N	Y	0	0	0	0	0	0	0	0	0	0	0	0	0	0	0	0	0	0	0	0	0	0	0	0
24-4-10/60448.0	3	1	A	0	0	1	N	U	0	0	0	0	0	0	0	0	0	0	0	0	0	0	0	0	0	0	0	0	0	0	0	0
24-4-10/60449.0	4	1	A	0	1	0	N	U	0	0	0	0	0	0	0	0	1	0	0	0	0	0	0	0	0	0	0	0	0	0	0	0
24-4-10/60450.0	IN	1	A	0	1	0	Y	U	0	0	0	0	0	0	0	0	0	0	0	0	0	0	0	0	0	0	1	0	0	0	0	0
24-4-10/60450.1	IN	1	J	0	0	1	Y	U	0	0	0	0	0	0	0	0	0	0	0	0	0	0	0	0	0	0	0	0	0	0	1	0
24-4-10/60451.0	1	1	A	0	1	0	Y	Y	0	0	0	0	0	0	0	0	0	1	1	1	0	0	0	0	1	0	0	0	0	0	0	0
24-4-10/60452.0	3	1	A	0	1	0	Y	Y	0	0	0	0	1	0	0	0	0	0	0	0	0	1	0	0	0	0	0	0	0	0	0	0
24-4-10/60453.0	3	1	A	1	0	0	Y	N	0	0	0	0	0	0	0	0	0	0	0	1	1	0	0	0	1	0	0	0	1	1	0	0
24-4-10/60454.0	2	1	A	1	0	0	N	Y	0	0	0	0	1	0	0	0	1	0	0	1	0	0	0	0	0	0	0	0	0	0	0	0
24-4-10/60455.0	1	1	A	0	1	0	N	Y	0	0	0	0	0	1	0	0	0	0	0	1	1	0	0	0	0	0	0	0	0	0	0	0
24-4-10/60456.0	IN	1	A	1	0	0	Y	Q	0	0	0	0	0	0	0	1	0	0	0	0	0	0	0	0	0	0	0	0	0	0	0	0
24-4-10/60457.0	IN	1	A	0	1	0	Y	Y	0	0	0	0	0	0	0	0	0	0	0	0	0	1	0	0	0	0	0	0	0	0	0	0
24-4-10/60458.0	IN	1	A	1	0	0	Y	Q	0	0	0	0	0	0	0	0	1	0	0	0	1	0	0	0	0	1	0	0	0	0	0	0
24-4-10/60459.0	4	1	A	0	0	1	N	U	0	0	0	0	0	0	0	0	0	0	0	0	0	0	0	0	0	0	0	0	0	0	0	0
24-4-10/60460.0	4	1	A	0	1	0	N	U	0	1	1	0	0	0	0	0	1	0	0	0	0	0	0	0	0	0	0	0	0	0	0	0
24-4-10/60461.0	3A	1	A	0	0	1	N	U	0	0	0	0	0	0	0	0	1	0	0	0	0	0	0	0	1	0	0	0	0	0	0	0
24-4-10/60462.0	BA	1	A	0	0	1	N	U	0	0	0	0	0	0	0	0	0	0	0	0	0	0	0	0	0	0	0	0	0	0	0	0
24-4-10/60463.0	IN	1	A	0	1	0	Y	Q	0	0	0	0	0	0	0	1	0	0	0	0	0	0	0	0	0	0	0	0	0	0	0	0
24-4-10/60464.0	4	1	A	1	0	0	N	U	0	0	0	0	0	0	0	0	0	0	0	0	0	0	0	0	0	0	0	0	0	0	0	0
24-4-10/60465.0	P	1	A	0	1	0	N	Q	0	0	0	0	0	0	0	0	1	0	0	0	0	0	0	0	0	0	0	0	0	0	0	0
24-4-10/60466.0	4	1	A	1	0	0	Y	Y	0	0	0	0	0	0	0	0	0	0	0	0	0	1	0	0	0	0	0	0	0	0	0	0
24-4-10/60467.0	P	1	A	1	0	0	N	U	0	0	0	0	0	0	0	0	0	0	0	0	0	1	0	0	0	0	1	0	0	0	0	0
24-4-10/60468.0	1	1	A	0	0	1	Y	U	0	0	0	0	0	0	0	0	0	0	0	0	0	0	0	0	0	0	0	0	0	0	0	0
24-4-10/60469.0	4	1	A	0	1	0	N	U	0	0	0	0	0	0	0	0	0	0	0	0	0	1	0	0	0	1	0	0	0	0	0	0
24-4-10/60470.0	4A	1	A	1	0	0	N	U	0	0	0	0	0	0	0	0	0	0	0	0	0	0	0	0	0	0	0	0	0	0	0	0
24-4-10/60471.0	5	1	A	0	1	0	Y	U	0	0	0	0	0	0	0	0	0	0	0	0	0	0	0	0	0	0	0	0	0	0	0	0
24-4-10/60472.0	IN	1	A	0	1	0	Y	Q	0	0	0	0	0	0	0	0	0	0	0	0	0	0	0	0	0	0	0	0	0	0	0	1
24-4-10/60473.0	1	1	A	1	0	0	Y	Y	0	0	0	0	1	0	0	0	1	0	0	0	0	0	0	0	0	0	0	0	1	0	0	0
24-4-10/60474.0	3	1	A	0	1	0	Y	U	0	0	0	0	0	0	0	0	0	0	0	0	1	1	0	0	0	0	0	0	0	0	0	1
24-4-10/60475.0	2A	1	A	0	1	0	Y	U	0	0	0	0	0	0	0	0	0	0	0	0	0	0	0	0	0	0	0	1	0	0	0	0
24-4-10/60476.0	BA	1	A	0	1	0	N	Q	0	0	0	0	0	0	0	0	1	0	0	0	0	0	0	0	0	0	0	0	0	0	1	0
24-4-10/60477.0	IN	1	J	0	1	0	N	Q	0	0	0	0	0	0	0	0	0	0	0	0	0	1	0	0	0	0	0	0	0	0	0	0
24-4-10/60477.1	IN	1	A	0	0	1	N	U	0	0	0	0	0	0	0	0	0	0	0	0	0	0	0	0	0	0	0	0	0	0	0	0
24-4-10/60479.0	B	1	A	0	1	0	N	Y	0	0	0	0	0	0	0	0	0	0	0	0	0	0	1	1	0	0	0	0	0	0	0	0
24-4-10/60480.0	P	1	A	0	1	0	N	Y	0	0	0	0	0	0	0	0	0	0	0	0	0	1	0	0	0	0	0	0	0	0	0	0
24-4-10/60481.0	P	1	A	0	1	0	N	Y	0	0	0	0	1	0	0	1	1	0	0	0	0	0	0	0	1	0	0	0	1	0	0	0
24-4-10/60482.0	2	1	A	1	0	0	Y	U	0	0	0	0	1	1	0	0	0	0	0	1	1	1	0	0	1	0	0	0	1	0	0	0
24-4-10/60483.0	3	1	A	1	0	0	Y	U	0	0	0	0	0	0	0	0	0	0	0	0	0	0	0	0	0	0	0	0	0	0	0	0
24-4-10/60483a	3	1	A	0	1	0	N	Y	0	0	0	0	0	0	0	0	0	0	0	0	0	0	0	0	0	0	0	0	0	0	0	0
24-4-10/60484.0	P	1	A	0	1	0	N	N	0	0	0	0	1	0	0	0	0	0	0	0	0	1	0	0	0	0	0	0	0	0	0	0
24-4-10/60485.0	1	1	A	1	0	0	N	U	0	1	1	0	1	0	0	0	0	0	0	0	0	0	0	0	1	0	0	0	0	0	0	0
24-4-10/60486.0	B	1	J	0	0	1	Y	U	0	0	0	0	0	0	0	1	0	0	0	0	0	1	0	0	0	0	0	0	0	0	0	0

Peabody number	CS	MNI	Age	Sex			AFO	Cr mod	Trauma				Infection				Arthritis					Physiological stress				Neurological			Tumor		Other path.	
				M	F	U			T c	T pc	T h	T d	I de	I di	I tr	I g	A sl	A m	A s	A sp	A o	EH	CO	PH	O	Sb sl	Sb m/s	O na	T bo	T o	OP d	OP c/pc
24-4-10/60486.1	B	1	A	0	0	1	Y	U	0	0	0	0	0	0	0	0	0	0	0	0	0	0	0	0	0	0	0	0	0	0	0	0
24-4-10/60487.0	4	1	A	0	1	0	Y	Y	0	0	0	0	0	0	0	0	0	0	0	0	0	0	0	0	0	0	0	0	0	0	0	0
24-4-10/60488.0	B	1	A	1	0	0	Y	U	0	0	0	0	0	0	0	1	1	0	0	1	0	0	1	0	0	0	0	0	0	0	0	0
24-4-10/60489.0	IN	1	A	0	1	0	N	Y	0	0	0	0	0	0	0	0	0	0	0	0	0	0	0	0	0	0	0	0	0	0	0	0
24-4-10/60490.0	1	1	A	1	0	0	N	U	0	1	1	0	1	1	0	0	1	0	0	1	1	0	0	0	1	0	0	0	0	0	0	1
24-4-10/60491.0	3	1	A	1	0	0	Y	U	0	0	0	0	0	0	0	0	0	0	0	0	0	0	0	0	0	0	0	0	0	0	0	0
24-4-10/60492.0	IN	1	J	0	0	1	Y	Y	0	0	0	0	0	0	0	0	0	0	0	0	0	0	1	0	0	0	0	0	0	0	0	0
24-4-10/60493.0	P	1	A	0	1	0	N	Q	0	0	0	0	0	0	0	0	0	0	0	0	0	0	0	0	0	0	0	0	0	0	0	0
24-4-10/60494.0	1	1	A	1	0	0	Y	Q	0	0	0	0	0	0	0	1	0	0	0	0	0	0	0	0	0	0	0	0	0	0	0	0
24-4-10/60495.0	IN	1	J	0	0	1	Y	Q	0	0	0	0	0	0	0	0	0	0	0	0	0	1	0	0	0	0	0	0	0	0	0	0
24-4-10/60496.0	3	1	A	0	1	0	Y	Y	0	0	0	0	0	0	0	0	0	0	0	1	0	0	0	0	0	0	0	0	0	0	0	0
24-4-10/60497.0	3	1	A	0	1	0	Y	N	0	0	0	0	0	0	0	1	0	0	0	0	0	1	0	0	0	0	0	0	0	0	0	0
24-4-10/60498.0	1A	1	A	0	0	1	Y	U	0	0	0	0	1	0	0	0	0	0	0	0	0	0	0	0	0	0	0	0	0	0	0	0
24-4-10/60499.0	1	1	A	0	0	1	N	U	0	0	0	0	0	0	0	0	0	0	0	0	0	0	0	0	0	0	0	0	0	0	0	0
24-4-10/60500.0	IN	1	A	0	1	0	Y	U	0	0	0	0	0	0	0	0	0	0	0	0	0	0	0	0	0	0	0	0	0	0	0	0
24-4-10/60501.0	1	1	A	0	1	0	N	Y	0	0	0	0	0	0	0	1	0	0	0	0	0	1	0	0	0	0	0	0	0	0	0	1
24-4-10/60502.0	IN	1	A	1	0	0	Y	N	0	0	0	0	1	0	0	0	0	0	0	0	0	1	0	0	0	0	0	0	0	0	0	0
24-4-10/60503.0	5	1	A	1	0	0	Y	U	0	0	0	0	1	0	0	0	1	0	0	0	0	0	0	0	0	0	0	0	0	0	0	0
24-4-10/60503a	5	1	A	0	0	1	Y	U	0	0	0	0	0	0	0	0	0	0	0	0	0	0	0	0	0	0	0	0	0	0	0	0
24-4-10/60504.0	2	1	A	0	0	1	Y	U	0	0	0	0	1	0	0	0	0	0	0	1	0	0	0	0	0	0	0	0	0	0	0	0
24-4-10/60505.0	3	1	A	0	1	0	Y	Q	0	0	0	0	0	0	0	0	0	0	0	0	1	1	1	0	0	0	0	0	0	0	0	0
24-4-10/60507.0	1A	1	A	0	0	1	Y	U	0	1	1	0	0	0	0	0	0	0	0	0	0	1	0	0	0	0	0	0	0	0	0	0
24-4-10/60508.0	BA	1	A	0	1	0	N	U	0	0	0	0	1	0	0	0	0	0	0	1	0	0	0	0	0	0	0	0	0	0	0	1
24-4-10/60509.0	1	1	J	0	0	1	Y	Y	0	0	0	0	0	0	0	0	0	0	0	0	0	1	1	0	0	0	0	0	0	0	0	0
24-4-10/60510.0	3	1	J	1	0	0	N	Y	0	0	0	0	0	0	0	0	0	0	0	0	0	0	1	0	0	0	0	0	0	0	0	0
24-4-10/60511.0	1	1	A	0	1	0	Y	U	0	0	0	0	0	0	0	0	0	0	0	0	0	0	0	0	0	0	0	0	0	0	0	0
24-4-10/60511.1	1	1	J	0	0	1	Y	U	0	0	0	0	0	0	0	0	0	0	0	0	0	0	0	0	0	0	0	0	0	0	0	0
24-4-10/60512.0	IN	1	A	0	1	0	Y	Y	0	0	0	0	1	0	0	0	1	0	0	0	0	0	0	0	1	0	0	0	0	0	0	0
24-4-10/60513.0	4	1	A	1	0	0	Y	N	1	0	1	0	0	0	0	0	0	1	0	1	0	1	0	0	0	1	0	0	0	0	0	0
24-4-10/60514.0	3	1	A	1	0	0	Y	Y	0	0	0	0	0	1	0	0	1	0	0	0	0	1	0	0	0	0	0	0	0	0	0	0
24-4-10/60515.0	3A	1	A	0	0	1	N	U	0	0	0	0	0	0	0	0	0	0	0	0	0	0	0	0	0	0	0	0	0	0	0	0
24-4-10/60516.0	3	1	A	0	1	0	N	U	0	0	0	0	0	0	0	0	0	0	0	0	0	1	0	0	0	0	0	0	0	0	0	1
24-4-10/60517.0	IN	1	A	0	0	1	N	U	0	1	1	0	0	0	0	0	0	0	0	0	0	1	0	0	0	0	0	0	0	0	0	0
24-4-10/60518.0	3	1	A	1	0	0	Y	N	0	1	1	0	1	0	0	0	0	1	0	1	0	0	0	0	0	0	0	1	0	0	0	0
24-4-10/60519.0	1A	1	A	1	0	0	N	Q	0	0	0	0	0	0	0	1	0	0	0	0	0	0	0	0	0	0	0	0	0	0	0	0
24-4-10/60520.0	4	1	A	0	1	0	Y	N	0	0	0	0	1	0	0	0	1	0	0	0	0	0	0	0	1	0	0	0	0	0	0	0
24-4-10/60521.0	1	1	A	0	1	0	N	Q	0	0	0	0	1	0	0	0	0	0	0	0	0	1	0	0	0	0	0	0	0	0	1	0
24-4-10/60522.0	3	1	A	0	1	0	N	Q	0	0	0	0	0	0	0	0	0	0	0	0	0	1	0	0	0	0	0	0	0	0	0	0
24-4-10/60523.0	3	1	J	0	0	1	N	Q	0	0	0	0	0	0	0	0	0	0	0	0	0	1	0	0	0	0	0	0	0	0	0	0
24-4-10/60523.1	3	1	J	0	0	1	N	U	0	0	0	0	0	0	0	0	0	0	0	0	0	1	0	0	0	0	0	0	0	0	0	0
24-4-10/60524.0	3	1	A	0	1	0	N	Q	0	0	0	0	0	0	0	1	0	0	0	0	0	1	0	0	0	0	0	0	0	0	0	1
24-4-10/60525.0	4A	1	A	0	1	0	N	U	0	0	0	0	1	0	0	0	0	1	0	0	0	0	0	0	0	0	0	0	0	0	0	0
24-4-10/60526.0	1A	1	A	0	1	0	Y	U	0	0	0	0	0	0	0	0	1	0	0	0	0	0	0	0	1	0	0	0	0	0	0	0
24-4-10/60527.0	3	1	A	1	0	0	Y	Y	0	0	0	0	0	0	0	0	1	0	0	0	0	0	0	0	0	0	0	0	0	0	0	0
24-4-10/60528.0	P	1	A	1	0	0	N	U	0	0	0	0	0	0	0	0	1	0	0	0	0	1	0	0	0	1	0	0	0	0	0	0

Peabody number	CS	MNI	Age	Sex			AFO	Cr mod	Trauma				Infection				Arthritis					Physiological stress				Neurological			Tumor		Other path.	
				M	F	U			T c	T pc	T h	T d	I de	I di	I tr	I g	A sl	A m	A s	A sp	A o	EH	CO	PH	O	Sb sl	Sb m/s	O na	T bo	T o	OP d	OP c/pc
24-4-10/60529.0	3	1	A	0	1	0	Y	Q	0	0	0	0	0	0	0	0	0	0	0	0	0	1	0	0	0	0	0	0	0	0	0	0
24-4-10/60530.0	IN	1	A	1	0	0	Y	U	0	0	0	0	1	0	0	0	0	1	0	1	0	1	0	0	0	0	0	0	0	0	0	0
24-4-10/60531.0	IN	1	J	0	0	1	Y	U	0	0	0	0	0	0	0	0	0	0	0	0	0	0	0	0	0	0	0	0	0	0	0	0
24-4-10/60532.0	1	1	A	0	1	0	Y	U	0	0	0	0	0	0	0	1	0	0	0	0	0	0	0	0	0	0	0	0	0	0	0	0
24-4-10/60532a	1	1	A	0	1	0	Y	U	0	0	0	0	0	0	0	0	0	0	0	0	0	0	0	0	0	0	0	0	0	0	0	0
24-4-10/60533.0	IN	1	A	0	1	0	Y	U	0	0	0	0	0	0	0	0	1	0	0	0	0	0	0	0	1	0	0	0	0	0	0	0
24-4-10/60534.0	2A	1	A	1	0	0	Y	Y	0	0	0	0	0	0	0	0	0	0	0	0	0	1	0	0	0	0	0	0	0	0	0	0
24-4-10/60535.0	3	1	A	0	0	1	Y	Q	0	0	0	0	0	0	0	0	0	0	0	0	0	0	0	0	0	0	0	0	0	0	0	0
24-4-10/60536.0	B	1	A	0	1	0	Y	Q	0	0	0	0	0	0	0	0	0	0	0	0	0	1	0	0	0	0	0	0	0	0	1	0
24-4-10/60536.1	B	1	J	0	0	1	Y	U	0	0	0	0	0	0	0	0	0	0	0	0	0	0	1	0	0	0	0	0	0	0	0	0
24-4-10/60537.0	3	1	A	0	0	1	Y	U	0	0	0	0	0	0	0	0	0	0	0	0	0	0	0	0	0	0	0	0	0	0	0	0
24-4-10/60538.0	3	1	A	0	1	0	Y	U	0	0	0	0	0	0	0	0	0	0	0	0	0	1	0	0	0	0	0	0	0	0	0	0
24-4-10/60539.0	1	1	A	1	0	0	Y	U	0	1	0	0	0	0	0	1	0	0	0	0	0	0	0	0	0	0	0	0	0	0	0	0
24-4-10/60540.0	3	1	A	0	0	1	N	U	0	0	0	0	0	0	0	0	0	0	0	0	0	0	0	0	0	0	0	0	0	0	0	0
24-4-10/60541.0	IN	1	A	1	0	0	Y	U	0	0	0	0	0	0	0	0	1	0	0	0	0	1	0	0	0	1	0	0	1	0	0	0
24-4-10/60542.0	IN	1	A	0	1	0	Y	N	1	0	1	0	1	0	0	1	0	0	0	1	0	0	0	0	0	0	0	0	0	0	0	0
24-4-10/60543.0	4	1	A	0	1	0	N	Y	0	0	0	0	1	0	0	1	1	0	0	0	0	1	0	0	0	0	0	0	1	0	0	0
24-4-10/60543.1	4	1	A	0	0	1	N	U	0	0	0	0	0	0	0	0	0	0	1	1	0	0	0	0	0	0	0	0	0	0	0	0
24-4-10/60544.0	IN	1	A	0	1	0	N	Y	0	0	0	0	0	0	0	0	0	0	0	0	0	1	0	0	0	0	0	0	0	0	0	0
24-4-10/60545.0	3	1	A	1	0	0	N	N	0	0	0	0	0	0	0	0	1	0	0	0	0	1	0	1	0	0	1	0	0	0	0	0
24-4-10/60545a	3	1	A	0	0	1	N	U	0	0	0	0	0	0	0	0	0	0	0	0	0	0	0	0	0	0	0	0	0	0	0	0
24-4-10/60546.0	4	1	A	0	1	0	Y	Q	0	0	0	0	0	0	0	0	0	0	0	0	0	1	0	0	0	0	0	0	1	0	1	0
24-4-10/60547.0	4	1	A	1	0	0	N	Q	0	0	0	0	0	0	0	0	0	0	0	0	0	1	0	0	0	0	0	0	0	0	0	0
24-4-10/60548.0	3	1	A	1	0	0	N	Q	0	0	0	0	0	0	0	0	1	0	0	0	0	1	0	0	0	0	0	0	0	0	0	0
24-4-10/60549.0	2	1	A	1	0	0	Y	Y	0	1	1	0	0	0	0	1	1	0	0	1	0	0	0	0	1	0	0	0	0	0	0	1
24-4-10/60549.1	2	1	A	0	0	1	Y	U	0	0	0	0	0	0	0	1	0	0	0	0	0	0	0	0	0	0	0	0	0	0	0	0
24-4-10/60550.0	B	1	A	0	1	0	Y	U	0	0	0	0	0	0	0	0	1	0	0	0	0	0	0	0	0	0	0	0	0	0	0	0
24-4-10/60551.0	IN	1	A	0	0	1	Y	U	0	0	0	0	0	0	0	0	0	0	0	0	0	0	0	0	0	0	0	0	0	1	0	0
24-4-10/60552.0	B	1	A	0	1	0	N	Q	0	1	1	0	0	0	0	0	0	0	0	0	0	0	0	0	0	0	0	0	0	0	0	0
24-4-10/60553.0	4	1	J	0	1	0	Y	Y	0	0	0	0	0	0	0	0	0	0	0	0	0	1	0	0	0	0	0	0	0	0	0	1
24-4-10/60554.0	3	1	A	1	0	0	N	U	0	1	1	0	0	0	0	0	0	0	0	0	0	0	0	0	0	0	0	0	0	0	0	0
24-4-10/60555.0	IN	1	A	0	1	0	Y	U	0	0	0	0	1	0	0	0	1	0	0	0	0	1	0	0	0	0	0	0	0	0	0	0
24-4-10/60556.0	4A	1	A	1	0	0	N	Y	0	0	0	0	1	0	0	1	0	1	0	0	0	1	0	0	0	0	0	0	0	0	0	1
24-4-10/60557.0	1	1	A	0	0	1	Y	U	0	0	0	0	0	0	0	0	1	0	0	0	0	0	0	0	0	0	0	0	0	0	0	0
24-4-10/60558.0	4	1	J	1	0	0	N	Q	0	0	0	0	0	0	0	0	0	0	0	0	0	1	0	0	0	0	0	0	0	0	0	0
24-4-10/60559.0	3	1	J	0	0	1	Y	Y	0	0	0	0	0	0	0	0	0	0	0	0	0	0	0	0	0	0	0	0	0	0	0	0
24-4-10/60560.0	IN	1	A	1	0	0	N	Y	1	0	1	0	0	0	1	0	1	0	0	0	0	1	0	0	0	1	0	0	0	0	0	0
24-4-10/60561.0	1	1	A	0	1	0	Y	Y	0	0	0	0	0	0	0	0	1	0	0	0	0	1	0	0	0	0	0	0	0	0	0	0
24-4-10/60562.0	B	1	A	1	0	0	Y	N	1	1	1	0	1	0	0	0	1	0	0	1	0	0	0	0	1	0	0	0	0	0	0	0
24-4-10/60563.0	B	1	A	0	1	0	N	U	0	0	0	0	0	0	0	0	1	0	0	0	0	0	0	0	0	0	0	0	0	0	0	0
24-4-10/60564.0	4	1	A	0	1	0	N	U	0	0	0	0	0	0	0	0	0	0	0	0	0	0	0	0	0	0	0	0	0	0	1	0
24-4-10/60565.0	3	1	A	0	1	0	Y	N	0	0	0	0	0	0	0	1	0	1	0	0	0	0	0	0	1	0	0	0	0	0	0	0
24-4-10/60566.0	5	1	J	0	1	0	Y	Y	0	0	0	0	0	0	0	0	0	0	0	0	0	0	0	0	0	0	0	0	0	0	0	0
24-4-10/60567.0	1A	1	A	0	0	1	N	U	0	0	0	0	0	0	0	0	1	0	0	0	0	0	0	0	1	0	0	0	0	0	0	0
24-4-10/60568.0	3	1	A	0	1	0	Y	Q	0	0	0	0	0	0	0	0	0	0	0	0	0	1	0	0	0	0	0	0	0	0	0	0

				Sex					Trauma				Infection				Arthritis					Physiological stress				Neurological			Tumor		Other path.	
Peabody number	CS	MNI	Age	M	F	U	AFO	Cr mod	T c	T pc	T h	T d	I de	I di	I tr	I g	A sl	A m	A s	A sp	A o	EH	CO	PH	O	Sb sl	Sb m/s	O na	T bo	T o	OP d	OP c/pc
24-4-10/60569.0	6	1	A	1	0	0	N	Y	1	1	1	0	0	0	1	0	0	1	0	1	0	1	0	0	0	0	0	0	0	0	0	0
24-4-10/60570.0	B	1	A	0	1	0	N	Y	0	0	0	0	0	0	0	0	0	0	0	0	0	1	0	0	0	0	0	0	0	0	0	0
24-4-10/60571.0	IN	1	A	0	1	0	N	Y	0	0	0	0	0	0	0	0	0	0	0	0	0	0	0	0	0	0	0	0	0	0	1	0
24-4-10/60572.0	5	1	A	0	1	0	N	Y	0	0	0	0	0	0	0	1	0	0	0	0	0	1	0	0	0	0	0	0	0	1	0	0
24-4-10/60573.0	1	1	A	1	0	0	N	U	0	0	0	0	0	0	0	0	1	0	0	0	0	0	0	0	0	0	0	0	0	0	0	0
24-4-10/60574.0	B	1	A	0	1	0	Y	Y	0	0	0	0	0	0	0	0	0	1	0	0	0	0	0	0	1	0	0	0	0	0	0	0
24-4-10/60575.0	2	1	J	0	1	0	Y	Q	0	0	0	0	0	0	0	0	0	0	0	0	0	1	0	0	0	0	0	0	0	0	0	1
24-4-10/60576.0	4	1	A	0	1	0	N	Q	0	0	0	0	0	0	0	0	0	0	0	0	0	0	0	0	0	0	0	0	0	0	0	0
24-4-10/60577.0	4A	1	A	1	0	0	Y	N	0	0	0	0	1	0	0	0	0	0	0	0	0	1	0	0	0	0	0	0	0	0	0	0
24-4-10/60578.0	1	1	A	0	1	0	Y	Q	0	0	0	0	0	0	0	1	0	0	0	0	0	0	0	0	0	0	0	0	0	0	0	0
24-4-10/60579.0	IN	1	A	1	0	0	Y	Q	0	0	0	0	0	0	0	0	0	0	0	0	0	1	0	0	0	0	0	0	0	0	0	0
24-4-10/60580.0	IN	1	A	0	1	0	Y	Y	0	1	1	0	0	0	0	0	1	0	0	0	0	0	0	0	1	0	0	0	0	0	0	0
24-4-10/60581.0	5	1	A	1	0	0	N	Q	0	0	0	0	1	0	0	0	0	0	0	0	0	0	0	0	0	0	0	0	0	0	0	0
24-4-10/60582.0	IN	1	A	0	1	0	Y	Q	0	0	0	0	1	0	0	0	1	0	0	0	0	0	0	0	0	0	0	0	0	0	0	0
24-4-10/60583.0	B	1	A	1	0	0	Y	Q	0	0	0	0	1	0	0	0	0	1	0	0	0	0	0	0	0	0	0	0	0	0	0	0
24-4-10/60584.0	P	1	J	0	0	1	Y	Q	0	0	0	0	0	0	0	0	0	0	0	0	0	0	0	1	0	0	0	0	0	0	0	0
24-4-10/60585.0	1	1	J	0	0	1	Y	Q	0	0	0	0	0	0	0	0	0	0	0	0	0	0	1	0	0	0	0	0	0	0	0	0
24-4-10/60586.0	IN	1	J	0	0	1	Y	Q	0	0	0	0	0	0	0	0	0	0	0	0	0	0	0	0	0	0	0	0	0	0	0	0
24-4-10/60587.0	1	1	J	0	0	1	Y	Q	0	0	0	0	0	0	0	0	0	0	0	0	0	0	0	0	0	0	0	0	0	0	0	0
24-4-10/60588.0	P	1	A	0	1	0	N	U	0	0	0	0	0	0	0	1	1	0	0	0	0	1	0	0	0	0	0	0	0	0	0	0
24-4-10/60589.0	1	1	A	1	0	0	Y	Q	0	0	0	0	0	0	0	0	1	0	0	0	0	1	0	0	0	0	0	0	0	0	0	0
24-4-10/60590.0	4	1	A	0	1	0	Y	Y	0	0	0	0	1	0	0	0	1	0	0	0	0	1	0	0	0	0	0	1	0	0	0	0
24-4-10/60591.0	1	1	A	0	1	0	Y	U	0	0	0	0	1	0	0	0	0	0	0	0	0	0	0	0	1	0	0	0	0	0	0	0
24-4-10/60592.0	1	1	A	1	0	0	Y	Y	0	0	0	0	1	0	0	0	0	0	0	0	0	0	0	0	0	0	0	0	0	0	0	0
24-4-10/60592.1	IN	1	A	0	0	1	Y	U	0	0	0	0	0	0	0	0	0	0	0	0	0	0	0	0	0	0	0	0	0	0	0	0
24-4-10/60593.0	6	1	A	1	0	0	N	Y	0	0	0	0	1	0	0	0	0	1	0	1	0	1	0	0	1	0	0	0	0	1	0	0
24-4-10/60594.0	4	1	A	0	1	0	Y	N	0	0	0	0	1	0	0	0	0	0	0	0	0	0	0	0	0	0	0	0	0	0	0	0
24-4-10/60595.0	4	1	A	1	0	0	Y	Q	0	0	0	0	0	0	0	0	0	0	0	0	0	0	0	0	0	0	0	0	1	0	0	0
24-4-10/60596.0	P	1	J	0	1	0	Y	Y	1	0	1	0	0	0	0	0	0	0	0	0	0	1	0	1	0	0	0	0	0	0	0	0
24-4-10/60597.0	1A	1	A	0	1	0	Y	Y	0	0	0	0	1	0	0	0	1	0	0	0	0	1	0	0	0	0	0	0	0	0	0	0
24-4-10/60598.0	1	1	A	1	0	0	N	Y	0	0	0	0	1	0	0	0	0	1	0	0	0	0	0	0	1	0	0	0	0	0	0	0
24-4-10/60599.0	1	1	A	1	0	0	Y	U	0	0	0	0	0	0	0	0	0	0	0	0	0	0	0	0	0	1	0	0	0	0	1	0
24-4-10/60600.0	BA	1	J	1	0	0	Y	Y	1	0	0	1	0	0	0	0	0	0	0	0	0	1	0	0	0	0	0	0	0	0	0	0
24-4-10/60601.0	IN	1	A	1	0	0	Y	Y	0	0	0	0	1	0	0	0	1	0	0	0	0	0	0	1	1	0	0	0	0	0	0	0
24-4-10/60602.0	1	1	A	1	0	0	Y	Y	0	0	0	0	1	0	0	0	0	0	1	1	0	0	1	0	0	0	0	0	0	1	0	0
24-4-10/60603.0	4	1	A	1	0	0	Y	U	1	0	1	0	0	0	1	0	1	0	0	0	0	0	0	0	0	0	0	0	0	0	0	0
24-4-10/60605.0	P	1	A	1	0	0	Y	Y	0	0	0	0	1	0	0	0	0	0	0	0	0	0	0	0	1	0	0	0	0	0	0	0
24-4-10/60606.0	2	1	J	0	0	1	Y	Q	1	0	0	0	0	0	0	0	0	0	0	0	0	0	0	0	0	0	0	0	0	0	0	0
24-4-10/60607.0	3	1	A	1	0	0	Y	Y	0	0	0	0	0	0	0	0	0	0	0	0	0	1	0	0	0	0	0	0	0	0	0	0
24-4-10/60608.0	P	1	A	0	1	0	N	Y	0	0	0	0	1	0	0	0	0	1	0	1	0	0	0	0	0	0	0	0	0	0	0	0
24-4-10/60609.0	P	1	A	0	1	0	Y	Y	0	0	0	0	1	0	0	0	1	0	0	0	0	0	0	0	0	0	0	0	0	0	0	0
24-4-10/60610.0	6A	1	A	0	1	0	N	Q	1	0	0	1	1	0	0	0	0	0	0	1	0	1	0	0	0	0	0	0	0	0	0	0
24-4-10/60611.0	B	1	A	0	1	0	Y	Y	0	0	0	0	0	0	0	0	0	0	0	0	0	0	0	0	0	0	0	0	0	0	0	0
24-4-10/60612.0	3	1	A	0	1	0	Y	Q	0	0	0	0	0	1	0	0	0	0	0	0	0	1	0	0	0	0	0	0	0	0	0	0
24-4-10/60613.0	1	1	A	0	1	0	N	U	0	0	0	0	0	0	0	1	1	0	0	0	0	0	0	0	0	0	0	0	0	0	0	0

Peabody number	CS	MNI	Age	Sex			AFO	Cr mod	Trauma				Infection				Arthritis					Physiological stress				Neurological			Tumor		Other path.	
				M	F	U			T c	T pc	T h	T d	I de	I di	I tr	I g	A sl	A m	A s	A sp	A o	EH	CO	PH	O	Sb sl	Sb m/s	O na	T bo	T o	OP d	OP c/pc
24-4-10/60614.0	4	1	A	0	1	0	Y	U	0	0	0	0	0	0	0	0	0	0	0	0	0	0	0	0	0	0	0	0	0	0	0	0
24-4-10/60615.0	6	1	A	0	1	0	Y	Y	0	0	0	0	1	0	0	0	0	0	0	0	1	1	1	0	0	0	0	0	1	0	0	0
24-4-10/60616.0	5	1	A	1	0	0	N	U	0	0	0	0	1	0	0	0	1	0	0	0	0	0	0	0	0	0	0	0	0	0	0	0
24-4-10/60616a	3A	1	A	1	0	0	Y	U	0	0	0	0	1	0	0	0	1	0	0	0	0	0	0	0	0	0	0	0	0	0	0	0
24-4-10/60617.0	3A	1	A	0	0	1	Y	U	0	0	0	0	1	0	0	0	0	0	0	0	0	0	0	0	0	0	0	0	0	0	0	0
24-4-10/60618.0	B	1	A	0	1	0	Y	U	0	0	0	0	0	0	0	0	0	0	0	0	0	0	0	0	0	0	0	0	0	0	0	0
24-4-10/60620.0	B	1	A	0	1	0	Y	Q	0	0	0	0	1	0	0	0	1	0	0	0	0	1	0	0	0	0	0	0	0	0	0	0
24-4-10/60621.0	2	1	J	0	0	1	N	Q	0	0	0	0	0	0	0	0	0	0	0	0	0	1	1	0	0	0	0	0	0	0	0	0
24-4-10/60622.0	4	1	J	0	0	1	N	U	0	0	0	0	0	0	0	0	0	0	0	0	0	0	0	0	0	0	0	0	0	0	0	0
24-4-10/60623.0	BA	1	A	0	1	0	Y	Q	0	1	0	0	0	0	1	0	0	1	0	1	0	1	0	0	0	0	0	0	0	0	0	1
24-4-10/N272.1	4	1	A	0	0	1	Y	U	0	0	0	0	0	0	0	0	0	0	0	0	0	0	0	0	0	0	0	0	0	0	0	0
24-4-10/N500.0	3	1	A	0	0	1	Y	U	0	0	0	0	0	0	0	0	0	0	0	0	0	0	0	0	0	0	0	0	0	0	0	0
25-3-10/60626.0	IN	1	J	0	1	0	N	Q	1	0	0	1	0	0	0	0	0	0	0	0	0	0	1	1	0	0	0	0	0	0	0	0
25-3-10/60627.0	IN	1	J	0	0	1	Y	U	0	0	0	0	0	0	0	0	0	0	0	0	0	0	0	0	0	0	0	0	0	0	0	0
25-3-10/60628a	CH	1	J	0	1	0	N	U	0	0	0	0	0	0	0	0	0	0	0	0	0	1	0	0	0	0	0	0	0	0	0	0
25-3-10/60628b	CH	1	J	0	0	1	N	U	0	0	0	0	0	0	0	0	0	0	0	0	0	0	0	0	0	0	0	0	0	0	0	0
25-3-10/60628c	CH	1	A	0	1	0	N	U	0	0	0	0	0	0	0	0	0	0	0	0	0	0	0	0	0	0	0	0	0	0	0	0
25-3-10/60628d	CH	1	A	1	0	0	N	U	0	0	0	0	0	0	0	0	0	0	0	0	0	0	0	0	0	0	0	0	0	0	0	1
25-3-10/60628e	CH	1	A	0	0	1	N	U	0	0	0	0	0	0	0	0	0	0	0	0	0	1	0	0	0	0	0	0	0	0	0	0
25-3-10/60628f	CH	1	A	0	0	1	N	U	0	0	0	0	0	0	0	0	0	0	0	0	0	0	0	0	0	0	0	0	0	0	0	0
25-3-10/60629.0	IN	1	J	0	0	1	N	U	0	0	0	0	0	0	0	0	0	0	0	0	0	0	0	0	0	0	0	0	0	0	0	0
25-3-10/60629a	IN	1	A	0	0	1	N	U	0	0	0	0	0	0	0	0	0	0	0	0	0	0	0	0	0	0	0	0	0	0	0	0
25-3-10/60630.0	3	1	A	0	1	0	N	N	0	0	0	0	0	0	0	0	0	0	0	0	0	1	0	0	0	0	0	0	0	0	0	0
25-3-10/60631.0	3	1	A	0	1	0	Y	Y	1	0	1	0	0	0	0	0	0	0	0	0	0	1	0	0	0	0	0	0	0	0	0	0
25-3-10/60631.1	3	1	A	0	0	1	Y	U	0	0	0	0	0	0	0	1	0	0	0	1	1	0	0	0	0	0	0	0	0	0	0	0
25-3-10/60632.0	IN	1	A	0	0	1	Y	U	0	0	0	0	0	0	0	0	0	0	0	0	0	0	0	0	0	0	0	0	0	0	0	0
25-3-10/60633.0	CH	1	A	1	0	0	N	Y	0	0	0	0	0	1	0	0	0	0	0	0	0	1	0	0	0	0	0	0	0	0	0	0
25-3-10/60634.0	CH	1	A	1	0	0	Y	Q	0	1	1	0	1	0	0	0	1	0	0	0	0	1	0	0	0	0	0	0	0	0	0	0
25-3-10/60651.0	CH	1	A	1	0	0	N	N	1	0	0	0	0	1	0	0	0	0	0	0	0	0	0	0	0	0	0	0	0	0	0	0
25-3-10/60652.0	CH	1	A	0	1	0	N	Y	0	0	0	0	0	0	0	0	0	0	0	0	0	0	0	0	0	0	0	0	0	0	0	0
25-3-10/60653.0	CH	1	A	0	1	0	N	Q	0	0	0	0	1	0	0	0	1	0	0	0	0	0	0	0	0	0	0	0	0	0	0	0
25-3-10/60654.0	CH	1	A	1	0	0	N	N	0	0	0	0	1	0	0	0	0	0	0	0	0	0	0	0	0	0	0	0	0	0	0	0
25-3-10/60655.0	1	1	A	1	0	0	N	Y	0	0	0	0	1	0	0	0	0	0	0	0	0	0	0	0	0	0	0	0	0	0	0	0
25-3-10/60656.0	3	1	A	0	1	0	N	Y	0	0	0	0	0	0	0	0	0	0	0	0	0	1	0	0	0	0	0	0	0	0	0	0
25-3-10/60657.0	3	1	A	0	1	0	Y	Y	0	0	0	0	0	0	0	0	0	0	0	0	0	1	0	0	0	0	0	0	0	0	0	0
25-3-10/60661.0	BA	1	A	0	1	0	Y	Y	0	0	0	0	1	0	0	0	1	0	0	1	0	0	0	0	0	0	0	0	1	0	0	0
25-3-10/60662.0	IN	1	A	1	0	0	N	Y	0	1	1	0	0	0	0	0	0	0	0	0	0	1	0	0	0	0	0	0	0	0	0	0
25-3-10/60663.0	CH	1	A	1	0	0	N	Q	1	1	1	0	0	0	0	1	0	0	1	1	1	0	0	0	0	0	0	0	0	0	0	0
25-3-10/60663.1	CH	1	A	0	0	1	N	U	0	0	0	0	0	0	0	0	0	0	0	1	1	0	0	0	0	0	0	0	0	0	0	0
25-3-10/60664.0	1A	1	A	0	1	0	N	U	0	0	0	0	0	0	0	0	0	0	0	1	0	0	0	0	0	0	0	0	0	0	0	1
25-3-10/60665.0	IN	1	A	1	0	0	Y	U	1	0	0	0	0	1	0	0	0	0	0	0	0	0	0	0	1	0	1	0	0	0	0	0
25-3-10/60666.0	3A	1	A	1	0	0	Y	U	0	1	1	0	0	0	1	0	1	0	0	1	0	0	0	0	1	0	0	0	0	0	0	0
25-3-10/60667.0	CH	1	A	1	0	0	N	Y	0	0	0	0	0	0	0	0	0	0	0	0	0	1	0	0	0	0	0	0	0	0	0	0
25-3-10/60668.0	IN	1	A	0	1	0	N	Y	0	0	0	0	1	0	0	0	0	0	0	1	0	0	0	0	1	0	0	0	0	0	0	0
25-3-10/60668.1	IN	1	A	0	0	1	N	U	0	1	0	1	0	0	0	0	1	0	0	0	0	0	0	0	0	0	0	0	0	0	0	0

Peabody number	CS	MNI	Age	Sex			AFO	Cr mod	Trauma				Infection				Arthritis					Physiological stress				Neurological			Tumor		Other path.	
				M	F	U			T c	T pc	T h	T d	I de	I di	I tr	I g	A sl	A m	A s	A sp	A o	EH	CO	PH	O	Sb sl	Sb m/s	O na	T bo	T o	OP d	OP c/pc
25-3-10/60668a	IN	1	A	1	0	0	N	Y	0	0	0	0	0	0	0	0	0	0	0	0	0	0	0	0	0	0	0	0	0	0	0	0
25-3-10/60670.0	IN	1	A	0	1	0	N	Y	0	0	0	0	0	0	0	0	1	0	0	0	0	0	0	0	0	0	0	0	0	0	0	0
25-3-10/60671.0	IN	1	A	0	1	0	Y	U	0	0	0	0	0	0	0	0	1	0	0	0	0	0	0	0	0	0	0	0	0	0	0	0
25-3-10/60672.0	CH	1	J	0	1	0	N	U	0	0	0	0	0	0	0	0	0	0	0	0	0	0	0	0	0	0	0	0	0	0	0	0
25-3-10/60673.0	IN	1	A	0	1	0	Y	Y	0	0	0	0	0	0	0	0	0	0	0	0	0	0	0	0	0	0	0	0	0	0	0	0
25-3-10/60673.1	IN	1	A	1	0	0	Y	U	0	0	0	0	0	0	0	0	0	1	0	0	0	0	0	0	1	0	0	0	0	0	0	0
25-3-10/60673.2	IN	1	A	1	0	0	Y	U	0	0	0	0	0	0	0	0	0	0	0	0	0	0	0	0	0	0	0	0	0	0	0	0
25-3-10/60674.0	IN	1	A	1	0	0	Y	Q	0	0	0	0	0	0	0	1	0	0	0	0	0	0	0	0	0	0	0	0	0	0	0	0
25-3-10/60674.1	IN	1	J	0	1	0	Y	U	0	0	0	0	0	0	0	0	0	0	0	0	0	0	0	0	0	0	0	0	0	0	0	0
25-3-10/60674.2	IN	1	A	0	1	0	Y	U	0	0	0	0	0	0	0	0	0	0	0	0	0	0	0	0	0	0	1	0	0	0	0	0
25-3-10/60675.0	IN	1	A	1	0	0	N	N	0	0	0	0	0	0	0	1	1	0	0	1	0	1	0	0	0	0	0	0	0	1	0	1
25-3-10/60676.0	IN	1	A	0	1	0	Y	U	0	0	0	0	0	0	0	0	0	0	0	1	0	0	0	0	0	0	0	0	0	0	0	0
25-3-10/60677.0	IN	1	A	1	0	0	N	Y	0	0	0	0	0	0	0	0	0	0	0	1	0	0	0	0	0	1	1	0	0	0	0	0
25-3-10/60677a	IN	1	A	0	1	0	N	Y	0	0	0	0	0	0	0	0	0	0	0	0	1	1	0	0	0	0	0	0	0	0	0	0
25-3-10/60678.0	IN	1	A	1	0	0	Y	U	0	0	0	0	0	0	0	0	1	0	0	0	0	0	0	0	0	0	0	0	0	0	0	0
25-3-10/60679.0	3	1	A	0	1	0	Y	Q	0	0	0	0	1	0	0	0	1	0	0	1	0	0	0	0	0	0	0	0	0	0	0	0
25-3-10/60679a	3	1	A	1	0	0	Y	Q	0	0	0	0	1	0	0	0	0	0	0	0	0	0	0	1	0	0	0	0	1	0	0	0
25-3-10/60680.0	CH	1	J	1	0	0	N	Q	0	0	0	0	0	0	0	0	0	0	0	0	0	0	0	0	0	0	0	0	0	0	0	1
25-3-10/60681.0	CH	1	A	0	1	0	N	Q	1	0	1	0	1	0	0	0	0	0	1	0	0	0	0	0	1	0	0	0	0	0	0	0
25-3-10/60683.0	CH	1	A	1	0	0	N	U	0	0	0	0	0	0	0	0	0	0	0	1	0	0	0	0	0	0	0	0	0	0	0	0
25-3-10/60683a	CH	1	A	0	1	0	N	U	0	0	0	0	0	0	0	0	0	0	0	0	0	0	0	0	0	0	0	0	0	0	0	0
25-3-10/60685.0	CH	1	A	0	1	0	N	N	0	0	0	0	1	0	0	0	0	0	0	0	0	1	0	0	0	0	0	0	0	0	0	0
25-3-10/60685.1	CH	1	A	0	0	1	N	U	0	0	0	0	0	0	0	0	0	0	0	0	0	0	0	0	0	0	0	0	0	0	0	0
25-3-10/60685.2	CH	1	A	0	0	1	N	U	0	0	0	0	0	0	0	1	0	0	0	0	0	0	0	0	0	0	0	0	0	0	0	0
25-3-10/60686.0	3A	1	A	1	0	0	Y	Y	0	0	0	0	0	0	0	1	0	0	0	0	0	0	0	0	0	0	0	0	0	0	0	0
25-3-10/60687.0	IN	1	A	1	0	0	N	Q	0	0	0	0	1	0	0	0	0	0	1	1	0	0	0	0	0	1	0	0	0	1	0	0
25-3-10/60688.0	IN	1	J	1	0	0	N	Q	0	0	0	0	0	0	0	1	0	0	0	0	0	1	0	0	0	1	0	0	0	0	0	0
25-3-10/60689.0	IN	1	A	0	1	0	Y	Q	0	0	0	0	0	0	0	0	0	0	0	1	0	0	0	0	0	0	0	0	0	0	1	0
25-3-10/60691.0	CH	1	A	1	0	0	Y	Q	1	1	0	0	1	0	1	0	0	1	0	1	0	0	0	0	0	0	0	1	0	0	0	0
25-3-10/60692.0	IN	1	J	0	0	1	N	Q	0	0	0	0	1	0	0	0	0	0	0	0	0	1	0	0	0	0	0	0	0	0	0	0
25-3-10/60693.0	6A	1	A	1	0	0	N	U	0	0	0	0	0	0	0	0	0	0	0	1	0	0	0	0	1	0	0	0	0	0	0	0
25-3-10/60694.0	IN	1	A	1	0	0	Y	Y	0	0	0	0	1	0	0	0	0	0	0	1	1	1	0	0	0	0	0	0	0	0	0	0
25-3-10/60695.0	1A	1	A	1	0	0	N	Q	0	1	1	0	0	0	1	0	0	0	1	1	0	0	0	0	0	0	0	0	0	1	0	0
25-3-10/60696.0	CH	1	A	0	1	0	N	Y	0	0	0	0	0	0	0	0	0	0	1	1	0	0	0	0	1	0	0	0	0	0	0	0
25-3-10/60697.0	IN	1	A	0	1	0	Y	Y	0	0	0	0	0	0	0	0	0	0	0	0	0	0	0	0	0	0	0	0	0	0	0	0
25-3-10/60698.0	IN	1	A	1	0	0	Y	Y	0	0	0	0	0	0	0	0	0	0	0	0	0	0	0	0	1	1	0	0	0	0	0	0
25-3-10/60699.0	IN	1	A	1	0	0	Y	Y	1	0	0	0	1	0	0	0	0	0	1	1	1	0	0	0	0	0	0	0	0	1	0	0
25-3-10/60700.0	IN	1	A	1	0	0	Y	Y	0	0	0	0	0	0	0	1	1	0	0	0	0	0	0	0	0	0	0	0	1	0	0	0
25-3-10/60701.0	1A	1	A	1	0	0	Y	Q	0	0	0	0	0	0	0	0	0	0	0	0	0	0	0	0	0	0	0	0	0	0	0	0
25-3-10/60702.0	5	1	A	0	0	1	Y	U	0	0	0	0	0	0	0	0	0	0	0	0	0	0	0	0	0	0	0	0	0	0	0	0
25-3-10/60703.0	IN	1	A	0	1	0	N	Q	0	1	1	0	1	0	0	0	0	0	0	0	0	0	0	1	0	1	0	0	0	0	0	0
25-3-10/60703a	IN	1	A	0	1	0	N	U	0	0	0	0	0	0	0	0	0	0	0	0	0	0	0	0	0	0	0	0	0	0	0	0
25-3-10/60704.0	CH	1	A	0	1	0	N	N	0	1	1	0	0	0	0	0	0	0	0	0	0	0	0	0	0	0	0	0	0	0	0	0
25-3-10/60704a	IN	1	A	0	1	0	N	Y	0	0	0	0	1	0	0	0	0	0	0	0	0	0	0	0	0	0	0	0	0	0	0	0
25-3-10/60706.0	CH	1	A	0	1	0	N	Q	0	0	0	0	0	0	0	1	0	1	0	1	0	0	0	0	0	0	0	0	0	0	0	0

Peabody number	CS	MNI	Age	Sex			AFO	Cr mod	Trauma				Infection				Arthritis					Physiological stress				Neurological			Tumor		Other path.	
				M	F	U			T c	T pc	T h	T d	I de	I di	I tr	I g	A sl	A m	A s	A sp	A o	EH	CO	PH	O	Sb sl	Sb m/s	O na	T bo	T o	OP d	OP c/pc
25-3-10/60708.0	CH	1	J	0	1	0	Y	Q	0	0	0	0	1	1	0	0	0	0	0	0	0	1	0	0	0	0	0	0	0	0	0	0
25-3-10/60709.0	2A	1	A	0	0	1	N	U	0	0	0	0	1	0	0	0	0	0	0	0	0	0	0	0	0	0	0	0	0	0	0	0
25-3-10/60710.0	IN	1	A	1	0	0	Y	Y	0	0	0	0	0	0	0	0	1	0	0	0	0	0	0	0	0	1	0	0	0	0	0	0
25-3-10/60711.0	CH	1	A	1	0	0	N	Q	1	1	1	1	0	0	0	0	0	0	0	0	1	0	0	0	0	1	0	1	0	0	0	0
25-3-10/60712.0	CH	1	J	0	0	1	N	Q	0	0	0	0	0	0	0	0	0	0	0	0	0	0	0	0	0	0	0	0	0	0	0	0
25-3-10/60713.0	3	1	A	0	1	0	N	U	0	0	0	0	1	0	0	0	0	0	0	0	0	0	0	0	0	0	0	0	0	0	0	0
25-3-10/60713.1	3	1	J	0	0	1	N	U	0	0	0	0	0	0	0	0	0	0	0	0	0	0	0	0	0	0	0	0	0	0	0	0
25-3-10/60714.0	CH	1	A	1	0	0	N	Y	0	0	0	0	1	0	0	0	0	0	1	1	0	0	0	0	1	0	0	0	0	0	0	0
25-3-10/60715.0	IN	1	A	0	1	0	Y	Y	0	0	0	0	0	0	0	0	1	0	0	0	0	1	0	0	0	0	0	0	0	0	0	0
25-3-10/60716.0	IN	1	J	1	0	0	Y	Y	0	0	0	0	0	0	0	0	0	0	0	0	0	0	0	0	0	0	0	0	0	0	0	0
25-3-10/60717.0	CH	1	A	1	0	0	N	Y	1	0	0	1	1	0	0	0	1	0	0	0	0	1	0	0	0	1	0	0	0	0	0	0
25-3-10/60718.0	CH	1	A	1	0	0	N	Q	1	0	1	0	0	0	1	0	0	0	0	0	0	0	0	1	0	0	0	0	0	0	1	0
25-3-10/60719.0	CH	1	J	0	1	0	N	Y	1	0	1	0	0	0	0	0	0	0	0	0	0	0	1	0	0	0	0	0	0	0	0	0
25-3-10/60720.0	CH	1	A	1	0	0	N	Q	0	0	0	0	1	0	0	0	0	0	0	0	0	0	0	0	0	0	0	0	0	0	1	0
25-3-10/60721.0	CH	1	A	1	0	0	N	Q	0	0	0	0	0	0	0	0	0	0	0	0	0	0	0	0	0	0	0	0	0	0	0	0
25-3-10/60722.0	IN	1	A	0	1	0	N	Y	0	0	0	0	0	0	0	0	0	0	0	1	0	1	0	0	1	0	0	0	0	0	0	0
25-3-10/60723.0	CH	1	A	0	1	0	N	Q	0	0	0	0	0	0	0	0	0	0	0	0	0	0	0	0	0	0	0	0	0	0	0	0
25-3-10/60724.0	CH	1	A	1	0	0	N	Q	0	0	0	0	0	0	0	0	0	0	0	0	0	0	0	0	0	0	0	0	0	0	0	0
25-3-10/60725.0	CH	1	A	1	0	0	N	U	0	0	0	0	0	0	0	0	1	0	0	0	0	0	0	0	0	1	0	0	0	0	0	0
25-3-10/60726.0	IN	1	J	0	0	1	Y	Q	0	0	0	0	0	0	0	0	0	0	0	0	0	0	1	0	0	0	0	0	0	0	0	0
25-3-10/60727.0	P	1	A	0	1	0	N	Q	0	0	0	0	1	0	0	0	0	0	1	1	0	0	0	0	1	1	0	0	1	0	0	0
25-3-10/60728.0	CH	1	A	0	1	0	N	U	0	1	1	0	0	0	0	0	1	0	0	0	0	0	0	0	0	0	0	0	0	0	0	0
25-3-10/60729.0	IN	1	A	1	0	0	N	U	0	1	1	0	0	0	0	1	1	0	0	1	0	0	0	0	0	0	0	0	0	0	0	0
25-3-10/60730.0	CH	1	J	0	0	1	N	Q	0	0	0	0	0	0	0	0	0	0	0	0	0	1	0	0	0	0	0	0	0	0	0	0
25-3-10/60731.0	IN	1	A	0	1	0	Y	Y	1	0	0	1	1	0	0	0	0	0	1	0	0	0	0	0	1	0	0	0	0	0	0	0
25-3-10/60732.0	IN	1	J	0	1	0	Y	U	0	0	0	0	0	0	0	0	0	0	0	0	0	0	0	0	0	0	0	0	0	0	0	0
25-3-10/60733.0	IN	1	A	1	0	0	Y	Q	0	0	0	0	0	0	0	0	0	0	0	1	0	1	0	0	0	0	0	0	0	0	0	0
25-3-10/60735.0	CH	1	J	1	0	0	N	Q	0	0	0	0	0	0	0	0	0	0	0	1	0	0	0	0	0	0	0	1	0	0	0	0
25-3-10/60735a	CH	1	J	0	0	1	Y	U	0	0	0	0	0	0	0	0	0	0	0	0	0	0	0	0	0	0	0	0	0	0	0	0
25-3-10/60735b	CH	1	A	1	0	0	Y	U	0	0	0	0	0	0	0	0	0	0	0	0	0	0	0	0	0	0	0	0	0	0	0	0
25-3-10/60736.0	5	1	A	0	1	0	N	Y	0	0	0	0	1	0	0	0	0	0	0	1	0	0	0	0	0	0	0	0	0	0	0	0
25-3-10/60737.0	CH	1	A	1	0	0	N	N	0	0	0	0	0	0	0	1	0	0	0	0	0	0	0	0	0	0	0	0	0	0	0	0
25-3-10/60738.0	3	1	A	0	1	0	N	Y	0	0	0	0	1	0	0	0	0	0	0	0	0	1	0	0	0	0	0	0	0	0	0	0
25-3-10/60739.0	IN	1	A	0	1	0	Y	U	0	0	0	0	0	0	0	0	1	0	0	1	0	0	0	0	1	0	0	0	0	0	0	0
25-3-10/60740.0	B	1	A	0	1	0	N	Q	0	0	0	0	0	0	0	0	0	0	0	0	0	0	0	0	1	0	0	0	0	0	0	0
25-3-10/60741.0	IN	1	A	1	0	0	Y	Q	0	0	0	0	0	0	0	0	0	0	0	0	1	1	0	0	0	0	0	0	0	0	0	0
25-3-10/60742.0	IN	1	J	1	0	0	N	U	0	0	0	0	0	0	0	0	0	0	0	0	0	0	0	0	0	0	0	0	0	0	0	0
25-3-10/60743.0	5	1	A	1	0	0	N	Q	1	0	1	0	1	0	0	0	0	0	1	0	1	0	0	0	0	1	0	0	0	0	0	0
25-3-10/60744.0	IN	1	A	0	1	0	N	Y	0	0	0	0	0	0	0	0	1	0	0	0	0	1	0	0	1	0	0	0	0	0	0	0
25-3-10/60745.0	B	1	A	1	0	0	Y	U	0	0	0	0	0	0	0	0	0	0	0	0	0	0	0	0	0	0	0	0	0	0	0	0
25-3-10/60746.0	IN	1	A	0	1	0	Y	Y	0	0	0	0	1	0	0	0	0	0	0	1	0	0	0	0	0	0	0	0	0	0	0	0
25-3-10/60747.0	CH	1	A	1	0	0	N	Q	0	0	0	0	0	0	0	0	0	0	0	0	0	1	0	0	0	0	0	0	0	0	0	0
25-3-10/60748.0	CH	1	A	1	0	0	N	Q	1	0	1	1	0	0	0	0	0	0	0	0	0	1	0	0	0	0	0	0	0	0	0	0
25-3-10/60749.0	CH	1	A	1	0	0	N	Q	1	0	1	0	0	0	0	1	0	0	0	0	0	1	0	0	0	0	0	0	0	0	0	0
25-3-10/60750.0	CH	1	A	0	1	0	N	Q	1	0	0	1	0	0	0	0	0	0	0	0	0	0	0	0	0	0	0	0	0	0	0	0

Peabody number	CS	MNI	Age	Sex			AFO	Cr mod	Trauma				Infection				Arthritis					Physiological stress				Neurological			Tumor		Other path.	
				M	F	U			T c	T pc	T h	T d	I de	I di	I tr	I g	A sl	A m	A s	A sp	A o	EH	CO	PH	O	Sb sl	Sb m/s	O na	T bo	T o	OP d	OP c/pc
25-3-10/60750b	CH	1	A	1	0	0	N	U	0	0	0	0	0	0	0	0	0	0	0	0	0	1	0	0	0	0	0	0	0	0	0	0
25-3-10/60751.0	CH	1	A	1	0	0	N	N	1	0	0	1	0	0	0	0	0	0	0	0	0	0	0	0	0	0	0	0	0	0	1	0
25-3-10/60752.0	3	1	A	0	1	0	N	Q	0	1	1	0	1	0	0	0	1	0	0	0	0	0	0	0	0	0	0	0	0	0	0	0
25-3-10/60752a	3	1	J	0	1	0	N	U	0	0	0	0	0	0	0	0	0	0	0	0	0	0	0	0	0	0	0	0	0	1	0	0
25-3-10/60753.0	CH	1	A	1	0	0	N	Y	0	0	0	0	0	0	0	0	0	0	0	0	0	0	0	0	0	0	0	0	0	0	0	0
25-3-10/60754.0	CH	1	A	1	0	0	Y	U	0	0	0	0	0	0	0	0	1	0	0	1	0	0	0	0	0	0	0	0	0	0	0	0
25-3-10/60754a	CH	1	A	0	0	1	Y	U	0	0	0	0	0	0	0	0	0	0	0	0	0	0	0	0	0	0	0	0	0	0	0	0
25-3-10/60756.0	IN	1	A	0	1	0	Y	U	0	1	1	0	0	0	0	0	1	0	0	0	0	0	0	0	1	0	0	0	0	0	0	0
25-3-10/60757.0	CH	1	A	1	0	0	N	Y	1	0	0	1	0	0	0	0	0	0	0	0	0	0	0	0	0	0	0	0	0	0	0	0
25-3-10/60758.0	CH	1	J	0	0	1	N	Q	1	0	0	1	0	0	0	0	0	0	0	0	0	0	0	0	0	0	0	0	0	0	0	0
25-3-10/60759.0	CH	1	A	0	1	0	N	N	0	0	0	0	1	0	0	0	0	0	0	0	0	0	0	0	0	0	0	0	0	0	0	0
25-3-10/60760.0	CH	1	J	0	1	0	N	Q	1	0	0	1	0	0	0	0	0	0	0	0	0	0	0	0	0	0	0	0	0	0	0	0
25-3-10/60761.0	CH	1	A	1	0	0	N	Q	1	0	0	0	0	0	0	0	0	0	0	0	0	0	0	0	0	0	0	0	0	0	0	0
25-3-10/60762.0	CH	1	A	0	1	0	N	U	1	0	0	1	0	0	0	0	1	0	0	0	0	0	0	0	0	0	0	0	0	0	0	0
25-3-10/60763.0	IN	1	A	1	0	0	Y	Y	1	0	1	0	0	0	0	0	0	0	0	0	0	0	0	0	0	0	0	0	1	0	0	0
25-3-10/60764.0	CH	1	A	1	0	0	N	Q	0	0	0	0	0	0	0	0	0	0	0	0	0	0	0	0	0	0	0	0	0	0	0	0
25-3-10/60765.0	CH	1	A	0	1	0	N	N	0	0	0	0	1	0	0	0	0	0	0	0	0	0	0	0	0	0	0	0	0	0	0	0
25-3-10/60766.0	CH	1	A	1	0	0	N	Q	0	0	0	0	0	0	0	0	0	0	0	0	0	0	0	0	0	0	0	0	0	0	0	0
25-3-10/60767.0	CH	1	A	0	1	0	N	Q	0	0	0	0	1	0	0	1	0	0	0	0	0	0	0	0	0	0	0	0	0	0	0	0
25-3-10/60768.0	CH	1	A	1	0	0	N	Q	0	0	0	0	0	0	0	1	0	0	0	0	0	0	0	0	0	0	0	0	0	0	0	0
25-3-10/60769.0	CH	1	A	1	0	0	N	Q	0	0	0	0	0	0	0	0	0	0	0	0	0	0	0	0	0	0	0	0	0	0	0	0
25-3-10/60770.0	CH	1	A	1	0	0	N	Y	0	0	0	0	0	0	0	1	0	0	0	0	0	0	0	0	0	0	0	0	0	0	0	0
25-3-10/60771.0	CH	1	A	0	1	0	N	U	0	0	0	0	0	0	0	0	0	0	0	0	0	0	0	0	0	0	0	0	0	0	0	0
25-3-10/60771.1	CH	1	J	0	0	1	N	U	0	0	0	0	0	0	0	0	0	0	0	0	0	0	0	0	0	0	0	0	0	0	0	0
25-3-10/60772.0	IN	1	A	0	1	0	Y	Y	0	0	0	0	0	0	0	0	0	0	0	0	1	0	1	0	0	0	0	0	0	0	1	0
25-3-10/60773.0	IN	1	A	1	0	0	Y	Q	0	0	0	0	0	0	0	0	1	0	0	0	0	1	0	0	1	1	0	0	0	0	0	0
25-3-10/60774.0	B	1	A	0	1	0	Y	Y	0	0	0	0	1	0	0	0	1	0	0	0	0	0	0	0	0	0	0	0	0	0	0	0
25-3-10/60775.0	IN	1	A	1	0	0	Y	Y	1	0	0	1	0	0	0	1	0	0	0	0	0	1	0	0	0	0	0	0	0	0	1	0
25-3-10/60777.0	IN	1	A	0	0	1	N	U	0	0	0	0	0	0	0	0	0	1	0	0	0	0	0	0	0	0	0	0	0	0	0	0
25-3-10/60778.0	IN	1	J	1	0	0	N	Y	0	1	1	0	0	0	0	0	0	0	0	0	0	0	0	0	0	0	0	0	0	0	0	0
25-3-10/60779.0	CH	1	A	0	1	0	N	Y	0	0	0	0	1	1	0	0	0	0	0	0	1	0	0	0	1	0	0	0	0	0	0	0
25-3-10/60781.0	IN	1	A	1	0	0	Y	Y	0	0	0	0	0	0	0	0	0	0	0	0	0	0	0	0	0	0	0	0	0	0	0	0
25-3-10/60781.1	IN	1	A	0	1	0	N	U	0	0	0	0	0	0	0	0	0	0	0	0	0	0	0	0	1	0	0	0	0	0	0	0
25-3-10/60782.0	CH	1	A	0	1	0	N	U	0	0	0	0	0	0	0	0	0	0	0	0	0	0	0	0	0	1	0	0	0	0	0	0
25-3-10/61122.0	CH	1	A	0	1	0	N	Y	1	0	0	1	1	0	0	0	0	0	0	0	0	0	0	0	0	0	0	0	0	0	0	0
25-3-10/61152.0	B	1	A	0	1	0	N	Y	0	0	0	0	1	0	0	0	0	0	0	0	0	1	0	0	0	0	0	0	0	0	0	0
25-3-10/61153.0	B	1	A	0	1	0	N	Q	0	0	0	0	1	0	0	1	0	0	0	0	0	1	0	0	0	0	0	0	0	0	0	0
25-3-10/61154.0	B	1	A	1	0	0	N	Q	0	0	0	0	0	0	0	0	0	0	0	0	0	0	0	0	0	0	0	0	0	0	0	0
25-3-10/61155.0	CH	1	A	0	1	0	N	Q	0	0	0	0	1	0	0	0	0	0	0	0	0	0	0	0	0	0	0	0	0	0	0	0
25-3-10/61172.0	B	1	J	0	0	1	N	Q	0	0	0	0	0	0	0	0	0	0	0	0	0	1	1	0	0	0	0	0	0	0	0	0
26-3-10/61179.0	B	1	A	0	1	0	N	U	0	0	0	0	0	0	0	0	0	0	0	1	0	1	0	0	0	0	0	0	0	0	0	0
26-3-10/61180.0	B	1	A	0	1	0	N	Q	0	0	0	0	0	0	0	0	0	0	0	0	0	0	0	0	0	0	0	0	0	0	0	0
26-3-10/61181.0	B	1	A	1	0	0	N	Q	0	0	0	0	0	0	0	0	1	0	0	1	1	0	0	0	0	0	0	0	0	0	0	1
26-3-10/61182.0	B	1	A	0	1	0	N	Y	0	0	0	0	0	0	0	0	0	0	0	0	0	0	0	0	0	0	0	0	0	0	0	0
26-3-10/61183.0	B	1	A	1	0	0	Y	U	0	0	0	0	0	0	0	0	1	0	0	0	0	0	0	0	0	0	0	0	0	0	0	0

				Sex					Trauma				Infection				Arthritis					Physiological stress				Neurological			Tumor		Other path.	
Peabody number	CS	MNI	Age	M	F	U	AFO	Cr mod	T c	T pc	T h	T d	I de	I di	I tr	I g	A sl	A m	A s	A sp	A o	EH	CO	PH	O	Sb sl	Sb m/s	O na	T bo	T o	OP d	OP c/pc
26-3-10/61184.0	B	1	A	0	1	0	Y	Y	0	0	0	0	0	0	0	0	0	0	0	0	0	0	0	0	0	0	0	0	0	0	0	0
26-3-10/61185.0	B	1	J	0	0	1	N	U	0	0	0	0	0	0	0	0	0	0	0	0	0	0	0	0	0	0	0	0	0	0	0	0
26-3-10/61186.0	B	1	J	0	0	1	N	U	0	0	0	0	0	0	0	0	0	0	0	0	0	0	0	0	0	0	0	0	0	0	0	0
26-3-10/61187.0	B	1	A	1	0	0	Y	U	0	0	0	0	0	0	0	0	0	0	0	0	0	1	0	0	0	0	0	0	0	0	0	0
26-3-10/61187.2	B	1	J	0	0	1	Y	U	0	0	0	0	0	0	0	0	0	0	0	0	0	1	0	0	0	0	0	0	0	0	0	0
26-3-10/61188.0	B	1	J	0	0	1	N	U	0	0	0	0	0	0	0	0	0	0	0	0	0	0	0	0	0	0	0	0	0	0	0	0
26-3-10/61191.0	B	1	A	0	0	1	N	U	0	0	0	0	0	0	0	0	0	0	0	0	0	0	0	0	0	0	0	0	0	0	0	0
26-3-10/61197.0	CH	1	J	1	0	0	N	Q	1	1	1	1	0	1	0	0	0	0	0	0	1	0	0	0	0	0	0	0	0	0	0	0
26-3-10/61198.0	B	1	A	0	1	0	N	Y	0	0	0	0	1	0	0	0	1	0	0	1	0	0	0	0	0	0	0	0	0	0	0	0
26-3-10/61199.0	B	1	J	0	0	1	N	U	0	0	0	0	0	0	0	0	0	0	0	0	0	0	0	0	0	0	0	0	0	0	0	0
26-3-10/61200.0	B	1	J	0	0	1	N	U	0	0	0	0	0	0	0	0	0	0	0	0	0	0	0	0	0	0	0	0	0	0	0	0
26-3-10/61201.0	B	1	A	0	0	1	N	U	0	0	0	0	0	0	0	0	0	0	0	1	0	0	0	0	0	0	0	0	0	0	0	0
26-3-10/61202.0	B	1	J	0	0	1	N	U	0	0	0	0	0	0	0	0	0	0	0	0	0	0	0	0	0	0	0	0	0	0	0	0
26-3-10/61203.0	B	1	A	1	0	0	Y	U	0	0	0	0	0	0	0	0	0	0	0	0	0	0	0	0	0	0	0	0	0	0	0	0
26-3-10/61204.0	B	1	A	1	0	0	N	U	0	0	0	0	0	0	0	0	0	0	0	0	0	1	0	0	0	0	0	0	0	0	1	0
26-3-10/61216.0	B	1	A	0	1	0	N	Q	0	0	0	0	0	0	0	0	1	0	0	1	0	0	0	0	0	0	0	0	0	0	0	0
27-1-10/60684.0	B	1	J	0	1	0	N	Q	0	0	0	0	0	0	0	0	0	0	0	0	0	0	0	0	0	0	0	0	0	0	0	0
27-1-10/61261.1	CH	1	J	1	0	0	N	U	0	0	0	0	1	0	0	0	0	0	0	0	0	1	0	0	0	0	0	0	0	0	0	0
27-1-10/61261.2	CH	1	A	0	1	0	N	U	0	0	0	0	0	0	0	0	0	0	0	0	0	0	0	0	0	0	0	0	0	0	0	0
27-1-10/61261.3	CH	1	A	1	0	0	N	U	0	0	0	0	0	0	0	0	0	0	0	0	0	0	0	0	0	0	0	0	0	0	0	0
27-1-10/61262.0	B	1	A	0	0	1	N	U	0	0	0	0	0	0	0	0	0	0	0	0	0	0	0	0	0	0	0	0	0	0	0	0
27-1-10/61265.0	B	1	J	0	0	1	Y	U	0	0	0	0	0	0	0	0	0	0	0	0	0	0	0	0	0	0	0	0	0	0	0	0
27-1-10/61266.0	B	1	A	0	0	1	Y	U	0	0	0	0	0	0	0	0	0	0	0	0	0	0	0	0	0	0	0	0	0	0	0	0
27-1-10/61268.0	B	1	J	0	0	1	N	Q	0	0	0	0	0	0	0	0	0	0	0	0	0	0	0	0	0	0	0	0	0	0	0	0
27-1-10/61269.0	B	1	J	0	0	1	N	U	0	0	0	0	0	0	0	0	0	0	0	0	0	0	0	0	0	0	0	0	0	0	0	0
27-1-10/61270.0	B	1	J	0	0	1	N	Q	0	0	0	0	0	0	0	0	0	0	0	0	0	0	0	0	0	0	0	0	0	0	0	0
27-1-10/61271.0	B	1	J	0	0	1	Y	U	0	0	0	0	0	0	0	0	0	0	0	0	0	0	0	0	0	0	0	0	0	0	0	0
27-1-10/61272.0	B	1	J	0	0	1	N	U	0	0	0	0	0	0	0	0	0	0	0	0	0	0	0	0	0	0	0	0	0	0	0	0
27-1-10/61276.0	B	1	A	1	0	0	N	Q	0	0	0	0	0	0	0	0	0	0	0	0	0	0	0	0	0	0	0	0	0	0	0	0
27-1-10/61277.0	B	1	A	0	0	1	N	U	0	0	0	0	0	0	0	0	0	0	0	0	0	0	0	0	0	0	0	0	0	0	0	0
27-1-10/61280.0	B	1	A	0	0	1	N	Q	0	0	0	0	0	0	0	0	0	0	0	0	0	0	0	0	0	0	0	0	0	0	0	0
27-1-10/61280a	B	1	A	0	0	1	N	U	0	0	0	0	0	0	0	0	0	0	0	0	0	1	0	0	0	0	0	0	0	0	0	0
27-1-10/61281.0	B	1	A	0	0	1	N	Q	0	0	0	0	0	0	0	0	1	0	0	0	0	0	0	0	0	0	0	0	0	0	0	0
27-1-10/61282.0	B	1	A	1	0	0	Y	U	0	0	0	0	0	0	0	0	1	0	0	0	0	0	0	0	1	0	0	0	0	0	0	0
27-1-10/61282.1	B	1	A	0	0	1	Y	U	0	0	0	0	0	0	0	0	0	0	0	0	0	0	0	0	0	0	0	0	0	0	0	0
27-1-10/61284.0	B	1	J	0	0	1	N	U	0	0	0	0	0	0	0	0	0	0	0	0	0	0	0	0	0	0	0	0	0	0	0	0
27-1-10/61285.0	B	1	J	0	0	1	Y	Q	0	0	0	0	0	0	0	0	0	0	0	0	0	0	0	0	0	0	0	0	0	0	0	0
27-1-10/61286.0	B	1	A	0	0	1	N	U	0	0	0	0	0	0	0	0	0	0	0	0	0	0	0	0	1	0	0	0	0	0	0	0
27-1-10/61287.0	B	1	A	1	0	0	N	U	0	0	0	0	0	0	0	0	0	0	0	1	0	0	0	0	0	0	0	0	0	0	0	0
27-1-10/61288.0	B	1	A	0	1	0	N	U	0	1	1	0	0	0	0	0	0	0	0	0	0	0	0	0	0	0	0	0	0	0	0	0
27-1-10/61289.0	CH	1	A	1	0	0	Y	Q	1	0	0	1	0	0	0	0	0	0	0	0	0	0	0	0	0	0	0	0	0	1	0	0
27-1-10/61290.0	B	1	A	0	1	0	N	Q	0	0	0	0	1	0	0	0	0	0	0	0	0	0	0	0	0	0	0	0	0	0	0	0
27-1-10/61291.0	B	1	A	0	1	0	N	Q	0	0	0	0	1	0	0	0	0	0	0	0	0	0	0	0	0	0	0	0	0	0	0	0
27-1-10/61293.0	B	1	J	0	0	1	N	Q	0	0	0	0	0	0	0	0	0	0	0	0	0	0	0	0	0	0	0	0	0	0	0	0
27-1-10/61294.0	B	1	J	0	0	1	N	Q	0	0	0	0	0	0	0	0	0	0	0	0	0	0	1	0	0	0	0	0	0	0	0	0

				Sex					Trauma				Infection				Arthritis					Physiological stress				Neurological			Tumor		Other path.	
Peabody number	CS	MNI	Age	M	F	U	AFO	Cr mod	T c	T pc	T h	T d	I de	I di	I tr	I g	A sl	A m	A s	A sp	A o	EH	CO	PH	O	Sb sl	Sb m/s	O na	T bo	T o	OP d	OP c/pc
27-1-10/61295.0	B	1	A	0	1	0	Y	Q	0	0	0	0	0	0	0	0	0	0	0	0	0	0	0	0	0	0	0	0	0	0	0	0
27-1-10/61295.1	B	1	J	0	0	1	Y	U	0	0	0	0	0	0	0	0	0	0	0	0	0	1	0	0	0	0	0	0	0	0	0	0
27-1-10/61296.0	B	1	A	0	0	1	Y	U	0	0	0	0	0	0	0	0	0	0	0	0	0	0	0	0	0	0	0	0	0	0	0	0
27-1-10/61297.0	B	1	J	0	0	1	N	U	0	0	0	0	0	0	0	0	0	0	0	0	0	0	0	0	0	0	0	0	0	0	0	0
27-1-10/61298.0	B	1	A	0	0	1	Y	U	0	0	0	0	0	0	0	0	0	0	0	0	0	0	0	0	0	0	0	0	0	0	0	0
27-1-10/61299.0	B	1	A	0	1	0	Y	U	0	0	0	0	0	0	0	0	0	0	0	0	0	0	0	0	0	0	0	0	0	0	0	0
27-1-10/61300.0	B	1	A	1	0	0	N	U	0	0	0	0	0	0	0	0	0	0	0	0	0	0	0	0	0	0	0	0	0	0	0	0
27-1-10/61304.0	B	1	A	1	0	0	N	Y	0	0	0	0	1	0	0	0	0	0	0	0	1	1	0	0	0	0	0	0	1	0	0	0
27-1-10/61309.0	B	1	J	0	0	1	Y	U	0	0	0	0	0	0	0	0	0	0	0	0	0	0	0	0	0	0	0	0	0	0	0	0
27-1-10/61310.0	B	1	A	0	1	0	N	Q	0	0	0	0	1	0	0	0	0	0	0	0	0	1	1	0	0	0	0	0	0	0	0	0
27-1-10/61312.0	B	1	J	0	1	0	Y	Y	0	0	0	0	0	0	0	0	0	0	0	0	0	1	0	0	0	0	0	0	0	0	0	0
27-1-10/61314.0	B	1	J	0	0	1	N	N	0	0	0	0	0	0	0	0	0	0	0	0	0	1	0	0	0	0	0	0	0	0	0	1
27-1-10/61315.0	B	1	A	1	0	0	Y	Y	0	0	0	0	0	0	0	0	1	0	0	0	0	1	0	0	0	0	0	0	0	0	0	0
27-1-10/61322.0	B	1	A	0	1	0	N	U	0	0	0	0	0	0	0	0	0	0	0	0	0	1	0	0	0	0	0	0	0	0	0	0
27-1-10/61323.0	B	1	A	0	0	1	N	U	0	0	0	0	0	0	0	0	0	0	0	0	0	1	0	0	0	0	0	0	0	0	0	0
27-1-10/61327.0	B	1	J	0	0	1	N	U	0	0	0	0	0	0	0	0	0	0	0	0	0	0	0	0	0	0	0	0	0	0	0	0
27-1-10/61328.0	B	1	A	0	0	1	Y	U	0	0	0	0	0	0	0	0	0	0	0	0	0	0	0	0	0	0	0	0	0	0	0	0
27-1-10/61329.0	B	1	J	0	0	1	N	U	0	0	0	0	0	0	0	0	0	0	0	0	0	0	0	0	0	0	0	0	0	0	0	0
27-6-10/60889.0	B	1	A	1	0	0	N	Y	0	0	0	0	1	0	0	0	0	0	0	0	0	0	0	0	0	0	0	0	1	0	0	0
27-6-10/60900.0	CH	1	J	0	0	1	N	N	0	0	0	0	0	0	0	0	0	0	0	0	0	0	0	0	0	0	0	0	0	0	0	0
27-6-10/60905.0	CH	1	J	0	1	0	N	Q	1	0	0	1	0	0	0	0	0	0	0	0	0	0	0	0	0	0	0	0	0	0	0	0
27-6-10/60906.0	B	1	A	0	1	0	N	Q	0	0	0	0	1	0	0	0	0	0	0	0	0	1	0	0	0	0	0	0	0	0	0	0
27-6-10/60910.0	B	1	A	1	0	0	N	U	1	0	1	0	0	0	0	0	0	0	0	0	0	0	0	0	0	0	0	0	0	0	0	0
29-2-10/60888.0	IN	1	A	0	1	0	Y	Y	0	1	1	0	0	0	0	0	1	0	0	1	0	0	0	0	1	0	0	0	0	0	0	0
29-2-10/60890.0	B	1	J	0	0	1	N	U	0	0	0	0	0	0	0	0	0	0	0	0	0	1	0	0	0	0	0	0	0	0	0	0
29-2-10/60891.0	B	1	J	0	0	1	N	Q	0	0	0	0	0	0	0	0	0	0	0	0	0	0	0	0	0	0	0	0	0	0	0	0
29-2-10/60892.0	B	1	J	0	0	1	Y	U	0	0	0	0	0	0	0	0	0	0	0	0	0	0	0	0	0	0	0	0	0	0	0	0
29-2-10/60893.0	B	1	J	0	0	1	Y	U	0	0	0	0	0	0	0	0	0	0	0	0	0	0	0	0	0	0	0	0	0	0	1	0
29-2-10/60894.0	IN	1	A	1	0	0	N	N	0	0	0	0	1	0	0	0	0	0	0	0	0	1	0	0	0	0	0	0	0	0	0	0
29-2-10/60895.0	IN	1	J	0	1	0	Y	Q	0	0	0	0	0	0	0	0	0	0	0	0	0	1	0	0	0	0	1	0	0	0	0	0
29-2-10/60896.0	B	1	J	0	0	1	Y	U	0	0	0	0	0	0	0	0	0	0	0	0	0	0	0	0	0	0	0	0	0	0	0	0
29-2-10/60897.0	B	1	J	0	0	1	Y	U	0	0	0	0	0	0	0	0	0	0	0	0	0	1	0	0	0	0	0	0	0	0	0	0
29-2-10/60898.0	IN	1	J	0	0	1	N	Q	0	0	0	0	0	0	0	0	0	0	0	0	0	0	0	0	0	0	0	0	0	0	0	0
29-2-10/60899.0	IN	1	J	0	0	1	N	U	0	0	0	0	0	0	0	0	0	0	0	0	0	0	1	0	0	0	0	0	0	0	0	0
29-2-10/60901.0	IN	1	J	1	0	0	Y	Q	0	0	0	0	0	0	0	0	0	0	0	0	0	1	0	0	0	0	0	0	0	0	0	0
29-2-10/60902.0	IN	1	A	1	0	0	N	Y	0	0	0	0	0	0	0	0	1	0	0	0	0	1	0	0	0	0	0	0	0	0	0	0
29-2-10/60903.0	B	1	A	0	1	0	N	Q	0	0	0	0	0	0	0	0	0	0	1	1	0	1	0	0	0	0	0	0	0	0	0	0
29-2-10/60907.0	IN	1	A	1	0	0	Y	Q	0	0	0	0	0	0	0	0	0	0	0	0	0	1	0	0	0	0	0	0	0	0	0	0
29-2-10/60908.0	IN	1	A	0	1	0	Y	U	0	0	0	0	0	0	0	0	0	0	0	0	0	0	0	0	0	0	0	0	0	0	0	0
29-2-10/60909.0	IN	1	A	1	0	0	Y	U	0	0	0	0	0	0	0	0	0	0	0	1	0	0	0	0	1	1	0	0	0	0	0	0
29-2-10/60911.0	IN	1	A	0	1	0	Y	Y	0	0	0	0	1	0	0	0	0	0	1	1	0	0	0	0	0	0	0	0	0	0	0	0
29-2-10/60912.0	IN	1	A	1	0	0	N	Q	0	0	0	0	1	0	0	0	0	0	1	0	0	1	0	0	0	0	0	0	0	0	0	0
29-2-10/60913.0	IN	1	J	0	0	1	N	Q	0	0	0	0	0	0	0	0	0	0	0	0	0	0	0	0	0	0	0	0	0	0	0	0
29-2-10/60914.0	IN	1	J	0	0	1	Y	U	0	0	0	0	0	0	0	0	0	0	0	0	0	0	0	0	0	0	0	0	0	0	0	0
29-2-10/60915.0	IN	1	A	1	0	0	Y	U	0	0	0	0	0	0	0	0	0	0	0	0	0	0	0	0	0	0	0	0	0	0	0	0

Peabody number	CS	MNI	Age	Sex			AFO	Cr mod	Trauma				Infection				Arthritis					Physiological stress				Neurological			Tumor		Other path.	
				M	F	U			T c	T pc	T h	T d	I de	I di	I tr	I g	A sl	A m	A s	A sp	A o	EH	CO	PH	O	Sb sl	Sb m/s	O na	T bo	T o	OP d	OP c/pc
29-2-10/60916.0	4	1	A	1	0	0	Y	U	0	0	0	0	1	0	0	0	0	0	0	0	0	0	0	0	0	0	0	0	0	0	0	0
29-2-10/60917.0	IN	1	J	0	0	1	Y	Y	0	0	0	0	0	0	0	0	0	0	0	0	0	0	0	0	0	0	0	0	0	0	0	0
29-2-10/60918.0	IN	1	A	1	0	0	N	Q	0	0	0	0	0	0	0	0	0	0	0	0	0	1	0	0	0	0	0	0	0	0	0	0
29-2-10/60919.0	IN	1	A	0	1	0	N	U	0	0	0	0	0	0	0	1	0	0	0	0	0	1	0	0	0	0	0	0	0	0	0	0
29-2-10/60921.0	IN	1	J	0	0	1	N	U	0	0	0	0	0	0	0	0	0	0	0	0	0	0	0	0	0	0	0	0	0	0	0	0
29-2-10/60922.0	IN	1	J	0	0	1	N	U	0	0	0	0	0	0	0	0	0	0	0	0	0	0	0	0	0	0	0	0	0	0	0	0
29-2-10/60923.0	IN	1	J	0	0	1	N	U	0	0	0	0	0	0	0	0	0	0	0	0	0	0	0	0	0	0	0	0	0	0	0	0
29-2-10/60924.0	IN	1	J	0	0	1	Y	Q	0	0	0	0	0	0	0	0	0	0	0	0	0	0	0	0	0	0	0	0	0	0	0	0
29-2-10/60925.0	3	1	J	0	0	1	N	Q	0	0	0	0	0	0	0	0	0	0	0	0	0	0	0	0	0	0	0	0	0	0	0	0
29-2-10/60926.0	IN	1	J	0	0	1	Y	U	0	0	0	0	0	0	0	0	0	0	0	0	0	0	0	0	0	0	0	0	0	0	0	0
29-2-10/60927.0	IN	1	J	0	0	1	Y	Q	0	0	0	0	0	0	0	0	0	0	0	0	0	0	0	0	0	0	0	0	0	0	0	0
29-2-10/60928.0	IN	1	J	0	0	1	Y	Q	0	0	0	0	0	0	0	1	0	0	0	0	1	0	0	0	0	0	0	0	0	0	0	0
29-2-10/60929.0	P	1	J	0	0	1	Y	Q	0	0	0	0	0	0	0	0	0	0	0	0	0	0	0	0	0	0	0	0	0	0	0	0
29-2-10/60930.0	IN	1	A	0	1	0	Y	Y	0	0	0	0	0	0	0	0	0	0	0	0	0	0	0	0	0	0	0	0	0	0	0	1
29-2-10/60931.0	IN	1	A	0	1	0	Y	Y	0	1	1	0	1	0	0	0	1	0	0	0	0	0	0	0	1	0	0	0	0	0	0	0
29-2-10/60932.0	IN	1	A	1	0	0	N	Y	0	0	0	0	1	0	0	0	1	0	0	1	0	1	0	0	0	1	0	0	0	0	0	0
29-2-10/60933.0	IN	1	A	0	1	0	N	Q	0	0	0	0	1	0	0	0	0	0	0	0	0	0	0	0	0	0	0	0	0	0	0	0
29-2-10/60934.0	IN	1	J	0	0	1	N	U	0	0	0	0	0	0	0	0	0	0	0	0	0	0	0	0	0	0	0	0	0	0	1	0
29-2-10/60935.0	IN	1	A	1	0	0	Y	Q	0	0	0	0	1	0	0	0	0	0	0	0	0	0	0	0	0	1	0	0	0	0	0	1
29-2-10/60936.0	IN	1	A	0	1	0	Y	Y	0	0	0	0	1	0	0	0	1	0	0	1	0	0	0	0	0	0	0	0	0	1	0	0
29-2-10/60937.0	IN	1	J	0	0	1	Y	U	0	0	0	0	0	0	0	0	0	0	0	0	0	0	0	0	0	0	0	0	0	0	0	0
29-2-10/60938.0	IN	1	J	0	0	1	Y	U	0	0	0	0	0	0	0	0	0	0	0	0	0	0	0	0	0	0	0	0	0	0	0	0
29-2-10/60939.0	IN	1	J	0	0	1	Y	Q	0	0	0	0	0	0	0	0	0	0	0	0	0	0	0	0	0	0	0	0	0	0	0	0
29-2-10/60940.0	IN	1	J	0	0	1	Y	U	0	0	0	0	0	0	0	0	0	0	0	0	0	0	0	0	0	0	0	0	0	0	0	0
29-2-10/60941.0	IN	1	J	0	0	1	N	U	0	0	0	0	0	0	0	0	0	0	0	0	0	0	0	0	0	0	0	0	0	0	0	0
29-2-10/60978.0	IN	1	A	1	0	0	N	Y	0	0	0	0	1	0	0	0	1	0	0	1	0	0	0	0	0	0	0	0	0	0	0	0
29-2-10/60979.0	IN	1	J	0	0	1	N	U	0	0	0	0	0	0	0	0	0	0	0	0	0	0	0	0	0	0	0	0	0	0	0	0
29-2-10/60980.0	IN	1	A	1	0	0	Y	Y	1	0	1	0	1	0	0	0	1	0	0	0	0	0	0	0	0	0	0	0	0	0	0	0
29-2-10/60981.0	IN	1	J	0	0	1	N	U	0	0	0	0	0	0	0	0	0	0	0	0	0	0	0	0	0	0	0	0	0	0	0	0
29-2-10/60983.0	P	1	A	0	1	0	Y	Y	0	0	0	0	0	0	0	0	0	0	0	0	0	0	1	0	0	0	0	0	0	0	0	0
29-2-10/60984.0	IN	1	A	0	1	0	Y	U	0	0	0	0	1	0	0	0	1	0	0	0	0	0	0	0	0	1	0	0	0	0	0	0
29-2-10/60985.0	3	1	J	0	0	1	Y	Q	0	0	0	0	0	0	0	0	0	0	0	0	0	0	0	0	0	0	1	0	0	0	0	0
29-2-10/60986.0	1	1	A	0	1	0	Y	Y	0	0	0	0	0	0	0	0	0	0	0	0	0	1	0	0	0	0	0	0	0	0	0	0
29-2-10/60987.0	IN	1	A	1	0	0	Y	Q	0	0	0	0	1	0	0	1	1	0	0	1	1	0	0	0	0	0	0	0	0	0	0	1
29-2-10/60988.0	IN	1	A	1	0	0	Y	U	0	0	0	0	1	0	0	0	1	0	0	0	0	1	0	0	0	0	0	0	0	0	0	0
29-2-10/60989.0	IN	1	A	1	0	0	N	Q	0	0	0	0	1	0	0	0	1	0	0	1	0	0	0	0	0	0	0	0	0	0	0	0
29-2-10/60990.0	IN	1	A	0	1	0	Y	Y	0	1	1	0	0	0	1	1	0	0	0	0	0	1	0	0	0	0	0	0	0	0	0	0
29-3-10/60993.0	IN	1	J	0	0	1	Y	Y	0	0	0	0	0	0	0	0	0	0	0	0	0	0	0	0	0	0	0	0	0	0	1	0
29-3-10/60994.0	IN	1	A	1	0	0	Y	Q	0	0	0	0	0	0	0	0	1	0	0	0	0	0	0	0	0	1	0	0	0	0	0	0
36-116-10/N1472.0	IN	1	A	0	1	0	Y	U	0	0	0	0	0	0	0	0	0	0	0	0	0	0	0	0	0	0	0	0	1	0	0	0
36-95-10/N1466.0	B	1	J	0	0	1	N	Y	0	0	0	0	0	0	0	0	0	0	0	0	0	0	0	0	0	0	0	0	0	0	0	0
45-1-10/N3986.2	IN	1	J	0	0	1	N	U	0	0	0	0	0	0	0	0	0	0	0	0	0	0	0	0	0	0	0	0	0	0	0	0
985-27-10/66196.0	IN	1	A	0	0	1	N	U	0	0	0	0	0	0	0	0	0	0	0	1	0	0	0	0	0	0	0	0	0	0	0	0
985-27-10/N10649.0	IN	1	A	0	0	1	N	U	0	0	0	0	0	0	0	0	0	0	0	0	0	0	0	0	0	0	0	0	0	0	0	0
985-27-10/N10650.0	IN	1	J	0	0	1	N	U	0	0	0	0	0	0	0	0	0	0	0	0	0	0	0	0	0	0	0	0	0	0	0	0

Peabody number	CS	MNI	Age	Sex			AFO	Cr mod	Trauma				Infection				Arthritis					Physiological stress				Neurological			Tumor		Other path.	
				M	F	U			T c	T pc	T h	T d	I de	I di	I tr	I g	A sl	A m	A s	A sp	A o	EH	CO	PH	O	Sb sl	Sb m/s	Ona	T bo	T o	OP d	OP c/pc
997-27-10/N10668.0	B	1	J	0	0	1	Y	U	0	0	0	0	0	0	0	0	0	0	0	0	0	0	0	0	0	0	0	0	0	0	0	0
997-27-10/N10669.0	B	1	A	0	0	1	N	U	0	0	0	0	0	0	0	0	0	0	0	0	0	0	0	0	0	0	0	0	0	0	0	0
997-27-10/N10670.0	B	1	A	0	0	1	N	U	0	0	0	0	0	0	0	0	0	0	0	0	0	0	0	0	0	0	0	0	0	0	0	0

Appendix D

Miscellaneous Additional Measurements and Other Demographic Notes for PMAE Numbers Listed in Appendixes A and B

Michèle E. Morgan

Key to Abbreviations	
di, dc, dm	Deciduous incisors, canines, molars
prox	Proximal
dist	Distal
med	Medial
I1, I2	Permanent incisors, first and second
P1, P2	Permanent premolars, first and second
M1, M2, M3	Permanent molars, first, second, and third
pcs	Pieces
max lgth	Maximum length
Limb elements	HM, humerus; RD, radius; UL, ulna; FM, femur; TB, tibia; FB, fibula; IL, ilium
PC	Postcrania

Peabody number	Measurements and notes
00-0-10/N9468.44	Not all dms erupted
00-0-10/N9468.46	Not all dms erupted
19-6-10/59828.0	Epiphyses fused but not obliterated, prox clavicle not fused
19-6-10/59877.1	Very large extremities, which are all that are present
19-6-10/59879.3	Epiphyseal fusion suggests 16–19 yrs
19-6-10/59894.0	Spheno-occipital just fused; very small individual
19-6-10/59968.0	Dist RD unfused; many long bones incompletely fused
19-6-10/59973.0	3 pcs pelvis not fused; M2s are erupted, M3 crowns fully formed
19-6-10/59974.0	Epiphyses fused but lines visible on most bones
19-6-10/59983.0	Spheno-occipital not fused; M3s just starting to erupt
19-6-10/59992.0	M2s erupting, lower P2s not yet in; max lgth (mm): HM 220, IL 101, TB 240
19-6-10/59996.0	Epiphyses fused except prox clavicle
19-6-10/60013.0	Prox clavicle not fused, iliac crest stage 1
19-6-10/60014.0	Max lgth left (mm): HM 169, RD 129, UL 143, FM 238, TB 195, FB 191, IL 87; max lgth right HM 172, RD 130, UL 145, FM 239, TB 196, FB 192, IL 87 (upper M2s smallish buds)
19-6-10/60020.0	Iliac crest stage 1; prox clavicle unfused
19-6-10/60022.0	M3 roots probably not complete
19-6-10/60023.0	PC all fused except prox clavicle, ischial tuberosity, iliac crest partially fused; M3s erupting
19-6-10/60024.0	Nearly edentulous
19-6-10/60025.0	Maxillary molars lost antemortem
19-6-10/60030.0	Edentulous, likely very old
19-6-10/60031.0	All molars lost antemortem
19-6-10/60037.0	PC: several appendicular bones incompletely fused
19-6-10/60043.0	Max lgth left (mm): HM 230, FM 330, TB 275; nearly all long bones unfused
19-6-10/60044.0	Prox clavicle not fused
19-6-10/60045.0	Max lgth right w/ epiphyses (mm): HM 285, FM 412, TB 330; wrist and knee unfused
19-6-10/60056.0	Max lgth right (mm): HM 166, FM 220
19-6-10/60067.0	Unfused epiphyses: prox clavicle, humeral head, iliac crest, dist FM, prox FB
19-6-10/60069.0	Max lgth right (mm): HM 168, RD 133, UL 145, FM 241, TB 200, FB 195, IL 81; permanent M1s fully erupted
19-6-10/60078.0	Max lgth left (mm): HM 125, FM 160, TB 136
19-6-10/60079.0	Max lgth left (mm): HM 165, RD 130, UL 145, IL 88
19-6-10/60097.0	Prox clavicle unfused, dist RD and UL incompletely fused, dist TB incompletely fused, iliac crest incompletely fused
19-6-10/60150.0	Max lgth left (mm): HM 216, RD 166, UL 179, FM 310, TB 257, FB 247, IL 105
19-6-10/60151.0	Max lgth right(mm): HM 176, RD 138, UL 153, FM 238, TB 203, FB 197, IL 88
19-6-10/60155.0	Max lgth left (mm): HM 188, RD 147, UL 159, TB 274(?), IL 93; max lgth right (mm): FB 232
19-6-10/60157.0	Max lgth left (mm): TB 312; max lgth right (mm): RD 220
19-6-10/60162.0	Max lgth left (mm): HM 181, RD 137, UL 152, FM 262, TB 211, FB 204, IL 84
19-6-10/60183.0	Max lgth right (mm): HM 169, RD 120, UL 135, FM 227, TB 187, FB 181, IL 87
19-6-10/60184.0	Max lgth left (mm): HM 111, FM 144, RB 122, IL 62
19-6-10/60185a	Max lgth left (mm): HM 77, RD 62, FM 93, TB 80, IL 43
19-6-10/60186.0	Max lgth left (mm): HM 63, FM 71, TB 60
19-6-10/60190.0	Max lgth left (mm): HM 230, RD 171, FM 304, TB 257, FB 252, IL 102; M2s in
19-6-10/60647.0	Long bone epiphyses and iliac crest just fused
19-6-10/60958.1	Max lgth left (mm): FM 150, HM 230
19-6-10/60964.0	Subadult, iliac and ischial crest stage 1 fusion
19-6-10/60965.0	Iliac and ischial crests unfused
19-6-10/60977.11	Humeral head diameter 39 mm
19-6-10/60998.0	Dm1s erupted; max lgth left (mm): TB 200
19-6-10/61111.0	Lower dm2s still present, M2s erupted, upper P2s erupting, 3 of 4 canines fully erupted
19-6-10/61113.0	M2s erupted, M3s unerupted
20-6-10/60204.0	Deciduous teeth just in, M1s unerupted; max lgth right (mm): FM 138, TB 113, IL 59
20-6-10/60206.0	M1s unerupted; max lgth left (mm): HM 173, UL 139, FM 231, TB 190, FB 188, IL 82
20-6-10/60208.0	M1s erupted; max lgth left (mm): HM 202, RD 157, UL 164, IL 90
20-6-10/60209.0	Max lgth right (mm): FM 222, TB 190, HM 164
20-6-10/60214.0	Max lgth right (mm): HM 264; epiphyseal fusion 10–15 yrs
20-6-10/60217.0	Max lgth left (mm): HM 220, RD 168, UL 184, FM 303, TB 249, FB 245, IL 108

Peabody number	Measurements and notes
20-6-10/60235.0	Prox clavicle not fused, all other long bones fused
20-6-10/60239.0	Iliac crest incompletely fused
20-6-10/60245.0	Max lgth right (mm): HM 222, UL 190, FM 307, IL >100
20-6-10/60263.0	Max lgth left (mm): HM 151, FM 216, TB 180, IL 78
20-6-10/60265.0	Clavicle not fused, other long bones barely fused; iliac crest and auricular surface suggest 18–24 yrs
20-6-10/60269.0	Dms erupted, M1s unerupted; max lgth left (mm): HM 152, RD 112, UL 120, FM 199, TB 162, FB 160, IL 82
20-6-10/60271.0	Iliac crest incompletely fused
20-6-10/60292.0	Iliac crest incompletely fused, some other long bones also incompletely fused
20-6-10/60304.0	Iliac crest unfused, many long bones stage 1 fusion
20-6-10/60305.0	PC epiphyses 14–15 yrs; max lgth left (mm): FM 373, TB 310
20-6-10/60311.0	Long bone epiphyseal lines still visible
24-4-10/60431.0	Max lgth left (mm): HM 227, UL 187, FM 315, TB 266, FB 259
24-4-10/60460.0	Vt/s1–2 not completely fused, but individual appears old
24-4-10/60479.0	Spheno-occipital fused, 1 M3 erupted
24-4-10/60509.0	8 dms, 2 Ms erupted, 4 Ms unerupted
24-4-10/60510.0	Epiphyseal fusion suggests 14–16 yrs
24-4-10/60511.1	8 dms, 4 Ms erupted, 3 Ms unerupted
24-4-10/60514.0	Spheno-occipital fused
24-4-10/60523.0	2 unerupted M3s
24-4-10/60531.0	4 dms, 1 M erupting
24-4-10/60535.0	All epiphyses just fused, lines prominent
24-4-10/60538.0	Long bone epiphyses just fused
24-4-10/60554.0	Prox clavicle just fusing
24-4-10/60559.0	3 molars still unerupted
24-4-10/60566.0	Many epiphyses just fusing; 18–20 yrs
24-4-10/60568.0	Upper M3s just erupting; prox clavicle just fusing
24-4-10/60575.0	M3s erupting; appendicular epiphyses just fused, not prox clavicle
24-4-10/60576.0	Cranial sutures suggest young adult
24-4-10/60582.0	Prox clavicle not fused; M3s just erupting; spheno-occipital fused
24-4-10/60584.0	M1s erupted
24-4-10/60585.0	Deciduous dentition present, M1s unerupted
24-4-10/60587.0	M1s erupted, Is erupting
24-4-10/60596.0	M3s erupted, iliac crest, dist UL and RD not fused
24-4-10/60599.0	Small male
24-4-10/60600.0	Spheno-occipital not fused
24-4-10/60606.0	8 dms erupted, M1s not erupted
24-4-10/60613.0	Max lgth (mm): HM 300
24-4-10/60621.0	M1s erupted
25-3-10/60626.0	Max lgth left and right (mm): IL 130; ischial tuberosity just fused, iliac crest unfused
25-3-10/60627.0	Unfused dist FM and prox HM, acetabulum fused but pubis-ilium line still visible, iliac crest not fused
25-3-10/60669.0	PC all fused except prox clavicle
25-3-10/60672.0	Epiphyseal union of appendicular elements suggests 14–16 yrs
25-3-10/60674.1	Epiphyses of dist RD and dist UL unfused
25-3-10/60680.0	Lower M3s just erupting; max lgth left (mm): RD 209, UL 229, FM 395, IL 124; max lgth right (mm): RD 210, FM 395, TB 289
25-3-10/60688.0	PC epiphyses of dist FM, RD, UL, and prox TB, FB, HM are all unfused; other appendicular epiphyses are stage 1 fusion
25-3-10/60691.0	Very robust individual
25-3-10/60708.0	PC epiphyses of dist FM, UL, RD, and iliac crest unfused, prox HM stage 2
25-3-10/60712.0	M1s fully erupted, I2s erupting
25-3-10/60716.0	Iliac crest stage 1, dist FM unfused
25-3-10/60726.0	M1s erupted
25-3-10/60730.0	M1s erupted
25-3-10/60752a	Epiphyses of prox HM, dist RD and UL are stage 1
25-3-10/60757.0	Spheno-occipital line visible
25-3-10/60771.1	4 dms, 4 Ms unerupted
25-3-10/60778.0	Prox HM and dist RD unfused, many appendicular epiphyses incompletely fused
25-3-10/61152.0	Different cranial shape

Peabody number	Measurements and notes
26-3-10/61200.0	RD and IL suggest 7.5–8.5 yrs
26-3-10/61202.0	Dm2s unerupted; max lgth left (mm): FM 130; max lgth right (mm): TB 108, IL 59
27-1-10/61265.0	Max lgth left (mm): HM 92, FM 112, TB 96, FB 93, IL 49
27-1-10/61268.0	Only lower dis erupted; max lgth left (mm): HM 86, RD 68, UL 77, TB 86
27-1-10/61269.0	Max lgth left (mm): HM 97, RD 74, UL 84
27-1-10/61270.0	Max lgth left (mm): RD 50, UL 56, FM 71, TB 62, FB 59, IL 33
27-1-10/61272.0	Dentition suggests 0.75 ± 0.25 yrs; max lgth right (mm): HM 82
27-1-10/61279.0	M1s unerupted
27-1-10/61284.0	Max lgth left (mm): IL 33; max lgth right (mm): HM 63, UL 61, FM 74
27-1-10/61285.0	PC all epiphyses unfused; max lgth right (mm): RD 157, UL 164, FM 270, TB 230, FB 220, IL 101
27-1-10/61294.0	Max lgth left (mm): RD 123, UL 137, FM 220, TB 187; max lgth right (mm): HM 164, FB 80, IL 83; M1s not erupted
27-1-10/61295.1	M3s not erupted
27-1-10/61297.0	4 dms, 6 Ms
27-1-10/61309.0	Iliac crest unfused
27-1-10/61312.0	Retained lower right dm2
27-1-10/61329.0	Max lgth right (mm): HM 99, UL 87, TB 105, FB 101, IL 56
27-6-10/60900.0	2 unerupted molars
27-6-10/60905.0	Spheno-occipital not fused, M3s not erupted
29-2-10/60890.0	All dms erupted, 1 M erupted
29-2-10/60891.0	Dm1s erupted; PC 0.5–1.5 yrs
29-2-10/60892.0	Dms all erupted, M1s not erupted
29-2-10/60893.0	Dms all erupted, M1s not erupted; max lgth right (mm): IL 60
29-2-10/60895.0	M3s just erupting; shoulder, wrist, all lower limb bones unfused
29-2-10/60896.0	M1s erupted, max lgth ilium 7.5–8.5 yrs
29-2-10/60897.0	M3s erupting
29-2-10/60898.0	Dm2s erupting
29-2-10/60899.0	Max lgth left (mm): TB 100; dc and dms unerupted
29-2-10/60901.0	Acetabulum fused but lines visible, prox RD and dist HM fused, all other limb elements unfused; max lgth left (mm): TB 300
29-2-10/60913.0	Dis erupting; max lgth right (mm): HM 80, RD 61, UL 69, FM 100, IL 42
29-2-10/60914.0	All deciduous dentition erupted, M1s not erupted
29-2-10/60917.0	M2s erupting; max lgth left (mm): HM 211, RD 160, FM 290, TB 240, FB 234, IL 101; all PC epiphyses unfused except ischio-pubis
29-2-10/60921.0	Mandible not fused; dental age based on tooth sockets 7 ± 2 mo in utero; max lgth left (mm): FM 65; max lgth right (mm): HM 57, RD 45, UL 52
29-2-10/60923.0	Dentition suggests 18 ± 6 mo
29-2-10/60924.0	Dentition suggests 7–9 mo in utero; max lgth right (mm): HM 55, RD 46, UL 52, FM 64, TB 57, FB 52, IL 29
29-2-10/60925.0	No teeth erupted; max lgth right (mm): HM 80, RD 65, UL 73, FM 93, TB 81
29-2-10/60926.0	Dentition suggests 9 ± 3 mo, all lower dms present; max lgth left (mm): HM 90, RD 70, UL 77, FM 120, TB 99, FB 94, IL 50
29-2-10/60927.0	All deciduous dentition erupted, M1s not erupted; max lgth left (mm): HM 103, RD 82, UL 89, FM 128, TB 107, IL 53
29-2-10/60928.0	All deciduous dentition erupted, M1s not erupted; max lgth left (mm): HM 123, FM 162, TB 136; max lgth right (mm): IL 62
29-2-10/60929.0	Only dis have erupted; max lgth right (mm): HM 98, RD 75, UL 82, FM 122, TB 103, FB 99, IL 50
29-2-10/60933.0	Stage 1 fusion spheno-occipital and prox HM; prox HM max diameter 38 mm, max lgth 311 mm
29-2-10/60934.0	All deciduous dentition erupted, upper M1s erupted, lower M1s emerging; max lgth left (mm): FM 232; max lgth right (mm): IL 86
29-2-10/60937.0	Dm1s just erupting; long bones suggest 0.5–1.5 yrs; ilia, right lower limb elements, left upper limb elements present
29-2-10/60938.0	All deciduous dentition erupted, no permanent dentition erupted; max lgth left (mm): FM 155, TB 125, IL 60; max lgth right (mm): HM 118, RD 91, UL 99;
29-2-10/60939.0	Dm2s erupting; max lgth left (mm): RD 88, UL 99, FM 144, TB 122, IL 68
29-2-10/60940.0	No teeth have erupted; max lgth right (mm): HM 73, RD 58, UL 63, FM 94, TB 77, IL 42
29-2-10/60941.0	All deciduous dentition erupted, M1s unerupted; max lgth left (mm): HM 138, UL 111, FM 181, TB 148, IL 68
29-2-10/60979.0	All deciduous dentition erupted, M1s unerupted
29-2-10/60981.0	Only 4 dms present; max lgth left (mm): IL <70
29-2-10/60984.0	FM head max diameter 41 mm
29-2-10/60985.0	M1s erupted, lower I2s erupted; max lgth left (mm): HM 178, RD 130, UL 147, FM 238, TB 193, FB 86, IL 87
29-2-10/60986.0	2 Ms unerupted; prox clavicle unfused; stage 1 fusion prox HM, dist RD, iliac crest, ischial crest, spheno-occipital
29-2-10/60990.0	Unfused prox clavicle; stage 1 fusion: dist UL, RD, prox HM, iliac and ischial crests

Peabody number	Measurements and notes
29-3-10/60993.0	Maxilla has no place for upper I2s, upper P1s; P2s and M1s erupted, lower P1s and M1s erupted, lower P2s erupting, upper and lower M2s erupting; max lgth left (mm): HM 253, FM 350, FB 291, IL 113+; all observable epiphyses unfused
997-27-10/N10668.0	Max lgth left (mm): TB 112
997-27-10/N10670.0	Very little wear on premolars

Appendix E

Hooton's Cranial, Postcranial, and Nonmetric Data with New Sex Estimations

Katherine E. Weisensee and Richard L. Jantz
Transcribed by Katherine Weisensee

PART 1: CRANIAL DATA

Key to Cranial Measurements and Definitions

The ID number is the catalogue number assigned by the Peabody Museum of Archaeology and Ethnology when the human remains were accessioned.

Abbrev.	Measurement (mm)	Description
GOL	Glabella-occipital length	Hold the left point of the caliper on glabella and move the right point up and down the occiput in the median line, watching the scale. Record the maximum measurement.
XCB	Maximum width	Take the greatest breadth perpendicular to the median sagittal plane wherever the greatest breadth occurs, avoiding the supra-mastoid crest and making due allowance for any warping outward of the temporal squama.
BBH	Basion-bregma height	From basion, the middle point on the anterior border of the foramen magnum to bregma, the meeting place of the coronal and sagittal sutures.
WFB	Minimum frontal diameter	Minimum breadth between the temporal crests on the frontal bone.
MNH	Menton-nasion height	Adjust mandible so that teeth are in occlusion and measure from nasion to gnathion (menton, or the middle point on the lower border of the mandibular symphysis). This measurement cannot be taken unless teeth occlude.
ANH	Alveon-nasion height	From nasion to alveon—the most inferior point of the mandible[a]
XDB	Maximum diameter bi-zygomatic	Greatest breadth between zygomatic arches, perpendicular to the median sagittal plane.

Abbrev.	Measurement (mm)	Description
BD	Bigonial diameter	Diameter between gonia (most external points of juncture of ascending ramus and horizontal ramus).
MAJ	Mean angle lower jaw	Place the mandible on the goniometer in natural position and raise the inclined plane so that it is tangent to the posterior edges of both ascending rami.
HS	Height of symphysis	Distance between gnathion and infra-dentale (point on alveolar border between middle incisor teeth of mandible).
IW	Bicondylar width	Diameter between most external points of mandibular condyles.
MAR	Minimum breadth of ascending ramus	Smallest distance between anterior and posterior borders of left ascending ramus. Taken perpendicular to height.
HAR	Height of ascending ramus	Place the mandible on the measuring board with the horizontal rami against the upright plane. Measure the height of the condyles with the square. Record the height of the left condyle.
CSL	Condylo-symphysial length	Place the mandible on the measuring board with the condyles tangent to the upright plane. Measure with the square to the most anterior point of the mental process.
OBH	Orbits: height (right and left)	The height is taken from the upper to the lower border in the middle of the orbit and perpendicular to the long axis.
OBB	Orbits: breadth (right and left)	The breadth is measured from dacryon to the middle of the external border (ectoconchion).
NLH	Nasal height	Height from nasion to subnasal point. Take the mean of the heights to lower borders of the nasal aperture on each side of the spine.
NLB	Nasal breadth	Maximum breadth of nasal aperture perpendicular to height.
BAL	Basion-alveon	From basion to alveon—the most inferior point on the mandible.[a]
BNL	Basion-nasion	The fixed caliper point is placed on nasion and the movable point adjusted to basion.
PXL	Palate-external length	Length taken from prosthion to the points tangent to the posterior edges of the alveolar borders.
PXW	Palate-external width	Maximum external breadth of palate on the outside of the alveolar borders. This is usually at the level of the second molars.
FOL	Foramen magnum length	The direct distance of basion from opisthion.
FOB	Foramen magnum breadth	The distance between the lateral margins of the foramen magnum at the point of greatest curvature.

[a] Measurement definitions were not published by Hooton but are likely to have been similar to measurements defined by Zobeck (1983) and Moore-Jansen and Jantz (1989).

ID number	Sex	GOL	XCB	BBH	WFB	MNH	ANH	XDB	BD	MAJ	HS	IW	MAR	HAR	CSL	OBH R	OBH L	OBB R	OBB L	NLH	NLB	BAL	BNL	PXL	PXW	FOL	FOB
59.801	F	162	143	140	91	125	78	137	100	116	37	123	38	60	109	36	35	37.5	40	53	25	97	103	58	68	34	28
59.802	F	171	136	130	95	115	71	130	94	122	33	113	35		106	32.5	33	42	40	49.5	27	95	97	54	61	35	30
59.803	F	157	154	134	92	113	67	130	93	111	35	116	35	66	99	34	35	41	40	45	22	91	92				
59.804	M				97				93	120	34	123	29	61	96											31	29
59.805	M	159	144	138	99	112	71	133	98	107	33	116	41	69	95	35	36	40	39	53	26	93	101	56	66		
59.806	M	157	152	136	99	113	68	142	100	114	34	126	39	63	103	35	35	41	41	47	24	95	102	54	63		
59.807	M		150		98	112	68		108	117	35	117	32.5	61	89	37		40		46	23			51	65		
59.811	F	154	147	138	90	112	71	132	88	114	35	113	38	64	93	32	33	38	37.5	46.5	25	90	91	53	66	31	27
59.812	I	166	147		96	125	79	139	105	116	37	120	36	72	92	38	36.5	40	42	53	25			55	65		
59.813	F	169	144	137	91	123	76	132	102	126	34	123	37	60	114	36	41	37	41	53	26.5	98	102			36	28
59.815	F	158	144	134	89	112	70	132	92	120	32	123	38	60	101	30	29	38.5	38.5	48	24	98	97	54	65	31	26
59.816	M	155	152	144	95	112	66	143	97	116	31	122	33	56	98	33	33	40.5	37	47	24.5			50	59	36	28
59.817	I	170	134	139	92	118	73	134	98	124	36	114	35	56	105	35	34	38	38	48	29	98	98	58	73		
59.818	F	159	141																								
59.819	F	179																									
59.820	I	147	139	137	91	113	68	130	94		36	121	31	53	97	35	35	38	39	48	24.5	89	92	53	63	33	28
59.822	I	184	132	132	96	121	73	131	98	123	34	118	36	51	103	35	35	37	38	53	24	97	105	53	67	36	29
59.823	M	158	152	150	103	120	69	144	113	111	38	125	35	71	102	34	34	41	40	47	27	99	98	56	71		
59.825	M	175	139		92				91	114	29	112	35	65	98												
59.826	I	161	151		96	117	73	146	101	116	34	124	39	64	99	36	36	39	40	48	27			56	65		
59.827	M	174	145	144	94	117	70	133			37		33			31	30.5	37	38	46	22	100	103	53	63		
59.828	I	162	143	140	91	113	68	127	96	127	35	103	36	56	102	34.5	34	37	37	45	24	96	96	55	64.5	35	27
59.829	F	157	133	134	92				99	126	32	112	37	51	102											32	29
59.831	I	171	142	136	95	122	75	142	109	118	36	125	38	62	101	35	36	39	37	52	26	96	102	56	66		
59.832	M	174	151	146	101	129	76	148	98	113	39		39	71	107	36.5	36	43	40	50.5	23	96	104	60	68		
59.839	M	158	143	139	97				132		38		36			35.5	34	40	41	52	24		103			34	26
59.840	I	164	151	136	94	112	71	148	102	125	31	125	32.5	57	109		34		40	51	26	102	106	52	61		
59.841	F	151	143	132	93	115		130	88	123	33	116	32	52	102	35	36.5	39	38.5	48	25		92		69	29	23
59.843	F	159	133		97	113	71				34		38	62		34	36	41		49	24.5			55	59.2		
59.844	M	163	149	142	92	127	75	144			36					36	35.5	40.5	39.5	54	23	99	103				
59.845	F	158	143	127	83	122	74.5	134	91	121	36	111	34.5	54	99	34.5	36	38	38.5	50	26	92	95			32	27
59.846	F	185	139	136	98			132					37			36	35	38	37.5	49	27					30	24
59.849	F	156	146	135	94	108	65	128	89	112	31	109	37	57	96	32.5	33.5	39	39	46	23.5	86	93	49	65	31	27
59.850	I	159	140	135	90	120	71		85				34	51						50	26	96	95	53	64.5		
59.851	M	178	151	150	96	122	77	148	111	119	37	130	37	66	113	38	38	41	42	56	26	97	106	55	70		
59.852	I	183	144	138	92		75	143								37	36	40	39	51	26	105	109				
59.855	F	161	150	143	92			133			34		38	55		34.5	36	41	40	47	27	102	102			35	25
59.857	F	170	145	136	94											37.5	37	37	40	51	24.5					33	25
59.858	I	173	139	139	95	119	72	141	106	116		126	37	67	108	34	34.5	40.5	41	52	27	100	103				
59.859	M	157	143		96		73	128								36.5	36.5	39	39	49	25			54	61		
59.860	I	166	133	127	93	100	59	129	93	122	28	111	37	52	100	29	30.5	39	38.5	43	26	93	95	49	64	32	25
59.862	M	161	140	142	96	123	77	140	98	119	35	126	37.5	59	112	36	35	42	42	54	27	101	105	59	67		
59.863	M	195	143	152	110						43		41	72										58	69	32	26
59.864	I	165		136	100	108	63				32		40	52		31.5	33.5	39	38							37	26
59.866	M	189	138	133	97	123	76	143	101	118	39	125	38	59	106	35	35	41	41	52	26.5	105	105	59	68	31	26
59.867	I	145	138	124	85	113	69	123			33		36	56		36	35.5	36	35.5	51	24	97	91	53	61	29	26
59.868	I	165	138		83	114	70		104	113	31		34	62		36	35	39	37	49	27						

ID number	Sex	GOL	XCB	BBH	WFB	MNH	ANH	XDB	BD	MAJ	HS	IW	MAR	HAR	CSL	OBH R	OBH L	OBB R	OBB L	NLH	NLB	BAL	BNL	PXL	PXW	FOL	FOB
59.870	M	173	144	146	98	130	77	137					39	65		36	35	39	39	52	26	102	105	58	68		
59.871	F	153	144	133	98		70	134								34.5	34.5	40	39	49	28	91	98	53	64	33	27
59.872	I	165	131	130	85			130		122		119	36	55	105	36	36	39	39.5	50	26		97			31	26
59.875	I	163	133	144	90	112	71	132					37	68		35	36	37.5	36	49	23.5	95	100	55	64		
59.876	M	179	135	132	96	129	81	144	106	120	37	130	36	60	111	38	38	41.5	39	57	25	99	105	56	68		
59.877	F	170	132	135	97	111	69			123	31	106	36	52	101	33	34	38	37	47	26	97	100	53	60	30	24
59.889	I	160	138	138	98			129								35	35	37	39	48	29					31	26
59.891	I	172	142	141	97	120	73	137	98	122	30	116	34	65	103	31	32	41	41	53	24	94	104	50	62		
59.892	I	167	132	136	92											35	35	40	38	48	26			57	59		
59.893	I	170	138	136	97											36.5	36	37.5	38	52	23						
59.898	I	167	135	136	89		69	129								34.5	33	36	37	49	21	94	101	52	61	33	29
59.901	I	178	135	135	94	104	69	134	91	130	30	106	30	62	101	33	32.5	37	35	51	22	90	99	50	66		
59.904	I	173	145	142	97																					33	29
59.905	I	181	144	140	104											35	35	41	42	52	28						
59.906	I	171	139	138	92	116	69	138	106	128	34	117	37	57	108	34.5	34.5	40	38.5	50	27.5	95	101	52	64		
59.907	M	145	135	146	92	124	79	143	108	126	35	122	39	66	106	35	37	41	37	55	26	103	105	58	65		
59.908	I	160	143	131	87	121	72	131	95	122	35	118	35	58	99	34	36	36	37	51	25	96	102	54	60	33	28
59.910	I	153	145	139	92	118	71	138	100	126	36	118	38	65	110	37	36.5	39	38	49	25	102	97	60	66		
59.911	F	164	140	139	99	117	74	130	95		34		36	56		36	36	43	41	53	27	103	108	56	61	33	25
59.912	F	149	139	131	83	100		130	96	120		118	35	58	92	35	35	41	41	48	28					35	27
59.915	M	158	144	141	92	126	74	143	109	126	38	126	41	58	117	34.5	35	41.5	40	52	26	104	101	57	67		
59.916	I	181	133	143	96																					34	26
59.917	F	160	136		94	104	65	124	87	120	31	110	35	57	95	34	33	39	38.5	47	21			49	62		
59.918	M	171	140	151	103	131	80	135	109	113	36	119	39	68	102	38	38	38	38.5	56	27	96	107	56	66		
59.919	M	163	136	144	94	117	70	131	96	118	36		32	68	97	35	37	37.5	37	45.5	23	99	102	54	61		
59.920	F	154	138	142	91	114	71	134		119	34		40	66	102	36	37	38	38	49	23	103	104	56	59		
59.923	F	165	134		87	110	65	133	93		34		34			33	34	39	40	45	27			57	65		
59.924	M	187	140		98	120	75	138	98	112	35	121	42	74	111	36.5	37.5	40	39	54	27			63	65		
59.925	F	157	143	140	89	120	71				37.5		34	52		34		38		51.5	27	96	101	56	66		
59.952	I	166	132	133	93						34		34									93	100			35	23
59.953	M	161	156	141	93	118	74		106	122	34	119	38	62	109	36	36	41	43	52	25	95	104	54	66		
59.954	M	164	147	144	97	118	67	143	102	117		126	38	67	108	33	32	39	39	52	27		98				
59.955	I	154	150	139	94		64	141								36	36	38	38	46	26	92	99	50	64		
59.957	M	180	134	137	93	122		138	106	118		116	36	68	103	34.5	36	39	39	50	29						
59.958	M	158	145	136	96				117		37		39							52	26					32	23
59.960	I	143	147		94	116	70	136	106	134	38	120	35	50	103	34	34	40	39	46	24			56	66		
59.962	M	167	148	144	98	125	75	136	90	117	34		37	63	110	36	36.5	41	40	49	23	96	103	53	63		
59.963	M	170	145	138	88	110	70	140	106	115		127	38	67	105	33	31	38	40	52	27	99	103				
59.965	F	146	145	132	96	106	62	134	92	122	31		35	49	93		36		37	48	28					32	26
59.966	M	163	157	143	94	113	69	154	105	115	29	126	38	63	103	36	35.5	39	39	50	24	93	102	53	72		
59.968	I	165	144	144	96	117	73	137	104	121	35	121	35	60	102	33	35	40	39	50	25	96	102	56	68		
59.969	F	153	145	127	89	115	69	135	93	123	36	117	35	50	100	35	35	36.5	37	45.5	27			58	66	30	27
59.970	I	168	153	144	94	126	78	144	113	117	37		35	65	104	36	37	42	40	54	26	95	101				
59.973	I	174	127		88	103	64	122	84	120	28	101	36	46	90	32	33	36	35	43	24						
59.974	M	154	143		99		68	136								36	36	40	40	50	26			57	65		
59.976	F	170	131	131	93	120	72	132	91	123	37	109	31	63	107	35.5	35	38	38	49	25	100	99	53	68		
59.977	I	161	141	134	98				91	130		113	37	55	105	37	35	39	38	45.5	28					28	23
59.978	F	160	146	142	97											37		39		50.5	26					32	31

ID number	Sex	GOL	XCB	BBH	WFB	MNH	ANH	XDB	BD	MAJ	HS	IW	MAR	HAR	CSL	OBH R	OBH L	OBB R	OBB L	NLH	NLB	BAL	BNL	PXL	PXW	FOL	FOB
59.979	F	150	145	133	88				93		34		36	54						49	24					34	28
59.980	F	164	133	135	88			132	98	114		126	39	61	99	35	35	38	37	45	25	94	95			32	28
59.981	M	157	143	146	89	113	68				35		32	60		32	32.5	36.5	37.5	48	26	99	105	57	62	33	29
59.982	M	180	132	139	91	120	74	142	101	113	36	119	38	62	105	36	35	41	41	54	24.5	100	106	53	66		
59.983	I	162	137	138	92	121	70	128	98	129	38	118	34	48	103	32	33.5	37	36	48	26	97	101	55	67	34	27
59.984	M	174	146	137	100	120	78	128	112	109		126	35	70	101	35	33	35	37	55.5	26	98	105	53			
59.985	F	177	123	151	97	112	65	136	91	110	34	118	37	74	96	34	34	38	38	46	24			53	62	33	27
59.986	F	165	140	133	93	113	67	132	93	116	36	111	42	64	106	31	32	38	37	46	26	102	99				
59.987	M	161	142		86	116	69	140	105	117	34		37	69	107	36	36	37	38	49	26			54	64		
59.989	F	151	147		89																						
59.990	M				91	124	74		116		37		31	58		33	33	39	39	52	27			50	67		
59.991	F	170	135	135	88				98	126	35	110	33	56	105											33	26
59.993	F	152	137	131	91	109	69	131	90	122	31	116	36	55	98	33	34	37.5	37	50	26	91	91	51	62	30	24
59.994	M	150	164		104	120	75	147			37					36	36	41.5	41	52.5	25			55	66		
59.995	F	162	142		96	110	67	126	85	115	30	106	38	62	97	33	35	40	38	46	22			54	59		
59.996	F	159	139	130	92	107	63	129	99	122	33	118	35.5	50	95	31	31	37	37	44.5	23	90	91	49	61	31	27
59.998	M	160	137	137	90	116	71	138	104	114	36		40	66	104	35.5	36	41	38	51.5	24	100	103	56	65	31	28
59.999	F	163	138	138	95	110	66	133			35		41	59		33	33	37	37	48	23	95	101	52	58		
60.000	I	161	154	148	99	110	65	144	104	119	34	127	37	57	100	35	34	41	39	49	25	98	105	52	67		
60.002	M	168	142	138	92	131	78	142			36		36			37	37	40	39.5	55	28	98	104	54	61		
60.003	M	175	140	135	100	119	74	137	97	119	34	117	39	59	107	36	36	42	40	52	27	98	99	55	63		
60.004	F	169	139	144	88	128	68	129	103	125	38	112	33	63	101	36	36	39	39	52	27	91	99	58	70		
60.005	F	159	143	143	97	114	71	131			38		33			35	35	39	38	47	24	101	97			29	25
60.006	F	161	142	138	98	116	72	129			36		34	57		35	36	39	39	50	26	90	97			30	23
60.007	I	162	139	129	82	111	67	133		128			33	46	88		37		38	48	29	86	94			32	26
60.008	F	156	142	131	85	116	71	128	97	122	34	114	37	57	99	35	36	37	37	51	25	93	97	50	60	32	27
60.009	F	152	136	133	90			130	85	123		113	34	51	97	33	33	38.5	38	47	25.5					28	27
60.010	M	166	150	129	96	109	67	137	103	111		120	40	67	100	32	32	40	38	46	30	96	100	58	66		
60.011	F	158	140	133	89	110	66	131	100	125	30		35	53	98	37	37	37	38	46.5	23	92	93	54	60	35	30
60.013	M	171	146	140	91	121	74	130	92	128	36	115	31	57	108	33	33	39	38	49	30	101	105	55	68	32	27
60.015	I	152	140	135	89	121	74	136	104	125	33	121	35	57	110	35	35	38	39	50	25	97	96	57	72	34	30
60.016	I	164	145	136	94	125	76	131	96	124	36	114	32	56	110	34	33	39	40	51	27	97	98	54	61	31	30
60.017	F	177	141	133	92	108	67	120			36		35	50		32	31	36	37	45	26			52	61		
60.018	M	176	145	146	103	123	75		112	116	35	131	33	64	102	36	36.5	42	41	54	25	95	106	53	70		
60.019	M	168	148	150	100	130	81	142	98		40		37			34	36	42	41	55	25	106	106	60	69		
60.021	F	150	140	134	90	107	66	131	96	125	32	119	36	51	101					46	24	88	89	48	60	30	26
60.023	F	166	133	148	94	115		135	95	120	35		41	61	108	34	35	37	39	48	24	105	103	56	68	35	25
60.024	M	173	136	133	87				103	127		117	31	53	108											34	26
60.025	I	156	142	138	98	118	72	136	94	118.5	35	124	35	62	104	37	36	40	40	50	25	94	102			27	28
60.026	M	182	134	140	100	120	75	140	103	114		118	33	66	98	34	34	43	41	52	26	95	105				
60.027	M	156	159		95	119	73	147	108	116	35	127	41	65	108	36	36	39	39.5	48	27			54	65		
60.029	F	163	142	135	90	116	69		91		35		38			33	35	39	39	50	26	93	99	50	62	31	24
60.031	I	167	134	128	86	106	68	128	85		31		35			36.5	35	37	37	48	24	94	98	56	61	28	24
60.032	I	157	146	131	91		68	131												50	23.5	89	94	50	60	32	27
60.033	I	155	145	138	96																						
60.034	I	154	145	141	92	115	70	132	90	119	30	119	31	61	102	34.5	35	39	38	48	27	97	101	54	67	32	27
60.036	F	158	133	136	89	109	71	130	93	120	33.5	119	37	57	89	36	36	41	39	49	24.5	96	96	54	63	33	28
60.037	M	158	156		103	125	78				36		35			36		44		54	25			56	66		

ID number	Sex	GOL	XCB	BBH	WFB	MNH	ANH	XDB	BD	MAJ	HS	IW	MAR	HAR	CSL	OBH R	OBH L	OBB R	OBB L	NLH	NLB	BAL	BNL	PXL	PXW	FOL	FOB
60.038	M	154	146		89	122	73	134	109	126	39	124	34	52	102	32		38						55	63		
60.039	M	159	155	156	105		77	148	111				39			34	34	41	41.5	54	26	102	106				
60.041	F	166	128	129	88	118	69	127	98	118	36	118	36	59	101	32.5	34	36.5	37	47	26	96	99	55	55		
60.042	I	155	144	142	112			140	110	113		121	38	66	98	36	37	40	39	48	24.5	102	108				
60.046	M	180	151	151	105	122	74	147	108	112	39	137	44	72	105	36	35	42	41	50	26	103	108	56	67	36	29
60.047	M	166	139	133	91	115	71	133			32		30	64			36		39			97	103	55	71	33	28
60.048	F	184	132		100				93	115	33	120	39	60	97					54	26			50	59		
60.049	F	157	138	135	84		70	130	96	116		121	39	60	103	36	36	38.5	39	50	21	93	100			31	24
60.050	M	173	149	139	96			133	98	121		128	36	61	107	36	35	42	40	50	26						
60.051	M	171	142	147	89	118	72	141	106	118	37	131	36	67	105	33	32	39	38.5	49	26	100	104	55	68		
60.052	I	169	131	136	93	118	70	132	96	128	36	114	34	53	105	34	33.5	38	38	47	26	93	99	54	64	31	25
60.053	I	164	132	134	88		73	131								35	36	40	38.5	51	26	98	102	52	66	32	24
60.055	F	161	145	133	91	110	65	128		126	30		34	55	98	33	34	39	38.5	45	28	97	97	53	63	28	26
60.057	F	156	140	134	92	114	68	132	84	118	34		37	60		34	33	39	38.5	46	26			55	61	31	25
60.058	I	171	151	142	104															55	26						
60.059	F	165		133	89	116	69	131			34.5					34	34	37	37	49	28	93	101	53	59	31	26
60.060	I	153	154	135	96	110	69	133	98	124	30	112	36	59	104	37	37	40	40	49	22	96	100	52	60	32	29
60.061	F	171	139	142	100	110	69	136	97	124		131	39	56	100	33	32.5	40	40	49	26	94	99	53	58	33	28
60.062	F	167	139	137	92	111	71	132			34		34			35	34	39	38	52	26	97	104	54	60	29	25
60.063	F	163	140	136	91	112		130	100		33		38													33	28
60.064	M	169	147	143	105	122	76	146	106	112	35	128	40	66	105	37	36	43	41	52.5	25	98	99	54	67		
60.065	M	165	134	135	93	121	77	142	102	113.5		126	37	63	102	38	36	39	40	56	25	97	99	56	68		
60.067	I	154	140	138	95	116	68	124	94		31		35			36	36	39	38	47	26	99	100	54	65	33	25
60.068	I	147	147		94		67									32.5	33	36.5	37	47	26			53	59		
60.070	F	148			93	117	74	140		127	37		32	56	107	33.5	35	40	39.5	50	26						
60.071	M	175	142	140	87	121	75	130		111	36	112	44	70	105	35	34	37	38	51	21	101	102	58	61		
60.073	F				94	119	72									38	38	41	38	48	26			51	61		
60.074	M	154	146	134	89	117	71	142	88	105	35.5	122	40	74	98	35	35.5	42	41	51	27	98	99	58	69		
60.075	M	180	125	133	91	118		133	97	126		121	37	55	110	32.5	32	39.5	38	50	27						
60.076	M	163	136	121	82	119	74	130	103	124	33	120	31	55	99	33	33.5	37	37.5	53	37.5	93	92	51	61		
60.080	I	166	152	142	97	120	75	142	98	114	35	125	34	60	103	36	38	42.5	42.5	55	26	99	104	55	67		
60.081	F	153	136		83	102	63	125		116	31		36.5	64	102	33	33	37	37	44.5	24			55	62		
60.082	F	159	138	133	98	118	73	127			35		32	58		37	36	39	37	51	25	100	100	56	60	34	26
60.083	F	163	136	137	95					129			38	54		35	35	37	37	49	27	101	102			33	25
60.084	F	153	139	132	96		65	132								32		38.5		47.5	23	92	95			29	28
60.085	M	168	149	145	92	121	72	144	111	119	36	119	36	64	111	35	35	40	40	52	25	93	101	57	65		
60.086	M	162	142	138	89		69	138								36	37	39	39.5	49	27	97	97			33	25
60.087	M	154	148	143	91	121	76		114		35		35	59		35	35.5	38	38	55	24	93	95	56	65		
60.088	M	176	150	140	95	125	81	145	113	118	38	126	36	58	102	38	37.5	41	42	57	23	100	109	57	67		
60.089	I	156	150		104	125	77	130	93	121	37		42	57	104	35.5	35.5	42	41	53	28			58	66		
60.090	M	171	150		109	112	70	161	109		36		39	66		37	35	40	42	48	29			59	68		
60.091	F	161	151		92				93	125.5	37		37	50	108												
60.092	M	175	149	139	96	120	73	136	95	127	35	125	36	61	104	34	34	39.5	39	53	26	99	102	59	66	33	26
60.093	F	178	140	139	98	116	68	130	92		34		35	56		35.5	36	41	42	49	26	95	101	55	65	34	26
60.094	F	161	136	131	86	104	63	130	88	116	29	108	34	55	102	33.5	33	37	38	47	26	97	97	53	62	30	25
60.095	F	162	139	137	93	115	71	127	95		29		32			37	38	39	38	52	24	85	93			35	30
60.096	M	165	150	144	90	127	78	142					39	78		34	34	40	39	50	27	100	101	58	64		
60.097	I	159	148	138	95	116	67	129		127	35		39	55	111		32		37	47	28	102	100	56	63	30	26

ID number	Sex	GOL	XCB	BBH	WFB	MNH	ANH	XDB	BD	MAJ	HS	IW	MAR	HAR	CSL	OBH R	OBH L	OBB R	OBB L	NLH	NLB	BAL	BNL	PXL	PXW	FOL	FOB
60.099	I	157	149	129	91	108	65	129	92	118	31	125	35	53	93	34	34	37	35	47	25	91	93	51	61	31	25
60.100	I	168	133	126	86	111	67	125	89	124	34	106	34	52	96	33	33	36	36	46	25	89	94	50	65	29	24
60.101	F	170		137	95	109	66									33	32	38	36	47	25	98	103	50	58		
60.102	M	175	143	138	105	121	74	140	109	114	36	121	38	68	102	34	35	41	40	50	28	96	99	52	66		
60.104	I	178	137	139	91			140								32.5	32.5	41	40	50	27						
60.105	I	183	135	134	93		74	133									35		40	51	28	101	101				
60.106	M	154	158	140	100	118	70	139	101	111	35	130	34	65		35	35	42	41	51	28	91	95	54	63		
60.107	M	160	143	148	94	115	67	134	97	119	35	117	38	61	105	35	35	38	38.5	49	26	93	98	53	66		
60.108	F	156	147	138	98	116	70	133	94	132	32	115	39	46	100	35	35	39	38	50	25	95	98	58	64	31	22
60.136	I				101																						
60.138	F	156	141		90																						
60.141	I	175	125	136	85				90		34		36	61												32	29
60.142	M	181	126	139	93		68	131																52	64	33	25
60.144	F	156	135	133	87																						
60.145	F	164	136		85																						
60.165	F	160	142			117	71		98	129	34		36	53	110	33		39		51	25.5			57	68		
60.175	F	165	142		103		72				36					33.5		38		49.5	26			55	63		
60.177	M				92	114	66	136			32		39	53		34.5	34	42	40	45	27			61	66		
60.179	M				96								35	51		35	34	36	38	45.5	23			54	64		
60.188	I	162	140	138	95		71	133								34.5	35	38.5	38.5	50	27.5	101	100	56	70	32	30
60.194	M	172	140	137	94		68	129	90	119	34	121	40	53	104	36	37	41	39	48	25	96	98	52		32	25
60.195	M	156	157			121	76		113	127	36	126	35	51	103					55	24.5			55	69		
60.196	I	166	145	134					92		32		35	50												33	25
60.198	M	176	140		95				94	122	36	116	36	54	109												
60.199	F	167			96	123			96	131	37	105	34.5	47	106	35	34	39	38	47.5	26			52	60		
60.200	M				93	116	67		98		33		35	63	96	33	33	39	38	47	25.5			49.5	62		
60.201	M	180	122		88																						
60.207	I	159	146	131	92		75													53	25			48	61	35	31
60.211	F	157	147	140	95		72									36	37	40.5	40	51	25	96	100			36	26
60.212	M	170	143	129					100	121		124	37	70	101											31	30
60.213	I	148	138	139	86																			57	62	31	26
60.218	M	177	145	139	98	125		138	100	118	38	128	36	64	108	34.5	34	41	41	49.5	25						
60.219	M	170			92	122	76	133	100	113.5	33	120	37	68	98	38	37	40	39	51	22			54	64		
60.220	M	170	141		99	114			104	112			33	64						49.5	23.5			55	62		
60.222	F	147	147		90	110	71	131	100		31		36			34	33	39.5	39.5	47	24						
60.225	I	180	121	134	88	107	66	128	100	117	35		33	60	91	33	33	38	37	43	26	91	97	51	65		
60.226	M	168	145		95	115	73		98	114	35	110	37	69	105	36	36	40	40	53.5	25						
60.227	F	176	131		97	112	67			130	33		36	52		33	33.5	40	39	48	27			54	67		
60.228	F	159	134	140	85	111	66	127		126	30	107	31	52	99	34.5	35	38	38	46.5	23	90	99			32	26
60.229	F								107		37		36	57										55	61		
60.230	M	177	133		96	121	73										35		42	54	28						
60.231	I	158	141	141	101	113	69	154	105	115	29	126	38	58	100	35	35	39	43	47	25	96	103				
60.234	I	177	120		88						37		38	52													
60.235	M	182	138		92				92		31		31														
60.236	M	181		141	95	118	73			110	31		42	62	105	36	35	42.5	42.5	54.5	26.5	92	102	51	65	35	27
60.237	I				97	126	77				38		38.5	76		35	34	42	44	51	26			56	70		
60.238	M	168	136	143	92	119	73	140	108	113	36	138	37	65	102	33.5	32.5	37	38	53.5	26	102	108	56	69	35	30
60.239	F	153	147	133		111	68	131	106		34		34	54		34	34	39	39	46	24	93	93	55	68	32	26

ID number	Sex	GOL	XCB	BBH	WFB	MNH	ANH	XDB	BD	MAJ	HS	IW	MAR	HAR	CSL	OBH R	OBH L	OBB R	OBB L	NLH	NLB	BAL	BNL	PXL	PXW	FOL	FOB
60.240	M	161	145	137	95											36		39.5								34	26
60.241	F	161	130	132	85	102	64	130			28		37	57		34	34	38	39	47	27	97	97	53	65	30	25
60.242	M	163	145	146	92			142	104	118		127	36	60	106	34	33	42	41	52.5	25	95	106			35	31
60.243	M	170	143	139	94		77				40		36			35		41		53	25	104	101				
60.248	F	169	130	135	90	121	74				36.5					35.5	36	39.5		50	26	103	103	58	64		
60.249	F	158	138	129	99	119	74				37					34.5	34	40	39.5	48	25	94	94	54	62		
60.252	F	160	147		93	119	69	129	104	123	36	115	39	58	105	33	32	39	38	48	24			53	62		
60.255	F	156	143	136	90		70	127		125			36	48		33	34	40	38	51	26.5	101	100	54	61	30	24
60.256	I	180	142	130	95	126	77		99	114	38	127	38	69		34	34	42	40.5	51	24	101	100	58	69		
60.257	F	157	147	140	89	106	65				33	116	35	48	99	34	33	38	37	45.5	24	89	93			33	29
60.258	M	166	136	131	90	118	70	132	98	120.5	35	125	36	60	108	33	33	39	38	49	25.5	96	99	55	73		
60.259	M	167	144	148	93	123	75		94	116	34		37	60	103	34	36	41	40	52	23			54	64	32	27
60.260	I	168	136	139	92	111	67	144	103	116	33	131	35	68	105	33.5	33.5	40	41	48	27	103	105	56	69		
60.261	M	182	127		87	115	73	126			31		37	68		35	35	38	35	51	26			52	65		
60.262	I	144	150	137	94	115	70		100	120	34.5		37.5	57		33	33	37	37	48.5	25	91	94	54	65	30	26
60.266	M									125	36	124	37	59	113									59	66	34	31
60.267	M	181	138	148	98		73	145	108	110	35	135	39	74	98					54	28		108			36	27
60.271	I	177	140		96	119	73	131	98	135	39	110	35	49	109	34	33.5	38	38	48	27			61	65		
60.274	M	178	136	133	90	121	70	138	94	116	37	118	38	61	106	33	33	38	37	49	28	100	101	53	67		
60.276	M	176	136		99	117	72		88	117	33		35	65	108	34	36	42	41	51	25			52	62		
60.277	M	182	139	133	92	117	70	138			36		39	59		33	33	42	42	46	26	96	101	57	62		
60.280	M	177	134		96	112	70	135		116	33		34	63	101	33.5	34	39.5	39	49	27			54	63		
60.281	F	172	132																								
60.282	I	156	137	134	94	111	67	130	88	123	34	120	38	55	103	34.5	34	37.5	38	48	24	95	97	48	62	30	23
60.283	M	176	126	134	89	121	71		94	119		112	39	71	107	32	33	37	38	48	26	96	101	55	62		
60.284	M						·		109	124	33	122	35	57	95									55	63		
60.286	I	182	146	146	103	135	84	147	112	118	41	128	39	68	110	37	36.5	43	42	54	24	97	110	54	68		
60.287	F	173	140		93				94	128	39	121	38	56	112												
60.291	F	170	141			109		127			33		37	51	96					49	26			56	66		
60.292	F	162	150	134	96	126	71				41		37	55		34		40				95	93			31	27
60.293	M	167	149	141	98	120	74	141	107		38		38	62		34.5	34	41	40	50	29	97	102	60	66		
60.297	I	162	139	137	96	119	72	131			37		39	59		35	34	43	41	46	28	99	100	56	64	31	24
60.301	I	172		136	90	115	70	126	100	121	32		35	59	107	35	33	38.5	39	51	27	107	108	59	63	33	23
60.302	F	148	146	141	99	114	73	133	99	125	32	121	34.5	59	107	34.5	36	39	40	51	27	92	97	53	65	32	30
60.303	F	166	143	137	94				92	122	34	128	38	50	104							99		56.5	66	33	28
60.307	F	169	133		91	113	68	123	86	120	33	111	33	54	196	34	36	37.5	36.5	48	22			53	57	30	22
60.309	F	163	148	143	99																					35	29
60.313	M	170	139	137	98	112	70	131	90	125	33	115	35	58	107	36	35	40	41	48	29	99	105	54		32	28
60.314	M	165	155	143	97	126	75	139	103	119	36	125	42	57	110	34	34	39	39	53	26	92	97	54	66		
60.315	M	183	141	140	100	121	75	142	90	114	37	118	40	70	108	36.5	37	41	40	53	24	103	103	57	69		
60.316	M	171	143	147	103			143								34	34.5	42	40	52.5	29.5		107				
60.317	I	174	145	140	92						37		42	61												35	30
60.318	M	173	146	138	95	125	77	141	102	116.5	37		39	64	108	35	36	41	40	53	28	99	100	58	66		
60.319	M	173	136	128	98	117	74	131	94	119	35	117	39	60	99	34	34	38.5	38	50	27	94	101	53	63		
60.320	M	140	137	133	83	107	67	122	94	125	32	111	36	54	98	34	33	36	37	48	24	94	94	53	60	30	26
60.342	I	162	143	136	94	114	71	133	98	127	34	125	33	52	102	34	36	39	36.5	50	27	101	101	52			
60.372	I	156	143	133	97																						
60.428	I	169	131	132	88	114	71			112			34							50	24			46	60	35	29

ID number	Sex	GOL	XCB	BBH	WFB	MNH	ANH	XDB	BD	MAJ	HS	IW	MAR	HAR	CSL	OBH R	OBH L	OBB R	OBB L	NLH	NLB	BAL	BNL	PXL	PXW	FOL	FOB
60.429	I	182	136	152	102	114	75	139					35	64		36	36	43	41	55	29	99	106			33	28
60.440	F	164	142	134	89	113	69	131			36					35	35	39	39	48	25.5	95	97	51	64	31	27
60.442	I	156	145	146	92	117	71	134		118	34		35.5	54	101	34	35	40	39.5	46.5	23	90	96	51	61		
60.445	M	163	150	136	101	108				117			37	60		35	35	40	40	50	24.5	88.5	95				
60.446	F	156	143	131	95		72	136		121		121	36	55	105	36	36.5	40	39.5	49	25	97	95				
60.451	I	151	152	137	89				104			130	33	54	99	35	35.5	41.5	40	49	25.5		95				
60.452	F	146	136	130		115	70	125			36		33.5	51	103	36		36.5		49	24	96	97	54	63	30	25
60.453	M	175	127	142	90		73	136								37	36	40.5	40	54	28	98	103				
60.454	M	175	142	144	105	121	75	136	106	108	37	116	37	77	105	35	35	43	42	50	25	97	102	55	67	33	28
60.455	F	149	150	134	99	104	64	140			28	115	36	34	100	34	34	41.5	39.5	44.5	26.5	94	92	52			
60.456	I	161	155	142	97	121	74	143		117	35		40	62	115	37	36.5	44	41	55	27	103	102	59	71	32	30
60.457	I	158	149	136	97	107	64	129			32		36	57		33	32	40	39	46	27	90	93	51		29	27
60.463	I	151	142	138	88		63						34.5	57		34.5	34		38	55	21.5	90	95	58		32	26
60.464	I	163	146	134	94	120	72	127	94	115	33	117	36.5	61	102					51	23			51	65	30	25
60.479	I	162	138	135	88	110	67	128	92		32	109	35	54	96	33	33	39	37.5	47.5	23.5	91	94	57	59	33	28
60.480	I	146	150	142	93	109	67	128	100	109	31		41	64		33	35	41.5	38	49	24	94	99	50	62		
60.483	M	148	147	135	92	116	75	125	97		34		36	48		36	35.5	41	40	50.5	22	92	93	51	55	30	25
60.484	M	171	135	138	93	122	71				37		38	60		35.5		37.5		50	26.5	99	95	57	71		
60.494	F	167	141	141	86		78									35.5	35	41	40	55	25.5	96	96	58	68	32	33
60.496	F	157	137	131	94	107	66	128	101		31		34	54		34	33	39	40	45	26	95	95				
60.497	F	157	142	139	91	109	69			123	31		32	53	95	33	36	39	37	46	26	96	95	53	62	31	25
60.501	I	149	140	129		112	65	130		119.5	32		40	56	108	35		40.5		48	27	101	95	54	62	30	25
60.502	M	173	143		95	122	76	137	93	115	35.5	123	35	64	100	32	33	39	40	53	25			56	67		
60.505	F	158	150	139	91	105	64	137	90	96	30	119	35	55	118	33	33	40	40	46	25	92	95	50	61	31	29
60.513	M	180	144		102		91	142		116	42	85	39	69		38	38	45	44	62	25			54	70		
60.514	M	168	141	133	94		80	137			40		36	64		34	32	41.5	41.5	55	26.5	100	101	58	65		
60.518	M	177	141	141	93								39	61		33	32.5	38	40	50.5	26	99	105				
60.520	F	178	139	130	90	118	73	140	103	125	39	118	37	61	109	31	31	40	41	47	29	102	99	59	70	31	25
60.521	F	175	134	127	98	116	67				33.5		34.5	57		33	33	39.5	38.5	44	21.5						
60.522	F	149	132	127	86	113	69	124		122	34		35	52	102	35	34	39	38	47	26	96	92	52	62		
60.524	M	152	144	138	96	118	74	125	96	130	34	106	36	51	101	36	36.5	40	39	53	27	92	94	50	62	33	26
60.529	F	169	128	135	99	108	66	125	95	123.5	32.5	108	38	52	99	36	34	39	40	46.5	27	95	98	49	61	30	24
60.534	I	169	147			120	76		101		35		37	62		34		39		52	25			57	68		
60.539	I	158	147	145	96	117	71	129			35		36	57		36.5	36	41	40	53	25	99	105	52	65		
60.542	F	181	128		97	115	70	137	100	128	36	124	37	52	105	35	35	40	40	49.5	29			54	66		
60.543	F	157	136	134		113	70		92	120	36		37	60		36		39.5		51	28	101	99	54	60	31	26
60.544	I	166	143	134	97		67	130								36.5	37	40	38	48	25	93	98	53	61	26	23
60.546	I	160	136	132	89		67	128								34	34	37.5	37.5	47	23	92	98	52	60	33	30
60.548	I	147	151	132	96	113	70	137			31		32.5	54		36	36	40.5	41	52	26		92	51	64		
60.554	I																							52	65		
60.556	M	158	140		88	122	76	130			35		37			38	38	41	39	53	24			52	65		
60.559	I	150	139		96		71	122		111	31.5		37	62		36	35	38	38	51	24			53	58		
60.560	M	153	154	139	95	118	73		106	110.5	34	124	40	63	101	39		40		51	24.5			52	62	35	30
60.561	F					111	69				32		34	59		35	35	40	39	47.5	25	94	89	53	64	33	27
60.562	M	163			92		76						38	58		33.5	34	38	37	52	24			53	60		
60.568	I	153	149	139	94	122	74	131	101		33		32	52		35	36	39	39.5	52	25.5	89	96	53	62	32	26
60.569	M	162	147		95	120	76	145	108	115	37	132	40	67	102	39	38	42	41	51.5	24.5			55	70		
60.570	M	177	126	123	95	109	65	128	96	115	28.5		40	60		35	36	42.5	41	48	27	96	100	50.5		34	27

ID number	Sex	GOL	XCB	BBH	WFB	MNH	ANH	XDB	BD	MAJ	HS	IW	MAR	HAR	CSL	OBH R	OBH L	OBB R	OBB L	NLH	NLB	BAL	BNL	PXL	PXW	FOL	FOB
60.571	F	156	138	137	92	112	70		88	122	30	102	37.5	52	102	36	36	39	39	49.5	22	92	98	48	63		
60.574	F	159	143	145	98			134					32	50	108	32	32	41	42	48	27		105			30	25
60.575	F	170	138	133	92	112	69	133	97	119	31.5		38	57		36	36	41	41	49.5	27	96	100	53	65	31	28
60.577	I	181	142		91																			58	61		
60.579	M	167	151			126	81		106	111	37		39	69	99		38	44	41	57	25			55	66		
60.580	F	162	143	141	95	110	68			119	34	120	34.5	56	103	35.5	35.5	41	41	50.5	26	92	102			34	27
60.581	M	167	144	137	96	116	71	141					40		104	36	33	41	39.5	50	27	97	101		65		
60.582	F	155	140	129	78	109	64				31		36	50		33.5	33	39.5	37	47	24	92	93				
60.583	I	160	150	143	92		78	147	98	113			36.5	62	102	36	36	41	41	57.5	26.5	97	104	58	70	33	30
60.590	F	167	145	141	97	120	74		104	121	36	121	34	55.5	105	36.5	36	41		50.5	28	101	103	54	67	35	26
60.593	I	164	142	136	100	122	75	139			33	128	35	58	104	35	34.5	42.5	42	51.5	28	93	97	53	65	32	24
60.594	F	168	129		96	108	64	125			34					33	32	39	38	44	26			52	60		
60.595	I	165	145	141	91	122	77	139	116		37		41	60	105	34.5	35	39.5	39.5	51	22.5	95	104	51	66	36	31
60.600	I	157	146	137	101	104	64.5	129	96	122	29	116	31.5	51	100	35	35	40	37	48	27	92	95	53	66	29	30
60.601	F	151	153	146										63	100		36		40	47	24			54	67		
60.605	I	185	138		98	126	80				34		39	69	105	37	35	41.5	41	55	27			56	67		
60.608	I	157	140		93		68	133			32		31														
60.609	F	150	146	139	94	115	73	131	97	120.5	32	114	33.5	60	101	36	36	40	39	52	25.5	89	95	51	63	31	26
60.610	F	173	141	135	95	116	74	127	95	122	36.5	111	42	55	107	34	37	39	39	50	28	98	100	57	65	33	26
60.611	F	166	134	133	91		71	129								32	33	41	41	51	26	92	94	52		35	29
60.612	I	168	136	129	105	121	74	133								38	36	44.5	42	48.5	27	100	101	56.5	63	32	27
60.615	F	158	139	128	92	127	77	130			37	106	32	55	105	38.5	37	41	39	52	24	85	94	51	60	30	27
60.620	F	178	148	140	97	120	74	138			37		40	57		33.5	33	41	41	50	25	103	102	58	67		
60.623	I	162	140		88	110	68	127	88	120.5	30.5	114	35	54	102	34	32	37	37	49	23			51	59		
60.624	M	158	159		90	115	74				34		34	48			37.5		39.5	52	23						

PART 2: POSTCRANIAL DATA

Key to Postcranial Measurements and Definitions

The ID number is the catalogue number assigned by the Peabody Museum of Archaeology and Ethnology when the human remains were accessioned. The data for each ID number continue from the left page to the right page. Parentheses in the following table indicate measurements that Hooton estimated.

Abbrev.	Measurement (mm)	Description
FOL	Femur length bicondylar	Length of the femur in natural position with the condyles resting against the vertical wall of the measuring board and the movable square tangent to the extremity of the femoral head.
FML	Femur maximum length	Greatest length from the internal condyle to the extreme point of the head on the measuring board.
XHD	Femur: diameter of head maximum	Maximum diameter taken with sliding caliper.
APD	Femur: subtrochanteric diameter anteroposterior	Sagittal diameter of the shaft below the lesser trochanter at the level of the greatest transverse diameter.
MLD	Femur: subtrochanteric diameter lateral	Transverse diameter perpendicular to anteroposterior diameter at subtrochanteric.
APS	Femur: middle shaft diameter anteroposterior	Diameter between the dorsal and ventral surfaces of the femur at the middle of the shaft, perpendicular to the ventral surface.
MLS	Femur: middle shaft diameter lateral	Transverse diameter perpendicular to anteroposterior diameter at middle shaft.
TML	Tibia: length maximum (minus spine)	Place the tibia with the malleolus against the vertical wall of the osteometric board and with the long axis of the bone parallel with the long axis of the board. Place the square against the anterior edge of the lateral condyle external to the tibial spine.
TAPS	Tibia: middle diameter anteroposterior	Mark the middle point on the tibial crest. Measure the anteroposterior diameter.
TMLS	Tibia: middle diameter lateral	Measure perpendicular to middle anteroposterior diameter.
APN	Tibia: nutrient foramen diameter anteroposterior	Anteroposterior diameter at level of base of nutritive foramen on external surface of bone.
MLM	Tibia: nutrient foramen diameter lateral	Measure perpendicular to nutritive foramen anteroposterior diameter.
BML	Fibula-maximum length	Maximum distance between the proximal and distal ends measured with an osteometric board.[a]
HML	Humerus: maximum length	Place the head against the fixed vertical of the board and adjust the movable upright to the distal end. Obtain the maximum length.[a]
MDA	Humerus: middle diameter anteroposterior	Diameter taken at exactly the midpoint, in the anteroposterior plane.[a]
MDM	Humerus: middle diameter lateral	Diameter taken at exactly the midpoint in the medial-lateral plane.[a]
MDH	Humerus: maximum diameter superior articular head	Taken from a point on the edge of the articular surface of the bone across the opposite side.[a]

Abbrev.	Measurement (mm)	Description
RML	Radius: maximum length	Maximum length from the head to tip of the styloid process.[a]
UML	Ulna: maximum length	Maximum length from the top of the olecranon process to the tip of the styloid process.[a]
PXB	Pelvis as a whole: maximum breadth	Maximum diameter outside of iliac crest.
PDII	Pelvis: distance between ischiatic spines	Posterior diameter (bispinous): distance between ischial spines.[a]
INH	Ossa innominata: height	Maximum height of the innominate bone between the iliac crest and the ischiopubic ramus measured with an osteometric board.
ILB	Ossa innominata: breadth	The distance from the anterio-superior iliac spine to the posterior--superior iliac spine.[a]
SAL	Sacrum: height	The distance from a point on the promontory positioned in the midsagittal plane to a point on the anterior border of the tip of the sacrum measured in the midsagittal plane.[a]
SMB	Sacrum: maximum breadth	Greatest breadth of sacrum at level of anterior projection of the auricular surfaces.

[a] Measurement definitions were not published by Hooton but are likely to have been similar to measurements defined by Zobeck (1983) and Moore-Jansen and Jantz (1989).

ID number	Sex	FOL R	FOL L	FML R	FML L	XHD R	XHD L	AP D R	AP D L	ML D R	ML D L	APS R	APS L	MLS R	MLS L	TM L R	TM L L	TAPS R	TAP S L	TML S R	TML S L	AP N R	AP N L	ML M R	ML M L
59.801	F	415	416	418	419	42	43	23	21	31	30	26	26	25	24	347		31	31	19	19	34.5	35	22.5	23
59.802	F	403	404	406	406	38	37	20	21	26	26	26	26	20	21	338	340	28	27	18	18	30	30	19	19
59.803	F	402	398	404	400	38	39	22.5	22	29	30.5	27.5	26	24	25	342	341	32	32	20	20	35	36	22	21
59.804	M	(410)	412	(413)	414	40	38	21	22	27	28	27	28	19	21	355	351	27	30	19	18	28	30	22	21
59.805	M	432	430	436	439	42	42	22	23	32	31	27	28	25	25	374	370	31	32	20	20	36	36	21	22
59.806	M	408	(408)	409	410	44	45	24	24	35	36	26	27	28	27	366		33	32	22	22	34	35	23	25
59.807	M	(408)	(410)	(410)	(411)	41	41	21	20	29	19	26	26	22	21										
59.809	F	420	418	425	423	38	38	22	22	30	31	26	26	23	24	352	354	26	27	19	19	31	30	22	23
59.810	I	418	415	420	416	42	41.5	22	22.5	30	30	28	26	23	23	350		27	29	18	19	31.5	30	19	20
59.811	F					37	36	22	21	30	31	24	23	23	24	331		27	26	18	18	29	30	19	20
59.813	F	420	426	424	428	42	41	22	21	32	33	26	25	26	27	346		29	29	19	18	32	32	19	18
59.814	M	380	380	380	383	35		19		24		24	20	18		312	316	23	24	19	19	27	28	21	21
59.815	I	389	391	392	395	39	39	21	21	28	29	24	25	23	24	324	319	29.5	27	17	17	30	30	18	17
59.816	F		405		406	40	45	22	20	33	32	23	24	23	24										
59.817	F	379	381	383	384	39	38	21	21	29	28	28	23	23	23	323	320	27	27	19	19	30	29	20	21
59.818	I	(395)	401		403	39	40	22.5	23	30	30	26	27	25	26		336	29	29	20	20	31	31	20	19
59.819	I	372		376		37		19	19	28	27	25	24	21	21		303		26		19		28		20
59.820	M	396	398	398	399		37	21	22	27	28	23	21	22	24	326	324	26	27	18	18	28	29	19	19
59.823	M	417	413	420	417	47	47	24	25.5	34	34.5	27	26	23	24	345	345	29	32	22	23	33	34	23	23
59.824	I		388		389	38	38	19	19	28	27	23	22	24	22.5		328	29	29	16	16.5	30	32	18	18
59.825	M	427	432	430	435	44	44	24.5	26	34	33	30	30	27	28	354		33	34	23	22	34	35	22	23
59.826	I	428	432	429	436	40	42	22	23	32	33	25	25	25	26	361	359	31	31	21	21	34	34	22.5	22.5
59.827	F	408	414	411	416	41	41	22	23	29.5	29	28	29	23	23	354	356	31	32	21	22	35	36	22	23
59.828	I			377		38	38.5	20	21	28	27	24	24	21	22										
59.829	M	389	397	393	400	36.5	36	18	19	27	26	23	23	20	20			28		17		28		18	
59.830	M	(417)	413	(417)	416		45	24	22	31	31	26	24	26	26	(348)	350	33	32	19	19	36	36	21	22
59.831	I	454	462	455	463	44	44	24	25	32.5	32	28	30	27	25	388	388	33	34	24	24	36	37	23	23
59.832	F	(465)		(470)		45	43	24	25	32	34	30	30	23	23.5			34	34	26	25.5	37	38	26	25.5
59.833	F	(435)	440	(441)	445	43	43	24	25	31	30	29	29	23	23										
59.834	M		355		357		36		22		27		24		22	288	285	26	27	17	17	28	29	18.5	19
59.835	F	404	403		406	37	37	18.5	19	28	27	23	22	20	22		342	28	27.5	17	17.5	31	31	17	18
59.837	F	437		443		43		22	23	32.5	33	25	26	23	22.5										
59.838	F		421		422	43	44	23	27	35	35	30	31	28	29	342		35	35	20	20	40	38.5	21	20
59.839	I	447	(442)	448	(445)	41	41	23	25	30	31	29	29	23	24		374		30		18		31		19
59.840	M	(434)	429	(438)		44	45	22	22	33	33		26		25	361	360	37	36	18	18	38	37	18	18
59.841	I	384		388		36	37	19	20	28	29	24	24	22	23	327	328	26	27	18	17	28	29	18	17
59.842	F	420		422		43	43	24	23.5	34	33	29	31	25	25	(345)		35		20		40		20	
59.843	F	397	400	397	404	40	38	24	26	32	30	29	29	24	24		336	28	30	20	21	32	33	22	21
59.844	I	429	430	(430)	432	43	43	25.5	26	32	30	28	30	25	23		365		34		22		38		22
59.845	M	397	397	399	400	40	40	21	22	29	31	26	25	24	25	340	341	30	28	17	18	32	30	18	20
59.846	I	407	408	408	410	39	38	23.5		31.5		25	26	26	26	348	344					28	26	22	22.5
59.849	M	377	379	384	381	37	37	21	20	29	29	24	24	22	21	311	310	27	28	18	18	30	29	21	19
59.851	M	431		434		45		22		35		28		26		(345)	348		35		19.5		39		
59.853	I	343	438	438	444	42	41	21	22	34	32	28	29	24	21	360	358	33	33	19	23	36	37	19.5	20
59.855	M					41		20	22	33	32	23	26	24	26		347	27	28	20	19.5	31	31.5	20	20
59.856	I	415	410	416	413	41	41	22	22	33	30	28	28	23	24										
59.857	I	390	(387)	392		39	36.5	20	21	30	29	25	24	23	22		323	29	28	19	18	31	29	20	18
59.858	M	469	473	471	476	49	49	28	28	36	36	31	32	27	28	392	396	34	35	22	21	41	43	25	26
59.859	F	404	407	408	411	43	42	22	23	32	31	28	28	24	24	342	339	34	33	26	23	36	36	26	25
59.860	I	384	386	386	390	39	39	21	22	31	30	22	22	21	21	314	321	26	26	17	17	29	28	18	18
59.862	I	424	428	429	432	44	45	22.5	22	32	30	37	37	23	21	362	361	32	31.5	18	20	33	34	19	19
59.863	M	430	433	432	424	45	44	25	25	32	33	31	31	26	27	373	373	36	37	24	24	38	39	26	25
59.865	F	409	416	412	418	43	43	23	24	33	34	28	28	24	25	342		33	32	21	22	38	37	20	21
59.866	I	413	413	415	416	47	46	25	24	31	31	27	28	25.5	26	342	339	35	38	21	21		41		22
59.867	I	371	364	373	371	35		17.5	17	29	27	23	21	20	19	317	310	26	24	18	17	28	28	18	19
59.870	I	454	(457)	457		45	45		24		33	30	29	26	23	386	377	36	34	21	20	40	37	21	21
59.871	I	399	402	401	404	38	38	23	24	31	30	28	28	24	24	338	344	30	31	18	18	32	33	19	19
59.872	I	391	392	397	398	40	40	20	21	25	28	28	26	23	23	331	338	29	33	19	19	32	34	20	19.5
59.874	I						46	27	29	33	35	32	34	27	27			32	34.5	25	26	36	37	26	27
59.876	I	445	445	448	449	48	49	26	27	31	33	32	32	24	23	390		35	35	21	19	38	40	22	21

ID number	Sex	BML R	BML L	HML R	HML L	MDA R	MDA L	MDM R	MDM L	MDH R	MDH L	RML R	RML L	UML R	UML L	PXB	PDII	INH R	INH L	ILB R	ILB L	SAL	SMB
59.801	F			302	(300)	21	20	15	15	43	42		237		(253)	267	89	195	192			95	110
59.802	F	332		294		15		21		37		224		245									
59.803	F	332		295	288	23	22	16	16	41	41	216	214	231	232	259		184	183	137	135	107	112
59.804	M			304		22	21	14	15	39	37	231	228		241								
59.805	M	362		300	311	21	21	17	17	43	43	244	245	267	267								
59.806	M	350		324	315	22.5	21	17.5	17	46	44		249										
59.807	M				299	21	21	14	15		43												
59.809	F	343	346	311	302	22	23	16	16	39	39					282		191	187	149	146	108	116
59.810	I	341	343	(312)	306	24	22	18	18	43	42		238	254									
59.811	F			278		22	22	19	20	38		217	216	231	226								
59.813	F		341	303		23	22	14	14	39			227		241								
59.814	M	(305)	307		275		20		14		37				225	235		178	182	132	134	109	114
59.815	I			273	270	21	21	14	14	39	40	212	208	232	227	276		(184)	184			89	114
59.816	F																						
59.817	F	317	316	277	277	21	21	14	14	37	37	211	208	233	228	266	95	187	186	143	145	(96)	122
59.818	I				283	22	22	16	16		37		225										
59.819	I																						
59.820	M	320		301	297	21	20.5	14	14	39	40	224				260	88	188	189		145	111	115
59.823	M	350	(349)	313	310	24	21	17	16	47	44	243	243	263	262	265	81	208	207	152	148	107	118
59.824	I	326		275	273	20	19	14	13	39	38			232	227							101	114
59.825	M			320		24		17		44													
59.826	I			316	311	23.5	22	14	14.5	40	40	239	236	257	253							98	122
59.827	F				290	21.5	20	17	15		42	234											
59.828	I			277		20		14		34.5		213											
59.829	M	316	317			19	19	13	12.5														
59.830	M	346			308		21		16		45		233		249								
59.831	I	370	369	334	334	21.5	21	16	15	46	44	253	251	272	271	269		211	210		146	114	114
59.832	F	387	(385)	339		22.5		16		43			261		274								
59.833	F				315	22	21	16	15		45												
59.834	M	280														243		(167)	161	123	126	91	105
59.835	F			397	391	20	18	13	14	39	39	(220)		241	236								
59.837	F				323		20		14		(43)												
59.838	F			310	318	22	20	16	16		45												
59.839	I		361	319		24		15		39													
59.840	M	353	356	317	318	20	19.5	16	15.5	44	44.5	246	251	264	267	266			219	148	148	109	113
59.841	I	318	316	287	280	22	22	14	13	37.5	38.5	217	213	235	229	252		178	182	133	134	82	112
59.842	F																						
59.843	F				276		23		17		38	219	217	240	235								
59.844	I				310		21		19		44												
59.845	M	334	335	308	301	21	21	16	16.5	38	36	230	228	250	246	253		192	190		140	114	118
59.846	I		330	304		24		16.5		36		226	222		239	265		197	199		149	114	122
59.849	M		305	273	268	21	21	15	14	37.5	38	205	208		226	256		177	177	(135)	140		
59.851	M		349	315	319	21	19.5	16	15	45	44.5		233		257								
59.853	I			312	312	23	22	17	17	43	42		252	276									
59.855	M																						
59.856	I			298	301	24	23	18	17	41	41	234	232	252	246	(257)		202	198	(154)	152	105	115
59.857	I				278	20	20	15	15	37	37	218	218										
59.858	M	391	389	347	346	30	25		17	52	50	256	258	270	269	285		218	214	165			
59.859	F				288	24	22	16	17		41	223		245								105	112
59.860	I	306	310	281	(280)	19	19	15	14.5	37		214											
59.862	I			315	312	21	21	15	16	46	45	247	244	269	266	261	87	204	206	150	147	111	117
59.863	M			323	326	25	23	19	18	46	45		247		262								
59.865	F			298		24	24	17	18	48													
59.866	I	336	333	303	300	21	20	16	16	46	46	220	217	240	238	271	75	205	204	144	144	102	114
59.867	I	311		274	269	20	20	14	13.5		34		203		218	235		179	176	129	124	114	106
59.870	I			(335)	331	21	21	16.5	15	45	44	251				288	67	215		157	158	110	126
59.871	I	334	342	304	301	22	22.5	15	15	39	40		234		254	262		190	194	142	144	101	111
59.872	I	327	324	290	291	23	22	13.5	14	39	39	221	222	235	236	261	110	195	195	140		122	120
59.874	I				327	24	24	18	19		(43)					292	(84)	213	213		163	104	112
59.876	I	378		325	321	22	21	15	15	48	47	252	246	271	265	285		213	215	156	159	111	124

ID number	Sex	FOL R	FOL L	FML R	FML L	XHD R	XHD L	AP D R	AP D L	ML D R	ML D L	APS R	APS L	MLS R	MLS L	TM L R	TM L L	TAPS R	TAP S L	TML S R	TML S L	AP N R	AP N L	ML M R	ML M L
59.877	I	357	357	358	358	37	37	21	21.5	28	28	25	25	23	24	298	296	27	27	19	19	28.5	28	19	19
59.907	I	457	459	460	462	47.5	47	27	26	32	32	30	30	26	25	372	372	34	35	20	22	37	39	21	21
59.908	M	390	390	395	394	38	38	19	19	29	29	22.5	21	22	22.5	314	316	26	26	27	27	27	27	17.5	18
59.910	I	377	384	379	386		38	22	22	30	32	26	25	23.5	23.5	322	320	31.5	30	19	18	33		21	
59.911	I	398	393	399	396	41	40	24	23	29	32	28	28	25	26		332		30		21		34		22
59.912	F	388	383	389	384	38	40	22	23	28.5	30	27	25	23	24	331	327	27	30	19	20	32	32	20	21
59.915	F	427	425	432	428	42	41	22	22	28	28	22	23	20	21	346	344	27	27	19	20	29	30	21	22
59.917	M	380	383	384	386	38	37	20	20	29	29	24	24	22	22	317	316	27	29	19	20	30	30	20.5	21
59.918	I	437	442	440	446	47.5	48	24	24	37	37.5	27.5	27	27.5	28	376	379	39.5	39	22	21	40	40	23	22
59.919	F	420	428	422	428	46	46	23	24	36	34	27	29	28	25	(358)	356	34	35	22	21	36	36	21.5	22
59.920	M	(394)	(391)		(397)	40		20	21	29	29	24	25	23	23	318		27	28	19	19	28	28	20	20
59.921	M		431		435	42	43	23	22.5	30	30	28	28	24	25	353	356	33	33	19	19	35	35	20	20
59.922	F		401		410		39		20		32		24		22.5	332		32		18		33		18	
59.923	F	394	398	397	400	39	39	21	22	29	29	25	25	22	22	318	320	28	29	17	17	31	31	19	19
59.924	M	422	419	423	423	49	49	24	23	34	35	26	25	29	29	351	349	35	33.5	20	21	35.5	35	22	23
59.925	F	(403)	398	(404)	403	(41)	41	24	23	30	28.5	26	26.5	22	23			29	31	20	19	32	34	20	21
59.953	I	447		447	448	43	42	22	21	33	33	28	26	23	25	369		36	32	20	20	36	33	21	20
59.954	M		427		432		46		22		29		25		22	355	351	31	30	20	20	33	31	21	22
59.956	M		420	423	424	44	44	19	19	33	32	23	23	23	21		361	31	30	17	18	32	31	19	20
59.957	I	404		406	412	44	45	23	23	32	31	27	28	25	24	358	357	31	32	22	23	34	35	26	25
59.958	M	408	411	409	410	42	42	23	23	33	32	27	27.5	26	25		345	32	33	19	19	36	35	18	18.5
59.962	M	431	435	433	437	45	45	26	25	31	32	27	30	27	25	359	356	33	32	19	21	38	36	22	23
59.963	I	437	436	440	439	44	44	25	25	32.5	34	31	32	24	25	375	374	34	34	22	21	37	37	21	21
59.964	M	364	367	365	370	38	38	20	19	27	28	23	22	22	22	301	302	27	26	16	16	30	29	18	18
59.965	M	406	405	408	409	39	40	22	23	29	29	27	27	22	22	343	342	32	31	21	23	32	33	22	23
59.966	F	458	459	460	463	46	46	23	24	26	35	28	29	28	27	382	381	36	35	18	19	37	37	20	21
59.967	M	451	452	454	454	47	46	23	23	37	36	30	30	29	26	382	380	34	35	21	21	38	41	21	21
59.969	I	352	354	356	358	41	40	22	21	31	31	24	24	26	26										
59.974	F	401	403	402	404	42	43	21	21	28.5	29	25	26	21	22										
59.975	I	450	446	455	454	42	41	22	22.5	33	32	28	30	24	25		389	32	31	23	22	36	35	23.5	24
59.976	I	400	402	404	406	40	40	23	21.5	30	30	25	25	24	24										
59.977	M	381	381	383	383	36.5	36	20	22	28	28	26	27	23	23	320	320	26	27	17	17.5	28	29	18	18.5
59.978	F	386	381	388	384	38	37	21	21	30	30	25	25	23	21.5	326	332	26	27	18	19	29	30	20	19
59.979	I	387	387	389	389	37	36	21	22	29	29	27	27	23	23	(314)		30	30	18	20	31	30	18	20
59.980	F	411	411	414	415	40	40	23	22	31	31	24	23.5	24	24.5	333	330	26	27	20	20	30	30	21	22
59.981	F	411	412	414	417	42	42	19.5	21	31	31.5	24	24	21	22.5	339	340	30	31.5	18	18	32	32	21	19
59.982	F	459	458	460	459	43	44	24	23.5	34	35	31	32	26	26.5	375	374	35	36	19	21	35	38	19	20
59.984	M	417	421	423	426	44	44	22	23	32	32	28	31	22	23	362		31	33	21	21	35		22	
59.985	M	398	400	401	403	42	41	22	22	30	31	24	23	23	25	334	333	33	31	24	21	35	35	21	20
59.986	I	393	399	395	400	37	36	21	20.5	30	31	24	26	23	24	332		29		17		30		17	
59.987	M	428	422	433	426	43	42	22	20.5	30	31	27	25	20	21	364	360	32	33	19	19	36	36	22	20
59.988	F	398		400		42		25		30.5		27.5		25		316	314	27	27	21	21	31	30.5	22	22
59.989	F	392	394	396	398	37	39	24	23	30	30	25	28	22	23		319		30		18		32		20
59.990	M		446		450		46	27.5	28	31.5	31.5	29.5	29	24	24	376	372	33	34	23	22	37	37	21.5	23
59.991	F		414		415	40	40	25	25.5	28	28	29	29	23	24										
59.993	M	372	376	372	379	37	36	22	22	29	29	26	25	23	22	303	301	30	27	19	19	33	30	20	20
59.994	F	439	442	442	446	45	45	23.5	24	33	35	30	28	24	24	361	358	36.5	36	20	20	38	40	21	22
59.995	F	381	385	381	385	37	36	22.5	20.5	29	30	25	24.5	23	23	(306)	307	26	29	20	19	28.5	30	20	20
59.996	M	369	371	369	372	36	36	22	21.5	29	29	21.5	20.5	23.5	23	303	305	28	26	17	18		28		19
59.997	F	404	409	405	411	42	41	20	21	29	29	26.5	26	21	21				30		19	31	31	23	21
59.998	F	427	428	429	431	47	45	23	23	34	34	27	26	27	28	354	347	34	35	20	20	37	37	21	21.5
59.999	M	383	387	383	391	41	39	20	22	30	30	24	25	24	24	313	320	27	28	18	17	31	30	21	20
60.000	F	424	422	425	425	43	42	24	25	33	34.5	27.5	29	27	27	359	357	34	33	26	23	36	37	24	24
60.001	I	396	402	398	403	38	38	23	25	28	29	25	25	21	22			28	27	18	18	30	29	18	19
60.002	M	469	477	473	482	47	47	30.5	30	33	33	38	36	28	27	400	406	34	35	25	26	40	39	26	27
60.003	M	416	416	419	419	42	40	22.5	24	32	31	27	26	25	26	344	343	30	31	21	22	33	33	22	23
60.004	F	408	410	411	414	40	39	21	23	28	28	27	27.5	23	23.5	344	338	28	29	19	19.5	31	31	21	20
60.005	F	382	388	385	388	41	40.5	20.5	20	31	30	23	23	23	23.5	324	300	26	27	20	20	30	30	22	22
60.006	F	387	389	391	391	38	37	20.5	21	30	30	26	24	23	24										
60.007	I	388	390		393			21	20	27	28	25	25	22	23			27	26	19	19	30	32	19	20
60.008	F	392		397		37		23	23	30	28	26	25	24	23										

ID number	Sex	BML R	BML L	HML R	HML L	MDA R	MDA L	MDM R	MDM L	MDH R	MDH L	RML R	RML L	UML R	UML L	PXB	PDII	INH R	INH L	ILB R	ILB L	SAL	SMB
59.877	I		292	268	267	21	22	15	15	37	37	193	190	212	209	251		181	184	138	136		
59.907	I		366	318	317	25.5	23	17	18	50	49					270	(90)	221	218	161	158	90	110
59.908	M	308	309	291	290	21	21	14.5	14	38	38	219	218	238	233	262	97	181	176	136	137	96	115
59.910	I			286	278	21.5	20.5	15	15	39	37	222	219		233								
59.911	I		322	286	280	23	21	17	15	40	40				236							101	121
59.912	F			288	287	22	20	14	15	38	38		216		231								
59.915	F			305						44			236									93	108
59.917	M		304	279	279	21	20	14	14	37	37	209	208		224								
59.918	I	363		327	321	24	23	17	15	51	48	255	252	274	268								
59.919	F			319	315	23	21	16.5	16	48.5	47				(265)								
59.920	M	307				22	21	15	15.5		41.5												
59.921	M		346	320	316	21	19	14	15	46	44	240	238	263	260								
59.922	F		330	293		22		15		39													
59.923	F	310	315	286	282	23	21	14.5	14	38.5	38		213		227	263		192	193	133	134	84	106
59.924	M	344	343		299	22	21	16	16	47	47.5		226			273		210				96	114
59.925	F																						
59.953	I	361		311	311	20	17	16	16	44	43	243		261	257								
59.954	M	343																					
59.956	M			298		21		15		46		239				291	80					106	115
59.957	I	350		297	299	23	21	18	18	46	45		238	259	256	274		203	206	151	154	100	118
59.958	M	345	342	301	301	23	19	16	15	44	45	242	239	260		282	86	196	197	150	148	106	113
59.962	M		348	(310)	306	21	19	17	16	45	45	243					87						
59.963	I	369	366	326	323	22	21	16	17.5	47		256	258	274	269		72	210		151		104	121
59.964	M			269	270	21	21	14	16				211										
59.965	M	339		290	289	20.5	21	15.5	15	38	38	231				279	(106)	194	195	147	148		
59.966	F	377	371	326	326	21	19	16	14	48	46	246		265									
59.967	M		376	334	334	24	22	17	17	46		249	249	266									
59.969	I			263	260	22	21	15	14		39	199	196	214	215								
59.974	F																						
59.975	I			330	327	22	20	16	15	42	42		255	279									
59.976	I			306	302	21	21	16	16	41	39	219	221	247									
59.977	M				290	20	21	14	15		38	219	219	236	235								
59.978	F	321	326	294	289	22	21	14	13	37	35	230	227		245	267		185	185	136	135	105	115
59.979	I			289	281	20	20	16	15.5	39	38												
59.980	F	326		300	290	22.5	23	15	16	40	38	218	217	231		271		185	192				114
59.981	F	325	329	302	297	20.5	18	16	15.5	43	44	225	225	244	244								
59.982	F	369	370	333	331	21	20	17	16	44	43	261	261		275	285	(78)	210	213	160	161		
59.984	M			301	297	24	21	17	15	43	42	242	243	261	260								
59.985	M	327	330	291	283	21	18	15	14	41	40	227	226	250	248								
59.986	I			281		24		15		38		220	219		232								
59.987	M	353		307	305	20	20	16	16	44	43	239	237	257	255								
59.988	F															261	(109)	193	192		145	107	118
59.989	F		312		283	21	21	14.5	14.5		37												
59.990	M	371	367	317		24		16		44		253		274	274	285	101	207	207	150	152	96	120
59.991	F													250	249	277		205	207		152	101	113
59.993	M			270	269	23	21	14	14	40	38												
59.994	F	351	353	316	314	23	20	17.5	16	45.5	45	246	242	263		275	85	204	203	147	150	135	117
59.995	F				283	22	23	15	14	35	35			233		257							
59.996	M		300	281	281		20.5	14	14	36	36	206	204		219	241	110	177	177	136	134		
59.997	F	330	330									229			247	(257)	77	192	193	142	(141)	103	109
59.998	F	347	345	311	315	24.5	21.5	16.5	16	45	45	236	238	251		290	86	206	207	150	152	104	120
59.999	M			282		21		15		39													
60.000	F	345	344	317	313	22	20	15	14	45	44		236		254	268	93		199		156	100	113
60.001	I		325	296	292	20	19	14	14		39	227	223	240	235								
60.002	M	392	393			26	24	20															
60.003	M	336	336	305	299	23	23	16	16	40	39	223	220	238	233								
60.004	F	333	330	313	308	23	22	15	14.5	41	39					271	116	194	194		145	97	122
60.005	F	314	314	294	290	23	24	14	14.5	41	41	214	210	233	230	258	115	188	191	141	141		
60.006	F			286	286	24	24	15	16	37	38		226		241	267	(87)	185	185	140	142		
60.007	I				293	23	22	14	15		38		211										
60.008	F			284	286	22	22	16	16	38	38				228							101	110

ID number	Sex	FOL R	FOL L	FML R	FML L	XHD R	XHD L	AP D R	AP D L	ML D R	ML D L	APS R	APS L	MLS R	MLS L	TM L R	TM L L	TAPS R	TAP S L	TML S R	TML S L	AP N R	AP N L	ML M R	ML M L
60.009	F	401	400	401	401	38	39	21	19	29	29	26.5	25	22	22	333		29		17.5		32	32	17.5	20
60.010	M	405	404	406	406	42	41	22	22	32	32	26	25	27	26	340		33	33	19	18	34	34	20	21
60.011	F	396	396		399	38	38	22	20	30	31	24	23	24.5	25	333	329	29	30	16	16.5	30	30	19	19.5
60.012	M	364	375	370	376	38		19	19	28	29	22	23	22	22	304	298	27	27	18	17		29		19
60.013	I		394		395	42	41	20.5	23	30	30	24	25	23	24	(340)	337	30	30	20	20	31	31	21	21
60.015	I	424	427	429	431	40	40	21	21	32	31	27.5	27	23	22	357	357	29	29	18	18	31	33	20	20
60.017	F	397		400		40		23	23	30	28	25	28	24	23	325	323	28	28	19	19	29	29	18	19
60.018	M	(417)	423	(422)	426	45	45	23	21	35	35	25.5	28	25	23	348	343	33	34	21	20	36		21	
60.019	M			453	454	45	45	22	25	36	36	28	30	24	26			35	36	23	22	38	39	25.5	24
60.020	F	402	403	406	405	38	38	20	23	27.5	29	24.5	27	21	22	344	344	27	28.5	20	19.5	29	30	21	20.5
60.021	F	378	375	384	384	40	39	19	19.5	30	30	22	22	23	23	319	317	36	37	18	18	28	28	20	21
60.022	M	423	424	427	427	43	43	22.5	22	30	30	29	29	23	22		350		32		19		35.5		21
60.023	I	392	392	397	394	40	39	21	22	30	30	25	25	21	23	340	336	27	30	19	19	31	31	21	21
60.024	M	449	453	454	456	46	46	26	26	30	30	31	33	24	24			35	35.5	21.5	21.5	40	39	21	21
60.025	M		383		388	39	38	20	21	29	30	24	24	23	24	317		26	27	18	18	29	30	19	18
60.026	F	447	443	448	446	46	46	25	24.5	35	35	29	30	27	28	373	367	36	36	22	21	38	39	22	22
60.027	I	396	399	400	401	42	42	22	22	32	32	24	23	24	26	337	335	33	34	18	18	35	35	19	18
60.028	I	414	411	416	413	41	42	22	22	32	31.5	26	27	26	26	338	340	33	34	18	18		34		17
60.029	I	414	412	418	417	38	38	21	22	30	29	25	26	23	22	337	337	29	28	17	19	30	31.5	19	21
60.030	I	443	432	446	433	44	44	24	24	31	30	31	30	24	23	357	362	30	33	21	23	35	37	23	24
60.036	F	369	371	371	375	37	36.5	21	20.5	28	28.5	24	26.5	21	21	321	325	27	26.5	18	19	29	28	21	20
60.037	M							24	23	29	29	27	27	23	23	362	363	29	31	20	21	33	34	20	22
60.038	M	396	393	401	397	40	40	21	20	29	30	24	24	25	23										
60.039	M	408		417		43	43	24	22	30	30	29	28	25	24	(347)	352	36	35.5	20	29	38	38	21	21
60.040	F	382	382	384	383	40	39	21	21	31	31	24	23	26	26	319		29	28	17	18	31	32	30	21
60.041	I	408	409	412	414	41	40	23	23	31	32	28	28	24	24	348	343	26	26	19	20	27	29.5	19	20
60.044	M	417	419	422	423	44	44	22	22.5	29	29	26	26	23	23	352	350	30	31	23	24	33	33	23	23
60.045	M	409	415	414	421	39	38	21	20	26	22	24	23	21.5	21	328	330	26	26	18	18	26.5	28	20	21
60.046	F	436	434	444	441		47	26	25	34	34	28	28	27	25	361	365	33	34	26	26	37	39	25	28
60.047	F	401	406	403	409	40	41	23	23	28	29	26	26	22.5	23		341	32	33	19	19	35	35	20	20
60.048	M	410	413	416	418	40	41	22	23	30	31	28	28	25	25	347	345	32	32	21	20	35	37	20	19
60.049	M	392	393	394	394	40	39	22	22	29	29	24	24	22	23	327	327	25.5	26	20	20.5	29	30	21	22
60.050	I		410		410	43	41	25	24	31	31.5	28	28	25	26		340		33		22		34		23
60.051	I	418	417	420	421	45	44	22	21.5	31	31.5	29.5	29	22	22	349		33		19		35		21	
60.054	F	381		382		37.5	38	19	19	28.5	27.5	24	25	22	21	324	321	26	27	19	20	30	31	20	21
60.055	F	368	365	370	369	35	36	19	19	27	26	23	22	20	19	308	303	26	26	17	16	29	28	18	17
60.057	I	402	401	403	402	40	39	22	22	28	29	26.5	27	24	25	331	326	30	30	22	21	30	31	21.5	21
60.058	F	389	390	391	391	37	39	20	21	27	28	23	24	22	23		359	31	29	19	19	33	31	21	21
60.059	I	(442)	(445)	445	446			21	20	30	31	26	25	25	24	369		30	27	20	20	31	32	21	22
60.060	F	405	406	406	409	39	38	19	19.5	32	31	22	23	25	22	325	325	25	24	18	19	27	26	20	20
60.061	F	400	399	405	403	43	42	23	21	32	31	25	26	24	24										
60.062	F	380	381	385	386		41	22	23	32	32	26	26	25	25	308	313	32	33	20	20	34.5	35	21	20
60.063	M	403	403	405	405	40	40	23	22	27	28	26	26	22.5	22.5	330	323	25	27	17	19	28	29.5	20	20
60.064	M	455	457	458	462	46	45	25	26	36	34	30	31	30	28	392	392	35	35	24	24.5	38	38.5	25	26
60.065	I	426	426	430	428	40	40	21	22	31	30	28	27	23	23	370	367	32	33	19	19	35	34	21	22
60.070	I	400	398	406	405	41	42	25	24	31	32	29	28	26	25	350		32	30	20	19	34	34	19	18
60.071	F	391	391	393	394	43	43	22	24	32.5	32	25.5	27	25	23	316	312	32	31.5	19	19	35	33	21	19
60.073	M	396	392	396	397	38.5	40	20	20	28	31	25	24	22	22	320	323	30	31	17	16	32	34	17.5	18
60.074	F	442	447	445	449	44	44	25	25	31	33	29	30	25	27	361	358	36	36	23	24	39	40	24	26
60.075	M	432	431	444	434	45	44	25	24	32	31	30	28	25	24	361		37	35	20	20.5	39	38.5	20	21
60.076	M	408	408	411	410	(42)	42	24	26	31	29	27.5	28	24	24			29	29	20	20	32	32.5	22	21
60.077	M	392	390	395	394	37	38	22	21	29	27	27	27	28	25	331	328								
60.080	I	449	456	452	459	44	44	24	24.5	31	30	28	29	25	24	389	385	32	33	22	22	34	33	23	24
60.081	F	428	425	430	429	39	38	22.5	22	28	29	28	28	25	25			29	30	22	21	33	32	23	22
60.082	F	400	400	405	405	38	38	20	21	30	31	25	25	22	23			26	28	17	18	30	30	19	19.5
60.083	F		382		383		38		21.5		30		25		24		322		31.5		19		34		19
60.084	F	388	388	388	389	40	40	21.5	23	30	31	27	26.5	24	24	312	313	28	29.5	19	20	30	31	20.5	20.5
60.085	M	442		443				24	24.5	33	32	30	32	25	25		373	33	34	20	20	35	36	21	21
60.086	M	430	429	434	432	42	42	24	24	34	35	31	33	26	26	373	373	32	35	22	22	36	37	23	22
60.087	M	400	399	402	401	(42)	43	23	26	31	34	25	27	24.5	26		334								
60.088	M	426	425	427	425	43	42	25	25	35	34.5	31	31	29	27	370	365	34	35	19	19	38	38	21	22

ID number	Sex	BML R	BML L	HML R	HML L	MDA R	MDA L	MDM R	MDM L	MDH R	MDH L	RML R	RML L	UML R	UML L	PXB	PDII	INH R	INH L	ILB R	ILB L	SAL	SMB
60.009	F					18	18	14	13	40	40												
60.010	M	336	335	294		23.5	22	16	16	44		224	221		242	251		195	194	144	139		
60.011	F		329	297	290	22	22	14.5	15	39	39			249		260	(114)	187	186	142	139	105	121
60.012	M			274	265	21		15		38			204										
60.013	I	330		301	300	21	21	15	14	44	42	231	231	255	252								
60.015	I	351	347	309	306	22	23	15	15	39	40	239	241	254	254	272	117	207	207	148	148	108	119
60.017	F	326	(318)	(298)		24		15		40													
60.018	M	340		299	301	25	21	18	16	48	46.5	234		254		280	90	199	201		157	95	112
60.019	M				321	25	22	16	15		45		256	273									
60.020	F		340	282	283	20.5	21	14	14	38	37	227	228	246	247								
60.021	F	310	309	284	275	22	22	14	14	41	41					251	103	192	193	141	143	104	116
60.022	M	346	342		301		20		17		44	233		247					201			120	115
60.023	I	333		293	286	22	21	15	14	40	39	232	229		246								
60.024	M									47	47		239										
60.025	M				272	21	20	15	15	37	37	219											
60.026	F	366	366	332	328	25	22	18	18	48	47	256	252	271									
60.027	I	327	325		282		19		15		43		216		236								
60.028	I	332	334	304	302	22	22	18	16	43	41												
60.029	I		329	317	312	23	23	14	14	40	39	230	230	248	247								
60.030	I				320		19		16		44			265	260								
60.036	F	318	321	(278)	279	22	21	15	14	36	35	212		227									
60.037	M	356	355	304		21	21	16	19	45	44	243		263									
60.038	M	331		298	299	21	18	15	15	42	42	229	226		243	257	88	195	195	143	143		
60.039	M			288		21	21	17	15	44			237										
60.040	F																						
60.041	I	341	336	299	297	24	23	16	15	40	41	231	230	251	248	270		180	201	138	139		
60.044	M	344	347	309	307	22	21	15	16	45	43	230	229	254		265	90	201	200	153	150	126	116
60.045	M			289	285	20	19.5	13	14	40.5	39					235	85	177		134		97	109
60.046	F	359	356			25	22	21	19	47		243			262	283	93	212	214	155	150	110	115
60.047	F	338	339	298	294	21	20	14	14	46	44	230	(227)	249	246								
60.048	M	334	339	293	290	21	20	15	14	41	40		228		246								
60.049	M	321	319	297	290	22	21.5	15	15	41	39	222	218	233		266	(102)	190		145		99	112
60.050	I				300	25	26	15	16		43	234	231	249	245								
60.051	I	342	344		314	23	20	17	15		43		244			242	71	197	197	138	135		
60.054	F											210											
60.055	F			268	267	23	22	15	15	37	37	198	198	218		242	97	174	173	131	128	83	101
60.057	I	321	320	284	284	22	21	15	14.5	40	39	221	216	239								71	109
60.058	F			286	277	20	20	14	14	40	39	208	209										
60.059	I																						
60.060	F			299	296	22	21	15	14	38	37	222		240		(207)	104	191	195	136		106	122
60.061	F			304	296	23	23	15	16	43	42	226	225	243	240	(267)		195	196	142		86	122
60.062	F	302	297	272	272	24	24	16	16	39	38		208	231	228	261	99	183	188	138	142	108	116
60.063	M			306	298	23	21	16	15.5	41	41	224	221	240	235	264		181	186	140	136		
60.064	M	386	387	325		24.5	23	19	16	48		255	255	275	272	283	90	211	214	157	160	94	119
60.065	I		359	320	314	20	21	16	16	42	43	243			269	258			199	149	147		
60.070	I																						
60.071	F	312		281	276	21	21	15.5	15	45	43	211			223	272	(92)	196	194	140	146	102	118
60.073	M				287	22	22	15	14	40	39		221	241		(266)	89	188	187		143		
60.074	F	352	345	312	307	25.5	24	18	18	45	43	246	243		258	282	78	214	213	148	151	106	113
60.075	M		358	311	308	24	21	18	19	46	44		243	265									
60.076	M			307	303	26	27	17	16	40	40	226			244								
60.077	M	322	318	275		21.5	21	16	16	35	36	218			227								
60.080	I	379	383	319	322	21	18	16	15	49	46	257	259	272	275	281	83	202	202	149	150	117	112
60.081	F			307	301	23	22	15	15	39	39			259		273		198	198	154			
60.082	F				304	22	22	15	14	38	37					(261)	(108)		189		140	106	118
60.083	F		312																				
60.084	F			284	276	23	23	14	14	38	38		202			272	(105)	184	185	139	142	89	116
60.085	M		367	324	317	23	20	17	17	47	46	248		264									
60.086	M	369	366		327	24	21	18	16		43	248	246	271	267								
60.087	M			292	289	22	21	15	15	42		223	225		242							105	116
60.088	M		364	315	313	21	20	16	15	43	42	245	243	261	259	272	(75)	200	201	144	143	96	112

ID number	Sex	FOL R	FOL L	FML R	FML L	XHD R	XHD L	AP D R	AP D L	ML D R	ML D L	APS R	APS L	MLS R	MLS L	TM L R	TM L L	TAPS R	TAPS L	TML S R	TML S L	AP N R	AP N L	ML M R	ML M L
60.089	I	374	374	379	377	36	37	20	19	29	28	24	24	24	23	318		27	27	16	17	30	31	17	17
60.090	M					43		23		29		31		23			369	34	33	24	24	37	35	25	25
60.091	F	398	401	399	406	39	40	20	21	30	30	24	24	24	24	334	333	28	31	20	17		32		19
60.092	M	426	423	432	432	44	44	24	24.5	31	31	30	30	24.5	25	360		33		22		36		23	
60.093	F	414	420	420	426	40	40	22	22	30	29	27	28	23	23	352		28	29	21	20	33	33	24	21
60.094	F	395		398	398	40	39	21.5	22	30	29	26	26	23	22	325		33	33	17	18	35	35.5	19	
60.095	F		391		393	39	39	19	20	28	29	24	26	21	23		327	30	32	16	16	30	31	16.5	17
60.096	M	445	446	447	449	45	44	24	24	35	34	30	29	29	26	381		38	35	22.5	22.5				
60.097	I	384	397	390	397	40	40	19	21	27	29	22	23	23	24	325	327	28	27		19	31		22	
60.098	I	375	380	376	382	38	38	19.5	20	31	30	22.5	23	23	23	309	304	27	27	19	19	28	28	21	21
60.099	I	411	417	414	420	39	40	22	23	25	23	26	26	23	22	324	330	27	27	19	18	32	32	22	21
60.100	F	384	387	387	388	35	35	20	20	30	30	24.5	25	22	23		315		28		17		28		18
60.101	M															339	330	30	32	21	17	35	34	21	19
60.102	I	459	455	461	459	46	44	24	24	32	35	29	27	25	24	401	402	33	33	23	24	36	34	25	26
60.103	I	405	404	408	405	44	43	20	20	31	30	25	25	25	25	343	343	28	29	20	20	32	32	22	21
60.106	M	435	430	438	432	44	44	23	23	31	33	28	27	24	23	357	352	33	33	18	17	36	36	20	18
60.107	M	431	427	433	432	42	43	22	21	33	32	28	26	25	23	350	349	28	28	20	21	31	29	22	21
60.108	F	369	369	374	375	36	37	20	20	29	27.5	23	24	22	21	306	306	25	25	19	20		26		22
60.133	I		455		461		49	26	26	34	36	31	31	27	26	396	391	34	35	23	23	37	38	25	25
60.135	F	421	423	428	428	42	42	22	23	29	29.5	26	27	22	23		344		29		19		31		20
60.138	I		395			39	38	22	22	33	33	27	26	24	24			31	30	18	18				
60.142	M	415	416	419	420	42	42	21	22	30	31	24	24	23	24		351	30	31	21	20	32	33	22	23
60.143	F	380	380	383	384	36	37	20	20	27	29	22	23	21	19.5		305		26		20		29		21
60.144	F	399	401	401	404	37.5	38.5	20	20	28	27	23	23.5	23	23	330	329	25	26.5	18	18	27	28	21	20
60.145	F		362		362	37	36	19	19	29	29	24	24	24	24	389	387	31	31	22	20	32	31	22	21
60.146	F	380	381	381	382	39	39	21	21	30	31	36	35	23	24	315	315	28	29	21	18	31	31	21	21
60.152	M		397		400		36	22	22.5	26	26	25	28	20	21	326	328	28	28	19	18	30	31	19	19
60.154	M	442	437	444	438	45	45	25		34		28	25	26	26	368		34		21		40		22	
60.165	I	390		396		39	38	22	22	31	31	25.5	26.5	23.5	23	329	323	30	31	20	19	32	35	20	18
60.168	M	399	405		406	40	40	22	24	32	31.5	27	26	24	24	333		30	30	18	17	34	32	19	19
60.170	M	374	373	380	381	38	37.5	20	20	31	30	24	23	22	23	318	319	25.5	25.5	17	17	28	28	19	17
60.171	I	385		389		38	39	22	23	28	27	25	25	22	22			29	29	20	19	31	29	20.5	20
60.172	M	412	414	414	415	41	40	22	23	32	31.5	26	27	25	23	345	346	30	31	20	20	33	35	22	22
60.174	F	362	368	362	368	39	38	19	20	31	31	24	25	22	24	311	299		32		19		34		20
60.175	M	397	396	398	397			26	26	32.5	31	28	27	25	25	331		34.5		23					
60.177	M		458		462		46		24		33		27		27		361		35		22		36		24
60.179	I			376	383	40		20	22.5	31.5	33	25	26	22	25			28	30	19	18	31	32	19	19
60.180	F	435	433	437	435	44	43	23	24	34	35	28	30	24	25.5	357	356	36	36	19	21	39	40	20	22
60.181	M	(450)					45	24	23	33	34	28	28	25	26.5	396	391	33	32	24	25	37	37	25	24
60.189	I	430	432	432	435	43	44	24	25	29	30	31	32	22	23	363	362	30	30	21	20	34	34	24	23
60.194	M	422	426	424	429	42	42	23	22	30	30	25	28	26	23	349	352	32	30	17	16	34	33	18	18
60.195	M	415		417		41		22		29		26		22			352		30		20		32		22
60.197	M	393		396		42		24		32		28		25											
60.198	F	437	442	439	445	43.5	43	23.5	24	33	33	30	31	25	26	363	356	37	37	20	19	39	38	21	20
60.199	I		357		(360)	37	36	18.5	20	27	28	23	24	22	23										
60.200	M	404	419	405	420	41	41.5	21	22	28	27	27	30	24	22	328	330	31	30	18	18	32.5	33	20	20
60.201	F	434	436	436	440	42	41	22.5	22	32	30	25	27	26	22		370	34	33	21	21	37	36	22	22
60.211	F	400	402	400	402	42	42	22.5	22.5	30	31	26	26.5	25	25.5	332	332	30.5	31	17	18	32	32	19	19
60.212	F	428	431	432	435	43	44	23	22	32	32	29	29	26	24	374		37	37	24	23	37	37	23	23
60.216	M	412	412	415	415	42	41	23	24	32	32	27	29	25	24.5	336	333	33	31.5	18	18	36	33.5	19	18
60.218	I	427	428	430	431	43	43	23	23	31	31	27	29	26	25	264	362	30	32	24	25	35.5	34	24	25
60.219	I	421	426	423	428	42	42	24	25	33	31	28	30	24	25	353	353	32	31	21	21.5	33	33.5	22	22
60.220	M	426	425	428	429		42	22	22	28	29	28	28	21	22	343	344	32	30	19	19	34	32	20	21
60.222	M	417	415	419	417	40	40	22	23	28.5	28	25	26	24	23	359	352								
60.224	I	418	421	419	422	43	42.5	24	24	30	30	32	31	23	23	353		35	35	20	19	38	39	21	21
60.225	M	377	377	379	379	39	39	20.5	20.5	31	31	25.5	26	24	25	303	303	28	28	20	20	31	31	22	21
60.226	F	424	421	426	423	43.5	44	25	27	33	34	30	32	26	25	355	358	36	36	22	23	41	42	23	25
60.227	M	398	403	401	405	39	40	22	22.5	29.5	29	27	28	23	23										
60.228	F	407	402	409	404	36	36	20	21	30	30	24	25	24	24.5	342	339	30	30	17	18	31	32	17	18
60.229	M					39		22	23.5	30	31	24	24.5	24.5	26	(316)		29	29	21	20	31.5	31	21	21
60.230	M	435	431	437	433	44	44	22	23	33	33	26	25	24	25			36.5	35	18	18	39	36	22	20

ID number	Sex	BML R	BML L	HM L R	HM L L	MDA R	MDA L	MD M R	MD M L	MDH R	MDH L	RML R	RML L	UML R	UML L	PXB	PDII	INH R	INH L	ILB R	ILB L	SAL	SMB
60.089	I			272		20		14		37		216	215	234		247	98	179	183		134	100	114
60.090	M	358		319	309	22	19	18	16	45		246					(84)					109	110
60.091	F	329			268		20		14		37.5		216										
60.092	M	355		313	313	21	19.5	17	16	43	42												
60.093	F	335		314	309	22	22	15	15	41		241	240	255		279		200		156			
60.094	F	322	320	290	287	25	23	16	15	41	37	226	225			269	109	191	193	146	146	105	116
60.095	F	319		295	294	21.5	20	13	13														
60.096	M			328	329	23	21	16	16	44	43		248		263	289		214	213			(117)	124
60.097	I			316		20		13		37		222		236									
60.098	I				273	22	21	14.5	14	39	38	208		226									
60.099	I			284	285	21	20	14	13	38	38												
60.100	F			279	273	21	21	15	15	34	34	209	207		220	258		178	180	135	134		
60.101	M				291	24	24	16	14		39												
60.102	I	387	391	318	320	23	21	15	15	45	45	254	254	274		272		205	206	152	151	112	119
60.103	I	334		294	294	21	20	16	15	44	43	225	226	246	241								
60.106	M	353	351	312	312	21	20	15	14	46	44		258			263	71	207	207	144	145	109	115
60.107	M	344	345	308	303	22	21.5	14	14.5	42	42.5					266	86	194	198	150	154	110	113
60.108	F			266	265	21.5	21	14	14	36	35	207	206	225	226	254	(105)	177	177	131	132	96	112
60.133	I	384	386													288		214	217	154	149	103	123
60.135	F		336	300	300	21	19	15	14	43	43		231		248	254	84	198	195	144		112	110
60.138	I			293	290	23	23	15	14	(38)	37			238		264	99	190	192	143	141	94	118
60.142	M			295	294	20	13	16	16	45	42											114	105
60.143	F			278	278	20	20	13	14	38	37	206	207	229	225	259	(92)	177	177	134	136	107	112
60.144	F			288		21	21	15	14	36	37					252			(187)			111	111
60.145	F			267	263	21	20	13.5	15	38	38												
60.146	F	313		283	278	21	21	14	14	38			207		222								
60.152	M																						
60.154	M			318	316	23	22	16	16	46	44												
60.165	I			292		23		15		39		217		235									
60.168	M					24		15		39	39	226											
60.170	M	314		281	271	21	22	15	14	38	38	219		237		266	88	190	188	145	144	98	119
60.171	I		311	292	287	21	21	15	14	39	39	222	219										
60.172	M	334											230		244	273		191		145			
60.174	F		298	273	271	24	23	14.5	16	41	40		210	226	225	243	101	179	180	132		77	114
60.175	M			286	284	27	27	17	17		42				236								
60.177	M		357		322		21		15.5		47		240		263								
60.179	I			282		25		16		43													
60.180	F			326		23.5	21	17.5	16	48		241	244			272	78	208	205	152	154	108	119
60.181	M			330		21		16		47						264	88	220	216	157			
60.189	I	356	358		301	21	19	15	16.5		44												
60.194	M	346	350	314	314	22	20	15	15	44	44	237	235	256		262	(82)	196	196	146	144	100	121
60.195	M			301		30		16		44			242										
60.197	M			289	286	22	23	17	16	43	42	224	224	241		268		203	205	153	152	108	106
60.198	F			306	308	22	20	16	16	44	42												
60.199	I			259	254	22	22.5	16	16	37	37	200											
60.200	M		317	287	287	22	19	15.5	16		41	228	226	241		260		196					
60.201	F	359			308	20	18	16	17		41		234									99	117
60.211	F			298	297	24	23	15	15	42		239				297	103	194	198	156	155	107	120
60.212	F	365		309	308	23	20	16	17	45	44	243			256	278		208	208	155	154	111	114
60.216	M			300	294	23	23	15	15	42	41	223	225										
60.218	I		353	301	301	24	23	17	17	45	45	236	237		250			199	197		144		
60.219	I			303	303	25	22	17	16		42		235	255	252								
60.220	M	325		303	306	21	21	15	15	45	45												
60.222	M	342		310	304	23.5	23	15	14.5	41.5	41		202		242	268		196		145	144	106	112
60.224	I																						
60.225	M			274	270	21	20.5	15	14	40	40	201			222	274		189	190	151	150		
60.226	F	345		311	301	23	21	19	18	43	42	246	246	260	257	227		209	209	150	150	85	120
60.227	M		325	292	284	23	22	16	16	40	38	229		246								96	121
60.228	F	333	334	294	284	23	23	14	14	38	37	227	221		236	251	(104)	183	184	138		104	110
60.229	M		301		282	25	24	15	15		41	211				280	(93)	189	191	146	147	99	118
60.230	M			306	306	21	21	16	15	45	(43)	233		251	249								

ID number	Sex	FOL R	FOL L	FML R	FML L	XHD R	XHD L	AP D R	AP D L	ML D R	ML D L	APS R	APS L	MLS R	MLS L	TM L R	TM L L	TAPS R	TAP S L	TML S R	TML S L	AP N R	AP N L	ML M R	ML M L
60.231	F	412	412		412	38	38	24	24	28.5	29.5	26	27	24	24										
60.232	F	492		495		47	48	27		32	28	35	34	25	25.5	415	413	35	35	25	25		39		
60.233	F	408	413	415	416	41	40	22	23	33	32	26	27	26	24	338		31	30	19	19	34	35	19	19
60.234	F	416	418	416	422	40	40	20	20	28	29.5	25	25	22	23	344	344	28	27	20	19	31	31	22	21
60.235	I	388	(390)	395	396	42	42	23	22	29	28	26	26	22	21	320		27		19		29		19	
60.236	F	(434)				44	46	24	25	32.5	30	31		26		372		37	36	18	17.5	42	40	19	18
60.238	M	470		472		45	44	23	25	34	33	29		26		394	392	32	32	23	23	36	35	23.5	23.5
60.239	M	376	372	380	377	37	37	21	22	28	27	25	24.5	21	22	308		25	27	20	19	28.5	28.5	20	20
60.240	I		411		412		41		24		30		28		25	339		31	31	18	18	33	35	19	19
60.241	M	398		395		38	38	22	20	28	27	26	26	23	23	326	329	27	27	17	16	30	30	17	17
60.242	I	437	440	439	441	44	43.5	24	24	31	33	33	32.5	23	23.5	368	372	34	35	22	21	38	38	23	23
60.243	M	418	419		427	43	41	25	23	29	30	28.5	29	22.5	22	358		32		21		36		23	
60.247	M	373	376	376	378	37	38	20	21	28	29	24	25	22.5	23	316	316	27	27	16	16	30	30	17	18
60.248	I	417	420	419	424	41	40	22	24	30	31	27	27	24	25	358		30	28	20	21	33	33	22	21
60.249	M	392		394		38	38	21		31.5		29		25		327		30	29	20	18	33	32	21	19
60.252	M	426	432	428	426	38	38	19	20	30	31	23.5	24	22	23					17	17	30	30	18	18
60.254	M		433				43		26		31		33		25		355		34		22		26		24
60.255	M		385		390		40		22		31.5		26		24	334		28	31	18	18	30		19	
60.257	F	386	387	389	391	39	38	21	20	30	29		26		24		319	32	30	17	17	30	31	18	17.5
60.258	I	415	416	418	418	39	39	20	21	30	29	26	27	22	23	339	338	28	29	16	16	30	31	17	17
60.259	M	424	430	430	434	39	40	23	22.5	28.5	28	28.5	29	21	21.5	363	359	27	28	19	20	30	31	20	20
60.260	M	418	414	421	417	43	42	23	24	31	31	29	29	24.5	24	346	348	34	33	22	21	37	36	23	21
60.261	I	399	405	405	410	40	41	20	21	31	30	24.5	25	23	22	352	360	30	30	20	19.5	32	33	21	21
60.266	F	412	409	413	410	43	44	23	25	31	31	27	28	24	24	349	345	34	33	22	22	36	37	24.5	23.5
60.267	F	442	445	445	448	48	49	24	24	36	34.5	31	31	25	25	371	370	33	34	20	20	38	38	22	23
60.268	F	447	451	453	455	48	48	26.5	28	36	34.5	29	30	28	26	392	387	40	38	25	25	43	45	24	24
60.271	M	404	407	409	413	38	40	20.5	21	29	28.5	25	25	21	22	341	341	28.5	28.5	17	17	31	30	20	18
60.272	I	392		398		37		20		26		22		20											
60.274	I	434		437		42		24		30		28		23		377		31		20		34		21	
60.275	F	413		419		44	44	24	24	31	31	27	30	25	22	339	343	33	33	17	17	38	37	19	18
60.276	F	416	416	417	418	42	42	24	23	31	30	28	29	23	22	(346)		34	35	21	21	39	38	21	22
60.277	F	417	422	419	424	43	43	26	27	31	31	29	30	25	25	359	347	33	32	20	20	34	35	20	22
60.278	F		427		432	44	44	22	22.5	32	31.5	28	28	25	23		366		31		22		36		24
60.279	M	421	419	424	420	43	42	24	25	31	30	31	32.5	23	23	364	363	35	35	17	17	37	39	18	18
60.280	M	(429)	(430)	(431)	(432)	40	40	23	24	30	31	28	28	23	24	358		30	29	19.5	19	32	33	21	21
60.281	M	(407)				40	39	23	23	31	30	26	26	25	25				30		20		34.5		21
60.282	M	385	384	389	388	37	36.5	22	23	30	29	26	26	25	22			28		19		31		19	
60.283	I	420		422		40		21.5	22	31	30	25	24.5	23	24	344		29	29	20	19.5	33	32	22	22
60.284	M	360	363	362	366	38	38	18	19	29	29	24	24	21	22	305		26	27	19	18	27	30	21	20
60.285	M	384	385	387	389	39	40	21	22	31	30	24	23.5	23	24	309		28	28	19	18		30		19
60.287	M		393		396		39		24		30		26		25		331		30		19		33		18
60.291	I	366	366	369	370	37	36	19	19	29	28	21	26.5	22	22		303	25	26	18	18	28	27	19	18
60.292	I	377	380	380	384	39	38	20	20.5	30	29	25	25	22	23	323		27	27	20	19	31	30	20	19
60.293	I	427	433	431	437	43	42	24.5	24	32	32	30	30	24.5	23		370	37	36.5	22	22	40	38.5	23	23.5
60.294	I	441		443		44	45	23	25	33	34	31	30	25	28	380		33	35	21	20	39	40	22	21
60.296	F	455	455	465	466	44	44	25	28	31	30	32	34	24	25	(377)	378	35	37	24.5	23	37	41	27	24
60.297	I	377	378	380	383	35	34	20	20	29	29	25	24.5	22	22	325	322	27	25.5	17	17	28	27	17	17
60.298	M	393	394	398	397	39	38	21	22	30	31	26	25	23	25	329	322	28	28	21	20	32	33	22	20
60.299	F	390	394	392	397	43	42.5	25	25	32	31	29	29	27	26	325	321	37	37	19	20	38	40	19	19
60.301	I	402		405		39		22		29						338	337	26.5	28	19	19	29.5	29.5	20.5	20.5
60.302	F	388	391	389	393	40	39	21	22	29	29	25	25	23	22	324	328	27	27	16	16	29	29	17	17
60.303	M	398	396	399	396	41	40	21	21	31	31	26	25	23	24	330	326	27.5	27	19	19	31	31	21	21
60.306	M	453	457	456	462	44	43	22	23	35	36	30	29	24	24	386	(383)	33.5	36	20	18	36.5	38	22	20
60.307	F	394	395	399	399	39	38	21	21	30	31	25	25	23	23		331	28	27	17	18	32	31	19	20
60.308	I	382	377	388	381	38	38	21.5	22	29	30	24.5	25	22	24	318		31	28	19	19	31	30	20	20
60.309	I	402	407	403	410	40	40	21.5	22	31.5	33	26	27	24	25		335	32	32	17	17	34	34	18	18
60.311	I	399	400	402	403	40	41	22	23	34	34	23	23	25	27	349		28		18		30.5		19	
60.312	I		398		399	37	37	22	22	28	29	25	26	21	22		333	27	28	16	17	30	30	17	17
60.313	I	459	461	462	463	44	44	24	25	34	34	27	29	25	25	373		36	36	21	21	38	39	21	21
60.314	I	418	418	422	422	44	43	22	21	31	30	26	28	26	23	366	366	32	33	19	19	35	35	21	21
60.315	M	410	409	413	411	43	43	23	25	30	31	30	31	24	23	346	347	33	32	22	23.5	34	33	22.5	24

ID number	Sex	BML R	BML L	HML R	HML L	MDA R	MDA L	MDM R	MDM L	MDH R	MDH L	RML R	RML L	UML R	UML L	PXB	PDII	INH R	INH L	ILB R	ILB L	SAL	SMB
60.231	F			299		22	21	14	14	39													
60.232	F			358	356	24	21	18	17	49	48	265		283									
60.233	F			(288)	(287)	22	22	14	14														
60.234	F		330	301	296	21	21	14	14	38.5	39												
60.235	I	317		278		20	20.5	16	15	41		229		246									
60.236	F	356		318	322	22	21.5	16	15	45.5	44				259								
60.238	M	380	384	326		24	22	17	18	43	42	257				285		207	207	150	151	112	126
60.239	M	293	294											228	224	256	106	177	178	129	133	99	115
60.240	I				301		24		15		42												
60.241	M	323														243	97	188	189	142			
60.242	I	357	363	316	314	22	19	15	16			243	243	263	262							89	124
60.243	M			303	305	21	19	18	16	42	41.5					254	(86)	198	196	146	140	112	98
60.247	M			266	268	20	21	18	18	37	36.5		204	225	222								
60.248	I	344		310	308	23	23	16	16	40	40	231			248								
60.249	M																						
60.252	M	360		325	324	21	21	14	14	39	38	238	237	258	257	254		196	198	139	141	111	123
60.254	M																						
60.255	M			292	288	24	22	15	15	40	39	224				242	105	193	192	134	142		
60.257	F	308		287	278	23	23	14.5	14	36	38					272		184	183	144	140	93	114
60.258	I	337	334	308	308	23	23	15	15	40	40		230	251	250	265		195	196	149	148	108	120
60.259	M	357		319	313	21	21	15	15	40.5	39	234	238		263								
60.260	M	336	338	397	399	23	21	20	17	46.5	46	230	231	249	247	272	96	200	202	145		100	113
60.261	I	347	357	292	292	22	20	15	14	42	41	230	226	247	248								
60.266	F	347		302	300	23	22	17	16.5	48	47	243		262	257								
60.267	F	368	371	332	330	23.5	23	18	18	50	48.5	251	251	270	268			220	220		159		
60.268	F	385		335		26	22	19	18			253	255		269								
60.271	M			290	288	21	22	16	16	39	38	229											
60.272	I			284	285	19	19	13	13	38	38		216		233	258		184	183	143	141		
60.274	I			321	320	22	20	17	17	43	41	250	249	266	263								
60.275	F	336	341	303	304	21	20.5	17	16		44		230		246	266	61	197	197		143	98	116
60.276	F		333		306		21		16		45			252		291	(80)	(217)	219	165		114	119
60.277	F	356	357	305	307	20.5	21	17	15	45	43	239	236			286		197	200		150	107	114
60.278	F	345		309	307	22	20	17	16	43	44	243	243	258	258								
60.279	M			305	303	21	20.5	16	14	45	45		242		257								
60.280	M				316		24		15		41				264								
60.281	M	334		291	293	22		15		41	40.5		219	240	241								
60.282	M	298		284	282	25	23	15	15	38	39	214											
60.283	I			299		20		16		43													
60.284	M	301		271		21		17		38		205				262	96	180	178	137	138	92	113
60.285	M	301		281	279	24.5	23.5	14	14	40.5	40	207	206			271		191	193	146	146	93	118
60.287	M			302		23		15		40													
60.291	I			270		21	21	15.5	14.5	37.5						241	(98)	175			133		
60.292	I	315	314	271		21		15.5		37		211		223									
60.293	I				317	24	23	17	16	46	43				268								
60.294	I			334	339	23	22	17.5	17.5	47	47		254			273			200		(148)		
60.296	F																						
60.297	I				273		22		14		36				232								
60.298	M	314	312	281	274	22	21.5	16	17	40	40	213	214	230	230	247		187	188	143	146	118	103
60.299	F			289		21	21.5	17	18	45													
60.301	I	332			291		23		14		38	227	226	245	243								
60.302	F		320	294	289	20	19.5	14	13	40.5	41	218	214	240		267		181	179		139	101	121
60.303	M	326	321	295	287	21	20.5	15	15	39	38	222	221		235	280	96	210	198	150	147	105	118
60.306	M	370	369	316	312	22	20	16	16	47	46	256	252	272	269	277	81	202	203	153	147		
60.307	F	329		296	287	23	21.5	14.5	14	39	38	227	222	244									
60.308	I			289	287	23	23	17	17	40	39	213			227		101					104	109
60.309	I	332	333	295	293	23	23	15	15	40	39	225		244	241	277		201	204	143			
60.311	I			290		22		13									115					105	114
60.312	I			291	288	19	20	14	15	37	35.5	216				262	(97)	191	191		137	98	111
60.313	I	364	367	337	331	22	19	17	16	47			253	268		279		208	208	150	152	105	124
60.314	I	358		312	312	22	20	18	17	44	43	248	248	268	266	259	90	207	212	148	146	110	117
60.315	M	334		303	301	23	21	17	18	45	45	234	234	252	251	260	(87)	199	203		147	107	111

ID number	Sex	FOL R	FOL L	FML R	FML L	XHD R	XHD L	AP D R	AP D L	ML D R	ML D L	APS R	APS L	MLS R	MLS L	TM L R	TM L L	TAPS R	TAP S L	TML S R	TML S L	AP N R	AP N L	ML M R	ML M L
60.316	M	(435)	436		439	43	43	24	25	33	33	27	28	26	26	360		33	32	21	21	35	34	22	22
60.318	F	460	458	463	461	43	43	25	25	37	37	29		27		401	393	37		22		39		23.5	
60.319	F	416	423	422	427	39	40	21	22	29	29	28	30	21	22	356	355	34	33	19	19	36	35	19	20
60.320	F		430		434	49	47	22	22	33	33	26	27	27	26	359		37		22		37		23	
60.426	I	407	417	408		42	42	23.5	24	30	31	27	28	23.5	24	354	353	33	31	20	22	36	33	20.5	22
60.436	M					40		23		30		28		23											
60.437	F	397	410	399	410	41	41	26.5	26	31	33	32	30	27	27.5		343	34	33.5	22	21	35	35	24	22.5
60.440	M		406		411		41	21	21	31	31	25	25.5	25	25.5		347	27	28	17.5	17	30	30	19	18
60.445	M		435		439		42		20.5		23		26		24										
60.446	M	390	388	398	392	40	41	24	24.5	30	31	26	25	26	25	316	313	30	30.5	20	18	33	32.5	20	19
60.451	F	389	388	390	390	39	40	22	21	28.5	29	27	27.5	22.5	23	310	312	26	26	18	18	28.5	29	20.5	19.5
60.452	F	379		385		39		20	21	28	27	24	25	22	23	306	305	26	25	20	19	29	27.5	20	20
60.453	F	458	455	461	459	43	43	23	25	34	33	27	29	26	25.5										
60.454	M	442	445	447	448	44	43	23	25	34	33	27	29	26	25.5	382	384	33	32	22	22	34.5	33.5	25	24.5
60.455	F	390	391	395	396	39	38	21	21	29	29		25		22	310		28.5		18		31.5		20	
60.458	I	440	440	445	445	45	45	23.5	23	31.5	31	27	27	24	24.5	376	372	34	34	22	21.5	37	36	23	23
60.466	I	436	435	438	436	43	42	23	22	34	32.5	28	32	28	25	369		36		21		38		21.5	
60.467	F	415	417	423	422	41.5	42	22	23	31	30	31	31	21.5	22		353		33	19			36		19.5
60.468	F	423	422	429	429	40	41	22.5	24	32	33	28	27	24	23	340	339	31.5	31	22	20	35	35	22	21.5
60.470	I	432	428	437	432	44	48	23.5	24	30	30	31.5	31	23.5	24	370	371	30	29	23	23	35	34	23	22.5
60.482	I	410	417	412	421	42		24	23	33		31	30	26	25	352	356	36	37	22	21	39	38	22	22
60.483	I	391	391		391	39	39	20.5	21	28.5	27	24	25	23	22										
60.484	I	(442)	441	(445)	446		43	24	24	32	33	27	27	25	27	365	361	32	34	24.5	22	34	34	29	25
60.491	M	419		420		44.5		23	24	32	32	30	30	24	23		345	33	29	23	22.5	35.5	33	25	25
60.494	I	397	397	398	(397)	40	41.5	22.5	22.5	33	32	24	24	27	27	325		28		18		30		19	
60.496	M	385	389	388	393	39	39	20	21	31	30	24	25	26	24	333	332	28.5		18		31.5		18.5	
60.497	F		373		376	36	37	19	20	29	29	23	23	22	23										
60.501	M	427	431	435	435	39	39	22	22	29	28	25	25	22	23.5	381	377	29	29	22	23	31	32	24	25
60.502	I	424	423	426	428	42	41	22	21.5	32	32	25	24	23	23	359	355	35	30	18	18	35.5	31	18	19.5
60.505	M	408	408	411	410	40	39	19	18.5	30	30	23	23	22	21	332		30		16		32		18	
60.511	M	407	412	408	414	37	37	21	20	28	27	25	25	22	21	348	346	29.5	30	18	18	32	32	21	19.5
60.513	F	472		474		47		25		32		31		26		423		34	33	24	21	37	36	25.5	22.5
60.514	F		406		409	41	41	22.5	23	25	26.5	25	26	20	21.5		336		31		18		33		19
60.518	F	417	420	421	423	43	43	22	21	32	31	28	27.5	25.5	23.5	350	358	32	33	20	19.5	36	34.5	21	21
60.520	I	402	403	405	403	40	40	22	23	32	32	26	27	26	24.5	338	339	31	30	19.5	21	33.5	33	19	20
60.521	M	397	400	400	401	37.5	37.5	22	23	26	26	26.5	28	21.5	22			27		17.5		30		18.5	
60.522	F	377		379		36		20		29		23		23		317		26		17		27.5		18	
60.524	M		425		428		42		25		31.5		29		24										
60.526	F	366	362	369	363	37	37	21	21	27	28	25	25	25	22.5	301	302	25.5	26	19	19	27.5	28.5	20	20
60.527	I	421	421	423	424	41	39	22	21.5	31	30	26.5	27	23	23	361	359	30	30	20	19	33	33	20	20
60.529	F	351	351	356	355	35	35	20	20	25	25	22.5	22	21	20	294	293	25	25	18	18	27.5	26.5	20	20
60.530	I	419	415	424	420	43	43	24	24	30	31	27	27	24	22.5	341	337	32	31	18.5	19	35	36	23	23
60.541	F	410	410	412	414	41	41	23	23	32	32	26	27	23.5	24	339	337	33.5	34	21	22	36.5	37.5	21.5	20.5
60.542	I	381	382	383	389	40	40	22	24	30.5	29.5	27	27.5	23	24	333	336	30	32	22.5	22	35	35.5	24	22
60.543	I	391	394	397	400	41	41	21	22	30	30	24	25	23	23	327	326	27	28.5	21.5	21	32	31	23	21
60.550	F		434		440	42	41	22.5	22	30.5	31	22	30	24	29										
60.551	I	420	423	423	425	41	41	24	24.5	29	30	31	32	24	23	356	353	33.5	35	21	19	37	37	23	22
60.554	I	424	424	426	425	43	43	23	23	33	33.5	30	29	24	25.5	359	359	35	34	18	18	37	38	19	18
60.555	F		408		410	39	40	24	24	32.5	34	26	29.5	24	24		337		31.5		21		34.5		22
60.556	F	375	374	378	376	40	40	21	23	28	28	27	27	22	22	317	315	30	32	20	19	33	35	22	20
60.560	F					45	44	23	23	35	33	27	26.5	28	27.5	346		34	34	20	18.5	39	35.5	22	20
60.561	I	377	380	380	384	38	38	18.5	19	25.5	26	22	22	21	21	315	313	25	26	18	17	26.5	25	19	18
60.562	F	419	418	422	421	43.5	43	23.5	23.5	32.5	32	25	25.5	27	25.5	350	349	31.5	32	19	20	37	37	21.5	22
60.564	F	392	395	395	397	40	40	21	21	30	29	24	24	24	23	325	324	28	28	20	20	31	32	20	20
60.568	I	381	378	383	382	37		21	21	27.5	28	22.5	22	21	22	315	313	28	28	17	17	30	30	18	19
60.569	M	420	421	423	424	46	46	25	26	32	32	29	30	30	28	351	356	34	34	21	22	39	37	22	21
60.570		415	421	417	427	41	41	23	23.5	29	29.5	28	28	23	23	342	344	28.5	29	20	20	32.5	30	21.5	31
60.571		383	388	387	390	37	37.5	19	21	29	29	25	25	22	22		321		29.5		17		30		18
60.573			400		401	39	38	21.5	20	28	28	28	26	22.5	21	345	344	30	30	19	19	32	33	19.5	19
60.574		374	377	377	381	40	40	23	22	31	31	25	25.5	24	25		307	28	29	19	19	30	30	21	21
60.575		401	400	403	402	40	40	23	23	29	29	26	29	23	24	327	326	30	31	17.5	17	37	33	21	18

ID number	Sex	BML R	BML L	HM L R	HM L L	MDA R	MDA L	MD M R	MD M L	MDH R	MDH L	RML R	RML L	UML R	UML L	PXB	PDII	INH R	INH L	ILB R	ILB L	SAL	SMB
60.316	M				316		22		16		44				264								
60.318	F	387	384	335	330	24	22	16	18	47	46	270	264	284		303		217	217	161		103	125
60.319	F	340		313		20	18	16	15	44		231	229	256	249								
60.320	F			319	308	22	18	17	16	49	46			266	262								
60.426	I																						
60.436	M															270	103	193		192		92	117
60.437	F	346		308	305	20	19	24	24	43	43.5	233	227	252		255	87	195		148	151	86	111
60.440	M			309	309	16	16	24	23	41	41		244			277		205	203	152		119	116
60.445	M		358																				
60.446	M			(281)	277	16	17	25	24		41		210		216								
60.451	F	307	306	298		14		22		38.5													
60.452	F	304		280	275	15	15	20	19							269	99	183	185			93	117
60.453	F																						
60.454	M	379	378		319	18.5	17.5	24	24	47	46	147			263								
60.455	F	317	317	294	288	15	15	24	22	37	37	216	219	232	234	272	106	196	188	150	150	98	113
60.458	I	368		309	305	19	19	23	23	45	45	247	245		263	268		205	207	146	145	113	114
60.466	I		368																				
60.467	F			315	317	(15)	15	19	17.5	45	43												
60.468	F																						
60.470	I		364	313		17.5	17	21.5	20	44		253		265									
60.482	I																						
60.483	I			318		18		26.5		46						265		202	200	144	142	94	112
60.484	I		351		301	16	16	25.5	25.5		41	229	225	249									
60.491	M			309		17	16	20	21	45	44												
60.494	I											240		215									
60.496	M				281		17		18		39					279		197	153	196	152	100	127
60.497	F			272		15		19									93	187		134			
60.501	M	357		308	305	16	15	22.5	22	40	39	239				261	(109)	193	194		141		
60.502	I				296	16	16	21	20	42	41												
60.505	M	324			288		13		16		41	221	221	239	239								
60.511	M	336	337																				
60.513	F		405		338	15.5	15	22	21	47	46			290				215		157		119	125
60.514	F			298	300	15	15	20	18.5	40	39.5					252	90	197	198	144	142	106	110
60.518	F	343	351	304	303	16	16.5	22	20	43.5		232	233	250	248	273		205	204	151	149	108	114
60.520	I			296	296	16	15.5	22	21	40	41		224		247								
60.521	M		327	298		16		22		39					254								
60.522	F	313		272		14		19		35													
60.524	M			308		18		23	42.5														
60.526	F																						
60.527	I			299	305	16	15.5	21	19.5	41	42	239											
60.529	F		285	262	254	15	14	21	21	35.5	34	195	192	209		236			168	127	128	83	101
60.530	I	334	332	304	302	16	15.5	22	20	47	46	235		250	250	262	77	199	200	146	416	97	109
60.541	F			303	302	24	21	19	22	45	44			333		263	(70)	191	193	145	146	104	111
60.542	I			285	273	18	19	26	26	41	39	214		233		255	(101)	190		146	147	83	110
60.543	I		321	278	276	18	16	23	23	40	41	220		236	(234)	246		(186)	189	(142)	144	113	110
60.550	F			313	307	15.5	16	25	24	39	40		250	248									
60.551	I		345									238	236										
60.554	I	363	358	321	317	21	14	16	19	44	43	241	236			267	76	213	213	153	151	102	114
60.555	F				306	17	17	23	23	43	41.5					273							
60.556	F			271		20		20.5		41		212		227		233		177	177	130	130	116	103
60.560	F	336	335	293	295	17	17.5	21.5	20	45	46		230	247		278		211	207	155	155	96	116
60.561	I	307	308											228									
60.562	F	344		297	300	16	21	22	22	44.5	44.5		232		246			200	200	144	141		
60.564	F			298	294	22	21	15	16	39	41												
60.568	I				280		23		18		37	216	215	236									
60.569	M	351	351										240	256	255								
60.570					298	16	15	22	22	42.5	40.5											102	111
60.571					272		16		19		41		211		230								
60.573				307	309	22	21	17	16	41	41	228	231	247								94	114
60.574						23	21	17	18.5	39	39	117											
60.575		320	322	291		21		15.5		39													

ID number	Sex	FOL R	FOL L	FML R	FML L	XHD R	XHD L	AP D R	AP D L	ML D R	ML D L	APS R	APS L	MLS R	MLS L	TM L R	TM L L	TAPS R	TAP S L	TML S R	TML S L	AP N R	AP N L	ML M R	ML M L
60.579		424	422	428	424	44	43	23.5	23	31	30.5	27.5	29	27	24										
60.580		386	389	390	393	39	38.5	21	21	23	27.5	25	25.5	22	22										
60.581		422	425	429		44	45	26	25.5	33	34	32	32	25	25		361	21	21	33	33	37	38	24	22
60.582		400	402	404	404	36	36	20	21	25.5	25	24	25.5	22	22		336		29.5		15		30		18
60.587		388	385	392	387	38.5	39	20	20.5	30		23.5	23		22										
60.589		430	435	431	440	43	43.5	21.5	21	31.5	31	25	26	23	21.5										
60.590		349	363	353	367	37	38	23.5	21.5	28	28	29	23.5	21	22.5	301	300	32	29	17.5	16.5	34	32	19	18
60.592		433	441	434	441	44	44	22.5	21	32	32	26.5	26	26	26										
60.593		422	429	427	432	41	40.5	22	22.5	35	34	27	29	26	25.5	356		32		18.5		34		21	
60.594		364	363	367	365	35.5	35.5	17	18.5	27	27	20	20.5	21	21.5	304	306.5	26	26.5	16	17	27	27.5	16	16
60.597		403	407	406	408	41	41	20	20	28	27.5	25.5	26	21.5	21.5		344	26.5	28	19	19		30		21
60.599		387	388	388	390	39	38	21	21	29	30	24	24.5	24	23	324	325	29	29	20	20	32	31	20	20
60.600		412	407	415	410	44	44	23	25	33	33	26.5	26	28	23.5	347	343	37	36	17	18	37	38	18	19
60.601		403	405	406	409	38	38	22	22	29	28	27	28	23	23	334	335	30	31	19	18	32	33	19	18
60.602		447	449	447	451	44	44	25	26	32.5	32	34	34	24	25		374	34	34.5	24	23	36	37	24.5	23
60.603		438	440	439	441	44	44.5	24.5	25	34	33	31	30	27	26		381	33	33	21	21	38	36	21	22
60.608			391		392		38		21		29		22		23		313		26		20		28.5		20
60.609		374	381	377	386	38	38	22	22	27	29	24	24	24	24.5	312	320	28.5	28	19.5	19.5	29	31	20.5	21
60.610		395	395	398	400	43	43	23.5	34	32	32	26	26	25	26	327	324	30	31	18	18.5	31.5	32	19	19
60.611						41	40	20	21	28	27.5	23.5	24	20.5	21.5	335		29	28	17	17.5	29.5	28	18	18
60.615		392	392	392	394	42	41	23	23	31	32	27.5	25.5	27	27	332	331	30	31	20	19	33	34	20	21.5
60.616		337	442	438	445	45	45	24.5	25	32	32.5	33	34	24	24	369	369	33	34	23	22	34	37	23	23
60.618		363	363	367	369	36	35	19	19	27.5	27	22	22	20.5	20.5	310	309	27	26	18	17	30	29	19.5	18
60.619		382	387	388	387	36	36	20	21	27	27.5	24	25	21	22	319	320	25.5	26	18	16.5	26	27	20	18
60.620		427	433	431	436	42	42	24	24.5	31	30	28	29	23.5	24.5	369	(366)	31	30.5	20	19	34.5	35	20	20
60.623			401		405		39	21	22	30	29	25	26	22	21	327	326	27	30	18	17	30	32	18	18.5
60.624		406		409		41		23		31		29		24			331	32	30	18	18.5	34	33	18	17.5
60.868		409	412	412		43	43	23.5	24	30	31	29	28	23.5	24										

ID number	Sex	BML R	BML L	HML R	HML L	MDA R	MDA L	MDM R	MDM L	MDH R	MDH L	RML R	RML L	UML R	UML L	PXB	PDII	INH R	INH L	ILB R	ILB L	SAL	SMB
60.579				307		20.5		18		47					255								
60.580				284	281	16	17	22.5	23	40	38	209		228									
60.581						17		23		46		241		258									
60.582		331	335	284	285	15	14	19	19		37	221	216	231	231								
60.587				281	279	15	15	22	21	39	38					265	101	189	191	141	136	101	(117)
60.589																							
60.590			292	270	269	15	15	22	20	40	39	208			224	254	120	190	191	(136)	143	94	111
60.592																							
60.593				308	303	15	16	23	24	41	40			254		282		188	189	143	149	94	116
60.594		292	300	261	259	12.5	12.5	18	17	34.5	34.5	199	198	222	218								
60.597			335	293	293	15	14	22	20		39					270			200		148		
60.599			317		283	16	15.5	20	18	40	41					253	82	185			136	112	87
60.600		339		298	295	22	20	17	16.5	44	43	236	233		247								
60.601		326	326	294	296	21.5	21	15	15.5	39	39	225	227	242	245								
60.602				320		24		23		46		246	247	260	261	276	74	215	215	160	159	99	120
60.603				327		20		24		49				272									
60.608					283				21		40		208			237	99	188	191			91	101
60.609		305	308	272	274	14	14	22	21	38	39	213	212	230	224	248	(114)	192	190	135	134	114	118
60.610		318	315		285	16	15	20	21	41	39.5	216	215	231		269	(106)	189	189			95	115
60.611			310	303	298	15	15	20.5	20		40			240									
60.615				306		16.5	16	20	20	42	41												
60.616		362	368	327	326	18	17	22.5	21	46	47	248	250	269									
60.618		301	301		263		15.5		21.5			228		210									
60.619		314	316	287		14.5		21		36.5		215				256	95	184	181	140	144	71	108
60.620				314		17		24		42	40.5	249		266	260	297							
60.623			313	295	294	15	15	21	21.5	39	38	222	221	236	235								
60.624																							
60.868																							

PART 3: NONMETRIC DATA

Key to Cranial Nonmetric Traits and Descriptions

The ID number is the catalogue number assigned by the Peabody Museum of Archaeology and Ethnology when the human remains were accessioned. The data for each ID number continue from the left page to the right page.

Abbrev.	Trait	Description
LBW	Lambdoid wormian	Scored presence of ossicles in suture, 1 = present, 0 = absent.
TPW	Temporo-parietal wormian (right and left)	Scored presence of ossicles in suture, 1 = present, 0 = absent.
TOW	Temporo-occipital wormian (right and left)	Scored presence of ossicles in suture, 1 = present, 0 = absent.
EPI	Epipteric (right and left)	Scored presence of ossicles in suture, 1 = present, 0 = absent.
BRG	Bregma	Scored presence of ossicles in suture, 1 = present, 0 = absent.
INC	Inca	Scored presence of ossicles in suture, 1 = present, 0 = absent.
CRN	Coronal	Scored presence of ossicles in suture, 1 = present, 0 = absent.
SAG	Sagittal (right and left)	Scored presence of ossicles in suture, 1 = present, 0 = absent.
PTE	Pterion	Scored shape of fronto-temporal articulation, 1 = K, 0 = H.
PTF	Parietal foramina (right and left)	Scored presence of foramen by side, 1 = present, 0 = absent.[a]
RMF	Retro-mastoid foramina (right and left)	Scored presence of foramen by side, 1 = present, 0 = absent.
SPI	Spade incisors	Presence of spade (shovel-shaped) incisors, 1 = present, 0 = absent.
UFM	Upper first molar pattern	Cusp pattern for upper first molars, 1 = 3 cusps, 0 = 4 cusps.
USM	Upper second molar pattern	Cusp pattern for upper second molars, 1 = 3 cusps, 0 = 4 cusps.
LFM	Lower first molar pattern	Cusp pattern for upper first molars, 1 = 4 cusps, 0 = 5 cusps.
LSM	Lower second molar pattern	Cusp pattern for upper first molars, 1 = 4 cusps, 0 = 5 cusps.
PLS	Palate shape	1 = parabolic, 2 = elliptical, 3 = hyperbolic, 4 = U-shaped.
MLF	Middle lacerate foramina	Presence of foramina, 1 = present, 0 = absent.
PLF	Posterior lacerate foramina	Presence of foramina, 1 = present, 0 = absent.
PCF	Postcondyloid foramina (right and left)	Presence of foramina, 1 = present, 0 = absent.
PSF	Pterygo-spinous foramina	Presence of foramina, 1 = present, 0 = absent.
DEH	Dehiscences (right and left)	Presence of dehiscences, 1 = present, 0 = absent.
IDF	Inferior dental foramina	Presence of foramina, 1 = present, 0 = absent.
MHR	Mylo-hyoid ridge	Presence of bridging, 1 = present, 0 = absent.
GNT	Genial tubercles	Presence of tubercles, 1 = present, 0 = absent.

[a] Measurement definitions were not published by Hooton but are likely to have been similar to measurements defined by Zobeck (1983) and Moore-Jansen and Jantz (1989).

ID number	LBW	R TPW	L TPW	R TOW	L TOW	R EPI	L EPI	BRG	INC	CRN	R SAG	L SAG	PTE	R PTF	L PTF	R RMF	L RMF
59.801	1	0	0	0	0	0	0	0	0	0	0	0	0	0	1	1	1
59.802	0	0	0	0	0	0	0	0	0	0	0	0	0	1	1	1	1
59.804	1	0	0	0	0	0	0	0	0	0	0	0	0	1	1	1	1
59.805	1	0	0	0	0	0	0	0	0	0	0	0	0	1	1	1	1
59.806		0	0	0	0	0	0	0	0	0	0	0		1	0	1	1
59.807	1	0	0			0	0	0	0	0	0	0	0	0	0		
59.808																	
59.809																	
59.810	1	0	1	0	0									1	0	1	1
59.811	1	0	0	0	0	0	0	0	0	0	0	0	0	1	1	1	1
59.812	1	0	0	0	0	1	1	0	0	0	0	0	0	0	0		
59.813	1	1	0	0	0	0	0	0	0	0	0	0	0	0	1	1	1
59.814																	
59.815	1	1	1	0	0	0	0	0	0	0	0	0	0	1	0	1	0
59.816	1	0	0	0	0	0	0	0	0	0	0	0	0	1	0	1	1
59.817	0	0	0	0	0	0	0	0	0	0	0	0	0	1	1	1	1
59.818	1													1	1		
59.820	0	0	0	0	0	0	0	0	0	0	0	0	0	0	1	1	1
59.822	1	0	0	0	1	0	0	0	0	0	0	0	0	0	1	0	1
59.823	1	0	0	0	0	0	0	0	0	0	0	0	0	1	0	1	1
59.825	0													1	0		
59.826	0	0	0	0	0	0	0	0	0	0	0	0	0	1	0	1	1
59.827	1	0	0	0	1	0	0	0	0	0	0	0	0	0	1	1	0
59.828	1	0	0	0	0	0	0	0	0	0	0	0	0	1	1	1	1
59.829	0	0	0	0	0	0	0	0	0	0	0	0	0	1	0	0	0
59.830	0	0	0	0	0	0	0	0	0	0	0	0	0	0	1	1	1
59.831	1	0	0	1	1	0	0	0	0	0	0	0	0	0	1	0	1
59.832													0	0	0	1	0
59.834																	
59.837													0	0	1	1	1
59.838													0	1	1		
59.839	1	0	0	0	0	0	0	0	0	0	0	0	0	1	1	0	1
59.840	0	0	0	0	0	0	0	0	0	0	0	0	0	1	1	1	1
59.841	1	0	0	0	0	0	0	0	0	0	0	0	0	1	1	1	1
59.842	1	0	0	0	0	0	0	0	0	0	0	0	0	0	1	1	1
59.843	0	0	0	0	0	0	0	0	0	0	0	0	0	1	0	1	1
59.845	0	0	0	0	0	0	0	0	0	0	0	0	0	1	1	1	1
59.846													0	1	1	1	1
59.849	1	0	0	0	0	0	0	0	0	0	0	0	0	1	1	1	1
59.850	0	0	0	0	0	0	0	0	0	0	0	0	0	0	0		
59.851													0	1	0	1	1
59.852	0	0	0	1	0	0	0	0	0	0	0	0	0	1	1	1	1
59.855													0	1	1	1	1
59.856													0				
59.857	1	0	0	0	0	0	0	0	0	0	0	0	0	0	1	1	1
59.859	1	0	0	0	0	0	0	0	0	0	0	0	0	1	0	1	1
59.860	1	0	0	0	0	0	0	0	0	0	0	0	0	0	0	0	1
59.862	1	0	0	0	0	0	0	0	0	0	0	0	0	1	1	1	1
59.863													0	1	1	1	1
59.864	0	0	0	0	0	0	0	0	0	0	0	0	0	1	1	0	1
59.866	0	1	0	0	0	1	1	0	0	0	0	0	0	1	1	1	1
59.867	1	0	0	1	0	0	0	0	0	0	0	0	0	1	1	1	1
59.868		0	0	0	0	0	0	0	0	0	0	0	0	0	0	1	1
59.870	0	0	0	1	0	0	0	0	0	0	0	0	0	1	1	1	1
59.871	1	0	0	0	0	0	1	0	0	0	0	0	0	1	0	1	1
59.872	1	0	0	0	0	0	0	0	0	0	0	0	0	0	0	1	1
59.875	1	0	0	0	0	0	0	0	0	0	0	0	0	1	0	1	1
59.876	1	0	0	0	0	0	0	1	0	0	0	0	0	1	1	1	1
59.877	0	0	0	0	0	0	0	0	0	0	0	0	0	1	1	1	1
59.889	0	0	0	0	0	0	0	0	0	0	0	0	0	0	0	1	1
59.891	1	0	0			0	0	0	0	0	0	0	0	1	0	0	0
59.892													0	0	0	1	1
59.893													0	0	0	1	1

ID number	SPI	UFM	USM	LFM	LSM	PLS	MLF	PLF	R PCF	L PCF	PSF	R DEH	L DEH	IDF	MHR	GNT
59.801						1	1	1	1	1	0	0	0	1	0	0
59.802	0			1	1	1	1	1	1	1	0	0	0	1	1	1
59.804	1					1		1	1	0		0	0	1	0	0
59.805		0	1	1	1	1	1				0	0	0	1	0	0
59.806						1	0	1	1	1	0	0	0	1	1	0
59.807	1	0	1	0	0	1					0	0	0	1	0	0
59.808	1					1								1	1	0
59.809	1	0	0	0	0	1								1	0	0
59.810		0	1									1	0	1	1	1
59.811	1	0	0	0	1	3	1	1	1	1	0	0	0	1	0	1
59.812	1					1					0	0	0	1	1	0
59.813							1	1	1	1	0	0	0	1	1	1
59.814	1	0	1			2										
59.815		0	1	1	0	1	1	1	1	1	1	0	0	1	1	1
59.816	1			0	1	4	1	1	1	1	0	0	0	1	0	0
59.817	1	0	0	0	0	3	1	1	1	1	0	0	0	1	0	0
59.818														1	0	0
59.820	1	0	0	0	1	1	1	1	1	1	0	0	0	1	0	0
59.822	1	0	0	0	0	4	1	1	1	1	0	0	0	1	0	0
59.823	1	0	0	0	1	1	1	1	1	1	0	0	0	1	0	1
59.825	0													1	1	0
59.826	0	0	0	0	1	4					0	0	0	1	1	1
59.827	0	0	1	1	1	1	1	1	1	0	0	0	0	0	0	0
59.828	1	0	0	0	0	4	1	1	1	1	1	0	0	1	0	0
59.829	1	0	0	0	0		1	1	1	1	0	0	0	1	0	0
59.830								1	1	1	0	0	0	1		1
59.831				0	0	4	1	1	1	1	0	0	0	1	0	0
59.832	0					4	1	1	1	1	0	0	0	1	0	0
59.834	1	0	1	0	0	1									0	0
59.837	1	0	0	0	0	4								1	1	1
59.838							1					0	0	1	1	0
59.839							1	1	1	1	1	1	0	1	1	1
59.840	1					1	1	1	1	1	0	0	0	1	1	0
59.841	1	0	1	0	1	4	0	1	1	0	0	0	0	1	1	0
59.842						4	1	1		1	0	0	0	1	0	0
59.843						4	0							1	0	0
59.845	0			0	0	4	1	1	1	1	0	0	1	1	1	0
59.846							1	1	1	1	0	0	0	1	1	0
59.849	1	0	1	0	1	1	0	1	1	1	0	0	0	1	0	0
59.850	1	0	0	0	0	1	1	1			0	1	1	1	0	0
59.851							1	1	1	1	0	0	0	1	1	0
59.852							1				0	0	0		0	0
59.855							1	1	1	0	0	0	0	1	0	0
59.856	1					1								1	0	0
59.857									1	0	0	0	1		1	0
59.859	1	0	1	0	0	4	1	1	1	1	0	1	1			
59.860	1	0	1	0	0	1	1	1	1	1	0	0	0	1	1	
59.862	0	0	1	0	0	4	1	1	1	1	0	0	0	1	0	1
59.863		0	0	0	0	1	1	1	1	1	0	0	1	1	0	0
59.864	0	0	0	0	1	1	0	1	1	1	0	0	0	1	0	1
59.866		0	0	0	0	1	1	1	1	1	0	0	0	1	1	1
59.867	1	0	0	0	0	4	1	1	1	1	0	0	0	1	0	0
59.868											1	1	0	1	1	0
59.870	1	0	0	0	0	1	1	1	1	1	0	0	0	1	1	0
59.871	1	0	0			1	1	1	1	1	0	1	1			
59.872						1		1	1	1	0	0	0	1	1	0
59.875	1					2	1	1	1	1	0	0	0	1	1	0
59.876		0	0	1	1	4	1	1	1	1	0	0	0	1	0	0
59.877						2	1	1	1	1	0	0	0	1	1	1
59.889							1	1	1	1	0	1	0			
59.891	0	0	1	1	1	1	1		1	1	0	0	0	1	1	0
59.892						4	1	1	1	1	0	0	0			
59.893							1	1	1	1	1	0	0			

ID number	LBW	R TPW	L TPW	R TOW	L TOW	R EPI	L EPI	BRG	INC	CRN	R SAG	L SAG	PTE	R PTF	L PTF	R RMF	L RMF
59.898	0	0	0	1	1	0	0	0	0	0	0	0	0	1	0	1	1
59.901	1	0	0	1	0	1	0	0	0	0	0	0	0	1	1	0	0
59.904	1					0	0	0	0	0	0	0		1	0		
59.905		0	0	0	0	0	0	0	0	0	0	0	0	0	0	1	1
59.906	1	0	0	0	0	0	0	0	0	0	0	0	1	1	0	1	1
59.907													0	1	1	1	1
59.908	1	0	0	0	0	0	0	0	0	0	0	0	0	1	0	1	1
59.910	1	0	0	0	0	1	1	0	0	0	0	0	0	1	1	1	1
59.911													0	0	0	1	1
59.912													0	1	0	1	1
59.915	0	0	1	0	1	0	0	0	0	0	0	0	0			1	1
59.916	1	1				0	0	1	0	0	0	0	0	0	0	1	1
59.917	1	0	0	1	1	0	0	0	0	0	0	0	0	1	1	1	1
59.918	1	0	0	0	1	0	0	0	0	0	0	0	0	1	1	1	1
59.919		0	0	0	0	0	1	0	0	0	0	0	1	1	1	1	1
59.920	1	0	0	0	0	0	0	0	0	0	0	0	1	1	1	1	1
59.921	1			1	1	0	0	0	0	0	0	0	0	1	0	1	1
59.922	1	0	0										0	1	0	1	1
59.923	0	0	0	0	1	0	0	0	0	0	0	0	1	1	1	1	1
59.924	1	0	0			0	0	0	0	0	0	0	1	1	0	1	1
59.925	0	0	0	1	1	1	1	0	0	0	0	0	1	1	1	1	1
59.952	1	1	0	1	0	0	0	0	0	0	0	0	0	1	1	1	1
59.953	1					0	0	0	0	0	0	0	0	1	1		
59.954	1	0	0	0	0	0	0	0	0	0	0	0	0	0	1	1	1
59.955	1	0	0	1	0	0	0	0	0	0	0	0	0	1	1	1	1
59.957	1	0	0	0	0	0	1	0	0	0	0	0	0	1	0	1	1
59.958																	
59.960	0	0	0			0	0	0	0	0	0	0	1	1	1	1	1
59.962		0	0	0	0	0	0	0	0	0	0	0	1	1	1	1	1
59.963	1	0	0	0	0	0	0	0	0	0	0	0	1	1	0	1	1
59.965	1	0	0	0	0	0	0	0	0	0	0	0	0	0	0	1	1
59.966	1	1	1	0	0	0	1	0	0	0	0	0	1	0	0	1	1
59.967		0	0										0	1	1		
59.968	0	0	0	0	0	0	0	0	0	0	0	0	0	0	1	1	1
59.969	0	0	0	0	0	0	0	1	0	0	0	0	1	0	1	1	0
59.973	1	0	0	0	0	0	0	0	0	1	0	0	0	1	1	1	1
59.974	0	0	0	0	0	0	0	0	0	0	0	0	0	0	1	1	1
59.976	1	0	0	1	1	0	0	0	0	0	0	0	1	1	1	1	0
59.977	0	0	0	0	0	0	0	0	0	0	0	0		1	0	0	0
59.978	1	0	0	0	0	0	0	0	0	0	0	0		0	0	1	0
59.979	1	0	0	0	0	0	1	0	0	0	0	0	0	0	0	1	1
59.980	0	0	0	0	0	0	0	0	0	0	0	0	0	1	1	1	1
59.981	1	0	0	0	0	0	0	0	0	0	0	0	1	1	1	1	1
59.982	0	0	0	0	0	0	0	0	0	0	0	0	0	1	0	1	1
59.983	1	0	0	0	0	0	0	0	0	0	0	0		0	1	1	1
59.984													0	1	0	1	1
59.985	1	0	0	1	0	0	0	0	0	0	0	0	0	1	1	1	1
59.986	0	0	0	0	0	0	0	0	0	0	0	0	0	1	0	1	0
59.987	1	0	0	0	0	0	0	0	0	0	0	0	0	1	1		
59.989	1	0	0	0	0	0	0	0	0	0	0	0	0	1	1	1	1
59.990													0	1	1		
59.991	0	0	0	0	0	0	0	0	0	0	0	0	0	1	1	1	1
59.993	1	0	0	0	0	0	0	0	0	0	0	0	0	0	1	1	0
59.994	0	0	0	0	0	0	0	0	0	0	0	0	0	1	1		
59.995	1	0	0	0	0	0	0	0	0	1	0	0	0	1	1	1	1
59.996	0	0	0	0	0	0	0	0	0	0	0	0	0			1	0
59.998	0	0	0	0	0	0	0	0	0	0	0	0	1	1	1	1	1
59.999	1	0	0	0	0	0	0	0	0	0	0	0	0	0	1	1	1
60.000	0	0	0	1	0	1	1	0	0	0	0	0	0	1	1	1	0
60.001													1	0	0		
60.002													0	0	0		
60.003	1	0	0	0	0	0	0	0	0	0	0	0	0	1	1	1	1
60.004	1	0	0	0	0	0	0	0	0	0	0	0	0	0	0	1	0

ID number	SPI	UFM	USM	LFM	LSM	PLS	MLF	PLF	R PCF	L PCF	PSF	R DEH	L DEH	IDF	MHR	GNT
59.898	0					4	1	1	1	1	0	0	0			
59.901	1	0	0	1	1	1	1		1	0	1	0	0	1	1	1
59.904							1	1	1	1	0	1	0			
59.905							1	1	1	1	1	1	0			
59.906						1	1	1	1	0	1	1	1	1	0	0
59.907	1	0	1	1	1	1	1	1	1	1	0	0	0	1	0	1
59.908	1	0	1	1	0	1	1	1	1	1	0	0	0	1	0	0
59.910	1	0	1	0	1	2	1	1	1	1	0	0	0	1	0	1
59.911							0	1	0	0	0	0	0	1	1	0
59.912							1	1	1	1	0	1	1	1	0	0
59.915	1	0	0	0	1	1	1	1	1	1	1	0	0	1	1	1
59.916							1	1			0	0	0			
59.917						1						1	1	1	0	0
59.918	1	0	0	0	0	4	1	1	1	1	0	0	0	1	0	0
59.919		0	0	0	0	1	1	1	1	1	0	1	1	1	1	0
59.920						4	1	1	0	0	0	0	0	1	1	0
59.921		0	1	0	1	1								1	0	0
59.922																
59.923	1					1	0	1	1	1		0	0	1	1	0
59.924						4						0	0	1	0	1
59.925	1	0	1	0	0	1	1	1	0	1	0	0	0	1	1	1
59.952							1	1	1	1	0	1	0	1	1	0
59.953	0	0	0	0	0	1	1	1	1	1	0	0	0	1	1	0
59.954		0	1			1	1	1	0	0	0	0	0	1	1	0
59.955						1	1	1	1	1	0	0	0			
59.957						1	1	1	1	1	1	0	0	1		0
59.958							1	1	1	0	0	0	0	1	0	0
59.960	1	0	0	0	0	3					0	0	0	1	1	1
59.962	0	0	1	1	1	1	1	1	1	1	0	0	0	1	1	0
59.963						1	1[a]	1	1	1	0	0	0	1	1	0
59.965						1		1	1	1		0	0	1	1	0
59.966	1					1	1	1	1	1	0	0	0	1	1	0
59.967														1	1	0
59.968	1	0	0	0	0	1		1	1	1	0	0	0	1	1	0
59.969	1	0	1	0	1	1		1	0	0	0	0	0	1	0	0
59.973						1					0	0	0	1	1	0
59.974	1	0	1			1					0	0	0			
59.976						2	1	1	1	1	0	0	0	1	1	0
59.977							1	1	1	1	0	0	0	1	1	0
59.978							1	1	1	1	0	0	0	1	1	0
59.979							1	1	0	1	0	1	1	1	0	0
59.980							1	1	1	1	1	0	0	1	1	0
59.981	1	0	0	0	1	4		1	1	1	0	1	1	1	1	0
59.982	0	0	0	0	1	1	1	1	0	0	1	0	0	1	0	0
59.983	1	0	0	0	1	1	1	1	1	1	0	0	0	1	1	0
59.984							1	1	1	1	0	0	0	1	1	1
59.985	0	0	0	0	0	1			1	1	1	0	0	1	0	0
59.986							1	1	1	1	1	0	0	1	1	1
59.987	1	0	0	0	1	1					0	0	0	1	1	1
59.989																
59.990	1	0	0	0	0	1	1	1			0	0	0	1	0	1
59.991						1	1	1	1	1	0	1	1	1	1	0
59.993						4	1	1	1	1	0	1	0	1	1	1
59.994	1	0	0	0	0	1					1	0	0		1	0
59.995	1	0	0	1	1	1					0	0	0	1	1	1
59.996	1	0	1	0	1	4	1	1	1	1	0	0	0	1	0	0
59.998	1	0	0	0	0	1	1	1	1	1	0	0	0	1	1	1
59.999						1	1	1	0	1	0	0	0	1	1	0
60.000		0	0	0	1	3	1	1	1	0	1	0	0	1	0	0
60.001							1	1	1	1	0	0	0	1	0	0
60.002						4	1	1			0	0	0		1	1
60.003	1	0	1	1	1	4	1	1	1	1	0	0	0	1	0	0
60.004	1	0	0	1	1	1	1	1	1	1	0	1	1	1	1	0

ID number	LBW	R TPW	L TPW	R TOW	L TOW	R EPI	L EPI	BRG	INC	CRN	R SAG	L SAG	PTE	R PTF	L PTF	R RMF	L RMF
60.005	1	0	1	0	0	0	0	0	0	0	0	0	1	1	1	1	1
60.006	1	0	0	0	0	0	0	0	0	0	0	0	0	1	1	1	1
60.007													1	0	1	1	1
60.008													0			1	1
60.009	0	0	0	0	0	0	0	0	0	0	0	0	1	1	1	1	1
60.010	0			0	0	0	0	0	0	0	0	0	0	1	0	1	1
60.011	1	0	0	0	1	0	0	1	0	0	0	0	0	1	1	1	1
60.012	1					0	0	0	0	0	0	0	0	0	0		
60.013	0	0	0	0	0	0	0	0	0	0	0	0	1	1	0	0	0
60.015	1	0	0	0	0	1	1	0	0	0	0	0	0	1	0	1	1
60.016	0	0	0	0	1	0	0	0	0	0	0	0	0	1	1	1	1
60.017	1	0	0	0	0	0	0	0	0	0	0	0	0	1	0	1	1
60.018	1	0	0	0	0	0	0	0	0	0	0	0	1	1	0		
60.019	1	0	0	0	0	0	0	0	0	0	0	0	0	1	0	1	1
60.020																	
60.021	0	0	0	1	1	0	0	0	0	0	0	0	1	1	1	1	1
60.022														1	1	1	1
60.023	1	0	0	0	0	0	0	0	0	0	0	0	0	1	1	1	1
60.025																1	1
60.026	0	0	0	0	0	0	0	0	0	0	0	0	0	1	0	1	1
60.027	1	0	0	1	1	0	0	0	0	0	0	0	0	1	0	1	1
60.028	1			0	0	0	0	0	0	0	0	0	1	0	1	1	1
60.029		0	0	0	0	0	0	0	0	0	0	0	1	1	0	0	1
60.031	0	0	0	0	0	0	0	0	0	0	0	0	0	1	1	1	1
60.032	1	0	0	1	1	0	0	0	0	0	0	0	1	1	1	1	1
60.033	1	0	0	0	0	0	0	0	0	0	0	0	0	1	1	1	1
60.034	0	0	0	1	0	0	0	0	0	0	0	0	0	0	1	1	1
60.036	0	0	0	1	0	0	0	0	0	0	0	0	0	1	1	1	1
60.038		0	0	0	0									1		1	1
60.039	0	0	0	1	0	0	0	0	0	0	0	0	1	1	0	1	1
60.040	0	0	0	0	0	0	0	0	0	0	0	0	0	0	0	1	1
60.041	1	0	0	0	1	0	0	0	0	0	0	0	1	0	0	0	1
60.042	1	0	0	1	0	0	0	0	0	0	0	0	1	0	0	1	1
60.044	0	0	0	0	0	0	0	0	0	0	0	0		0	0	1	1
60.045	1	0	0	0	0	0	1	0	0	0	0	0	0	1	1	1	1
60.046	0	0	0	0	0	0	0	0	0	0	0	0	1	0	1	1	1
60.047	0	0	0	1	1	0	0	0	0	0	0	0	1	1	1	1	1
60.048	0	0	0	0	0	0	0	0	0	0	0	0	1	0	1	1	1
60.049	0	0	0	0	0	0	0	0	0	0	0	0	1	0	1	1	1
60.051		0	0	0	0	0	0	0	0	0	0	0	1	0	1	1	1
60.052	1	0	0	1	0	0	0	0	0	0	0	0	0	1	1	1	1
60.053	0	0	0	0	0	0	1	0	0	0	0	0	1	1	0	1	1
60.055	0	0	0	0	0	0	0	0	0	0	0	0	0	0	1	1	1
60.057	1	0	0	0	0	0	0	0	0	0	0	0	1	1	1	1	1
60.058		0	0	0	0	0	0	0	0	0	0	0	0	0	1	1	1
60.059	1			0	0	0	0	0	0	0	0	0	1	1	1	1	1
60.060	1	0	0	0	0	0	0	0	0	0	0	0	1	1	0	1	0
60.061	0	0	0	0	0	0	0	0	0	0	1	1	1	0	1	1	1
60.062	1			1	0								0	1	1	1	1
60.063	1	1	1	0	0	0	0	0	0	0	0	0	0	1	1	1	1
60.064	1	0	0	0	0	0	0	0	0	0	0	1	0	1	1	1	1
60.065	1	0	0	0	1	0	0	0	0	0	0	0	1	0	1	1	1
60.067	1	0	0	0	0	0	0	0	0	0	0	0	1	0	0	1	0
60.068	0	0	0	1	0	0	0	0	0	0	0	0	1	0	1	1	1
60.070													0			1	1
60.071	1	0	0	0	0	0	0	0	0	0	0	0	0	0	1	1	1
60.073	1												0	0	0		
60.074	1	0	0	0	0	0	0	0	0	0	0	0	0	0	0	1	1
60.075													0	1	0	0	0
60.080	1	0	0	1	1	0	1	0	0	0	0	0	0	1	1	1	1
60.081	1	1	0	0	0	0	0	0	0	0	0	0	0	0	0	1	1
60.082	0	0	0	0	0	0	0	0	0	0	0	0		0	0	1	1
60.083		0	0	0	0	0	0	0	0	0	0	0	0	1	1	1	1

ID number	SPI	UFM	USM	LFM	LSM	PLS	MLF	PLF	R PCF	L PCF	PSF	R DEH	L DEH	IDF	MHR	GNT
60.005							1	1	1	1	0	0	0	1	1	1
60.006							1	1	0	1	0	1	1	1	1	0
60.007	1						1	1	1	1	0	0	0	1	0	0
60.008						4	1	1	1	1	0	0	0	1	1	0
60.009							1	1	1	1	0	0	0	1	0	0
60.010						1	1	1	1	1	0	0	0	1	1	0
60.011	1	0	1	0	1	4	1	1	1	1	0	0	0	1	0	0
60.012	1					1	1	1			0	0	0	1	0	0
60.013	1	0	1	0	1	1	1	1	1	1	0	1	1	1	1	0
60.015	1	0	0	0	0	1	1	1	1	1	1	0	0	1	1	0
60.016	1					4	1	1	1	1	0	1	1	1	1	1
60.017						1	1	1				1	1	1	1	1
60.018						3	1	1	1	1	0	0	0	1	1	1
60.019	1	0	0	0	1	1	1	1	1	1	0	0	0	1	1	0
60.020						2								1	0	
60.021	0	0	1	0	0	2	1	1	1	1	0	0	0	1	1	1
60.022																
60.023	0	0	0	0	0	1	1	1	1	0	0	0	0	1	1	1
60.025							1	1	1	1	0	0	0	1	1	0
60.026						1	1	1	1	1	1	0	0	1	1	1
60.027	1	0	1	0	0	1					0	0	0	1	0	1
60.028								1	1	1		0	1	1	1	1
60.029	1	0	1	1	1	1	1	1	1	1	0	0	0	1	1	0
60.031							1	1	1	1	1	0	0	1	1	0
60.032	0					1	1	1	1	1	0	0	0			
60.033							1				0	0	0	1	1	0
60.034	1			0	1	1	1	1	1	1	0	0	1	1	0	0
60.036		0	0	1	1	1	1	1	1	0	0	1	1	1	0	0
60.038	1	0	1	0	1	4			1	1		0	0	1	0	0
60.039							1	1	1	0	0	0	0	1	1	1
60.040						4	1	1			0	1	1	1	0	0
60.041	1	0	0	1	1	1	1	1	1	1	0	0	0	1	0	0
60.042							1	1	1	1	0	0	0	1	0	0
60.044	1	0	1	0	0	1		1	1	1		0	0	1	0	0
60.045	1	0	1	0	0	4	1	1	1	1	0	0	0	1	1	1
60.046						4	1	1	1	1	0	1	1	1	1	0
60.047	1	0	1	0	0	3	1	1	1	0	0	0	0	1	1	1
60.048						4								1	0	1
60.049							1	1	1	1	0	0	0	1	1	0
60.051		0	0	0	1	1	1	1	1	1	1	0	0	1	0	0
60.052	1	0	0	0	0	1	1	1	1	1	0	0	0	1	0	0
60.053	1					1	1	1	0	1	1	0	0			
60.055	1	0	1	0	0	4	1	1	1	1	0	1	1	0	1	0
60.057						4	1	1	1	1	0	0	1	1	0	0
60.058						4	1	1	1	1	1	0	0			
60.059	1					1	1	1	0	0	0	0	0	1	1	0
60.060						1	1	1	1	0	0	1	1	1	1	0
60.061							1	1	1	1	0	0	0	1	1	1
60.062						1	1	1	1	1	0	0	0	1	1	0
60.063							1	1	1	1	0	0	0	1	1	0
60.064	1					1	1	1	1	1	0	0	1	1	1	0
60.065						1	1	1	1	1	0	0	0	1	1	1
60.067	1	0	0	0	0	4	1	1	1	1	0	0	0	1	0	0
60.068						4						0	0			
60.070														1	1	1
60.071		0	0	0	1	4	1	1	0	1	0	0	0	1	1	1
60.073	1					2						0	0		0	1
60.074	1	0	0	0	0	1	1	1	1	1	0	0	1	1	0	0
60.075						4	1	1	1	1	0	0	0	1	1	1
60.080	1	0	0	0	1	3	1	1	1	1	1	0	0	1	1	0
60.081	1	0	0	0	0	3					0	0	0	1	1	0
60.082	1					4	1	1	1	1	0	0	0	1	0	0
60.083							1	1	1	1	0	0	0	1	0	0

ID number	LBW	R TPW	L TPW	R TOW	L TOW	R EPI	L EPI	BRG	INC	CRN	R SAG	L SAG	PTE	R PTF	L PTF	R RMF	L RMF
60.084	1	0	0	0	0	0	0	0	0	0	0	0	1	1	1	1	1
60.085	1	0	0	0	0	0	0	0	0	0	0	0	0	1	1	1	1
60.086	0	0	0	0	0	0	0	0	0	0	0	0	0	0	0	1	1
60.087	1	0	0			0	0	0	0	0	0	0	0	1	1		
60.088				0	0	0	0	0	0	0	0	0	0	1	0	1	1
60.089	1	0	0	0	0	0	0	0	0	0	0	0	1	1	1	1	1
60.090	0	0	0	0	0	0	0	0	0	0	0	0	1	0	0	1	1
60.091	1	0	0	0	0	0	0	0	0	0	0	0	0	0	0	0	0
60.092	0	0	0	1	1	0	1	0	0	0	0	0	0	0	1	1	1
60.093	0	0	0	0	0	0	0	0	0	0	0	0	0	0	0	0	0
60.094	0	0	0	0	0	0	0	0	0	0	0	0	0	1	1	1	1
60.095	1	0	0	1	1	0	0	0	0	0	0	0	0	1	1	1	1
60.096	1	0	0	1	1	0	0	0	0	0	0	0	0	1	0	1	1
60.097	1	0	0	0	0	0	0	0	0	0	0	0	0	0	0	1	1
60.098	0	0	0	0	0	0	0	0	0	0	0	0	0	1	1		
60.099	1	0	0	0	1	0	0	0	0	0	0	0	0	1	0	1	1
60.100	0	0	0	1	0	0	0	0	0	0	0	0	0	0	1	1	1
60.101	1	0	0	0	0	0	0	0	0	0	0	0	1	0	1	1	1
60.102	1	0	0	0	0	0	0	0	0	0	0	0	1	1	1		
60.103	0	0	0	0	0									0	1	0	0
60.105	1	0	0			0	0	0	0	0	0	0	0	0	0	1	0
60.106	1	0	0					1									
60.107	0	0	0	0	0	0	0	0	0	0	0	0	1	1	0	1	1
60.108	1	0	0	0	1	0	0	0	0	0	0	0	1	0	1	1	1
60.138	1	0	0	0	0	0	0	0	0	0	0	0		1	0	1	1
60.141	1												0				
60.142	0	0	0			0	0	0	0	0	0	0	0	1	1		
60.144	1												1	1	0		
60.145	0	0	0	0	0	0	0	0	0	0	0	0	0	1	1	1	1
60.165	1													0	0		
60.175	1	0	0	0	0	0	0	0	0	0	0	0	0	0	0		
60.177													0				
60.179																	
60.188	1	0	0	0	0	0	0	0	0	0	0	0	0	1	1	1	1
60.194	1	0	0	0	0	0	0	0	0	0	0	0	0	0	0	1	1
60.195	1	0	0	1	1	0	0	0	0	0	0	0	0	0	0	1	1
60.196	0	0	0	0	0	0	0	0	0	0	0	0	0	1	1	1	1
60.198	1	0	0			0	0	0	0	0	0	0	0	0	0		
60.199													0	1	1		
60.200	1												0	1	1		
60.201	1	0	0			0	0	0	0	0	0	0	0	1	0		
60.207	0	0	0	0	0	0	0	0	0	0	0	0	0	1	1	1	1
60.210	1	0	0			0	0	0	0	0	0	0		0	1		
60.211	0	0	0			0	0	0	0	0	0	0	0	0	1	1	0
60.212	1	0	0	0	0	0	0	0	0	0	0	0	0	0	0	1	1
60.213	1	0	0	0	0	0	0	0	0	0	0	0	0	0	1	1	1
60.216	1					0	0	0	0	0	0	0	0	0	1		
60.218	0	0	0	1	1	0	0	0	0	0	0	0	0	1	1	1	1
60.219	1	0	0	0	1	0	0	0	0	0	0	0	0	1	1	1	1
60.220													0	1	1		
60.222	0	0	0	1	1	0	0	0	0	0	0	0	0	1	1		
60.225	1	0	0	1	1	0	0	0	0	0	0	0		0	0		
60.226								1					0	1	1	1	1
60.227	1												0				
60.228													1	1	0		
60.229													0	1	1	1	1
60.230													0	0	0	1	1
60.231	1	0	0	1	0	0	0	0	0	0	0	0	0	0	0	1	1
60.234	1													0	1		
60.235	0	0	0	1	0	0	0	0	0	0	0	0	0	1	1	1	
60.236	0	0	0	0	0	0	0	0	0	0	0	0	0	1	1		1
60.237	0												0	1	1		
60.238	0	0	0	0	0	0	0	0	0	0	0	0	0	1	0	1	1

ID number	SPI	UFM	USM	LFM	LSM	PLS	MLF	PLF	R PCF	L PCF	PSF	R DEH	L DEH	IDF	MHR	GNT
60.084							1	1	1	1	0	1	1	1	0	0
60.085	1	0	0	0	0	1	1	1	1	1	0	0	0	1	1	0
60.086							1	1	1	1	0	0	0			
60.087	1	0	0	0	0	1	1	1			0	0	0	1	1	0
60.088						4	1	1	1	1	0	0	0	1	1	1
60.089	1					4	1				0	0	0	1	0	0
60.090						1					0	1	1	2	1	0
60.091						1					0	0	0	1	0	1
60.092	1	0	0	0	1	4	1	1	1	1	0	0	1	1	1	0
60.093	1	0	0	0	0	4	1	1	1	1	0	0	0	1	1	1
60.094	1	0	0	1	1	3		1	1	1	0	1	1	1	1	0
60.095							1	1	1	1	0	1	1	1	1	1
60.096						4	1	1	1	1	0	0	0	1	1	1
60.097	1	0	0	0	0	4	1	1	1	1	0	0	0	1	0	0
60.098	1	0	1	0	0	1					0	0	0	1	0	0
60.099	1	0	1	1	1	1	1	1	1	1	0	0	0	1	1	0
60.100	1	0	0	0	1	1	1	1	1	1	0	0	0	1	1	0
60.101						4						0	0	1	1	1
60.102	1	0	0	0	0	3	1	1	1	1	0	0	0	1	0	0
60.103	1	0	0	0	0	2						0	0			
60.105							1	1	0	1	1	0	0			
60.106	1					2	1	1	1	1	0	0	0	1	0	0
60.107	1	0	0	0	0	1	1	1				0	1	1	0	1
60.108	1	0	0	1	1	4	1	1	0	1	1	1	1	1	0	1
60.138									1	1		1	1	1	1	0
60.141	1					4					1	1	0	1	0	0
60.142	1	0	1			1			1	1	0	1	1			
60.144	1	0	1	0	0	1					0	0	0	1	0	0
60.145												0	0	1	1	1
60.165	1	0	0	0	1	1					0	0	0	1	1	1
60.175		0	1	1	0	1						0	0		0	0
60.177	1	0	0	0	1	4								1	1	1
60.179	1	0	1	1	1	4						1	1			
60.188		0	0			3	1	1	1	1	0	0	0			
60.194						4	1	1	1	1	0	0	0	1	1	0
60.195	1	0	1	0	1	1					0	0	0	1	0	0
60.196							1	1	0	1	0	0	1	1	1	0
60.198						4					0	0	0	1	1	1
60.199	1	0	0	0	1	4					0	0	0	1	1	1
60.200	1					1								1	1	0
60.201	1	0	1	1	1	4						0	0	1	1	0
60.207						1	1	1	1	1		0	0	0	0	0
60.210						1			1	1		0	0	1	0	0
60.211							1	1	1	0	0	0	0			
60.212							1	1	1	1	0	0	0	1	0	0
60.213	1	0	1	0	0	1	1	1	1	0	0	0	0	1	0	0
60.216						4					0	0	0	0	1	1
60.218						1	1	1	1	1	1	0	0	1	1	1
60.219						1					1	0	0	1	1	0
60.220						4						0	0	1	1	1
60.222	0	0	0	0	1	1					0	0	0	1	0	1
60.225						1	1	1			0	0	0	1	0	0
60.226														0	0	0
60.227	1	0	1	0	1	1					0	0	0	1	1	1
60.228	1						1	1	1	0	0	0	0	1	1	0
60.229						1					0	0	0	1	1	1
60.230											0	0	0	0	0	0
60.231							1	1	1	1	1	0	0	1	0	0
60.234						4						0	0	1	1	1
60.235	1	0	0	0	1	1			1	0		0	0	1	1	1
60.236						3	1	1	1	1	1	0	1	1	1	1
60.237														1	0	
60.238						4	1	1	1	0	1	0	0	1	1	0

ID number	LBW	R TPW	L TPW	R TOW	L TOW	R EPI	L EPI	BRG	INC	CRN	R SAG	L SAG	PTE	R PTF	L PTF	R RMF	L RMF
60.239	1	0	0	1	1	0	0	1	0	0	0	0	0	0	0	1	1
60.240	0	0	0			0	0	0	0	0	0	0	0	0	1		
60.241	1	1	1	0	0	0	0	0	0	0	0	0	0	1	0	1	1
60.242	0	0	0	0	0	0	0	0	0	0	0	0	0	0	0	1	1
60.243	1												0	1	0	1	0
60.248	1	0	0	0	0	0	0	0	0	0	0	0	0	0	0	1	1
60.249	1	0	0	0	0	0	0	0	0	0	0	0	0	1	1	1	0
60.252	1	0	0	0	0	0	0	0	0	0	0	0	0			1	1
60.255													0	1	0	1	1
60.256	0	1	1	0	1	0	0	0	0	0	0	0	0	1	0	1	1
60.257	0	0	0	0	0	0	0	0	0	0	0	0	0	0	1	1	1
60.258	0	0	0	0	0	0	0	0	0	0	0	0	0	0	0	1	1
60.259	1	0	0	0	0	0	1	0	0	0	0	0	0	0	0		
60.260														1	1	1	1
60.261		0	1	0	1	0	0	0	0	0	0	0	0	0	1	0	1
60.262	1	0	0	0	0	0	0	0	0	0	0	0	0	0	0	1	1
60.271	0	0	0	0	0	0	0	0	0	0	0	0	0	0	0	0	1
60.274	1	0	0	0	0	0	1	0	0	0	0	0	0	1	1	1	1
60.275	1	0	0	0	0	0	0	0	0	0	0	0	0	1	1	1	1
60.276		0	0	0	0	0	0	0	0	0	0	0	0	1	1	1	
60.277	0	0	0	0	0	1	1	0	0	0	0	0	0	1	0	1	1
60.278													0				
60.280	0	0	0	0	0	0	0	0	0	0	0	0	1	1	1	1	1
60.281	1	0	0	1	1	0	0	0	0	0	0	0		1	1		
60.282	0	0	0	0	0	0	1	0	0	0	0	0	0	1	1	1	1
60.283	1	0	0	0	0	0	0	0	0	0	0	0	0	1	0	1	1
60.284																	
60.286	0	0	0	0	0	0	0	0	0	0	0	0	0	0	0	1	1
60.287	0					0	0	0	0	0	0	0		0	0		
60.291	0	0	0			0	0	0	0	0	0	0		0	0		
60.292	0	0	0	0	0	0	0	0	0	0	0	0	1	1	0	1	1
60.293	1	0	0	0	1	0	0	0	0	0	0	0	0	1	0	1	1
60.297	1	0	0			0	0	0	0	0	0	0	1	0	1	1	1
60.301																	
60.302	1	0	0	0	0	0	1	0	0	0	0	0	0	0	0	1	1
60.303	1	0	0	0	0	1	0	0	0	0	0	0	1	0	0	1	0
60.307	1	0	0	0	0	0	0	0	0	0	0	0	1	0	1	1	1
60.309	1												1	0	0	0	0
60.313	0	0	0	0	0	0	0	0	0	0	0	0	0	1	1	1	1
60.314	1	0	0	0	0	1	1	0	0	0	0	0	0	0	0	1	1
60.315	1	1	0	0	0	0	1	0	0	0	0	0	0	1	0	1	1
60.316	1	0	0	0	0	0	0	0	0	0	0	0	0	1	1	1	1
60.317	1	0	0	0	0	0	0	0	0	0	0	0	0	1	1	1	1
60.318	1	0	0	0	0	0	0	0	0	0	0	0	0	1	0	1	1
60.319	1	0	0	0	1	0	0	0	0	0	0	0	0	1	0		
60.320	0	0	0	0	0	0	0	0	0	0	0	0	0	0	0	1	1
60.342	0	0	0	0	1	0	0	0	0	0	0	0	0	0	0	1	1
60.453													1	1	0	1	1
60.494				0	0	0	0	0	0	0	0	0	1	0	0	1	1
60.514		0	0	0	0	0	0	0	0	0	0	0	1	1	1	1	1
60.520	1	0	0	0	0	1	1	0	0	0	0	0	1	1	1	1	1
60.521	1	0	0	0	1	0	0	0	0	0	0	0	0	0	0	1	1
60.534	1	0	0	0	0	0	0	0	0	0	0	0		1	1	1	1
60.556	1	0	0			0	0	0	0	0	0	0	1	1	0	1	1
60.581	1	0	0			1	0	0	0	0	0	0	0	1	1	1	1
60.583														1	0	1	1
60.610	1	0	0	0	0	0	0	0	0	0	0	0	1	1	1	1	1
60.612	1	0	0	1	1	0	0	0	0	0	0	0	1	1	1	1	1

ID number	SPI	UFM	USM	LFM	LSM	PLS	MLF	PLF	R PCF	L PCF	PSF	R DEH	L DEH	IDF	MHR	GNT
60.239	1	0	0	0	0	1	1	1	1	1	0	0	0	1	1	0
60.240							1	1	1	1	1	0	0			
60.241						1	1	1	1	1	0	1	1	1	1	1
60.242							1	1	1	1	1	1	1	1	1	0
60.243							1	1	1	0	0	0	0	1	0	0
60.248	1					4			1	1		1	0	1	1	0
60.249	1					1	1				1	1	1		1	0
60.252	1	0	1	0	1	1					0	0	0	1	1	1
60.255						4	1	1	1	0	0	0	0	1	0	0
60.256	1	0		0	1	1	1	1	1	1	0	0	0	1	1	0
60.257							1	1	1	1	0	1	1	0	1	1
60.258	1					1	1	1	1	1	0	0	0	1	1	0
60.259	1	0	1	1	1	1			1	1		0	0	1	1	0
60.260	0					1	1	1	1	1	1	0	0	1	0	0
60.261	1	0	0			1					0	0	0	1	0	0
60.262	1	0	1	0	1	2	1	1	1	1	0	1	1	1	0	0
60.271	1	0	0	0	0	4					1	0	1	1	1	0
60.274						1		1	1	1	1	1	0	1	1	1
60.275	1					2					0	0	0	1	0	1
60.276						4						0	0	1	0	1
60.277	1	0	0	0	1	4	1	1	1	1	0	0	0	1	0	0
60.278												0	0	1	0	0
60.280	1					4					0	0	0	1	1	0
60.281						1					0	0	0			
60.282						1	1	1	1	1	0	0	0	1	1	0
60.283	1	0	0	1	1	1	1	1	1	1	0	0	0	1	1	0
60.284	1					2						0	0	1	1	0
60.286	1					1	1	1	1	1	1	1	1	1	1	0
60.287														1	1	0
60.291	1	0	0			4					0	0	0	1	0	0
60.292	1	0	1	0	1	1	1	1	1	1	0	0	0	1	0	1
60.293	1	0	1	0	1	1	1	1	1	1	0	0	0	1	1	0
60.297						1	1	1	1	1	0	1	1	1	0	0
60.301						4	1	1	1	1	0	0	0	1	1	0
60.302		0	0		1	4					0	0	0	1	0	0
60.303	1	0	0	0	0	1	1	1	1	1	1	0	0	1	0	0
60.307	1					4	1	1	1	1	0	0	0	1	1	1
60.309							1	1	1	1	0	0	1	1	1	0
60.313						4	1	1	1	1	0	0	0	1	1	0
60.314	1	0	0	0	0	1	1	1	1	1	1	0	0	1	1	0
60.315	0					1	1	1	1	1	1	0	0	1	1	1
60.316						1	1	1	1	1	1	0	0	1	0	1
60.317							1	1	1	0	0	0	0	1	0	0
60.318						4	1	1	1	1	0	0	0	1	0	0
60.319						1	1	1	1	1	0	0	0	1	0	0
60.320	0	0	0	0	0	1	1	1	1	1	0	0	1	1	1	0
60.342											1	0	0	0	1	0
60.453							0	1	1	1	0	0	0			
60.494	1	0	0			1	1	1	1	1	0	0	0			
60.514	1	0	0		0	1	1	1	1	0	0	0	0	1	1	0
60.520	1					4	1	1	1	1	0	0	0	1	1	1
60.521	1					4	1	1	1	0	1	0	0	1	0	0
60.534	1	0	0	0	1	1						0	0	1	1	1
60.556	1	0	1	1	1	1					0	0	0	1	0	0
60.581						3	1	1	1	1	0	0	0	1	0	1
60.583						1	0	1	1	0	1	0	0	1	1	0
60.610	0	0	1	1	1	2	1	1	1	1	1	0	0	1	0	0
60.612	1	0	1				1	1	1	1	1	1	1			

Appendix F

Dental Nonmetric Data for Pecos Pueblo

Christy G. Turner

Key to Column Heads (Dental Nonmetric Traits)	
1	Winging I1
2	Shoveling I1
3	Double shoveling I1
4	Interruptus groove I2
5	Tuberculum dentale I2
6	Mesial ridge C
7	Distal accessory ridge C
8	Uto-Aztecan P1
9	Hypocone M2
10	Cusp 5 M1
11	Carabelli cusp M1
12	Parastyle M3
13	Enamel extension M1
14	Root number P1
15	Root number M2
16	M3 (presence = 0)
17	Lingual cusp p2
18	Y groove m2
19	Cusp number m1
20	Cusp number m2
21	Deflecting wrinkle m1
22	Trigonid crest m1
23	Protostylid m1
24	Cusp 7 m1
25	Tomes root p1
26	Root number c
27	Root number m1
28	Root number m2
29	Odontome P1, P2, p1, p2

Key to Scoring (Procedures and Abbreviations Following Turner et al. 1991, as Cited in Chapter 6)	
Abs	Absent
Red	Reduced
M	Mesial-lingual border
D	Distal–lingual border
U	Unobservable
X	Cusps 1 and 4 are in contact
Y	Cusps 2 and 3 are in contact
+	Cusps 1–4 are in contact
NN	None noted

Peabody number	Dental nonmetric trait																											
	1	2	3	4	5	6	7	8	9	10	11	12	13	14	15	16	17	18	19	20	21	22	23	24	25	26	27	28	29
19-6-10/59807.0	U	U	U	0	0	0	3	0	1	0	0	0	0	U	2	0	8	X	5	5	U	U	0	0	NN	1	2	1	0
19-6-10/59809.0	3	3	M,D	0	0	0	3	0	3	1	3	0	3	1	U	0	0	+	5	5	2	+	1	1A	NN	1	2	U	0
19-6-10/59811.0	3	3	M	0	0	U	U	0	U	U	U	0	3	1	1	0	U	X	U	U	U	U	0	0	NN	1	2	1	0
19-6-10/59814.0	U	U	U	U	U	0	2	0	2	U	U	0	1	1	U	Red	U	U	U	U	U	U	U	U	U	U	U	U	0
19-6-10/59816.0	3	6	0	0	0	U	U	0	U	U	1	U	3	1	1	Abs	U	X	U	U	U	0	0	0	5	1	2	2	0
19-6-10/59817.0	3	5	M	0	0	0	U	0	3.5	U	U	0	1	1	2	Red	U	X	U	>4	U	U	U	0	NN	1	2	2	0
19-6-10/59820.0	3	5	M,D	M	5–	0	U	0	3.5	U	U	0	3	1	2	0	2	+	5	>4	U	U	0	0	NN	U	2	2	U
19-6-10/59821.0	3	3	0	M	0	0	–	0	3	1	0	0	3	1	U	0	0	Y	5	5	2	0	0	0	NN	1	2	2	0
19-6-10/59822.0	3	3	M	D	0	0	U	0	3.5	0	U	0	3	1	U	0	2	X	5	>4	U	U	0	0	NN	1	3	2	0
19-6-10/59823.0	3	4	0	0	0	0	U	0	3	0	U	0	0	1	3	Red	0	X	5	5	0	0	0	0	NN	U	U	U	0
19-6-10/59824.0	1B	3	0	0	0	0	0	0	2	2	1	U	3	1	3	Peg	U	U	U	U	U	U	U	U	U	U	U	U	0
19-6-10/59826.0	3	U	U	U	U	U	U	0	3.5	2	U	0	3	1	2	0	3	X	6	>4	U	U	0	3	NN	1	2	2	U
19-6-10/59828.0	3	3	M,D	0	5	0	U	0	3.5	1	1	0	1	U	3	0	5	X	6	6	U	U	2	0	NN	U	2	2	0
19-6-10/59829.0	U	U	U	0	0	U	U	U	3.5	U	0	U	U	1	U	0	2	X	U	>4	3	0	0	0	NN	U	2	1	+
19-6-10/59834.0	1A	3	M,D	0	1	0	U	0	3.5	1	1	0	0	1	1	0	0	X	6	>4	U	0	0	0	5	1	2	2	0
19-6-10/59837.0	1A	5	M	0	0	0	U	U	3.5	U	U	0	0	1	1	0	2	Y	U	>4	U	U	0	0	5	1	2	2	0
19-6-10/59847.0	U	U	U	U	U	0	2	0	3.5	0	1	U	0	1	3	0	2	X	5	U	2	0	1	0	5	1	2	2	0
19-6-10/59848.0	1B	3	0	0	0	U	U	0	3.5	0	0	U	0	U	U	0	U	X	5	4	0	0	0	0	NN	1	2	2	0
19-6-10/59849.0	3	3	M,D	D	5–	0	U	0	3.5	U	U	0	3	1	U	0	0	X	5	5	U	U	0	0	NN	U	2	1	0
19-6-10/59850.0	1B	5	M,D	D	5–	U	U	U	3.5	0	3	0	3	1	1	0	U	X	5	5	U	0	1	0	NN	1	2	1	0
19-6-10/59859.0	3	3	M,D	M	5	0	1	0	3.5	U	U	0	1	1	2	0	U	U	U	U	U	U	U	U	U	U	U	U	0
19-6-10/59871.0	1B	2	0	D,M	0	0	U	0	3.5	U	4	0	0	1	3	0	U	U	U	U	U	U	U	U	U	U	U	U	0
19-6-10/59900.0	U	U	U	U	U	0	0	0	4	0	3	U	0	U	U	0	U	U	U	U	U	U	U	U	U	U	U	U	0
19-6-10/59906.0	U	U	U	U	U	U	U	0	3	0	5	U	3	1	3	0	U	+	5	5	3	0	0	0	NN	1	2	2	0
19-6-10/59907.0	3	5	M	M,D	5	0	3	0	2	U	1	0	3	1	2	Red	0	X,+	5	5	U	0	0	0	NN	U	2	2	0
19-6-10/59910.0	3	3	0	M	0	0	U	0	3	0	1	0	3	1	3	0	U	U	U	U	U	U	U	U	U	U	U	U	0
19-6-10/59914.0	1A	4	M	U	U	0	U	0	3	U	4	0	1	1	3	0	0	X	6	5	U	0	1	0	NN	1	2	2	0
19-6-10/59915.0	3	3	M,D	M	1	0	3	0	3.5	0	1	0	3	1	U	0	2	X	6	5	U	0	1	0	NN	U	2	2	0
19-6-10/59918.0	3	3	M	M	0	0	U	0	3.5	U	0	U	1	U	3	0	0	X	5	>4	U	U	1	0	NN	1	2	2	0
19-6-10/59955.0	U	U	U	U	U	U	U	U	3.5	0	U	0	0	1	3	Peg	U	U	U	U	U	U	U	U	U	U	U	U	U
19-6-10/59956.0	3	3	0	0	0	0	U	0	3	0	U	0	0	1	3	0	0	+	5	5	U	0	0	0	NN	1	2	2	0
19-6-10/59961.0	U	3	0	U	U	U	U	U	3.5	0	0	U	0	U	U	U	U	U	5	U	3	+	0	4	U	U	2	U	U
19-6-10/59964.0	3	4	0	D	5–	0	0	0	1	0	0	U	3	1	1	0	0	X	5	5	U	0	1	0	NN	1	2	1	0
19-6-10/59968.0	3	3	0	0	5–	0	U	0	3.5	0	2	0	3	1	1	0	0	Y,X	5	5	0	+	1	0	NN	U	2	2	0
19-6-10/59973.0	U	U	U	U	U	0	1	0	3.5	0	1	0	0	1	2	0	2	+	6	5	2	0	2	0	NN	1	2	2	0
19-6-10/59974.0	3	3	M,D	D	1	0	2	0	1	0	1	0	2	1	1	0	U	U	U	U	U	U	U	U	U	U	U	U	0
19-6-10/59981.0	3	2	0	D	1	0	4	0	3	0	3	0	0	1	3	0	3	Y,X	6	5	2	0	0	0	NN	U	3	2	0
19-6-10/59983.0	3	5	M,D	D	0	0	2	0	3.5	0	4	0	1	1	2	0	9	X	U	5	U	0	3	0	NN	1	2	2	0
19-6-10/59987.0	1B	3	0	0	0	0	U	0	3	0	6	0	0	U	U	0	3	X	U	5	0	0	0	0	NN	U	2	1	0
19-6-10/59990.0	1A	5	0	0	5–	0	U	0	3	2	U	0	1	1	3	Red	2	X	5	5	3	0	0	1A	NN	U	2	2	0
19-6-10/59992.0	3	2	0	0	0	0	0	0	3.5	0	U	U	1	U	3	U	1	X	5	5	0	0	0	0	NN	U	2	2	0
19-6-10/59996.0	1B	3	0	0	0	0	0	0	2	0	2	U	0	1	3	0	0	Y	5	5	3	0	1	0	NN	U	2	2	0

Peabody number	Dental nonmetric trait																											
	1	2	3	4	5	6	7	8	9	10	11	12	13	14	15	16	17	18	19	20	21	22	23	24	25	26	27	28	29
19-6-10/59998.0	3	2	M	0	1	0	U	0	3.5	0	1	U	0	2	3	0	0	X,+	5	5	U	0	0	0	5	1	2	2	0
19-6-10/60013.0	3	3	0	0	0	1	0	0	0	0	0	0	0	1	U	0	U	U	U	U	U	U	U	U	U	U	U	U	0
19-6-10/60023.0	1B	2	0	0	5–	0	3	0	3.5	0	2	0	3	U	U	0	3	X	6	5	0	0	0	0	NN	U	2	U	0
19-6-10/60029.0	1B	3	0	0	0	U	U	0	U	0	1	0	0	1	1	Red	0	X	6	>4	U	0	1	0	NN	1	2	1	0
19-6-10/60043.0	3	3	0	M	0	U	U	0	2	2	0	U	1	1	3	0	0	X	6	6	2	0	3	0	NN	U	2	U	0
19-6-10/60045.0	3	5	0	0	0	1	0	U	3	0	0	0	3	1	1	0	0	+	5	5	3	0	0	0	NN	1	2	2	0
19-6-10/60052.0	1B	3	U	U	U	0	U	0	3.5	2	1	0	1	1	1	0	8	Y,X	6	>4	U	0	1	0	NN	U	2	1	0
19-6-10/60055.0	1A	5	M	D	5	0	U	0	U	U	3	0	3	1	3	0	0	X	U	>4	U	U	0	U	NN	U	2	2	0
19-6-10/60056.0	U	U	M	U	U	U	U	U	3	0	0	U	0	U	U	0	U	X	5	5	0	+	0	0	U	U	2	U	U
19-6-10/60067.0	3	3	0	M	0	0	0	0	3.5	U	2	0	0	1	U	0	0	X	5	5	0	0	1	1A	NN	1	2	2	0
19-6-10/60077.0	3	3	0	M	0	0	1	0	2	U	0	U	0	1	3	0	2	X	U	5	U	U	U	U	NN	1	2	2	0
19-6-10/60079.0	U	U	0	U	U	U	U	U	0	0	0	U	0	U	U	U	U	X	5	5	0	0	0	0	U	U	2	U	U
19-6-10/60097.0	3	2	M	M	0	0	2	0	3.5	0	2	0	3	U	3	0	2	X	6	5	0	+	U	1A	NN	U	2	1	0
19-6-10/60102.0	3	3	0	0	0	0	U	0	3	U	3	U	3	U	3	Peg	U	X	U	>4	U	U	0	0	NN	U	2	2	0
19-6-10/60106.0	1B	3	0	0	5–	U	U	0	3.5	U	U	U	3	1	3	0	2	X	5	5	U	0	1	0	NN	1	2	2	0
19-6-10/60107.0	3	4	M,D	0	5	0	3	0	3.5	2	5	0	0	1	U	0	4	X, +	6	>4	U	0	2	0	NN	1	2	2	0
19-6-10/60143.0	3	3	M,D	M	5–	0	1	0	3	0	2	0	1	1	2	0	2	X	U	5	2	0	0	0	NN	1	2	1	0
19-6-10/60149.0	3	2	M	0	0	U	U	U	0	0	0	U	0	U	U	Abs	U	X	5	4	2	0	0	0	U	U	2	U	U
19-6-10/60150.0	3	4	0	U	0	0	3	0	2	0	1	U	0	1	2	0	U	X, +	5	5	3	0	1	0	NN	1	2	2	0
19-6-10/60151.0	U	U	U	0	0	U	U	U	3	0	0	U	0	U	U	U	U	X	5	5	3	0	0	0	U	U	2	U	U
19-6-10/60152.0	1A	3	M	D	0	0	2	0	3	U	1	0	3	1	3	0	0	+	5	5	U	U	0	0	NN	1	2	2	0

Appendix G

Carbon and Nitrogen Stable Isotope Values for Pecos Pueblo Human Remains

Margaret Schoeninger

Peabody number	Sample number	$\delta^{13}C$‰	$\delta^{15}N$‰	Yield (%)	C:N ratio	Previous time period	Corrected time period	Sex	Age (years)
19-6-10/60077.0	849	−8.0	9.0	10.0	2.6	Trade	Pre-trade	F	18–20
19-6-10/59965.0	846	−7.9	9.3	7.3	3.4	Trade	Pre-trade	F	35–39
19-6-10/60013.0	851	−7.9	8.2	7.8	2.7	Trade	Pre-trade	F	20–24
20-6-10/60261.0	867	−7.9	8.7	11.3	2.9	Trade	Pre-trade	M	35–45
24-4-10/60602.0	855	−7.6	8.9	8.6	2.7	Trade	Pre-trade	M	50–59
24-4-10/60454.0	866	−7.5	9.8	9.7	3.0	Pre-trade	Pre-trade	M	45–49
24-4-10/60561.0	860	−7.4	9.0	7.2	3.1	Trade	Pre-trade	F	25–34
24-4-10/60490.0	850	−7.4	9.0	8.3	3.2	Historic	Pre-trade	M	35–39
20-6-10/60199.0	847	−6.7	10.0	9.4	2.7	Trade	Pre-trade	F	25–29
20-6-10/60216.0	904	−8.6	8.2	6.3	2.9	Trade	Trade	F	60+
20-6-10/60228.0	888	−8.2	9.5	6.8	3.1	Historic	Trade	F	35–44
19-6-10/59834.0	875	−8.1	9.9	11.5	3.1	Pre-trade	Trade	F	22–29
19-6-10/59987.0	863	−7.7	8.2	7.2	3.4	Trade	Trade	M	22–29
20-6-10/60279.0	887	−7.7	8.2	9.0	3.1	Historic	Trade	M	50–59
19-6-10/59996.0	870	−7.7	8.9	16.8	3.2	Pre-trade	Trade	F	18–19
20-6-10/60211.0	869	−7.7	10.0	—	3.2	Pre-trade	Trade	F	50–59
20-6-10/60313.0	876	−7.6	9.0	7.1	2.6	Historic	Trade	M	40–49
20-6-10/60212.0	903	−7.5	9.0	6.2	2.9	Pre-trade	Trade	M	50–59
20-6-10/60242.0	906	−7.5	8.6	14.1	2.9	Trade	Trade	M	45–55+
19-6-10/60041.0	878	−7.3	8.3	8.1	3.1	Pre-trade	Trade	F	25–34
19-6-10/59823.0	907	−7.3	8.8	14.5	3.0	Pre-trade	Trade	M	20–29
19-6-10/59838.0	862	−7.2	9.3	6.3	3.1	Trade	Trade	M	45–55+
20-6-10/60225.0	899	−7.2	8.8	6.1	2.9	Historic	Trade	F	50–59
20-6-10/60302.0	871	−7.0	7.8	8.2	3.4	Historic	Trade	F	30–39
20-6-10/60201.0	874	—	9.2	9.6	3.4	Pre-trade	Trade	M	30–39
19-6-10/60064.0	881	−8.5	9.5	5.4	3.2	Historic	Historic	M	40–44
20-6-10/60229.0	920	−8.3	8.9	7.3	3.2	Historic	Historic	F	45–59
19-6-10/59977.0	922	−8.3	10.1	6.4	2.7	Historic	Historic	F	50–59
19-6-10/60061.0	2653	−8.2	9.3	6.6	2.7	Trade	Historic	F	50–59

Peabody number	Sample number	$\delta^{13}C$‰	$\delta^{15}N$‰	Yield (%)	C:N ratio	Previous time period	Corrected time period	Sex	Age (years)
19-6-10/59957.0	917	−7.9	8.0	6.8	3.2	Pre-trade	Historic	M	50–59
19-6-10/59866.0	2651	−7.9	8.4	9.1	2.6	Historic	Historic	M	27–34
19-6-10/59805.0	921	−7.9	9.4	—	3.0	Historic	Historic	M	35–44
24-4-10/60593.0	856	−7.7	9.0	8.6	3.1	Trade	Historic	M	35–39
19-6-10/60065.0	933	−7.6	—	5.5	2.9	Historic	Historic	M	40–44
20-6-10/60194.0	942	−7.2	8.2	11.4	2.9	Pre-trade	Historic	M	50–59
19-6-10/59918.0	915	−7.2	10.4	12.2	3.3	Pre-trade	Historic	M	25–34
19-6-10/59982.0	2648	−7.1	8.0	11.9	2.8	Pre-trade	Historic	M	35–45
19-6-10/59871.0	2649	−7.1	9.5	10.1	2.7	Pre-trade	Historic	F	25–29
24-4-10/60566.0	923	−7.1	9.0	8.9	2.6	Pre-trade	Historic	F	18–20
19-6-10/59877.0	2647	−7.0	10.0	12.2	2.8	Trade	Historic	F	30–39
19-6-10/59998.0	929	−8.1	8.7	5.2	3.0	Historic	Indeterminate	M	20–24
19-6-10/59817.0	909	−7.9	9.3	7.4	2.8	Historic	Indeterminate	F	35–44
19-6-10/59864.0	861	−7.9	10.0	5.7	3.1	Historic	Indeterminate	F	25–29
19-6-10/59846.0	868	−7.9	10.0	5.6	2.9	Trade	Indeterminate	F	40–49
24-4-10/60542.0	852	−7.8	8.4	13.5	2.8	Trade	Indeterminate	F	35–39
24-4-10/60456.0	854	−7.6	9.6	9.1	3.3	Historic	Indeterminate	M	30–34
19-6-10/59803.0	859	−7.5	9.9	9.2	2.8	Pre-trade	Indeterminate	F	35–49
24-4-10/60502.0	914	−7.4	8.9	10.0	3.1	Pre-trade	Indeterminate	M	30–39
20-6-10/60318.0	2650	−7.4	9.8	9.4	2.8	Historic	Indeterminate	M	40–44

Appendix H

Bone Structural Data for the Combined Pecos Sample

Christopher Ruff

Key to Column Heads	
ID	Catalogue number assigned by the Peabody Museum of Archaeology and Ethnology when the human remains were accessioned
Sex	1 = male, 2 = female (only over 15 years of age)
Age	Determined to within 1 year in juveniles, to within 5 years in adults 20–39 years, and to within 10 years in adults 40–59 years (age midpoints given)
MLen	Maximum bone length (mm) including epiphyses (lengths only approximate in some younger juveniles)
Len'	Length', interarticular length (mm) (see Ruff 2002 for definitions)
TA	Total subperiosteal area (mm^2)
CA	Cortical area (mm^2)
MA	Medullary (endosteal) area (mm^2)
%CA	(CA/TA) x 100
Ix	Second moment of area about mediolateral axis (anteroposterior bending rigidity) (mm^4)
Iy	Second moment of area about anteroposterior axis (mediolateral bending rigidity) (mm^4)
Imax	Maximum second moment of area (maximum bending rigidity) (mm^4)
Imin	Minimum second moment of area (minimum bending rigidity) (mm^4)
J	Polar second moment of area (average bending/torsional rigidity) (mm^4)
Zp	Polar section modulus (average bending/torsional strength) (mm^3)

ID number	Sex	Age	MLen	Len'	TA	CA	MA	%CA	Ix	Iy	Imax	Imin	J	Zp
Femur														
59803	2	44	402	379	456	252	203	55.4	15063	12570	16354	11280	27634	2251
59805	1	32	438	410	501	371	130	74.0	20029	17932	21103	16858	37961	2960
59807	1	16	406	389	383	301	82	78.6	11807	11263	13724	9347	23071	1955
59815	2	27	393	370	428	343	85	80.1	13796	14522	15168	13150	28318	2415
59817	2	37	382	360	386	305	81	79.0	11078	11876	12054	10900	22954	2027
59822	1	17	416	384	434	281	153	64.8	14910	11941	15453	11398	26851	2106
59823	1	22	418	392	444	317	128	71.2	15951	13638	17236	12353	29589	2466
59845	2	32	398	376	444	352	91	79.4	14636	16364	19387	11613	31000	2490
59846	2	54	409	393	476	316	159	66.5	16055	16788	17276	15567	32843	2591
59847	2	16	383	363	311	194	116	62.6	6297	7137	8055	5379	13434	1224
59848		10	329	305	236	136	100	57.7	3811	3510	3826	3495	7321	785
59856	2	37	411	388	498	403	95	81.0	20707	19113	24927	14892	39819	3063
59862	1	32	427	410	442	363	79	82.1	14828	16158	19140	11846	30986	2561
59864	2	18	368	352	287	207	80	72.0	5967	6273	6556	5684	12240	1243
59873	1	16	443	413	440	298	142	67.7	15744	12359	16874	11229	28103	2294
59909		10.5	315	297	211	156	54	74.2	3486	3175	3673	2988	6661	781
59912	2	54	388	374	448	297	151	66.2	16380	12729	17077	12032	29109	2446
59918	1	27	445	422	544	388	157	71.2	21251	21998	22017	21232	43249	3252
59923	2	32	395	375	425	290	135	68.3	13823	12271	14290	11804	26094	2289
59973		13	365	341	213	154	59	72.3	3855	2896	3908	2843	6751	778
59981	1	22	415	387	397	297	101	74.7	12743	10876	12845	10774	23619	2104
59996	2	22	370	355	371	309	62	83.3	9072	12523	12630	8965	21595	1995
59998	1	22	427	403	479	367	112	76.7	14892	20446	21830	13508	35338	2745
60003	2	32	417	394	447	315	132	70.6	16637	13223	16696	13164	29860	2468
60005	2	44	383	365	375	266	109	70.8	10330	10492	10549	10273	20822	1893
60010	1	32	404	383	468	372	95	79.6	15790	18392	19737	14446	34183	2825
60013	2	18	386	365	400	264	136	66.0	10762	12005	12142	10626	22768	1976
60023	1	18	392	367	361	259	102	71.7	9421	10105	10773	8753	19526	1680
60026	1	54	444	418	550	310	240	56.3	19888	19327	21417	17797	39214	2842
60041	2	27	411	384	452	340	112	75.3	17745	13896	18674	12967	31641	2552
60044	1	22	420	393	443	317	126	71.6	15737	13317	15758	13296	29054	2457
60045	2	17	413	387	371	257	114	69.3	10127	9975	10849	9252	20101	1795
60049	2	54	393	374	382	235	146	61.7	10283	9710	10484	9509	19993	1793
60057	2	44	402	387	435	328	107	75.5	14070	14793	16329	12534	28863	2371
60063	2	37	403	385	411	281	130	68.4	13051	11638	13875	10813	24688	2142
60064	1	37	461	436	602	445	157	73.9	29463	25111	29653	24921	54574	3836
60069		7	263	239	143	103	41	71.7	1493	1508	1576	1425	3001	441
60075	1	54	434	410	505	312	194	61.7	19714	16366	22070	14011	36081	2754
60078		4	170	161	88	35	53	39.6	311	495	495	311	806	141
60086	1	37	432	414	544	382	162	70.3	23604	20776	26501	17880	44381	3324
60107	1	19	429	402	411	300	111	73.1	13170	12122	13575	11718	25293	1934
60109	1	27	433	409	493	360	133	73.0	21363	15968	23565	13766	37331	2911
60140		8	257	233	109	65	44	59.8	860	727	883	704	1587	248
60142	1	18	414	341	406	232	174	57.1	10245	11243	11324	10164	21488	1783
60143	2	19	381	351	306	226	81	73.7	7071	6948	7701	6317	14018	1316
60150		11.5	345	314	255	153	102	60.0	4171	4496	4699	3967	8666	947
60151		8	264	246	205	117	88	57.1	2658	2802	2831	2629	5460	658
60157	1	16	432	403	366	265	101	72.5	11397	8900	12195	8102	20297	1897
60180	1	44	432	415	553	361	192	65.3	22055	21261	24362	18954	43316	3227
60183		6	251	233	147	92	54	62.9	1518	1434	1522	1429	2951	420
60190		12	335	305	197	124	74	62.6	2631	2692	2842	2482	5324	645
60194	1	37	427	402	451	296	155	65.6	12986	16847	18372	11461	29833	2406
60201	1	27	435	414	467	337	130	72.2	15917	17117	20646	12388	33034	2627
60212	1	44	434	411	513	379	135	73.8	22143	17841	23375	16609	39984	3088
60213	2	17	411	390	443	259	185	58.3	11861	14289	15198	10952	26150	2051

ID number	Sex	Age	MLen	Len'	TA	CA	MA	%CA	Ix	Iy	Imax	Imin	J	Zp
60216	2	54	414	396	500	247	253	49.4	15734	14307	16504	13536	30040	2379
60217		9	336	310	241	163	78	67.5	3935	4330	4361	3903	8264	939
60218	1	37	430	405	547	379	169	69.2	23449	20775	24988	19237	44225	3331
60242	1	54	439	418	567	368	199	64.9	30560	17324	31795	16089	47884	3580
60252	2	32	426	405	367	297	70	80.9	9895	11004	11580	9318	20898	1896
60259	2	27	434	408	447	360	87	80.5	20427	12761	21866	11322	33188	2743
60260	1	37	418	389	483	299	184	61.9	18277	14481	18638	14120	32758	2569
60266	1	32	408	386	481	344	137	71.5	19759	14864	20036	14587	34623	2726
60267	1	44	440	418	520	391	129	75.2	23739	17696	23993	17443	41436	3069
60269		5	215	199	141	76	65	53.6	1174	1292	1298	1168	2466	354
60270		9	300	280	189	113	76	59.7	2883	2071	2885	2068	4953	576
60271	2	22	412	385	409	309	100	75.5	13980	11340	14415	10905	25320	2261
60275	1	32	420	394	580	406	174	70.1	27000	21785	28143	20643	48786	3849
60298	2	27	398	371	376	269	107	71.6	12047	9448	12052	9443	21495	1833
60302	2	37	387	371	431	314	117	73.0	13370	14461	15743	12087	27830	2399
60308	2	27	378	354	446	333	113	74.6	14830	15401	17762	12470	30232	2562
60309	2	44	410	389	484	246	238	50.8	14384	14429	14642	14171	28813	2314
60313	1	44	461	420	540	391	149	72.5	24029	20491	28901	15620	44521	3280
60314	1	27	422	396	474	353	121	74.4	18370	15733	20260	13842	34102	2790
60316	1	44	439	416	500	332	168	66.3	16918	18932	20523	15327	35850	2731
60446	2	54	393	367	433	290	143	66.9	13246	13905	14035	13116	27151	2185
60450	2	22	377	355	344	277	66	80.7	8591	9725	10279	8036	18315	1782
60454	1	54	447	415	528	344	184	65.2	23586	17141	26011	14716	40727	2962
60458	1	27	444	416	457	341	117	74.5	17417	14726	17606	14537	32143	2556
60596	2	16	415	392	376	289	88	76.7	12605	9534	12985	9154	22139	1913
60620	2	37	436	409	480	352	129	73.2	19174	15712	19975	14912	34887	2663
60663	1	54	420	397	505	385	120	76.3	18605	20600	23306	15899	39205	3027
60680		14.5	430	405	376	290	85	77.3	13164	8902	13692	8374	22066	1944
60685	2	22	375	355	286	231	55	80.6	6707	6158	7166	5699	12865	1255
60688	1	17	441	403	426	315	111	74.0	13065	14333	16229	11170	27399	2192
60732	1	18	400	376	388	278	110	71.7	13101	10431	14702	8830	23532	1945
60744	2	44	399	376	462	306	156	66.3	14791	15807	17061	13538	30599	2524
60772	2	22	383	357	318	245	73	77.0	8522	7114	8818	6819	15637	1526
60930	2	32	351	329	310	232	78	75.0	7052	7559	7577	7034	14611	1429
60935	1	22	405	379	407	305	101	75.1	16173	10234	17098	9309	26407	2272
60998		7	270	250	154	103	51	66.8	1585	1724	1726	1583	3309	456
Tibia														
59803	2	44	345	322	415	204	211	49.2	16834	8062	18706	6189	24895	1953
59805	1	32	375	353	452	337	115	74.5	24935	11273	27148	9059	36207	2897
59807	1	16	353	331	436	283	154	64.7	17034	12779	21135	8677	29812	2289
59815	2	27	317	296	302	215	87	71.3	10341	5446	12063	3724	15787	1382
59817	2	37	322	301	316	203	112	64.4	9742	5871	10773	4840	15613	1379
59822	1	17	355	334	367	215	152	58.5	10948	9750	15275	5423	20698	1690
59823	1	22	349	326	414	289	125	69.8	16391	11384	19105	8669	27774	2124
59845	2	32	344	321	365	268	97	73.3	15067	7496	16468	6095	22563	1928
59846	2	54	346	327	385	210	175	54.6	14321	6565	14498	6388	20886	1678
59847	2	16	332	313	267	153	114	57.4	5613	5408	8238	2783	11021	1062
59848		10	275	261	231	111	120	48.0	3650	3562	5437	1774	7211	749
59856	2	37	337	315	396	279	118	70.3	18147	9936	21257	6826	28083	2220
59862	1	32	365	341	385	280	104	72.9	15301	11186	20503	5984	26487	2098
59864	2	18	312	299	288	193	95	67.0	7919	5159	8651	4427	13078	1248
59873	1	16	374	355	409	219	190	53.5	12947	10789	16881	6855	23736	1884
59909		10.5	262	250	197	123	74	62.6	4210	1911	4393	1728	6121	682
59912	2	54	331	309	385	175	210	45.5	12047	6683	13888	4842	18730	1542
59918	1	27	380	356	537	362	175	67.5	35163	17117	42246	10033	52279	3526
59923	2	32	320	301	327	222	105	68.0	10132	7940	13373	4700	18073	1545

ID number	Sex	Age	MLen	Len'	TA	CA	MA	%CA	Ix	Iy	Imax	Imin	J	Zp
59973		13	300	285	193	132	62	68.1	4913	1529	4953	1488	6441	712
59981	1	22	344	320	377	263	114	69.7	19487	6531	20759	5259	26018	2238
59996	2	22	307	289	301	241	60	80.0	8661	7129	11040	4750	15790	1439
59998	1	22	356	332	418	288	130	68.8	19708	11668	24573	6803	31376	2368
60003	2	32	347	327	411	243	168	59.2	16142	8458	16963	7638	24601	1914
60005	2	44	326	304	390	199	191	51.1	12759	7674	14083	6349	20432	1808
60010	1	32	337	317	371	249	122	67.2	15025	10877	21294	4607	25901	2080
60013	2	18	339	314	402	221	182	54.8	11188	11813	16378	6623	23001	1913
60023	1	18	340	319	346	194	152	56.0	10263	7646	13219	4690	17909	1578
60026	1	54	370	347	454	243	211	53.5	22358	9660	25088	6930	32018	2251
60041	2	27	350	330	322	213	109	66.2	9938	6275	11105	5108	16213	1416
60044	1	22	352	328	437	246	191	56.3	14833	11915	17466	9282	26748	2038
60045	2	17	331	310	299	190	109	63.5	7810	5861	9610	4062	13672	1296
60049	2	54	330	308	354	186	168	52.6	9944	6474	10516	5902	16418	1453
60057	2	44	332	312	471	242	229	51.4	17716	11263	19496	9483	28979	2318
60063	2	37	335	311	332	203	129	61.1	9350	7243	11675	4918	16593	1495
60064	1	37	397	376	500	331	169	66.2	25772	13819	28451	11139	39590	2680
60069		7	225	218	131	94	38	71.3	1552	1189	1887	854	2741	403
60075	1	54	360	338	447	194	254	43.3	17462	8156	19398	6219	25617	1850
60078		4	145	140	74	29	45	39.6	294	276	333	237	570	107
60086	1	37	377	353	460	251	210	54.5	20358	10319	22046	8631	30677	2199
60107	1	19	352	331	375	268	108	71.4	12228	9728	14269	7686	21955	1778
60109	1	27	377	355	456	323	133	70.8	25306	12180	29458	8028	37486	2726
60140		8	205	200	115	73	42	63.2	1209	711	1212	708	1920	288
60142	1	18	355	335	407	195	212	47.9	14248	7745	15952	6042	21994	1756
60143	2	19	303	283	298	184	114	61.7	9047	4892	9423	4516	13939	1225
60150		11.5	280	268	237	137	100	57.9	5362	2657	5362	2657	8019	840
60151		8	226	215.5	250	118	132	47.2	4448	3182	5111	2518	7629	807
60157	1	16	355	334	392	232	159	59.3	16883	6249	16920	6212	23132	1840
60180	1	44	360	337	462	224	238	48.5	21744	8671	23453	6962	30415	2220
60183		6	208	201	155	90	65	57.9	1668	1590	2070	1188	3258	439
60190		12	280	265	207	111	96	53.6	3435	2157	3454	2137	5591	639
60194	1	37	356	333	338	209	129	61.9	12637	6371	14887	4120	19007	1597
60201	1	27	373	354	437	283	154	64.7	21984	9261	23758	7486	31244	2371
60212	1	44	375	350	672	264	407	39.4	33581	17533	38286	12828	51114	3556
60213	2	17	344	321	365	181	183	49.8	9852	7090	11160	5782	16942	1412
60216	2	54	340	318	452	122	330	27.0	12863	5345	14200	4009	18209	1474
60217		9	274	260	216	136	80	63.0	3778	3520	5372	1926	7298	806
60218	1	37	368	344	562	330	232	58.7	30040	14827	30620	14248	44868	3361
60242	1	54	371	346	530	350	180	66.1	31111	14585	34426	11270	45696	3366
60252	2	32	366	346	287	220	67	76.8	9298	5343	11007	3633	14640	1340
60259	2	27	363	342	360	266	94	73.9	12246	8981	14772	6455	21227	1826
60260	1	37	349	324	422	222	200	52.7	15183	9708	17861	7029	24890	1896
60266	1	32	347	328	498	337	161	67.6	25586	16485	30475	11596	42071	3181
60267	1	44	373	348	425	308	117	72.4	21820	11025	25554	7291	32845	2447
60269		5	182	174	125	64	61	51.1	1203	825	1342	685	2027	291
60270		9	250	245	181	95	86	52.6	2699	1841	3254	1286	4540	523
60271	2	22	344	324	321	209	112	65.2	10932	6008	12721	4219	16940	1516
60275	1	32	347	322	404	278	127	68.7	18949	10521	23654	5816	29470	2367
60298	2	27	326	308	383	240	143	62.6	14171	8134	14879	7426	22305	1886
60302	2	37	331	309	313	220	93	70.3	11009	5292	12090	4211	16301	1556
60308	2	27	312	291	362	226	137	62.3	13018	7543	15012	5548	20560	1717
60309	2	44	336	312	327	98	229	30.0	7367	4011	9246	2132	11378	970
60313	1	44	378	356	448	293	155	65.3	25077	9080	26322	7834	34156	2516
60314	1	27	369	347	395	280	115	71.0	18526	9713	22348	5891	28239	2259
60316	1	44	362	336	485	279	206	57.5	23105	12180	26206	9079	35285	2704
60446	2	54	315	292	378	218	159	57.8	14989	6979	16755	5213	21968	1747

ID number	Sex	Age	MLen	Len'	TA	CA	MA	%CA	Ix	Iy	Imax	Imin	J	Zp
60450	2	22	324	304	304	211	92	69.6	9906	6139	12334	3711	16045	1503
60454	1	54	389	366	503	307	196	61.1	26141	11939	27410	10669	38079	2826
60458	1	27	379	354	471	320	151	67.9	22996	14007	27596	9407	37003	2696
60596	2	16	349	330	362	242	120	66.9	12330	7809	13491	6667	20158	1745
60620	2	37	370	352	430	284	146	66.1	19304	10387	21758	7934	29692	2366
60663	1	54	348	321	382	258	124	67.5	16069	8281	17985	6364	24349	1976
60680		14.5												
60685	2	22	315	296	253	182	71	72.1	6983	3685	7862	2806	10668	1219
60688	1	17	362	344	409	290	119	70.9	16920	9926	18956	7890	26846	2474
60732	1	18	328	309	317	252	65	79.5	9489	6910	10478	5922	16400	1477
60744	2	44	330	308	359	235	124	65.5	14755	6526	15909	5372	21281	1800
60772	2	22	307	292	298	186	111	62.6	9218	5085	10121	4182	14303	1318
60930	2	32	291	273	267	165	102	62.0	7022	4290	7893	3420	11313	1112
60935	1	22	344	323	431	310	121	72.0	24457	10659	27782	7334	35116	2770
60998		7	220	210	139	99	40	71.6	2145	974	2160	959	3119	441
Humerus														
59803	2	44	293	289	258	104	153	40.5	3408	3748	4348	2803	7156	769
59805	1	32	312	306	280	182	98	65.0	6317	5213	6795	4741	11530	1227
59807	1	16	297	293	221	183	37	83.2	3830	3870	4460	3250	7700	856
59815	2	27	272	270	227	139	88	61.3	3601	3964	4573	2986	7564	836
59817	2	37	276	272	236	152	83	64.6	3986	4438	4977	3430	8424	955
59822	1	17	286	283	206	131	75	63.7	3600	2490	3860	2240	6090	706
59823	1	22	313	311	279	184	95	65.9	5794	6128	7537	4381	11922	1229
59845	2	32	306	303	225	147	77	65.5	3396	4144	4528	3020	7541	877
59846	2	54	302	299	261	114	147	43.7	3907	3907	4528	3283	7814	829
59847	2	16	280	277	175	101	74	57.7	1970	2340	2820	1490	4310	507
59848		10	239	263	118	78	40	65.9	945	1100	1200	846	2045	299
59856	2	37	301	299	274	162	112	59.2	4430	5925	6062	4293	10355	1049
59862	1	32	311	309	249	195	54	78.3	4652	5481	5961	4178	10133	1039
59864	2	18	267	263	182	115	67	63.1	2390	2530	3220	1700	4920	595
59873	1	16	302	298	221	126	96	56.7	3020	3570	3880	2710	6590	722
59909		10.5	219	217	123	82	42	66.2	1120	1260	1620	758	2380	344
59912	2	54	287	284	229	98	131	42.8	2791	3191	3556	2424	5983	667
59918	1	27	325	321	305	196	109	64.3	6440	7481	8397	5504	13921	1361
59923	2	32	285	283	247	145	102	58.6	3986	4764	5526	3214	8750	928
59973		13	248	245	159	95	63	60.2	1680	1820	2210	1290	3500	465
59981	1	22	296	293	213	153	60	71.8	3214	3862	3907	3157	7076	830
59996	2	22	280	276	212	150	62	70.6	3624	3964	5168	2401	7587	841
59998	1	22	308	305	281	171	110	60.7	5313	5905	6206	5011	11219	1111
60003	2	32	298	293	249	126	122	50.8	3260	4977	5078	3169	8237	834
60005	2	44	293	289	234	105	130	44.6	2803	3873	4302	2366	6676	706
60010	1	32	292	289	249	139	111	55.6	4775	3975	5660	3077	8750	902
60013	2	18	299	297	228	125	103	54.8	3030	4020	4590	2450	7050	762
60023	1	18	284	281	210	139	71	66.4	3190	3490	4330	2350	6680	740
60026	1	54	326	316	270	146	124	54.1	4528	5045	5269	4302	9572	992
60041	2	27	295	288	239	141	98	59.0	4065	4483	5626	2906	8548	868
60044	1	22	307	303	276	171	106	61.8	5563	5213	6642	4134	10776	1111
60045	2	17	284	281	190	122	68	64.2	2420	2760	3100	2080	5180	609
60049	2	54	289	286	237	128	110	53.8	3616	3738	4327	3027	7354	780
60057	2	44	283	280	220	133	87	60.4	3214	3828	4370	2654	7042	798
60063	2	37	304	289	236	135	101	57.2	3658	4077	4719	3009	7734	862
60064	1	37	324	320	326	222	104	68.2	7691	8320	9484	6517	16011	1479
60069		7	181	179	70	49	20	71.1	407	329	467	269	736	146
60075	1	54	310	303	268	137	131	51.1	4235	5000	5559	3658	9234	959
60078		4	132	130	74	22	52	30.3	228	211	238	201	439	83
60086	1	37	325	321	247	168	78	68.3	4820	4449	5336	3930	9269	1008

ID number	Sex	Age	MLen	Len'	TA	CA	MA	%CA	Ix	Iy	Imax	Imin	J	Zp
60107	1	19	301	293										
60109	1	27	312	308	311	223	88	71.6	8222	6629	9493	5358	14850	1459
60140		8	180	178	65	39	26	60.4	328	261	352	238	589	123
60142	1	18	294	289	229	125	104	54.4	3160	3730	4290	2600	6890	745
60143	2	19	276	273	185	113	72	61.1	2120	2960	3380	1700	5080	598
60150		11.5	233	230	127	73	54	57.3	1070	1060	1070	1060	2130	323
60151		8	187	185	125	51	73	41.1	711	948	948	711	1659	244
60157	1	16	308	303	198	144	54	72.8	3270	2690	3580	2370	5960	725
60180	1	44	324	320	268	134	134	50.1	4697	4325	5257	3748	9021	928
60183		6	180	178	79	43	36	55.0	397	408	463	342	805	151
60190		12	244	241	121	75	46	62.0	1040	1030	1230	835	2070	315
60194	1	37	314	307	231	125	106	54.2	3465	3601	4031	3032	7066	770
60201	1	27	307	303	202	128	74	63.3	2665	3453	3612	2504	6118	689
60212	1	44	307	304	234	147	86	63.0	4167	3805	4674	3294	7972	869
60213	2	17	299	296	210	106	105	50.2	2750	2800	3440	2110	5550	632
60216	2	54	293	288	247	91	157	36.7	2814	3362	3658	2516	6176	640
60217		9	232	229	99	67	32	67.9	663	759	763	659	1422	236
60218	1	37	301	296	272	160	112	58.7	5022	5369	5861	4528	10392	1047
60242	1	54	312	311	235	131	104	55.9	3612	3760	3873	3499	7372	817
60252	2	32	323	319	219	156	62	71.4	3248	4584	4966	2860	7832	873
60259	2	27	317	313	239	156	83	65.2	3794	4876	5336	3328	8670	950
60260	1	37	297	292	241	150	92	62.0	4302	4111	4910	3499	8413	932
60266	1	32	301	296	272	167	105	61.4	5626	5089	6406	4302	10716	1066
60267	1	44	330	325	293	183	110	62.5	6351	6373	7205	5257	12724	1235
60269		5	158	156	79	30	49	37.7	307	327	397	238	634	114
60270		9	206	204	99	57	42	57.5	701	583	718	566	1284	210
60271	2	22	287	283	209	143	66	68.5	2814	3975	4077	2711	6789	780
60275	1	32	303	298	229	157	72	68.4	3997	4009	4584	3419	8006	890
60298	2	27	281	278	242	143	100	58.9	3794	4977	5716	3032	8771	923
60302	2	37	293	288	203	131	72	64.6	2734	3556	4009	2274	6290	736
60308	2	27	285	281	243	141	102	58.0	3794	4652	5358	3077	8445	880
60309	2	44	292	289	232	55	178	23.4	1670	1868	1950	1588	3539	373
60313	1	44	336	330	274	158	116	57.5	4865	5548	6451	3952	10413	1065
60314	1	27	311	307	235	153	82	65.1	4291	3862	4573	3578	8153	906
60316	1	44	315	309	274	169	105	61.8	5325	5425	6061	4685	10750	1097
60446	2	54	271	269	245	100	145	40.9	3066	3442	3635	2883	6508	709
60450	2	22	283	276	201	144	57	71.5	2700	3918	4212	2412	6618	793
60454	1	54	319	315	272	175	98	64.1	5157	5693	5850	5000	10850	1121
60458	1	27	305	301	259	185	75	71.2	5604	5425	7127	3884	11029	1134
60596	2	16	291	287	177	122	54	69.3	2120	2660	2990	1790	4780	572
60620	2	37	314	310	264	156	107	59.3	4910	5134	5861	4178	10044	994
60663	1	54	309	307	257	171	86	66.4	5738	4348	6073	3997	10086	1029
60680		14.5	286	283	201	131	71	64.9	2800	2990	3320	2470	5790	687
60685	2	22	275	271	162	110	52	68.0	1892	2285	2768	1401	4177	537
60688	1	17	318	316	214	148	66	69.0	3220	3530	4070	2670	6750	774
60732	1	18	287	284	209	136	72	65.3	3170	3240	4040	2370	6410	726
60744	2	44	289	285	252	125	127	49.5	3465	4663	4876	3237	8127	831
60772	2	22	271	268	202	133	69	65.8	2734	3453	3533	2654	6187	739
60930	2	32	261	257	191	123	69	64.1	2516	2997	3385	2124	5513	672
60935	1	22	288	284	208	141	67	67.8	3191	3624	4077	2723	6815	764
60998		7	184	177	69	44	25	63.7	310	372	386	296	682	134
Radius														
59803	2	44	215	205	102	46	56	45.4	518	692	694	517	1210	193
59805	1	32	243	232	99	75	24	76.1	747	817	820	744	1564	265
59807	1	16	231	222	85	63	22	74.6	455	686	686	455	1141	187
59815	2	27	211	202	90	70	21	77.0	590	798	883	505	1388	239

ID number	Sex	Age	MLen	Len'	TA	CA	MA	%CA	Ix	Iy	Imax	Imin	J	Zp
59817	2	37	212	202	85	75	10	87.9	590	692	801	481	1282	234
59822	1	17	225	217	74	52	22	70.8	339	510	514	335	849	151
59823	1	22	243	229	128	99	29	77.6	1217	1530	1705	1043	2747	395
59845	2	32	229	214	93	71	22	76.8	586	826	843	570	1412	237
59846	2	54	221	210	98	58	40	59.3	475	930	938	467	1405	227
59847	2	16	215	207	75	50	25	66.6	311	511	514	308	822	142
59848		10	178	171	59	39	20	65.7	194	317	320	192	512	102
59856	2	37	231	218	104	78	26	75.1	708	1071	1087	692	1780	303
59862	1	32	243	232	84	78	6	92.4	531	726	794	464	1258	236
59864	2	18	202	192	64	51	13	79.3	324	338	402	260	662	128
59873	1	16	241	232	98	60	38	61.5	538	876	898	515	1413	215
59909		10.5	166	159	54	39	16	71.1	179	289	297	171	468	96
59912	2	54	216	205	73	41	32	56.3	325	356	356	325	681	138
59918	1	27	253	242	121	93	28	76.6	1014	1413	1424	1012	2427	375
59923	2	32	212	202	67	49	18	73.5	295	393	424	264	688	124
59973		13	197	189	51	38	13	74.1	158	265	265	157	422	91
59981	1	22	223	211	85	72	13	84.7	570	638	642	568	1208	217
59996	2	22	203	194	80	73	7	91.2	450	656	667	439	1106	206
59998	1	22	234	222	94	70	24	74.1	541	970	985	528	1511	246
60003	2	32	218	205	93	68	25	73.3	588	857	862	584	1445	240
60005	2	44	214	201	100	54	47	53.6	495	920	920	495	1415	218
60010	1	32	220	203	85	68	17	79.5	516	654	677	492	1169	224
60013	2	18	230	219	81	64	17	78.6	497	529	578	448	1026	193
60023	1	18	229	220	78	58	20	74.4	413	520	521	413	934	168
60026	1	54	251	240	110	73	36	66.9	774	1086	1119	741	1860	300
60041	2	27	229	218	96	77	19	80.3	676	868	923	620	1543	271
60044	1	22	230	215	101	77	24	76.1	727	844	850	722	1571	265
60045	2	17	220	210	70	56	15	79.1	286	523	527	282	809	154
60049	2	54	217	206	97	62	35	63.8	580	765	778	566	1345	230
60057	2	44	215	206	82	61	21	74.0	535	520	592	463	1055	194
60063	2	37	223	213	91	63	27	69.8	551	726	733	544	1277	221
60064	1	37	255	244	129	100	29	77.3	1100	1705	1705	1099	2805	398
60069		7	143	138	33	26	7	77.8	64	117	120	61	181	48
60075	1	54	243	232	105	64	41	61.3	719	846	859	706	1565	253
60078		4												
60086	1	37	245	233	99	73	26	73.6	653	922	935	639	1574	268
60107	1	19	232	215	96	71	25	73.8	559	888	893	554	1447	241
60109	1	27	243	230	113	87	27	76.5	966	1083	1176	872	2049	320
60140		8	151	145	30	18	12	61.5	45	82	82	45	128	38
60142	1	18	225	217	79	57	23	71.6	411	576	576	411	987	166
60143	2	19	207	195	67	54	13	80.7	284	468	468	284	752	137
60150		11.5	178	171	52	34	17	66.3	149	257	259	147	406	90
60151		8	147	142	54	26	29	47.3	165	187	193	159	352	74
60157	1	16	237	228	80	54	26	67.1	418	519	519	418	937	163
60180	1	44	243	232	98	64	34	65.1	592	847	856	585	1440	230
60183		6	129	124	36	23	13	64.9	83	104	104	83	187	50
60190		12	180	173	45	36	9	79.5	133	186	192	127	319	79
60194	1	37	236	223	108	80	29	73.5	831	956	1010	778	1787	307
60201	1	27	233	224	73	55	18	74.8	367	487	509	345	855	168
60212	1	44	238	228	91	64	27	70.8	500	875	879	495	1375	226
60213	2	17	224	217	91	58	33	63.5	479	696	708	467	1175	202
60216	2	54	223	212	90	43	47	48.3	447	569	592	424	1016	172
60217		9	181	174	43	30	14	68.4	114	165	168	111	279	70
60218	1	37	237	223	98	78	19	80.3	694	892	916	670	1585	253
60242	1	54	242	231	98	68	30	69.2	674	811	819	667	1486	265
60252	2	32	236	227	80	67	13	84.2	500	574	580	495	1074	193
60259	2	27	243	233	97	84	13	87.1	638	976	979	636	1614	270

ID number	Sex	Age	MLen	Len'	TA	CA	MA	%CA	Ix	Iy	Imax	Imin	J	Zp
60260	1	37	230	216	95	70	26	73.0	602	864	873	592	1466	255
60266	1	32	242	226	112	87	25	77.5	821	1295	1307	813	2116	315
60267	1	44	250	236	109	91	18	83.3	949	1045	1061	933	1994	318
60269		5	119	114	28	15	13	54.1	44	60	60	44	104	30
60270		9												
60271	2	22	228	218	81	68	14	83.0	397	748	749	396	1145	202
60275	1	32	229	216	83	67	17	80.0	505	637	691	448	1141	211
60298	2	27	214	203	100	82	18	82.3	616	1236	1236	614	1852	284
60302	2	37	217	206	79	67	12	84.6	445	596	596	445	1040	194
60308	2	27	211	198	92	63	29	68.4	495	844	855	484	1339	217
60309	2	44	223	212	102	30	71	30.0	294	591	592	293	885	130
60313	1	44	252	237	106	68	38	64.4	665	1248	1248	660	1912	268
60314	1	27	242	233	92	76	16	82.7	597	840	856	581	1437	248
60316	1	44	246	235	103	75	28	72.6	717	1004	1054	666	1720	282
60446	2	54	210	199	98	52	46	53.1	484	807	816	474	1290	219
60450	2	22	217	205	70	63	7	89.8	424	465	490	400	889	169
60454	1	54	247	234	110	82	28	74.3	789	1185	1193	780	1974	326
60458	1	27	245	233	105	91	14	86.9	762	1190	1198	753	1952	317
60596	2	16	211	203	68	56	12	82.2	287	495	501	282	783	144
60620	2	37	250	235	110	73	37	66.5	792	1096	1096	792	1889	279
60663	1	54	236	223	104	80	24	76.8	778	1107	1175	708	1885	280
60680		14.5	225	215	75	54	21	71.8	399	453	463	390	853	168
60685	2	22	197	187	59	49	10	82.7	221	358	359	219	579	124
60688	1	17	234	225	88	65	23	74.1	482	723	726	479	1205	209
60732	1	18	215	203	67	54	13	80.2	320	393	402	311	713	133
60744	2	44	211	202	90	58	32	64.5	595	636	720	511	1230	207
60772	2	22	208	198	78	61	17	78.2	470	533	561	442	1003	184
60930	2	32	196	188	70	58	11	84.0	310	547	552	305	857	176
60935	1	22	234	224	77	65	11	85.3	446	569	579	435	1015	191
60998		7	145	140	29	25	4	87.2	57	82	83	56	139	41
Ulna														
59803	2	44	230	205	102	46	56	45.4	518	692	694	517	1210	193
59805	1	32	261	232	99	75	24	76.1	747	817	820	744	1564	265
59807	1	16	249	222	85	63	22	74.6	455	686	686	455	1141	187
59815	2	27	232	202	90	70	21	77.0	590	798	883	505	1388	239
59817	2	37	232	202	85	75	10	87.9	590	692	801	481	1282	234
59822	1	17	244	217	74	52	22	70.8	339	510	514	335	849	151
59823	1	22	263	229	128	99	29	77.6	1217	1530	1705	1043	2747	395
59845	2	32	249	214	93	71	22	76.8	586	826	843	570	1412	237
59846	2	54	238	210	98	58	40	59.3	475	930	938	467	1405	227
59847	2	16	235	207	75	50	25	66.6	311	511	514	308	822	142
59848		10	190	171	59	39	20	65.7	194	317	320	192	512	102
59856	2	37	245	218	104	78	26	75.1	708	1071	1087	692	1780	303
59862	1	32	266	232	84	78	6	92.4	531	726	794	464	1258	236
59864	2	18	216	192	64	51	13	79.3	324	338	402	260	662	128
59873	1	16	257	232	98	60	38	61.5	538	876	898	515	1413	215
59909		10.5	178	159	54	39	16	71.1	179	289	297	171	468	96
59912	2	54	231	205	73	41	32	56.3	325	356	356	325	681	138
59918	1	27	273	242	121	93	28	76.6	1014	1413	1424	1012	2427	375
59923	2	32	227	202	67	49	18	73.5	295	393	424	264	688	124
59973		13	214	189	51	38	13	74.1	158	265	265	157	422	91
59981	1	22	244	211	85	72	13	84.7	570	638	642	568	1208	217
59996	2	22	217	194	80	73	7	91.2	450	656	667	439	1106	206
59998	1	22	251	222	94	70	24	74.1	541	970	985	528	1511	246
60003	2	32	233	205	93	68	25	73.3	588	857	862	584	1445	240
60005	2	44	232	201	100	54	47	53.6	495	920	920	495	1415	218

ID number	Sex	Age	MLen	Len'	TA	CA	MA	%CA	Ix	Iy	Imax	Imin	J	Zp
60010	1	32	241	203	85	68	17	79.5	516	654	677	492	1169	224
60013	2	18	251	219	81	64	17	78.6	497	529	578	448	1026	193
60023	1	18	244	220	78	58	20	74.4	413	520	521	413	934	168
60026	1	54	269	240	110	73	36	66.9	774	1086	1119	741	1860	300
60041	2	27	242	218	96	77	19	80.3	676	868	923	620	1543	271
60044	1	22	253	215	101	77	24	76.1	727	844	850	722	1571	265
60045	2	17	235	210	70	56	15	79.1	286	523	527	282	809	154
60049	2	54	232	206	97	62	35	63.8	580	765	778	566	1345	230
60057	2	44	233	206	82	61	21	74.0	535	520	592	463	1055	194
60063	2	37	239	213	91	63	27	69.8	551	726	733	544	1277	221
60064	1	37	274	244	129	100	29	77.3	1100	1705	1705	1099	2805	398
60069		7	151	138	33	26	7	77.8	64	117	120	61	181	48
60075	1	54	264	232	105	64	41	61.3	719	846	859	706	1565	253
60078		4												
60086	1	37	265	233	99	73	26	73.6	653	922	935	639	1574	268
60107	1	19	245	215	96	71	25	73.8	559	888	893	554	1447	241
60109	1	27	261	230	113	87	27	76.5	966	1083	1176	872	2049	320
60140		8	164	145	30	18	12	61.5	45	82	82	45	128	38
60142	1	18	243	217	79	57	23	71.6	411	576	576	411	987	166
60143	2	19	224	195	67	54	13	80.7	284	468	468	284	752	137
60150		11.5	187	171	52	34	17	66.3	149	257	259	147	406	90
60151		8	159	142	54	26	29	47.3	165	187	193	159	352	74
60157	1	16	256	228	80	54	26	67.1	418	519	519	418	937	163
60180	1	44	257	232	98	64	34	65.1	592	847	856	585	1440	230
60183		6	139	124	36	23	13	64.9	83	104	104	83	187	50
60190		12	192	173	45	36	9	79.5	133	186	192	127	319	79
60194	1	37	255	223	108	80	29	73.5	831	956	1010	778	1787	307
60201	1	27	246	224	73	55	18	74.8	367	487	509	345	855	168
60212	1	44	255	228	91	64	27	70.8	500	875	879	495	1375	226
60213	2	17	244	217	91	58	33	63.5	479	696	708	467	1175	202
60216	2	54	239	212	90	43	47	48.3	447	569	592	424	1016	172
60217		9	192	174	43	30	14	68.4	114	165	168	111	279	70
60218	1	37	250	223	98	78	19	80.3	694	892	916	670	1585	253
60242	1	54	262	231	98	68	30	69.2	674	811	819	667	1486	265
60252	2	32	256	227	80	67	13	84.2	500	574	580	495	1074	193
60259	2	27	262	233	97	84	13	87.1	638	976	979	636	1614	270
60260	1	37	246	216	95	70	26	73.0	602	864	873	592	1466	255
60266	1	32	261	226	112	87	25	77.5	821	1295	1307	813	2116	315
60267	1	44	269	236	109	91	18	83.3	949	1045	1061	933	1994	318
60269		5	126	114	28	15	13	54.1	44	60	60	44	104	30
60270		9												
60271	2	22	247	218	81	68	14	83.0	397	748	749	396	1145	202
60275	1	32	245	216	83	67	17	80.0	505	637	691	448	1141	211
60298	2	27	231	203	100	82	18	82.3	616	1236	1236	614	1852	284
60302	2	37	239	206	79	67	12	84.6	445	596	596	445	1040	194
60308	2	27	225	198	92	63	29	68.4	495	844	855	484	1339	217
60309	2	44	244	212	102	30	71	30.0	294	591	592	293	885	130
60313	1	44	267	237	106	68	38	64.4	665	1248	1248	660	1912	268
60314	1	27	266	233	92	76	16	82.7	597	840	856	581	1437	248
60316	1	44	263	235	103	75	28	72.6	717	1004	1054	666	1720	282
60446	2	54	225	199	98	52	46	53.1	484	807	816	474	1290	219
60450	2	22	235	205	70	63	7	89.8	424	465	490	400	889	169
60454	1	54	263	234	110	82	28	74.3	789	1185	1193	780	1974	326
60458	1	27	263	233	105	91	14	86.9	762	1190	1198	753	1952	317
60596	2	16	231	203	68	56	12	82.2	287	495	501	282	783	144
60620	2	37	261	235	110	73	37	66.5	792	1096	1096	792	1889	279
60663	1	54	254	223	104	80	24	76.8	778	1107	1175	708	1885	280

ID number	Sex	Age	MLen	Len'	TA	CA	MA	%CA	Ix	Iy	Imax	Imin	J	Zp
60680		14.5	245	215	75	54	21	71.8	399	453	463	390	853	168
60685	2	22	213	187	59	49	10	82.7	221	358	359	219	579	124
60688	1	17	249	225	88	65	23	74.1	482	723	726	479	1205	209
60732	1	18	231	203	67	54	13	80.2	320	393	402	311	713	133
60744	2	44	232	202	90	58	32	64.5	595	636	720	511	1230	207
60772	2	22	232	198	78	61	17	78.2	470	533	561	442	1003	184
60930	2	32	215	188	70	58	11	84.0	310	547	552	305	857	176
60935	1	22	256	224	77	65	11	85.3	446	569	579	435	1015	191
60998		7	155	140	29	25	4	87.2	57	82	83	56	139	41

Appendix I

Results of Multiple Comparisons Tests for Pelvic and Nonpelvic Variables

Robert G. Tague

The multiple comparisons tests show whether one sample has a significantly different mean or mean rank from another sample in pairwise comparisons for Pecos Pueblo (PP), Haida (H), Indian Knoll (IK), Libben (L), blacks (B), and whites (W). Sexes are evaluated separately. In each pairwise comparison, the abbreviation for the sample with the significantly higher mean or mean rank is given. The abbreviation NS = nonsignificant.

Variable	PP:H	PP:IK	PP:L	PP:B	PP:W	H:IK	H:L	H:B	H:W	IK:L	IK:B	IK:W	L:B	L:W	B:W
Females															
Anteroposterior diameter of inlet	H	IK	NS	B	W	NS	NS	NS	NS	NS	NS	NS	B	W	NS
Anteroposterior diameter of outlet	H	NS	NS	B	NS	H	NS	H	H	NS	B	NS	NS	NS	NS
Transverse diameter of inlet	NS	NS	NS	PP	NS	NS	NS	H	NS	NS	IK	NS	L	NS	W
Transverse diameter of outlet	NS	IK	NS	NS	NS	NS	NS	H	NS	NS	IK	NS	NS	NS	W
Posterior sagittal diameter of inlet	H	NS	NS	B	W	NS	NS	NS	NS	NS	NS	NS	NS	NS	NS
Posterior sagittal diameter of outlet	H	NS	NS	B	NS	NS	NS	NS	NS	NS	B	NS	NS	NS	NS
Anterior sagittal diameter of inlet	H	NS	NS	B	W	NS	H	NS	NS	NS	NS	NS	NS	NS	NS
Anterior sagittal diameter of outlet	NS	NS	NS	NS	NS	NS	NS	NS	NS	NS	NS	NS	NS	NS	NS
Breadth of sacrum at inlet	NS	NS	NS	PP	NS	NS	NS	H	NS	L	NS	NS	L	NS	W
Straight length of sacrum	NS	NS	NS	NS	NS	NS	NS	NS	NS	NS	NS	NS	NS	NS	NS
Angulation of sacrum	NS	NS	L	NS	NS	H	NS	NS	NS	L	NS	NS	NS	NS	NS
Depth of pelvis	H	IK	L	B	W	NS	NS	NS	NS	NS	NS	W	NS	NS	NS
Subpubic angle	H	IK	L	NS	NS	NS	NS	H	NS	NS	IK	IK	L	NS	NS
Area of inlet	H	IK	NS	B	W	NS	NS	NS	NS	NS	NS	NS	NS	NS	W
Area of outlet	H	NS	NS	NS	NS	NS	NS	H	NS	NS	NS	NS	NS	NS	NS
Femoral length	NS	IK	L	B	W	IK	L	B	W	L	B	NS	NS	NS	B
Femoral head diameter	H	NS	L	B	W	H	NS	NS	NS	L	B	W	B	W	NS
Clavicular length	NS	NS	L	B	NS	H	NS	NS	NS	L	B	W	NS	L	B

Variable	PP:H	PP:IK	PP:L	PP:B	PP:W	H:IK	H:L	H:B	H:W	IK:L	IK:B	IK:W	L:B	L:W	B:W
Males															
Anteroposterior diameter of inlet	H	IK	L	B	W	NS	NS	NS	NS	NS	IK	NS	NS	NS	W
Anteroposterior diameter of outlet	H	NS	L	B	NS	H	NS	NS	H	L	B	NS	NS	NS	NS
Transverse diameter of inlet	H	NS	L	PP	NS	H	NS	H	NS	L	IK	W	L	NS	W
Transverse diameter of outlet	NS	NS	L	NS	W	NS	NS	H	NS	NS	NS	NS	L	NS	W
Posterior sagittal diameter of inlet	H	NS	L	NS	W	H	NS	H	NS	NS	NS	NS	L	NS	W
Posterior sagittal diameter of outlet	NS	NS	NS	B	NS	NS	NS	NS	NS	NS	B	NS	NS	NS	NS
Anterior sagittal diameter of inlet	H	NS	L	B	W	NS	NS	NS	NS	NS	NS	W	NS	NS	NS
Anterior sagittal diameter of outlet	NS	NS	NS	NS	NS	NS	NS	H	NS	NS	NS	NS	L	NS	W
Breadth of sacrum at inlet	H	NS	L	PP	W	NS	NS	H	NS	NS	IK	NS	L	NS	W
Straight length of sacrum	H	NS	NS	NS	NS	H	H	H	H	NS	NS	NS	NS	NS	NS
Angulation of sacrum	NS	NS	NS	B	W	NS	NS	B	W	NS	B	W	B	W	NS
Depth of pelvis	H	NS	L	B	W	H	H	H	NS	NS	B	W	NS	W	W
Subpubic angle	NS	IK	L	B	NS	IK	NS	NS	NS	IK	IK	IK	NS	L	NS
Area of inlet	H	IK	L	NS	W	NS	NS	H	NS	NS	IK	NS	L	NS	W
Area of outlet	H	NS	L	NS	W	NS	NS	NS	NS	NS	NS	NS	L	NS	NS
Femoral length	NS	NS	L	B	W	NS	L	B	W	L	B	W	NS	NS	NS
Femoral head diameter	H	NS	L	B	W	H	NS	B	W	L	B	W	B	W	NS
Clavicular length	H	PP	L	B	NS	H	NS	NS	NS	L	B	W	NS	L	NS

Appendix J

Means for Six Pelvic Variables for Pecos Pueblo and Other Samples

Robert G. Tague

Key to Columns 2, 3, 5, and 9	
Sample number	The same sample number is used for females and males from the same population in a study. The sample number is used in figure 10.3.
Citation number	The citation number listed for each source refers to the following: (1) Acosta-Sison and Calderon 1919; (2) Adadevoh, Hobbs, and Elkins 1989; (3) Allen 1944; (4) Chen et al. 1982; (5) Chun et al. 1964; (6) Davivongs 1963; (7) Emmons 1913; (8) Garson 1881–1882; (9) Greulich, Thoms, and Twaddle 1939; (10) Hanson 1936; (11) Heyns 1944; (12) Heyns 1946; (13) Holland et al. 1982; (14) Howells and Hotelling 1936; (15) Hrdlička 1916; (16) İşcan and Cotton 1985; (17) Jordaan 1976; (18) K'o et al. 1957; (19) Kolawole, Adamu, and Evans 1978; (20) Krukierek 1951; (21) Lau, Leung, and Li 1998; (22) Litten 1925; (23) Manahan, Marquez, and Mangay 1954;

	(24) Martin 1866; (25) Martin and Saller 1959; (26) Mengert 1938; (27) Mengert 1948; (28) Nicholson 1939; (29) Nicholson 1945; (30) Nicholson and Allen 1946; (31) Noriega and Gutiérrez 1895; (32) Ogedengbe et al. 1993; (33) Orford 1934; (34) Pan 1929; (35) Paterson 1893; (36) Ross 1975; (37) Runge 1889; (38) Russell and Richards 1971; (39) Scheyer 1934; (40) Segebarth-Orban 1980; (41) Shordania 1932; (42) Sibley, Armelagos, and Van Gerven 1992; (43) Sibthorpe and Allbrook 1958; (44) Slome 1929; (45) Stewart, Cowan, and Philpott 1979; (46) Tague, present study, table 10.4; (47) Thoms 1915; (48) Thoms 1941; (49) Thoms 1946; (50) Thoms and Greulich 1940; (51) Thoms, Foote, and Friedman 1939; (52) Torpin 1951; (53) Warren 1897; (54) Wei, Chen, and Chen 1956; (55) Young and Ince 1940.
Conjugate diameter of inlet	Some researchers are not specific about whether they are reporting the obstetrical or anatomical conjugate diameter of inlet. Key: (a) = anatomical conjugate; (b) = unspecified whether obstetrical or anatomical conjugate.
Index of inlet shape	The index of inlet shape = (obstetrical conjugate diameter of inlet/transverse diameter of inlet)100. Some researchers report this index; for others, I computed the index from summary statistics.

Sample	Sample number	Citation number	Obstetrical conjugate diameter of inlet (mm), mean	Conjugate diameter of inlet (mm), mean	Anteroposterior diameter of outlet (mm), mean	Transverse diameter of inlet (mm), mean	Straight length of sacrum (mm), mean	Index of inlet shape, mean
Females								
North America								
Native American								
Pecos Pueblo	1	46	91.0	94.3 (a)	114.0	133.5	96.2	68.5
Arkansas and Louisiana	2	15	107.4			133.3	102	81.4
Amerindian[a]	3	7	103.9		116.1	129.9	98.5	80.1
Amerindian	4	25					110.0	
Haida	5	46	112.0	115.6 (a)	134.8	134.7	100.2	83.0
Indian Knoll	6	46	108.2	111.4 (a)	114.5	134.0	99.1	81.2
Lenape	7	15	110			130	99	84.5
Libben	8	46	98.0	101.1 (a)	121.2	134.3	100.0	72.7
Southern Utah	9	15	101			131		
Southwest, Pueblos	10	14		102.4 (a)	115.4	132.2	97.5	
Southwest and Mexico	11	15	107.5			129	104	83.1
United States								
Unspecified	12	27		116 (b)	121	125		
Unspecified	13	10			111.0			
Black	14	52	110.0			119.0		
Black	15	46	110.4	114.3 (a)	122.9	122.9	96.0	90.3
Black	16	49	113.9			119.0		95.7
Black	17	52		108.6 (b)		117.8		
Black	18	16		110.8 (a)		120.77		
White	19	52	110.3			128.8		
White	20	46	110.8	115.3 (a)	118.7	134.0	99.6	83.1
White	21	51	116			123.4		94.0
White	22	9	117.5			123.9		94.8
White	23	48	118.0			124.0		95.2
White	24	52		112.8 (b)		125.9		
White	25	16		116.29 (a)		132.96		
White	26	47			105.45			
White	27	26			114			
Mexico	28	31		106.2 (b)	85	124.0		
Mexico	29	31		108.0 (b)	95	130.0		
Mexico	30	52		116.8 (b)		127.5		
Africa								
Egypt								
Naqada	31	53					98.1	
Ethiopia	32	36	106		115	119	101	87.5

Sample	Sample number	Citation number	Obstetrical conjugate diameter of inlet (mm), mean	Conjugate diameter of inlet (mm), mean	Anteroposterior diameter of outlet (mm), mean	Transverse diameter of inlet (mm), mean	Straight length of sacrum (mm), mean	Index of inlet shape, mean
Ethiopia	33	36	113		117	119	104	94.8
Ghana	34	2	101.1					
Nigeria	35	32	112.0		121.0			
Nigeria	36	19		110.7 (a)	112.6		102.8	
South Africa	37	44		104.8 (b)		105.2	96.6	
Black	38	17	111.94			121.1		92.52
Bantu	39	12	105.9			113.4		93.4
Bantu	40	11	107.6		116.5	119.6	99.7	90.4
Bantu	41	33		104.0 (b)		114.2		
Bush	42	33		113.2 (b)		106.8		
Shona	43	45	115		120	118	105	97.5
Zulu	44	45	104		118	116	97	89.7
Sudan								
Kulubnarti	45	42	103		111	116		88.8
Uganda								
Bantu	46	43	99.7		110.3	112	95	87.6
Bantu	47	43	103		111	118	97	87.3
Asia								
China	48	4	116.5	118.2 (b)		124.5		93.6
China	49	54	117.1		112.6	125.9		93.0
China	50	21	119		109	122		97.5
China	51	39		108.5 (b)				
China	52	18		116 (b)	118	123	118	
Hong Kong	53	5	115.1		115.2	125.5		91.7
India	54	34		101.3 (a)		119.1		
Andaman Islands	55	8	99.2		100.6	102.8		96.3
Andaman Islands[b]	56	35					90.1	
Japan	57	25		107 (b)	111	121	94.0	
Japan	58	25					103.5	
Aino	59	25		111 (b)	118	129	95.0	
Malaysia	60	24		115 (b)		117		
Philippines	61	23	118.1		117.5	123.5		95.6
Philippines	62	1			110.04			
Russia	63	41		108 (b)				
Russia	64	37		110 (b)		132		
Australia								
Aborigine	65	8	108.6		107.6	118.2		92.5
Aborigine	66	25		113 (b)		123		
Aborigine	67	24		114 (b)		126		
Aborigine	68	6					88.12	

Sample	Sample number	Citation number	Obstetrical conjugate diameter of inlet (mm), mean	Conjugate diameter of inlet (mm), mean	Anteroposterior diameter of outlet (mm), mean	Transverse diameter of inlet (mm), mean	Straight length of sacrum (mm), mean	Index of inlet shape, mean
Aborigine	69	25					94.0	
Europe								
European	70	8	106.6		116.0	133.0		80.4
European	71	25		106 (b)		135		
European	72	25					101.0	
Belgium/France	73	40		112.4 (a)		130.1		
France	74	24		110 (b)		135		
Germany	75	22		104 (b)		140		
Germany	76	24		107 (b)		136		
Germany	77	24		112 (b)		135		
Poland	78	20	107.5			135.9		79.3
United Kingdom								
England	79	38	113		116	129		87.6
England	80	28	116.4		130.1	132.3		88.3
England	81	55	118.3	122.4 (a)	119.7	130.6		90.8
England	82	29	120.19			131.3		91.81
England	83	30	120.97		132.91	131.41		92.32
England	84	24		103 (b)		137		
Northern Ireland	85	13	112			127		88.9
Wales	86	19		121.5 (a)	119.3		113.2	
South America								
Peru	87	7	104.1		117.6	127.2	99.1	81.5
Other								
New Zealand	88	3	126.35		126.61	133.64		95.44
Males								
North America								
Native American								
Pecos Pueblo	1	46	90.6	94.3 (a)	107.5	126.0	100.9	71.6
Arkansas and Louisiana	2	15	104		130		109.5	79.8
Haida	5	46	102.7	106.4 (a)	117.0	131.6	111.9	78.4
Indian Knoll	6	46	103.1	106.7 (a)	108.6	125.1	104.4	82.3
Lenape	7	15	106			121	107	87.8
Libben	8	46	101.8	106.8 (a)	116.0	132.4	104.7	77.4
Southern Utah	9	15	100			124	108	80.6
Southwest, Pueblos	10	14		96.5 (a)	107.5	125.1	103.1	
Southwest and Mexico	11	15	97			123	107	78.7
United States								
Black	15	46	96.5	101.7 (a)	113.2	117.7	99.8	81.9
Black	18	16		100.97 (a)		111.61		

Sample	Sample number	Citation number	Obstetrical conjugate diameter of inlet (mm), mean	Conjugate diameter of inlet (mm), mean	Anteroposterior diameter of outlet (mm), mean	Transverse diameter of inlet (mm), mean	Straight length of sacrum (mm), mean	Index of inlet shape, mean
White	20	46	101.9	107.2 (a)	111.2	129.7	102.7	79.0
White	89	50	115.5			118.2		97.7
White	25	16		109.47 (a)		124.08		
Africa								
Egypt								
Naqada	31	53					97.45	
South Africa	37	44		86.6 (b)		97.2	90.2	
Bantu	39	12	103.3			112.6		91.7
Bantu	41	33		102.7 (b)		111.4		
Bush[c]	42	33		91.0 (b)		96.1		
Asia								
India								
Andaman Islands[b]	56	35					93.6	
Japan	57	25		103 (b)	103	120	99.0	
Japan	58	25					101.6	
Aino	59	25		103 (b)	109	121	99.0	
Australia								
Aborigine	66	25		107 (b)		115		
Aborigine	68	6					96.52	
Aborigine	69	25					97.0	
Aborigine[b]	90	35					98.7	
Europe								
European	71	25		104 (b)		130		
European	72	25					105.0	
Belgium/France	73	40		105.9 (a)		125.8		
Poland	78	20	98.1			129.7		75.72
United Kingdom								
England	81	55	115.1	119.6 (a)	119.4	120.8		95.3
Northern Ireland	85	13	102			118		86.5

[a] Emmons (1913) provided data on individual specimens. I did not use those data (1) based on "an approximate measurement . . . (or a) high or low value" (Emmons 1913:48) and (2) for obstetrical conjugate diameter of inlet if either a false or double promontory was present. For anteroposterior diameter of outlet and sacral length, I used only individuals with five sacral vertebrae.

[b] I used only individuals with five sacral vertebrae.

[c] Summary statistics are based on data for individuals presented by Orford (1934), and these statistics differ from those in her table 1.

Appendix K

Pecos Pueblo Sex Assignments by Tague and Hooton and Measurements by Tague

Robert G. Tague

The ID number is the catalogue number assigned by the Peabody Museum of Archaeology and Ethnology when the human remains were accessioned.

Angulation of sacrum and subpubic angle are in degrees. All other measurements are in millimeters. L = left, R = right.

Additional notes regarding data:

Twenty nine individuals were not included in this study or previous studies of pelvic size by me. These are noted by an asterisk following the ID number in the first column of the appendix. They were excluded from these studies for various reasons: probably not adult (ID 59822, 60304, 60484, 60558); could not determine age (ID 59962); adult, but could not determine age (ID 59838, 59839, 60309); not certain of sex or age (ID 59925, 60308, 60437, 60620); pelvic bones did not articulate well (ID 60303); and no pelvic data (ID 59828, 59849, 59852, 59918, 59922, 60013, 60101, 60227, 60230, 60266, 60276, 60278, 60319, 60454, 60456, 60611).

The curved length of anterior space of inlet (fig. 10.2d, B–H) can be computed as follows:

> Curved length of anterior space of inlet = curved length of linea terminalis – curved length of posterior space of inlet (fig. 10.2d, B–L and H–L, respectively).

The curved length of anterior space of inlet was directly measured for four individuals because posterior space of inlet could not be measured: ID 60158 = 101.6 mm (R); ID 60480 = 101.6 mm (R); ID 60663 = 109.2 mm (R); and ID 60888 = 105.4 mm (R).

The transverse diameter of midplane is not listed, because I could take this measurement on only three individuals: ID 60051 = 68.5 mm; ID 60902 = 75.1 mm; and ID 60558 = 82.8 mm. The difference in sample size of transverse diameter of midplane between Tague 1989, 1992, and Appendix K is that I computed this diameter for the former two studies, using formulas for triangles, for individuals with data on the anteroposterior diameter, anterior space, and posterior space of the midplane.

PART 1: SEX ASSIGNMENTS AND METRICS FOR ANTEROPOSTERIOR AND TRANSVERSE DIAMETERS OF PELVIS AND SOME VARIABLES OF SACRUM

ID number	Sex assignment by Tague	Sex assignment by Hooton	Antero-posterior diameter of inlet: Anatomical conjugate	Antero-posterior diameter of inlet: Obstetrical conjugate	Antero-posterior diameter of midplane	Antero-posterior diameter of outlet	Bi-iliac diameter	Transverse diameter of inlet	Transverse diameter of outlet	Sacrum, maximum breadth	Sacrum, breadth at inlet	Sacrum, straight length	Sacrum, curved length
59801	Male	Male			106.8	101.9	263	122.4	111.9	109.4	103.6		
59803	Female	Male	107.0	103.9	119.5	113.4	258	135.4	112.6			93.4	105.4
59805	Male	Male									109.1		
59809	Female	Male			105.8	101.2		132.1					
59814	Female	Female									103.7		
59815	Female	Female			123.4	116.2	271	140.1		113.5	109.7		
59817	Female	Male											
59818	Female	Female											
59822*	Male	Male	91.2	88.4	100.6	95.9		114.3	102.3				
59823	Male	Male	102.3	99.6	121.4	117.2	264	129.0	95.8	117.3	109.8	104.1	110.5
59824	Female	Female										99.3	107.3
59828*	Female	Female											
59829	Female	Female										106.4	115.6
59833	Male	Male											
59834	Female	Female								105.5		88.5	99.1
59838*	Male	Male								115.4			
59839*	Female	Indeter-minate											
59841	Female	Female	78.8	74.2	110.4	103.0	248	135.5	123.2	111.9	110.8	80.2	99.1
59845	Female	Female								118.0	108.8	111.7	116.8
59846	Female	Male	111.6	108.6	138.1	133.5	263	131.1	110.7	113.6	105.6	90.7	100.3
59849*	Female	Female											
59852*	Male	Male											
59856	Female	Male											
59857	Female	Female											
59861	Female									115.4		98.3	106.7
59862	Male	Male											
59865	Male	Male											
59866	Male	Male											
59867	Female	Female											
59870	Male	Male											
59871	Female	Female											
59874	Male	Male								110.0	105.8		
59876	Male	Male	95.7	93.0	128.1	125.7	280	138.6	98.3	121.9	114.0	109.0	116.2
59877	Female	Female											
59891	Male	Male											

ID number	Sex assignment by Tague	Sex assignment by Hooton	Antero-posterior diameter of inlet: Anatomical conjugate	Antero-posterior diameter of inlet: Obstetrical conjugate	Antero-posterior diameter of midplane	Antero-posterior diameter of outlet	Bi-iliac diameter	Transverse diameter of inlet	Transverse diameter of outlet	Sacrum, maximum breadth	Sacrum, breadth at inlet	Sacrum, straight length	Sacrum, curved length
59907	Male	Male											
59908	Female	Female											
59912	Female	Female											
59913	Female												
59918*	Male	Male											
59919	Male	Male										112.2	115.6
59921	Male	Male								116.8			
59922*	Female	Female											
59923	Female	Female											
59924	Male	Male											
59925*	Female	Male											
59954	Male	Male											
59956	Male	Male	96.2	92.9	118.4	115.8		134.2		114.5	110.7	103.7	109.2
59957	Male	Male										98.1	104.1
59958	Male	Male											
59962*	Male	Male											
59963	Male	Male								119.7	103.7	104.0	108.6
59965	Female	Female								114.2	109.9		
59972	Male												
59977	Female	Female											
59978	Female	Female											
59980	Female	Female											
59982	Male	Male											
59985	Male	Male											
59986	Female	Male											
59987	Male	Male											
59988	Female	Female	94.8	92.0	120.0	114.5	257	133.6	125.0	116.6	108.5	103.8	
59990	Male	Male	86.7	83.0	108.3	100.9	283	135.0	100.5	118.8	110.2	94.5	111.8
59991	Female	Female											
59994	Male	Male											
59995	Female	Female											
59996	Female	Female											
59997	Male	Male											
59998	Male	Male											
60003	Female	Male										111.6	118.7
60004	Female	Male											
60005	Female	Female										92.9	102.2
60006	Female	Female	92.0	90.1	100.0	99.3	262	125.9	104.9			103.7	108.0

ID number	Sex assignment by Tague	Sex assignment by Hooton	Antero-posterior diameter of inlet: Anatomical conjugate	Antero-posterior diameter of inlet: Obstetrical conjugate	Antero-posterior diameter of midplane	Antero-posterior diameter of outlet	Bi-iliac diameter	Transverse diameter of inlet	Transverse diameter of outlet	Sacrum, maximum breadth	Sacrum, breadth at inlet	Sacrum, straight length	Sacrum, curved length
60008	Female	Female										97.6	102.9
60009	Female	Female											
60010	Male	Male	84.2	78.4			249	113.8			94.8		
60011	Female	Female											
60013*	Female	Female											
60017	Female	Female											
60018	Male	Male											
60019	Male	Male											
60020	Female	Female											
60021	Female	Female											
60022	Male	Male											
60025	Female	Female											
60026	Male	Male											
60028	Male	Male											
60038	Male	Male										101.5	108.0
60039	Male	Male	87.2	84.5				121.0			107.2		
60040	Female	Male										109.4	115.6
60041	Female	Female	85.9	82.4			273	134.8			114.8		
60042	Male	Male										104.9	114.9
60044	Male	Male											
60046	Male	Male											
60049	Female	Female	89.0	85.7	114.6	109.8	264	141.2		111.8	110.2	96.9	106.7
60050	Female	Male											
60051	Male	Male	89.0	85.8	105.0	104.0		118.2	76.4	106.8	94.6	103.5	109.2
60052	Female	Female											
60055	Female	Female											
60057	Female	Female										71.8	93.3
60060	Female	Female	99.4	96.4	123.6	120.4		138.9		120.8	113.6	104.4	113.0
60061	Female	Male	95.2	91.4	124.5	117.1		131.6	91.3	121.6	112.2	84.8	100.3
60062	Female	Female			115.1	114.0	255	131.0	111.7	114.2	100.3	105.0	110.5
60063	Female	Female											
60064	Male	Male			115.7	108.0	282	132.8	99.2	118.3	112.2		
60065	Male	Male									101.5		
60066	Female											87.7	102.9
60071	Male	Male	77.0	74.7	96.0	95.7	273	131.5	96.6	117.5	110.9	99.8	104.1
60074	Male	Male	108.0	103.9	111.4	105.1	282	122.6	98.2	111.8	103.0	103.5	113.7
60077	Female	Female								116.1	115.2	89.4	97.8
60082	Female	Female								116.7	107.0	103.6	113.7

ID number	Sex assignment by Tague	Sex assignment by Hooton	Antero-posterior diameter of inlet: Anatomical conjugate	Antero-posterior diameter of inlet: Obstetrical conjugate	Antero-posterior diameter of midplane	Antero-posterior diameter of outlet	Bi-iliac diameter	Transverse diameter of inlet	Transverse diameter of outlet	Sacrum, maximum breadth	Sacrum, breadth at inlet	Sacrum, straight length	Sacrum, curved length
60083	Female	Female											
60084	Female	Female	84.2	78.6	115.8	108.4	270	136.4	108.1	115.4	114.0	86.2	104.8
60086	Male	Male										97.8	108.0
60088	Male	Male			108.2	103.3	268	122.5	79.1		103.4		
60089	Female	Female	85.0	83.2	105.9	99.6		130.0		112.2	104.7	99.0	109.9
60092	Male	Male											
60093	Female	Female											
60094	Female	Female	92.8	89.2	120.7	117.4	266	133.3	117.8	115.2	107.3	102.0	109.9
60096	Male	Male								124.4	115.0		
60100	Female	Female	86.6	84.2			257	127.5					
60101*	Female	Male											
60102	Male	Male	100.1	95.8	111.3	106.6	268	123.7	99.4	118.7	104.7	96.6	103.5
60106	Male	Male	98.2	95.6	108.6	104.2	261	119.9	91.2	113.6	104.2	107.4	114.9
60108	Female	Female	92.2	89.8	127.8	129.4	251	129.8	124.8	112.4	106.2	94.5	99.1
60109	Male											94.9	102.9
60133	Male	Male											
60134	Female										110.7		
60138	Female	Male			125.2	119.8	262	138.2	111.3				
60142	Male	Male										110.6	114.3
60143	Female	Female	87.1	85.8	104.8	101.8	257	124.8	103.1			104.2	111.1
60147	Female												
60154	Male	Male											
60158	Female		92.7	89.6	113.0	111.2		137.0		117.3	106.4	98.4	102.9
60160	Female												
60169	Female		92.6	90.2	116.5	112.2	262	132.0				94.0	105.4
60172	Female	Female								116.0	113.0		
60173	Female	Female	82.2	79.6	106.4	102.3		129.8				89.1	98.4
60174	Female	Female								115.0	110.4	74.2	95.2
60176	Male												
60179	Female	Female											
60180	Male	Male	97.6	93.6	110.7	106.2	268	131.2	98.7	117.6	107.3	106.8	114.9
60181	Male	Male											
60189	Male	Male											
60194	Male	Male	93.2	89.2	110.8	107.0	260	122.7	95.6	120.6	104.2	99.0	104.1
60197	Male	Male	101.2	97.4	115.1	109.0	265	123.2	97.5	105.4	96.0	106.1	114.9
60198	Male	Male											
60200	Male	Male											
60201	Male	Male								116.5		97.6	107.3

ID number	Sex assignment by Tague	Sex assignment by Hooton	Antero-posterior diameter of inlet: Anatomical conjugate	Antero-posterior diameter of inlet: Obstetrical conjugate	Antero-posterior diameter of midplane	Antero-posterior diameter of outlet	Bi-iliac diameter	Transverse diameter of inlet	Transverse diameter of outlet	Sacrum, maximum breadth	Sacrum, breadth at inlet	Sacrum, straight length	Sacrum, curved length
60202	Female												
60210	Female	Female	94.0	90.3	116.0	110.8	281	146.4	112.3	115.2	109.3	102.1	114.9
60211	Female	Female			129.0		301	140.9	116.0	120.0	113.1		
60212	Male	Male										109.3	113.0
60216	Female	Indeter-minate											
60218	Male	Male										88.8	100.3
60219	Male	Male											
60222	Female	Female									107.4	104.2	111.1
60224	Male	Male											
60225	Female	Indeter-minate								119.4			
60226	Male	Male	95.1	91.2	118.2	109.4	277	131.8	89.6	118.8	107.8	82.6	99.1
60227*	Female	Female											
60228	Female	Female											
60229	Female	Female									107.6		
60230*	Male	Male											
60232	Male	Male								121.6	99.7		
60233	Female	Female											
60238	Male	Male	90.2	87.4	116.4	117.9		138.0	110.8			110.2	115.6
60239	Female	Female											
60241	Female	Female										78.1	93.3
60244	Male												
60247	Female	Female											
60251	Female		86.6	81.9	115.6	111.7		129.8		112.2	107.6	95.2	111.8
60252	Female	Female								120.2	116.7	108.8	118.7
60254	Male	Male											
60257	Female	Female	91.4				270	136.2		114.5	113.2	93.4	102.9
60258	Female	Male	107.4	102.7	127.8	125.4	263	144.6	110.2	119.7	116.0	107.0	113.0
60259	Female	Female										104.3	118.1
60260	Male	Male										97.6	109.2
60266*	Male	Male											
60267	Male	Male											
60272	Female	Female	98.0	95.8			256	133.8	128.8		102.5		
60275	Male	Male										96.7	100.3
60276*	Male	Male											
60277	Male	Male								114.6		105.2	109.2
60278*	Male	Male											
60281	Female	Female											

ID number	Sex assignment by Tague	Sex assignment by Hooton	Antero-posterior diameter of inlet: Anatomical conjugate	Antero-posterior diameter of inlet: Obstetrical conjugate	Antero-posterior diameter of midplane	Antero-posterior diameter of outlet	Bi-iliac diameter	Transverse diameter of inlet	Transverse diameter of outlet	Sacrum, maximum breadth	Sacrum, breadth at inlet	Sacrum, straight length	Sacrum, curved length
60283	Male	Male											
60284	Female	Female								112.7	105.2	90.1	98.4
60285	Female	Female	94.8	92.4	117.8	110.4	269	136.6	124.5		112.1	91.4	108.6
60286	Male	Male								118.6	103.6	109.8	114.3
60290	Female		100.2	97.0				126.2	105.6	110.1	104.2		
60291	Female	Female	81.4	78.8	111.4	107.2		131.6	110.7	107.9	105.9		
60293	Male	Male											
60296	Male	Male										104.2	114.3
60297	Female	Female										69.3	
60298	Female	Female	98.0	95.5	123.3	119.2	247	130.4	122.2	117.5	103.9	102.0	110.5
60302	Female	Female	90.8	87.8	118.4	115.8	265	140.8		119.8	114.7	100.8	110.5
60303*	Female	Female								117.4	111.3	104.0	113.0
60304*	Female												
60306	Male	Male										95.0	101.0
60308*	?Female?	Female								108.0	101.2	101.6	108.0
60309*	Female	Male										88.6	99.1
60311	Female	Female										101.9	111.8
60312	Female	Female	98.1	96.5	111.8	104.9	262	128.2				96.2	108.0
60313	Male	Male								123.0	115.9	101.8	113.7
60314	Male	Male	88.6	83.6	116.8	114.4	256	128.8		116.8	110.5	108.1	115.6
60315	Male	Male								111.4		105.7	112.4
60318	Male	Male	98.3	93.2	116.6	112.7	301	137.8		123.8	115.4	101.0	108.0
60319*	Male	Male											
60428	Female	Female										98.0	108.0
60436	Female	Male	89.0	85.1	119.4	114.7	264	140.8	119.7	116.4	114.0	91.4	101.6
60437*	?Female?	Male									106.6	85.8	106.7
60451	Female	Indeter-minate											
60452	Female	Female											
60454*	Male	Male											
60455	Female	Female			124.4	124.5	270	135.9	119.4	113.4	111.8		
60456*	Male	Male											
60457	Female	Female											
60458	Male	Male	99.0	96.6	115.3	112.4	269	128.4	105.2	111.8	105.8	111.1	119.4
60480	Female	Female	100.5		109.9			129.0		112.4			
60481	Female												
60482	Male	Male										101.4	106.7
60483	Male	Male								110.1	108.5	91.6	106.7

ID number	Sex assignment by Tague	Sex assignment by Hooton	Antero-posterior diameter of inlet: Anatomical conjugate	Antero-posterior diameter of inlet: Obstetrical conjugate	Antero-posterior diameter of midplane	Antero-posterior diameter of outlet	Bi-iliac diameter	Transverse diameter of inlet	Transverse diameter of outlet	Sacrum, maximum breadth	Sacrum, breadth at inlet	Sacrum, straight length	Sacrum, curved length
60484*	Female	Male								118.2	109.3		
60488	Male												
60496	Female	Female	105.9	103.6	127.7	125.1		138.0		126.6	111.6	97.5	106.0
60511	Female	Female								124.2			
60513	Male	Male										116.7	
60514	Male	Male	98.4	95.4	111.3	111.7	251	122.2	107.0			100.6	102.9
60518	Male	Male								113.9	108.0		
60525	Female												
60526	Female	Female								108.4		91.5	104.1
60527	Male	Male											
60528	Male	Male											
60529	Female	Female											
60530	Male	Male	94.2		110.6	106.6	260	114.8	89.0	108.5	96.1	91.2	97.8
60533	Female	Female			122.3	115.6		131.2			107.3	77.0	90.2
60541	Male	Male									100.6		
60542	Female	Female								111.9	105.3		
60543	Female	Female	97.2	94.4	114.2	107.4	248	125.9		109.2	104.4	109.6	120.0
60545	Male	Male										112.5	118.1
60550	Female	Female											
60554	Male	Male								113.5			
60555	Female	Female									111.4		
60556	Male	Male	85.8	83.2	93.5	96.3	233	110.2	85.8	102.4	91.2	108.5	111.1
60558*	Male	Male	85.3	82.6				122.4	107.1				
60560	Male	Male								116.0		93.6	107.3
60562	Male	Male											
60563	Female	Female	104.9		114.4	109.8		147.9		124.3	119.0	107.8	114.3
60573	Male	Male										92.5	99.7
60574	Female	Female											
60578	Female	Female								118.5	114.9		
60588	Female	Female					257	134.0		117.8	114.0		
60590	Female	Female	88.6	85.4	120.0	116.2	252	135.6	131.7	110.7	108.2	92.9	103.5
60591	Female												
60594	Female	Female											
60597	Female	Male	106.7	101.0			265	138.2		114.8	112.3		
60599	Male	Male										85.4	97.2
60601	Female	Female								108.7	104.2	107.5	111.8
60602	Male	Male								120.5	111.7		
60608	Female	Female	93.0	90.0	116.7	109.2		128.3		106.0	101.3	90.0	101.6

ID number	Sex assignment by Tague	Sex assignment by Hooton	Antero-posterior diameter of inlet: Anatomical conjugate	Antero-posterior diameter of inlet: Obstetrical conjugate	Antero-posterior diameter of midplane	Antero-posterior diameter of outlet	Bi-iliac diameter	Transverse diameter of inlet	Transverse diameter of outlet	Sacrum, maximum breadth	Sacrum, breadth at inlet	Sacrum, straight length	Sacrum, curved length
60609	Female	Indeterminate	105.4	101.5	114.6	113.4	249	129.2	122.9	118.3	102.8	104.1	109.2
60610	Female	Male											
60611*	Female	Male											
60612	Female	Male								110.0			
60614	Female	Female										97.3	109.2
60616	Male												
60620*	?Female?	Male											
60633	Male		113.6	110.2	122.1	111.5	297	135.5	96.7	125.8	114.0	115.5	128.3
60634	Male		110.0	105.0	109.4	103.6	270	128.4	101.8	119.9	105.4	108.8	115.6
60647	Female												
60662	Male		98.4	94.2				130.7		120.4	114.4		
60663	Male		82.5	77.4	100.4	90.3	271	127.8	102.2			85.2	102.2
60664	Female											80.8	102.2
60665	Male										99.2	104.8	112.4
60669	Male		85.6	82.0	102.7	98.2	252	115.9		103.4	99.3	96.8	104.8
60671	Female									109.6	109.6	93.1	102.2
60674	Female		89.8	88.6	121.6	121.3	264	138.9	113.4	107.1	106.1	106.9	114.3
60675	Male		97.4	93.4	111.5		274	128.7	96.6		109.2		
60676	Female												
60677	Male		104.8	100.8	115.6	107.9	273	122.3	97.0	110.8	107.3	90.9	105.4
60678	Male										108.4		
60682	Male												
60685	Female												
60687	Male		94.0	90.2	122.8	119.4	271	133.1	81.9	124.8	112.8	103.0	114.3
60689	Male												
60691	Male		96.4	92.7	112.2	110.8	283	132.8	101.1	116.8	107.2	104.4	110.5
60693	Male												
60695	Male										90.8		
60696	Female												
60703	Female									117.7		101.8	109.2
60706	Female		97.6	95.0			262	122.1	103.3	107.5	97.4		
60713	Female		101.2	99.0	123.3	117.2	268	137.3		122.4	120.3	95.0	106.0
60717	Male									112.0	101.8	91.5	100.3
60722	Female											88.8	102.2
60725	Male												
60727	Female											104.4	114.3
60729	Male		88.4	83.2	107.0	105.0		129.4		113.9	105.0	109.6	119.4

ID number	Sex assignment by Tague	Sex assignment by Hooton	Antero-posterior diameter of inlet: Anatomical conjugate	Antero-posterior diameter of inlet: Obstetrical conjugate	Antero-posterior diameter of midplane	Antero-posterior diameter of outlet	Bi-iliac diameter	Transverse diameter of inlet	Transverse diameter of outlet	Sacrum, maximum breadth	Sacrum, breadth at inlet	Sacrum, straight length	Sacrum, curved length
60733	Male												
60736	Female		89.6	87.6	112.0	110.5		118.4	107.6	110.2	105.5	94.5	101.0
60737	Male		97.0	92.4	110.4	105.7	256	115.0	91.0	110.2	97.5	93.7	101.6
60739	Female												
60740	Female												
60741	Male		89.6	86.1	102.2	95.4	256	120.2	86.5	116.5	103.4	93.5	103.5
60743	Male				118.2	111.5	258	124.7	85.5	116.0	107.3	102.5	113.0
60744	Female		102.8	97.7	144.8	143.8		134.8	128.6	114.4	112.8	102.7	109.2
60754	Male		92.9		114.6	114.6	268	111.3	100.6			88.2	94.6
60755	Female												
60773	Male											96.4	99.1
60779	Female												
60782	Female		90.6	88.7	113.8	110.0	264	135.0	128.2	110.5	108.6	109.9	121.9
60888	Female		97.6	89.7		115.5	256	125.6				89.6	
60902	Male		88.8	84.8	117.2	113.4	262	128.8	84.2	107.7	103.7	109.5	119.4
60903	Female									112.8			
60909	Male											94.8	106.7
60911	Female												
60912	Male												
60930	Female		88.8				238	129.7			97.1		
60931	Female												
60935	Male		84.4	80.3	101.4	97.3	247	125.8	104.1		100.7	92.5	101.0
60984	Female									120.3	117.1	84.4	99.1
60986	Female											97.5	108.0

PART 2: METRICS FOR ANTERIOR SPACES OF PELVIS AND SOME VARIABLES OF SACRUM AND POSTERIOR SPACE OF INLET

ID number	Sacrum, angulation, L	Sacrum, angulation, R	Anterior space of inlet, L	Anterior space of inlet, R	Anterior space of midplane, L	Anterior space of midplane, R	Anterior space of outlet, straight length, L	Anterior space of outlet, straight length, R	Anterior space of outlet, curved length, L	Anterior space of outlet, curved length, R	Posterior space of inlet, straight length, L	Posterior space of inlet, straight length, R
59801	54.05	58.92	91.0	92.1			92.6	93.5	102.9	104.8	21.0	23.2
59803			99.5				90.9	90.8	96.5	97.8	29.6	31.2
59805												
59809		64.20		94.8				81.7		92.1	20.6	18.3
59814		75.26						80.2		88.9		
59815	72.71			98.9							33.0	31.3
59817	74.08						86.6	83.3	92.1	90.2		
59818	70.22						88.9		94.0			
59822*	59.49	56.48	84.0	85.8			76.9	76.5	88.9	89.5	19.7	21.1
59823	57.84	63.27		103.7			85.0	85.9	94.0	95.2	20.6	20.3
59824							87.0		95.2			
59828*												
59829												
59833												
59834												
59838*												
59839*												
59841	73.92	72.72	92.0	86.8			88.0	88.7	94.6	97.8	26.0	29.0
59845	83.29	77.74			92.2		86.6		95.2			
59846	79.90	74.12		107.4			96.3	94.8	102.9	102.9	21.3	24.9
59849*												
59852*												
59856	76.30	68.16						87.8		95.2		
59857							84.3					
59861												
59862	56.36	57.15						97.0		109.2		
59865	64.81						89.2		98.4			
59866							83.1		97.8			
59867	62.09						80.7		88.3			
59870												
59871												
59874	60.49							90.9		101.6		
59876	64.87	61.50	105.2	104.7			94.1	92.4	104.1	102.9	22.5	25.0
59877		76.52						83.4		92.1		
59891	63.82	62.50					86.0	87.2	99.7	102.2		

ID number	Sacrum, angulation, L	Sacrum, angulation, R	Anterior space of inlet, L	Anterior space of inlet, R	Anterior space of midplane, L	Anterior space of midplane, R	Anterior space of outlet, straight length, L	Anterior space of outlet, straight length, R	Anterior space of outlet, curved length, L	Anterior space of outlet, curved length, R	Posterior space of inlet, straight length, L	Posterior space of inlet, straight length, R
59907												
59908	70.80	75.35					84.6	84.2	91.4	91.4		
59912							93.0		100.3			
59913												
59918*												
59919							87.3					
59921												
59922*												
59923												
59924		68.06				88.2		86.2		98.4		
59925*												
59954												
59956			98.5	96.4							28.0	28.4
59957		71.72				92.0		84.2		94.0		
59958		57.65						85.0		96.5		
59962*												
59963								87.8		99.7		
59965		79.10						85.6		92.1		
59972	55.85				91.5		90.0		98.4			
59977												
59978	61.55	57.91						88.4		97.2		
59980								88.1				
59982												
59985							85.8		94.0			
59986												
59987		58.45										
59988		77.39		95.6				90.2		102.2	31.5	28.3
59990	64.92	61.09	91.3	92.2			86.8	84.0	97.8	95.2	25.4	22.0
59991		80.15						93.2		101.6		
59994		53.19					91.6	90.6	100.3	101.6		
59995												
59996												
59997	64.38	60.73			81.0		80.1	80.8	87.6	89.5		
59998	66.69	64.27					88.6	91.0	99.1	108.6		
60003	70.77	74.90										
60004	69.85	78.39			96.9	101.1	92.3	90.6	99.7	99.1		
60005		78.23						93.2		101.0		
60006	64.73	59.25	87.1	88.7			82.3	83.9	88.3	91.4	22.4	25.0

ID number	Sacrum, angulation, L	Sacrum, angulation, R	Anterior space of inlet, L	Anterior space of inlet, R	Anterior space of midplane, L	Anterior space of midplane, R	Anterior space of outlet, straight length, L	Anterior space of outlet, straight length, R	Anterior space of outlet, curved length, L	Anterior space of outlet, curved length, R	Posterior space of inlet, straight length, L	Posterior space of inlet, straight length, R
60008								85.6		92.1		
60009								88.5		97.2		
60010			82.1	84.9			79.2		86.4		22.7	20.8
60011	71.13	73.79					90.8					
60013*												
60017		63.95										
60018	58.03	58.92					87.3	89.6	95.9	99.7		
60019								93.1		108.0		
60020												
60021		66.53						87.3		96.5		
60022												
60025												
60026							87.3	89.3	101.0	105.4		
60028												
60038	62.12							77.4		87.6		
60039			94.8	92.0							19.0	22.9
60040					86.6		79.9		87.6			
60041			100.5	95.7				92.9		103.5	20.0	19.9
60042												
60044	48.69							88.0		101.6		
60046	63.31	57.06					90.2		102.9			
60049		63.70		94.0				91.7		100.3	28.3	29.0
60050	68.22						95.4		101.0			
60051	67.28	63.47	87.0	88.8	83.5	85.6	79.9	81.3	87.6	92.7	19.2	17.8
60052		60.24						84.3		92.7		
60055	64.38	65.28			88.4	80.9	85.0	77.4	93.3	86.4		
60057						94.8		86.8		93.3		
60060			103.8	100.2							28.4	28.3
60061	64.47	72.49	97.6	100.4			86.3	88.5	94.0	96.5	24.6	24.6
60062	63.11	62.76						87.5			31.6	32.7
60063												
60064	63.77	73.00	97.3	95.2		92.5	87.4	87.2	101.0	102.2	24.0	25.4
60065	60.37	56.66					88.1		98.4			
60066		63.33										
60071			87.0	88.2			79.5	79.5	87.0	88.3	19.4	21.0
60074	59.65			90.7		88.6	83.9	82.0	95.2	93.3	24.6	28.3
60077												
60082												

ID number	Sacrum, angulation, L	Sacrum, angulation, R	Anterior space of inlet, L	Anterior space of inlet, R	Anterior space of midplane, L	Anterior space of midplane, R	Anterior space of outlet, straight length, L	Anterior space of outlet, straight length, R	Anterior space of outlet, curved length, L	Anterior space of outlet, curved length, R	Posterior space of inlet, straight length, L	Posterior space of inlet, straight length, R
60083	72.28						89.5		96.5			
60084	60.13	68.95	99.3	98.7			87.4	84.0	96.5	93.3	21.2	21.1
60086	64.52	66.39					85.2		93.3			
60088	57.55	59.08	91.0	88.8			83.5	84.4	94.6	96.5	26.0	24.1
60089	67.42		86.8	87.6			84.1		91.4		28.8	28.4
60092	60.63	61.38						86.0		100.3		
60093												
60094	63.63	68.09	92.6	91.9			90.3	88.4	99.1	97.2	30.2	32.0
60096							95.0		102.9			
60100			85.5	88.2							30.6	26.4
60101*												
60102		65.40	94.6	94.3			85.5	85.0	97.2	99.1	23.7	20.6
60106	62.79	55.76	91.5	91.1			87.7	83.6	99.7	96.5	23.4	26.4
60108	77.02	78.60	93.0	94.8			87.4	85.5	94.0	92.7	22.8	29.3
60109		66.33						82.4		93.3		
60133								93.2		105.4		
60134												
60138	73.44			97.6			89.0	87.6	97.2	95.9	26.7	26.7
60142												
60143	66.04	66.15	86.4	85.0			81.1	80.4	90.2	88.9	24.3	28.3
60147		66.93										
60154	69.72						88.4					
60158		65.97	97.5	95.3				87.4		98.4	26.7	
60160							97.5	94.2	106.0	105.4		
60169			93.7	91.6			82.7				29.6	27.9
60172												
60173			91.7	93.6							19.0	18.2
60174	79.16						86.2		90.2			
60176		60.96			88.3		86.4		92.7			
60179		63.04										
60180	65.32		99.4	95.4			94.0	92.2	104.1	104.1	19.4	20.5
60181												
60189												
60194	60.94		93.3	94.2			86.7	83.9		97.8	20.1	19.0
60197	54.78	56.57	90.1	92.9			86.6	88.5	94.6	100.3	29.7	28.4
60198								91.0		106.7		
60200	58.75											
60201												

ID number	Sacrum, angulation, L	Sacrum, angulation, R	Anterior space of inlet, L	Anterior space of inlet, R	Anterior space of midplane, L	Anterior space of midplane, R	Anterior space of outlet, straight length, L	Anterior space of outlet, straight length, R	Anterior space of outlet, curved length, L	Anterior space of outlet, curved length, R	Posterior space of inlet, straight length, L	Posterior space of inlet, straight length, R
60202		61.58					88.8	91.6				
60210	72.83	67.04	101.7	102.3		93.0	91.4	89.0	97.2	95.2	31.8	31.0
60211	67.23	63.37	96.4	99.7			99.7	99.6	107.3	108.6	31.6	30.5
60212												
60216								91.3		97.2		
60218	63.42						85.4	83.4	96.5	99.1		
60219	49.73	54.01					91.4		100.3			
60222		72.46						95.9		106.7		
60224	66.20						81.9		91.4			
60225		65.74										
60226		68.67	92.6	94.9			82.3	82.7	88.3	91.4	28.7	27.6
60227*												
60228	71.09						91.3		98.4			
60229	59.90							87.9		94.6		
60230*												
60232												
60233		63.23										
60238	57.71	57.92	100.3	99.9			98.0	95.9	104.1		21.0	19.8
60239	72.49				88.4		89.0		95.2			
60241		74.79										
60244							88.4	89.7	99.7	104.1		
60247		70.53						82.7				
60251		63.81	93.8	95.2		83.0		79.5		85.1	31.2	27.0
60252	72.75						92.0	88.2	98.4	95.9		
60254	62.51						87.6		96.5			
60257			100.0								25.3	27.6
60258	69.32	72.18	103.0	104.9			89.5	81.7	97.8	89.5	30.6	33.6
60259		69.76										
60260							93.0		106.0			
60266*												
60267												
60272	69.37	70.57	95.5	93.0			88.3	87.4	97.2	97.8	30.7	34.9
60275					86.6		82.5	83.7		94.0		
60276*												
60277								88.2				
60278*												
60281	60.53	61.89										
60283							87.3		92.7			

ID number	Sacrum, angulation, L	Sacrum, angulation, R	Anterior space of inlet, L	Anterior space of inlet, R	Anterior space of midplane, L	Anterior space of midplane, R	Anterior space of outlet, straight length, L	Anterior space of outlet, straight length, R	Anterior space of outlet, curved length, L	Anterior space of outlet, curved length, R	Posterior space of inlet, straight length, L	Posterior space of inlet, straight length, R
60284	68.56	67.81						86.8		96.5		
60285	72.05	61.59	97.2	98.8			90.6	88.8	99.1	99.1	24.8	25.6
60286	53.86	51.41			94.2		89.1		99.7			
60290		76.04		100.9				84.2		90.2	26.3	26.3
60291	64.22	67.73	91.6	90.4	84.0		80.8	87.0		94.6	23.6	26.9
60293								93.0		109.2		
60296							91.1		104.1			
60297	81.82	80.38					83.4	88.6	88.9	95.2		
60298	66.61	71.97	95.2				88.3	84.2	95.2	94.0	31.2	34.4
60302			92.8	100.0							33.4	26.5
60303*							88.1	89.2	98.4	104.1		
60304*												
60306	56.44	58.01			94.8		92.9		100.3			
60308*	64.92	67.78										
60309*												
60311	70.29							91.0		102.2		
60312				94.4				85.3			24.6	28.2
60313	63.49	68.20						88.5		101.6		
60314			96.8	96.2				91.4		102.9	21.3	20.6
60315	55.99	61.85			85.8		89.0		99.1			
60318			100.4	102.3							21.8	24.4
60319*												
60428	67.93	64.08					89.4		102.2			
60436	73.51	71.59	96.4				92.0		99.7		26.3	23.6
60437*	67.45	72.82					94.5		101.6			
60451	72.41	74.31					95.1		100.3			
60452	64.43	69.50										
60454*												
60455	63.52	73.29		100.4			94.9	89.6	101.0	97.2	31.8	29.8
60456*												
60457												
60458		64.15	95.9		93.0		92.8	93.6		104.8	22.9	20.0
60480			92.9	94.1							23.2	
60481		67.44					93.3	89.4	99.7	96.5		
60482							85.4		91.4			
60483		68.67						88.0		100.3		
60484*		70.78										
60488							91.6	92.0	103.5	106.7		

ID number	Sacrum, angulation, L	Sacrum, angulation, R	Anterior space of inlet, L	Anterior space of inlet, R	Anterior space of midplane, L	Anterior space of midplane, R	Anterior space of outlet, straight length, L	Anterior space of outlet, straight length, R	Anterior space of outlet, curved length, L	Anterior space of outlet, curved length, R	Posterior space of inlet, straight length, L	Posterior space of inlet, straight length, R
60496		63.54	103.8	102.7				89.1		95.9	30.8	30.0
60511												
60513								90.3		100.3		
60514	58.00	62.19	94.2	94.4			83.9	82.9	95.2		21.0	24.0
60518	53.49	59.05					87.6		97.2			
60525								87.4				
60526												
60527												
60528								82.6				
60529												
60530	75.41		87.8	91.8				81.2		95.2	21.5	22.1
60533			99.0								29.7	28.9
60541							86.8					
60542	66.12	69.41						85.7		93.3		
60543			88.2	87.0							27.7	33.0
60545		61.77						85.8		100.3		
60550												
60554	60.54	60.66			92.1		86.3		97.2			
60555												
60556	53.81	49.99	81.3	82.8				76.8		90.8	20.6	21.8
60558*	67.56	66.84	88.2	89.0	86.8	85.2	82.7	84.2	92.7		23.0	20.8
60560	58.69						92.4		102.9			
60562	56.22	59.83					80.6	82.8				
60563		66.31		101.4				93.7		103.5	33.0	32.6
60573								85.8		97.8		
60574	67.41						84.0		87.6			
60578												
60588			97.2	94.3							26.3	30.8
60590	68.88		94.9	90.2	99.1		92.8	91.0	101.6	101.6	31.0	33.5
60591												
60594												
60597	58.58		104.5				90.8				31.4	30.6
60599	68.51	62.90						75.2		83.8		
60601	62.11	68.82						91.8		97.8		
60602	66.29	69.60					93.3		103.5			
60608	65.10			93.1							26.0	23.0
60609		65.05	96.6	95.0			86.9	85.8	94.0	95.2	34.6	35.0
60610												

ID number	Sacrum, angulation, L	Sacrum, angulation, R	Anterior space of inlet, L	Anterior space of inlet, R	Anterior space of midplane, L	Anterior space of midplane, R	Anterior space of outlet, straight length, L	Anterior space of outlet, straight length, R	Anterior space of outlet, curved length, L	Anterior space of outlet, curved length, R	Posterior space of inlet, straight length, L	Posterior space of inlet, straight length, R
60611*												
60612												
60614												
60616												
60620*	71.30											
60633	66.77	57.87	105.4	105.6		100.4	96.9	94.7	108.0	106.7	30.1	30.6
60634	57.24	59.55	96.6	99.3	86.6		87.2	89.4	99.7	102.9	31.1	28.8
60647												
60662			96.6	101.8							20.7	18.2
60663				97.6			82.0	83.9	90.8	95.2	19.2	
60664	77.10	73.45						86.8		94.0		
60665								90.7		102.2		
60669	60.10		82.8	82.8			76.7				17.9	24.1
60671	74.40	78.36			91.2		83.2		88.3			
60674	63.59	62.23	93.5	93.3			91.4		99.1		29.7	31.8
60675			93.4	92.0		84.2	84.1	81.3	95.2	93.3	24.2	24.7
60676												
60677	60.92	61.14	99.1	98.8			86.0	87.1	97.8	100.3	23.0	23.9
60678		56.25				88.2		90.1		100.3		
60682		66.88						83.2		101.0		
60685	66.64	68.90					87.8	83.4	94.0	91.4		
60687				94.4			91.3	92.2	98.4	101.6	24.6	28.0
60689	56.24						94.0	96.9	105.4	109.9		
60691		58.52		91.8				94.1		105.4	27.2	28.1
60693							86.1	83.6	94.0	96.5		
60695	64.82							90.0		103.5		
60696												
60703		72.03						90.2		98.4		
60706	68.25	72.19	96.6	98.4			88.3	85.5	94.0	93.3	27.3	25.3
60713	78.23		104.1	101.2							29.2	29.2
60717												
60722	79.75	82.51					92.8	91.0	101.6	100.3		
60725												
60727		69.83						91.4		97.8		
60729			92.6	91.7							25.0	27.2
60733												
60736	61.62	61.77	94.2	89.0	90.3		89.8	87.2	95.2	95.2	20.8	22.8
60737	63.12	63.17	87.9	87.8			82.3	81.9	92.1	92.7	23.6	22.1

ID number	Sacrum, angulation, L	Sacrum, angulation, R	Anterior space of inlet, L	Anterior space of inlet, R	Anterior space of midplane, L	Anterior space of midplane, R	Anterior space of outlet, straight length, L	Anterior space of outlet, straight length, R	Anterior space of outlet, curved length, L	Anterior space of outlet, curved length, R	Posterior space of inlet, straight length, L	Posterior space of inlet, straight length, R
60739							87.9	86.0	97.2	94.6		
60740							91.8		97.2			
60741	68.01		91.0	90.5			79.6	78.8	90.8			12.6
60743							92.6	91.4	102.2	102.2	24.2	24.6
60744	72.72	76.56	100.6	97.9	103.5		96.4	94.8	102.2	101.6	32.0	31.7
60754	66.68		84.1		86.8		81.2		94.6		25.8	
60755	68.29	67.09					97.4	99.2	102.9	106.7		
60773	61.37				84.8		79.2		88.9			
60779	55.27						90.6		97.8			
60782	67.02	69.27		91.4				87.0		95.9	25.0	27.3
60888				99.8								
60902	54.37	57.69	96.0	93.8	90.2	87.8	89.9	85.6	101.0	97.8	24.6	27.3
60903							86.2		92.1			
60909					91.2							
60911	72.54	75.99						101.4		108.6		
60912					88.8	88.2						
60930			91.2	91.4		87.4		83.3		90.8	26.2	30.4
60931	63.09	67.99					86.8		96.5			
60935	60.58	65.80	88.2	87.5	84.1		82.6	82.2	91.4	95.2	24.5	26.4
60984	77.17	74.69						92.1		101.6		
60986	64.70				89.0		83.2		90.8			

PART 3: METRICS FOR LINEA TERMINALIS, PELVIC DEPTH, AND SOME VARIABLES OF POSTERIOR SPACES OF PELVIS

ID number	Posterior space of inlet, curved length, L	Posterior space of inlet, curved length, R	Posterior space of midplane, L	Posterior space of midplane, R	Posterior space of outlet, L	Posterior space of outlet, R	Linea terminalis, straight length, L	Linea terminalis, straight length, R	Linea terminalis, curved length, L	Linea terminalis, curved length, R	Depth of pelvis, L	Depth of pelvis, R
59801	21.0	22.9			66.8	70.8	102.7	103.9	120.0	121.3	95.1	100.6
59803	30.5	31.8			82.9	80.8	113.5		134.6		101.5	102.7
59805												
59809	20.3	17.8				71.3	105.0		120.0			95.4
59814								102.0		121.3		90.2
59815	33.0	31.8			86.6			113.3		139.1	96.4	
59817							111.4		131.4		100.0	
59818							116.2		139.7		102.9	
59822*	20.3	21.0			71.7	67.8	94.4	98.6	114.3	114.3	99.0	101.9
59823	19.7	19.7			76.4	77.5		115.8		129.5	107.2	106.7
59824												
59828*												
59829												
59833							106.8		127.6			
59834								104.1		122.6		
59838*												
59839*								121.4		137.8		
59841	25.4	29.2			76.9	82.5	105.0	98.4	122.6	123.2	92.2	92.7
59845							112.5		133.4		100.9	102.4
59846	21.6	24.1			86.6	88.6		122.4		137.2	98.2	98.8
59849*												
59852*												
59856											105.4	107.4
59857												
59861							106.2		125.1			
59862								108.0		128.3	100.4	101.4
59865											99.9	
59866							100.1	98.8	114.3	111.8	101.9	
59867							96.0		106.0		87.4	
59870							117.2	121.8	137.2	142.2		
59871							111.4	105.6	134.0	124.5		
59874								113.7		130.8	106.8	106.8
59876	22.9	24.1			79.7	83.0	116.0	117.0	131.4	134.6	110.3	110.9
59877								109.1		123.8		95.2
59891							112.6	114.1	128.9	126.4	107.8	107.8
59907							105.8	105.5	127.0	129.5		
59908							102.6	103.3	125.1	120.6	91.4	89.8

ID number	Posterior space of inlet, curved length, L	Posterior space of inlet, curved length, R	Posterior space of midplane, L	Posterior space of midplane, R	Posterior space of outlet, L	Posterior space of outlet, R	Linea terminalis, straight length, L	Linea terminalis, straight length, R	Linea terminalis, curved length, L	Linea terminalis, curved length, R	Depth of pelvis, L	Depth of pelvis, R
59912							109.1		131.4		96.5	
59913								110.2				
59918*												
59919							106.6		131.4		101.0	105.1
59921							103.4		123.8			
59922*												
59923								114.9		139.1		
59924								110.2		128.3		109.4
59925*								113.8		129.5		
59954							103.3		130.8			
59956	28.6	28.6					114.8	109.6	134.0			
59957								101.8		115.6		100.3
59958								103.4		125.1		99.7
59962*							107.0	107.2				
59963								115.2		131.4	107.5	102.0
59965								117.6		134.6	98.8	98.4
59972							102.0		119.4		98.4	
59977							110.8		129.5			
59978								116.6		137.2	98.2	104.6
59980								112.1		130.8		99.8
59982							112.8		132.7			
59985												
59986								104.8		124.5		
59987							101.5		117.5			95.0
59988	31.8	28.6			91.2	86.8		108.8			103.4	103.2
59990	25.4	22.2			72.7	71.7	104.3	102.7	123.2	120.0	102.8	99.6
59991								122.6		147.3		104.6
59994							102.4	101.4	121.9	122.6	102.9	100.9
59995							101.8	106.2	121.3	124.5		
59996												
59997							103.2	103.6	116.8	117.5	101.3	100.4
59998							109.9	110.8	137.8	132.1	99.4	101.0
60003											102.0	100.8
60004								116.3		133.4	102.3	102.9
60005								116.6		131.4		96.4
60006	22.9	24.8			67.0	70.5	95.5	99.6	118.7	121.3	95.2	95.3
60008												
60009								112.2		134.0	94.6	98.0
60010	22.9	21.0					94.9	96.9	109.9	111.8	93.0	

ID number	Posterior space of inlet, curved length, L	Posterior space of inlet, curved length, R	Posterior space of midplane, L	Posterior space of midplane, R	Posterior space of outlet, L	Posterior space of outlet, R	Linea terminalis, straight length, L	Linea terminalis, straight length, R	Linea terminalis, curved length, L	Linea terminalis, curved length, R	Depth of pelvis, L	Depth of pelvis, R
60011							111.3	111.0	132.7	137.2	101.5	100.8
60013*												
60017												103.4
60018											91.0	90.3
60019								112.1		133.4		109.6
60020							110.9		132.1			
60021							116.8		138.4			100.8
60022							108.5		129.5			
60025								113.3		133.4		
60026							110.1	111.0	130.8	132.1	101.8	101.3
60028							108.5		135.3			
60038								105.1		121.9	94.8	97.8
60039	18.4	22.9					106.0	108.1	120.6			
60040												
60041	19.7	20.3					112.2	102.2	126.4	123.8		91.8
60042											99.6	
60044								107.0		124.5	99.4	98.9
60046							110.8		125.1		99.2	103.2
60049	28.6	29.2				76.8		104.8		133.4		99.4
60050							118.1		144.8		101.0	
60051	19.0	17.8	47.2	44.5	64.0	59.8	96.0	98.0	109.9	110.5	99.8	97.8
60052							101.5	101.4	118.7	123.2		92.9
60055							103.1	107.4	122.6	125.1	91.6	91.6
60057								109.4		125.7		98.2
60060	29.2	28.6					118.2	115.2	137.2	135.3		
60061	24.8	24.1			72.7	73.7	111.5	113.9	130.8	132.1	101.6	100.2
60062	31.1	32.4			81.3	80.4					97.4	102.8
60063							109.8		128.9			
60064	23.5	24.8			71.7	77.1	111.5	109.3	130.2		105.4	105.5
60065							105.7		122.6		91.4	95.9
60066												99.6
60071	19.0	21.6			68.0	67.2	94.9	96.3	113.0	116.8	95.1	97.1
60074	24.8	28.6		54.5	64.1	77.4		108.3		127.0	102.2	107.8
60077								104.5		121.9		
60082							104.6		123.2			
60083							115.0		135.9		103.2	
60084	21.6	21.0			74.0	71.1	107.3	108.2	131.4	127.0	98.0	94.4
60086							104.6		118.1		101.4	97.8
60088	26.0	24.1			57.6	55.6	106.9	104.7	122.6	122.6	104.1	101.3

ID number	Posterior space of inlet, curved length, L	Posterior space of inlet, curved length, R	Posterior space of midplane, L	Posterior space of midplane, R	Posterior space of outlet, L	Posterior space of outlet, R	Linea terminalis, straight length, L	Linea terminalis, straight length, R	Linea terminalis, curved length, L	Linea terminalis, curved length, R	Depth of pelvis, L	Depth of pelvis, R
60089	29.2	28.6			71.6		100.6	101.3	123.2	123.8	94.5	
60092								113.1		132.7	107.3	106.0
60093							119.3		141.6			
60094	30.5	32.4			80.9	85.4	108.3	107.0	129.5	132.7	100.0	101.1
60096							113.0		138.4		106.7	
60100	30.5	26.0					99.6	101.2	127.0	123.8		
60101*												
60102	23.5	20.3			76.2	80.0	108.4	107.6	127.0	123.2	106.0	102.8
60106	23.5	26.7			68.0	66.6	107.7	106.5	122.6	125.1	106.8	108.9
60108	22.9	29.2			101.6	100.4	103.4	111.0	123.8	129.5	94.3	98.0
60109								110.8		124.5		98.9
60133							111.8	116.3	123.2	128.3		106.0
60134								126.4		144.1		
60138	27.3	26.7			92.0	81.0		108.6		131.4	96.1	99.6
60142							90.2		102.2			
60143	24.1	28.6			69.0	68.0	98.8	102.6		119.4	91.3	95.1
60147							125.4		146.7			110.2
60154							114.2	106.2	131.4	125.1	111.2	
60158	27.3					82.0	107.9		130.8			93.6
60160							115.1	119.8	143.5		107.0	112.1
60169	29.8	28.6			77.6		108.0	104.0	130.8	129.5	101.4	
60172							114.0		134.6			
60173	18.4	17.1					99.6	102.5	116.8	118.1		
60174							110.8		126.4		94.9	88.5
60176												101.8
60179												95.8
60180	19.7	19.7			66.8	72.1	106.1	106.9	128.3	121.9	100.3	99.5
60181												
60189								110.5		127.0		
60194	20.3	19.0			70.8	71.5	103.8	104.3	118.7	120.0	97.1	94.1
60197	29.8	28.6			68.2	63.0	105.8	107.3	129.5	133.4	98.7	96.5
60198												
60200											96.9	
60201												
60202							119.6	124.8	144.8	143.5	109.2	110.0
60210	31.8	30.5		64.0	78.8	78.8	114.4	114.8	141.0	141.0	102.4	101.8
60211	31.8	30.5					111.0	114.8	134.6	135.9	99.0	100.0
60212												
60216								124.6		145.4		106.2

ID number	Posterior space of inlet, curved length, L	Posterior space of inlet, curved length, R	Posterior space of midplane, L	Posterior space of midplane, R	Posterior space of outlet, L	Posterior space of outlet, R	Linea terminalis, straight length, L	Linea terminalis, straight length, R	Linea terminalis, curved length, L	Linea terminalis, curved length, R	Depth of pelvis, L	Depth of pelvis, R
60218							95.6	94.3	109.2		97.2	97.9
60219							105.6		119.4		102.9	99.0
60222								111.2		141.0		105.2
60224							106.0		122.6		102.2	105.3
60225											94.2	100.2
60226	29.2	27.9			77.6	77.7	108.0	110.0	129.5	129.5	102.4	103.4
60227*												
60228							113.2	107.6	132.1	134.6	95.1	
60229								105.8		128.9	98.7	95.8
60230*												
60232											121.3	
60233												103.6
60238	21.0	19.0			78.0	74.5	111.6	110.3	127.0	127.0	104.8	107.0
60239								110.7		127.6	96.0	
60241												92.0
60244							109.2		126.4		106.4	103.1
60247								105.3		123.2		88.3
60251	31.8	27.3		62.0		78.2	111.4	109.3	128.3	126.4		96.7
60252							119.8	119.7	143.5	143.5	107.2	109.2
60254							108.5		125.1		96.2	
60257	24.8	27.3					111.4		134.6			
60258	31.1	33.7			84.8	87.8	118.0	122.6	142.9	145.4	96.1	98.7
60259							120.0		141.0			101.5
60260							102.2		119.4		97.8	
60266*												
60267												
60272	31.1	34.9					112.0	109.1	132.7	134.6	100.9	100.8
60275							102.7		118.1		96.4	
60276*												
60277								108.1		128.3	100.9	
60278*												
60281											102.4	99.9
60283												
60284								98.7		118.7	94.2	93.8
60285	24.8	25.4			87.2	80.3	107.4	112.0	130.8	132.1	99.4	100.2
60286											110.1	109.0
60290	25.4	25.4						118.7		130.8	100.0	98.6
60291	22.9	27.3	60.4		74.7	77.0	102.2	102.2	121.9	125.1	93.6	93.2
60293												

ID number	Posterior space of inlet, curved length, L	Posterior space of inlet, curved length, R	Posterior space of midplane, L	Posterior space of midplane, R	Posterior space of outlet, L	Posterior space of outlet, R	Linea terminalis, straight length, L	Linea terminalis, straight length, R	Linea terminalis, curved length, L	Linea terminalis, curved length, R	Depth of pelvis, L	Depth of pelvis, R
60296							117.2		131.4		97.6	
60297							110.6	109.3	124.5	123.8	95.0	97.6
60298	31.1	34.3			89.5	90.5	112.4		133.4		98.8	97.4
60302	33.7	25.4					108.1	115.0	134.0	133.4		
60303*							119.4	117.4	144.8	145.4	108.0	107.6
60304*												
60306							114.0	110.3	123.2	123.2	98.3	100.2
60308*							102.7		127.6		98.9	95.4
60309*								117.7		141.6		
60311								117.6		137.8	97.0	99.0
60312	25.4	29.2				84.9		108.9		130.2		103.6
60313								114.4		135.9	103.7	107.2
60314	21.6	21.0				69.3	107.8	107.9	125.1	123.2		99.8
60315							100.0		118.7		94.6	90.7
60318	22.9	24.8					112.7	114.1	130.2	137.2		
60319*												
60428							101.5		123.8		90.6	90.9
60436	26.0	23.5			83.5	82.3	106.6		130.2		94.8	96.2
60437*							116.6		139.7		100.0	101.4
60451							118.8		137.8		102.5	102.6
60452											91.4	94.2
60454*												
60455	31.8	29.8			79.8	86.8		116.9		136.5	95.6	95.0
60456*												
60457								110.4		131.4		
60458	23.5	19.7	58.7		72.0	72.5	107.7		126.4		108.2	104.9
60480	22.9						104.2		122.6			
60481								102.8		123.8		98.7
60482							105.9	105.3		118.1	100.8	
60483								110.0		122.6	93.8	94.6
60484*												104.6
60488											104.6	
60496	31.1	30.5				91.0	120.8	119.0	141.6	139.7		103.2
60511								112.1		135.3		
60513								115.5		132.7	111.1	
60514	21.0	23.5			83.2	84.7	104.2	107.2	120.0	123.8	100.3	102.2
60518											102.4	103.7
60525								115.6		137.2		
60526											94.8	

ID number	Posterior space of inlet, curved length, L	Posterior space of inlet, curved length, R	Posterior space of midplane, L	Posterior space of midplane, R	Posterior space of outlet, L	Posterior space of outlet, R	Linea terminalis, straight length, L	Linea terminalis, straight length, R	Linea terminalis, curved length, L	Linea terminalis, curved length, R	Depth of pelvis, L	Depth of pelvis, R
60527								104.6		124.5		
60528												
60529							102.4		116.2			
60530	21.0	21.6	60.6		81.6	72.1	100.0	104.5	116.2	120.6	97.1	101.0
60533	29.8	28.6				86.0	116.2		133.4			97.5
60541							105.4		119.4		96.4	94.0
60542								113.4		128.9	100.5	98.1
60543	27.9	33.0					102.7	108.4	126.4	125.1		
60545												96.9
60550								106.3		130.8		
60554							113.2		130.8		109.3	110.2
60555							107.8		126.4			
60556	21.0	22.9			62.4	60.6	93.6	93.4	108.0	112.4	91.4	94.6
60558*	22.9	20.3					100.9	100.0	117.5	116.8	99.4	95.0
60560							117.9		135.9		103.6	
60562								113.9		132.1	106.9	108.1
60563	33.7	33.0				76.0		117.6		142.9		102.4
60573								100.9		115.6		97.0
60574							106.4		123.2		94.1	
60578								113.2		131.4		
60588	25.4	30.5					112.6	111.1	131.4	132.7		
60590	30.5	33.0	67.4		84.6	91.8	111.3	108.4	130.2	127.6	94.9	
60591							111.0	108.2	134.6			
60594							96.7		123.2			
60597	31.8	31.1					121.2		141.6		104.8	
60599								93.0		106.0	88.9	90.2
60601								114.2		130.8	101.0	99.8
60602							121.2		137.2		107.7	106.4
60608	25.4	23.5			74.5			102.4		124.5	93.6	
60609	34.3	34.9			92.1	89.0	116.5	116.5	137.8	135.3		103.3
60610							113.4	113.0	133.4	134.0		
60611*												
60612								113.9		137.2		
60614								108.8		130.8		
60616								100.7		113.7		
60620*								129.2		147.3	107.6	
60633	30.5	30.5		49.7	70.4	70.8	123.7	125.0	146.7	147.3	116.1	119.8
60634	31.8	29.2			71.0	71.5	116.0	115.6	134.0	134.6	107.6	109.0
60647												

ID number	Posterior space of inlet, curved length, L	Posterior space of inlet, curved length, R	Posterior space of midplane, L	Posterior space of midplane, R	Posterior space of outlet, L	Posterior space of outlet, R	Linea terminalis, straight length, L	Linea terminalis, straight length, R	Linea terminalis, curved length, L	Linea terminalis, curved length, R	Depth of pelvis, L	Depth of pelvis, R
60662	21.0	17.8					108.0	113.4	126.4	129.5		
60663	18.4				67.8	60.4					93.9	
60664								117.5		141.0	104.6	100.9
60665								110.5		123.2	102.9	103.0
60669	17.1	24.1			63.8		93.1	98.1	107.3	114.3	94.4	
60671							106.4		127.0		99.6	100.6
60674	29.8	31.8			84.0	78.8	103.8	105.0	130.8	133.4	95.9	94.8
60675	24.8	24.8		59.0			106.9	105.2	125.7	125.1	103.2	101.4
60676							116.4		134.6			
60677	22.9	23.5			66.6	70.7	115.1	114.7	128.9	130.2	102.5	108.0
60678								110.5		132.1	99.2	102.3
60682								108.1		121.9		98.5
60685							99.9	105.1	122.6	124.5	87.8	90.9
60687	24.8	27.9			65.5	69.4		111.8		129.5	107.1	107.7
60689							114.2	116.4	135.9	131.4	106.7	103.7
60691	27.3	27.9			70.1	66.4		106.4		128.3	105.0	102.8
60693								113.2				107.6
60695								97.4		118.1	99.8	94.0
60696												
60703								109.0		132.1		92.5
60706	27.3	24.8					112.3	113.3	130.8	131.4	98.4	97.4
60713	28.6	28.6			86.4		121.4	120.1	139.1	137.2	103.4	
60717								102.1		118.7	91.4	91.5
60722							120.2	118.2		138.4	101.2	103.1
60725								112.5		133.4		
60727								115.8		127.0		93.1
60729	25.4	27.9					103.7	104.1	123.2	118.1		
60733											104.6	
60736	21.0	22.2	55.2		69.0	75.6	108.4	103.6	121.9	118.1	88.9	89.2
60737	22.9	22.2			75.0	70.8	101.4	101.7	119.4	116.8	102.7	101.6
60739											90.8	91.8
60740												
60741	12.7	14.6			68.2	65.8	97.4	98.5			97.4	98.8
60743	24.1	24.1			65.6	67.4					107.5	105.6
60744	31.8	31.8	81.3		101.6	107.3	119.8	116.4	139.1	138.4	102.0	100.7
60754	26.0		57.8		83.9	76.9	103.3		116.8		100.2	
60755							128.0	125.9		147.3	98.4	96.6
60773							101.0		116.8		93.3	
60779							120.1			144.1	105.2	

ID number	Posterior space of inlet, curved length, L	Posterior space of inlet, curved length, R	Posterior space of midplane, L	Posterior space of midplane, R	Posterior space of outlet, L	Posterior space of outlet, R	Linea terminalis, straight length, L	Linea terminalis, straight length, R	Linea terminalis, curved length, L	Linea terminalis, curved length, R	Depth of pelvis, L	Depth of pelvis, R
60782	25.4	26.7			84.4	86.6		105.1		124.5	94.0	96.1
60888												
60902	24.8	27.3	52.6	54.4	61.5	62.5	108.0	109.2	126.4	129.5	102.4	102.5
60903							109.0		133.4		100.6	
60909												
60911								137.3		155.6	117.7	117.0
60912							107.3	105.7	130.2	126.4		
60930	26.7	31.1					102.0	105.0	123.2	128.9		95.0
60931							104.3		131.4		97.4	96.5
60935	24.1	26.0	55.6		76.4	72.1	101.2	100.7	121.9		98.6	96.9
60984								127.0		146.7	102.1	104.0
60986							106.2		125.7		103.3	

PART 4: METRICS FOR PUBIC HEIGHT, SUBPUBIC ANGLE, FEMORAL AND CLAVICULAR LENGTHS, AND FEMORAL HEAD DIAMETER

ID number	Height of pubis, L	Height of pubis, R	Subpubic angle, L	Subpubic angle, R	Femoral length, L	Femoral length, R	Femoral head diameter, L	Femoral head diameter, R	Clavicular length, L	Clavicular length, R
59801	34.4	31.6	29.00	32.50		416	42.8		153	
59803	35.8		43.50	47.50	399		38.6			
59805						434				
59809	33.8				423	422	38.4			
59814		33.4			381		35.4	34.8	124	
59815		38.2		41.00		390		39.2	130	
59817	36.4	35.2	46.00	35.00		381		38.2	134	134
59818	32.1						39.7	38.8		
59822*	37.2	37.1	37.50	42.00			41.6	42.3	135	135
59823		41.7	35.00	31.00	415		46.4			148
59824			47.00		387		37.7	38.0		129
59828*					379		38.6	38.0		
59829					398	391				
59833										
59834				37.00	355	345		36.7		
59838*		33.7					43.8			
59839*						447	41.8			
59841	33.5	31.4	40.00	44.00		386	36.0	35.2	140	139
59845	32.1		47.00			397		39.9	139	134
59846		34.6	36.00	45.50		406		38.4	144	144
59849*					379	381	36.7			
59852*								45.0		
59856				38.00						
59857			41.00				38.2		131	
59861			51.50		391	391	39.8			128
59862		33.1		26.00	429		44.8		150	151
59865						411	44.0	44.0		
59866	36.4	38.0	33.00	32.50	412	412	45.6	46.0		136
59867	28.5		51.00							
59870				28.00						
59871			45.00	44.00	402	399	38.6	37.6	145	
59874		39.6		31.00			45.7			
59876	37.6	37.4	30.00	27.00	446	445	48.6	47.8		
59877		34.8			355	355	37.1	36.2		
59891			33.00	28.00	456	452	45.6	47.2	148	147
59907				37.50		459		46.8	163	165
59908	37.6	34.8	34.00	40.50	392	392	38.2	38.6		136

ID number	Height of pubis, L	Height of pubis, R	Subpubic angle, L	Subpubic angle, R	Femoral length, L	Femoral length, R	Femoral head diameter, L	Femoral head diameter, R	Clavicular length, L	Clavicular length, R
59912	34.3		37.75		383		38.6			
59913				53.00		416				
59918*						438		47.7		
59919							46.0			141
59921			30.00		432		42.6	42.2	158	153
59922*						407		38.3		
59923					398					
59924		38.6		34.50	421		48.2			
59925*				44.00	401	403	41.1			
59954			36.00		430		45.4			
59956							44.6	44.6		
59957		37.9		36.00			45.6		152	
59958		40.5		26.00	408	407	41.5	41.0	146	145
59962*										
59963		33.1			434		44.1		148	148
59965		35.2		42.00	407	406	40.0	39.7	140	140
59972	33.9		24.00		412	410	45.4	45.6		
59977					380	379				
59978		34.1		46.50	391	385	38.5	38.5	140	
59980				46.00	412	411		39.8	130	
59982			26.00		457	458	44.8	43.0		
59985			41.00		401		41.6			
59986					397		36.0	36.7		
59987						430	42.6	42.7		
59988		40.9		48.00		397		41.6		
59990	37.4	37.5	30.00	26.00	448		46.6			
59991							41.0	39.7		
59994	42.0	39.2	28.00	28.00	442	440	44.8	44.3	152	151
59995						379		36.4		
59996				47.50		366				
59997	28.6	28.4	29.00					41.7		
59998	40.4	36.0	35.00	34.50	429		46.2			
60003					417		40.0			138
60004		37.0	46.00	37.00	412	409		39.8	138	135
60005		33.0		41.50	390		40.7		131	
60006	32.2	31.9	42.50	39.50		388	38.2	38.7	129	118
60008				46.50				36.8		
60009		36.2		37.50	399	400	39.4	38.6	136	
60010	32.2		26.00			403		41.3		
60011					398		38.4	37.7		

ID number	Height of pubis, L	Height of pubis, R	Subpubic angle, L	Subpubic angle, R	Femoral length, L	Femoral length, R	Femoral head diameter, L	Femoral head diameter, R	Clavicular length, L	Clavicular length, R
60013*					393					
60017						397		40.2		
60018			29.50	26.00	423	419	45.8	44.5	147	
60019		39.0								149
60020			55.00		402			38.3		131
60021	33.2		40.50	40.00	380		39.7			125
60022			35.00		425	424	43.8	43.2	157	151
60025				40.50	386		39.2	38.7		
60026	38.9	38.2	38.00			444		45.5		153
60028					410	413	41.7	40.4		139
60038		32.4		32.00	395	399	40.1	40.2	148	146
60039					415				143	
60040			43.00		382	381	39.6	39.1		
60041		36.5	43.00	44.00	413		40.1			
60042					426					
60044		31.7		35.00	421				155	154
60046	37.2		41.00		438		47.2			
60049		32.9		42.50	392		40.0		132	
60050	38.5		39.00		409		41.8			143
60051	31.3	31.5	30.00	29.00	418	418	44.1	44.8		
60052		32.5	41.50	38.00	388	388	38.2			
60055	33.1	40.0	48.50	49.50	368	368	35.4	35.7		
60057		37.0		43.50	401		39.2			
60060				40.00	406	403		38.6		131
60061	33.2	35.0	38.50	39.50	402		42.2		139	131
60062					385		41.5			
60063					404		40.0		133	123
60064	39.5	39.0	38.50	40.50		457		45.2	154	151
60065	30.3		33.00		427		39.5			
60066								37.6		
60071	36.0	36.3	29.50	29.50	392	390	43.2	42.6	140	
60074		35.8	32.00	31.00	448	443	44.5	44.3	153	
60077						393				
60082					402	402	38.0	36.8	133	
60083			50.00							
60084	34.8	37.0	41.50	38.50		386		40.5	132	129
60086	37.4		32.00		430		42.0		158	159
60088	37.6	35.4	32.00	29.00	423	425	42.3	42.8	145	149
60089	33.8		44.50		374	375	37.0		129	
60092		35.4		30.00	429	429	44.4	43.8	149	

ID number	Height of pubis, L	Height of pubis, R	Subpubic angle, L	Subpubic angle, R	Femoral length, L	Femoral length, R	Femoral head diameter, L	Femoral head diameter, R	Clavicular length, L	Clavicular length, R
60093					423	418				
60094	36.2	35.0	50.50	42.00		395	39.2			141
60096	36.4		28.00		445	444	44.4			
60100			50.00	48.00	386	385	35.3	35.4	125	126
60101*					395	393	39.1	38.9	132	
60102	35.0	35.6	33.50	31.50	456	458	44.6	45.2	147	
60106	33.1	34.9	32.00	29.00	436	428	43.6	43.6		
60108	33.8	33.0	42.50	45.50	372		37.2			
60109		31.1		31.00		435		43.8		149
60133		32.6	34.00	32.00	457					
60134					415	410	38.2	38.1		
60138		36.1	41.50	42.00				38.9		
60142										
60143	31.7	35.5		50.00	380					
60147			40.50					39.8		
60154					441	435	44.5			
60158			47.00	48.50				41.3		
60160	38.4						41.2		147	
60169					373		41.0	38.7		
60172			36.00		412	412	40.8	40.5		
60173				43.00				37.8		
60174	36.6		37.00		367		39.7	39.7		
60176										
60179								40.6	147	
60180	32.8	34.2	26.00	27.00		432		43.7		
60181			39.00				45.3			
60189				33.00	433		45.0			
60194	36.0	36.0	30.00	29.00		422		42.8		
60197	41.0	38.5	29.00	33.00		394		42.0		144
60198				33.50	444	438				
60200										
60201			34.00		438		41.4			
60202			42.00							
60210	35.2	36.0	35.00	39.50			39.1	39.2		140
60211	40.8	40.1	42.00	47.50	401		44.2			
60212						430		43.2		147
60216		38.4		34.50	414		41.6			134
60218	34.4	36.1	35.50	34.50		427		43.2		147
60219	36.9		30.50		427	422				
60222		35.6			415		39.7			

ID number	Height of pubis, L	Height of pubis, R	Subpubic angle, L	Subpubic angle, R	Femoral length, L	Femoral length, R	Femoral head diameter, L	Femoral head diameter, R	Clavicular length, L	Clavicular length, R
60224	35.4		31.00				43.1	42.7		
60225						375				127
60226	43.2	43.1	30.00	24.00	422	425	44.4	43.9	146	149
60227*					404	400	39.9		139	
60228	31.9		38.00	35.50		408		36.3		
60229		36.5		33.00				40.0		
60230*					431	435				
60232								46.9		
60233								41.1		
60238	34.1	32.5	30.00			470			154	
60239			52.50		376	379		36.8		
60241			47.25			394	38.2	37.8	134	130
60244	36.2		31.00	26.00			45.4		163	158
60247		31.8			376		37.3			
60251	32.4	35.1	42.00	37.50		364				
60252	34.0	35.8	54.00		433				151	149
60254	38.8		30.00				47.1			
60257					390					
60258	35.1	43.8	46.83		417	417	39.8	39.4	146	
60259	37.9		46.00			429				
60260	36.8		31.00		413		42.7		145	
60266*						411		43.1		
60267			26.00	23.00	444		48.4		157	155
60272	37.1	36.8	54.50	51.50		397		37.3	136	129
60275	38.2	38.1		30.00			43.7		143	139
60276*					415	414	41.8	42.5	150	151
60277				30.00	420		43.1			
60278*										
60281							39.7			
60283						420		39.8		
60284		34.2		47.50	364	362	38.5	37.5	128	124
60285	33.9	35.8	50.00	44.00		386		39.4		
60286			33.00						168	170
60290		39.9		39.50			38.9			
60291	37.0	30.9		41.00	369					
60293				28.00	434		42.6			
60296	37.4				463		43.5		158	
60297	37.3	34.8	31.00	30.00		379		34.8		
60298	36.1		42.00	37.00	396		38.9			130
60302			44.50	34.50	391		39.3			

ID number	Height of pubis, L	Height of pubis, R	Subpubic angle, L	Subpubic angle, R	Femoral length, L	Femoral length, R	Femoral head diameter, L	Femoral head diameter, R	Clavicular length, L	Clavicular length, R
60303*	36.5	36.4	33.50	34.00	397	397	40.6	40.7	132	134
60304*							39.3			
60306	37.4				459	450				155
60308*						385		36.6		
60309*				37.00		401		40.0		136
60311		33.0			399					
60312		37.1		47.50	398		38.0	37.5		
60313		38.1		25.00		460		43.9	156	
60314	36.6	37.4		34.00		420		44.2		
60315	35.7		29.00		409		43.7			155
60318				32.00	459	460	42.8	43.1	160	
60319*					426	420	40.2			146
60428	32.0		53.00				39.2		126	
60436	32.8		43.50	41.50				40.1		
60437*	36.9		41.00		410	398	42.0	41.8		
60451	34.5		34.00		389		40.9	38.8	138	
60452						383				
60454*					447				149	145
60455		34.1	42.50	40.50	395	394	37.8	39.3	132	136
60456*							45.3			
60457										
60458	33.8		31.50	28.00		444		44.9	156	155
60480			45.50	41.50						
60481		37.8	45.00	43.00				39.7		
60482	35.9	37.4				412		42.4	149	
60483		33.6		34.00		435		42.6		
60484*					444					
60488			23.00	27.00						
60496		42.8	38.00	43.75	393	386	39.8	39.2		
60511				41.25						
60513		36.6		26.00		473		47.0	170	
60514	34.4	34.0	37.00		407		42.0			
60518			28.00		422				146	
60525		37.4		39.50		456				
60526										
60527					423		40.0			
60528				31.00						
60529						354			124	122
60530		32.7		30.00		424		43.6	139	139
60533										

ID number	Height of pubis, L	Height of pubis, R	Subpubic angle, L	Subpubic angle, R	Femoral length, L	Femoral length, R	Femoral head diameter, L	Femoral head diameter, R	Clavicular length, L	Clavicular length, R
60541	32.6						42.3			
60542		43.5		41.50		382		40.5		
60543					400		42.2			
60545					427	428	42.8	42.2	150	154
60550					438					
60554	33.8		27.50		425		44.0		147	
60555							42.1			
60556		27.9		30.00		377		40.0		
60558*	34.6	35.5	39.00	28.00			41.5	40.8		
60560	34.8		28.00				44.2			150
60562			35.00	35.00	421		43.4		149	
60563		37.6	47.00	47.50					148	143
60573		32.9		24.00	417		42.2		151	155
60574	37.4		46.50		381	376				
60578				43.00			39.2			
60588				45.50	410	404	38.8	39.7		
60590	39.7	35.2	49.50	44.50		352		37.5	131	
60591			43.50	39.50	392	382	37.6	36.4		
60594			40.50		365					
60597	36.5				408	405	41.2			
60599		33.7		30.50	390	388	38.6	39.6		
60601		35.2		43.00	408	405	38.0	38.6	137	
60602	40.6		28.00		450		44.4			
60608			41.00	34.00			38.6		127	
60609	34.8	36.4	41.50	43.50		376		38.8	134	129
60610			42.50	36.50	399		43.4			132
60611*										
60612							41.5			
60614										
60616										
60620*						430		42.7		
60633	36.1	38.0	29.00	30.00					153	161
60634	38.0	35.2	32.00	35.00		428		46.6	145	146
60647				46.00	426	426		40.9		
60662			26.50	26.50						
60663			36.00			419		45.7	143	142
60664		39.2		49.00				41.8		
60665		36.7		27.00			43.8	43.8		
60669	34.0	35.9	39.00	40.00		420		42.4	145	
60671	37.2		47.00			380				

ID number	Height of pubis, L	Height of pubis, R	Subpubic angle, L	Subpubic angle, R	Femoral length, L	Femoral length, R	Femoral head diameter, L	Femoral head diameter, R	Clavicular length, L	Clavicular length, R
60674	34.6		52.50	40.50		404		40.0	129	127
60675	35.0	33.7	33.00					42.7	145	
60676			36.50							
60677	43.0	42.1	31.00	26.00	440		44.8			
60678		32.5		23.00		443		47.3		
60682		40.6		38.00						
60685	31.4	30.7	49.50	44.00	378				125	
60687		35.2	24.00	21.50	407					
60689	35.7	36.6	32.00	26.00						
60691		33.5		31.00	441	441	46.9		155	161
60693			33.00	34.50						
60695		35.3		27.00	408				145	
60696			30.00		402		41.6		128	128
60703		29.5		44.00		398		40.8		136
60706	32.4	33.2	37.00	36.00		373		37.7		125
60713			46.50							
60717					410	412	42.7	42.1		
60722	36.9	39.6		48.00		420		41.4		
60725					429	431	45.8			
60727		34.2			406	400		40.2		
60729			26.00	29.00						
60733						447		45.1		
60736	29.5	30.1	39.50	42.00	376	380	35.6			
60737	35.1	35.2	35.00	29.00						
60739			39.25	27.00				36.6		
60740					388					
60741	32.7	33.1	29.00	27.00	385		45.7	46.0	152	
60743			22.00	20.00	451	451	44.2	43.6	164	143
60744	36.7	37.2	42.50	47.50		397		41.4		
60754	38.2		35.00						151	
60755	34.6	34.4	46.25	40.25	392	390	39.1	40.0		
60773	31.1									
60779	38.1				423	426	42.0	41.7	141	134
60782		32.4	46.00	42.50	397	397	39.8	39.9		
60888				38.00					137	136
60902	37.3	38.2	27.00	32.00		437		43.6		
60903	37.1		37.00							
60909			31.00						148	154
60911		37.4		39.25		404		43.2		136
60912				25.00		418		42.0		

ID number	Height of pubis, L	Height of pubis, R	Subpubic angle, L	Subpubic angle, R	Femoral length, L	Femoral length, R	Femoral head diameter, L	Femoral head diameter, R	Clavicular length, L	Clavicular length, R
60930		31.0		41.00	349		36.2			118
60931	36.0					372	36.5	37.2		
60935	33.5	33.2	36.00	36.00		405		41.5		
60984		36.8		46.50			41.1	41.4		
60986	35.6				384	389		38.5	135	133

References

Abbott, S., E. Trinkaus, and D. B. Burr
1996 "Dynamic Bone Remodeling in Later Pleistocene Fossil Hominids." *American Journal of Physical Anthropology* 99:585–601.

Acosta-Sison, H., and F. Calderon
1919 "Pelvimetry and Cephalometry among Filipino Women and Newborn Babies." *Philippine Journal of Science* 14:253–271.

Adadevoh, S. W. K., C. Hobbs, and T. E. Elkins
1989 "The Relation of the True Conjugate to Maternal Height and Obstetric Performance in Ghanaians." *International Journal of Gynecology and Obstetrics* 28:243–251.

Adams, E. B.
1954 *Bishop Tamarón's Visitation of New Mexico, 1760.* Historical Society of New Mexico, Albuquerque.

Adams, E. B., and Fray A. Chavez, eds.
1956 *The Missions of New Mexico, 1776: A Description by Fray Francisco Atanasio Dominguez with Other Contemporary Documents.* University of New Mexico Press, Albuquerque.

Adams, E. C.
1991 *The Origin and Development of the Pueblo Katsina Cult.* University of Arizona Press, Tucson.

Adams, E. C., and A. I. Duff, eds.
2004 *The Protohistoric Pueblo World, A.D. 1275–1600.* University of Arizona Press, Tucson.

Adler, M. A., and H. Dick
1999 *Picuris Pueblo through Time: Eight Centuries of Change at a Northern Rio Grande Pueblo.* William P. Clements Center for Southwestern Studies, Southern Methodist University, Dallas.

Adler, M. A., T. V. Pool, and R. D. Leonard
1996 "Ancestral Pueblo Population Aggregation and Abandonment in the North American Southwest." *Journal of World Prehistory* 10(3):375–438.

Ahlstrom, R. V. N., C. Van West, and J. S. Dean
1995 "Environmental and Chronological Factors in the Mesa Verde–Northern Rio Grande Migration." *Journal of Anthropological Archaeology* 14:125–142.

Allen, E. P.
1944 "Radiological Pelvimetry: A New Zealand Study." *New Zealand Medical Journal* 43: 116–129.

Allen, R.
1992 "The Use of Shellfish and Shell Beads at Santa Cruz Mission." *Pacific Coast Archaeological Society Quarterly* 28(2):18–34.

Ambrose, S., and L. Norr
1993 "Experimental Evidence for the Relationship of the Carbon Stable Isotope Ratios of Whole Diet and Dietary Protein to Those of Bone Collagen and Carbonate," in *Prehistoric Human Bone: Archaeology at the Molecular Level,* J. B. Lambert and G. Grupe, eds., pp. 1–38. Springer-Verlag, Berlin.

Anderson, G. C.
1999 *The Indian Southwest, 1580–1830: Ethnogenesis and Reinvention.* University of Oklahoma Press, Norman.

Arriaza, B., M. Allison, and E. Gerszten
1988 "Maternal Mortality in Pre-Columbian Indians of Arica, Chile." *American Journal of Physical Anthropology* 77:35–41.

Bailey, V.
1913 "Life Zones and Crop Zones of New Mexico," in *North American Fauna,* pp. 1–100. U.S. Bureau of Biological Survey, no. 35. Department of Agriculture, Washington, D.C.

Bandelier, A. F.
1881 "Report on the Ruins of the Pueblo of Pecos." Papers of the Archaeological Institute of America, American Series, vol. 1, pp. 37–133. A. Williams and Co., Boston.

Barnes, E.
1994 *Development Defects of the Axial Skeleton in Paleopathology.* University of Colorado Press, Niwot.

Bass, S., P. D. Delmas, G. Pearce, E. Hendrich, A. Tabensky, and E. Seeman
1999 "The Differing Tempo of Growth in Bone Size, Mass, and Density Is Region Specific." *Journal of Clinical Investigation* 104:795–804.

Baugh, T. G.
1984 "Southern Plains Societies and Eastern Frontier Pueblo Exchange during the Protohistoric Period," in *Collected Papers in Honor of Harry L. Hadlock,* N. L. Fox, ed., pp. 157–166. Papers of the Archaeological Society of New Mexico 9. Albuquerque.

Beck, L.
2006 "Kidder, Hooton, Pecos, and the Birth of Bioarchaeology," in *Bioarchaeology: The Contextual Analysis of Human Remains,* J. E. Buikstra and L. A. Beck, eds., pp. 83–94. Elsevier, Amsterdam.

Beck, T., A. Looker, C. Ruff, H. Sievanen, and H. Wahner
2000 "Structural Trends in the Aging Femoral Neck and Proximal Shaft: Analysis of NHANES III DXA Data." *Journal of Bone and Mineral Research* 15:2297–2304.

Beck, T. J., C. B. Ruff, A. C. Looker, and F. A. Mourtada
1996 "Ethnic and Sex Differences in Age Trends of Proximal Femur Geometry from NHANES III Hip Dxa Data." *Journal of Bone and Mineral Research* Suppl. 1:61.

Bello, S. M., A. Thomann, M. Signoli, O. Dutour, and P. Andrews
2006 "Age and Sex Bias in the Reconstruction of Past Population Structures." *American Journal of Physical Anthropology* 129:24–38.

Bennett, K. A.
1973 *The Indians of Point of Pines, Arizona; A Comparative Study of Their Physical Characteristics.* Anthropological Papers of the University of Arizona 23. University of Arizona Press, Tucson.

Bernard, R. M.
1952 "The Shape and Size of the Female Pelvis." *Transactions of the Edinburgh Obstetrical Society* 59:1–16.

Bernardini, W.
1998 "Conflict, Migration, and the Social Environment: Interpreting Architectural Change in Early and Late Pueblo IV Aggregation," in *Migration and Reorganization: The Pueblo IV Period in the American Southwest,* K. Spielmann, ed., pp. 91–105. Arizona State University Anthropological Research Papers 51. Tempe.

Berry, D. R.
1985 "Dental Paleopathology of Grasshopper Pueblo, Arizona," in *Health and Disease in the Prehistoric Southwest,* C. F. Merbs and R. J. Miller, eds., pp. 253–274. Arizona State University Anthropological Research Papers 34. Tempe.

Boutton, T. W., T. A. Harrison, and B. N. Smith
1980 "Distribution of Biomass of Species Differing in Photosynthetic Pathway along an Altitudinal Transect in Southeastern Wyoming Grassland." *Oecologia* 45:287–298.

Bowers, B.
1966 "Field Report: 1964 Excavations of the Davis and Larson Sites, near Mobridge,

South Dakota." *Plains Anthropologist* 12: 204–205.

Brew, J. O.
1937 "The First Two Seasons at Awatovi." *American Antiquity* 3(2):122–137.

Brewer-Carias, C., S. A. LeBlanc, and J. V. Neel
1976 "Genetic Structure of a Tribal Population, the Yanamama Indians, 13, Dental Microdifferentiation." *American Journal of Physical Anthropology* 44(1):5–14.

Bridges, P. S.
1985 "Changes in Long Bone Structure with the Transition to Agriculture: Implications for Prehistoric Activities." Ph.D. dissertation, Department of Anthropology, University of Michigan, Ann Arbor.
1989 "Changes in Activities with the Shift to Agriculture in the Southeastern United States." *Current Anthropology* 30:385–394.

Bridges, P. S., J. H. Blitz, and M. C. Solano
2000 "Changes in Long Bone Diaphyseal Strength with Horticultural Intensification in West-Central Illinois." *American Journal of Physical Anthropology* 112(2):217–238.

Brock, S. L. and C. B. Ruff
1988 "Diachronic Patterns of Change in Structural Properties of the Femur in the Prehistoric American Southwest." *American Journal of Physical Anthropology* 75:113–127.

Brown, D. N.
1999 "Picuris Pueblo in 1890: A Reconstruction of Picuris Social Structure and Subsistence Activities," in *Picuris Pueblo through Time: Eight Centuries of Change at a Northern Rio Grande Pueblo,* M. A. Adler and H. Dick, eds., pp. 19–41. William P. Clements Center for Southwestern Studies, Southern Methodist University, Dallas.

Buccimazza, S. S., C. D. Molteno, T. T. Dunne, and D. L. Viljoen
1994 "Prevalence of Neural-Tube Defects in Cape-Town, South Africa." *Teratology* 50(3):194–199.

Buikstra, J. E., and D. H. Ubelaker, eds.
1994 *Standards for Data Collection from Human Skeletal Remains: Proceedings of a Seminar at the Field Museum of Natural History.* Arkansas Archaeological Survey Research Series 44. Fayetteville.

Burr, D. B., C. B. Ruff, and D. D. Thompson
1990 "Patterns of Skeletal Histologic Change through Time: Comparison of an Archaic Native American Population with Modern Populations." *Anatomical Record* 226:307–313.

Caldwell, W. E., and H. C. Moloy
1933 "Anatomical Variations in the Female Pelvis and Their Effect in Labor with a Suggested Classification." *American Journal of Obstetrics and Gynecology* 26:479–505.
1938 "Anatomical Variations in the Female Pelvis: Their Classification and Obstetrical Significance." *Proceedings of the Royal Society of Medicine* 32:1–30.

Cameron, C. M.
1995 "Migration and Movement of Southwestern Peoples." *Journal of Anthropological Archaeology* 14:104–124.

Caperton, T.
1981 "An Archaeological Reconnaissance of the Gran Quivira Area," in *Contributions to Gran Quivira Archeology,* A. Hayes, ed., pp. 3–11. National Park Service Publications in Archeology 17. Washington, D.C.

Carter, D. R.
1978 "Anisotropic Analysis of Strain Rosette Information from Cortical Bone." *Journal of Biomechanics* 11:199–202.

Cavagnaro, J. B.
1988 "Distribution of C3 and C4 Grasses at Different Altitudes in a Temperate Arid Region of Argentina." *Oecologia* 76:273–277.

Cavalli-Sforza, L., P. Menozzi, and A. Piazza
1994 *The History and Geography of Human Genes.* Princeton University Press, Princeton, New Jersey.

Cavalli-Sforza, L. L., E. Minch, and J. L. Mountain
1992 "Coevolution of Genes and Languages Revisited." *Proceedings of the National Academy of Sciences* 89:5620–5624.

Chamberlin, M.
2008 "Evaluating the Cultural Origins of Complexity in the Ancestral Pueblo World." Ph.D. dissertation, School of Human Evolution and Social Change, Arizona State University, Tempe.

Chen, H.-Y., Y.-P. Chen, L.-S. Lee, and S.-C. Huang
1982 "Pelvimetry of Chinese Females with Special Reference to Pelvic Type and Maternal Height." *International Surgery* 67:57–62.

Chun, D., P. S. Kan, H. C. Ho, and C. P. Wong
1964 "The Hong Kong Chinese Female Bony Pelvis and Its Influence on Labour: A Radiographic and Clinical Study of 1,005 Women." *Journal of Obstetrics and Gynaecology of the British Commonwealth* 71:556–564.

Cleveland, W. S.
1979 "Robust Locally Weighted Regression and Smoothing Scatterplots." *Journal of the American Statistical Association* 74:829–836.

Cordell, L. S.
1997 *Archaeology of the Southwest.* Academic Press, San Diego.
1998 *Before Pecos: Settlement Aggregation at Rowe, New Mexico.* Maxwell Museum of Anthropology, Anthropological Papers 6. University of New Mexico Press, Albuquerque.

Cordell, L., D. Doyel, and K. Kintigh
1993 "Processes of Aggregation in the Prehistoric Southwest," in *Themes in Southwest Prehistory,* G. Gumerman, ed., pp. 109–134. School of American Research Press, Santa Fe, New Mexico.

Cordell, L. S., and D. D. Fowler, eds.
2005 *Southwest Archaeology and the Twentieth Century.* University of Utah Press, Salt Lake City.

Cosgrove, H. S., and C. B. Cosgrove
1932 *The Swarts Ruin: A Typical Mimbres Site in Southwestern New Mexico.* Papers of the Peabody Museum of American Archaeology and Ethnology, vol. 15, no. 1. Peabody Museum of Archaeology and Ethnology, Harvard University, Cambridge, Massachusetts.

Crown, P. L.
1994 *Ceramics and Ideology.* University of New Mexico Press, Albuquerque.

Crown, P. L., J. D. Orcutt, and T. A. Kohler
1996 "Pueblo Cultures in Transition: The Northern Rio Grande," in *The Prehistoric Pueblo World,* A.D. *1150–1350,* M. Adler, ed., pp. 188–204. University of Arizona Press, Tucson.

Cunningham, F. G., N. F. Gant, K. J. Leveno, L. C. Gilstrap III, J. C. Hauth, and K. D. Wenstrom
2001 *Williams Obstetrics.* 21st ed. McGraw-Hill, New York.

Cyron, B. M., W. C. Hutton, and J. D. G. Troup
1976 "Spondylolytic Fractures." *Journal of Bone and Joint Surgery* 58-B(4):462–466.

Dahlberg, A. A.
1956 "Materials for the Establishment of Standards for Classification of Tooth Characteristics, Attributes, and Techniques in Morphological Studies of the Dentition." Ph.D. dissertation, Department of Anthropology, University of Chicago.

Davivongs, V.
1963 "The Pelvic Girdle of the Australian Aborigine: Sex Differences and Sex Determination." *American Journal of Physical Anthropology* 21:443–455.

Deagan, K.
1985 "Spanish-Indian Interaction in Sixteenth-Century Florida and Hispaniola," in *Cultures in Contact,* W. Fitzhugh, ed., pp. 281–318. Smithsonian Institution Press, Washington, D.C.
2001 "Dynamics of Imperial Adjustment in Spanish America: Ideology and Social Integration," in *Empires: Perspectives from Archaeology and History,* S. Alcock, T. D'Altroy, K. Morrison, and C. Sinopoli, eds., pp. 179–194. Cambridge University Press, Cambridge.

Dean, J. S.
1996 "Demography, Environment, and Subsistence Stress," in *Evolving Complexity and Environmental Risk in the Prehistoric Southwest,* J. A. Tainter and B. B. Tainter, eds., pp. 25–56. Addison-Wesley, Reading, Massachusetts.

Dean, J. S., W. H. Doelle, and J. D. Orcutt
1994 "Adaptive Stress: Environment and Demography," in *Themes in Southwest Prehistory,* G. J. Gumerman, ed., pp. 53–86. School of American Research Press, Santa Fe, New Mexico.

Dean, J. S., D. M. Meko, and T. W. Swetnam
1994 *Tree Rings, Environment, and Humanity: Processes and Relationships.* RADIOCARBON, Department of Geosciences, University of Arizona, Tucson.

Defrise-Gussenhoven, E.
1967 "Generalized Distance in Genetic Studies." *Acta Genetica et Medicae Gemellologiae* 17:275–288.

DeNiro, M. J., and S. Epstein
1978 "Influence of Diet on Distribution of Carbon Isotopes in Animals." *Geochimica et Cosmochimica Acta* 42(5):495–506.

Di Peso, C. C.
1974 *Casas Grandes: A Fallen Trading Center of the Gran Chichimeca.* Amerind Foundation Publications 9. Northland Press, Flagstaff, Arizona.

Dobyns, H. F.
1989 "Native Historic Epidemiology in the Greater Southwest." *American Anthropologist* 91(1):171–174.
1992 "Native American Trade Centers as Contagious Disease Foci," in *Disease and Demography in the Americas,* J. W. Verano and D. H. Ubelaker, eds., pp. 215–222. Smithsonian Institution Press, Washington, D.C.

Dozier, E.
1970 *The Pueblo Indians of North America.* Holt, Reinhart, and Winston, New York.

Duff, A.
1998 "The Process of Migration in the Late Prehistoric Southwest," in *Migration and Reorganization: The Pueblo IV Period in the American Southwest,* K. A. Spielmann, ed., pp. 31–52. Arizona State University Anthropological Papers 51. Tempe.

Elliott, M.
1995 *Great Excavations: Tales of Early Southwestern Archaeology, 1888–1939.* School of American Research Press, Santa Fe, New Mexico.

Elliott, M. L.
1982 *Large Pueblo Sites near Jemez Springs, New Mexico.* Santa Fe National Forest, Cultural Resources Report 3. U.S. Department of Agriculture, Forest Service, Southwestern Region, Santa Fe, New Mexico.

El-Najjar, M. Y.
1976 "Maize, Malaria, and the Anemias in the Pre-Columbian New World." *Yearbook of Physical Anthropology* 20:329–337.

Emmons, A. B.
1913 "A Study of the Variations in the Female Pelvis, Based on Observations Made on 217 Specimens of the American Indian Squaw." *Biometrika* 9:34–57.

Erickson, G. M., J. Catanese III, and T. M. Keaveny
2002 "Evolution of the Biomechanical Material Properties of the Femur." *Anatomical Record* 268(2):115–124.

Evans, S.
1998 "Toltec Invaders and Spanish Conquistadors," in *Studies in Culture Contact: Interaction, Culture Change, and Archaeology,* J. G. Cusick, ed., pp. 335–357. Center for Archaeological Investigations, Southern Illinois University, Carbondale.

Ezzo, J. A.
1991 "Dietary Change at Grasshopper Pueblo, Arizona: The Evidence from Bone Chemistry Analysis." Ph.D. dissertation, Department of Anthropology, University of Wisconsin, Madison.
1992 "Dietary Change and Variability at Grasshopper Pueblo, Arizona." *Journal of Anthropological Archaeology* 11:219–289.
1993 *Human Adaptation at Grasshopper Pueblo: Social and Ecological Perspectives.* Archaeological Series 4. International Monographs in Prehistory, Ann Arbor, Michigan.

Feik, S. A., C. D. Thomas, R. Bruns, and J. G. Clement
2000 "Regional Variations in Cortical Modeling in the Femoral Mid-Shaft: Sex and Age Differences." *American Journal of Physical Anthropology* 112(2):191–205.

Feik, S. A., C. D. L. Thomas, and J. G. Clement
1996 "Age Trends in Remodeling of the Femoral Midshaft Differ between the Sexes." *Journal of Orthopaedic Research* 14:590–597.

Feldesman, M. R., and R. L. Fountain
1996 "'Race' Specificity and the Femur/Stature Ratio." *American Journal of Physical Anthropology* 100:207–224.

Feldesman, M. R., J. G. Kleckner, and J. K. Lundy
1990 "Femur/Stature Ratio and Estimates of Stature in Mid- and Late-Pleistocene Fossil Hominids." *American Journal of Physical Anthropology* 83:359–372.

Ferretti, J. L., R. F. Capozza, G. R. Cointry, S. L. Garcia, H. Plotkin, M. L. Alvarez Filgueira, and J. R. Zanchetta
1998 "Gender-Related Differences in the Relationship between Densitometric Values of Whole-Body Bone Mineral Content and Lean Body Mass in Humans between 2 and 87 Years of Age." *Bone* 22(6):683–690.

Fiedel, S. J.
2004 "The Kennewick Follies: Peopling of the Americas." *Journal of Anthropological Research* 60(1):75–110.

Floberg, J., P. Belfrage, M. Carlsson, and H. Ohlsen
1986 "The Pelvic Outlet: A Comparison between Clinical Evaluation and Radiologic Pelvimetry." *Acta Obstetricia et Gynecologica Scandinavica* 65:321–326.

Ford, R. I.
1972 "Barter, Gift, or Violence: An Analysis of Tewa Intertribal Exchange," in *Social Exchange and Interaction,* E. N. Wilmsen, ed., pp. 21–46. Anthropological Papers 62. Museum of Anthropology, University of Michigan, Ann Arbor.
1987 "The New Pueblo Economy," in *When Cultures Meet: Remembering San Gabriel del Yunque Oweenge,* pp. 73–91. Sunstone Press, Santa Fe, New Mexico.

Ford, R. I., A. H. Schroeder, and S. L. Peckham
1972 "Three Perspectives on Puebloan Prehistory," in *New Perspectives on the Pueblos,* A. Ortiz, ed., pp. 19–40. University of New Mexico Press, Albuquerque.

Fowles, S.
2004 "Tewa versus Tiwa: Northern Rio Grande Settlement Patterns and Social History, A.D. 1275–1540," in *The Protohistoric Pueblo World, A.D. 1275–1600,* E. C. Adams and A. I. Duff, eds., pp. 17–25. University of Arizona Press, Tucson.

France, D. L.
1988 "Osteometry at Muscle Origin and Insertion in Sex Determination." *American Journal of Physical Anthropology* 76(4):515–526.

Francis, J. R.
1999 "Temporal Trends in Mission Populations: A Comparison of Pathological Frequencies and Long Bone Length at Mission San Juan de Capistrano, San Antonio, Texas." M.A. thesis, Department of Anthropology, University of Texas, San Antonio.

Frankenberg, S. R., and L. W. Konigsberg
2006 "A Brief History of Paleodemography from Hooton to Hazards Analysis," in *Bioarchaeology: The Contextual Analysis of Human Remains,* J. E. Buikstra and L. A. Beck, eds., pp. 227–261. Academic Press, Boston.

Gage, T.
1989 "Bio-Mathematical Approaches to the Study of Human Variation in Mortality." *Yearbook of Physical Anthropology* 32:185–214.

Garn, S. M.
1970 *The Earlier Gain and the Later Loss of Cortical Bone.* Charles C. Thomas, Springfield, Ill.
1972 "The Course of Bone Gain and the Phases of Bone Loss." *Orthopedic Clinics of North America* 3:503–520.

Garn, S. M., C. G. Rohmann, B. Wagner, and W. Ascoli
1967 "Continuing Bone Growth throughout Life: A General Phenomenon." *American Journal of Physical Anthropology* 26:313–318.

Garner, M. E.
1918 "Pelvimetry and Cephalometry of Chinese." *Chinese Medical Journal* 32:121–125.

Garson, J. G.
1881–1882 "Pelvimetry." *Journal of Anatomy and Physiology* 16:106–134.

Genoves, S.
1967 "Proportionality of the Long Bones and Their Relation to Stature among Mesoamericans." *American Journal of Physical Anthropology* 26:67–78.

Gilbert, B. M., and T. W. McKern
1973 "A Method for Aging the Female Os Pubis." *American Journal of Physical Anthropology* 38(1):31–38.

Gill, G. W., and S. Rhine, eds.
1990 *Skeletal Attribution of Race: Methods for Forensic Anthropology.* Maxwell Museum of Anthropology, Anthropological Papers 4. University of New Mexico Press, Albuquerque.

Givens, D. R.
1992 *Alfred Vincent Kidder and the Development of Americanist Archaeology.* University of New Mexico Press, Albuquerque.

Glassow, M.
1980 *Prehistoric Agricultural Development in the Northern Southwest.* Ballena Press, Socorro, New Mexico.

Goldstein, M.
1953 "Some Vital Statistics Based on Skeletal Material." *Human Biology* 25:3–12.

Gompertz, B.
1825 "On the Nature of the Function Expressive of the Law of Human Mortality, and on a New Mode of Determining the Value of Life Contingencies." *Philosophical Transactions of the Royal Society of London* (Series A) 115:513–585.

Goodman, A.
1997 "Racializing Kennewick Man." *Anthropology Newsletter* 38(10):3.

Gower, J. C.
1972 "Measures of Taxonomic Distance and Their Analysis," in *The Assessment of Population Affinities in Man,* J. S. Weiner and J. Huizinga, eds., pp. 1–24. Clarenden Press, Oxford.

Graves, W. M., and S. L. Eckert
1998 "Decorated Ceramic Distributions and Ideological Developments in the Northern and Central Rio Grande Valley, New Mexico," in *Migration and Reorganization: The Pueblo IV Period in the American Southwest,* K. Spielmann, ed., pp. 263–283. Arizona State University Archaeological Research Papers 51. Tempe.

Greulich, W. W., H. Thoms, and R. C. Twaddle
1939 "A Study of Pelvic Type and Its Relationship to Body Build in White Women." *Journal of the American Medical Association* 112:485–493.

Guthe, C.
1918 "Report on the Packing and Shipping of the Skeletons, Spring 1918." Manuscript on file, Hooton Papers, Peabody Museum of Archaeology and Ethnology Archives, Cambridge, Massachusetts.

Haapasalo, H., P. Kannus, H. Sievanen, M. Pasanen, K. Uusi-Rasi, A. Heinonen, P. Oja, and I. Vuori
1996 "Development of Mass, Density, and Estimated Mechanical Characteristics of Bones in Caucasian Females." *Journal of Bone and Mineral Research* 11(11):1751–1760.

Haas, J., and W. Creamer
1996 "The Role of Warfare in the Pueblo III Period," in *The Prehistoric Pueblo World,* A.D. *1150–1350,* M. A. Adler, ed., pp. 205–213. University of Arizona Press, Tucson.
1997 "Warfare among the Pueblos: Myth, History, and Ethnology." *Ethnohistory* 44(2): 235–261.

Habicht-Mauche, J.
1988 "An Analysis of Southwestern Style Utility Ware Ceramics from the Southern Plains in the Context of Protohistoric Plains-Pueblo Interaction." Ph.D. dissertation, Department of Anthropology, Harvard University, Cambridge, Massachusetts.
1991 "Evidence for the Manufacture of Southwestern-Style Culinary Ceramics on the Southern Plains," in *Farmers, Hunters, and Colonists,* K. A. Spielmann, ed., pp. 51–70. University of Arizona Press, Tucson.
2000 "Pottery, Food, Hides, and Women: Labor, Production, and Exchange within the Protohistoric Plains-Pueblo Frontier Economy," in *The Archaeology of Regional Interaction in the American Southwest,* M. Hegmon, ed., pp. 209–231. University of Colorado Press, Boulder.

Habicht-Mauche, J., A. A. Levendosky, and M. J. Schoeninger
1994 "Antelope Creek Phase Subsistence: The Bone Chemistry Evidence," in *Skeletal Biology in the Great Plains: Migration, Warfare, Health, and Subsistence,* D. W. Owsley and R. L. Jantz, eds., pp. 291–304. Smithsonian Institution Press, Washington, D.C.

Hackett, C. W.
1937 *Historical Documents Relating to New Mexico, Nueva Vizcaya, and the Approaches Thereto, to 1773,* vol. 3. Carnegie Institution of Washington, Washington, D.C.

Hackett, C. W., and C. Shelby
1942 *Revolt of the Pueblo Indians of New Mexico and Otermin's Attempted Reconquest, 1680–1682.* University of New Mexico Press, Albuquerque.

Hale, K., and D. Harris
1979 "Historical Linguistics and Archaeology," in *Handbook of North American Indians,* vol. 9, *Southwest,* A. Ortiz, ed., pp. 170–177. Smithsonian Institution, Washington, D.C.

Hammett, J. E.
1992 "Ethnohistory of Aboriginal Landscapes in the Southeastern United States." *Southern Indian Studies* 41:1–50.
1997 "Interregional Patterns of Land Use and Plant Management in Native North America," in *People, Plants, and Landscapes: Studies in Paleoethnobotany*, K. J. Gremillion, ed., pp. 195–216. University of Alabama Press, Tuscaloosa.

Hanson, S.
1936 "The Transversely Contracted Midpelvis with Particular Reference to Forceps Delivery." *American Journal of Obstetrics and Gynecology* 32:385–398.

Hare, P. E., M. L. Fogel, T. W. Stafford, A. D. Mitchell, and T. C. Hoering
1991 "The Isotopic Composition of Carbon and Nitrogen in Individual Amino Acids Isolated from Modern and Fossil Proteins." *Journal of Archaeological Science* 18(3):277–292.

Harris, A. H.
1999 "The Nonavian Fauna from Picuris Pueblo," in *Picuris Pueblo through Time: Eight Centuries of Change at a Northern Rio Grande Pueblo*, M. A. Adler and H. Dick, eds., pp. 127–139. William P. Clements Center for Southwestern Studies, Southern Methodist University, Dallas.

Haury, E. W.
1936 *The Mogollon Culture of Southwestern New Mexico.* Medallion Papers 20. Gila Pueblo, Globe, Arizona.

Hawkey, D. E.
1998 "Out of Asia: Dental Evidence for Microevolution and Affinities of Early Populations from India/Sri Lanka." Ph.D. dissertation, Department of Anthropology, Arizona State University, Tempe.

Hayes, A. C.
1974 *The Four Churches of Pecos.* University of New Mexico Press, Albuquerque.
1981 *Mound 7 Excavations at Las Humanas.* Publications in Archaeology 16. U.S. Department of the Interior, National Park Service, Santa Fe, New Mexico.

Head, G. N., and J. D. Orcutt, eds.
2002 *From Folsom to Fogelson: The Cultural Resources Inventory Survey of Pecos National Historic Park.* National Park Service, U.S. Department of the Interior, Washington, D.C.

Heyns, O. S.
1944 "A Study of the Bantu Female Pelvis." *Journal of Anatomy* (London) 78:151–166.
1946 "The Superiority of the South African Negro or Bantu as a Parturient." *Journal of Obstetrics and Gynaecology of the British Empire* 53:405–428.
1947 "A Preliminary Report on the Pelvic Brim Index." *Journal of Obstetrics and Gynaecology of the British Empire* 54:39–64.

Hiernaux, J.
1985 "A Comparison of the Shoulder-Hip-Width Sexual Dimorphism in Sub-Saharan Africa and Europe," in *Human Sexual Dimorphism*, J. Ghesquiere, R. D. Martin, and F. Newcombe, eds., pp. 191–206. Taylor and Francis, Philadelphia.

Hobson, K. A., and R. G. Clark
1992 "Assessing Avian Diets Using Stable Isotopes, 2: Factors Influencing Diet-Tissue Fractionation." *The Condor* 94:189–197.

Holden, J.
1955 "A Preliminary Report on Arrowhead Ruin." *El Palacio* 62:102–119.

Holden, W. C.
1932 "Excavations at Tecolote during Summer of 1931." *Bulletin of the Texas Archaeological and Paleontological Society* 4:25–28.

Holland, E. L., G. W. Cran, J. H. Elwood, J. H. M. Pinkerton, and W. Thompson
1982 "Associations between Pelvic Anatomy, Height and Year of Birth of Men and Women in Belfast." *Annals of Human Biology* 9:113–120.

Holland, T. D., and M. J. O'Brien
1997 "Parasites, Porotic Hyperostosis, and the Implications of Changing Perspectives." *American Antiquity* 62(2):183–193.

Holman, D.
2000 mle: A Programming Language for Building Likelihood Models. Version 2, http://faculty.washington.edu/~holman/mle.

Hooton, E. A.
1930 *The Indians of Pecos Pueblo: A Study of Their Skeletal Remains.* Yale University Press, New

Haven, Connecticut [= Papers of the Southwestern Expedition 4, Department of Archaeology, Phillips Academy, Andover, Massachusetts].
1935 "Development and Correlation of Research in Physical Anthropology at Harvard University." *Proceedings of the American Philosophical Society* 75(6):499–516.
1954 *Up from the Ape.* Macmillan, New York.

Hoppa, R. D., and J. W. Vaupel, eds.
2002 *Paleodemography: Age Distributions from Skeletal Samples.* Cambridge University Press, Cambridge.

Houdek, D., J. E. Buikstra, and C. Stojanowski
2000 "Osteological Analysis," in *Never Anything So Solemn: An Archeological, Biological, and Historical Investigation of the Nineteenth-Century Grafton Cemetery,* J. E. Buikstra, J. A. O'Gorman, and C. Sutton, eds., pp. 91–114. Kampsville Studies in Archeology and History, vol. 3. Center for American Archeology, Kampsville, Illinois.

Howard, R.
1959 "Comments on the Indians' Water Supply at Gran Quivira National Monument." *El Palacio* 66:85–91.

Howells, W. W.
1973 *Cranial Variation in Man: A Study by Multivariate Analysis of Patterns of Differences among Recent Human Populations.* Papers of the Peabody Museum of Archaeology and Ethnology, vol. 67. Harvard University, Cambridge, Massachusetts.

Howells, W. W., and H. Hotelling
1936 "Measurements and Correlations on Pelves of Indians of the Southwest." *American Journal of Physical Anthropology* 21:91–106.

Howland, M. R., L. T. Corr, S. M. Young, V. Jones, S. Jim, N. J. van der Merwe, A. D. Mitchell, and R. P. Evershed
2002 "Expression of the Dietary Isotope Signal in the Compound-specific δ^{13}C Values of Pig Bone Lipids and Amino Acids." *International Journal of Osteoarchaeology* 13(1–2): 54–65.

Hrdlička, A.
1916 *Physical Anthropology of the Lenape or Delawares, and of the Eastern Indians in General.* Bulletin of the Bureau of American Ethnology, no. 62. Smithsonian Institution, Washington, D.C.
1935 "The Pueblos, with Comparative Data on the Bulk of the Tribes of the Southwest and Northern Mexico." *American Journal of Physical Anthropology* 20(3):235–460.

Hsieh, Y. F., A. G. Robling, W. T. Ambrosius, D. B. Burr, and C. H. Turner
2001 "Mechanical Loading of Diaphyseal Bone in Vivo: The Strain Threshold for an Osteogenic Response Varies with Location." *Journal of Bone and Mineral Research* 16(12):2291–2297.

Hui, S. L., L. Zhou, R. Evans, C. W. Slemenda, M. Peacock, C. M. Weaver, C. McClintock, and C. C. Johnston Jr.
1999 "Rates of Growth and Loss of Bone Mineral in the Spine and Femoral Neck in White Females." *Osteoporosis International* 9(3): 200–205.

Humphreys, S. B.
1971 "The Skeletal Biology of Eighteenth-Century Coahuiltecan Indians from San Juan Capistrano Mission, San Antonio, Texas." Master's thesis, Department of Anthropology, Southern Methodist University, Dallas.

Ince, J. G. H., and M. Young
1940 "The Bony Pelvis and Its Influence on Labour: A Radiological and Clinical Study of 500 Women." *Journal of Obstetrics and Gynaecology of the British Empire* 47:130–190.

İşcan, M. Y., and T. S. Cotton
1985 "The Effect of Age on the Determination of Race from the Pelvis." *Journal of Human Evolution* 14:275–282.

Ivey, J. E.
1994 "The Greatest Misfortune of All: Famine in the Province of New Mexico, 1667–1672." *Journal of the Southwest* 36(1):76–100.

Jackes, M.
2000 "Building the Bases for Paleodemographic Analysis: Adult Age Determination," in *Biological Anthropology of the Human Skeleton,* M. A. Katzenberg and S. R. Saunders, eds., pp. 417–466. Wiley-Liss, New York.

Jantz, L. M., R. L. Jantz, N. P. Herrmann, C. S. Sparks, K. E. Weisensee, and D. V. Kopp
2002 *Archaeological Investigations at the Last Spanish Colonial Mission Established on the Texas*

Frontier: Nuestra Señora del Refugio (41RF1), Refugio County, Texas, vol. 2, *Osteological Analyses.* Archaeological Survey Report 315. Center for Archaeological Research, University of Texas, San Antonio.

Jantz, R. L., and D. W. Owsley
2001 "Variation among Early North American Crania." *American Journal of Physical Anthropology* 114:146–155.

Jarcho, J.
1933 *The Pelvis in Obstetrics: A Practical Manual of Pelvimetry and Cephalometry, Including Chapters on Roentgenological Measurement.* Paul B. Hoeber, New York.

Jarvinen, T. L., P. Kannus, and H. Sievanen
2003 "Estrogen and Bone: A Reproductive and Locomotive Perspective." *Journal of Bone and Mineral Research* 18(11):1921–1931.

Jarvinen, T. L., I. Pajamaki, H. Sievanen, T. Vuohelainen, J. Tuukkanen, M. Jarvinen, and P. Kannus
2003 "Femoral Neck Response to Exercise and Subsequent Deconditioning in Young and Adult Rats." *Journal of Bone and Mineral Research* 18(7):1292–1299.

Jelinek, A. J.
1967 *A Prehistoric Sequence in the Middle Pecos Valley, New Mexico.* Anthropological Papers 31. Museum of Anthropology, University of Michigan, Ann Arbor.

Jim, S., S. H. Ambrose, and R. P. Evershed
2004 "Stable Carbon Isotopic Evidence for Differences in the Dietary Origin of Bone Cholesterol, Collagen and Apatite: Implication for Their Use in Palaeodietary Reconstruction." *Geochimica et Cosmochimica Acta* 68(1): 61–72.

Johnston, F. E.
1961 "Sequence of Epiphyseal Union in a Prehistoric Kentucky Population from Indian Knoll." *Human Biology* 33:66–81.

Johnston, F. E., and E. S. Watts
1969 "Endosteal Deposition of Bone at the Midshaft of the Second Metacarpals of Adolescent Females." *Anatomical Record* 163:67–70.

Jones, H. H., J. D. Priest, W. C. Hayes, C. C. Tichenor, and D. A. Nagel
1977 "Humeral Hypertrophy in Response to Exercise." *Journal of Bone and Joint Surgery* 59A:204–208.

Jones, O. L. Jr.
1966 *Pueblo Warriors and Spanish Conquest.* University of Oklahoma Press, Norman.

Jordaan, H. V. F.
1976 "The Determinants of Pelvic Brim Morphology in the Female." *South African Medical Journal* 50:772–778.

Jurmain, R. D.
1977 "Stress and the Etiology of Osteoarthritis." *American Journal of Physical Anthropology* 46:353–366.
1980 "The Pattern of Involvement of Appendicular Degenerative Joint Disease." *American Journal of Physical Anthropology* 53:143–150.
2001 "Paleoepidemiological Patterns of Trauma in a Prehistoric Population from Central California." *American Journal of Physical Anthropology* 115:13–23.

Kaldahl, E. J., S. Van Keuren, and B. J. Mills
2004 "Migration, Factionalism, and the Trajectories of Pueblo IV Period Clusters in the Mogollon Rim Region," in *The Protohistoric Pueblo World, A.D. 1275–1600,* E. C. Adams and A. I. Duff, eds., pp. 85–94. University of Arizona Press, Tucson.

Kelly, J. H.
1984 *The Archaeology of the Sierra Blanca Region of Southeastern New Mexico.* Anthropological Papers 74. Museum of Anthropology, University of Michigan, Ann Arbor.
1991 "An Overview of the Capitan North Project," in Mogollon V, compiled by P. Beckett, pp. 166–176. COAS Publishing and Research, Las Cruces, New Mexico.

Kent, S.
1986 "Influence of Sedentism and Aggregation on Porotic Hyperostosis and Anaemia: A Case Study." *Man* 21(4):605–636.

Kern, K. F.
2006 "T. Wingate Todd: Pioneer of Modern American Physical Anthropology." *Kirtlandia* 55:1–42.

Kessell, J. L.
1979 *Kiva, Cross, and Crown: The Pecos Indians and New Mexico, 1540–1840.* National Park Service, U.S. Department of the Interior, Washington, D.C.
1987 *Kiva, Cross, and Crown: The Pecos Indians and New Mexico, 1540–1840.* 2nd ed. National

Park Service, U.S. Department of the Interior, Washington, D.C.

Key, P., and R. Jantz
1990 "Statistical Assessment of Population Variability: A Methodological Approach." *American Journal of Physical Anthropology* 82(1):53–59.

Kidder, A. V.
1924 *An Introduction to the Study of Southwestern Archaeology, with a Preliminary Account of the Excavations at Pecos.* Yale University Press, New Haven, Connecticut [= Papers of the Southwestern Expedition 1, Department of Archaeology, Phillips Academy, Andover, Massachusetts].
1929 "Instructions for Lab of Anthropology Rockefeller Students, 1929." Notes on file, Laboratory of Anthropology, Santa Fe, New Mexico.
1931 *The Pottery of Pecos,* vol. 1, *Dull Paint Wares.* Yale University Press, New Haven, Connecticut [= Papers of the Southwestern Expedition 5, Department of Archaeology, Phillips Academy, Andover, Massachusetts].
1932 *The Artifacts of Pecos.* Yale University Press, New Haven, Connecticut [= Papers of the Southwestern Expedition 6, Department of Archaeology, Phillips Academy, Andover, Massachusetts].
1958 *Pecos, New Mexico: Archaeological Notes.* Papers of the Robert S. Peabody Foundation for Archaeology 5. Phillips Academy, Andover, Massachusetts.
1962 *An Introduction to the Study of Southwestern Archaeology.* Yale University Press, New Haven, Connecticut.

Kidder, A. V., and M. A. Kidder
1917 "Notes on the Pottery of Pecos." *American Anthropologist* 19:325–360.

Kidder, A. V., and A. O. Shepard
1936 *The Pottery of Pecos,* vol. 2, *The Glaze-Paint, Culinary and Other Wares.* Yale University Press, New Haven, Connecticut [= Papers of the Southwestern Expedition 7, Department of Archaeology, Phillips Academy, Andover, Massachusetts].

King, G., and T. Y. Teh
1937 "Obstetrical Criteria in North China, II: The Weights and Measurements of the Mature New-born Child." *Chinese Medical Journal* 52:501–506.

Knaut, A. L.
1995 *The Pueblo Revolt of 1680.* University of Oklahoma Press, Norman.

K'o, Y.-K., L. Yueh, W.-M. Li, F.-K. Chou, and C.-Y. Hou
1957 "Measurements of the Chinese Female Pelvis." *Chinese Medical Journal* 75:656–660.

Kohler, T. A., and C. van West
1996 "The Calculus of Self-Interest in the Development of Cooperation: Sociopolitical Development and Risk among the Northern Anasazi," in *Evolving Complexity and Environmental Risk in the Prehistoric Southwest,* J. A. Tainter and B. B. Tainter, eds., pp. 169–196. Addison-Wesley, Reading, Massachusetts.

Kolawole, T. M., S. P. Adamu, and K. T. Evans
1978 "Comparative Pelvimetric Measurements in Nigerian and Welsh Women." *Clinical Radiology* 29:85–90.

Konigsberg, L.
1990 "Analysis of Prehistoric Biological Variation under a Model of Isolation by Geographic and Temporal Distance." *Human Biology* 62(1):49–70.

Konigsberg, L., and S. Frankenberg
1992 "Estimation of Age Structure in Anthropological Demography." *American Journal of Physical Anthropology* 89(2):235–256.
2002 "Deconstructing Death in Paleodemography." *American Journal of Physical Anthropology* 117:297–309.

Konigsberg L., and S. Ousley
1995 "Multivariate Quantitative Genetics of Anthropometric Traits from the Boas Data." *Human Biology* 67:481–498.

Kontulainen, S., H. Sievanen, P. Kannus, M. Pasanen, and I. Vuori
2002 "Effect of Long-Term Impact-Loading on Mass, Size, and Estimated Strength of Humerus and Radius of Female Racquet-Sports Players: A Peripheral Quantitative Computed Tomography Study between Young and Old Starters and Controls." *Journal of Bone and Mineral Research* 17(12): 2281–2289.

Krogman, W. M.
1962 *The Human Skeleton in Forensic Medicine.* Charles C. Thomas, Springfield, Illinois.

Krogman, W. M., and M. Y. İşcan
1986 *The Human Skeleton in Forensic Medicine.* 2nd ed. Charles C. Thomas, Springfield, Illinois.

Krukierek, S.
1951 "The Sexual Differences of the Human Pelvis." *Gynaecologia* 132:92–110.

Kucklelman, K. A.
2002 "Thirteenth-Century Warfare in the Central Mesa Verde Region," in *Seeking the Center Place,* M. Varien and R. Wilshusen, eds., pp. 233–253. University of Utah Press, Salt Lake City.

Kuhlbusch, T. A., J. M. Lobert, P. J. Crutzen, and P. Warneck
1991 "Molecular Nitrogen Emissions from Denitrification during Biomass Burning." *Nature* 351:135–137.

Lang, R. W., and A. H. Harris
1984 *The Faunal Remains from Arroyo Hondo Pueblo, New Mexico: A Study in Short-term Subsistence Change.* School of American Research Press, Santa Fe, New Mexico.

Lanyon, L., and T. Skerry
2001 "Postmenopausal Osteoporosis as a Failure of Bone's Adaptation to Functional Loading: A Hypothesis." *Journal of Bone and Mineral Research* 16(11):1937–1947.

Larsen, C. S.
1981 "Functional Implications of Postcranial Size Reduction on the Prehistoric Georgia Coast, U.S.A." *Journal of Human Evolution* 10:489–502.

Larsen, C. S., and C. B. Ruff
1994 "Stresses of Conquest in Spanish Florida: Structural Adaptation and Change Before and After Contact," in *In the Wake of Contact: Biological Responses to Conquest,* C. S. Larsen and G. R. Milner, eds., pp. 21–34. Wiley and Sons, New York.

Larson, D., H. Neff, D. A. Graybill, J. Michaelson, and E. Ambos
1996 "Risk, Climatic Variability, and the Study of Southwestern Prehistory: An Evolutionary Perspective." *American Antiquity* 61(2):217–241.

Lau, T. K., C. M. Y. Leung, and C. Y. Li
1998 "A Study of the Use of Radiological Pelvimetry in a Chinese Population." *Acta Obstetricia et Gynecologica Scandinavica* 77:41–44.

LeBlanc, S. A.
1998 "Settlement Consequences of Warfare during the Late Pueblo III and Pueblo IV Periods," in *Migration and Reorganization: The Pueblo IV Period in the American Southwest,* K. A. Spielmann, ed., pp. 115–135. Arizona State University Anthropological Papers 51. Tempe.
1999 *Prehistoric Warfare in the American Southwest.* University of Utah Press, Salt Lake City.
2000 "Regional Interaction and Warfare in the Late Prehistoric Southwest," in *The Archaeology of Regional Interaction,* M. Hegmon, ed., pp. 41–70. University Press of Colorado, Boulder.

Lee-Thorp, J. A., J. C. Sealy, and N. J. van der Merwe
1989 "Stable Carbon Isotope Ratio Differences between Bone Collagen and Bone Apatite, and Their Relationship to Diet." *Journal of Archaeological Science* 16:585–599.

Leonard, K.
2000 "Directionality and Exclusivity of Plains-Pueblo Exchange during the Protohistoric Period." M.A. thesis, Department of Anthropology, Arizona State University, Tempe.

Levine, F.
1999 *Our Prayers Are in This Place: Pecos Identity over the Centuries.* University of New Mexico Press, Albuquerque.
2004 "Surviving Extinction," in *Identity, Feasting, and the Archaeology of the Greater Southwest,* B. J. Mills, ed., pp. 93–106. University of Colorado Press, Boulder.

Levine, F., and A. LaBauve
1997 "Examining the Complexity of Historic Population Decline: A Case Study of Pecos Pueblo, New Mexico." *Ethnohistory* 44(1):75–112.

Levine, F., M. Norcini, and M. Foster
1994 "An Ethnographic Overview of Pecos National Historical Park." Unpublished summary of a contract of the National Park Service, Southwest Region, Santa Fe, New Mexico.

Lewontin, R.
1972 "The Apportionment of Human Diversity," in *Evolutionary Biology,* vol. 6, T. Dobzhan-

sky, M. Hecht, and W. Steere, eds., pp. 381–398. Appleton-Century-Crofts, New York.

Lintz, C.
1984 "The Plains Villages: Antelope Creek," in *Prehistory of Oklahoma,* R. B. Bell, ed., pp. 325–346. Academic Press, New York.
1986 "Historical Development of a Culture Complex: The Basis for Understanding Architectural Misconceptions of the Antelope Creek Focus." *Plains Anthropologist Memoir* 21:111–128.
1991 "Texas Panhandle–Pueblo Interactions from the Thirteenth through the Sixteenth Century," in *Farmers, Hunters, and Colonists: Interaction between the Southwest and the Southern Plains,* K. A. Spielmann, ed., pp. 89–106. University of Arizona Press, Tucson.

Lipe, W. D.
1995 "The Depopulation of the Northern San Juan: Conditions in the Turbulent 1200s." *Journal of Anthropological Archaeology* 14: 143–169.
2002 "Social Power in the Central Mesa Verde Region," in *Seeking the Center Place,* M. Varien and R. Wilshusen, eds., pp. 203–232. University of Utah Press, Salt Lake City.

Litten, L.
1925 "Über die Normalmasse des it Weichteilen Ausgekleideten Weiblichen Beckens." *Zentralblatt für Gynäkologie* 49:817. [As cited in Jarcho 1933.]

Lorenz, J. G., and D. G. Smith
1996 "Distribution of Four Founding MtDNA Haplogroups among Native North Americans." *American Journal of Physical Anthropology* 101:307–323.

Lovejoy, C. O.
1985 "Dental Wear in the Libben Population: Its Functional Pattern and Role in the Determination of Adult Skeletal Age at Death." *American Journal of Physical Anthropology* 68:47–56.

Lovejoy, C. O., R. S. Meindl, T. R. Pryzbeck, T. S. Barton, K. G. Heiple, and D. Kotting
1977 "Paleodemography of the Libben site, Ottawa County, Ohio." *Science* 198:291–293.

Lovell, N. C.
1989 "Test of Phenice's Technique for Determining Sex from the Os Pubis." *American Journal of Physical Anthropology* 79:117–120.
2000 "Paleopathological Description and Diagnosis," in *Biological Anthropology of the Human Skeleton,* M. A. Katzenberg and S. R. Saunders, eds., pp. 211–242. Wiley-Liss, New York.

Mackey, J. C.
1980 "Appendix G: Arroyo Hondo Population Affinities," in *The Arroyo Hondo Skeletal and Mortuary Remains,* by A. M. Palkovich. School of American Research Press, Santa Fe, New Mexico.

MacLaughlin, S. M., and M. F. Bruce
1990 "The Accuracy of Sex Identification in European Skeletal Remains Using the Phenice Characters." *Journal of Forensic Sciences* 35:1384–1392.

Manahan, C. P., C. Marquez, and A. S. Mangay
1954 "A Roentgenologic Study of the Filipino Female Pelvis." *American Journal of Obstetrics and Gynecology* 68:228–235.

Mann, R. W., and S. P. Murphy
1990 *Regional Atlas of Bone Disease: A Guide to Pathologic and Normal Variation in the Human Skeleton.* Charles C. Thomas, Springfield, Illinois.

Mantel, N.
1967 "The Detection of Disease Clustering and a Generalized Regression Approach." *Cancer Research* 27:209–220.

Martin, C.
1866 "Beckenmessung an Verschiedenen Menschenracen." *Monatsschrift für Geburtshilfe und Frauenheilkunde* 28:23. [As cited in Jarcho 1933.]

Martin, D. L., N. J. Akins, A. H. Goodman, H. W. Toll, and A. C. Swedlund
2001 *Totah: Time and the Rivers Flowing. Excavations in the La Plata Valley,* vol. 5, *Harmony and Discord: Bioarchaeology.* Museum of New Mexico, Santa Fe.

Martin, R., and K. Saller
1959 *Lehrbuch der Anthropologie,* vol. 2. Gustav Fischer Verlag, Stuttgart.

Martin, R. B., and P. J. Atkinson
1977 "Age and Sex-Related Changes in the Structure and Strength of the Human Femoral Shaft." *Journal of Biomechanics* 10:223–231.

Martinson, E. A.
2002 "Reassessing the Etiology of Cribra Orbitalia and Porotic Hyperostosis: A Case Study of the Chiribaya of the Osmore Drainage, Peru." Ph.D. dissertation, Department of Anthropology, University of New Mexico, Albuquerque.

Matson, R. G., and B. Chisholm
1991 "Basketmaker II Subsistence: Carbon Isotopes and Other Dietary Indicators from Cedar Mesa." *American Antiquity* 56(3): 444–459.

McGuire, R. H., E. C. Adams, B. A. Nelson, and K. A. Spielmann
1994 "Drawing the Southwest to Scale: Perspectives on Macroregional Relations," in *Themes in Southwest Prehistory,* G. J. Gumerman, ed., pp. 239–265. School of American Research Press, Santa Fe, New Mexico.

McKern, T. W., and T. D. Stewart
1957 *Skeletal Age Changes in Young American Males, Analyzed from the Standpoint of Identification.* Headquarters Quartermaster Research and Development Command, Technical Report EP-45. Natick, Massachusetts.

McKusick, C. R.
1981 "The Faunal Remains of Las Humanas," in *Contributions to Gran Quivira Archaeology,* A. Hayes, ed., pp. 39–65. Publications in Archaeology 17. U.S. Department of the Interior, National Park Service, Santa Fe, New Mexico.

Meindl, R. S., C. O. Lovejoy, R. P. Mensforth, and L. Don Carlos
1985 "Accuracy and Direction of Error in the Sexing of the Skeleton: Implications for Paleodemography." *American Journal of Physical Anthropology* 68:79–85.

Mengert, W. F.
1938 "Pelvic Measurements of 4,144 Iowa Women." *American Journal of Obstetrics and Gynecology* 36:260–267.
1948 "Estimation of Pelvic Capacity." *Journal of the American Medical Association* 138:169–174.

Mera, H. P.
1934 *A Survey of the Biscuit Ware Area in Northern New Mexico.* Laboratory of Anthropology Technical Series Bulletin 6. Santa Fe, New Mexico.
1935 *Ceramic Clues to the Prehistory of North Central New Mexico.* Laboratory of Anthropology Technical Series Bulletin 8. Santa Fe, New Mexico.

Merbs, C. F.
1983 *Patterns of Activity-Induced Pathology in a Canadian Inuit Population.* National Museum of Man, Mercury Series, Archaeological Survey of Canada, Paper 119. National Museums of Canada, Ottawa.

Miles, A. E.
1963 "The Dentition in the Assessment of Individual Age in Skeletal Material." *Dental Anthropology* 5:191–208.

Milner, G. R., J. W. Wood, and J. L. Boldsen
2000 "Paleodemography," in *Biological Anthropology of the Human Skeleton,* M. A. Katzenberg and S. R. Saunders, eds., pp. 467–497. Wiley-Liss, New York.

Mobley, C.
1980 "Demographic Structure of Pecos Indians: A Model Based on Life Tables." *American Antiquity* 45(3):518–530.

Moore-Jansen, P. H., and R. L. Jantz
1989 *Data Collection Procedures for Forensic Skeletal Material.* Report of Investigations 48. Department of Anthropology, University of Tennessee, Knoxville.

Morley, S. G.
1910 *The South House at Puye, New Mexico.* Papers of the School of American Archaeology 7. Archaeological Institute of America, Santa Fe, New Mexico.

Nagurka, M. L., and W. C. Hayes
1980 "An Interactive Graphics Package for Calculating Cross-Sectional Properties of Complex Shapes." *Journal of Biomechanics* 13:59–64.

Natural Resources Conservation Service
1999 "Wild Turkey *(Meleagris gallopavo)*." Fish and Wildlife Management Leaflet 12. U.S. Department of the Interior, Madison, Wisconsin.

Nei, M., and A. Roychoudhury
1974 "Genetic Variation within and between the Three Major Races of Man, Caucasoids, Negroids, and Mongoloids." *American Journal of Human Genetics* 26:421–443.

Nelson, B. A., D. L. Martin, A. C. Swedlund, P. R. Fish, and G. J. Armelagos
1994 "Studies in Disruption: Demography and Health in the Prehistoric Southwest," in *Understanding Complexity in the Prehistoric Southwest,* G. J. Gumerman and M. Gell-Mann, eds., pp. 59–112. Addison-Wesley, Reading, Massachusetts.

Nelson, N. C.
1914 *Pueblo Ruins of the Galisteo Basin, New Mexico.* Anthropological Papers of the American Museum of Natural History, vol. 15, no. 1. New York.

Newman, R. W.
1957 *A Comparative Analysis of Prehistoric Skeletal Remains from the Lower Sacramento Valley.* California Archaeological Survey Report 39. University of California, Berkeley.

Nichol, C. H.
1990 "Dental Genetics and Biological Relationships of the Pima Indians of Arizona." Ph.D. dissertation, Department of Anthropology, Arizona State University, Tempe.

Nicholson, C.
1939 "The Interpretation of Radiological Pelvimetry." *Journal of Obstetrics and Gynaecology of the British Empire* 45:950–984.
1945 "The Two Main Diameters at the Brim of the Female Pelvis." *Journal of Anatomy* (London) 79:131–135.

Nicholson, C., and H. S. Allen
1946 "Variation in the Female Pelvis." *Lancet* 251: 192–195.

Noriega, T., and M. Gutiérrez
1895 "Apuntamientos para el estudio comparativo de la pelvis mexicana y la europea, y consecuencias prácticas á que da lugar la especial conformación de la primera." *Transactions of the Pan-American Medical Congress,* part 1:889. [As cited in Jarcho 1933.]

Ogedengbe, O. K., O. Okikiolu, R. I. E. Macrae, and O. F. Giwa-Osagie
1993 "Erect Lateral Pelvimetry and the Outcome of Labor in Lagos." *Journal of the National Medical Association* 85:41–44.

O'Neill, M. C., and C. B. Ruff
2004 "Estimating Human Long Bone Cross-sectional Geometric Properties: A Comparison of Noninvasive Methods." *Journal of Human Evolution* 47: 221–235.

Orcutt, J. D.
2002 "Demography," in *From Folsom to Fogelson: The Cultural Resources Inventory Survey of Pecos National Historical Park,* vol. 1, G. N. Head and J. D. Orcutt, eds., pp. 221–236. National Park Service, U.S. Department of the Interior, Santa Fe, New Mexico.

Orford, M.
1934 "The Pelvis of the Bush Race." *South African Journal of Science* 31:586–610.

Ortner, D. J., and W. G. J. Putschar
1981 *Identification of Pathological Conditions in Human Skeletal Remains.* Smithsonian Institution Press, Washington, D.C.

Owsley, D. W., and K. S. Bruwelheide
2008 "Cranial Trauma in Historic Period Central and Northern Plains Indians," in *Skeletal Biology and Bioarchaeology of the Northwestern Plains,* G. W. Gill and R. L. Weathermon, eds., pp. 212–225. University of Utah Press, Salt Lake City.

Owsley, D. W., K. S. Bruwelheide, D. Hunt, R. Clark, M. K. Spradley, N. Suro, M. A. Cashion Lugo, J. Romero Palanco, M. Perez, M. Rosario, and Fray V. Diaz
2002 "Forensic Identification of Ramón Power y Giralt: Osteological Documentation Related to the Power Project." Technical report for the Proyecto Power Comisión Investigadora. Manuscript on file, Department of Anthropology, National Museum of Natural History, Washington, D.C.

Owsley, D. W., and R. L. Jantz
1989 *A Systematic Approach to the Skeletal Biology of the Southern Plains, from Clovis to*

Comanchero: Archaeological Overview of the Southern Great Plains. Arkansas Archaeological Survey Research Series 35. Fayetteville.

Owsley, D. W., R. L. Jantz, S. Novak, D. Kollmann, and J. Francis
1999 "Documentation of Human Skeletons from San Juan Capistrano." Technical report for the Center for Archaeological Research, University of Texas, San Antonio.

Oxorn, H.
1980 *Human Labor and Birth.* 4th ed. Appleton-Century-Crofts, New York.

Palkovich, A. M.
1983 "A Comment on Mobley's 'Demographic Structure of Pecos Indians.'" *American Antiquity* 48(1):142–149.
1994 "Historic Epidemics of the American Pueblos," in *In the Wake of Contact: Biological Responses to Conquest,* C. S. Larson and G. R. Milner, eds., pp. 87–95. Wiley-Liss, New York.

Pan, N.
1929 "Measurements of the Pelvis in Hindu Females." *Journal of Anatomy* 63:263–266.

Paterson, A. M.
1893 "The Human Sacrum." *Scientific Transactions of the Royal Dublin Society* 5:123–204.

Pauwels, F.
1980 "Principles of Construction of the Lower Extremity: Their Significance for the Stressing of the Skeleton of the Leg," in *Biomechanics of the Locomotor Apparatus,* F. Pauwels, ed., pp. 193–204. Springer-Verlag, Berlin.

Peabody Museum of Archaeology and Ethnology and Robert S. Peabody Museum of Archaeology
1998 "Inventory of Human Remains and Associated/Unassociated Funerary Objects for Five Sites in the Pecos Valley: Pecos Pueblo (LA625), Dick's Ruin (LA276), Forked Lightning (LA672), Loma Lothrop (LA277), Rowe (LA108)." Manuscript on file, Peabody Museum of Archaeology and Ethnology, Harvard University.

Pecos Burial Cards
n.d. Field burial cards on file, Peabody Museum of Archaeology and Ethnology Archives, Harvard University, Cambridge, Massachusetts.

Pendall, E., J. L. Betancourt, and S. W. Leavitt
1999 "Paleoclimatic Significance of δD and $\delta^{13}C$ Values in Piñon Pine Needles from Packrat Middens Spanning the Last 40,000 years." *Palaeogeography, Palaeoclimatology, Palaeoecology* 147:53–72.

Petersen, H.
2000 "On Statistical Methods for Comparison of Intrasample Morphometric Variability: Zalavar Revisited." *American Journal of Physical Anthropology* 113(1):79–84.

Phenice, T. W.
1969 "A Newly Developed Visual Method of Sexing the Os Pubis." *American Journal of Physical Anthropology* 30:297–301.

Piper, C.
2002 "The Morphology of Prehispanic Cradleboards: Form Follows Function," in *Children in the Prehistoric Puebloan Southwest,* K. A. Kamp, ed., pp. 41–70. University of Utah Press, Salt Lake City.

Plog, S., and J. Solometo
1997 "The Never-Changing and the Ever-Changing: The Evolution of Western Pueblo Ritual." *Cambridge Archaeological Journal* 7:161–182.

Powell, M.
2002 "Ceramics," in *From Folsom to Fogelson: The Cultural Resources Inventory Survey of Pecos National Historical Park,* vol. 1, G. N. Head and J. D. Orcutt, eds., pp. 237–304. National Park Service, U.S. Department of the Interior, Santa Fe, New Mexico.

Powell, M., and K. Benedict
2002 "Ceramic Chronology," in *From Folsom to Fogelson: The Cultural Resources Inventory Survey of Pecos National Historic Park,* G. N. Head and J. D. Orcutt, eds., pp. 59–98. National Park Service, U.S. Department of the Interior, Santa Fe, New Mexico.

Ramenofsky, A.
1996 "The Problem of Introduced Infectious Diseases in New Mexico: A.D. 1540–1680." *Journal of Anthropological Research* 52(2):161–184.

Reff, D. T.
1992 "Contact Shock in Northwestern New Spain, 1518–1764," in *Disease and Demography in the Americas,* J. Verano and D. H.

Ubelaker, eds., pp. 265–276. Smithsonian Institution Press, Washington, D.C.
1993 "Alternative Explanation of Subsistence Change during the Early Historic Period at Pecos Pueblo." *American Antiquity* 58(3):563–564.

Reid, J. J., and S. M. Whittlesey
1999 *Grasshopper Pueblo: A Story of Archaeology and Ancient Life.* University of Arizona Press, Tucson.

Reinhard, K. J.
1988 "Cultural Ecology of Prehistoric Parasitism on the Colorado Plateau as Evidenced by Coprology." *American Journal of Physical Anthropology* 77:355–366.
1990 "Archaeoparasitology in North America." *American Journal of Physical Anthropology* 82(2):145–163.

Reiter, P.
1938 *The Jemez Pueblo of Unshagi, New Mexico.* Monographs of the School of American Research, nos. 5–6. University of New Mexico Press, Albuquerque.

Relethford, J.
1991 "Genetic Drift and Anthropometric Variation in Ireland." *Human Biology* 63:155–165.

Relethford, J., and H. Harpending
1994 "Craniometric Variation, Genetic Theory, and Modern Human Origins." *American Journal of Physical Anthropology* 95:249–270.

Resnick, D., and G. Niwayama
1978 "Intravertebral Disk Herniations: Cartilaginous Schmorl's Nodes." *Radiology* 126:57–65.

Reynolds, E.
1930 "Appendix 2: A Study of the Pecos Pelves with Special Reference to the Brim," in *The Indians of Pecos Pueblo: A Study of Their Skeletal Remains,* by E. A. Hooton, pp. 374–386. Yale University Press, New Haven, Connecticut [= Papers of the Southwestern Expedition 4, Department of Archaeology, Phillips Academy, Andover, Massachusetts].

Ricklis, R. A.
1996 *The Karankawa Indians of Texas: An Ecological Study of Cultural Tradition and Change.* University of Texas Press, Austin.

Riley, M. M., J. L. Halliday, and J. M. Lumley
1998 "Congenital Malformations in Victoria, Australia, 1983–95: An Overview of Infant Characteristics." *Journal of Paediatrics and Child Health* 34(3):233–240.

Robbins, D. R., K. R. Rosenberg, and C. B. Ruff
1989 "Activity Patterns in Late Middle Woodland, Delaware." *American Journal of Physical Anthropology* 78:290–291.

Roche, M. B., and G. G. Rowe
1952 "The Incidence of Separate Neural Arch and Coincident Bone Variations." *Journal of Bone and Joint Surgery* 34-A(2):491–494.

Rose, M. R., J. S. Dean, and W. J. Robinson
1981 *The Past Climate of Arroyo Hondo, New Mexico Reconstructed from Tree Rings.* School of American Research Press, Santa Fe, New Mexico.

Rosenberg, K. R., Z. Lü, and C. B. Ruff
2006 "Body Size, Body Proportions and Encephalization in a Middle Pleistocene Archaic Human from Northern China." *Proceedings of the National Academy of Sciences* 103:3552–3556.

Ross, S. M.
1975 "A Radiological Survey of the Female Pelvis in Ethiopia." *Ethiopian Medical Journal* 13:153–159.

Ruff, C. B.
1981 "A Reassessment of Demographic Estimates for Pecos Pueblo." *American Journal of Physical Anthropology* 54:147–151.
1984 "Allometry between Length and Cross-Sectional Dimensions of the Femur and Tibia in *Homo sapiens sapiens.*" *American Journal of Physical Anthropology* 65:347–358.
1987 "Sexual Dimorphism in Human Lower Limb Bone Structure: Relationship to Subsistence Strategy and Sexual Division of Labor." *Journal of Human Evolution* 16:391–416.
1989 "New Approaches to Structural Evolution of Limb Bones in Primates." *Folia Primatologia* 53:142–159.
1991 *Aging and Osteoporosis in Native Americans from Pecos Pueblo, New Mexico: Behavioral and Biomechanical Effects.* Garland Publishing, New York.
1994a "Biomechanical Analysis of Northern and Southern Plains Femora: Behavioral Implications," in *Skeletal Biology in the Great*

Plains: A Multidisciplinary View, D. W. Owsley and R. L. Jantz, eds., pp. 235–245. Smithsonian Institution Press, Washington, D.C.

1994b "Morphological Adaptation to Climate in Modern and Fossil Hominids." *Yearbook of Physical Anthropology* 37:65–107.

1995 "Biomechanics of the Hip and Birth in Early *Homo.*" *American Journal of Physical Anthropology* 98:527–574.

1998 "Evolution of the Hominid Hip," in *Primate Locomotion: Recent Advances,* E. Strasser, A. Rosenberger, H. McHenry, and J. Fleagle, eds., pp. 449–469. Plenum Press, Davis, California.

1999 "Skeletal Structure and Behavioral Patterns of Prehistoric Great Basin Populations," in *Prehistoric Lifeways in the Great Basin Wetlands: Bioarchaeological Reconstruction and Interpretation,* B. E. Hemphill and C. S. Larsen, eds., pp. 290–320. University of Utah Press, Salt Lake City.

2000a "Biomechanical Analyses of Archaeological Human Skeletal Samples," in *Biological Anthropology of the Human Skeleton,* M. A. Katzenburg and S. R. Saunders, eds., pp. 71–102. Alan R. Liss, New York.

2000b "Body Mass Prediction from Skeletal Frame Size in Elite Athletes." *American Journal of Physical Anthropology* 113:507–517.

2000c "Body Size, Body Shape, and Long Bone Strength in Modern Humans." *Journal of Human Evolution* 38:269–290.

2002a "Long Bone Articular and Diaphyseal Structure in Old World Monkeys and Apes, I: Locomotor Effects." *American Journal of Physical Anthropology* 119:305–342.

2002b "Variation in Human Body Size and Shape." *Annual Review of Anthropology* 31: 211–232.

2003a "Growth in Bone Strength, Body Size, and Muscle Size in a Juvenile Longitudinal Sample." *Bone* 33:317–329.

2003b "Ontogenetic Adaptation to Bipedalism: Age Changes in Femoral to Humeral Length and Strength Proportions in Humans, with a Comparison to Baboons." *Journal of Human Evolution* 45:317–349.

Ruff, C. B., and W. C. Hayes

1982 "Subperiosteal Expansion and Cortical Remodeling of the Human Femur and Tibia with Aging." *Science* 217:945–948.

1983a "Cross-Sectional Geometry of Pecos Pueblo Femora and Tibiae: A Biomechanical Investigation. I. Method and General Patterns of Variation." *American Journal of Physical Anthropology* 60:359–381.

1983b "Cross-Sectional Geometry of Pecos Pueblo Femora and Tibiae: A Biomechanical Investigation. II. Sex, Age, and Side Differences." *American Journal of Physical Anthropology* 60:383–400.

1984a "Age Changes in Geometry and Mineral Content of the Lower Limb Bones." *Annals of Biomedical Engineering* 12:573–584.

1984b "Bone Mineral Content in the Lower Limb: Relationship to Cross-Sectional Geometry." *Journal of Bone and Joint Surgery* 66A:1024–1031.

1988 "Sex Differences in Age-Related Remodeling of the Femur and Tibia." *Journal of Orthopaedic Research* 6:886–896.

Ruff, C. B., B. H. Holt, and E. Trinkaus

2006 "Who's Afraid of the Big Bad Wolff? Wolff's Law and Bone Functional Adaptation." *American Journal of Physical Anthropology* 129:484–498.

Ruff, C. B., and C. S. Larsen

2001 "Reconstructing Behavior in Spanish Florida: The Biomechanical Evidence," in *Bioarchaeology of Spanish Florida: The Impact of Colonialism,* C. S. Larsen, ed., pp. 113–145. University Press of Florida, Gainesville.

Ruff, C. B., and F. P. Leo

1986 "Use of Computed Tomography in Skeletal Structure Research." *Yearbook of Physical Anthropology* 29:181–195.

Ruff, C. B., H. M. McHenry, and J. F. Thackeray

1999 "The 'Robust' Australopithecine Hip: Cross-Sectional Morphology of the Sk 82 and 97 Proximal Femora." *American Journal of Physical Anthropology* 109:509–521.

Ruff, C. B., W. W. Scott, and A. Y.-C. Liu

1991 "Articular and Diaphyseal Remodeling of the Proximal Femur with Changes in Body Mass in Adults." *American Journal of Physical Anthropology* 86:397–413.

Ruff, C. B., E. Trinkaus, and T. W. Holliday

1997 "Body Mass and Encephalization in Pleistocene *Homo.*" *Nature* 387:173–176.

Ruff, C. B., E. Trinkaus, and B. Holt

2000 "Lifeway Changes as Shown by Postcranial Skeletal Robustness." *American Journal of Physical Anthropology* Suppl. 30:266.

Ruff, C. B., E. Trinkaus, A. Walker, and C. S. Larsen
1993 "Postcranial Robusticity in *Homo*, I: Temporal Trends and Mechanical Interpretation." *American Journal of Physical Anthropology* 91: 21–53.

Ruff, C. B., A. Walker, and E. Trinkaus
1994 "Postcranial Robusticity in *Homo*, III: Ontogeny." *American Journal of Physical Anthropology* 93:35–54.

Runge, G.
1889 "Das Russische Weibliche Becken in Anthropologischer Beziehung." *Zeitschrift für Geburtshilfe und Gynäkologie* 16:131. [As cited in Jarcho 1933.]

Russell, J. G. B. , and B. Richards
1971 "A Review of Pelvimetry Data." *British Journal of Radiology* 44:780–784.

Sandness, K. L., and K. J. Reinhard
1992 "Vertebral Pathology in Prehistoric and Historic Skeletons from Northeastern Nebraska." *Plains Anthropologist* 37(141): 299–309.

Sando, J. S.
1982 *Nee Hemish: A History of Jemez Pueblo.* University of New Mexico Press, Albuquerque.
1992 *Pueblo Nations: Eight Centuries of Pueblo Indian History.* Clear Light Publishers, Santa Fe, New Mexico.

Saunders, R.
1998 "Forced Relocation, Power Relations, and Cultural Contact in the Missions of La Florida," in *Studies in Culture Contact,* J. Cusick, ed., pp. 402–429. Occasional Paper 25. Center for Archaeological Investigations, Southern Illinois University, Carbondale.

Scheuer, L., and S. Black
2000 *Developmental Juvenile Osteology.* Academic Press, San Diego.

Scheyer, H. E.
1934 "Anthropological and Roentgenological Observations on the Pelves of Chinese Women." *Chinese Medical Journal* 48:1228–1234.

Schoenau, E., C. M. Neu, F. Rauch, and F. Manz
2001 "The Development of Bone Strength at the Proximal Radius during Childhood and Adolescence." *Journal of Clinical Endocrinology and Metabolism* 86:613–618.

Schoeninger, M. J.
1995a "Stable Isotope Studies in Human Evolution." *Evolutionary Anthropology* 4(3):83–98.
1995b "Dietary Reconstruction in the Prehistoric Carson Desert: Stable Carbon and Nitrogen Isotopic Analyses," in *Bioarchaeology of the Stillwater Marsh: Prehistoric Human Adaptation in the Western Great Basin,* C. Larsen and R. L. Kelly, eds., pp. 96–106. Anthropological Papers of the American Museum of Natural History 77. New York.

Schoeninger, M. J., and M. J. DeNiro
1984 "Nitrogen and Carbon Isotopic Composition of Bone-Collagen from Marine and Terrestrial Animals." *Geochimica et Cosmochimica Acta* 48(4):625–639.

Schoeninger, M. J., K. Spielmann, K. Moore, and C. Kellner
2003 "Diet at Gran Quivira." Paper presented at the annual meeting of the Society for American Archaeology, Milwaukee, Wisconsin.

Schoeninger, M. J., N. J. van der Merwe, K. Moore, and J. A. Lee-Thorp
1990 "Decrease in Diet Quality between the Prehistoric and the Contact Period on St. Catherine's Island, Georgia," in *Postcontact Biocultural Adaptation of Native American Populations on St. Catherine's Island, Georgia,* C. Larsen, ed., pp. 78–93. Anthropological Papers of the American Museum of Natural History 68. New York.

Scholes, F. V.
1936–1937 "Church and State in New Mexico, 1610–1650." *New Mexico Historical Review* 11(1): 9–71, (2):145–178, (3):283–294, (4):297–343; 12(1):78–106.
1937–1941 "Troublous Times in New Mexico, 1659–1670." *New Mexico Historical Review* 12(2): 134–174, (4):380–452; 13(1):63–84; 15(3):249–268, (4):369–417; 16(1):15–40, (2):184–205, (3):313–327.

Schroeder, A. H.
1968 "Birds and Feathers in Documents Relating to Indians of the Southwest," in *Collected Papers in Honor of Lyndon Lane Hargrave,* A. H. Schroeder, ed., pp. 95–114. Papers of the Archaeological Society of New Mexico 1. Museum of New Mexico Press, Santa Fe.

1979a "Pueblos Abandoned in Historic Times," in *Handbook of North American Indians,* vol. 9, *Southwest,* A. Ortiz, ed., pp. 236–254. Smithsonian Institution, Washington, D.C.
1979b "Pecos Pueblo," in *Handbook of North American Indians,* vol. 9, *Southwest,* A. Ortiz, ed., pp. 430–437. Smithsonian Institution, Washington, D.C.

Schurr, M., and M. J. Schoeninger
1995 "Associations between Agricultural Intensification and Social Complexity: An Example from the Prehistoric Ohio Valley." *Journal of Anthropological Archaeology* 14(3):315–339.

Schwarcz, H. P.
2000 "Some Biochemical Aspects of Carbon Isotopic Paleodiet Studies," in *Biogeochemical Approaches to Paleodietary Analysis,* S. Ambrose and M. A. Katzenberg, eds., pp. 189–210. 5th ed. Kluwer Academic/Plenum Publishers, New York.

Scott, G. R.
1973 "Dental Morphology: A Genetic Study of American White Families and Variation of Living Southwest Indians." Ph.D. dissertation, Department of Anthropology, Arizona State University, Tempe.

Scott, G. R., and C. G. Turner II
1997 *The Anthropology of Modern Human Teeth: Dental Morphology and Its Variation in Recent Human Populations.* Cambridge University Press, Cambridge.

Segebarth-Orban, R.
1980 "An Evaluation of the Sexual Dimorphism of the Human Innominate Bone." *Journal of Human Evolution* 9:601–607.

Selker, F., and D. R. Carter
1989 "Scaling of Long Bone Fracture Strength with Animal Mass." *Journal of Biomechanics* 22:1175–1183.

Senior, L., and L. J. Pierce
1989 "Turkeys and Domestication in the Southwest: Implications from Homol'ovi III." *Kiva* 54(3):245–259.

Shafer, H. J.
2003 *Mimbres Archaeology at the NAN Ranch Ruin.* University of New Mexico Press, Albuquerque.

Sheridan, T. E.
1992 "The Limits of Power: The Political Ecology of the Spanish Empire in the Greater Southwest." *Antiquity* 66:153–171.

Shields, G. F., A. M. Schmeichen, and B. L. Frazier
1993 "MtDNA Sequences Suggest a Recent Evolutionary Divergence for Beringian and Northern North American Populations." *American Journal of Human Genetics* 53:549–562.

Shordania, J.
1932 "Die Bedeutung der Äusseren Pelvimetrie für die Diagnose des Engen Beckens." *Archiv für Gynäkologie* 148:89. [As cited in Jarcho 1933.]

Sibley, L. M., G. J. Armelagos, and D. P. Van Gerven
1992 "Obstetric Dimensions of the True Pelvis in a Medieval Population from Sudanese Nubia." *American Journal of Physical Anthropology* 89:421–430.

Sibthorpe, E. M., and D. B. Allbrook
1958 "A Radiological Survey of the Female Pelvis in Uganda: With Addendum on the Measurements of Specimens of the Female Bony Pelvis." *Journal of Obstetrics and Gynaecology of the British Empire* 65:600–605.

Simmons, M.
1979 "History of Pueblo-Spanish Relations to 1821," in *Handbook of North American Indians,* vol. 9, *Southwest,* A. Ortiz, ed., pp. 178–193. Smithsonian Institution, Washington, D.C.

Slome, D.
1929 "The Osteology of a Bushman Tribe." *Annals of the South African Museum* 24:33–60.

Smith, B. H.
1991 "Standards of Human Tooth Formation and Dental Age Assessment," in *Advances in Dental Anthropology,* M. A. Kelley and C. S. Larsen, eds., pp. 143–168. Wiley-Liss, New York.

Smith, R. W., and R. R. Walker
1964 "Femoral Expansion in Aging Women: Implications for Osteoporosis and Fractures." *Science* 145:156–157.

Smith, W., R. B. Woodbury, and N. F. S. Woodbury
1966 *The Excavation of Hawikuh by Frederick Webb Hodge: Report of the Hendricks-Hodge Expedition, 1917–1923.* Museum of the American Indian, Heye Foundation, New York.

Smithgall, E. B., F. E. Johnston, R. M. Malina, and M. A. Galbraith
1966 "Developmental Changes in Compact Bone Relationships in the Second Metacarpal." *Human Biology* 38:141–151.

Smouse, P., and J. Long
1992 "Matrix Correlation Analysis in Anthropology and Genetics." *Yearbook of Physical Anthropology* 35:187–214.

Snead, J. E., W. Creamer, and T. Van Zandt
2004 "'Ruins of Our Forefathers': Large Sites and Site Clusters in the Northern Rio Grande," in *The Protohistoric Pueblo World, A.D. 1275–1600,* E. C. Adams and A. I. Duff, eds., pp. 26–34. University of Arizona Press, Tucson.

Snow, D. H.
1981 "Protohistoric Rio Grande Pueblo Economics: A Review of Trends," in *The Protohistoric Period in the North American Southwest, A.D. 1450–1700,* D. R. Wilcox and W. B. Masse, eds., pp. 354–377. Arizona State University Anthropological Research Papers 24. Tempe.
1984 "Prologue to Rio Grande Prehistory," in *Collected Papers in Honor of Harry L. Hadlock,* N. L. Fox, ed., pp. 125–132. Papers of the Archaeological Society of New Mexico 9. Albuquerque.
1991 "Upland Prehistoric Maize Agriculture in the Eastern Rio Grande and Its Peripheries," in *Farmers, Hunters, and Colonists: Interaction between the Southwest and the Southern Plains,* K. Spielmann, ed., pp. 71–88. University of Arizona Press, Tucson.

Specker, B., and T. Binkley
2003 "Randomized Trial of Physical Activity and Calcium Supplementation on Bone Mineral Content in 3- to 5-Year-Old Children." *Journal of Bone and Mineral Research* 18(5):885–892.

Speth, J.
2004 *Life on the Periphery: Economic Change in Late Prehistoric Southeastern New Mexico.* Memoir 37. Museum of Anthropology, University of Michigan, Ann Arbor.

Spielmann, K. A.
1983 "Late Prehistoric Exchange between the Southwest and Southern Plains." *Plains Anthropologist* 28(102):257–272.
1989 "Colonists, Hunters, and Farmers: Plains-Pueblo Interaction in the Seventeeth Century," in *Columbian Consequences,* vol. 1, *Archaeological and Historical Perspectives on the Spanish Borderlands West,* D. H. Thomas, ed., pp. 101–113. Smithsonian Institution Press, Washington, D.C.
1991a *Interdependence in the Prehistoric Southwest: An Ecological Analysis of Plains-Pueblo Interaction.* Garland Publishing, New York.
1991b "Coercion or Cooperation? Plains-Pueblo Interaction in the Protohistoric Period," in *Farmers, Hunters, and Colonists: Interaction between the Southwest and the Southern Plains,* K. A. Spielmann, ed., pp. 36–50. University of Arizona Press, Tucson.
1996a "Pueblo III Settlement Trends among the Rio Abajo and Eastern Border Pueblos," in *The Prehistoric Pueblo World, A.D. 1150–1350,* M. Adler, ed., pp. 177–187. University of Arizona Press, Tucson.
1996b "Exchange and Evolution: Themes in the Study of Plains-Pueblo Relations," in *Interpreting Southwestern Diversity: Underlying Principles and Overarching Patterns,* P. R. Fish and J. J. Reid, eds., pp. 35–40. Arizona State University Anthropological Research Papers 48. Tempe.
1998 "Ritual Influences on the Development of Rio Grande Glaze A Ceramics," in *Migration and Reorganization: The Pueblo IV Period in the American Southwest,* K. A. Spielmann, ed., pp. 253–262. Arizona State University Anthropological Research Papers 51. Tempe.
2004 "Clusters Revisited," in *The Protohistoric Pueblo World, A.D. 1275–1600,* E. C. Adams and A. I. Duff, eds., pp. 137–143. University of Arizona Press, Tucson.

Spielmann, K., and E. A. Angstadt-Leto
1992 "Hunting, Gathering, and Health in the Prehistoric Southwest," in *Evolving Complexity and Environmental Risk in the Prehistoric Southwest,* J. A. Tainter and B. B. Tainter, eds., pp. 79–106. Addison-Wesley, Reading, Massachusetts.

Spielmann, K., M. J. Schoeninger, and K. Moore
1990 "Plains-Pueblo Interdependence and Human Diet at Pecos Pueblo, New Mexico." *American Antiquity* 55(4):745–765.

Stanislawski, M. B.
1983 "The American Southwest as Seen from Pecos." Manuscript on file, National Park Service, Southwest Regional Office, Santa Fe, New Mexico.

Stearns, M. L.
1981 *Haida Culture in Custody: The Masset Band.* University of Washington Press, Seattle.

Steuter, A. A., E. M. Steinauer, G. L. Hill, P. A. Bowers, and L. L. Tieszen
1995 "Distribution and Diet of Bison and Pocket Gophers in a Sandhills Prairie." *Ecological Applications* 5(3):756–766.

Stevenson, M. C.
1901–1902 *The Zuni Indians: Their Mythology, Esoteric Fraternities, and Ceremonies.* 23rd Annual Report of the Bureau of American Ethnology. Smithsonian Institution, Washington, D.C.

Stewart, K. M.
1973 "Mohave Warfare," in *The California Indians: A Source Book,* 2nd ed., R. F. Heizer and M. A. Whipple, eds., pp. 431–444. University of California Press, Berkeley.

Stewart, K. S., D. B. Cowan, and R. H. Philpott
1979 "Pelvic Dimensions and the Outcome of Trial Labour in Shona and Zulu Primigravidas." *South African Medical Journal* 55:847–851.

Stewart, T. D.
1956 "Examination of the Possibility that Certain Skeletal Characters Are Predisposed to Defects in the Lumbar Neural Arches." *Clinical Orthopaedics* 8:44–60.

Stock, J., and S. Pfeiffer
2001 "Linking Structural Variability in Long Bone Diaphyses to Habitual Behaviors: Foragers from the Southern African Later Stone Age and the Andaman Islands." *American Journal of Physical Anthropology* 115:337–348.

Stodder, A. L. W.
1989 "Bioarchaeological Research in the Basin and Range Region," in *Human Adaptations and Cultural Change in the Greater Southwest,* A. H. Simmons, A. L. W. Stodder, D. D. Dykeman, and P. A. Hicks, eds., pp. 167–190. Arkansas Archaeological Survey Research Series 32. Fayetteville.
1994 "Bioarchaeological Investigations of Protohistoric Pueblo Health and Demography," in *In the Wake of Contact,* C. S. Larsen and G. R. Milner, eds., pp. 97–108. Wiley-Liss, New York.
1996 "Paleoepidemiology of Eastern and Western Pueblo Communities in Protohistoric and Early Historic New Mexico," in *Bioarchaeology of Native American Adaptation in the Spanish Borderlands,* B. Baker and L. Kealhofer, eds., pp. 148–176. University Press of Florida, Gainesville.

Stodder, A. L. W., and D. L. Martin
1992 "Health and Disease in the Southwest before and after Spanish Contact," in *Disease and Demography in the Americas,* J. W. Verano and D. H. Ubelaker, eds., pp. 55–74. Smithsonian Institution Press, Washington, D.C.

Stodder, A. L. W., D. L. Martin, A. H. Goodman, and D. T. Reff
2002 "Cultural Longevity and Biological Stress in the American Southwest," in *The Backbone of History: Health and Nutrition in the Western Hemisphere,* R. H. Steckel and J. C. Rose, eds., pp. 481–505. Cambridge University Press, Cambridge.

Stuart, D. E., and R. P. Gauthier
1981 *Prehistoric New Mexico: Background for Survey.* New Mexico Historic Preservation Bureau, Santa Fe, New Mexico.

Stuart-Macadam, P.
1985 "Porotic Hyperostosis: Representative of a Childhood Condition." *American Journal of Physical Anthropology* 66:391–398.

Suchey, J. M., D. V. Wiseley, R. F. Green, and T. T. Noguchi
1979 "Analysis of Dorsal Pitting in the Os Pubis in an Extensive Sample of Modern American Females." *American Journal of Physical Anthropology* 51:517–539.

Sumner, D. R., B. Mockbee, and K. Morse
1985 "Computed Tomography and Automated Image Analysis of Prehistoric Femora." *American Journal of Physical Anthropology* 68: 225–232.

Sutherland, L. D., and J. M. Suchey
1991 "Use of the Ventral Arc in Pubic Sex Determination." *Journal of Forensic Sciences* 36:501–511.

Systat: Statistics, V. E.
1990 SYSTAT, Inc., Evanston, Illinois.

Tague, R. G.
1986 "Obstetric Adaptations of the Human Bony Pelvis." Ph.D. dissertation, Department of Anthropology, Kent State University, Kent, Ohio.
1988 "Bone Resorption of the Pubis and Preauricular Area in Humans and Nonhuman Mammals." *American Journal of Physical Anthropology* 76:251–267.
1989 "Variation in Pelvic Size between Males and Females." *American Journal of Physical Anthropology* 80:59–71.
1991 "Commonalities in Dimorphism and Variability in the Anthropoid Pelvis, with Implications for the Fossil Record." *Journal of Human Evolution* 21:153–176.
1992 "Sexual Dimorphism in the Human Bony Pelvis, with a Consideration of the Neandertal Pelvis from Kebara Cave, Israel." *American Journal of Physical Anthropology* 88: 1–21.
1994 "Maternal Mortality or Prolonged Growth: Age at Death and Pelvic Size in Three Prehistoric Amerindian Populations." *American Journal of Physical Anthropology* 95(1): 27–40.
2000 "Do Big Females Have Big Pelves?" *American Journal of Physical Anthropology* 112:377–393.

Tague, R. G., and C. O. Lovejoy
1986 "The Obstetric Pelvis of A.L. 288–1 (Lucy)." *Journal of Human Evolution* 15:237–255.

Tanner, J. M., P. C. R. Hughes, and R. H. Whitehouse
1981 "Radiographically Determined Widths of Bone Muscle and Fat in the Upper Arm and Calf from Age 3–18 Years." *Annals of Human Biology* 8:495–517.

Tarpy, C.
2000 "Pueblo Ancestors Return Home." *National Geographic* 198(5):118–125.

Teeri, J. A., and L. G. Stowe
1976 "Climatic Patterns and the Distribution of C4 Grasses in North America." *Oecologia* 23:1–12.

Thoms, H. K.
1915 "A Statistical Study of the Frequency of Funnel Pelves and the Description of a New Outlet Pelvimeter." *American Journal of Obstetrics and Diseases of Women and Children* 72:121–132.
1939 "Routine Roentgen Pelvimetry in 600 Primiparous White Women Consecutively Delivered at Term." *American Journal of Obstetrics and Gynecology* 37:101–106.
1941 "The Clinical Application of Roentgen Pelvimetry and a Study of the Results in 1,100 White Women." *American Journal of Obstetrics and Gynecology* 42:957–975.
1946 "A Discussion of Pelvic Variation and a Report on the Findings in 100 Negro Women." *American Journal of Obstetrics and Gynecology* 52:248–254.
1947 "The Role of Nutrition in Pelvic Variation." *American Journal of Obstetrics and Gynecology* 54:62–73.

Thoms, H. K., W. R. Foote, and I. Friedman
1939 "The Clinical Significance of Pelvic Variations: A Dimensional Study of the Upper, Mid, and Lower Pelvis in 200 White Primiparous Women." *American Journal of Obstetrics and Gynecology* 38:634–642.

Thoms, H., and W. W. Greulich
1940 "A Comparative Study of Male and Female Pelves." *American Journal of Obstetrics and Gynecology* 39:56–62.

Tieszen, L. L.
1994 "Stable Isotopes on the Plains: Vegetation Analyses and Diet Determinations," in *Skeletal Biology in the Great Plains: Migration, Warfare, Health, and Subsistence,* D. W. Owsley and R. L. Jantz, eds., pp. 261–282. Smithsonian Institution Press, Washington, D.C.

Tieszen, L. L., and T. W. Boutton
1989 "Stable Carbon Isotopes in Terrestrial Ecosystems Research," in *Stable Isotopes in Ecological Research,* P. W. Rundel, J. R. Ehleringer, and K. A. Nagy, eds., pp. 167–195. Springer-Verlag, New York.

Tieszen, L. L., and T. Fagre
1993a "Carbon Isotopic Variability in Modern and Archaeological Maize." *Journal of Archaeological Science* 20:25–40.
1993b "Effect of Diet Quality and Composition on the Isotopic Composition of Respiratory CO_2, Bone Collagen, Bioapatite, and Soft Tissues," in *Prehistoric Human Bone: Archaeology at the Molecular Level,* J. B. Lambert and G. Grupe, eds., pp. 121–156. Springer-Verlag, Berlin.

Timoshenko, S. P., and J. M. Gere
1972 *Mechanics of Materials.* Van Nostrand Reinhold, New York.

Torpin, R.
1951 "Roentgenpelvimetric Measurements of 3,604 Female Pelves, White, Negro, and Mexican, Compared with Direct Measurements of Todd Anatomic Collection." *American Journal of Obstetrics and Gynecology* 62: 279–293.

Trager, G. L.
1967 "The Tanoan Settlement of the Rio Grande Area: A Possible Chronology," in *Studies in Southwestern Linguistics,* D. H. Hymes and W. E. Bittle, eds., pp. 335–350. Mouton, The Hague.

Trigg, H. B.
1999 "The Economy of Early Colonial New Mexico, A.D. 1598–1680: An Investigation of Social Structure and Human Agency Using Archaeological and Documentary Data." Ph.D. dissertation, Department of Anthropology, University of Michigan, Ann Arbor.

Trinkaus, E.
1976 "The Evolution of the Hominid Femoral Diaphysis during the Upper Pleistocene in Europe and the Near East." *Zeitschrift für Morphologie und Anthropologie* 67:291–319.
1981 "Neanderthal Limb Proportions and Cold Adaptation," in *Aspects of Human Evolution,* C. B. Stringer, ed., pp. 187–224. Taylor and Francis, London.
1989 "Olduvai Hominid 7 Trapezial Metacarpal 1 Articular Morphology: Contrasts with Recent Humans." *American Journal of Physical Anthropology* 80:411–416.

Trinkaus, E., S. E. Churchill, B. Holt, and C. B. Ruff
2003 "Patterns of Diaphyseal Cross-Sectional Geometry between Central and Western European Early/Middle Upper Palaeolithic Humans," in *Changements biologiquese et culturels en Europe de la fin du Paléolithique Moyen au Néolithique,* J. Bruzek, B. Vandermeersch, and M. D. Garralda, eds., pp. 75–86. Laboratoire d'Anthropologie des Populations du Passé, Université de Bordeaux, Talence.

Trinkaus, E., S. E. Churchill, and C. B. Ruff
1994 "Postcranial Robusticity in *Homo,* II: Humeral Bilateral Asymmetry and Bone Plasticity." *American Journal of Physical Anthropology* 93:1–34.

Trinkaus, E., and C. E. Hilton
1996 "Neandertal Pedal Proximal Phalanges: Diaphyseal Loading Patterns." *Journal of Human Evolution* 30:399–425.

Trinkaus, E., and C. B. Ruff
1989 "Diaphyseal Cross-Sectional Morphology and Biomechanics of the Fond-de-Forêt 1 Femur and the Spy 2 Femur and Tibia." *Bulletin de la Societé Royale Belge d'Anthropologie et de Préhistoire* 100:33–42.
1999 "Diaphyseal Cross-Sectional Geometry of Near Eastern Middle Paleolithic Humans: The Femur." *Journal of Archaeological Science* 26:409–424.

Trinkaus, E., C. B. Ruff, F. Esteves, J. M. S. Coelho, M. Silva, and M. Mendonça
2002a "The Lower Limb Remains," in *Portrait of the Artist as a Child: The Gravettian Human Skeleton from the Abrigo do Lagar Velho and Its Archaeological Context,* J. Zilhão and E. Trinkaus, eds., pp. 435–465. Instituto Português de Arqueologia, Lisbon.
2002b "The Upper Limb Remains," in *Portrait of the Artist as a Child: The Gravettian Human Skeleton from the Abrigo do Lagar Velho and Its Archaeological Context,* J. Zilhão and E. Trinkaus, eds., pp. 466–488. Instituto Português de Arqueologia, Lisbon.

Trotter, M., R. R. Peterson, and R. Wette
1968 "The Secular Trend in the Diameter of the Femur of the American Whites and Negroes." *American Journal of Physical Anthropology* 28:65–74.

Turner, C. G. II
1984 "Advances in the Dental Search for Native American Origins." *Acta Anthropogenetica* 8:23–62.
1987 "Affinity and Dietary Assessment of Hohokam Burials from the Site of La Ciudad, Central Arizona," in *La Ciudad: Specialized Studies in the Economy, Environment, and Culture of La Ciudad,* J. E. Kisselberg, G. E. Rice, and B. L. Shears, eds., pp. 215–230. Anthropological Field Studies, Arizona State University, Tempe.
1990 "The Major Features of Sunadonty and Sinodonty, Including Suggestions about East Asian Microevolution, Population History, and Late Pleistocene Relationships with

Australian Aboriginals." *American Journal of Physical Anthropology* 82:295–317.
1993 "Southwest Indian Teeth." *National Geographic Research and Exploration* 9(1):32–53.
1999 "The Dentition of Casas Grandes with Suggestions on Epigenetic Relationships among Mexican and Southwestern U.S. Populations," in *The Casas Grandes World,* C. F. Schaafsma and C. L. Riley, eds., pp. 229–233. University of Utah Press, Salt Lake City.
2002 "Teeth, Needles, Dogs, and Siberia: Bioarchaeological Evidence for the Colonization of the New World." *Memoirs of the California Academy of Sciences* 27:123–158.

Turner, C. G. II, and J. D. Irish
1989 "Further Assessment of Hohokam Affinity: The Classic Period Population of the Grand Canal and Casa Buena Sites, Phoenix, Arizona," in *Archaeological Investigations at the Grand Canal Ruins: A Classic Period Site in Phoenix, Arizona,* D. R. Mitchell, ed., pp. 775–792. Publications in Archaeology 12. Soil Systems, Phoenix, Arizona.

Turner, C. G. II, C. H. Nichol, and G. R. Scott
1991 "Scoring Procedures for Key Morphological Traits of the Permanent Dentition: The Arizona State University Dental Anthropology System," in *Advances in Dental Anthropology,* M. A. Kelley and C. S. Larsen, eds., pp. 13–31. Wiley-Liss, New York.

Tyson, R. A.
1977 "Historical Accounts as Aids to Physical Anthropology: Examples of Head Injury in Baja California." *Pacific Coast Archaeological Society Quarterly* 13:52–58.

Ubelaker, D. H.
1988 "North American Indian Population Size, A.D. 1500 to 1985." *American Journal of Physical Anthropology* 77(3):289–294.
1989 *Human Skeletal Remains: Excavation, Analysis, Interpretation.* 2nd ed. Taraxacum, Washington, D.C.
1999 *Human Skeletal Remains: Excavation, Analysis, Interpretation.* 3rd ed. Taraxacum, Washington, D.C.

Upham, S.
1992 "Population and Spanish Contact in the Southwest," in *Disease and Demography in the Americas,* J. Verano and D. Ubelaker, eds., pp. 223–236. Smithsonian Institution Press, Washington, D.C.

Van Gerven, D. P., G. L. Armelagos, and M. R. Bartley
1969 "Roentgenographic and Direct Measurement of Femoral Cortical Involution in a Prehistoric Mississippian Population." *American Journal of Physical Anthropology* 31: 23–38.

Vehik, S.
2002 "Conflict, Trade, and Political Development on the Southern Plains." *American Antiquity* 61(1):37–64.

Vogel, J. C.
1978 "Isotopic Assessment of the Dietary Habits of Ungulates." *South African Journal of Science* 74:298–301.

Volkman, S. K., A. T. Galecki, D. T. Burke, M. R. Paczas, M. R. Moalli, R. A. Miller, and S. A. Goldstein
2003 "Quantitative Trait Loci for Femoral Size and Shape in a Genetically Heterogeneous Mouse Population." *Journal of Bone and Mineral Research* 18:1497–1505.

Wahlen, M.
1994 "Carbon Dioxide, Carbon Monoxide, and Methane in the Atmosphere: Abundance and Isotopic Composition," in *Stable Isotopes in Ecology and Environmental Science,* K. A. M. Lajtha and R. H. Michener, eds., pp. 83–113. Blackwell Scientific, Oxford.

Walker, P. L.
1985 "Anemia among Prehistoric Indians of the American Southwest," in *Health and Disease in the Prehistoric Southwest,* C. F. Merbs and R. J. Miller, eds., pp. 139–162. Arizona State University Anthropological Research Papers 34. Tempe.
1989 "Cranial Injuries as Evidence of Violence in Prehistoric Southern California." *American Journal of Physical Anthropology* 80:313–323.

Walrath, D.
2003 "Rethinking Pelvic Typologies and the Human Birth Mechanism." *Current Anthropology* 44:5–31.

Ward, R. H., A. Redd, D. Valencia, B. Frazier, and S. Paabo
1993 "Genetic and Linguistic Differentiation in the Americas." *Proceedings of the National Academy of Sciences* 90:10663–10667.

Ware, J. A., and E. Blinman
2000 "Cultural Collapse and Reorganization: The Origin and Spread of Pueblo Ritual

Sodalities," in *The Archaeology of Regional Interaction,* M. Hegmon, ed., pp. 381–409. University Press of Colorado, Boulder.

Warren, E.
1897 "An Investigation on the Variability of the Human Skeleton: With Especial Reference to the Naqada Race Discovered by Professor Flinders Petrie in his Explorations in Egypt." *Philosophical Transactions of the Royal Society of London* (Series B) 189: 135–227.

Washburn, S. L.
1948 "Sex Differences in the Pubic Bone." *American Journal of Physical Anthropology* 6:199–207.

Webb, W. S.
1974 *Indian Knoll.* University of Tennessee Press, Knoxville.

Wei, P. Y., H. Y. Chen, and Y. P. Chen
1956 "A Roentgenological Study of the Chinese Female Pelvis." *American Journal of Obstetrics and Gynecology* 72:1015–1020.

Weisensee, K.
2001 "Pecos Revisited: A Modern Analysis of Earnest Hooton's *The Indians of Pecos Pueblo.*" M.A. thesis, Department of Anthropology, University of Tennessee, Knoxville.
2008 "Reassessing Sex in the Pecos Pueblo Population: New Analyses of Hooton's Data," in *Reanalysis and Reinterpretation in Southwestern Bioarchaeology,* A. Stodder, ed., pp. 185–196. Anthropological Research Papers 59. Arizona State University Press, Tempe.

Weiss, K.
1972 "On the Systematic Bias in Skeletal Sexing." *American Journal of Physical Anthropology* 37(2):239–250.

Wendorf, F., and E. K. Reed
1955 "An Alternative Reconstruction of Northern Rio Grande Prehistory." *El Palacio* 62:131–173.

Wetterstrom, W.
1986 *Food, Diet, and Population at Prehistoric Arroyo Hondo Pueblo, New Mexico.* School of American Research Press, Santa Fe, New Mexico.

White, T. D.
1991 *Human Osteology.* Academic Press, San Diego.

Wilcox, D. R.
1981 "Changing Perspectives in the Protohistoric Pueblos, A.D. 1450–1700," in *The Protohistoric Period in the North American Southwest,* D. R. Wilcox and W. B. Masse, eds., pp. 378–409. Arizona State University Anthropological Research Papers 24. Tempe.

Wilcox, D. R., and J. Haas
1994 "The Scream of the Butterfly: Competition and Conflict in the Prehistoric Southwest," in *Themes in Southwest Prehistory,* G. Gumerman, ed., pp. 211–238. School of American Research Press, Santa Fe, New Mexico.

Willey, G. R., and J. Sabloff, eds.
1993 *A History of American Archaeology.* W. H. Freeman, New York.

Williams, P. L.
1995 *Gray's Anatomy.* 38th British ed. Churchill Livingstone, New York.

Wilshusen, R. H.
2002 "Estimating Population in the Central Mesa Verde Region," in *Seeking the Center Place,* M. Varien and R. Wilshusen, eds., pp. 101–120. University of Utah Press, Salt Lake City.

Wilson, J. P.
1973 "Quarai." *El Palacio* 78:14–28.

Winship, G. P.
1896 *The Coronado Expedition, 1540–1542.* Fourteenth Annual Report of the Bureau of Ethnology, 1892–1893, Part 1, pp. 329–613. Washington, D.C. Reprint, Rio Grande Press, Chicago, 1964.

Wolff, J.
1892 *Das Gesetz der Transformation der Knochen.* A. Hirchwild, Berlin.

Woo, S. L. Y., S. C. Kuei, D. Amiel, M. A. Gomez, W. C. Hayes, F. C. White, and W. H. Akeson
1981 "The Effect of Prolonged Physical Training on the Properties of Long Bone: A Study of Wolff's Law." *Journal of Bone and Joint Surgery* 63A:780–787.

Wood, J., D. Holman, K. O'Conner, and R. Ferrell
2002 "Mortality Models for Paleodemography," in *Paleodemography: Age Distributions from Skeletal Samples,* R. Hoppa and J. Vaupel, eds., pp. 129–168. Cambridge University Press, Cambridge.

Wood, J. W., G. R. Milner, H. C. Harpending, and K. M. Weiss
1992 "The Osteological Paradox: Problems of Inferring Prehistoric Health from Skeletal Samples." *Current Anthropology* 33(4):343–370.

Worth, J. E.
2001 "The Ethnohistorical Context of Bioarchaeology in Spanish Florida," in *Bioarchaeology of Spanish Florida,* C. S. Larsen, ed., pp. 1–21. University Press of Florida, Gainesville.

Young, M., and J. G. H. Ince
1940 "A Radiographic Comparison of the Male and Female Pelvis." *Journal of Anatomy* (London) 74:374–385.

Zimmerman, M. R., and M. A. Kelley
1982 *Atlas of Human Paleopathology.* Praeger, New York.

Zobeck, T. S.
1983 "Postcraniometric Variation among the Arikara." Ph.D. dissertation, Department of Anthropology, University of Tennesee, Knoxville.

Contributors

Lane Anderson Beck
Arizona State Museum
University of Arizona

Karin S. Bruwelheide
Department of Anthropology
National Museum of Natural History
Smithsonian Institution

Patricia Capone
Peabody Museum of Archaeology and Ethnology
Harvard University

Richard L. Jantz
Department of Anthropology
University of Tennessee–Knoxville

Corina M. Kellner
Department of Anthropology
Northern Arizona University

Steven A. LeBlanc
Peabody Museum of Archaeology and Ethnology
Harvard University

Katherine Moore
Department of Anthropology
University of Pennsylvania

Michèle E. Morgan
Peabody Museum of Archaeology and Ethnology
Harvard University

Douglas W. Owsley
Department of Anthropology
National Museum of Natural History
Smithsonian Institution

Christopher Ruff
Center for Functional Anatomy and Evolution
Johns Hopkins University School of Medicine

Margaret Schoeninger
Department of Anthropology
University of California at San Diego

Katherine Spielmann
School of Human Evolution and Social Change
Arizona State University

Robert G. Tague
Department of Geography and Anthropology
Louisiana State University

Christy G. Turner
School of Human Evolution and Social Change
Arizona State University

Katherine E. Weisensee
Department of Sociology and Anthropology
Clemson University